Two v

Digital Image Analysis of Microbes

MODERN MICROBIOLOGICAL METHODS

Series Editor Michael Goodfellow, *Department of Agricultural and Environmental Science, University of Newcastle, Newcastle upon Tyne, UK*

Molecular Biological Methods for *Bacillus* (1990)
Edited by Colin R. Harwood and Simon M. Cutting

0 471 92393 1

Nucleic Acid Techniques in Bacterial Systematics (1991)
Edited by Erko Stackebrandt and Michael Goodfellow

0 471 92906 9

Chemical Methods in Prokaryotic Systematics (1994)
Edited by Michael Goodfellow and Anthony G. O'Donnell

0 471 94191 3

Digital Image Analysis of Microbes
Imaging, Morphometry, Fluorometry and Motility Techniques and Applications (1998)
Edited by Michael H.F. Wilkinson and Frits Schut

0 471 97440 4

Digital Image Analysis of Microbes

Imaging, Morphometry, Fluorometry and Motility Techniques and Applications

edited by

M.H.F. Wilkinson

Centre for High Performance Computing,
University of Groningen,
Groningen, The Netherlands

and

F. Schut

Microscreen BV,
Centre for Microbial Detection and Identification Technology,
Groningen, The Netherlands

JOHN WILEY & SONS

Chichester · New York · Weinheim · Brisbane · Singapore · Toronto

Baffins Lane, Chichester,
West Sussex PO19 1UD, England

National 01243 779777
International (+44) 1243 779777
e-mail (for orders and customer service enquiries): cs-books@wiley.co.uk
Visit our Home Page on http://www.wiley.co.uk
or http://www.wiley.com

Commissioned in the UK on behalf of John Wiley & Sons, Ltd by Medi-Tech Publications, Storrington, West Sussex RH20 4HH, UK

Other Wiley Editorial Offices

John Wiley & Sons, Inc., 605 Third Avenue,
New York, NY 10158-0012, USA

WILEY-VCH Verlag GmbH, Pappelallee 3,
D-69469 Weinheim, Germany

Jacaranda Wiley Ltd, 33 Park Road, Milton,
Queensland 4064, Australia

John Wiley & Sons (Asia) Pte Ltd, 2 Clementi Loop #02-01,
Jin Xing Distripark, Singapore 129809

John Wiley & Sons (Canada) Ltd, 22 Worcester Road,
Rexdale, Ontario M9W 1L1, Canada

Library of Congress Cataloging-in-Publication Data

Digital image analysis of microbes : imaging, morphometry, fluorometry, and motility techniques and applications / edited by M.H.F. Wilkinson, F. Schut.
p. cm. — (Modern microbiological methods)
Includes bibliographical references and index.
ISBN 0-471-97440-4 (alk. paper)
1. Microbiology—Technique. 2. Image processing—Digital techniques. 3. Confocal microscopy. 4. Fluorescence microscopy. 5. Fluorimetry. I. Wilkinson, M.H.F. II. Schut, F. III. Series.
QR68.D54 1998
579′.028—dc21 97-44412
CIP

British Library Cataloguing in Publication Data

A catalogue record for this book is available from the British Library

ISBN 0-471-97440-4

Typeset in 10/12 Palacio from the authors' disks by Mathematical Composition Setters Ltd, Salisbury, Wiltshire
Printed and bound in Great Britain by Bookcraft (Bath) Ltd, Midsomer Norton.
This book is printed on acid-free paper responsibly manufactured from sustainable forestry, in which at least two trees are planted for each one used for paper production.

For Judith Wilkinson, and all ME/CFS sufferers.

Contents

Contributors

T. Agger *Centre for Process Biotechnology, Department of Biotechnology, Technical University of Denmark, DK-2800 Lyngby, Denmark*

Professor R. Amann *Max-Planck-Institut für marine Mikrobiologie, Arbeitsgruppe Molekulare Ökologie, Bremen, Germany*

Dr M.R. Barer *Department of Microbiology, The Medical School, University of Newcastle, Newcastle upon Tyne, NE2 4HH, UK*

Dr F.R. Boddeke *Pattern Recognition Group, Faculty of Applied Physics, Delft University of Technology, Lorentzweg 1, 2628 CJ Delft, The Netherlands*

Dr M. Carlsen *Centre for Process Biotechnology, Department of Biotechnology, Technical University of Denmark, DK-2800 Lyngby, Denmark*

Dr G. Cercignani *Laboratori di Biochimica, Dipartimento di Fisiologia e Biochimica, Università di Pisa, Via S. Maria 55, 56126 Pisa, Italy*

R. Grewal *Department of Microbiology, The Medical School, University of Newcastle, Newcastle upon Tyne, NE2 4HH, UK*

Dr C.-T. Huang *Department of Microbiology, Montana State University, Bozeman, MT 59717, USA*

A. Hunt *Adaptrix, 221 Eastbourne Avenue, Gateshead, NE8 4NL, UK*

Dr P. Hutzler *GSF-Forschungszentrum für Umwelt und Gesundheit, Institut für Pathologie und biomedizinische Bildanalyse, Oberschleissheim, Germany*

Dr G.J. Jansen *Department of Medical Microbiology, University of Groningen, Hanzeplein 1, 9713 GZ Groningen, The Netherlands*

Dr J.R. Lawrence *National Hydrology Research Institute, 11 Innovation Blvd, Saskatoon, Saskatchewan, Canada S7N 3H5*

Dr S. Lucia *Istituto di Biofisica del CNR, Via S. Lorenzo 26, 56127 Pisa, Italy*

Professor G.A. McFeters *Department of Microbiology, Montana State University, Bozeman, MT 59717, USA*

Dr B.C. Meijer *Regional Laboratory for Public Health, Van Ketwich-Verschuurlaan 59, 9721 SW Groningen, The Netherlands*

Dr T.R. Neu *UFZ, Sektion Gewässerforschung, Am Biederitzer Busch 12, 39114 Magdeburg, Germany*

Professor J. Nielsen *Centre for Process Biotechnology, Department of Biotechnology, Technical University of Denmark, DK-2800 Lyngby, Denmark*

Dr A.C. Peters *Food Safety Research Group, University of Wales Institute, Cardiff, CF3 7XR, UK*

Dr D. Petracchi *Istituto di Biofisica del CNR, Via S. Lorenzo 26, 56127 Pisa, Italy*

Dr M.-N. Pons *Laboratoire des Sciences du Génie Chimique, CNRS-ENSIC-INPL, Nancy, France*

Dr J.B.T.M. Roerdink *Department of Mathematics and Computing Science, University of Groningen, PO Box 800, 9700 AV Groningen, The Netherlands*

Dr F. Schut *Microscreen BV, Centre for Microbial Detection and Identification Technology, Zernikepark 8, 9747 AN Groningen, The Netherlands*

Dr M.E. Sieracki *Bigelow Laboratory for Ocean Sciences, McKown Pt Rd, W. Boothbay Harbor, ME 04575, USA*

Dr J. Slavik *Institute of Physiology, Czech Academy of Sciences, Videnska 1083,* CZ-142 20 Prague 4, Czech Republic

Dr A. Spohr *Centre for Process Biotechnology, Department of Biotechnology, Technical University of Denmark, DK-2800 Lyngby, Denmark*

Dr P.S. Stewart *Department of Microbiology, Montana State University, Bozeman, MT 59717, USA*

Dr D. Sudar *Life Sciences Division MS 74-157, Lawrence Berkeley National Laboratory, 1 Cyclotron Road, Berkeley, CA 94720, USA*

Dr L.V. Thomas *School of Pure and Applied Biology, University of Wales College of Cardiff, Cardiff, CF1 3TL, UK*

Professor (emerit) D. van der Waaij *Emeritus Professor of the Department of Medical Microbiology, University of Groningen, Hansplein 1, 9713 GZ Groningen, The Netherlands*

Dr L.J. van Vliet *Pattern Recognition Group, Faculty of Applied Physics, Delft University of Technology, Lorentzweg 1, 2628 CJ Delft, The Netherlands*

Dr C.L. Viles *School of Information and Library Science, University of North Carolina, Chapel Hill, North Carolina, USA*

Dr H. Vivier *Laboratoire des Sciences du Génie Chimique, CNRS-ENSIC-INPL, Nancy, France*

Dr M. Wagner *Lehrstuhl für Mikrobiologie, Technische Universität, München, Germany*

Dr A.S. Whiteley *Department of Microbiology, The Medical School, University of Newcastle, Newcastle upon Tyne, NE2 4HH, UK*

Dr M.H.F. Wilkinson *Centre for High Performance Computing, University of Groningen, PO Box 800, 9700 AV Groningen, The Netherlands*

Dr J.W.T. Wimpenny *School of Pure and Applied Biology, University of Wales College of Cardiff, Cardiff, CF1 3TL, UK*

Dr G.M. Wolfaardt *Department of Applied Microbiology and Food Science, University of Saskatchewan, Saskatoon, Saskatchewan, Canada S7N 5A8*

Professor I.T. Young *Pattern Recognition Group, Faculty of Applied Physics, Delft University of Technology, Lorentzweg 1, 2628 CJ Delft, The Netherlands*

Series Preface

The science of microbiology owes its existence as well as its underlying principles to the talent and practical prowess of pioneers such as Van Leeuwenhoek, Pasteur, Koch and Beijerinck. Interest in microbiology has recently increased quite significantly given the exciting developments in genetics and molecular biology and the growth of microbial technology. There was a time when most microbiologists were acquainted with many of the techniques used in microbiology. It is, however, now becoming increasingly difficult for research workers to keep abreast of the bewildering range of techniques currently used in microbiological laboratories. This problem is compounded by the fact that scientists in any one field increasingly need to apply techniques developed in other scientific disciplines.

The series 'Modern Microbiological Methods' aims to identify specialist areas in microbiology and provide up-to-date methodological handbooks to aid microbiologists at the laboratory bench. The books will be directed primarily towards active research workers but will be structured so as to serve as an introduction to the methods within a speciality for graduate students and scientists entering microbiology from related disciplines. Protocols will not only be described but difficulties and limitations of techniques and questions of interpretation fully discussed.

In summary, this series of books is designed to help stimulate further developments in microbiology by promoting the use of new and updated methods. Both authors and the editor-in-chief will be grateful to hear from satisfied or dissatisfied users so that future books in the series can benefit from the informed comment of practitioners in the field.

MICHAEL GOODFELLOW

Preface

Many microbiological methods are aimed at some form of counting of micro-organisms: colonies on agar plates, bacteria or yeast on microscopic slides, etc. Many of these methods are very straightforward technically, but also excruciatingly boring. Skilled technicians in laboratories the world over spend a lot of their time doing work that any patient simpleton could do, namely counting small things against a more-or-less smooth background. This is one reason why the computer, patient simpleton *par excellence*, is being recruited into microbiology, for the express purpose of relieving the tedium by doing the evaluation of images automatically. Of course, this is by no means the only reason to introduce digital image analysis to microbiology, where it has now certainly come to stay, nor does it explain why it has made its entry in the past few decades.

The fact that microbiology has always needed optical aids such as microscopes to study the subject matter has meant that advances in optical systems and recording techniques have had a large impact on this field of science. The recent developments of powerful and affordable computers on the one hand, and electronic imaging devices on the other have led to the widespread availability of affordable digital image processing systems for microscopy. These systems allow rapid quantification of many parameters which could hitherto only be described qualitatively. This is the main reason why digital image processing and analysis are rapidly becoming important tools in many areas of microbiology. Applications have been reported in the fields of medicine, food hygiene, environmental microbiology (in soil and aquatic environments) and biotechnology (fermentation), using a great variety of optical systems, cameras and image processing hardware and software. This development is certainly a boon to microbiology, although, as in the case of any new technique, many new kinds of problems have been encountered. Many of these centre around the surprise felt by microbiologists when it turns out that a task which is trivial to the couple of pounds of image processing 'wetware' we carry on our shoulders is by no means trivial to implement in software. Novices in the field often fail to see that computer vision is certainly one of the most computationally intensive, and arguably one of the most difficult, fields of computer science. Having said that, the images found in many microbiological settings are far simpler than those in the world around us. Therefore, many of the simpler computer vision techniques are quite adequate in many microbiological applications.

Successful development of image processing applications in microbiology, or any other biomedical field, is a highly multidisciplinary undertaking. Obviously, it requires a proper understanding of the biomedical problem to be studied, to ensure that the relevant parameters are measured. Secondly, it requires understanding of computer vision. But besides these, knowledge of optical systems, the quantum and wave nature of light, opto-electronics of cameras, mechanisms of fluorescence and range of available fluorescent probes etc., are all relevant to image analysed microscopy. It is no coincidence that the image processing work at the Department of Medical Microbiology of the University of Groningen, where both of us did our first work in image processing, was started by a medical doctor with a BSc in physics. Such multi-talented people are quite rare, so it is not surprising that most successful groups involve both microbiologists and computer scientists, and often physicists and biochemists as well. In this book we strive to bring together knowledge from many disciplines, and to show what the importance of each of these is within the field of digital image analysis in microbiology. We hope this allows individual researchers to learn at least the basics of each of the disciplines involved. However, this book must be seen as an encouragement of, not a replacement for, multidisciplinary teamwork.

OUTLINE

This volume focuses mainly on the theory and application of detection, morphometry, fluorometry and motility measurement techniques applied to bacteria, fungi, yeasts and protozoa. Regarding imaging systems, the main focus is on optical microscopy, since the light microscope is very much the workhorse of microbiology. Although fascinating examples of three-dimensional reconstruction of organelle or even protein structure from electron micrographs have been reported, we feel that this would be more suitable for a volume on protein science or microbial structure. Other new microscopic techniques such as atomic force microscopy are also beginning to be used within the field of microbiology, and are dealt with briefly, comparing the pros and cons of these emerging techniques with classical and confocal optical microscopy. One chapter on agar plate applications has been included, since the techniques used in this field are very closely related to those used in optical microscopy. Each of the chapters includes a brief discussion of future developments, to stimulate readers into exploring new fields.

The book is divided into three parts. The first and shortest part is mainly technical, dealing with imaging and optical systems, the physics of fluorescence, calibration techniques, the theory of image restoration, segmentation and mathematical morphology. The aim of this first part is to provide enough theoretical and technical background for proper understanding of the applications described in the second and third parts of the volume. In Chapter 1 the basics of image analysis are dealt with, and the terminology explained. Chapter 2 deals with image detectors: charge-coupled devices (CCD), image intensifiers and video cameras. Various image errors and sources of noise typical to some or all detectors are reviewed, along with image restoration methods to deal with them.

Chapter 3 discusses different types of optical microscopy, and in particular methods used to correct image aberrations by digital image processing. In Chapter 4 the power and problems of fluorescence microscopy are discussed. A whole chapter is devoted to this, because many applications described in the latter sections use fluorescence in one form or another. In Chapter 5 guidelines on the organization of software development are discussed, and especially how to prevent software obsolescence (which is an increasing problem as hardware platforms come and go). Chapter 6 is devoted to image segmentation or object recognition. Both manual and automatic methods are reviewed, and the problems encountered when working with objects close in size to the optical limit of resolution are discussed.

The second part focuses on applications to single-celled organisms of methods expounded in the first part. Each of the chapters highlights particular techniques, using specific applications as illustrations. First of all, enumeration and sizing of bacteria are described (Chapter 7). These methods include volume estimation from two-dimensional shape, which may be used to measure biomass and biovolume in fermentors and aquatic and terrestrial ecosystems. The microscopic techniques are compared to enumeration and sizing by flow cytometry, which is in a sense the 'arch rival' of image analysed microscopy. Chapter 8 deals with morphometry of yeast, and in particular methods to determine numbers of dividing or budding cells.

In Chapter 9, statistically optimized morphometry methods applied to complex microfloras such as the human and murine intestinal microflora are described. Single-colour fluorometry is dealt with in Chapter 10. Quantitative immunofluorescence and fluorescence *in situ* hybridization (FISH), which allows detection and enumeration of phylogenetic groups (species, genera) of microbes, are used as illustration. A comparison with flow cytometry is also presented. Chapter 11 discusses the use of densitometry and fluorometry combined with dyes that allow measurement of cellular activity, such as respiration and enzyme activity.

In Chapter 12 microspectrofluorometry, or multiple colour fluorometry is discussed. This has been applied to measuring pH changes in, for example, sporulating bacteria and changes in Ca^{2+} concentration in motile amoeba. Finally, in Chapter 13, motility measurements, both in real time on free swimming and tethered bacteria, and clustering as a result of chemotaxis or phototaxis, are presented.

The third part deals with multicellular organisms and clusters of single-celled organisms (colonies, biofilms, floc or soil communities). Its first topic is the application of morphometry to filamentous organisms such as fungi (Chapter 14). These methods are of importance to biotechnology, as many filamentous organisms are used in the production of, for example, antibiotics. Next, biofilm formation and analysis by classical and confocal microscopic techniques are discussed. Microbial biofilms form in a variety of circumstances, and the growth and effects have been studied on implants, catheters, etc. The techniques to study these film communities include cryosectioning combined with two-dimensional microscopy (Chapter 15), or optical sectioning using the three-dimensional imaging abilities of the confocal microscope (Chapter 16). Either technique may be used with any of the single- and multiple-colour fluorescent staining techniques of Part II. Chapter 17 deals with FISH applied to three-dimensional communities (flocs in, for example, activated sludge, soil communities etc.), studied by confocal microscopy, which may yield insight into the species distribution throughout a three-dimensional ecosystem.

Finally, in Chapter 18, colony counting and sizing, and more advanced agar plate techniques used for susceptibility testing are dealt with. The latter include drug susceptibility testing using more objective and therefore more reproducible methods of suppression zone measurement, and determination of environmental susceptibilities using single- and multiple-gradient plates.

ACKNOWLEDGEMENTS

Like any undertaking in image processing in microbiology, the creation of this book has been the result of multidisciplinary teamwork. The first team member we would like to thank is Sue Horwood, of Meditech Publishers, who first proposed the idea of a book on computer analysis of microbes. Without her initiative and assistance we would probably not have drawn up the proposal for this book in the first place. We would also like to thank Sally Betteridge, Sarah Lock and Lisa Tickner at John Wiley & Sons, for their assistance throughout the project, and the many reviewers of our proposal for their criticism, which greatly improved the original proposal, and for their encouragement. We have benefited greatly from the expertise of many others, in particular the authors, whom we would like to thank for their excellent and timely contributions. We would also like to thank our former colleagues Drs B.C. Meijer, H.Z. Apperloo-Renkema, G.J. Jansen, G.W. Welling and Professor D. van der Waaij, and all others at the image processing group of the Department of Medical Microbiology, University of Groningen. It was there that both of us became involved in this fascinating field.

Finally, we would like to thank Drs V. Rusch and K. Zimmermann, and Wilma Jung, of the Institute for Microbiology and Biochemistry, Herborn-Dill, Germany, for their financial and moral support of the aforementioned image processing group. Without their assistance little of that work would have been possible.

MICHAEL H.F. WILKINSON
FRITS SCHUT
November 1997

Part I

Technical Background

1

An Introduction to Digital Image Processing

Jos B.T.M. Roerdink
University of Groningen, Groningen, The Netherlands

1.1 INTRODUCTION

Digital image processing is the process of manipulating images by computer with the purpose of extracting useful information about the objects which appear in the image, and is a multidisciplinary field involving elements of optics, electronics, mathematics and computer engineering. As such, it suffers from an extensive but often rather imprecise jargon. It is the purpose of this chapter to collect a number of basic concepts in order to introduce the reader to the fundamentals of digital image processing, and serve as a reference for other chapters in this book. It should be realized that a certain amount of mathematical machinery is unavoidable in this field. The mathematical level of the exposition in this chapter conforms to that in the average textbook on digital image processing: elementary linear algebra (matrix calculus) and analysis, i.e. mainly convolution and Fourier transforms. An exception is Section 1.6, where a knowledge of basic set theory is assumed.

Because of its introductory nature, this chapter treats many aspects only in a very global way. Also, recent developments such as processing of three-dimensional images, parallel image processing, special purpose hardware or high resolution display systems are not discussed. The same holds for new methodologies such as wavelet transforms for image compression or colour image processing. The reader is referred to Castleman (1996), Gonzalez and Woods (1992), Rosenfeld and Kak (1982), Haralick and Shapiro (1992) or Jain (1989) for a more complete treatment of the fundamentals of digital image processing.

The organization of this chapter is as follows. In Section 1.2 a global sketch of the basic concepts and terminology is given. Section 1.3 introduces the basics of linear filtering, with its associated concepts of convolution and point-spread function. Section 1.4 gives examples of image enhancement by point operations (e.g. contrast stretching) or spatial filtering (smoothing, sharpening etc.). Image restoration – that is, the process of undoing degradations of the image that occurred during image acquisition (blurring, noise) – is presented in Section 1.5. Nonlinear image filtering is described in Section 1.6, where a concise introduction to mathematical morphology is presented. The remaining sections describe a number of specific

Digital Image Analysis of Microbes: Imaging, Morphometry, Fluorometry and Motility Techniques and Applications. Edited by M.H.F. Wilkinson and F. Schut.

image processing tasks, such as segmentation (Section 1.7) and description or feature extraction (Section 1.8).

1.2 BASIC CONCEPTS

In its most simple form, a digital image processing system contains an input device, or image sensor, to acquire the image, a computer upon which to process the image, and an output device (Figure 1.1; see also Section 5.2). As an input device one may use a camera or an image digitizer (see Chapter 2). Output may go to a computer display or to hardcopy devices such as printers or plotters. Other components involve communication and image storage. Here we are mostly concerned with the computer processing aspects.

1.2.1 Digitization

An image is a spatial representation of an object or a two- or three-dimensional scene. Usually the image is represented as a function $(x, y) \rightarrow f(x, y)$, where the domain of (x, y) values is a discrete grid of small rectangular regions called picture

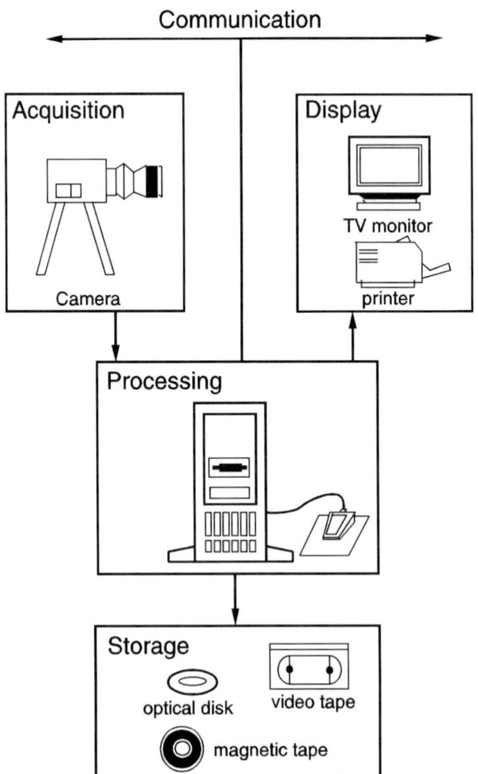

Figure 1.1 A digital image processing system.

elements or *pixels*. The values $f(x, y)$ are called grey levels: if n bits per pixel are available, the number of possible grey levels is 2^n. Eight-bit images, for which grey levels range from 0 to 255, are very common.

The process of converting the values of the continuous luminance distribution reaching the image sensor to a numerical form readable by a computer is called *digitization*, or *analog-to-digital conversion*. This involves sampling of the image brightness at each pixel location and quantization of the value within the allowed grey level range. A simple approach is to use uniform spatial sampling and equally spaced grey levels. More advanced techniques use adaptive or non-uniform sampling and quantization.

1.2.2 Digital imaging

Many different areas involve the formation and manipulation of images by computer, such as image processing, image analysis and computer graphics (see Table 1.1). *Image processing* is concerned with *transforming* images: the input of the processing is an image; the output is, again, an image. In *image analysis* we want to obtain *quantitative* data from the images, such as the number of cells in a microscopic image, the diameter of blood vessels in a cardiac scan etc.; in this case, input consists of one or more images, whereas the output contains numeric or symbolic data. Finally, in *computer graphics* the goal is image synthesis: starting from data derived from a mathematical description of objects, images are constructed on a computer display.

1.2.3 Classification of image operations

Assuming that a digitized image is available in computer memory, various operations can be performed on the image, depending on the application. These operations can be classified with respect to their mathematical properties, or with respect to the goal of the operation (or set of operations). A first classification considers the correspondence between the pixels of input and output images.

(a) *Point operation*: the grey level at each pixel of the output image f_{out} depends only on the grey level of the corresponding pixel in the input image f_{in}:

$$f_{\text{out}}(x, y) = \mathcal{O}(f_{\text{in}}(x, y))$$

where $\mathcal{O}$ denotes the image operation. Examples are contrast stretching or histogram equalization (see Section 1.4.1). *Algebraic* operations, in which the output image is formed by pointwise addition, subtraction, multiplication or

Table 1.1 Subfields in digital imaging.

	Image out	Data out
Image in	Image processing	Image analysis
Data in	Computer graphics	Data analysis

division of two input images, also belong in this category. An example is subtraction of two images for shading correction (Section 2.4.7).

(b) *Local operation*: the grey level at each pixel (x, y) of the output image depends on the grey levels of pixels which are contained in a small neighbourhood $\mathcal{B}(x, y)$ of the pixel (x, y) in the input image:

$$f_{\text{out}}(x, y) = \mathcal{O}(\{f_{in}(x', y') : (x', y') \in \mathcal{B}(x, y)\})$$

See Section 1.4.2 for examples.

(c) *Global operation*: the grey level at each pixel of the output image depends on the grey levels of all pixels in the input image. An example of such an operation is a (discrete) Fourier transform of the input image (see Section 1.3).

(d) *Geometric operation*: this involves a spatial transformation of the image such as scaling, translation, rotation or perspective projection (e.g. used for distortion correction, Section 3.5.5).

1.2.4 Connectivity

The above spatial relationships between pixels were expressed by using the concept of the neighbourhood of a pixel. Although the notion of neighbourhood is intuitively obvious, a more precise formulation is needed for the case of digital images, which are defined on discrete grids. In this context the notion of *connectivity* is essential. When, on a rectangular pixel grid, only the pixels to the north, south, east and west of a pixel are considered as the neighbours of this pixel, we speak of *4-connectivity*. When in addition the diagonal pixels are considered as neighbours, we speak of *8-connectivity*. We also say that the neighbours of a pixel are *adjacent* to that pixel. This is illustrated in Figure 1.2.

A set M of pixels is called *connected* if and only if for every pair of pixels $p, q \in M$ there exists a path between p and q which only passes through pixels of M, where a path only moves between neighbouring pixels. A *connected component* is a

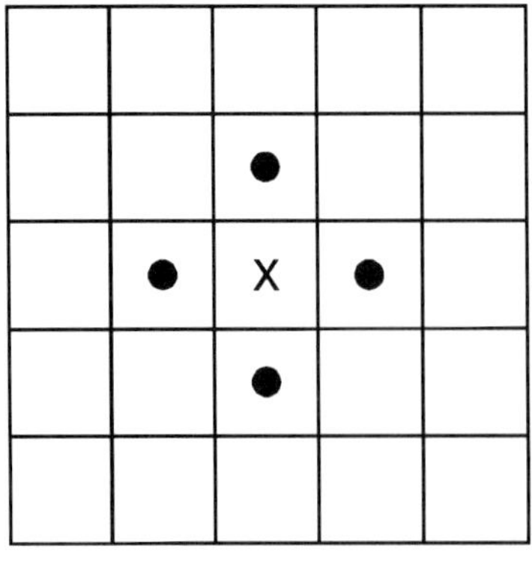

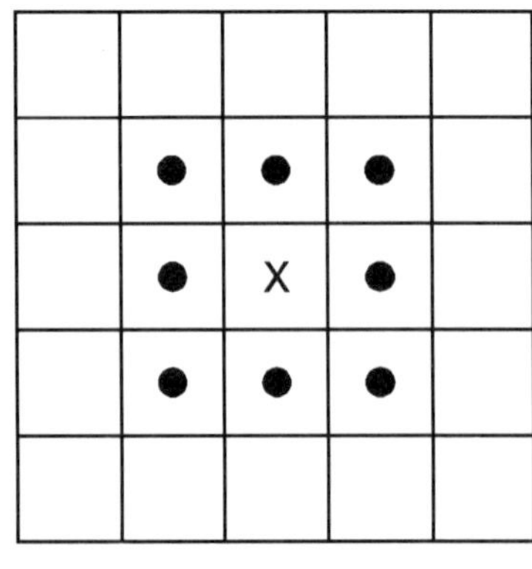

Figure 1.2 Pixels that are (a) 4-connected and (b) 8-connected to the centre pixel x.

non-empty connected set of pixels of maximal size. Note that the result of applying this definition depends on the type of neighbour relation used (4-connectivity or 8-connectivity). For binary images, where pixels have a value of 1 or 0, connected components can be defined for both foreground (1-pixels) and background (0-pixels). The *border* of a connected component *C* of 1-pixels is the set of pixels of *C* which are a neighbour of 0-pixels. We can also introduce (4- or 8-) connectivity for the 0-pixels, and define the connected components of these pixels too. To distinguish both component types, let us speak about 0-components and 1-components. However, one has to be careful in the choice of connectivity for both types of pixels. It can be shown that if *C* is a 4-connected 1-component and *D* is an adjacent 8-connected 0-component, then either *C* surrounds *D* or *D* surrounds *C* (Rosenfeld, 1970). This remains the case when we use 4-connectivity for the 0-pixels and 8-connectivity for the 1-pixels, but not when we use 4-connectivity (or 8-connectivity) for *both* 0- and 1-pixels.

Extraction of connected components is a step often used in segmentation or description (see Sections 1.8 and 6.7.7). Efficient algorithms exist to perform this task based on some form of label propagation, for example the two-pass algorithm (Rosenfeld and Kak, 1982).

1.2.5 Methodology

In digital image processing one may distinguish the following three levels:

(a) *Low level*: the first step after image acquisition is pre-processing to obtain elementary or *atomic* features (representation in terms of *iconic* data).
(b) *Intermediate level*: grouping of atomic features (segmentation, description, feature extraction).
(c) *High level*: recognition and interpretation of objects (representation in terms of *symbolic* data).

Let us expand a little upon the individual steps involved in a digital image processing system (Figure 1.3):

(a) *Image formation*: this is the first step, and a vital one. The quality of further processing depends in a crucial way upon the quality of the sensor, illumination conditions, digitization etc.
(b) *Pre-processing*: the assumption is that the image contains an informative part and variations which one wants to suppress. The pre-processing step, also called *conditioning*, encompasses enhancement, image restoration, noise suppression, background normalization etc.
(c) *Segmentation*: the assumption is that the image has a spatial structure. A first step is labelling, where we try to organize the pixels into sets of correlated points. Examples are thresholding, edge detection and corner detection. Next comes *grouping*; for example, connected component detection, segmentation on the basis of grey values, linking edges into borders. Note that there is a change of logical data structure here: from pixels into pixel *sets*.

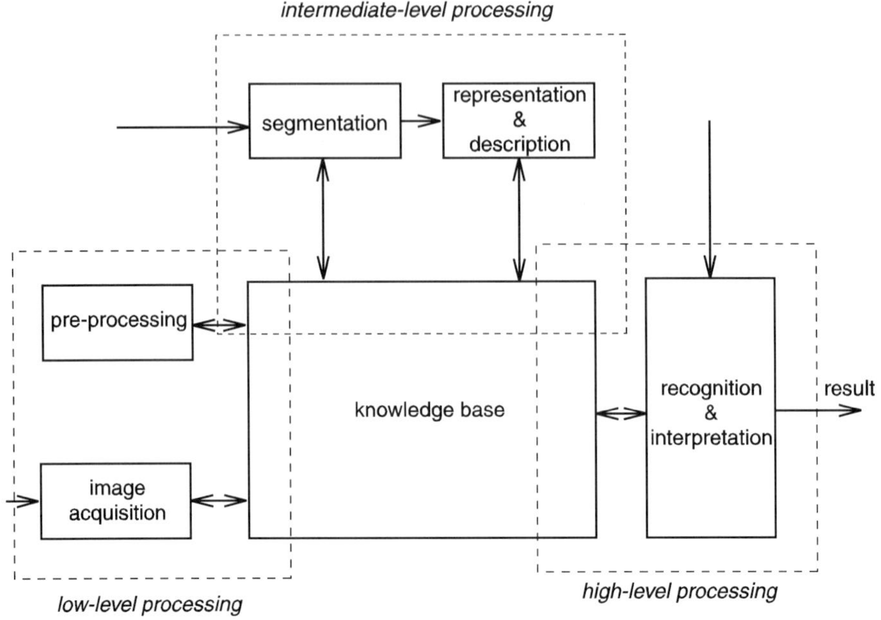

Figure 1.3 Phases in digital image processing.

(d) *Description*: in this step, also called *feature extraction*, pixel sets are grouped into lists of properties: area, centre of gravity, number of holes, curvature, relation to other groups.
(e) *Interpretation*: find the meaning of the extracted properties in terms of real-world properties. Examples are template matching, statistical pattern recognition (matching vectors of image properties to vectors of object properties) and structural pattern recognition (matching vectors together with relations between them).

In the rest of this chapter we give a more precise definition and description of the different phases of image processing outlined above.

1.3 LINEAR FILTERING

Many of the operations commonly used for image enhancement involve the application of linear filters. There is a well-developed mathematical discipline underlying this, called linear system theory, which is an essential ingredient of linear signal and image processing. In this section we give the fundamentals of this theory, including a discussion of sampling, convolution, Fourier transforms and filter design. For simplicity we first discuss one-dimensional signals, which are functions $f(t)$ of a single parameter, where t denotes the time. This is easily extended to images, which are – in the two-dimensional case – signals $f(x, y)$ of two spatial variables.

1.3.1 Linear shift-invariant systems

Two basic assumptions are made about the system that transforms signals to signals (or, for that matter, images to images). The first one is additivity: operating on the sum of two input signals gives the same result as operating on the two input signals separately and adding the results. A similar property holds with respect to scaling the intensity of the signal: first scaling and then operating on the signal gives the same result as first performing the signal operation and then scaling the result. A second basic assumption concerns time shifts: it makes no difference whether the system operates on a time-shifted signal or on the original signal followed by a time shift. (In the case of two-dimensional images there are two shift parameters, one for the horizontal and one for the vertical direction.) The usefulness of introducing systems with these two basic properties comes from the fact that such systems form a good approximation to many optical systems used in practice (see Sections 2.2.1 and 3.2.2). The above leads to the following precise definition:

A linear shift-invariant (LSI) system $\mathcal{O}$ maps an input signal $f(t)$ to an output signal $g(t)$, denoted as $f(t) \xrightarrow{\mathcal{O}} g(t)$, such that the following two conditions hold:

(a) *Linearity*: if $f_1(t) \xrightarrow{\mathcal{O}} g_1(t)$, and $f_2(t) \xrightarrow{\mathcal{O}} g_2(t)$, then, for arbitrary constants a, b:

$$af_1(t) + bf_2(t) \xrightarrow{\mathcal{O}} ag_1(t) + bg_2(t)$$

(b) *Shift invariance*: if $f(t) \xrightarrow{\mathcal{O}} g(t)$, then for each time shift T:

$$f(t - T) \xrightarrow{\mathcal{O}} g(t - T)$$

1.3.1a Signals

The output of an LSI system is given by the *convolution* of the input with a function $k(t)$ (called a *kernel*) which is characteristic for that system:

$$g(t) = (k * f)(t) = \int_{-\infty}^{\infty} k(t - s)f(s)\ \mathrm{d}s \tag{1.1}$$

If the input is an impulse, i.e. a delta function at time 0, then the output is precisely $k(t)$. Therefore, the filter kernel $k(t)$ is usually called the *impulse response function*.

The one-dimensional Fourier transform LSI systems can be conveniently described by using Fourier transforms. The Fourier transform of a function $f(t)$, denoted by $F(u)$, is defined by:

$$F(u) = \int_{-\infty}^{\infty} \mathrm{e}^{-i2\pi ut} f(t)\ \mathrm{d}t \tag{1.2}$$

For discrete and finite data the Fourier transform (1.2) is replaced by the forward discrete Fourier transform (DFT), which transforms a signal represented by the vector $[f(0), f(1), ..., f(N-1)]$ into a transformed vector $[F(0), F(1), ..., F(N-1)]$ by:

$$F(u) = \frac{1}{\sqrt{N}} \sum_{x=0}^{N-1} f(x) e^{-i2\pi ux/N} \qquad u = 0, 1, ..., N-1$$

The original signal can be recovered from its discrete Fourier transform by the inverse discrete Fourier transform (IDFT):

$$f(x) = \frac{1}{\sqrt{N}} \sum_{x=0}^{N-1} F(u) e^{i2\pi ux/N} \qquad x = 0, 1, ..., N-1$$

The DFT can be implemented efficiently by the fast Fourier transform (FFT) algorithm.

Introducing the Fourier transforms F, G and K of f, g and k, we can write the convolution equation (1.1) in the form:

$$G(u) = K(u)F(u)$$

So in the Fourier domain, convolution becomes multiplication, which is the basis of computer implementation using the FFT: first compute the DFT of k and f, then perform the multiplication of the Fourier transforms K and F, and finally compute the inverse DFT of G to obtain g. The function $K(u)$, which is the Fourier transform of the impulse response function, is called the *transfer function* of the linear system.

1.3.1b Images

In the two-dimensional case, the convolution takes the form:

$$g(x, y) = (k * f)(x, y) = \int_{-\infty}^{\infty} \int_{-\infty}^{\infty} k(x - x', y - y') f(x', y') \, dx' \, dy' \qquad (1.3)$$

Now $k(x, y)$ is often referred to as the *point-spread function* (PSF). In the case of digital images, the convolution integral (1.3) becomes a summation:

$$g(x, y) = \sum_{m=0}^{M-1} \sum_{n=0}^{N-1} k(x - m, y - n) f(m, n) \qquad (1.4)$$

for $x = 0, 1, ..., M-1$, $y = 0, 1, ..., N-1$. Examples of discrete convolution for image filtering are given in Section 1.4.2.

The two-dimensional Fourier transform For the case of images we use a two-dimensional DFT:

$$F(u, v) = \frac{1}{\sqrt{NM}} \sum_{x=0}^{M-1} \sum_{y=0}^{N-1} f(x, y) e^{-i2\pi(ux/M + vy/N)}$$

$$u = 0, 1, ..., M-1, \; v = 0, 1, ..., N-1$$

with inverse

$$f(x, y) = \frac{1}{\sqrt{NM}} \sum_{x=0}^{M-1} \sum_{y=0}^{N-1} F(u, v) e^{i2\pi(ux/M + vy/N)}$$

$$x = 0, 1, ..., M-1, \; y = 0, 1, ..., N-1$$

Introducing the two-dimensional Fourier transforms F, G and K of f, g and k, we can again write the convolution equation (1.4) in the form of a product:

$$G(u, v) = K(u, v)F(u, v)$$

1.3.2 Filter design

Linear filters are commonly classified with respect to their frequency domain characteristics (Figure 1.4). *Lowpass* filters suppress high frequencies, and can therefore be used to suppress noise. The simplest case is the box filter k, for which $K(u) = 1$ for $-W \leqslant u \leqslant W$. Here W is the cut-off frequency. *Bandpass* or *bandstop* filters pass or suppress energy, respectively, if the frequency u is within a given window of size 2Δ.

Highpass filters suppress low frequencies. This is useful for edge detection, for example (see Section 1.4.2b). An example is the difference-of-Gaussians (DOG) filter, whose transfer function is given by:

$$K(u) = Ae^{-u^2/\alpha_1^2} - Be^{-u^2/\alpha_2^2} \qquad A \geqslant B, \; \alpha_1 > \alpha_2$$

1.3.3 Sampling

Suppose the Fourier transform of a signal f is only non-zero in the interval $[-W, W]$. Such a signal is called *bandlimited*, and W is called the *bandwidth*. The famous sampling theorem of Shannon says that bandlimited signals can be represented by a sampling series:

$$f(x) = \sum_n f(n\Delta)\mathrm{sinc}_W(x - n\Delta) \tag{1.5}$$

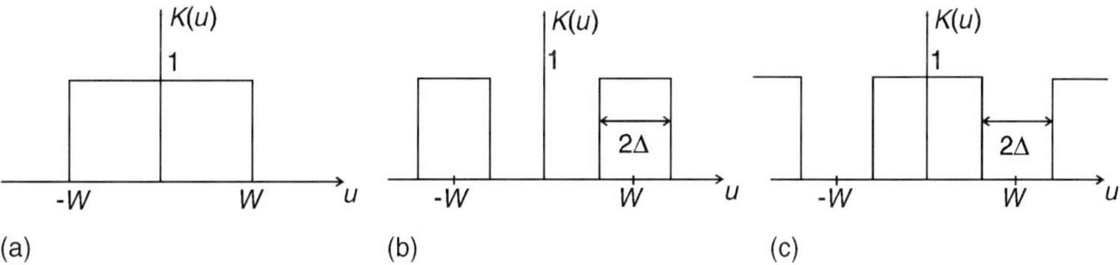

Figure 1.4 Ideal transfer functions: (a) lowpass, (b) bandpass and (c) bandstop.

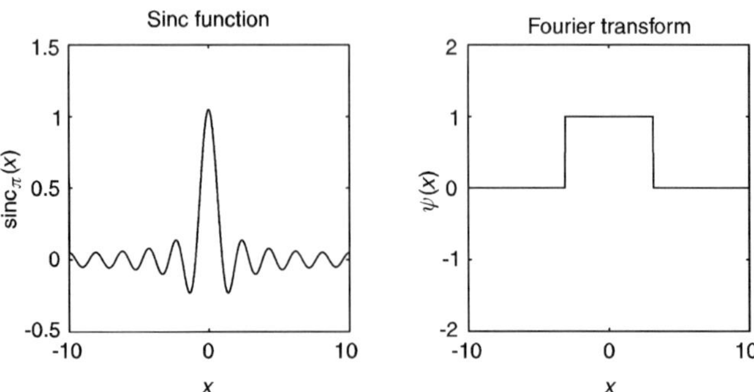

Figure 1.5 Sinc function and its Fourier transform ψ for $W=1/2$.

where the sinc function is defined by $\text{sinc}_W(x) = \sin(2\pi Wx)/(2\pi Wx)$ for $x \neq 0$, and $\text{sinc}_W(0) = 1$. Here $\Delta := 1/(2W)$ is the sampling interval and $f_s := 1/\Delta$ the corresponding sampling frequency. The Fourier transform ($\mathscr{F}$) of the sinc function is the box function (see Figure 1.5): $\mathscr{F}(\text{sinc}_W)(u) = \Delta$ for $|u| < W$ and zero elsewhere. Equation (1.5) says that f is completely determined by its values at equidistant points $0, \pm\Delta, \ldots$. Note that a function with bandwidth W is certainly bandlimited on $[-W', W']$ with $W' > W$. So we can always use a smaller sampling step $\Delta' < \Delta$ in (1.5). The minimal sampling frequency required is f_s: this is called the *Nyquist frequency*. Note that the minimal frequency $f_s = 2W$ is twice the maximum frequency W contained in the signal f.

If one undersamples the signal by using a sampling frequency lower than the Nyquist frequency, distortions in the sampled signal appear; this is called *aliasing*. In the case of images visible artefacts appear, which are called Moiré patterns. Unfortunately, real signals or images are defined on a bounded domain, which implies that they cannot be strictly bandlimited. However, by a judicious choice of digitizing parameters the so-called aliasing error can be reduced to an acceptable level. A further discussion of aliasing and its effects can be found in Sections 7.2.2 (in two dimensions) and 16.4.2 (in three dimensions).

1.4 IMAGE ENHANCEMENT

Image enhancement is the process of improving image quality so that the result is more suitable for a specific application. Examples are point operations such as contrast stretching, or spatial filtering for edge enhancement, noise suppression, smoothing and sharpening.

1.4.1 Point operations

Point operations are often used to improve the visual appearance of an image. As defined in Section 1.2, a point operation is distinguished by the fact that the value at each pixel of the output image depends only on the value of the corresponding pixel in the input image:

$$f_{\mathrm{out}}(x, y) = \mathcal{O}(f_{\mathrm{in}}(x, y))$$

where $\mathcal{O}$ denotes the *grey scale transformation* (GST) function. This can be a linear or nonlinear function. We give two examples.

1.4.1a Contrast stretching

If the features of interest occupy only a small range of the available grey levels, one may use a point operation so that the output image spans a larger range. This is called contrast stretching. For example, let the input image f_{in} have minimum and maximum grey levels m and M, respectively $(M > m \geqslant 0)$. Then the following operation stretches the image to full grey level range [0..255]:

$$f_{\mathrm{out}}(x, y) = \frac{255}{M - m}(f_{\mathrm{in}}(x, y) - m)$$

See Figure 1.6B for an example, and Sections 16.4.2 and 18.3.7 for applications.

1.4.1b Histogram equalization

The *histogram* $h(m)$ of a digital image f is the number of times the grey value m occurs in the image. Now suppose we want to apply a point operation to the image f which produces a flat histogram in the output, i.e. on the average an equal number of pixels at each grey level. So if the image has N pixels and grey level range $[0..M]$, the output image will have M/N pixels at each grey level (see Figure 1.6C). It can be shown that the grey scale transformation that achieves this histogram equalization is given by $\mathcal{O}(f(x, y)) = M\mathsf{P}(f(x, y))$ where:

$$\mathsf{P}(\ell) = \frac{1}{N}\sum_{m=0}^{\ell} h(m)$$

is the normalized cumulative histogram function.

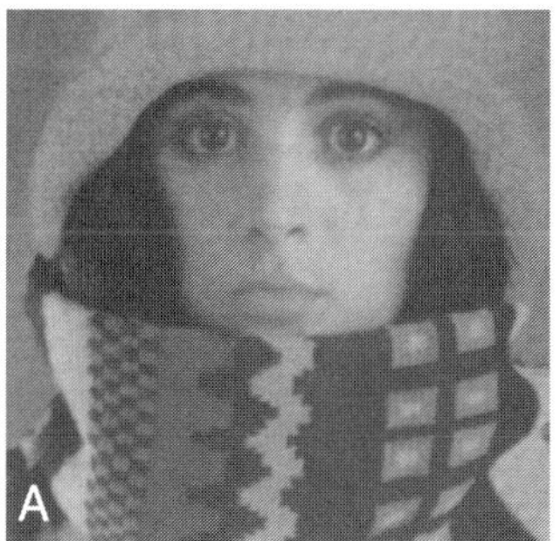

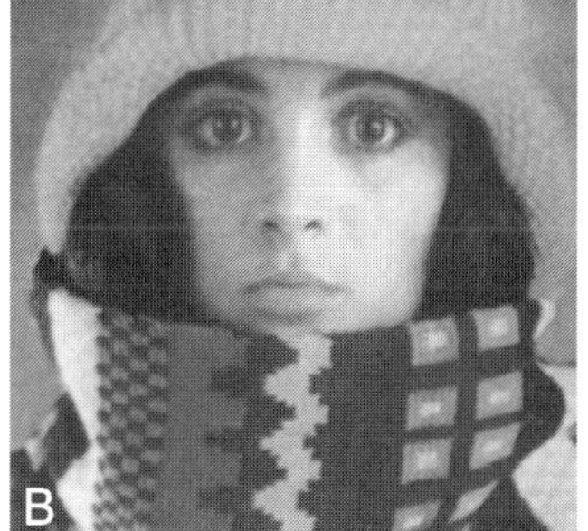

Figure 1.6 (A) Input image. (B) Contrast stretch of (A). (C) Histogram equalization of (A).

1.4.2 Spatial filtering

1.4.2a Smoothing filters

To suppress noise one may use lowpass filtering by local averaging within a small neighbourhood, or *mask*, surrounding each pixel. This can be implemented by convolution filtering (see Section 1.3). For example, take as the mask a 3×3 neighbourhood, with $k(m, n) = 1/9$ for $|m| \leqslant 1, |n| \leqslant 1$ in (1.4); this is called *uniform filtering* (see Figure 1.7B). This may be extended to more general filter kernels, such as a Gaussian function.

To obtain noise reduction without blurring of the edges, one may use *rank-order*, or *percentile*, filters based upon ranking the pixel values within the mask surrounding the centre pixel (see Figure 1.7C). An example is *median filtering*: sort the set of pixel values and assign the median (the value m such that half the values in the set are less than m and half are greater than m) to the centre pixel. This class of filters is nonlinear and therefore more difficult to analyse than linear filters based on convolution (see also Section 6.4.1 for a comparison of uniform and median filtering). More examples of such nonlinear filters are provided by the morphological filters (see Section 1.6).

1.4.2b Sharpening filters

The goal of sharpening is to enhance fine details such as edges in an image that has been blurred during image acquisition. This can be performed by using highpass filters based upon spatial differentiation. Some of the most well known examples are the following (see Figure 1.11):

(a) *Laplace operator*: a digital implementation of Laplacian filtering uses the discrete convolution kernels shown in Figure 1.8.
(b) *Roberts operator*: when $f(x, y)$ is the input image, this operator produces an output $g(x, y)$ given by:

$$g(x, y) = \sqrt{\{f(x, y) - f(x+1, y+1)\}^2 + \{f(x+1, y) - f(x, y+1)\}^2}$$

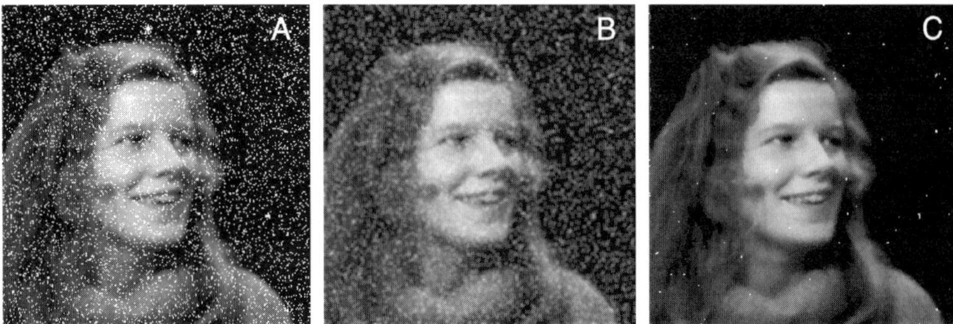

Figure 1.7 Smoothing filters: (A) original, (B) uniform filter and (C) percentile filter.

0	-1	0
-1	4	-1
0	-1	0

-1	-1	-1
-1	8	-1
-1	-1	-1

Figure 1.8 Laplacian convolution kernels.

(c) *Sobel operator*: each point of the input image is convolved with two kernels, shown in Figure 1.9, one for detecting horizontal edges and the other for vertical edges. The output of the operator is the maximum of the two convolutions; one may also take the square root of the sum of the squares (see Section 6.7.4).
(d) *Prewitt operator*: this works in exactly the same way as the Sobel filter, except for the convolution kernels, which are now as shown in Figure 1.10.

1.4.3 Frequency domain filtering

Enhancement can also be performed in the frequency domain: compute the Fourier transform of the input image, multiply the result by a filter transfer function, and take the inverse Fourier transform. For example, to suppress noise one may use lowpass filtering (see Section 1.3.2), which attenuates the high-frequency components. On the other hand, edge sharpening is obtained by highpass filtering (Figure 1.11).

-1	-2	-1
0	0	0
1	2	1

-1	0	1
-2	0	2
-1	0	1

Figure 1.9 Kernels for the Sobel operator.

-1	-1	-1
0	0	0
1	1	1

-1	0	1
-1	0	1
-1	0	1

Figure 1.10 Kernels for the Prewitt operator.

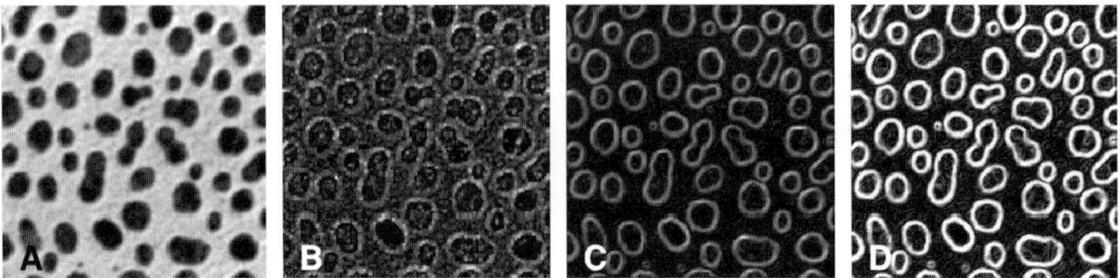

Figure 1.11 Effect of sharpening filters: (A) original, (B) Laplace filter, (C) Roberts filter and (D) Sobel filter.

1.5 IMAGE RESTORATION

By image restoration we mean the process of removing or reducing image degradations which occur during image formation. Important degrading factors are: (i) *blurring* produced by the optical system or object motion during acquisition, and (ii) *noise* from electronic and optical devices.

1.5.1 Linear degradation model

A simple model assumes linear degradation of the ideal image $f(x, y)$ described by a convolution kernel $k(x, y)$ and additive noise $n(x, y)$, resulting in the degraded image $w(x, y)$:

$$w(x, y) = (k * f)(x, y) + n(x, y) \tag{1.6}$$

Restoration now has to undo the degradation due to convolution and noise.

1.5.2 Linear restoration model

Using a linear model for restoration as well (for this reason, one often speaks of *deconvolution*), we compute an estimate $\tilde{f}(x, y)$ of the ideal image by using a convolution kernel $r(x, y)$:

$$\tilde{f}(x, y) = (r * w)(x, y)$$

The complete process of degradation and restoration is represented graphically in Figure 1.12.

A simplified restoration problem results if we consider images of finite size, say $N \times N$. In that case we can work with matrix-vector calculus (Andrews and Hunt, 1977). The restoration problem (1.6) then has the form

$$\mathbf{w} = \mathbf{Kf} + \mathbf{n} \tag{1.7}$$

where $\mathbf{w}$, $\mathbf{f}$, $\mathbf{n}$ are column vectors of length N^2, and $\mathbf{K}$ is an $N^2 \times N^2$ matrix representing the convolution. To give an example, for an image of size 512×512 the

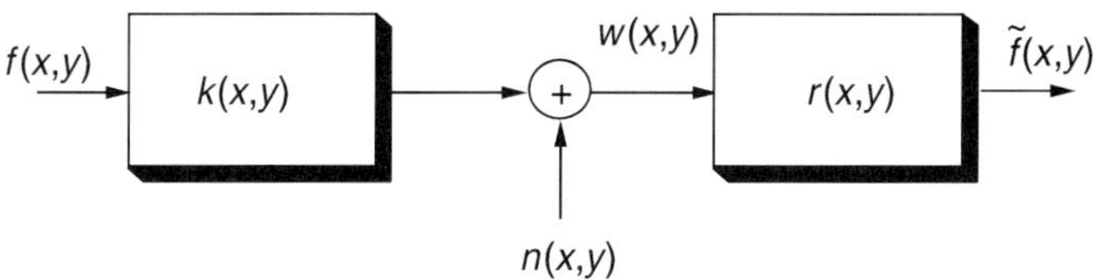

Figure 1.12 Linear image restoration model.

matrix **K** contains over 60 000 million entries, but in practice it is sparse, i.e. many entries are zero. For shift-invariant systems, the matrix **K** has a periodic structure, which can be exploited to achieve efficient computation using the FFT (see Section 1.3).

1.5.3 Constrained least squares restoration

A number of well-known restoration filters is captured by considering the following *constrained least squares restoration* model. First, the estimated solution vector $\tilde{\mathbf{f}}$ should satisfy (1.7); this gives the constraint that the norms of $\mathbf{w} - \mathbf{K}\tilde{\mathbf{f}}$ and $\mathbf{n}$ are equal:

$$\| \mathbf{w} - \mathbf{K}\tilde{\mathbf{f}} \|^2 = \| \mathbf{n} \|^2 \tag{1.8}$$

Second, one may subject the solution $\tilde{\mathbf{f}}$ to additional constraints; this can be realized by introducing a matrix $\mathbf{Q}$ such that the term:

$$\| \mathbf{Q}\tilde{\mathbf{f}} \|^2 \tag{1.9}$$

is minimized. To take both contributions into account, we want to minimize:

$$\mathscr{E}(\tilde{\mathbf{f}}) = \| \mathbf{Q}\tilde{\mathbf{f}} \|^2 + \gamma^{-1}(\| \mathbf{w} - \mathbf{K}\tilde{\mathbf{f}} \|^2 - \| \mathbf{n} \|^2) \tag{1.10}$$

where γ is a constant that determines the relative weight of the terms (1.8) and (1.9). The solution of (1.10) is:

$$\tilde{\mathbf{f}} = (\mathbf{K}^t\mathbf{K} + \gamma\mathbf{Q}^t\mathbf{Q})^{-1}\mathbf{K}^t\mathbf{w} \tag{1.11}$$

where the superscript t denotes transposition of vectors or matrices. Special cases are the following:

(a) *Pseudoinverse filter*: take $\mathbf{Q} = \mathbf{I}$, the identity matrix. Thus we minimize the norm of $\tilde{\mathbf{f}}$ subject to the constraint (1.8). The solution (1.11) then reduces to:

$$\tilde{\mathbf{f}} = (\mathbf{K}^t\mathbf{K} + \gamma\mathbf{I})^{-1}\mathbf{K}^t\mathbf{w}$$

An alternative formulation of this result makes use of the discrete Fourier transforms $K(u, v)$, $W(u, v)$, $F(u, v)$ of $k(x, y)$, $w(x, y)$, $f(x, y)$ in (1.6). The estimate $\tilde{F}(u, v)$ of $F(u, v)$ according to the pseudoinverse filter has the form:

$$\tilde{F}(u, v) = [|K(u, v)|^2 + \gamma]^{-1} K^*(u, v) W(u, v) \tag{1.12}$$

with $K^*(u, v)$ the complex conjugate of $K(u, v)$.

(b) *Inverse filter*: for $\gamma = 0$ the equation for the pseudoinverse filter reduces to the 'inverse' filter:

$$\tilde{\mathbf{f}} = (\mathbf{K}^t\mathbf{K})^{-1}\mathbf{K}^t\mathbf{w}$$

or, in terms of Fourier transforms,

$$\tilde{F}(u, v) = \frac{W(u, v)}{K(u, v)} = F(u, v) + \frac{N(u, v)}{K(u, v)} \tag{1.13}$$

where $N(u, v)$ is the Fourier transform of the noise term $n(x, y)$ in (1.6). In practice, inverse filtering is often useless, the reason being that the matrix $\mathbf{K}$ may be (almost) singular. Equivalently, this means that the Fourier transform $K(u, v)$ has values that are zero or very small at certain points (u, v) in the frequency plane, leading to amplification of the noise term in (1.13). An example is given below.

(c) *Parametric Wiener filter*: assume $\mathbf{f}$ and $\mathbf{n}$ to be random vectors with zero mean and covariance matrices $\mathbf{R}_f = \mathbb{E}(\mathbf{f}\mathbf{f}^t)$ for the signal and $\mathbf{R}_n = \mathbb{E}(\mathbf{n}\mathbf{n}^t)$ for the noise.

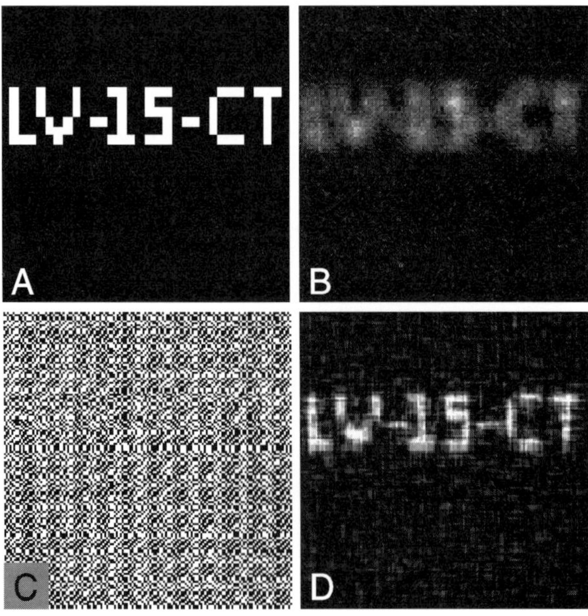

Figure 1.13 Effect of restoration filters: (A) original, (B) degraded image, (C) inverse filtering and (D) pseudoinverse filtering ($\gamma = 0.01$).

Let $\mathbf{Q}$ be the noise-to-signal ratio:

$$\mathbf{Q} = \left(\frac{\mathbf{R}_n}{\mathbf{R}_f}\right)^{\frac{1}{2}}$$

Then the solution (1.11) reads:

$$\tilde{\mathbf{f}} = (\mathbf{K}^t\mathbf{K} + \gamma\mathbf{R}_f^{-1}\mathbf{R}_n)^{-1}\mathbf{K}^t\mathbf{w}$$

For shift-invariant blur and stationary signal and noise, this filter is the parametric Wiener filter, which for $\gamma = 1$ reduces to the ordinary Wiener filter.

As an example, we present in Figure 1.13 the results of applying inverse and pseudoinverse filters to an image degraded by uniform blurring and Gaussian noise. As can be seen, the inverse filter is useless, because of noise amplification. The pseudoinverse filter, however, leads to satisfactory results.

1.6 MATHEMATICAL MORPHOLOGY

Mathematical morphology was developed at the Paris School of Mines as a set-theoretical approach to image analysis (Matheron, 1967; Serra, 1982). For a first introduction, see, for example, Giardina and Dougherty (1988), Haralick *et al.* (1987) or Serra and Vincent (1992). The method works by translating small subsets B, called *structuring elements*, of various forms and sizes over the image plane and recording the locations where certain relations between the image and the translated structuring element are satisfied. This can be compared to the neighbourhood operations of linear filtering (see Section 1.4.2), the main difference being that the operations are now *nonlinear*. Morphological operations are usually performed in a sequence, which is chosen in such a way that the end result is in a form suitable for quantitative measurements. First we look at some of the basic transforms for the case of binary images. After that, the case of grey value images is considered.

1.6.1 Morphology for binary images

Binary images have only two values, say, zero or one (black and white). This means that we can describe such images in terms of *sets* in the plane: all white pixels form a set X, and all black pixels the complement of X in the plane. The theory of mathematical morphology not only works for discrete images, but for continuous ones as well, with a unified formulation. In the continuous case one considers images as subsets of the Euclidean space $E = \mathbb{R}^2$, in the discrete case as subsets of the discrete grid $E = \mathbb{Z}^2$.

We add a brief reminder about sets and the basic set operations. A set X with elements $x_1, x_2, x_3, \ldots$ is denoted by $X = \{x_1, x_2, x_3, \ldots\}$. The expression $x \in X$ means that x is an element of X. The set Y consisting of those elements of X satisfying some condition C, is written $Y = \{x \in X : C\}$. For example, if $X = \{-2, -1, 0, 1, 2\}$, the elements of X that are positive can be selected as follows: $Y = \{x \in X : x > 0\} = \{1, 2\}$.

If X and Y are subsets of a certain universal set E (say, the plane), the union $X \cup Y$ contains all elements that occur in X or in Y (possibly in both), the intersection $X \cap Y$ contains all elements that occur both in X and Y, the set difference $X \backslash Y$ contains those elements of X that are not an element of Y, and the complement X^c contains all elements of E that are not in X, i.e. $X^c = E \backslash X$.

1.6.1a *Dilation and erosion*

Let A be a set in the plane (the structuring element). We write $x + y$ for the vector sum of x and y, and $-x$ for the reflected vector of x with respect to the origin. If we translate (i.e. shift) A along the vector h, we write the result as A_h, so:

$$A_h = \{a + h : a \in A\}$$

Also, if we reflect all points of a set A, we get the *reflected set* $\check{A}$, i.e.:

$$\check{A} = \{-a : a \in A\}$$

Now we can define the following elementary algebraic operations between two sets X and A:

(a) *Minkowski addition*:

$$X \oplus A = \{x + a : x \in X, a \in A\} = \bigcup_{a \in A} X_a = \bigcup_{x \in X} A_x$$

(b) *Minkowski subtraction*:

$$X \ominus A = \bigcap_{a \in A} X_{-a}$$

When the set A is fixed, the mapping $X \mapsto \delta_A(X) = X \oplus A$ is called the *dilation* with structuring element A, and the mapping $X \mapsto \varepsilon_A(X) = X \ominus A$ is called the *erosion* with structuring element A. There is a simple geometrical interpretation of these operations, expressed by the following formulas:

(a) *Dilation*:

$$X \oplus A = \{x \in E : (\check{A})_x \cap X \neq \varnothing\}$$

(b) *Erosion*:

$$X \ominus A = \{x \in E : A_x \subseteq X\}$$

In words, the dilation of X by A is the collection of points h such that $\breve{A}$ after translation over the vector h 'hits' the set X. Similarly, the erosion of X by A is the collection of points h such that after shifting A over the vector h it 'fits' into X. For the discrete case, an example for the dilation is given in Figure 1.14, and for the erosion in Figure 1.15. Pixels with value 1 (1-pixels) are represented by dots; pixels with value 0 by blanks. The position of the origin in the structuring element is indicated by the symbol $\ulcorner$ (reflecting the fact that we use a row–column orientated coordinate system).

Dilation and erosion are *translation-invariant*: if the origin of X is shifted over a vector h, followed by a dilation or erosion, the result is the same as first dilating or eroding the set, followed by translation. For example:

$$X_h \oplus A = (X \oplus A)_h$$

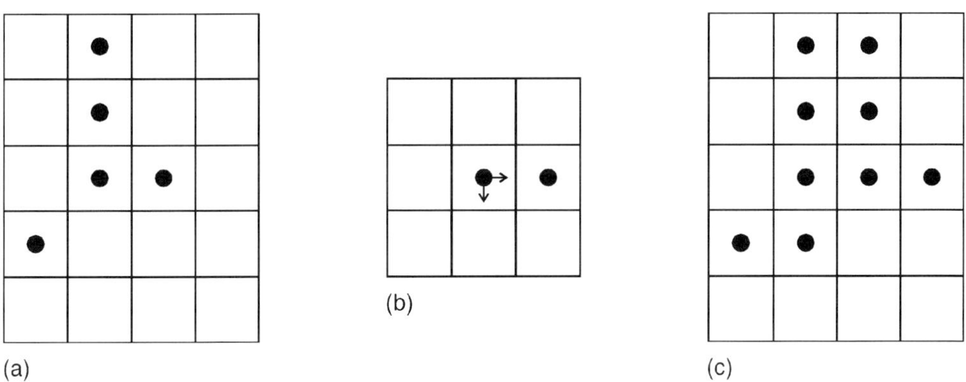

Figure 1.14 (a) Binary image X. (b) Structuring element A. (c) Dilation of X by A. Pixels with a value of 1 (1-pixels) are represented by dots, pixels with a value of 0 by blanks, and the position of the origin in the structuring element is indicated by the symbol $\ulcorner$.

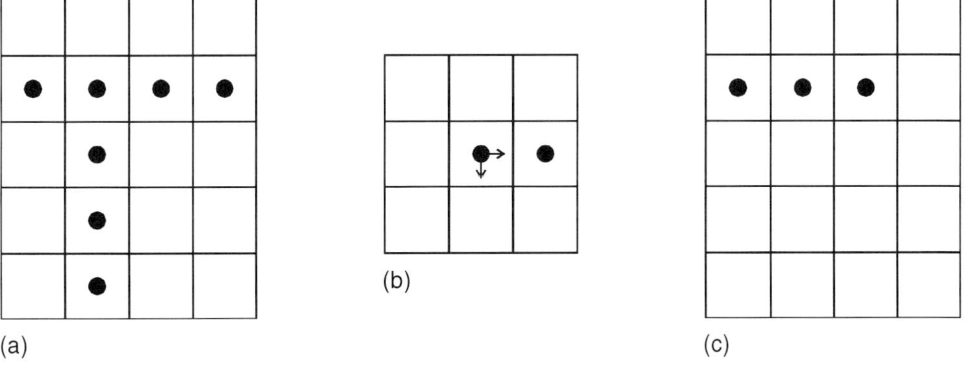

Figure 1.15 (a) Binary image X. (b) Structuring element A. (c) Erosion of X by A.

There exists a *duality relation* with respect to set-complementation:

$$X \oplus A = (X^c \ominus \check{A})^c$$

That is, dilating an image by A gives the same result as eroding the background by $\check{A}$, followed by taking the complement. Applications of binary dilation and erosion can be found in Sections 8.3.2c and 10.6.3.

1.6.1b The hit-or-miss transform

The hit-or-miss transform of an image X by a composite structuring element (A^1, A^2) is the collection of points h such that the shifted set A_h^1 fits into X and the shifted set A_h^2 has no overlap with X. Expressed in an equation:

$$\begin{aligned} X \circledast (A^1, A^2) &= \{h \in E : A_h^1 \subseteq X, A_h^1 \subseteq X^c\} \\ &= (X \ominus A^1) \cap (X^c \ominus A^2) \end{aligned}$$

This is a form of 'template matching'. This transformation can be used to detect special points in an image. An example is given in Figure 1.16, which shows an application where north-east pixels are detected.

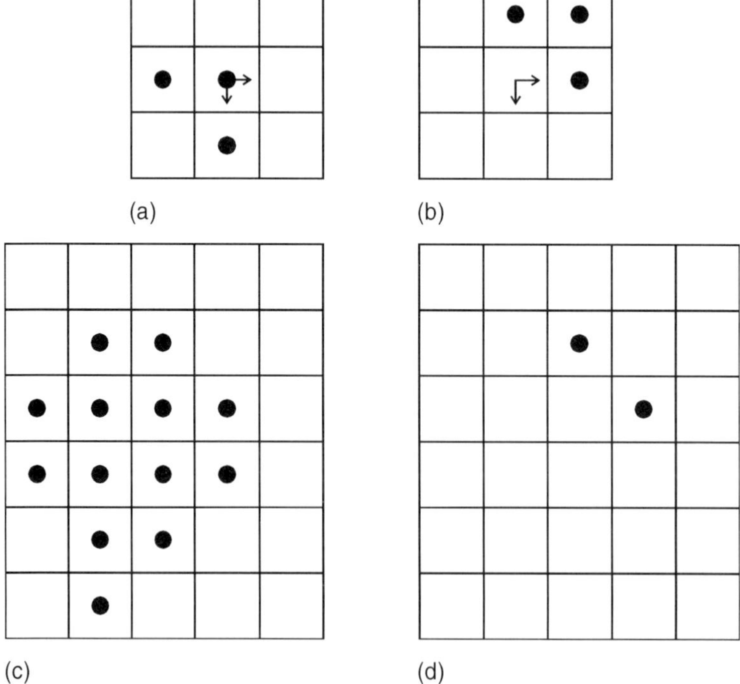

Figure 1.16 (a) Structuring element A^1. (b) Structuring element A^2. (c) Binary set X. (d) Hit-or-miss transform of X.

Erosion, or the hit-or-miss transform, can also be used to compute the genus or the number of regions in a binary image (Haralick and Shapiro, 1992; see also Section 1.8.2b).

1.6.1c *Openings and closings*

Opening and closing by a structuring element A are obtained by an erosion followed by a dilation, and conversely:

(a) *Opening*:

$$X \circ A = (X \ominus A) \oplus A$$

(b) *Closing*:

$$X \bullet A = (X \oplus A) \ominus A$$

It can be shown that the opening is the union of all shifted copies of the structuring element A which are included in the set X. An opening smooths contours, cuts narrow bridges, and removes small islands and sharp corners; a closing fills narrow channels and small holes. For the discrete case, an example is given in Figure 1.17. It is easy to see that opening and closing are translation-invariant, just like dilation and erosion. A translation of the structuring element A, however, does not affect the outcome.

1.6.1d *Thinnings and thickenings*

Also much used are the *thinning* and *thickening* of a set X by $B = (B^1, B^2)$ where B^1 and B^2 should be disjoint:

(a) *Thinning*:

$$X \circ B = X \backslash (X \circledast B) = X \cap (X \circledast B)^c$$

(b) *Thickening*:

$$X \odot B = X \cup (X \circledast B)$$

1.6.1e *Distance transforms*

On the discrete grid we use a *digital* distance function, which depends on the type of connectivity: if 4-connectivity is used, each of the four neighbours is by definition at distance 1 from the centre pixel. This is called the city-block distance. We note that other digital metrics have been considered involving so-called *chamfer distances*, which approximate the Euclidean distance very well (Borgefors, 1984) (see Section 8.3.2c for an application).

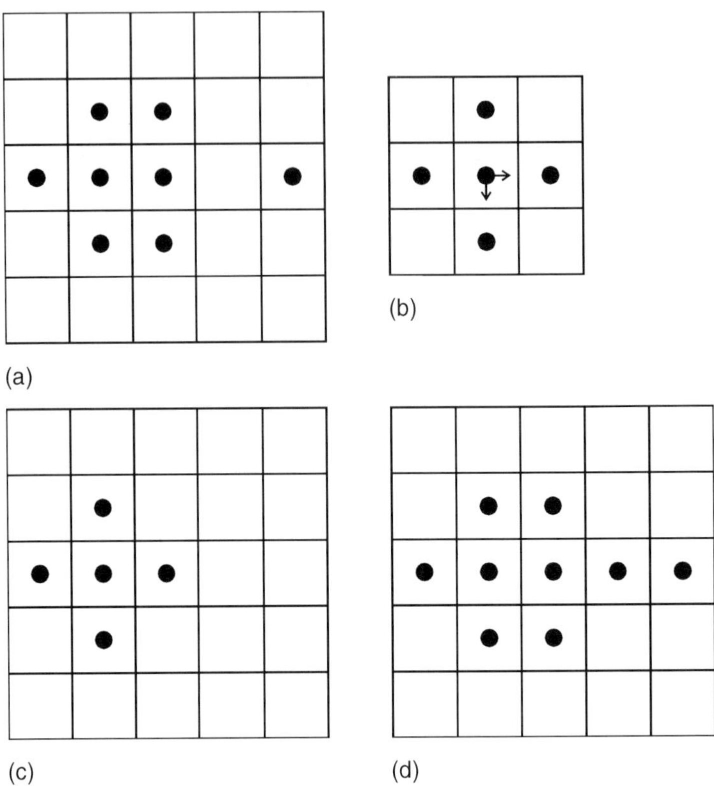

Figure 1.17 (a) Binary image X. (b) Structuring element A. (c) Opening of X by A. (d) Closing of X by A.

Paths on the grid are formed by steps in the horizontal and vertical directions (each of length 1). The length of a path is the sum of the lengths of the individual steps. The distance $d(x, y)$ between two points x and y on the grid is then the minimum of the lengths of all allowed paths from x to y on the grid. We also define the distance $d(x, A)$ between a point x and a set A as the minimum of the distances between x and points of A.

Let A be a discrete binary image. The *distance transform* of A is the grey value function $D : E \longrightarrow \mathbb{N}$ which associates to point x the distance of x to the complement of A: $D(x) = d(x, A^c)$. Notice that $D(x) = 0$ for points $x \in A^c$. The distance transform can be computed by a two-pass procedure (Rosenfeld and Pfaltz, 1968), consisting of a forward scan (top-down, left-to-right), which takes only steps to the left or upwards into account, followed by a reverse scan (bottom-up, right-to-left), which considers steps to the right and downwards. An example is shown in Figure 1.18.

1.6.1f Skeletonization

For purposes of shape description and efficient coding it is desirable to extract features from binary images that represent only the 'backbone' or skeleton of the set.

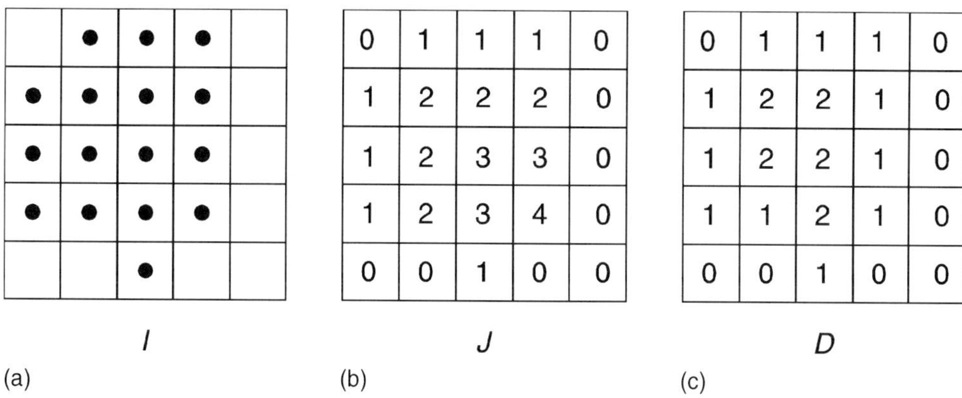

Figure 1.18 (a) Digital image I. (b) Forward scan output image J. (c) City-block distance transform of I produced by reverse scan of J.

Roughly speaking, these are connected lines within the binary image which are in the middle of the set and 'as thin as possible'. Applications of skeletonization can be found in Chapter 14.

For a binary image A one can define the so-called SKIZ ('skeleton by influence zones'), which is a skeleton of the background. Let:

$$A = \bigcup_{n=1}^{N} C_n$$

with $C_1, C_2, ..., C_N$ the connected components of A (see Section 1.2.4). The zone of influence Z_n of C_n is the set of points closer to C_n than to any other component:

$$Z_n = \{x : d(x, C_n) < d(x, C_m), m \neq n\}$$

The points that do not belong to any Z_n (points that are equidistant to at least two components) form the SKIZ S_z of A, or the skeleton of the background A^c:

$$S_z(A) = \left(\bigcup_{n=1}^{N} Z_n \right)^c$$

See Figure 1.19 for an example.

1.6.2 Morphology for grey value images

Morphological operations for grey value images can be defined in analogy with the binary case. The basic idea is to consider a grey value image as a landscape in three-dimensional space, where the vertical dimension denotes grey value.

We can define morphological transformations acting on the image f by introducing the concept of *umbrae* ('shadows') (Sternberg, 1986). For a given grey value image f, the umbra of f is the set $U(f)$ consisting of all points below the graph of f (when one

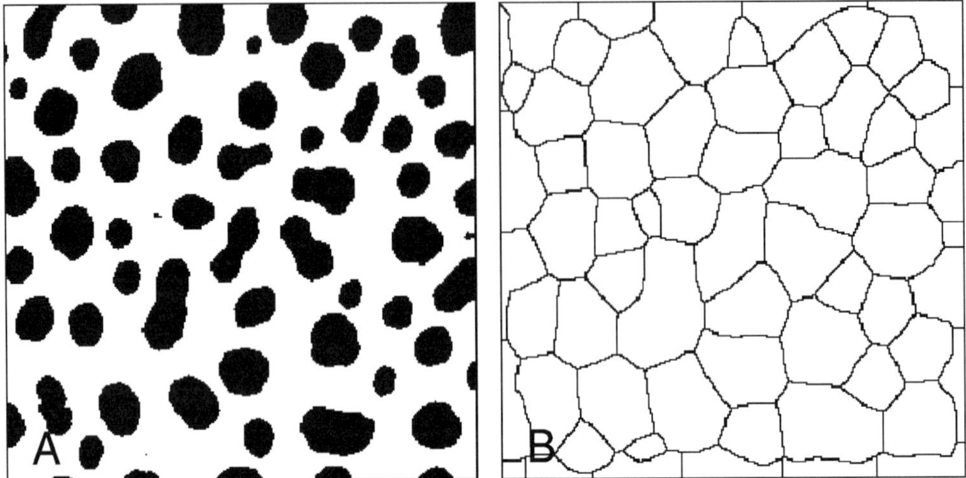

Figure 1.19 Skeletonization: (A) binary image and (B) SKIZ of (A).

imagines a light source above the landscape, the umbra is made up of all points 'in the shadow'). Now one may apply a binary morphological operation to the set $U(f)$, which yields a transformed set. One can again associate a grey value image to this set by looking at the *top surface* of the transformed set. More details can be found in Serra (1988) or Haralick and Shapiro (1992). For a mathematical treatment, see, for example, Heijmans (1994). Here we confine ourselves to a presentation of some of the most important morphological operations for grey value images.

1.6.2a Grey value dilation and erosion

Let f and k be two grey value images with domain F and K, respectively. The *grey value dilation* $f \oplus k$ and the *grey value erosion* $f \ominus k$ of f by k are the functions defined by:

$$(f \oplus k)(x) = \max_{y \in K,\; x-y \in F} [f(x-y) + k(y)]$$

$$(f \ominus k)(x) = \min_{y \in K,\; x+y \in F} [f(x+y) - k(y)]$$

That is, $f \oplus k$ equals the maximum of $f(x-y) + k(y)$ when y runs over the set K, and similarly for $f \ominus k$.

A special case occurs for so-called *flat* structuring functions, which have a constant value of zero on a domain K. Then we get flat grey value dilation and erosion, which simply replace each pixel by the maximum or minimum in a neighbourhood defined by a structuring element K:

$$(f \oplus k)(x) = \max_{y \in K,\; x-y \in F} [f(x-y)]$$

$$(f \ominus k)(x) = \min_{y \in K,\; x+y \in F} [f(x+y)]$$

For this reason these operations are called *minimum* and *maximum* filters.

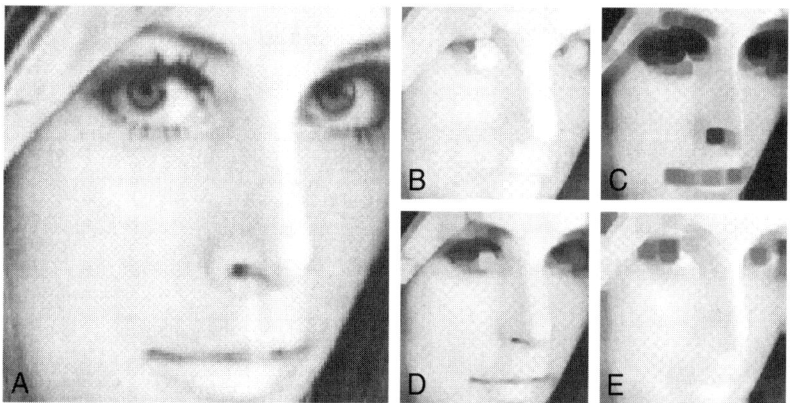

Figure 1.20 (A) A grey value image. (B–E) (Reduced by a factor of 2): (B) grey value dilation, (C) grey value erosion, (D) grey value opening and (E) grey value closing.

1.6.2b Grey value opening and closing

By taking products of grey value dilation and erosion we can construct grey value openings and closings just as in the binary case. For flat structuring functions with domain K, the *grey value opening* $f \circ K$ and the *grey value closing* $f \bullet K$ are defined by:

$$(f \circ K)(x) = \max_{z \in K} \min_{y \in K} f(x - z + y)$$

$$(f \bullet K)(x) = \min_{z \in K} \max_{y \in K} f(x + z - y)$$

The grey value opening and closing filters are also called *min–max* and *max–min* filters, respectively. Opening eliminates peaks, closing valleys. Figure 1.20 shows the result of grey value dilation, erosion, opening and closing for a grey value image.

If we apply grey value opening with a flat structuring element to an image f and subtract the result from f, we get Meyer's top-hat transform (Serra, 1982). This transform extracts the peaks from a grey value image (see Sections 6.4.2, 6.7.2 and 14.2).

1.7 IMAGE SEGMENTATION

The process of isolating the important objects in the image from the background is called image segmentation (Haralick and Shapiro, 1985). More precisely, image segmentation can be defined as the process of partitioning an image into disjoint regions. Each region should be uniform and homogeneous with respect to some property, such as grey value or texture, and differ significantly from neighbouring regions.

The following subdivision of image segmentation methods can be made:

(a) *Region based approach*: assign pixels to regions or objects.
(b) *Boundary based approach*: locate boundaries between regions.
(c) *Edge based approach*: identify edge pixels and link them to form boundaries.

A review of image segmentation techniques in microbiology is given in Chapter 6.

1.7.1 Region based approaches

1.7.1a Thresholding

Let $f(x, y)$ be a grey value image. Let t be a fixed grey value called the 'threshold'. Then we define a binary image f_t by thresholding as follows:

$$f_t(x, y) := \begin{cases} 1 & \text{if } f(x, y) \geqslant t \\ 0 & \text{if } f(x, y) < t \end{cases}$$

In this method, pixels above a certain value t – the 'threshold' – are declared to be 'object'; those below the threshold 'background'. This will work well when grey levels of both objects and background are relatively uniform, but distinct. When the background grey level is not constant one may use a threshold that varies over the image: this is called *adaptive thresholding*.

An important problem is to determine the threshold automatically. This can be done by considering the histogram of the grey values of the image (see Section 1.4.1b). If the histogram is bimodal, i.e. has two peaks, an obvious choice for the threshold is a value at the local minimum in the valley between the two maxima (see Figure 1.21; see also Section 6.7.3).

When the histogram does not have a convenient bimodal shape, one may determine a global threshold statistically, by minimizing *within group variance* (Haralick and Shapiro, 1992). This means splitting the set of grey values into two disjoint subsets V_1 and V_2 in such a way that the weighted sum of the variances of the groups V_1 and V_2 (which are measures of homogeneity) is minimal.

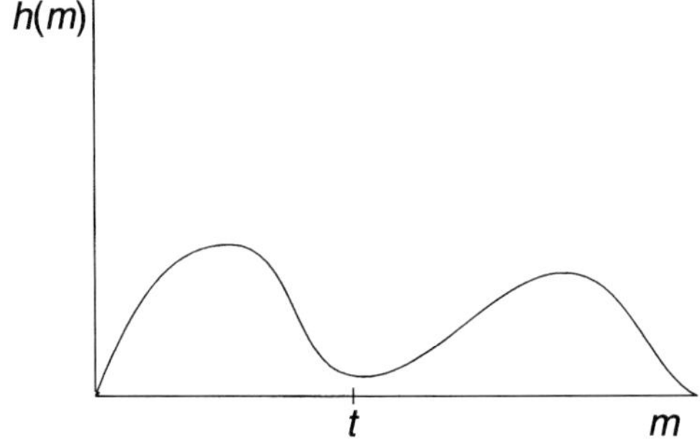

Figure 1.21 Choice of the threshold t for a bimodal histogram.

1.7.1b Watershed segmentation

The standard approach to image segmentation in morphological image processing (see Section 1.6) is by the watershed transform originally proposed by Lantuéjoul (Serra, 1982; Vincent and Soille, 1991; Beucher and Meyer, 1993). The intuitive idea underlying this method is that of flooding a landscape or topographic relief with water, by piercing holes at the positions of local minima and immersing the landscape into a lake. Basins will fill with water starting at local minima, and at points where water coming from different basins would meet, dams are built. When the water level has reached the highest peak in the landscape, the process is stopped. The set of dams thus obtained partitions the region into 'catchment basins' separated by dams. These dams (more precisely, their projections on the horizontal plane) are called watershed lines or simply *watersheds*.

A recursive algorithm for computing the watershed transform was given by Vincent and Soille (1991). The basic structure of the algorithm is a loop in which the image is thresholded at successive grey levels. In every iteration the current basins belonging to the minima are extended by their influence zones within the binary image obtained by thresholding at the current grey level (compare with the computation of the SKIZ).

Other approaches exist based on a definition of watersheds in terms of distance functions (Meyer, 1994; Najman and Schmitt, 1994; Meijster and Roerdink, 1996).

To separate the regions in a grey value image, one usually first computes the derivative of the image using some edge detector, before applying the watershed. In practice, the watershed transform suffers from severe *over-segmentation*; that is, the creation of many small basins in regions with many local minima. Therefore, one usually has to use some pre-processing method to reduce the number of local minima, for example by first applying some smoothing filter.

As an application, we show in Figure 1.22 segmentation of images of coffee beans (see Vincent, 1990). The goal is to segment the image in non-overlapping regions, each containing exactly one coffee bean. The most important steps to achieve this are the following. First segment the image in foreground and background by simple thresholding, and remove some small holes inside the beans (Figure 1.22B). Note that using the connected components in this binary image directly will not work, because some of the beans overlap. To separate these overlapping beans we first apply a distance transform to the inverted image of Figure 1.22B. This will yield a single maximum in the centre of the isolated coffee beans, and two (or more) maxima in overlapping beans (Figure 1.22C). Then invert the distance transformed image so that maxima become minima, and use these minima as points for starting a watershed algorithm. The resulting watershed lines are shown in Figure 1.22D. Finally, we show in Figure 1.22E the original image with the watershed lines superimposed.

1.7.1c Region growing, split-and-merge

A merging method starts with many small regions subdividing the image. In each region one computes specific properties which represent membership to objects. Next, boundaries are merged if their strength, defined according to certain criteria,

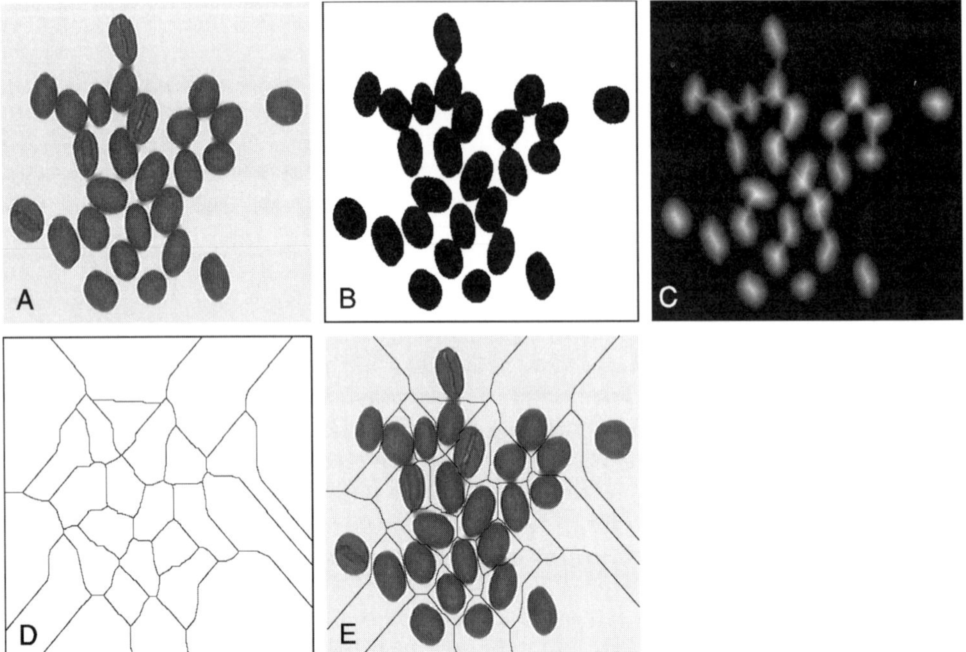

Figure 1.22 Segmenting an image of coffee beans: (A) original image, (B) original after thresholding and small hole removal, (C) distance transform of the inverse of (B), (D) watershed lines of the inverse of (C) and (E) original image with watershed lines superimposed.

is below a given threshold; that is, if the corresponding regions are similar enough. This process is repeated until no more merging takes place.

One can also start with the entire image as the initial region, and successively split regions into sub-regions if they are not homogeneous enough. Since this method tends to produce sub-optimal boundaries, a combination of splitting and merging operations, called the split-and-merge strategy, is often used.

1.7.2 Boundary based approaches

Boundary techniques attempt to find the edges directly by computing points with high gradient magnitude.

1.7.2a Boundary tracking

After taking the gradient of the initial image, the pixel with the highest value is located, and a boundary tracking process is started at this pixel. Then, new boundary points are computed iteratively by searching in the local neighbourhood of the current point for the point – not yet chosen as a boundary point – with maximum grey level.

If several candidate points are available, one chooses arbitrarily. To combat noise, the image can be smoothed before tracking, or one may use a larger local window within which searching takes place. An application of boundary tracking can be found in Section 13.4.5.

1.7.2b Laplacian filtering

This filter was mentioned above as a sharpening filter for image enhancement (see Section 1.4.2b). The Laplacian is a second derivative operator ∇^2 defined by:

$$\nabla^2 f(x, y) = \left(\frac{\partial^2}{\partial x^2} + \frac{\partial^2}{\partial y^2}\right) f(x, y)$$

which is linear and shift-invariant; its transfer function is zero at the origin in frequency-space. Therefore, after Laplacian filtering the image will have zero average grey value. Laplacian filtering of an image will produce a zero-crossing at an edge (Figure 1.23). In the presence of noise, filters such as these, which are based upon differentiation, lead to very unstable results.

One solution is first to convolve the image with a smoothing kernel, such as a Gaussian, prior to differentiation. All derivatives of the image undergo the same Gaussian smoothing process, which is equivalent to convolving the image with derivatives of a Gaussian. The *Canny edge detector* computes the first derivatives of the Gaussian-smoothed image. Edges are identified as locations where the gradient magnitude has a maximum.

Another commonly used edge detector is the *Marr–Hildreth* operator, which uses second derivatives; the convolution kernel $K_\sigma(x, y)$ is the Laplacian-of-Gaussian (LoG), with σ the standard deviation of the Gaussian kernel:

$$K_\sigma(x, y) = -\frac{2}{\pi\sigma^4}\left(1 - \frac{r^2}{2\sigma^2}\right) e^{-r^2/2\sigma^2} \qquad (r^2 = x^2 + y^2)$$

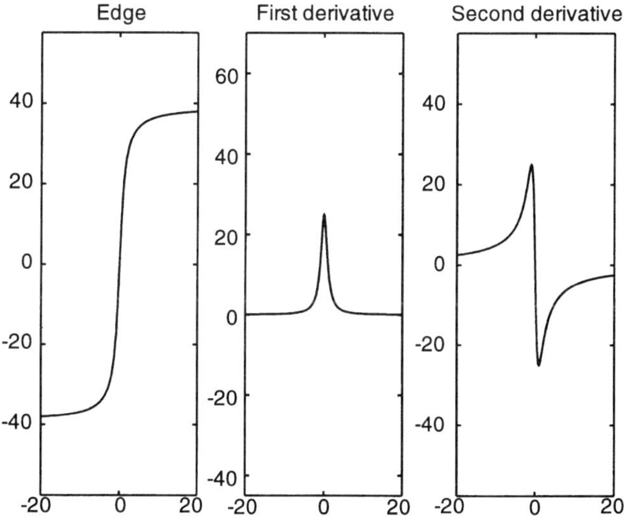

Figure 1.23 Derivatives of an edge.

Edges of f are identified as zero-crossings of $K_\sigma f$. The LoG is a 'Mexican hat'-shaped function (Figure 1.24). An application to images of bacteria can be found in Section 6.7.5.

1.7.3 Edge based approaches

These approaches consists of two phases. The first is to determine for each pixel whether it is on the boundary of an object using some appropriate criterion. The result is called an *edge image*. This will not result in closed contours in general. To obtain these, a second step is required in which edge points are linked.

1.7.3a Detecting edge points

Edge points are determined by looking at the magnitude and direction of the gradient of the input image. Most methods use local convolution with a set of directional derivative masks, as in the case discussed above in Section 1.7.2b. Other examples of edge operators were presented in Section 1.4.2.

1.7.3b Edge linking

For strong edges and low noise, the edge image may be thresholded and a thinning applied (Section 1.6.1d) to the resulting binary image until closed contours are obtained with thickness of one pixel. In general the edge image will contain gaps which must be closed. When the gaps are small, some heuristic search procedure for

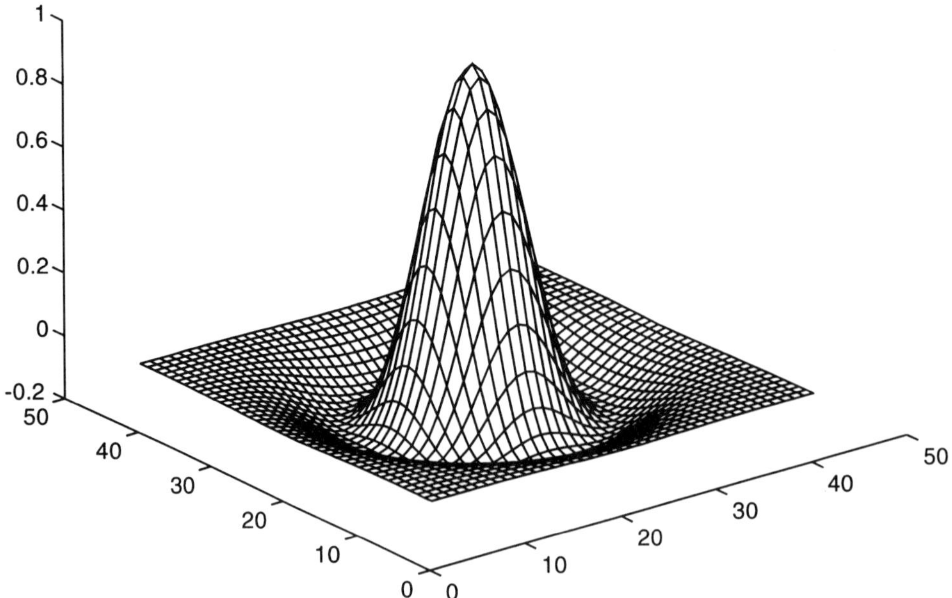

Figure 1.24 Surface plot of the Laplacian-of-Gaussian function.

finding other pixels to link to may be used. For sparse edge points, curve fitting to form boundaries may be used. Straight-line fitting can be performed conveniently by using the *Hough transform*, which maps a straight line $y = mx + b$ to polar coordinates ρ, θ by the relation $\rho = x \cos \theta + y \sin \theta$. Consider a set of edge points (x_i, y_i) lying on a straight line with parameters ρ_0, θ_0. Each point is represented in ρ, θ space by a sinusoidal curve, but all these curves go through a common point ρ_0, θ_0. Thus, a straight line through a number of edge points can be found by looking for a local maximum of the histogram representing the points in ρ_0, θ_0 space. Note that the Hough transform can also be used to detect other shapes, e.g. circles (see Section 8.3.1).

1.8 DESCRIPTION AND MEASUREMENT

After image segmentation the resulting collection of regions is usually represented and described in a form suitable for higher level processing. In this section we outline the most important representations based on shape or texture. In the next stage, image descriptors based on the chosen representation are presented. A desirable property of these descriptors is that they should be insensitive to changes in size, translation or rotation. The actual measurement of features in digital images makes use of many of the techniques discussed above, such as linear or morphological image operators.

1.8.1 Representations

Roughly speaking, representations can be given in terms of the boundary or the interior of the regions (Castleman, 1996, chapter 11).

1.8.1a Boundary representations

Chain codes are strings of integers used to represent a boundary by a connected sequence of straight-line segments of specified length and direction. The direction of each line segment is coded using a numbering scheme, which may be adapted to the connectivity used. The accuracy of the straight-line representation depends on the spacing of the sampling grid.

A digital boundary can also be approximated by a polygon, possibly with minimum length. Computing such rubber-band approximations directly within the grey value image has recently become popular under the name of active contour models, or *snakes* (Kass *et al.*, 1988).

1.8.1b Region representations

An important shape representation of regions is obtained by computing a *skeleton*, based on the medial axis (Blum, 1973), thinning algorithms (Section 1.6.1d) or morphological skeletons (see Section 1.6.1f). The goal of skeletonization is reduction of information while keeping the essential characteristics of the image (see Section 14.2.2).

1.8.2 Descriptors

1.8.2a Boundary descriptors

Simple descriptors of boundaries are length, diameter (maximal distance between points on the contour) or curvature. Length can be approximated simply by counting the number of pixels along the contour. Curvature can be computed from the chain code. *Shape numbers* of a boundary can be defined in terms of its chain code (Gonzalez and Woods, 1992).

Fourier descriptors are based upon the representation of the boundary as a sequence $s(k) = x(k) + iy(k)$, $k = 0, 1, \ldots, N - 1$ where $(x(k), y(k))$ are coordinates of points on the contour. Taking a discrete Fourier transform of the vector $s(0), s(1), \ldots, s(N - 1)$ yields N complex coefficients called the Fourier descriptors of the boundary. Approximations of the boundary are obtained by only using $M < N$ of the Fourier coefficients and performing an inverse discrete Fourier transform.

The *fractal dimension* of a (compact) set can be computed as follows (Barnsley 1988). Let $\mathcal{N}(A, \varepsilon)$ denote the smallest number of closed balls of radius $\varepsilon > 0$ needed to cover the set A. Then:

$$D = \lim_{\varepsilon \to 0} \left(\frac{\log(\mathcal{N}(A, \varepsilon))}{\log(1/\varepsilon)} \right)$$

is called the fractal dimension of A. The fractal dimension of a boundary can be estimated using morphological operations on real images through the concept of Minkowski dimension, which is based upon the Minkowski addition (Serra, 1982, chapter 5).

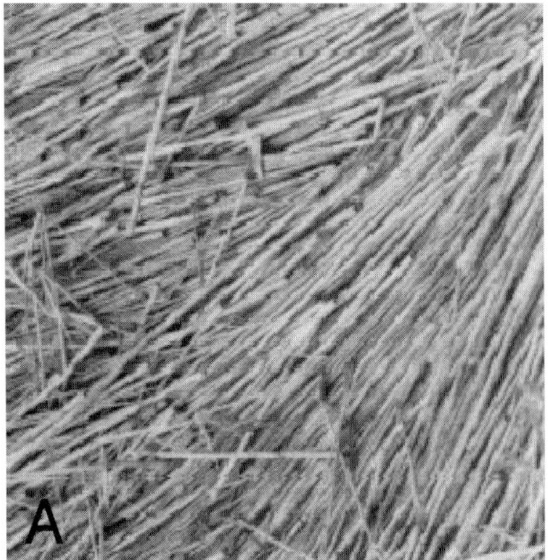

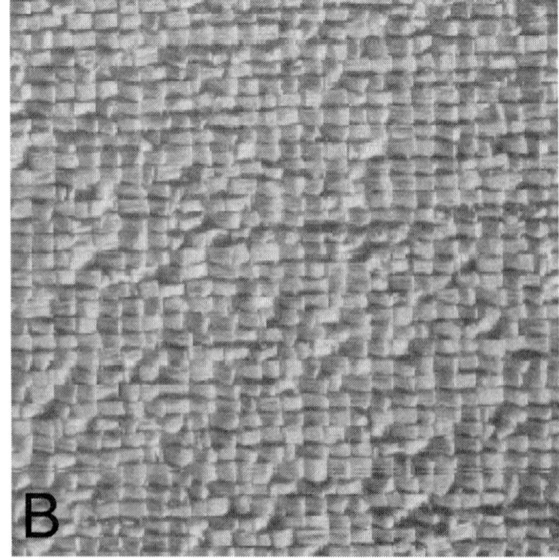

Figure 1.25 Textures: (A) straw and (B) raffia.

1.8.2b Regional descriptors

Simple descriptors based upon regions are area, perimeter or derived quantities such as perimeter squared divided by area, which is a *circularity measure* (Castleman, 1996, section 19.3). Such measures are applied throughout this volume (see Sections 7.5, 8.3, 9.2, 14.2 and 18.3.9).

Topological descriptors are invariant to a large class of local deformations. Examples are number of connected components, number of holes, or the genus or Euler number (number of connected components minus the number of holes). These quantities can be measured digitally by using erosions or hit-or-miss transforms as introduced in Section 1.6.1; see also Haralick and Shapiro (1992).

Texture of a region refers to the spatial distribution of discrete grey value variations, described in terms of uniformity, coarseness, regularity and directionality. Some examples are given in Figure 1.25.

The three main approaches used to describe texture are: (i) statistical, (ii) structural, viewing a texture as an arrangement of texture primitives, and (iii) spectral, using the Fourier transform to detect global periodicities.

Moments of an image $f(x, y)$ are integrals over its domain with respect to polynomial weight functions:

$$M_{m,n} = \iint x^m y^n f(x, y)\, \mathrm{d}x\, \mathrm{d}y, \qquad m, n = 1, 2, \ldots$$

Invariant moments are combinations of several $M_{m,n}$ which are invariant to translation, rotation and scale-change.

1.9. SUMMARY

In this chapter we have introduced the basics of digital image processing, concentrating on the main concepts and terminology used in the field. Aspects of both linear and nonlinear image processing have been presented, such as image enhancement, spatial filtering, restoration, morphological image processing, segmentation and description.

Although many aspects have been treated only in a very global way, it is hoped that this chapter will provide to the reader sufficient background to understand the basis of the image processing techniques to be found in the later chapters of this volume.

REFERENCES

Andrews, H.C. and Hunt, B.R. (1977) *Digital Image Restoration.* Prentice Hall: Englewood Cliffs, NJ.

Barnsley, M.F. (1988) *Fractals Everywhere.* Academic Press: New York.

Beucher, S. and Meyer, F. (1993) The morphological approach to segmentation: the watershed transformation. In *Mathematical Morphology and Image Processing* (Dougherty, E.R., ed.), pp. 433–81. Marcel Dekker: New York.

Blum, H. (1973) Biological shape and visual science (part I). *J. Theor. Biol.* **38**: 205–87.

Borgefors, G. (1984) Distance transformations in arbitrary dimensions. *Comp. Vis. Graph. Image Proc.* **27**: 321–45.

Castleman, K.R. (1996) *Digital Image Processing*. Prentice Hall: Englewood Cliffs, NJ.

Giardina, C.R. and Dougherty, E.R. (1988) *Morphological Methods in Image and Signal Processing*. Prentice Hall: Englewood Cliffs, NJ.

Gonzalez, R.C. and Woods, R.E. (1992) *Digital Image Processing*. Addison-Wesley: Reading, MA.

Haralick, R.M. and Shapiro, L.G. (1985) Survey: image segmentation techniques. *Comp. Vis. Graph. Image Proc.* **29**: 100–32.

Haralick, R.M. and Shapiro, L.G. (1992) *Computer and Robot Vision*. Addison-Wesley: Reading, MA.

Haralick, R.M., Sternberg, S.R. and Zhuang, X. (1987) Image analysis using mathematical morphology. *IEEE Trans. Pattern Anal. Mach. Intell.* **9**: 532–50.

Heijmans, H.J.A.M. (1994) *Morphological Image Operators*, vol. 25 of *Advances in Electronics and Electron Physics*, Supplement. Academic Press: New York.

Jain, A.K. (1989) *Fundamentals of Digital Image Processing*. Prentice Hall: Englewood Cliffs, NJ.

Kass, M., Witkin, A. and Terzopoulos, D. (1988) Snakes: active contour models. *Int. J. Comp. Vis.* **1**: 321–31.

Matheron, G. (1967) *Eléments pour une Théorie des Milieux Poreux*. Masson: Paris.

Meijster, A. and Roerdink, J.B.T.M. (1996) Computation of watersheds based on parallel graph algorithms. In *Mathematical Morphology and its Applications to Image and Signal Processing* (Maragos, P., Shafer, R.W. and Butt, M.A., eds), pp. 305–21. Kluwer: Dordrecht.

Meyer, F. (1994) Topographic distance and watershed lines. *Sign. Proc.* **38**: 113–25.

Najman, L. and Schmitt, M. (1994) Watershed of a continuous function. *Sign. Proc.* **38**: 99–112.

Rosenfeld, A. (1970) Connectivity in digital pictures. *J. Ass. Comp. Mach.* **17**: 146–60.

Rosenfeld, A. and Kak, A. (1982) *Digital Picture Processing*, vols. I and II. Academic Press: New York.

Rosenfeld, A. and Pfaltz, J. (1968) Distance functions on digital pictures. *Patt. Recogn.* **1**: 33–61.

Serra, J. (1982) *Image Analysis and Mathematical Morphology*. Academic Press: New York.

Serra, J. (1988) *Image Analysis and Mathematical Morphology*, vol. 2: *Theoretical Advances*. Academic Press: New York.

Serra, J. and Vincent, L. (1992) An overview of morphological filtering. *Circuits Systems Sign. Proc.* **11**: 47–108.

Sternberg, S.R. (1986) Grayscale morphology. *Comp. Vis. Graph. Image Proc.* **35**, 333–55.

Vincent, L. (1990) Algorithmes Morphologiques a Base de Files d'Attente et de Lacets. Extension aux Graphes, PhD thesis, Ecole Nationale Supérieure des Mines de Paris, Fontainebleau.

Vincent, L. and Soille, P. (1991) Watersheds in digital spaces: an efficient algorithm based on immersion simulations. *IEEE Trans. Pattern Anal. Mach. Intell.* **13**: 583–98.

2

Image Detectors for Digital Image Microscopy

Lucas J. Van Vliet[1], Frank R. Boddeke[1], Damir Sudar[2] and Ian T. Young[1]
[1]*Delft University of Technology, Delft, The Netherlands*
[2]*Lawrence Berkeley National Laboratory, California, USA*

2.1 INTRODUCTION

Digital image microscopy is more than putting a camera on top of a microscope to produce pretty pictures for reproduction or archiving. Its goal is to analyse a wide variety of 'analog' quantities from digitized data as accurately and precisely as possible. Recent advances in molecular biology and biochemistry have made it possible to tag selectively specific parts of cells or cellular constituents. For example, using fluorescence *in situ* hybridization (FISH), specific target sequences of DNA molecules can be fluorescently labelled. For very small sequences ($\sim$1k base pairs) these signals are very localized and extremely weak. To facilitate imaging we need state-of-the-art instrumentation and image analysis software. An item that is frequently overlooked is the image sensor. This chapter focuses on the properties of charge-coupled device (CCD) based image sensors ranging from video cameras to scientific CCD cameras and their applicability in low light level quantitative fluorescence microscopy. Modern image sensors are based on solid-state technology. Hence we limit ourselves to CCD-based camera systems. Similar to photomultipliers (point imaging) and intensified vidicon tubes (scanning), three generations of image intensifiers for CCD cameras are widely available.

This chapter is neither a comprehensive market survey nor a buyer's guide. As of this writing, sensor systems such as those discussed in this chapter can be bought from a variety of manufacturers and vendors. This chapter may help you decide which properties are useful for your application in fluorescence microscopy. In Section 2.2 we introduce the duality of light, as waves and as particles. Section 2.3 describes the various types of CCD image sensors. The basic properties of CCD cameras that characterize the performance of the sensor are presented in Section 2.4. Section 2.5 deals with image intensifiers that can be combined with any type of CCD sensor for various applications.

Digital Image Analysis of Microbes: Imaging, Morphometry, Fluorometry and Motility Techniques and Applications. Edited by M.H.F. Wilkinson and F. Schut.

More details about techniques for CCD camera characterization that are also used in this chapter can by found in Mullikin *et al.* (1994) and Van Vliet *et al.* (1997). Other reading material includes an introduction to image sensors, image formation and image processing (Castleman, 1996), principles of fluorescence microscopy (Inoue, 1986; Young, 1989), a general description of scientific CCD cameras (Aikens *et al.*, 1989) and an introduction to image intensifiers (Csorba, 1985).

2.2 LIGHT: WAVES AND PARTICLES

A wide variety of CCD cameras are used in light microscopy as image sensors. Before exploring the various properties of CCD cameras, we will explain the physical descriptions of light that are needed to understand the process of image formation and image acquisition. Physics teaches us two descriptions of light: as waves and as particles. Both descriptions of light yield fundamental limits to the quality of the image that can be observed. A state-of-the-art camera neither reduces the optical resolution below the diffraction limit, nor does it add a substantial amount of noise on top of the photon noise.

2.2.1 Wave description of light

The wave description of light allows us to explain diffraction. Diffraction limits the optical resolution of a microscope. An ideal lens is a lens without aberrations (see Section 3.5). Such a lens allows us to model the imaging process as a linear shift-invariant (LSI) system followed by a pure magnification system. The definitions and consequences of LSI systems are discussed in Castleman (1996) and Young (1989); see also Section 1.3.1a). The LSI system can be fully characterized by its impulse response: the point-spread function (PSF). Each point source in the object plane is replaced by a scaled and translated PSF in the image plane. The PSF depends on the size and shape of the lens aperture (usually specified as numerical aperture, NA) and the wavelength of light (λ) being imaged. A circular aperture yields a circularly symmetric PSF or impulse response, also called an Airy disc (Figure 2.1). Each impulse response has a corresponding optical transfer function (OTF) – they form a Fourier transform pair (see Section 1.3.1b). The OTF describes how the spatial frequencies in the image are modified as they pass through the optical system.

A uniformly illuminated field (which has spatial frequency $f = 0$) passes through an optical system unaltered: $\text{OTF}(0) = 1$. Any object can be thought of as a linear superposition of spatial frequencies, which pass through the optical system with reduced amplitude. The OTF of the lens is bandlimited, i.e. there exists a highest frequency f_c ($f_c = 2\text{NA}/\lambda$) above which the OTF equals zero. Thus the lens acts as a lowpass filter with cutoff frequency f_c. A higher cutoff frequency yields a 'crisper' image. The cutoff frequency is proportional to NA/λ. For example, an oil immersion lens with $\text{NA} = 1.25$ and green light with a wavelength of 500 nm yields a cutoff frequency of $f_c = 5.0$ cycles per micrometre.

If we are to work with a digitized (sampled) image, then the Nyquist sampling theory requires the number of image samples per micrometre to be greater than twice the cutoff frequency. Satisfying this condition guarantees that the analog

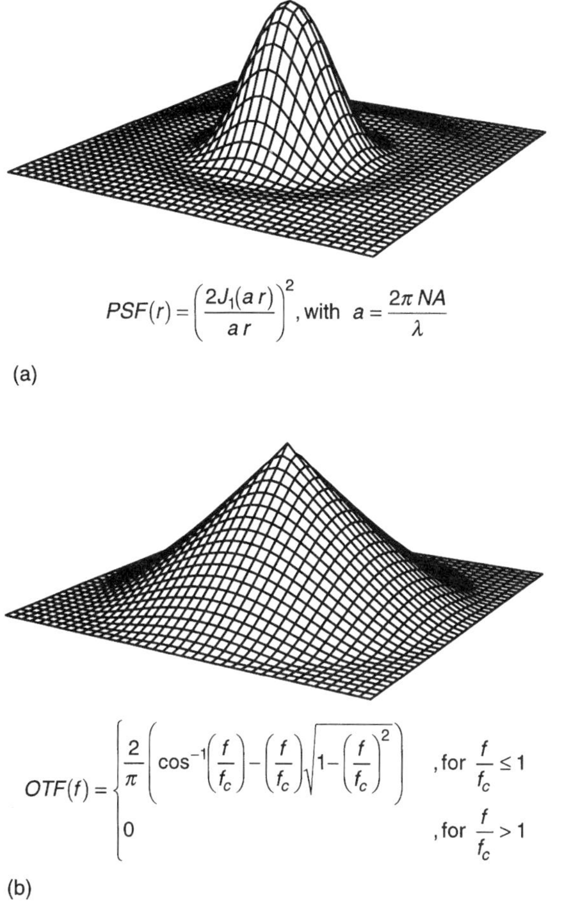

Figure 2.1 (a) The PSF or Airy disc and (b) the OTF of diffraction limited optics.

image (also visible through the eyepieces of the microscope) is accurately represented by the image samples. For the above example with a cutoff frequency of 5.0 cycles per micrometre, the Nyquist sampling rate is 10 samples per micrometre (requiring an image scale of 0.1 μm/pixel). The combination of microscope and camera needs to satisfy this spatial sampling rate. The overall optical magnification divided by the pixel size needs to be larger than the Nyquist rate. This favours CCDs with small pixels over CCDs with large pixels.

2.2.2 Quantum nature of light

The quantum nature of light explains the noisy images that often occur in low light level situations. Light can also be considered as a series of particles called photons. Each photon carries a certain amount of energy, $E = h\nu = hc/\lambda$, where h is Planck's constant from quantum mechanics and c the speed of light. Scientific CCD cameras are sensitive enough to detect, store and count individual incoming photons per pixel.

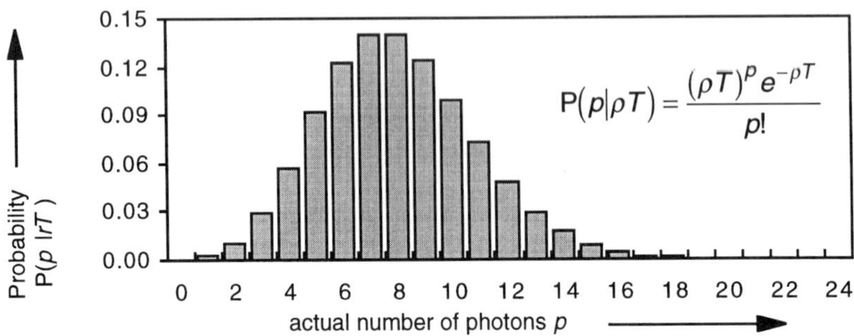

Figure 2.2 Poisson distribution for an expected number of photons $\rho T = 8$.

Photon production by any light source is a statistical process governed by the laws of quantum physics. The source emits photons at random time intervals. Consequently, the number of photons detected in a fixed observation interval will obey Poisson statistics. The probability distribution for counting p photons in an observation window of T seconds and with a photon flux of ρ in photons per seconds is displayed in Figure 2.2 for $\rho T = 8$. However, instead of the expected number of photons (ρT), each observation will find a number p with a probability given by $\mathrm{P}(p \mid \rho T)$. The average of a large number of observations will approximate the expected photon production ρT.

2.3 INTRODUCTION TO CCD CAMERAS

CCD cameras can be divided roughly into two groups: video cameras and scientific cameras. Video cameras deliver a continuous analog output signal that conforms to a specified standard. The video signal can be displayed using a video monitor or digitized by any compatible frame grabber. Scientific cameras deliver a single high-quality digital image on command. They do not obey any standard and are available with a wide variety of options. These options include pixel sizes, image dimensions, readout rate, integration time, cooling, quality grade and dynamic range. The use of these options will be clarified in this chapter.

2.3.1 From photon to electron

A CCD camera is a semiconductor device that acts as a transducer between incoming light and electrical charge. It consists of a rectangular array of photosensitive elements called pixels. The fixed spatial organization avoids geometric distortions such as pincushioning. Also, assuming proper lens configuration, there is no vignetting near the corners of the CCD. After production each CCD chip receives a quality mark which depends on the number of 'bad' pixels and other blemishes.

In order to understand how the camera functions, we have to work with the quantum description of light. An incident photon of sufficient energy can release an electron from the semiconductor's valence band into its conduction band by creating a so-called electron–hole pair. The freed electrons (called photoelectrons) are collected in potential wells (Figure 2.3). After collecting electrons for a certain time the image is ready for readout. CCD readout is very important and occurs differently for scientific and video cameras.

2.3.2 Scientific CCD cameras

Scientific CCD cameras employ full frame CCD elements (Figure 2.4). The entire pixel array consists of photosensitive elements. After a specified integration time, all charge is shifted towards the serial register. Consequently, a shutter is needed to block all light during image readout (which may take a 'long' time (seconds) for slow-scan CCD cameras). Pixel by pixel the charge is amplified and transformed into an electrical signal. This electrical signal is then converted into a discrete

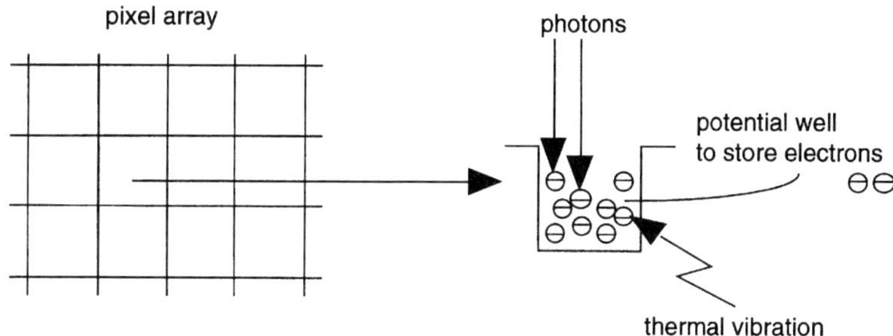

Figure 2.3 A CCD consists of an array of potential wells to store electrons.

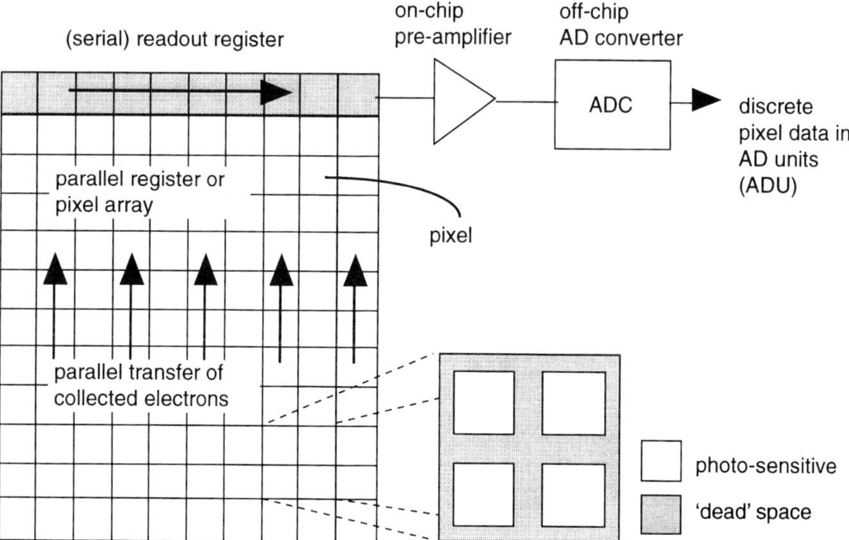

Figure 2.4 A full frame CCD as employed in scientific CCD cameras.

number by an analog-to-digital converter (ADC). In scientific cameras, there exists a one-to-one relationship between the CCD's potential wells and the pixels of a digitized image. Another aspect of scientific CCDs is their high fill factor. A fill factor of 100% means that the entire pixel surface on the CCD is photosensitive, i.e. there is no dead space between adjacent pixels.

2.3.3 Video cameras

Video cameras offer a continuous stream of video images that conform to a standard. Every complete image is composed of two consecutive fields. The odd fields contain the odd image lines and the even fields the even image lines. Unfortunately, video standards vary across the world, i.e. NTSC in the USA, PAL in most of Europe and SECAM in France. Typical field rates, formats and pixel rates are listed in Table 2.1.

The CCD elements used in video cameras have a special architecture to permit a continuous high-speed image readout. These architectures are frame-transfer and line-transfer CCDs (Figure 2.5). After an integration time of 20 ms (PAL) a field is transferred from a photosensitive area to a masked area. This happens almost instantaneously. During the next 20 ms this field is read out at the specified pixel rate while at the same time the next field is being imaged. Modern video cameras allow on-chip integration in integer multiples of 20 ms (PAL). After charge amplification at the end of the serial register, the analog electrical signal is put into video format. A frame grabber samples and quantifies the video signal. In contrast to scientific cameras, there is no longer a one-to-one relationship between a potential well on the CCD and a pixel in the digitized image.

2.3.4 Colour CCD cameras

The CCD sensor is a monochrome imaging device, i.e. photoelectrons induced by light of different wavelengths are indistinguishable. To create a colour image the incoming light needs to be selected before it hits the CCD. Two methods are frequently used. The first uses a single CCD chip in which each pixel is covered by a single colour filter. The spatial distribution of coloured pixels may vary: colour striping or a pseudo-random colour mosaic. Note that each colour channel is sparsely sampled and that roughly one-third of the incoming photons reaches the CCD. The second method uses photon sorting to separate the incoming colour image into three colour channels: red, green and blue. Each channel is imaged onto its own CCD. Three CCD colour cameras with photon sorting do not sacrifice spatial resolution and do not waste any of the incoming photons. The distribution of colour over the three colour channels is shown in Figure 2.6.

Table 2.1 Specification of video formats.

Videl type	Field rate (Hz)	Format (H $\times$ V)	Pixel rate (MHz)
NTSC	60	$(4/3 \times 1)$ 525	11
PAL/SECAM	50	$(4/3 \times 1)$ 625	13

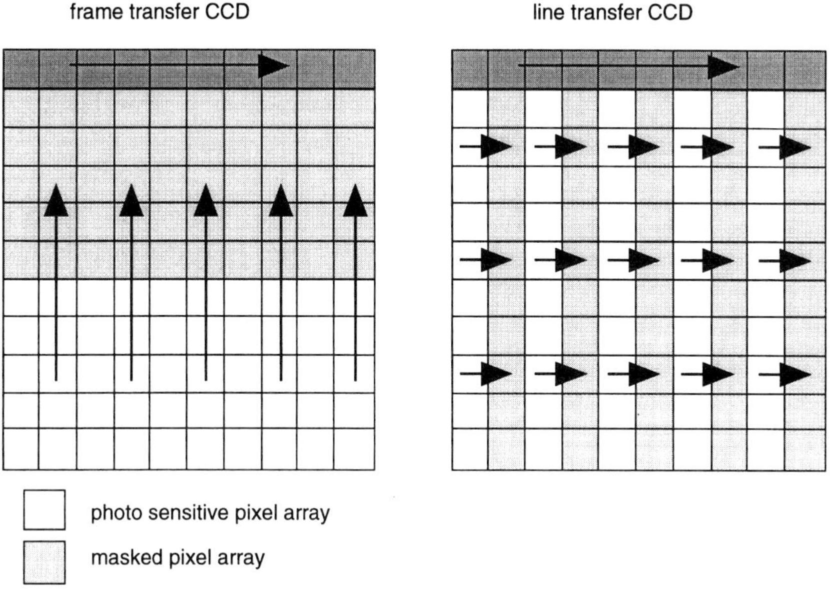

Figure 2.5 Frame transfer and line transfer CCDs consist of photosensitive and masked pixels. The pixel shape in line transfer CCDs can be such that the pixel spacing is equal in both directions.

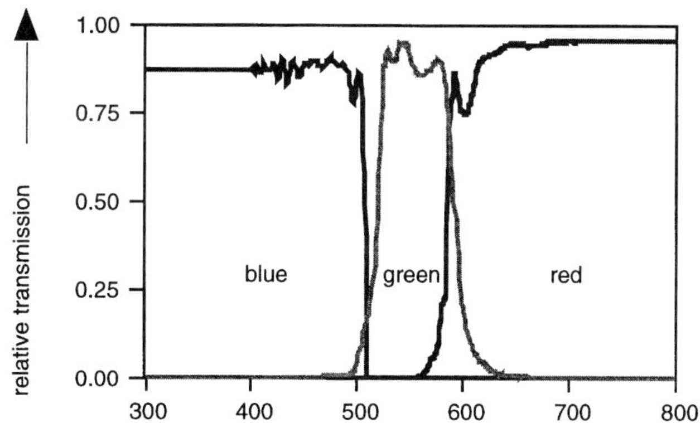

Figure 2.6 Relative transmission of colour filters in RGB cameras.

2.4 PROPERTIES OF CCD CAMERAS

A variety of properties are important for applications in quantitative fluorescence microscopy. To measure the amount of fluorescence, the camera should have a linear response. If weak signals require long integration times the camera needs cooling to suppress dark current. Why does my camera have a low readout rate? How many photons contribute to a single output unit? Is my camera photon limited? All of these questions, and more, are addressed below.

2.4.1 Noise sources

All acquired images will be contaminated by noise from a variety of sources. Noise is a stochastic phenomenon that can neither be compensated for nor eliminated, as opposed to systematic distortions such as shading or some forms of image blur. The noise sources that play a role in scientific CCD cameras are: photon noise, thermal noise (dark current and hot pixels), readout noise (amplifier noise, on-chip electronic noise and so-called KTC noise) and quantization noise.

Some of these noise sources can be made negligible by proper electronic design and careful operating conditions. One of them – photon noise – can never be eliminated and thus forms the limiting case when all other noise sources have become negligible compared to this one.

2.4.1a Photon noise

Photon noise is unavoidable and is caused by a fundamental law of nature – the quantum nature of light. The number of photoelectrons (n) generated by incident photons inherits the photon's statistical properties, i.e. Poisson distribution. The Poisson distribution has a fixed relationship between its expected value and its variance, $\mathrm{E}(n) = \mathrm{var}(n)$. Even if the photon noise were the only noise source, the signal-to-noise ratio (SNR) would still be finite. SNR improves slowly with increasing photon (photoelectron) counts, i.e. more light or longer integration time. Thus the ideal SNR becomes:

$$\mathrm{SNR}_{\mathrm{photon}} = 10 \log(n) \text{ dB} \tag{2.1}$$

It is important to remember that photon noise is not independent of the signal, not additive, and not Gaussian, thus violating three properties often used to optimize image processing methods. The maximum SNR is limited by the well capacity. The well capacity is proportional to the pixel area and, with current technology, a photoelectron density given by about 700 $e/\mu\mathrm{m}^2$. A chip with small pixels (6.8 μm × 6.8 μm) has a maximum SNR of 45 dB, whereas a chip with large pixels (23 μm × 23 μm) yields a maximum SNR of 56 dB.

2.4.1b Thermal noise: dark current and hot pixels

Thermal noise or dark current refers to the creation of electron–hole pairs due to thermal vibration. These thermal electrons cannot be distinguished from photoelectrons. Dark current is a stochastic process and yields a Poisson distribution for the number of thermal electrons generated in a fixed time interval. The production rate of thermal electrons is an exponentially increasing function of temperature. Dark current reduces the dynamic range of a pixel and adds a substantial amount of noise.

Thermal noise can be greatly reduced by cooling the CCD chip (Figure 2.7). The dark current reduces by a factor of two for every 6 °C reduction of temperature. Cooling to −40 °C can be achieved using Peltier elements, which themselves need to be cooled by air or liquids such as ethylene glycol. Cooling below 4 °C requires a

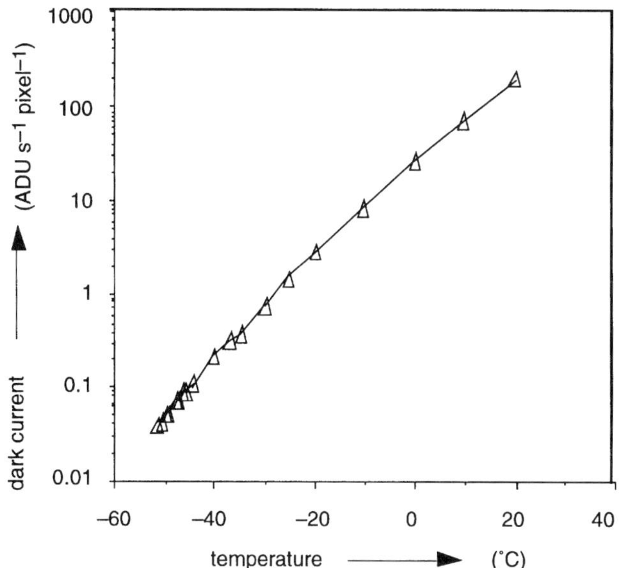

Figure 2.7 Dark current measurements. Average dark current (in ADU per second per pixel) as a function of temperature. The pixel size is 23 μm × 23 μm and each ADU level corresponds to about 90 thermal electrons.

vacuum around the CCD chip to avoid condensation. Air-cooled cameras with operating temperatures around 4 °C should not be used in areas with a high humidity. Note that an air-conditioned room typically has a very low humidity. Some CCD chips can be operated in a special accumulation mode called multiphase pinning (MPP). This technique may reduce the average dark current significantly in exchange for a smaller potential well for storing electrons.

Due to impurities in the CCD's silicon layer, some pixels suffer severely from dark current. They build up thermal electrons at a much faster rate (often up to a hundredfold) than the majority of pixels. After a few seconds of integration on a non-cooled camera, a dark image looks like an image with stars on a clear night (see Table 2.2 and Figure 2.8). Cooling the CCD also reduces that impact of hotspots or hot pixels by the same amount as the average dark current.

Some video cameras do on-board dark current subtraction. They estimate the average dark current using a strip of masked pixels. This average is subtracted from the output signal. Note that subtraction of the average dark current reduces neither the dark current noise nor the 'hot' pixels.

2.4.1c Readout noise: on-chip electronics, pre-amplifier and KTC noise

This noise originates in the process of reading the signal from the sensor. It is caused by the CCD's on-chip electronics and strongly depends on the readout rate. For extremely low readout rates the noise has a $1/f$ character, where f is the readout rate. For moderate readout rates the readout noise is minimal and approximately constant. Scientific cameras usually operate in this range (20 kHz, 500 kHz). For

Table 2.2 Some statistics of dark images (shutter closed) acquired using a non-cooled CCD camera with MPP to reduce dark current. The hot pixels have a dark current of roughly 20 times the mode of the ordinary 'cold-pixel' dark current distribution. The values below are averages over 10 images of image minima ('cold' pixel), image modes and image maxima ('hot' pixels).

Integration time (s)	ADU value of 'cold' pixels	ADU value of mode	ADU value of 'hot' pixels
0	88	103	119
1	90	106	149
5	100	114	459
10	114	138	815
50	159	193	1863

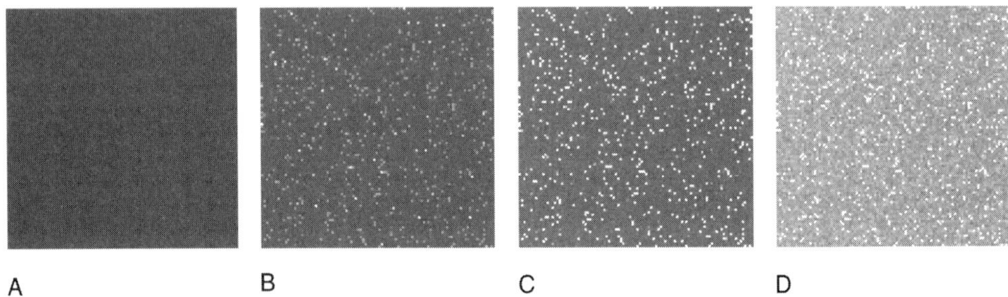

Figure 2.8 Hot pixel measurements. Dark images of (A) 1, (B) 5, (C) 10 and (D) 50 seconds of integration time acquired using a non-cooled CCD camera with MPP to reduce dark current. In all images, black refers to zero and white refers to 255. All pixel values higher than 255 have been clipped.

high readout rates the readout noise increases and becomes a significant component of the overall noise. The readout noise is additive, Gaussian distributed and independent of the signal. It is therefore expressed by its standard deviation (root mean square or RMS value) in number of electrons. At low readout rates the readout noise may be as low as a few electrons. At high readout rates (video speed $\sim$13 MHz) it can be as high as a few hundred electrons per pixel. Figure 2.9 shows the histogram of intensity levels for a dark image (shutter closed) with zero integration time and a readout rate of 4 MHz. Readout noise is the only noise source that contributed to this image.

Readout noise originates from several on-chip sources, the pre-amplifier and the field effect transistor (FET). Pre-amplifier noise is generally negligible in well-designed electronics. So-called KTC noise (associated with the gate capacitor of an FET) can be eliminated almost completely by proper design of the ADC.

2.4.1d Quantization noise

Quantization noise is inherent in the quantization of the pixel amplitudes into a finite number of discrete levels by the ADC. The ADC converts the amplitude of an

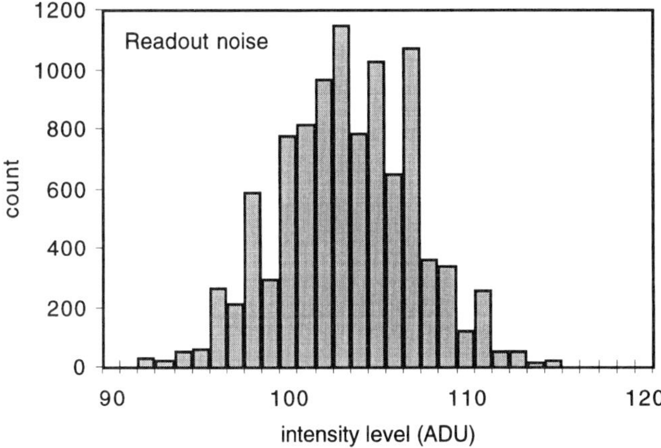

Figure 2.9 Distribution of readout noise for a 12 bit CCD camera. The readout rate is 4 MHz. The readout noise with an RMS error of 3.9 ADU (= 39 e^-) sits on top of an offset of 103 ADU. Each ADU corresponds to about 10 electrons.

electronic signal into a binary representation, a pixel value. The associated round-off errors are called quantization noise. Ideally, this noise is additive, uniformly distributed [−0.5, +0.5] and independent of the signal. The SNR for quantization noise is $\mathrm{SNR}_{qn} = 6b + 11$ dB, where b is the number of bits. Scientific CCD cameras use a high quality ADC with 8–16 bits. Quantization noise is very small and is usually ignored. A summary of the noise sources is presented in Table 2.3.

2.4.2 Linearity of photometric response

The pixel values (in ADC units (ADU), or grey levels) should be linearly proportional to the number of captured electrons and thus to the amount of incoming light. Because a (continuous) CCD readout with zero integration time consists solely of readout noise, a small offset (typically around 50–100 ADUs for a 12-bit digital signal) prevents the clipping of the signal at the lowest value. A proper setting of the electronic gain guarantees that the ADC's dynamic range stays within the linear working range of the CCD. In Figure 2.10 we show that this excellent linearity is present over the entire range of light levels. Note that video cameras are equipped with auto gain control and gamma correction. These options enhance the image to please the human eye. However, they spoil any quantitative analysis. These functions need to be switched off in order to obtain a linear response.

2.4.3 Signal-to-noise ratio

As described above all images are contaminated by noise. Table 2.3 states that all noise sources can be effectively suppressed or avoided except photon noise. The SNR of a state-of-the-art CCD camera should therefore be photon-noise limited. Photon noise is the only noise source that depends on the signal. Due to the properties of its Poisson distribution, the pixel variance, var(I), increases linearly

Table 2.3 Summary of noise sources.

Noise	Distribution	Dependent on	SNR	Remarks
Photon	Poisson	Signal	$\sqrt{n}$ $10\log(n)$ dB	Unavoidable, SNR increases with signal
Thermal	Poisson	Temperature and integration time	$\sqrt{n_t}$ $10\log(n_t)$ dB	Effectively suppressed by cooling
Readout	Gaussian, additive	Readout rate	rms = 5–10 e^- at [20,500] kHz	Noise increases rapidly with readout rate
Quantization	Uniform, additive	Number of bits in ADC	$2^b\sqrt{12}$ $11+6b$ dB	Negligible for ADC with $b>8$ bits

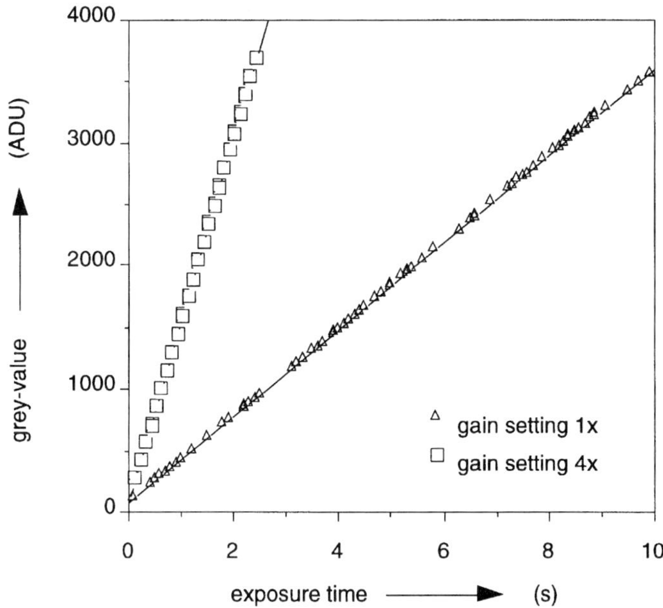

Figure 2.10 Linearity of photometric response for two different settings of the (electronic) camera gain. Data was obtained with a scientific CCD camera equipped with a 12 bit ADC. The pixel values (in ADU) are shown as a function of the integration time for a scientific CCD camera cooled to −40 °C.

with the pixel value, l, as is shown for the camera in Figure 2.11. The slope is equal to the camera gain:

$$\frac{\mathrm{var}(l)}{\bar{l}} = \frac{g^2 \cdot \mathrm{var}(n)}{g \cdot \bar{n}} = g \tag{2.2}$$

The SNR of this slow-scan (500 kHz) cooled CCD camera is photon limited over almost the entire range of light levels, as we can see from the slope of Figure 2.11, which is linear over the entire range of light levels. Note that we have measured the variance of the signal at individual pixels.

Measuring the image variance instead of the pixel variance yields a completely different result. The pixel variation is much larger than the Poisson noise of individual pixels. The pixel variation is caused by a difference in response of the individual CCD wells to the same amount of light. This is clearly demonstrated in Figure 2.12. Here the image SNR before calibration never exceeds 30 dB. To achieve photon limited behaviour, we need to calibrate the image. Image calibration corrects the image for differences in pixel response (see Section 2.4.7).

Video cameras are not photon limited. For the low output values in particular the noise can be substantially higher than the photon noise. SNR measurements for video cameras can produce values that are higher than the theoretical maximum, but at the expense of image sharpness. The conversion to and from video employs lowpass filters along the image lines. This reduces the noise in exchange for a lower spatial resolution.

Figures 2.11 and 2.12 show a higher SNR when the image intensity in ADUs increases. This is only true for a fixed setting of the electronic gain, i.e. a fixed conversion factor from photoelectrons into ADUs. Increasing the electronic gain not only amplifies the signal, but also the inherent Poisson noise. Figure 2.13 shows two images of the same DAPI-stained metaphase spread. The average intensity in ADUs is identical. The high quality picture was taken with a standard gain setting and an integration time of 2 seconds. The low quality picture was taken with a 16 times larger electronic gain, but an integration time of only 1/8 second. The product of electronic gain and integration time was kept constant.

2.4.4 Sensitivity

Unfortunately, not all photons that reach the CCD are converted into electrons. CCD sensors, like many other sensors, including the human eye, are not equally sensitive to all wavelengths of light (see Figure 2.14). The spectral sensitivity for CCD sensors is called the *quantum efficiency* (QE). The QE denotes the probability that a photon of a certain wavelength will create a photoelectron. The QE for front-illuminated CCDs is virtually zero for UV light below 400 nm and reaches its maximum for IR around 1000 nm. Due to the sensitivity of the CCD sensor to IR, an IR blocking filter is put in the light path. An IR blocking filter greatly enhances the contrast. The absence of such a filter results in images with a very high background, which are perceived as 'foggy'. Special attention is needed for users of IR fluorophores such as Cy5 and Cy7. Many standard IR blocking filters block their emission light as well. These dyes require an IR blocking filter with a higher cutoff wavelength to allow near IR excitation in exchange for a slightly higher background signal.

To increase the sensitivity for low wavelengths, the CCD surface must be coated with a very thin layer of a so-called wavelength converter. This layer absorbs UV and emits light at a higher wavelength. Back-illuminated CCDs offer superior spectral sensitivity compared to the normal front-illuminated CCDs. Apart from the QE, there are two ways to describe the sensitivity of cameras. Both are based on photoelectrons rather than photons:

(a) *Absolute sensitivity*: the minimum number of detectable photoelectrons is called the absolute sensitivity. Here we are not limited by photon noise, but by readout noise. Even the very best cameras have about 5 electrons RMS readout noise. To ensure detectability of a signal it should be three times larger than the RMS readout noise. A cooled intensified CCD camera allows detection of single photoelectrons in an extremely low photon flux. Cooling needs to be such that the thermal electron flux is negligible compared to the detected photoelectron flux.

(b) *Relative sensitivity*: the number of photoelectrons needed to add a single brightness level after analog-to-digital conversion is called relative sensitivity. The relative sensitivity is the inverse of the camera's gain. This quantity can easily be measured for photon-limited CCD cameras. The brightness level $l = gn$, where g is the gain and n the number of photoelectrons. The variance in brightness is $\text{var}(l) = g^2\text{var}(n) = g^2n$. Thus, after eliminating any offsets, $g = \text{var}(l)/l$. The CCD gain is roughly proportional to the ratio of the well

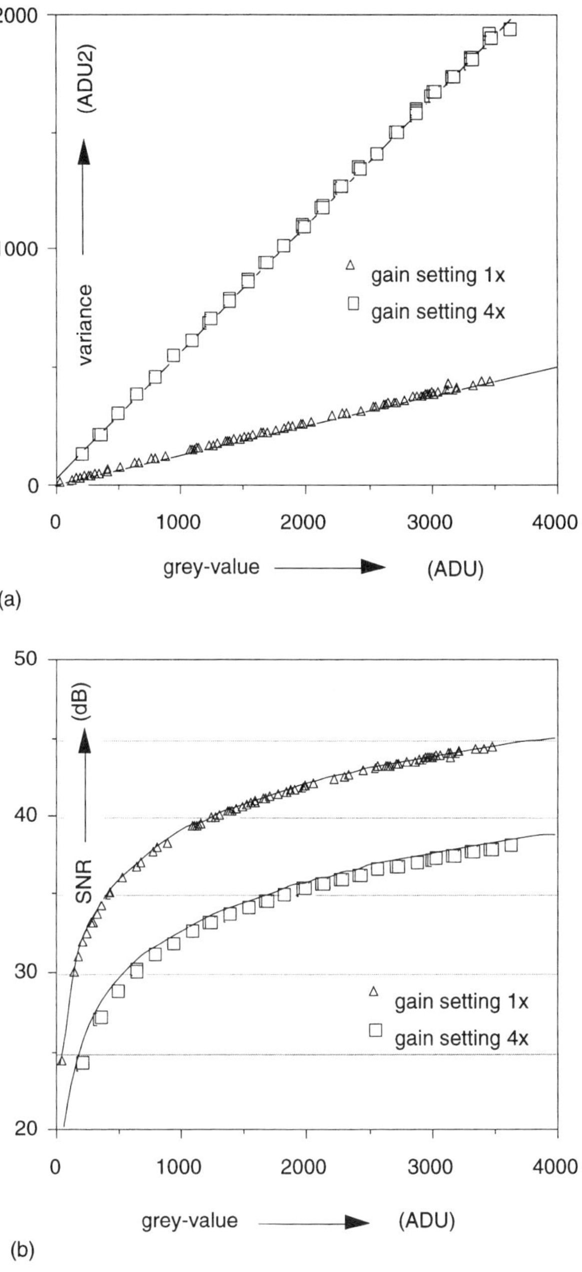

Figure 2.11 SNR of a slow-scan (500 kHz) cooled (−40 °C) CCD camera for two different settings of the electronic gain. (a) The pixel variance as a function of pixel value. (b) The SNR as a function of pixel value.

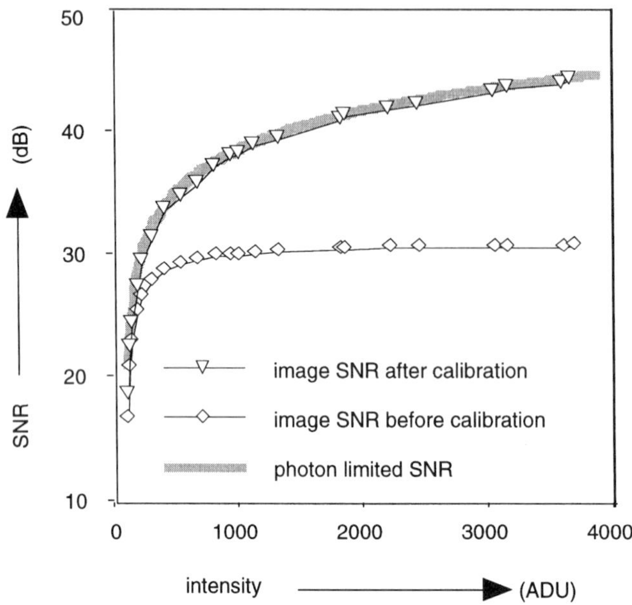

Figure 2.12 SNR before and after calibration, to correct for differences between pixels.

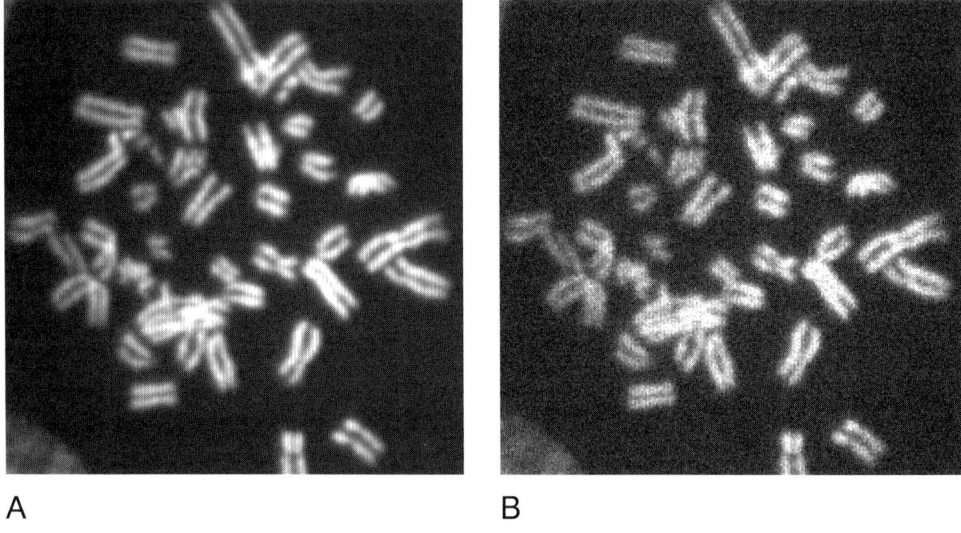

Figure 2.13 A DAPI-stained metaphase spread captured with different settings of the electronic gain and integration time: (A) gain is $1\times$ and integration time is 2 s, (B) gain is $16\times$ and integration time is $1/8$ s.

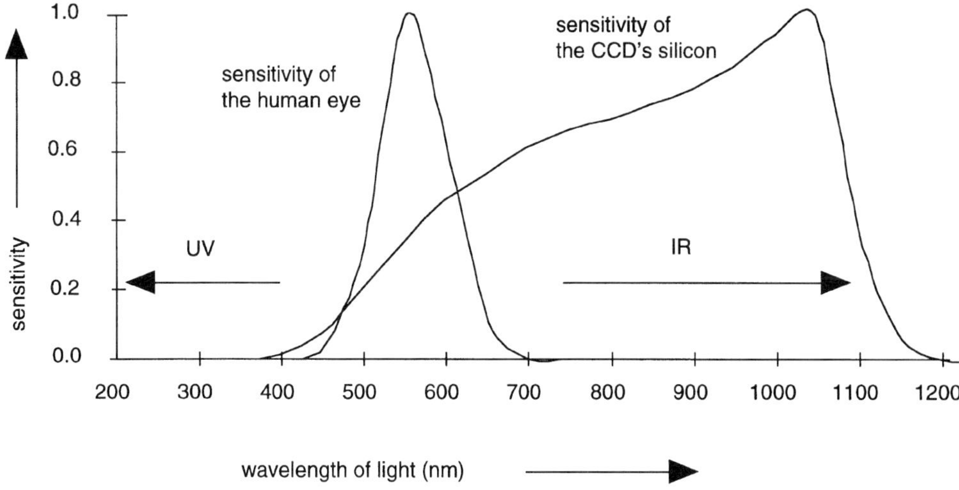

Figure 2.14 Sensitivity of CCD's silicon (QE) versus the sensitivity of the human eye. Note that the QE of a back-illuminated CCD comes close to silicon's QE. The QE for a front-illuminated CCD is roughly half that of silicon's QE.

capacity and the total output range. The first depends on the pixel size and the second on the number of output bits. Some examples are listed in Table 2.4.

Finally, when comparing video to scientific cameras, video cameras have a lower sensitivity than scientific cameras due to the absence of a special coating, more 'dead' space between the pixels, and a built-in IR filter with a smooth slope into the 'red' wavelengths.

2.4.5 Dynamic range and blooming

Dynamic range is the maximum signal divided by the camera's noise floor. For cooled, slow-scan CCD cameras the noise floor is the readout noise. For non-cooled cameras one should be aware that dark current not only adds uncertainty, but its average value also reduces the dynamic range. Ideally, the number of output bits should be just large enough to accommodate the full dynamic range. Excess bits will

Table 2.4 Relative sensitivity ($S = g^{-1}$) for various CCD cameras.

Pixel size (μm × μm)	S (e^-/ADU)	ADC bits
23.0 × 23.0	90.9	12
22.0 × 22.0	9.7	16
6.8 × 6.8	7.9	12
11.0 × 5.5	130	8 (video camera)

only be filled with noise. Use of too few bits reduces the relative sensitivity, and ultimately does not allow the detection of very weak signals, i.e. signals just above the readout noise.

Very bright signals may not only saturate the camera output, but also spill electrons to adjacent wells. This flooding of electrons to neighbouring pixels is called *blooming*. Some CCDs have a higher resistance against blooming in exchange for a lower well capacity and lower quantum efficiency.

2.4.6 Spatial frequency response

The spatial frequency response (SFR) quantifies the camera response as a function of spatial frequency. An ideal response is flat (equal to 1) for all spatial frequencies. This can only be achieved by point sampling. However, a CCD pixel has a photosensitive surface that may be as large as the entire pixel (fill factor= 1). This corresponds to a sinc shape response, as depicted in Figure 2.15.

The overall SFR (camera $\times$ optics) should be compared to the optical transfer function (OTF) for sampling at the Nyquist rate. The camera should not significantly reduce the overall response below the OTF. Other sources that may decrease the spatial frequency response are:

(a) mechanical vibration of the camera during image integration (forced air cooling using a fan mounted on the camera head)
(b) leakage of charge (smearing) during CCD readout
(c) conversion to and from video

Scientific cameras usually do not suffer from any of these problems and do not significantly reduce the optical resolution for proper sampling. Video cameras employ a lowpass filter along the image lines. This may cause some additional image blur and thus a lower spatial frequency response.

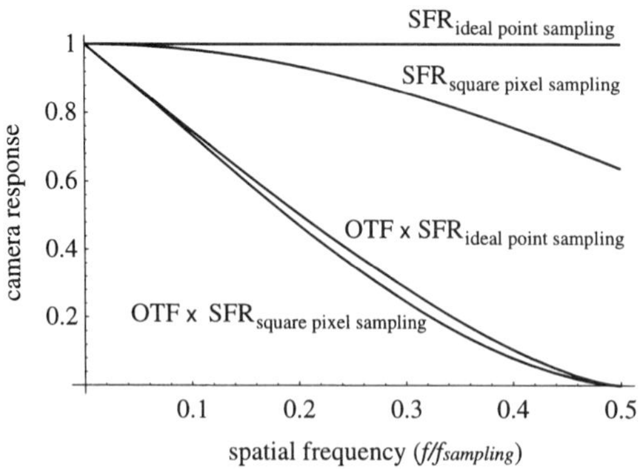

Figure 2.15 Spatial frequency response. Point sampling yields a constant response, whereas a rectangular pixel reduces the higher frequencies. The overall response of the optics and a square-pixel camera is similar to the overall response of an 'ideal' diffraction limited system.

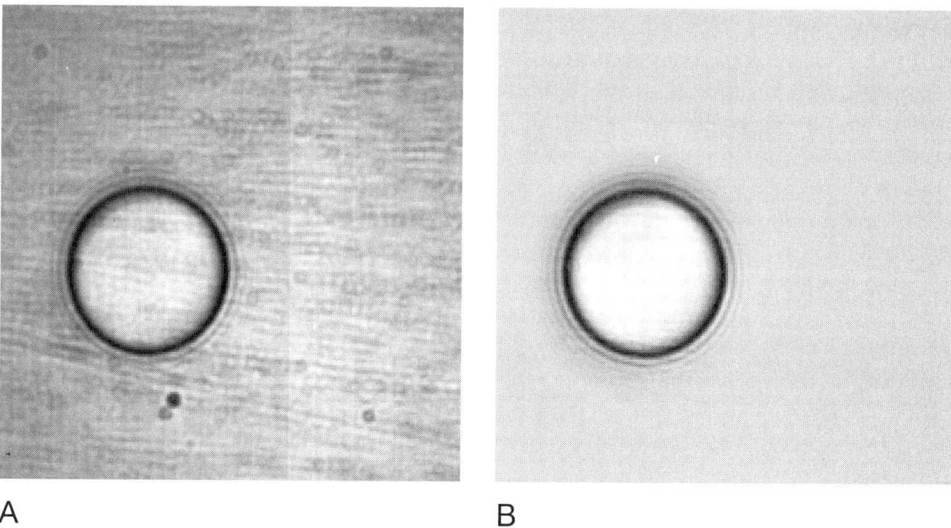

Figure 2.16 Brightfield image of a latex sphere (A) before and (B) after image calibration.

2.4.7 Image calibration: flat-field correction

Image calibration or flat-field correction corrects a variety of systematic errors such as pixel variation, 'hot' pixel artefacts, non-homogeneous illumination, shading and dirt on glass surfaces. In addition to the image I, it requires two additional images of the same exposure time: a dark image I_{dark} (shutter closed) and a blank field image I_{blank} (illuminated image without objects). The corrected image $I_{corrected}$ becomes:

$$I_{corrected}(x, y) = \frac{I(x, y) - I_{dark}(x, y)}{I_{blank}(x, y) - I_{dark}(x, y)} K \qquad (2.3)$$

Equation (2.3) is a point operation that needs to be computed using floating-point numbers, rather than integers. The first term performs the normalization whereas the second term, K, scales the image. The effect of image calibration is depicted in Figure 2.16.

2.5 IMAGE INTENSIFIERS

We have shown that scientific CCD cameras are almost perfect image sensors, i.e. their SNR is photon limited and spatial blurring is negligible. In low light level imaging applications, a longer exposure time (cooled CCD element) permits accumulation of photoelectrons to obtain a higher SNR. Image intensifiers are typically used in very low light level imaging applications when there is a time constraint on the maximum exposure time, i.e. fast autofocusing and dynamic studies. Under these circumstances CCD sensors are seriously degraded by their readout noise, whereas image intensifiers amplify the incoming photoelectrons before the camera degrades the image with readout noise. As a consequence the imaging system is photon limited for lower light levels. Ultimately, an intensified CCD (ICCD) can be used for photon counting. Other applications of image intensifiers include use of the unique gain modulation capabilities of second-generation image intensifiers, which allow ultra-short exposure times (down to nanoseconds) or sinusoidal modulation up to the hundred megahertz range.

An image intensifier can be mounted in front of a CCD camera, as shown in Figure 2.17. An input image is projected onto a photocathode, which converts a detected photon into one electron. Each electron is then accelerated and/or multiplied before it hits a phosphorus screen. The screen converts electrons back into photons. Depending on its energy, one electron can create multiple photons. The output image on the phosphorus screen is an intensified copy of the input image projected on the photocathode. The space between the photocathode and phosphorus screen needs to be a vacuum to avoid ionization of the air and thus unwanted electrons. An image intensifier not only amplifies the image, but also converts the spectrum of input wavelengths (colour image) into the spectrum of the phosphorus screen. The spectral sensitivity depends on the type of photocathode. The spectrum of the phosphorus screen may be chosen to match the sensitivity of the CCD camera.

In general the input and output windows of intensifiers are made of glass (anti-veiling glare glass or AVG) or fibre optics. Fibre optics are made of stacked optical fibres and guide the image – each fibre can be seen as a pixel of the fibre optic stack – so that the input image is projected onto the output of the stack. Using fibre optics it is possible to (de)magnify an image (tapered fibre optics) or rotate it over a certain angle (twisted fibre optics). It is also possible to project a flat image onto a spherical

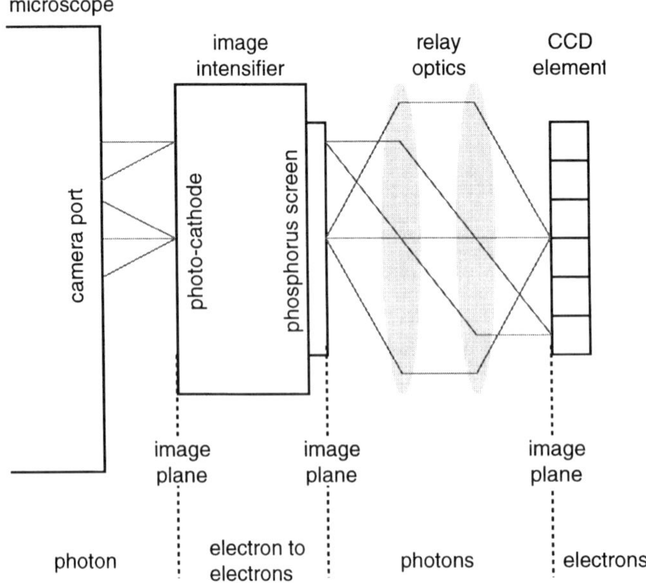

Figure 2.17 Intensified imaging setup: image intensifier, relay optics and CCD camera.

surface just by making the output surface of the fibre optics spherical. This is particularly useful for electrostatic focused first-generation image intensifiers, as described below.

In most cases the output image of the image intensifier is projected onto the CCD camera by relay optics. In some cases the CCD chip is directly mounted onto the fibre optic output of the intensifier.

Three main architectures of image intensifiers can be distinguished: first-, second- and third-generation image intensifiers. A detailed description is given in the appendix and they are summarized in Table 2.5. High spatial resolution and high gain are conflicting constraints. The characteristics of an image intensifier not only depend on the generation type, but also the materials of which the photocathode and the phosphorus screen are made. In general the characteristics of image intensifiers are determined by:

(a) gain ranges from 10 to 10^6
(b) spectral sensitivity depending on the material of the photocathode
(c) spatial resolution depending on the focusing mechanism, micro-channel plates (MCP) and the phosphor screen
(d) noise characteristics

These characteristics are described in the following sections.

2.5.1 Intensifier gain and linearity of photometric response

The photometric response of an image intensifier is linear up to saturation when the intensifier is not controlled by an auto-gain-control (AGC) circuit. Figure 2.18 shows the response of an ICCD camera for various gain settings (MCP voltages; see appendix) of a second-generation image intensifier.

The total gain of an image intensifier is determined by the spectral sensitivity or quantum efficiency of the photocathode, the electron multiplication factor (if an MCP is present) and the efficiency of the phosphorus screen, which depends on the energy of the electrons hitting the screen.

The gain is often specified as luminance gain (candela per square metre per lux). This luminance gain is a photometric measure specifying the amplification as perceived by the human eye. Transformation of the luminance gain into a photon-per-photon gain requires knowledge about the spectral distribution of the incoming light, the spectral sensitivity of the photocathode, the emission spectrum of the phosphorus screen, and the spectral sensitivity of the eye.

Table 2.5 Types of image intensifiers and their characteristics.

Intensifier generation/type	Resolution	Gain	Remarks
First/proximity focusing	+++	+	No geometric distortion
First/electrostatic focusing	++	++	Geometric distortion at border
Second/micro-channel plates	+	+++	Small physical size, fast gating
Third/(MCP/no MCP)	+/+++	+++/+	Higher red/near IR sensitivity

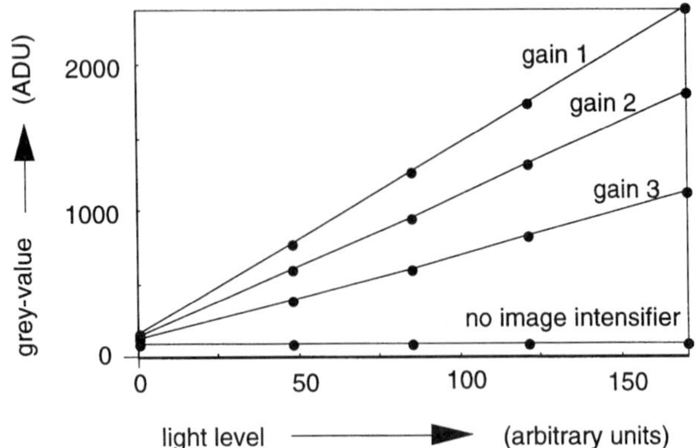

Figure 2.18 Photometric response of an image intensifier.

2.5.2 Signal-to-noise ratio

Every amplifier (including image intensifiers) not only amplifies the signal but also adds noise. Intensifier noise is mainly caused by the discrete nature of the electron amplification process. Each photon entering the image intensifier will induce a large number of photons leaving the phosphorus screen. The variation in the gain for each incoming photon and the correlation in time of the output photons are the source of the noise.

A single input photon yields a burst of output photons. The departure times of these output photons are strongly correlated in time. As a consequence, the Poisson process at the input of the intensifier yields a filtered Poisson process at the output, which has an increased variance compared to ordinary Poisson-distributed noise. Reducing the correlation between the output photons induced by a single input photon (i.e. if they are randomly spread over a large period of time) yields a stochastic process that approximates a normal Poisson process. A very slow phosphorus screen greatly reduces the correlation and thus reduces the noise in exchange for a much lower time resolution (the effective exposure time is longer).

Like Poisson statistics, filtered Poisson noise will also cause the pixel value variance of an ICCD image sensor to increase linearly with the pixel value. This is demonstrated in Figure 2.19. The CCD readout noise is the dominant white noise source. Therefore the variance lines in Figure 2.19 of the CCD and ICCD intersect at the point with pixel value equal to zero at which the variance is equal to the readout noise. For longer exposure times – which are exceptional for ICCD systems – the image intensifier might add some dark current noise. Figure 2.20 shows the SNR calculated from the same data. Note that the SNR is plotted as a function of camera response. The amount of light required for one ADU for the ICCD is much less than for the CCD sensor.

Figures 2.19 and 2.20 are not in favour of the ICCD sensor. The true application for an image intensifier is in low light situations, as can be seen in Figure 2.21. For extremely low light levels the intensified system yields a higher SNR than the non-intensified CCD camera.

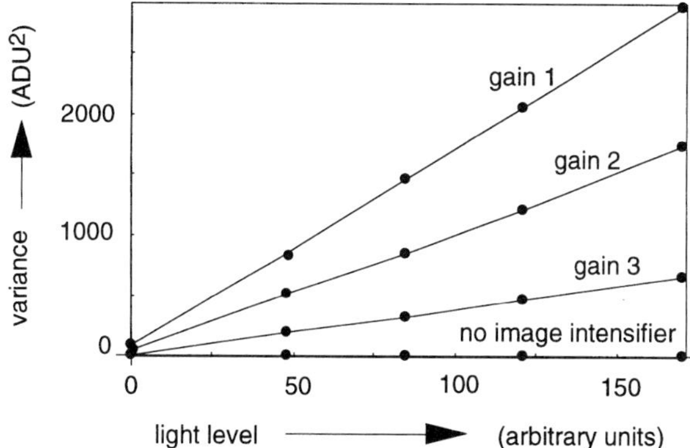

Figure 2.19 The pixel value variance as a function of the amount of light, for a CCD and an ICCD image sensor.

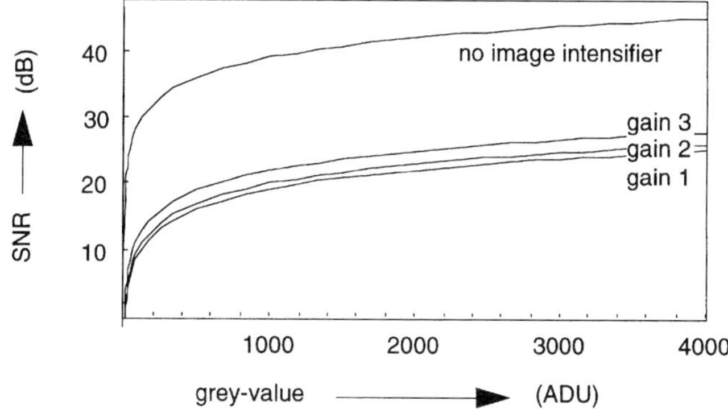

Figure 2.20 The SNR as a function of the pixel value for a CCD and an ICCD image sensor. Note that the amount of light required for one ADU for the ICCD is much less than for the CCD sensor.

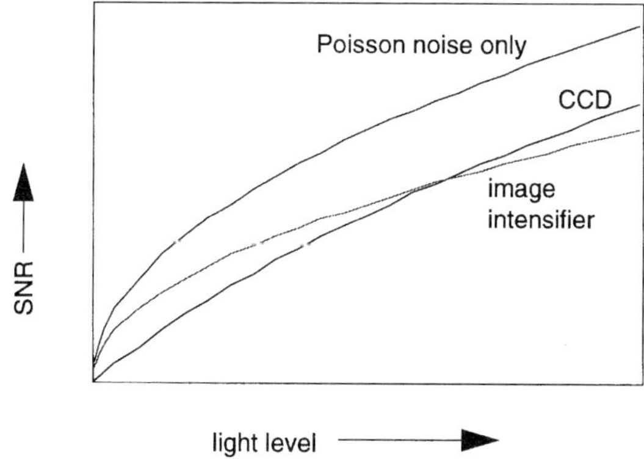

Figure 2.21 The SNR as a function of the amount of light, for a CCD and an ICCD image sensor for extremely low light levels.

2.5.3 Spectral sensitivity

The spectral sensitivity or quantum efficiency of the photocathode is an important factor. In an ICCD sensor, the relative sensitivity of the photocathode is equivalent to the quantum efficiency in 'standard' CCD cameras. It denotes the probability that a photon is detected. The quantum efficiency of an image intensifier is generally much lower than that of a CCD sensor. It depends strongly on the material of the photocathode. Figure 2.22 shows the quantum efficiency as a function of the wavelength for a few common photocathode materials (S20, S25 and gallium arsenide).

The sensitivity of a photocathode is often specified in photometric units describing the response of a photocathode in terms of milliamperes current generated for light with a certain colour temperature measured in lumen. It is again very difficult to transform these photometric quantities into radiometric quantities.

2.5.4 Spatial resolution

The spatial resolution of image intensifier systems is limited by the grain size and thickness of the phosphorus screen and the diameter of the MCP channels. Given the spatial resolution of the image intensifier together with a possible magnification or demagnification with the relay optics, one can determine what resolution a CCD camera must have in order to be suitable for a given intensifier type. CCDs with relatively large pixels match the low resolution of the current intensifiers. Intensifiers with a fibre optic output window show a 'chicken-wire' structure superimposed on the acquired image which becomes clearly visible for very bright output images. This 'chicken-wire' structure originates from the fibre bundles (which consists of very fine fibres) packed in a hexagonal grid.

2.5.5 Special applications of image intensifiers

The most common application of image intensifiers is imaging in low light level situations where CCD cameras suffer from readout noise. Although the image

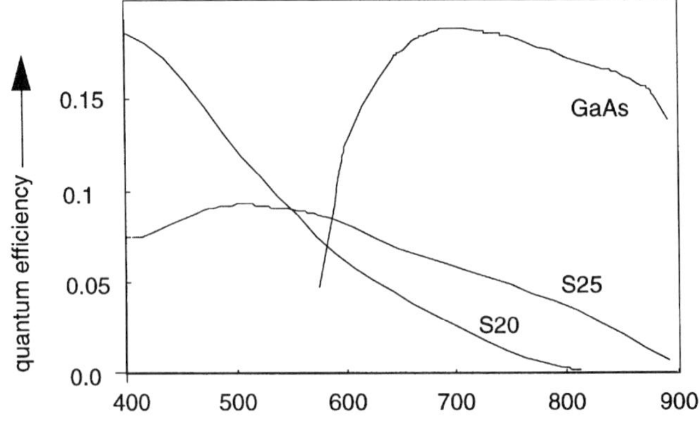

Figure 2.22 Quantum efficiency for a few common photocathode types.

intensifier overcomes the problem of readout noise, images acquired with an image intensifier suffer from gain noise and filtered Poisson noise (which is worse than Poisson noise).

In very low light level situations (i.e. a few photons per frame) it is possible to use image intensifiers and reach Poisson limited imaging. This technique is called photon counting (Wick, 1987). The intensified CCD is used as a single photon detector. Each time a photon enters the intensifier, a burst of photons is detected by the CCD camera. The centre of mass of this burst is calculated in software and marked in a digital image in computer memory. An image of the scene is gradually built up in this way, frame-by-frame. Poisson limited imaging is retrieved while the variation in gain is eliminated due to the fact that single photon bursts (of different numbers of output photons) are detected and only the position is used.

The gain of a second-generation image intensifier can be modulated in time without deforming the image. Gating is a form of gain modulation that is used in ultra-high speed photography. Incoming light is only imaged when the gate is open, during a very short time (down to nanoseconds).

A sinusoidally modulated gain of an image intensifier is used in fluorescence lifetime imaging microscopy (FLIM; Lakowicz, 1992). In this application the lifetime of the excited state of fluorescence molecules (in the nanosecond range) is imaged according to a homodyne or heterodyne detection scheme (see also Section 4.4). This nanosecond phenomenon becomes a stationary process. Modulation of the gain at several tens of megahertz is feasible using ordinary second-generation image intensifiers.

2.6 CONCLUSIONS

This chapter shows how the performance of state-of-the-art scientific CCD cameras is limited by the laws of nature. The optimal resolution is determined by ideal diffraction limited optics that create an image at the surface of the CCD chip. The fixed square sampling grid allows equidistant sampling. Each sampling element consists of a square photosensitive tile, which causes negligible image blur if sampled at the Nyquist rate. Improvement of the resolution beyond the diffraction limit requires sophisticated image restoration algorithms (Van Kempen *et al.*, 1997). The most important practical points are summarized below:

(a) CCD arrays with many small pixels are very useful in quantitative microscopy. Small pixels allow sampling at the Nyquist rate. The size of the CCD array should be slightly smaller than the microscope's tube diameter to permit a large field of view.
(b) All images are contaminated by noise. Photon noise cannot be avoided and is caused by the quantum nature of light. High quality cameras yield photon-limited signal-to-noise ratios over almost the entire range of output signals.
(c) The readout rates of scientific cameras can be set to moderate levels (2–4 MHz) for most applications. Only for extremely weak signals is the readout noise a serious threat to the overall signal-to-noise ratio and requires slow to very slow readout rates. By contrast, video cameras offer high readout rates in exchange for a much higher readout noise.

(d) Cooling virtually eliminates dark current and the influence of hot pixels. Peltier cooling (air or liquid) may lower the operating temperature to −40 °C. This is needed for applications that require integration times of more than 3–5 seconds. Slow-scan CCD cameras always need cooling to suppress dark current during image readout.
(e) The relative sensitivity of scientific CCD cameras can be extremely high (around 2 photoelectrons per ADU level) and limited by the CCD's quantum efficiency. In general, video cameras are not very sensitive. Note that image brightness can also be increased by using better optics, that is, lenses that, for a given magnification, have the highest possible NA.
(f) Image intensifiers can be used in combination with any CCD camera. They multiply the detected photoelectrons in exchange for a lower spatial resolution and intensifier noise. In special cases, in particular when short exposures are required, they perform better than a stand-alone CCD camera.

CCD cameras are available from a variety of manufacturers and have been used successfully for more than a decade in many applications in cell biology.

ACKNOWLEDGEMENTS

This work was partially supported by the Royal Dutch Academy of Arts and Sciences (KNAW), the Dutch Foundation for Technical Sciences (STW) project 2987, Lambert Instruments (Leutingewolde, The Netherlands), the Resource for Molecular Cytogenetics (US DOE DE-AC03-76SF00098), the Netherlands Organization for Scientific Research (NWO) grant 900-538-040, the Concerted Action for Automated Molecular Cytogenetics Analysis (CA-AMCA), the Human Capital and Mobility Project FISH, and the Rolling Grants programme of the Foundation for Fundamental Research in Matter (FOM).

REFERENCES

Aikens, R.S., Agard, D.A. and Sedat, J.W. (1989) Solid-state imagers for microscopy. In *Methods in Cell Biology*, vol. 29: *Fluorescence Microscopy of Living Cells in Culture* (Taylor, L.S. and Wang, Y.-L., eds), Part A, pp. 291–313. Academic Press: San Diego, CA.

Castleman, K.R. (1996) *Digital Image Processing*, 2nd edn. Prentice Hall: Englewood Cliffs, NJ.

Csorba, I.P. (1985) *Image Tubes*. Howard W. Sams & Co: Indianapolis, IN.

Inoue, S. (1986) *Video Microscopy*. Plenum Press: New York.

Lakowicz, J.R. (1992) Fluorescence lifetime sensitive imaging generates cellular images. *Laser Focus World* May, 60–2.

Mullikin, J.C., Van Vliet, L.J., Netten, H., Boddeke, F.R., Van der Feltz, G.W. and Young, I.T. (1994) Methods for CCD camera characterization. *SPIE* **2173**: 73–84.

Van Kempen, G.M.P., Van Vliet, L.J., Verveer, P. and Van der Voort, H.T.M. (1997) A quantitative comparison of image restoration techniques for confocal microscopy. *J. Microscopy* **185**: 354–65.

Van Vliet, L.J., Sudar D. and Young, I.T. (1997) Digital fluorescence imaging using cooled CCD array cameras. In *Cell Biology: A Laboratory Handbook*, 2nd edn (Celis, J.E., ed.). Academic Press: New York.

Wick, R.A. (1987) Quantum-limited imaging using microchannel plate technology. *Appl. Opt.* **26**: 3210–18.

Young, I.T. (1989) Image fidelity: characterizing the imaging transfer function. In *Methods in Cell Biology*, vol. 30: *Fluorescence Microscopy of Living Cells in Culture* (Taylor, L.S. and Wang, Y.-L., eds), Part B, pp. 1–45. Academic Press: San Diego, CA.

APPENDIX: IMAGE INTENSIFIERS

The properties of image intensifiers depend on their architecture. We distinguish three main types: first-, second- and third-generation image intensifiers.

A.1 First-generation image intensifiers

The first-generation image intensifier applies an electric field to accelerate the electrons from the photocathode to the phosphorus screen (Figure 2.23). An approximately uniform electric field is created by applying a voltage between the photocathode and the phosphorus screen. This field will force the electrons towards the phosphorus screen. Unfortunately, upon creation in the photocathode, the electrons have a velocity in a random direction. Accelerating the electrons in the direction of the phosphorus screen does nothing to the initial lateral velocity, which causes blurring. To reduce blurring, first-generation image intensifiers use proximity focusing or electrostatic focusing.

Proximity focusing puts the two plates close together. In this way the electrons have a short time to deviate. As a result of this small distance, the voltage that can be applied over the plates is limited and so is the gain.

Electrostatic focusing uses charged deflectors to create a complex electric field, which focuses the electrons from the photocathode onto the phosphorus screen. To accomplish minimal geometric distortion of the image, the photocathode and phosphorus screen are spherical surfaces and use fibre optics to convert from a flat image to the spherical surface of the photocathode, and vice versa for the phosphorus screen. Although this type of image intensifier can have a much higher gain compared to the proximity-focused intensifier with the same spatial resolution, it usually suffers from geometric distortions at the image borders. Electrostatic focusing also allows (de)magnification of the image in the intensifier.

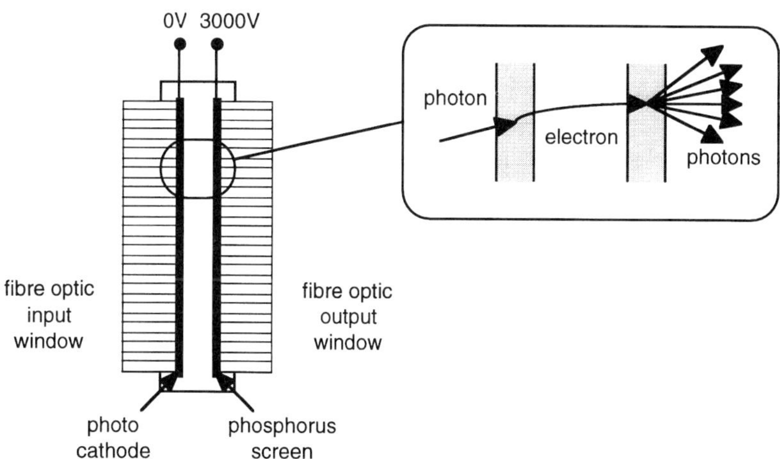

Figure 2.23 Typical first-generation proximity-focused image intensifier.

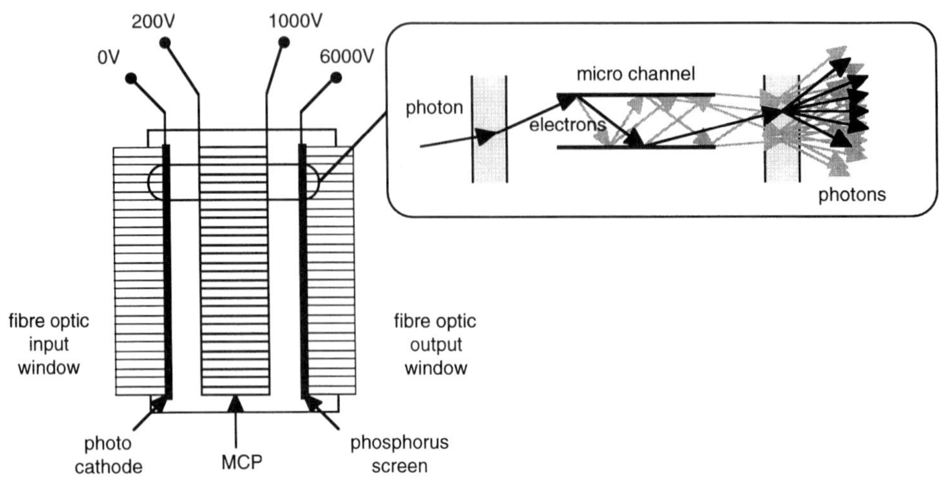

Figure 2.24 Typical second-generation image intensifier.

A.2 Second-generation image intensifiers

In a second-generation image intensifier the problem of focusing and high gain is solved by electron multiplication inside a micro-channel plate (MCP). An MCP consists of hexagonally stacked hollow tubes with a typical diameter of 10–15 μm (Figure 2.24). The electrons created in the photocathode are proximity focused onto the MCP by the first electric field. When a voltage is applied over the MCP, each channel acts as an electron multiplier: an electron entering a channel is forced through the channel by the electric field.

The electron will hit the inner resistive surface of the channel (due to its initial velocity towards this surface), creating multiple secondary electrons. The process is repeated with these secondary electrons, giving a multiplier effect. Many electrons leave the end of a channel as result of a single incident electron. These electrons are accelerated in the last electric field between the MCP and the phosphorus screen (proximity focusing) so that each electron hitting the phosphorus screen creates a number of photons. The resolution is limited by the dimensions of the MCP.

A.3 Third-generation image intensifiers

Third-generation image intensifiers have a special photocathode made from a single crystal of gallium arsenide. GaAs photocathodes are much more sensitive to light in the red to near infrared (NIR) part of the spectrum. Third-generation image intensifiers can be made as first-generation proximity focused or second-generation MCP type intensifiers. The photocathode has to be flat (GaAs photocathodes are single crystals), therefore third-generation image intensifiers cannot apply electrostatic focusing.

3

Optical Systems for Image Analysed Microscopy

Michael H.F. Wilkinson
University of Groningen, Groningen, The Netherlands

3.1 INTRODUCTION

Most of the applications described in this book involve an optical microscope in some shape or form. Since their first use in microbiology by Antonie van Leeuwenhoek in the seventeenth century, microscopes have become more and more powerful, and more and more sophisticated. For users of modern microscopes with Köhler illumination, plan-apochromatic objectives, wide angle, high eyepoint ocular lenses etc., it is really amazing what excellent work was done by this early researcher, with just a single carefully ground and polished sphere of glass, set between two brass plates. Many early researchers made or designed their own instruments, and they probably had a more thorough understanding, or at least working knowledge, of all the aberrations and distortions of their instruments than many, if not most, microscope users today. Because of the ease of use and the excellent image quality of today's microscopes, the residual aberrations and distortions are so small that they go unnoticed by most users. However, if a microscope is used in conjunction with quantitative image analysis, it is necessary to have a good understanding of the basic principles of image formation, and of the strengths and weaknesses of the type of microscope used. Residual distortions and aberrations must be measured and, if possible, corrected digitally.

This chapter aims to provide sufficient background knowledge of the optical systems used in image analysed microscopy for the proper understanding of further chapters. The emphasis is on understanding which types of aberrations exist, and which can be corrected digitally in quantitative work. Apart from the well-known techniques, such as phase contrast, Nomarski differential interference contrast, epifluorescence and confocal microscopy, a number of special microscopic techniques is described, to illustrate the wealth of techniques available to the optical microscopist. The list is far from complete, since proper treatment would require at least a book the size of this one. For more information the reader is referred to one of many excellent textbooks (Wilson and Sheppard, 1984; Taylor and Wang, 1989; Wang and Taylor, 1989; Russ, 1990; Herman and LeMasters, 1992).

Digital Image Analysis of Microbes: Imaging, Morphometry, Fluorometry and Motility Techniques and Applications. Edited by M.H.F. Wilkinson and F. Schut.

3.2 THE BASIC PRINCIPLES OF OPTICAL MICROSCOPY

3.2.1 Components of a microscope and basic optical pathway

Unlike van Leeuwenhoek's original single-lens design, the modern, finite tube length microscope is basically a two-lens design (Figure 3.1), even though in a practical design a dozen or more lenses may be involved. First, a short focal length objective lens creates a real, magnified, primary image at a specified distance (the optical tube length, usually 160 mm). The microscopist observes this real image using a second lens, the eyepiece or ocular. What the microscopist sees is a magnified, virtual, secondary image at some distance behind the ocular. In this type of design, the microscope's objective lens is specifically corrected for the exact *optical* tube length, and hence for a particular working distance.

The *optical* distance d_{opt} through a medium relates to the *physical* distance d_{phys} that light travels as:

$$d_{\text{opt}} = n \cdot d_{\text{phys}} \tag{3.1}$$

where n is the index of refraction of the intervening medium (1.00 for air). One problem of the fixed tube optical length design is that it is difficult to insert any auxiliary optics, e.g. filters, into the optical pathway, since these change the optical tube length by changing the index of refraction. To correct this problem, a collimator lens, which turns the converging rays from the objective into parallel rays, must be

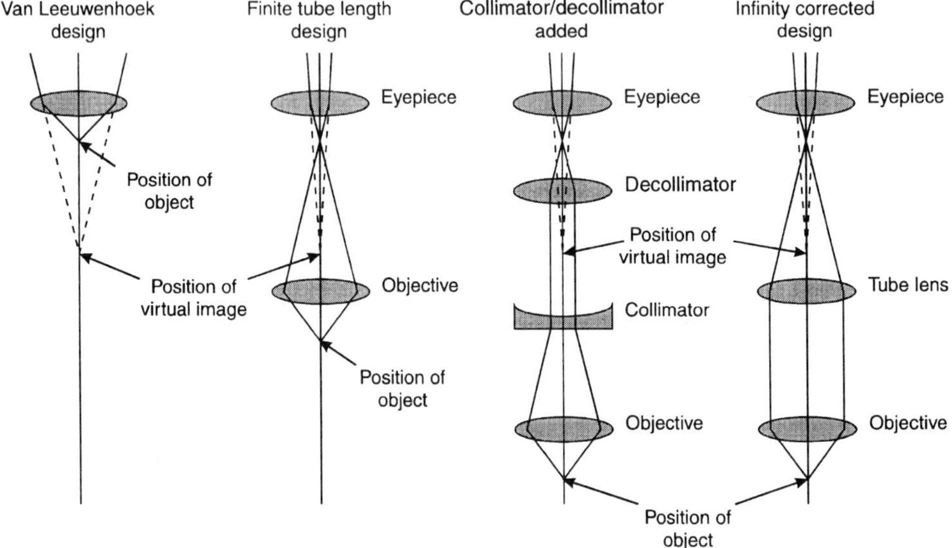

Figure 3.1 Four stages of microscope evolution. Leftmost is the original design of van Leeuwenhoek, with just a single lens; later microscopes used the familiar finite tube length design, featuring two (usually compound) lenses. Next, to allow insertion of, for example, filters into the optical pathway, a collimator/decollimator pair is needed. Rightmost is the modern infinity corrected system.

placed below the inserted optics. This collimator produces a virtual image at infinity, which must be turned back into a real image by a decollimating lens above the inserted optics. Because the virtual image created by the collimator is at infinity, any optical system that changes the optical path length in the intervening space between collimator and decollimator has no effect on the position of the real image or the working distance of the objective. Thus, the performance of the objective is not compromised. The disadvantage of this system is that adding extra lenses leads to a loss of image brightness and contrast, due to absorption of light in the glass, internal reflections at the glass–air interfaces etc.

One solution to this problem is the use of *infinite* optical tube lengths (or infinity corrected optics), which are used in all of the latest models of research microscopes. The infinite tube length design (Figure 3.1) is basically a three-lens design. An objective lens creates a virtual image at infinity. This is converted to a real image by a second lens, the tube lens. As in the case of finite tube length, the eyepiece creates a magnified virtual image of this real image. Again, because the virtual image (created by the objective this time) is at infinity, any change in optical pathway between objective and tube lens has no effect on the performance.

Infinity corrected optics are not necessarily superior to finite tube length optics, unless auxiliary optics such as filters are to be used. In this case, the reduction of the number of air–glass interfaces results in a clearer, brighter image. Nonetheless, infinity corrected microscopes are often superior to older research microscopes, but this may have more to do with general improvements in optical manufacture and the widening range of optical glasses (e.g. ultra low dispersion glasses) than the infinity corrected optical system itself. In any case, almost all research microscopes have a host of filters, dichroic mirrors and other devices that are routinely inserted into the optical path, so infinity correction is bound to pay off.

3.2.2 Resolution limit and the importance of proper illumination

As discussed in Section 2.2.1, due to the wave nature of light, the image of a point source of light cannot be a perfect point. Diffraction at the aperture of the lens causes the point source to be imaged as a small circular disc of finite diameter, surrounded by a series of diffraction rings, or fringes. This diffraction pattern is usually referred to as the Airy disc. The first fringe (nearest to the centre) is usually the only one sufficiently bright to be noticed. The *Rayleigh criterion* for the resolution limit can be expressed as:

$$r_0 = \frac{0.61\lambda}{\mathrm{NA}} \tag{3.2}$$

where λ is the wavelength of light, and NA the numerical aperture. This is related to the physical radius r of the lens aperture and its working distance d as:

$$\mathrm{NA} = n\,\frac{r}{\sqrt{r^2 + d^2}} = n\ \sin\theta_{\mathrm{max}} \tag{3.3}$$

where n is the index of refraction of the medium between lens and specimen, and θ_{max} the maximum opening angle of the lens (see Figure 3.2). Thus, the numerical

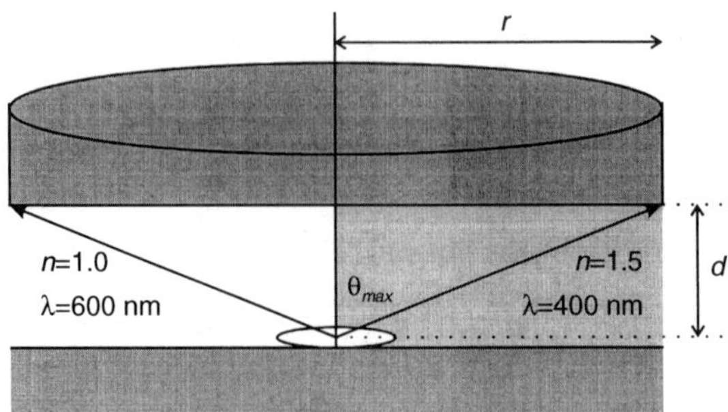

Figure 3.2 The numerical aperture (NA) of a lens depends on the opening angle θ_{max}, which can be computed from the radius r of the aperture and effective working distance d, and the refractive index n of the medium between the object and the lens, due to the decreased wavelength in a high refractive index medium.

aperture is always smaller than the index of refraction of the medium, i.e. NA < 1 in air, NA < 1.3 in water and NA < 1.5 in oil.

It should be noted that only the illuminated part of the aperture contributes to the image formation, and therefore to the image sharpness. If the object emits light itself, e.g. by fluorescence, all angles α will be emitted, so the entire aperture of the lens is illuminated. However, in the case of brightfield microscopy, light rays traverse the object, rather than being emitted. In this case, if no rays enter the object at an angle α, none will enter the objective at that angle. In other words, a part of the aperture will not be illuminated. This will lead to a lower cutoff frequency and decreased resolution. Properly aligned Köhler illumination guarantees that the entire aperture is illuminated, ensuring optimal resolution (Figure 3.3). If the aperture stop of the condenser is misaligned, the point-spread function (PSF) will lose its ideal radial symmetry; if the aperture stop is too small, the PSF becomes broader, reducing resolution.

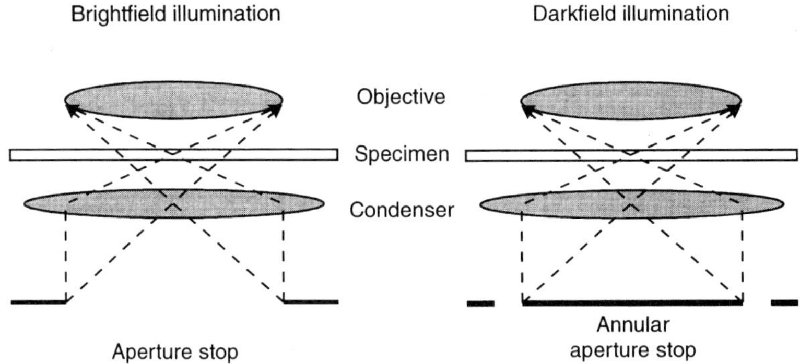

Figure 3.3 Brightfield and darkfield illumination. In brightfield illumination the aperture stop of the condenser is imaged onto the aperture of the objective lens, ensuring optimal resolution. In darkfield illumination, the aperture stop is annular, allowing none of the light from the light source to enter the objective directly.

3.3 CLASSICAL OR WIDE FIELD LIGHT MICROSCOPES

3.3.1 Brightfield and darkfield illumination

Brightfield illumination is the standard form of illumination on most light microscopes. To obtain contrast, objects must differ in light absorption. In optical parlance these are called *amplitude objects*, i.e. objects which alter the amplitude of light waves passing through them. The image intensity may be expressed as:

$$I(x, y) = I_0 e^{-\tau(x,y)} \tag{3.4}$$

in which $\tau(x, y)$ is (for thin specimens):

$$\tau(x, y) = \int_0^{z_{max}} \varepsilon(x, y, z)\,dz \tag{3.5}$$

where z_{max} is the thickness of the specimen, and $\varepsilon(x, y, z)$ is the extinction coefficient at each point in the specimen. Microbes must usually be stained in some way to show up in brightfield images, since the extinction coefficients for visible light are low in most microbes. When using such dyes, $\varepsilon(x, y, z)$ is proportional to the dye concentration, and $\tau(x, y)$ is proportional to the amount of dye along the optical path through the specimen in each part of the image, which can then be determined simply:

$$\tau(x, y) = \ln I_0 - \ln I(x, y) \tag{3.6}$$

Assuming a uniform dye distribution we have:

$$\tau = a_{dye} z_{max} C_{dye} \tag{3.7}$$

where a_{dye} is the molar absorption constant, and C_{dye} is the dye concentration. Equation (3.6) is known as Lambert–Beer's (or Beer–Lambert's) Law; (3.7) is known as Beer's law.

One problem with this densitometric approach to dye quantification is the lack of linearity of the method. For very high $\tau(x, y)$, a small absolute error in the intensity measurement translates to a huge error in the concentration. Conversely, at low dye concentrations, the extinction by the dye is hard to distinguish from that of the cells or tissue.

A further drawback of many traditional dyes for brightfield microscopy is that they are toxic to microbes at useful concentrations for densitometry. Studying living cells requires low concentrations of dyes, or no dye at all. Furthermore, some micro-organisms are notoriously difficult to stain.

Thus there is a need for imaging modes for *phase objects*, i.e. objects that change only the phase of light passing through them in a different way from the surrounding medium. One of the first techniques used to observe unstained specimens by light microscopy was darkfield illumination (Figure 3.3). None of the light from the condenser enters the objective, unless it is scattered or refracted by an object in the optical pathway.

3.3.2 Phase-contrast illumination

Phase-contrast illumination was developed by Zernike in 1935 (Longhurst, 1973) as a further technique to observe unstained biological specimens. In phase-contrast microscopy differences in refractive index (phase delays) are translated into differences in image intensity. The principle is demonstrated in Figure 3.4.

An annular aperture stop in the back focal plane of the condenser is imaged onto an annular phase plate in the back focal plane of the objective. In the absence of phase objects, all undiffracted light passes through the annular region of reduced (or increased) phase delay. Light diffracted by objects passes through the thicker (or thinner) region of the phase plate. Interference of the diffracted and undiffracted light creates the image contrast.

3.3.3 Nomarski differential interference contrast

Nomarski differential interference contrast (DIC) also relies on differences in index of refraction in the specimen. Here, however, two birefringent prisms are inserted in the optical pathway. Birefringent materials have different indices of refraction for two linear polarization directions at right angles to each other. This property is used in the Nomarski set-up to shear the wave fronts of polarized light in say the X and Y planes (Figure 3.5). The first prism is inserted below the condenser, where it produces such a shear, which is precisely compensated by the shear introduced by the second prism, located above the objective. Though there are now phase differences between X and Y polarized light from any phase object in the field of view, these remain invisible, until a polarizer is used. Varying the angle between polarizer and the prisms' X-axis varies the contrast.

Nomarski DIC cannot be used to image birefringent materials, which is of more concern to materials scientists than to microbiologists. Reflected mode Nomarski DIC is also possible and has been used to image biofilm development on metal surfaces for research into fouling (Keevil and Walker, 1992).

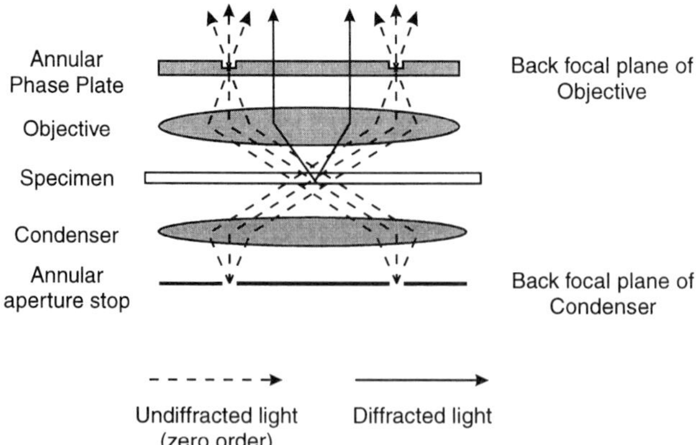

Figure 3.4 Phase contrast according to Zernike. Light diffracted by objects receives a different phase delay in the annular phase plate than the undiffracted light. Interference of the diffracted and undiffracted light creates the image contrast.

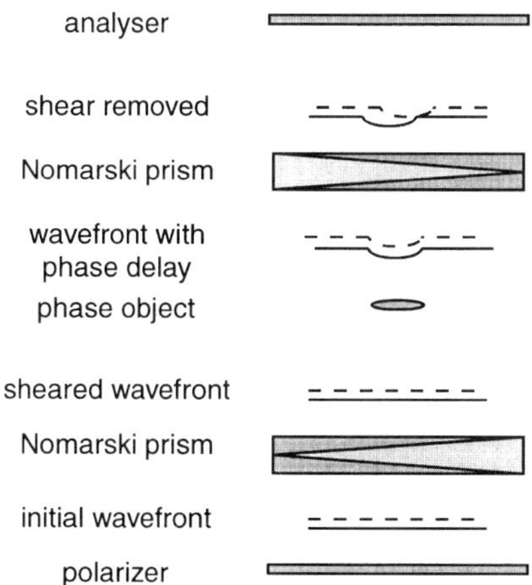

Figure 3.5 Nomarski DIC. Light entering the microscope from below is first polarized linearly at an angle to the x and y axes (the optical axis is z). A birefringent prism below the object (normal microscope) displaces (shears) the wavefronts of incoming light in the x (solid) and y (dashed) directions. A phase object in the optical path then introduces phase delays, distorting the wavefronts. After a second Nomarski prism, the shear is compensated, and the wavefront distortions in the x and y polarized beams no longer coincide, leading to interference if viewed with a second polarizer (or analyser).

3.3.4 Hoffman modulation contrast

Like both previous types, the Hoffman modulation contrast microscope is used to image unstained transparent objects (Hoffman, 1977). It converts phase gradients rather than differences in the image into intensity changes. Assume we have an illuminated slit aperture in the back focal plane of the condenser (Figure 3.6). After passing through the condenser, the (parallel) rays of light then pass through the specimen, and are focused onto the focal plane of the objective. If no phase gradients are present, the slit will be imaged at a specific location on this plane. If we introduce a prism-like object into the parallel rays between the condenser and objective, most light passing through it will be refracted (not diffracted!). A second slit image will then appear in the focal plane of the objective, at a location determined only by the phase gradient (angle of refraction), not by the position of the object in the object plane. If we introduce a graded filter into this plane, parts of objects with different gradients pass through different parts of the filter, and phase gradients are effectively transformed into intensity fluctuations.

Any brightfield microscope can be modified to Hoffman modulation contrast by the addition of (i) an off-axis slit aperture, partly covered by a polarizer below the condenser, (ii) a rotatable polarizer below this slit, and (iii) a modulator filter, which consists of a narrow dark strip, a narrow, grey (15% transmission) strip and a large clear region. The clear half of the slit below the condenser is imaged onto the grey area of the modulator.

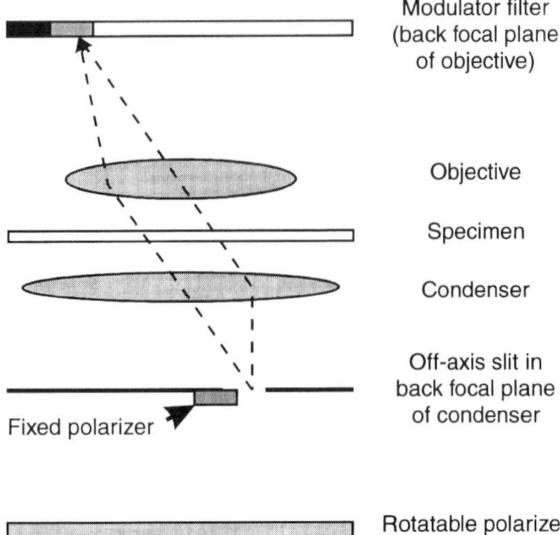

Figure 3.6 The principle of the Hoffman modulation contrast microscope. If no phase gradients are present, all incident light from the off-axis slit passes through the grey part of the modulator filter. Phase gradients cause the light passing through the object to pass through either the black (where it is absorbed) or transparent parts of the modulator filter.

Unlike Nomarski DIC and phase contrast, Hoffman modulation contrast does not rely on interference, and does not have a tendency to produce interference fringes or halos around objects.

3.3.5 Polarization microscopy

The basic polarization microscope is a simple modification of the brightfield microscope, requiring only the addition of two linear polarization filters: one below (the polarizer) and one above (the analyser) the specimen. Unlike the previous contrast techniques, the method relies on differences in optical activity, rather than phase shifts. Optically active substances have different refractive indices for right-handed and left-handed circularly polarized light. Linearly polarized light may be considered to be a superposition of right-handed and left-handed circularly polarized beams of light of equal intensity, and with the phase difference between them determining the orientation of the plane of polarization. Therefore, if a beam of linearly polarized light passes through an optically active medium, the phase difference between the constituent circularly polarized beams changes, and the plane of polarization of the emerging beam is rotated. If the planes of polarization of the polarizer and analyser are aligned, and no optically active substances are in between, a featureless, bright image will be the result. If objects do show optical activity, light passing through them will be partially blocked by the analyser. Conversely, if the polarizer and analyser are at right angles, a featureless, dark image will be the result, unless optical activity in some part of the specimen allows some of the light to pass through the analyser.

This technique is not commonly used in microbiology, due to the small path differences through microbiological specimens.

3.3.6 Epifluorescence microscopes

One way to increase contrast and improve sensitivity to low concentrations of dyes is by using fluorescence (see Chapter 4). Since fluorescent substances emit light at a different (usually longer) wavelength than the exciting light, colour filters can separate the two, enhancing the contrast of the image. Ideally, a fluorescence microscope should illuminate the specimen with high intensity light of the optimal excitation wavelength, while preventing the exciting light from reaching the observer, especially when using ultraviolet illumination. These two design goals have been achieved in a very elegant way through the use of the epifluorescence microscope illumination system developed by Ploem (1967). The epifluorescence microscope is therefore probably the most commonly used fluorescence microscope.

The basic emission and excitation light paths are shown in Figure 3.7. The first part of the excitation light path is at right angles to the optical axis of the microscope. Light from the excitation light source is first collimated and focused on the back focal plane of the objective lens. It is then passed through an excitation filter and bounced off a dichroic mirror mounted in the optical axis of the microscope, at an angle of 45°. This dichroic mirror works as a long pass filter, reflecting light with wavelengths shorter than a particular cutoff. The higher energy part of the excitation light is reflected down towards the specimen. The objective thus doubles as a condenser. By illuminating the specimen from the top, most of the excitation light is automatically removed from the light returning up through the microscope, as the reflectance of most transparent specimens in biology and medicine is low. The emitted fluorescence and reflected excitation light pass back through the objective. When the dichroic mirror is reached, the higher energy light

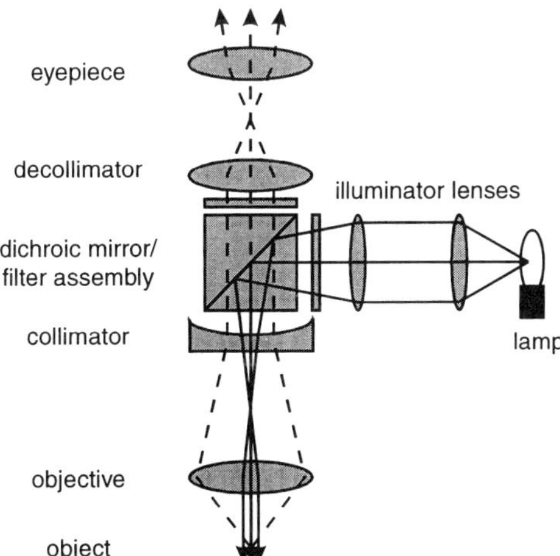

Figure 3.7 Illumination (solid) and emission pathways in epifluorescence microscopy (see text for details).

is again reflected, filtering most of the remaining excitation light from the light reaching the observer. Usually, extra emission filters are mounted above the dichroic mirror. To keep the optical length of the tube at 160 mm, a collimator/decollimator lens pair is mounted above and below the dichroic and filter assembly. Microscopes with infinity optics do not need these lenses. Using the objective as a condenser has the advantage that only a very small area of the specimen is illuminated, and that the illumination is intense, especially with high NA immersion objectives.

Other methods of fluorescent illumination, such as brightfield and darkfield, require more filters to eliminate the short wavelengths, and may have problems with thick specimens, since the bottom is illuminated and fluoresces most strongly, but the emission is observed from the top, so the light may be partially absorbed in the top parts of the specimen. A more detailed discussion of fluorescence microscopy is given in Chapter 4.

3.3.7 Three-dimensional imaging using wide field microscopy

Although confocal microscopes are probably the most commonly used three-dimensional light microscopes, classical microscopes may also be used to obtain three-dimensional images in two ways: (i) by tilting the specimen, creating stereo images (Willis *et al.*, 1993), and (ii) by digital optical section microscopy (Agard *et al.*, 1989). Both techniques require considerable computing power to reconstruct the three-dimensional image from either (i) two or more images taken at different angles through specimens, or (ii) multiple images taken at different depths of focus in the specimen. In the latter case, deconvolution of the highly complex three-dimensional PSF is performed for reconstruction. This process is hampered by the fact that the PSF is very extended in the z-direction. Besides, the three-dimensional PSF may differ significantly from the theoretical shape, which means that it must be determined experimentally for most reconstruction methods. A new deconvolution technique, called blind deconvolution (Holmes *et al.*, 1995), may be used to circumvent the latter problem. Given the small size of bacteria, the missing information in spatial frequencies in the z-direction may limit the practical value of this technique in microbiology.

3.4 CONFOCAL MICROSCOPES

3.4.1 The principle

The confocal scanning laser microscope (CSLM) was first developed by Brakenhof *et al.* (1979). Since then, this type of instrument has had a large impact on all fields of biology and medicine in which classical fluorescence microscopy was used before. Numerous modes of confocal microscopy exist (Wilson and Sheppard, 1984), but this discussion focuses on confocal fluorescence microscopy, which is by far the most popular imaging mode.

In CLSM light from a laser is focused onto the specimen, forming (ideally) a diffraction limited spot of intense exciting light on the specimen. The fluorescence from this spot is imaged onto a pinhole aperture, which will only allow the light from the plane of focus to pass unattenuated. The light passing through the pinhole is consequently imaged onto a detector, or multiple detectors, after passing through wavelength-sorting beam splitters. By scanning either the stage or (more practically) mirrors used to image the spot on the specimen, an image is formed. The practical upshot of this is twofold: (i) the image is formed only of light coming from a narrow region around the plane of focus, and (ii) the PSF of the CSLM is the product of the PSFs of the illumination and imaging systems. Since these systems are usually one and the same lens, this yields a confocal PSF equal to the square of the classical PSF (disregarding the wavelength dependence of the PSF). This improves the lateral resolution by some 20–60% (Brakenhoff *et al.*, 1989).

The ability to exclude most light from out-of-focus parts of the specimen means that three-dimensional images can be formed by stacking a series of images taken at different depths of focus (see Chapters 16 and 17, and Plates 5–7). Due to the relatively small extent of the PSF of a CSLM in the z-direction, deconvolution is not essential to form a three-dimensional image of reasonable quality, although significant improvements are possible when they are used (Van Kempen *et al.*, 1997). Even if just a single two-dimensional image is obtained, exclusion of out-of-focus light leads to greatly improved contrast.

Chapters 16 and 17 deal with many practical aspects of confocal microscopy in microbiology. More information on the principles and practice of confocal microscopy can be found in Pawley (1995); a further review can be found in Laurent *et al.* (1994). White *et al.* (1996) discuss a number of pitfalls in quantitating the results of three-dimensional fluorescence microscopy, in particular distortions caused by differences in index of refraction within the specimen.

3.4.2 Multiple photon excitation

Two-photon and three-photon excitation systems are a relatively recent development (Denk *et al.*, 1990). Unlike any other fluorescence technique used in microscopy, multiple photon excitation uses light of a wavelength longer than the emission wavelength to excite fluorescence. Using light of a longer wavelength requires two or more photons to strike the fluorochrome simultaneously, to supply sufficient energy for excitation. The probability that n photons hit a target simultaneously is proportional to the intensity raised to the nth power. Suppose we illuminate a thick specimen with a conical beam focused on a spot at $z = 0$. At this spot the radius of the beam is $r_0 > 0$, determined by the diffraction limit. Disregarding extinction, the radius of the beam above and below the plane of focus becomes (roughly):

$$r_{\text{beam}}(z) = r_0 + \alpha\,|\,z\,| \tag{3.8}$$

where α is the tangent of the half opening angle of the illuminating cone of light. The intensity as a function of z is simply:

$$I(z) = \frac{I_0}{\pi r_{\text{beam}}^2(z)} = \frac{I_0}{\pi(r_0 + \alpha\,|\,z\,|)^2} \tag{3.9}$$

In the case of n-photon excitation, the effective illumination is:

$$I_{\text{eff}}(z) = I^n(z) = \frac{I_0^n}{\pi^n (r_0 + \alpha |z|)^{2n}} \tag{3.10}$$

Since the illuminating intensity is far higher in the plane of focus, the level of excitation outside this plane falls off very rapidly (see Figure 3.8).

Now suppose we build an image by scanning the specimen in x and y. Suppose each sample in the $z = 0$ plane is illuminated for a time t_0. The total illumination time as a function of z becomes:

$$t_{\text{ill}}(z) = \pi (r_0 + \alpha |z|)^2 t_0 \tag{3.11}$$

During each scan, the total effective irradiance per scan as a function of z is (again disregarding extinction in the specimen):

$$I_{\text{eff, total}}(z) = t_{\text{ill}}(z) I^n(z) = \frac{t_0 I_0^n}{\pi^{n-1} (r_0 + \alpha |z|)^{2n-2}} \tag{3.12}$$

Note that for single photon excitation ($n = 1$), this is independent of z, so the entire specimen is bleached (Figure 3.8). All of this leads to three effects: (i) the extent of the PSF in the z-direction becomes smaller, (ii) photobleaching of the fluorochrome and photo-damage to the specimen are reduced, and (iii) the depth of penetration into the specimen is increased, since the parts of the specimen above the plane of focus (in epi-illumination) are not excited and do not therefore cause extinction. An added bonus is that no (expensive) UV laser or UV-corrected optics are needed to excite blue fluorochromes. A pulsed laser, yielding very intense short pulses (in the

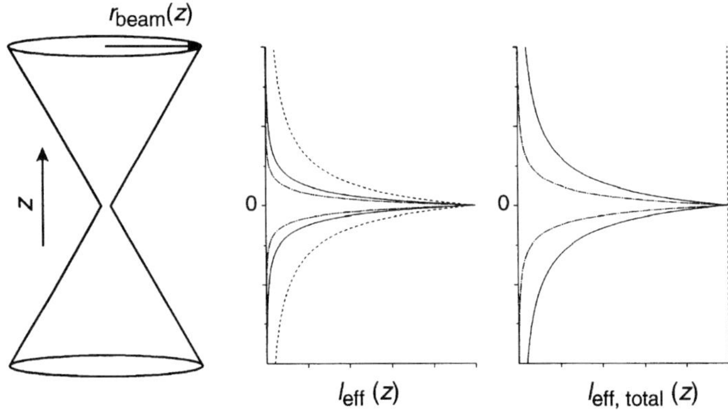

Figure 3.8 Multiple photon excitation. When a conical beam illuminates a spot at $z = 0$, the effective illumination I_{eff} falls off far more rapidly as a function of z for two-photon (solid curve) and three-photon (dash-dot) than single-photon (dashed) excitation. Per scan of an entire optical section, the total effective irradiance $I_{\text{eff,total}}$ is independent of z for single-photon excitation, but falls off dramatically for two- and three-photon excitation.

order of 100 femtoseconds) and with considerably higher power (25–50 mW instead of 3–5 mW) is needed (Denk *et al.*, 1990).

Because the excitation itself is limited to a specific plane, observation of the fluorescence is usually not performed confocally, or with the confocal pinhole in front of the detector fully open (Summers *et al.*, 1993).

3.5 DEPARTURES FROM THE LINEAR, SHIFT-INVARIANT MODEL

The ideal microscope can be modelled as a linear shift-invariant (LSI) system, followed by a pure magnification system (Young, 1989). This is of some importance to image analysis, as LSI systems have properties that may be of use in image restoration (Section 1.3.1). Linearity is not often violated in optics as it is in detector systems (Chapter 2), but shift invariance is, due to numerous aberrations. Aberrations come in two basic types: chromatic and achromatic. The former are due to varying indices of refraction as a function of wavelength in glass, and can only be detected when light of different wavelengths is used. The latter apply to all wavelengths equally, and can therefore be seen when using monochromatic light from a laser. Some of these are independent of the angle of incidence, and are therefore captured within the definition of the LSI model. An example is spherical aberration, which is due to the fact that rays at the edge of the lens are focused slightly closer to the lens than rays closer to the optical axis. In the following sections several aberrations are discussed which do depend on angle of incidence and therefore violate shift invariance. In general, errors of this type increase with the angle of incidence, so the problems they pose increase with the field of view of the camera or eyepiece used. Some, but by no means all, of these defects can be corrected digitally. However, digital correction never eliminates the problems completely, and highest grade optical systems are always to be preferred.

3.5.1 Coma and astigmatism

These two achromatic aberrations cause the PSF of the optical system to vary as a function of angle of incidence. Coma increases linearly with the angle of incidence, whereas astigmatism increases with the square of the angle of incidence. If the resolving power of the system, as measured by any of the methods in Young (1989) using monochromatic light, varies with the distance to the optical axis, and refocusing does not improve the edge resolution, one or both of these errors is present. By fitting the resolution limit as a function of the angle of incidence to a parabola, the magnitude of both aberrations may be measured.

3.5.2 Chromatic aberration

Chromatic aberrations can have two distinct effects: (i) different wavelengths are imaged at different positions along the optical axis (axial chromatic aberration), and (ii) different wavelengths have different magnifications (lateral chromatic aberration). Both types of error cause blurring of the image, which can be reduced by using narrow wavelength ranges for imaging, or more complicated optical design. Achromatic and apochromatic objectives are corrected for two or three wavelengths, respectively, but either may show residual aberrations when used in critical applications. Most objectives are only corrected for the visual range of wavelengths (400–700 nm). Increasing use of deep red and near infrared fluorochromes, and the use of (UV excited) blue fluorochromes means that there is a need for objectives with an extended wavelength range. These are available, but are costly.

Digital image processing can in principle restore these errors if several more or less monochromatic images in different wavelength bands of the same field of view can be obtained separately (Waggoner *et al.*, 1989; Bruins *et al.*, 1994).

In two-dimensional microscopy, lateral and axial chromatic aberrations require separate methods of treatment to correct them. Scale changes due to lateral aberration can be corrected by geometric procedures (Sections 1.2 and 3.5.6). Axial chromatic aberrations can be corrected by measuring the shift of the focal plane. By altering focus using precise stage control in the *z*-direction, or accurate autofocus (Section 3.7.2), images in each wavelength band could be obtained in the appropriate plane of focus. I am not aware of any application of these techniques in microbiology. In three-dimensional microscopy, both axial and lateral chromatic aberration can be considered as geometric errors, and can therefore be corrected in a single step (Section 3.5.6).

3.5.3 Image misalignments in multiple colour fluorescence microscopy

Apart from scale changes due to residual chromatic aberrations, images obtained at different wavelengths through different filtersets may be shifted in position (Plate 1). These shifts have two components: (i) a fixed part due to prismatic errors in the filters, and (ii) a random part caused by mechanical disturbance of the microscope during the process of switching filters. The latter part can be minimized by automation of the switching process, thus minimizing any (variable) shake caused by human operators. The former can be measured and corrected with geometric procedures (see Plate 1(a, b) and Section 3.5.6). Errors of this type have been reduced greatly by improved optical quality of filters in recent years. Plate 1(a) shows the alignment errors for an older research microscope (ca. 1975). Quite clearly the errors are unacceptable, and digital correction is required (Plate 1(b)). Plate 1(c) shows alignment errors for a newer research microscope (ca. 1995). Quite clearly, the alignment of images taken through different filters is far better, yet not perfect, as can be seen in Plate 1(d), where the green and blue components have been subtracted. Even such small misalignments can cause problems in critical applications (see Sections 10.5.1 and 11.4.2g), and correction becomes necessary (Waggoner *et al.* 1989; Bruins *et al.*, 1994).

One way to counter such problems is through the use of multi-band filtersets. These filtersets allow simultaneous observation of two or three fluorochromes. Originally intended mainly for visual inspection and colour photography, they have been found suitable for quantitative work (Lowy, 1995). Observation of single fluorochromes through such filtersets can be achieved by inserting extra bandpass filters in the excitation pathway. Switching these does not affect image position. A possible danger is increased cross-talk, i.e. fluorescence of one fluorochrome 'leaking' into the other pass bands, compared to the use of separate filtersets for each fluorochrome, but this does not appear to be a large problem in practice (Lowy, 1995).

3.5.4 Distortion

If the position (x', y') in the image plane is no longer a linear function of object position (x, y) distortion of the image occurs. Another way of looking at distortion is to say that the magnification varies with image coordinates if distortion is present. Pincushion distortion occurs when the magnification increases with the distance to the optical axis, whereas barrel distortion occurs when the magnification decreases with the distance to the optical axis. Jericevic *et al.* (1989) give a detailed description of how to measure and correct for distortion using digital image processing (see also Section 3.5.6).

Global correction of distortion is essential if information is derived from the global geometry of the image, i.e. if distances of the order of the width of the image are to be measured. When the objects of interest are small compared to the image width, and the relative positions of the objects are not of any interest, the resulting errors may be negligible.

3.5.5 Curvature of field

When imaging a flat object, such as a slide, the image plane should be just that: a plane. However, in many optical systems the image plane is curved. Visually, this curvature may not be apparent, since the eye can accommodate very rapidly while scanning the field of view. When using flat detectors in cameras, whether digital or photographic, it may be a serious problem. When the centre is in focus, the edges are not, causing the effective PSF to vary with the distance from the optical axis. However, unlike any other aberration, the edges can be focused properly, but only when the centre is out of focus. There are no practical digital correction methods in two-dimensional imaging, other than limiting the useful field of view.

When using three-dimensional imaging, this problem becomes one of geometric distortion: each object is imaged at a z-position depending on its true z-position and the distance to the optical axis. Therefore, this problem can be corrected using methods described in the next subsection.

3.5.6 Correcting geometric errors

Correcting all geometric errors in images, whether distortion, chromatic errors or curvature of field in three dimensions, may be performed by geometric transformation of the image. All geometric errors (in two dimensions) cause a point which should have been imaged at (x, y) to be imaged at (x', y'). Mathematically this can be expressed as:

$$\mathbf{x}' = \mathcal{G}(\mathbf{x}) \tag{3.13}$$

where we use vector notation $\mathbf{x} = (x, y)$ (or $\mathbf{x} = (x, y, z)$ in three dimensions) for convenience, and $\mathcal{G}(\mathbf{x})$ denotes the geometric distortion which maps $\mathbf{x}$ to $\mathbf{x}'$. The aim of geometric correction is first to determine the nature of the mapping, and then to invert this mapping:

$$f(\mathbf{x}) = g(\mathcal{G}(\mathbf{x})) \tag{3.14}$$

where f is the corrected image and g the distorted image. In the case of shifts and linear scale changes, the transformation $\mathcal{G}(\mathbf{x})$ is a linear one:

$$\mathbf{x}' = \mathcal{G}(\mathbf{x}) = a\mathbf{x} + \mathbf{b} \tag{3.15}$$

where a is the scale change and $\mathbf{b}$ the shift. More generally, the mapping $\mathcal{G}(\mathbf{x})$ can be modelled as a polynomial function.

Determining this polynomial function is fairly straightforward using fiducial markers, which may be either internal or external to the image. Fiducial markers are objects that occur in all colour components in the case of correction of chromatic and image misalignment errors, or which have a known distribution in the image (e.g. a grid in the case of distortion correction in monochrome images). Using external markers means that a separate slide containing the markers is used to measure the mapping function $\mathcal{G}(\mathbf{x})$. This function is then applied to any number of images recorded subsequently with the same instrument and settings. This can only be done if the geometric errors are stable in time. Distortion and chromatic errors usually are, but image misalignments need not be due to mechanical instabilities. Internal markers are objects in the image itself of which it is known that they have a fixed position in the specimen, and that they are visible in all wavelength bands under consideration.

Once a suitable set of fiducial markers has been selected, their positions in both the distorted image and the reference image (in which there is no distortion) are determined, usually using the centre of gravity as a measure for location (Waggoner *et al.*, 1989). Performing a linear, or polynomial, least squares fit of the position ($\mathbf{x}'$) in the distorted image, as a function of the 'true' position ($\mathbf{x}$) in the reference image yields an approximation of the mapping function $\mathcal{G}(\mathbf{x})$. The image can then be transformed by applying (3.14).

Applying this approximation of $\mathcal{G}(\mathbf{x})$ using (3.14) to digital images requires some further thought. Consider the image in Plate 1(a). Let us assume that the green component has the correct geometry, and that the blue and red need correction. We must therefore compute which position $\mathbf{x}'$ in the red (or blue) image corresponds to each position $\mathbf{x}$ centred on a pixel in the green image. In general, each $\mathbf{x}'$ will not be centred on a pixel in the red (blue) image. The simplest approach is to use the nearest pixel position instead. This can lead to a somewhat blocky appearance of the result, especially if a number of corrections are done sequentially (Figure 3.9). A better approach is to use the pixels surrounding the position $\mathbf{x}'$ in the distorted image (four in two dimension, eight in three dimensions) and perform a bilinear (trilinear in three dimensions) interpolation. Suppose our distorted coordinate pair (x', y') lies between (n, m), $(n + 1, m)$, $(n, m + 1)$, and $(n + 1, m + 1)$. We first interpolate in the x-direction:

$$g(x', m) = (n + 1 - x')g(n, m) + (x' - n)g(n + 1, m) \tag{3.16a}$$

$$g(x', m + 1) = (n + 1 - x')g(n, m + 1) + (x' - n)g(n + 1, m + 1) \tag{3.16b}$$

After this we interpolate in the y-direction:

$$f(x, y) = g(x', y') = (m + 1 - y')g(x', m) + (y' - m)g(x', m + 1) \tag{3.16c}$$

This correction, using a linear distortion model determined using external fiducial markers, was used to obtained Plate 1(b). The red and green images are now aligned properly. Equations (3.16a–c) represent a digital form of (3.14) and can easily be extended to three dimensions, by adding an extra interpolation.

One final caveat concerns the image borders, where data may be missing in one or more colour plane. In this case, the image must be cropped to exclude such areas.

3.5.7 Uneven illumination and vignetting

Any decrease in the transmission of light from object to image plane with increasing angle of incidence is called vignetting. Apart from transmission losses between object and image, the illumination of the object might also fall off with increasing angle of incidence, due to vignetting in the illumination pathway. These errors are unavoidable: even in theory, light transmission falls off as the square of the cosine of the angle of incidence. Fortunately, these errors are easily corrected by digital image processing, together with errors such as dust in the optical pathway, spatial variations of the sensitivity of the camera (Section 2.4.7) etc. All of these errors together are called *shading*; the correction process is called deshading, shading correction or flat-field correction.

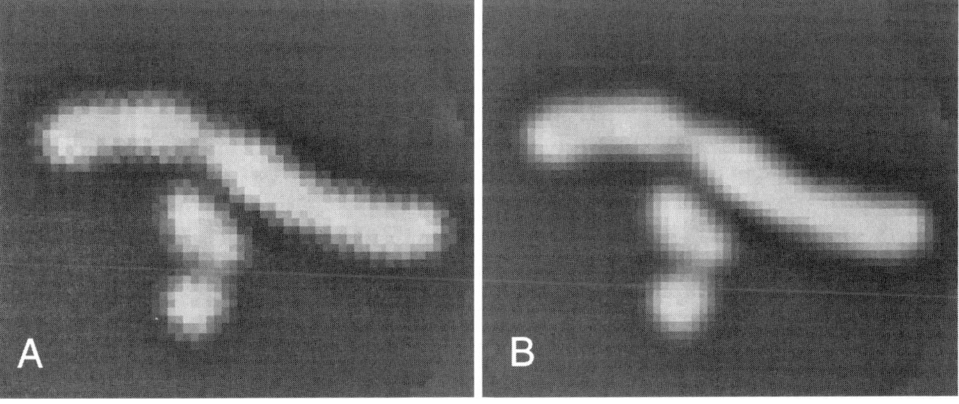

Figure 3.9 Geometric image transform (rotation over 30°) using (A) nearest neighbour approximation, and (B) bilinear interpolation. The images have been enlarged to show pixel structure. Interpolation clearly provides a better approximation than the nearest neighbour approach.

3.6 SPECIAL TECHNIQUES

3.6.1 Total internal reflectance fluorescence microscopy

Total internal reflectance fluorescence (TIRF) microscopy (Axelrod, 1989) attempts to provide a single, very thin optical section of a fluorescently stained specimen. A stained specimen mounted on a slide is illuminated by light of the appropriate excitation wavelength from below, in such a way that total internal reflectance occurs at the boundary between glass and specimen (Figure 3.10). In that case an exponentially decreasing 'evanescent wave' penetrates a very short distance (in the order of a wavelength or less) into the specimen. Put another way, some photons 'tunnel' through the barrier at the interface, and may therefore excite fluorochromes on the part of the surface of the specimen that is attached to the glass. Since the bulk of the cell is not illuminated, background fluorescence may be reduced greatly. Obviously, the method is limited to the study of the surface of cells.

3.6.2 Fluorescence correlation spectroscopy

Fluorescence correlation spectroscopy (FCS; Magde *et al.*, 1974) is a technique to determine the motility of fluorochromes within the specimen. The principle is simple: a small volume of the specimen is illuminated (usually with a laser) and the fluorescence from that volume is measured as a function of time. If the number of fluorochromes in the volume is constant on average, the signal should be constant, with variations caused by two factors: (i) photon noise, and (ii) fluctuations caused by fluorochromes moving in and out of the volume element, due to diffusion, bulk flow etc. The first form of noise is spectrally white: there is no correlation between measurements at time t and $t + \Delta t$ for any Δt. By contrast, the other type of noise shows increasing correlation with decreasing Δt. Movements of molecules in and out of the volume element take time, so long-period variation is likely to be larger than short-period. In theory, as Δt approaches zero, the autocorrelation function $G(\Delta t)$ of the fluorescence signal $F(t)$ approaches the variance of the 'molecular'

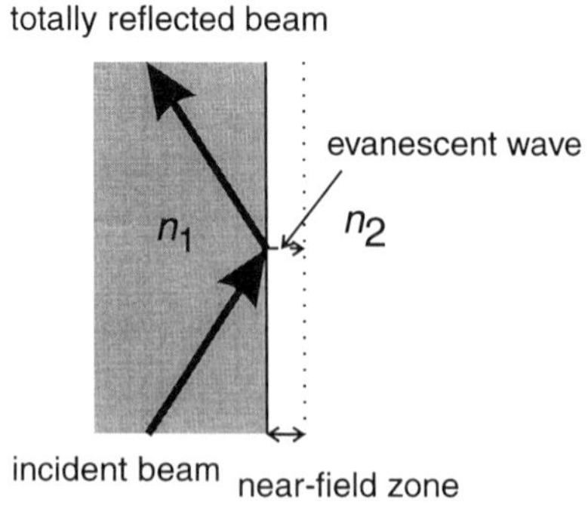

Figure 3.10 Optical sectioning in TIRF microscopy. Light incident at an angle larger than the critical angle $\theta_c = \arcsin(n_2/n_1)$ at the interface between media of high (n_1) and low (n_2) refractive indices reflects of the boundary, yet an exponentially decaying, evanescent wave penetrates a small distance into the specimen, exciting fluorescence only there.

noise, which yields the number of fluorescent entities (which may be clusters of multiple labelled molecules):

$$\lim_{\Delta t \to 0} \frac{G(\Delta t)}{\bar{F}} = \frac{1}{N} \tag{3.17}$$

where $\bar{F}$ is the mean fluorescence and N the number of fluorescent entities. The autocorrelation function $G(\Delta t)$ is defined as:

$$G(\Delta t) = \int_{-\infty}^{\infty} F(t)F(t + \Delta t)\, dt \tag{3.18}$$

When M discrete samples have been obtained, (3.11) takes the form of a finite sum. Autocorrelation functions can be computed very efficiently using the fast Fourier transform (FFT, Section 1.3.1). By fitting a theoretical curve (e.g. Gaussian; see Petersen *et al.*, 1993) the limit in (3.17) can be taken and the number of clusters determined. Computing the variance of the signal directly is of no use, since this is dominated by photon noise. FCS can also be used to determine diffusion rates, membrane fluidity etc.

Petersen *et al.* (1986, 1993) describe two related techniques, called *scanning fluorescence correlation spectroscopy*, and *image correlation spectroscopy*, in which the spatial, rather than temporal, autocorrelation functions are obtained. The first works by scanning a laser over the specimen at a constant rate and determining the fluctuations in the detected signal. The second acquires images, and computes the two-dimensional autocorrelation of grey levels as a function of x, and y. The small size of bacteria may limit the use of these latter techniques in microbiology, since a central assumption is that the mean concentration of fluorescent molecules is constant over a large area.

3.6.3 Fluorescence recovery after photobleaching

Fluorescence recovery after photobleaching (FRAP) is another technique to assess fluorochrome motility (Wolf, 1989). After measuring the fluorescence intensity in a volume of the specimen, intense spot illumination (by a laser) is used to destroy the fluorochromes within it. After this, the time required for the fluorescence to return to its original strength (or some defined percentage) is determined. The recovery speed is a measure of the diffusion rate. This technique is discussed in more detail in Chapter 4.

3.6.4 Delayed fluorescence microscopy

An epifluorescence microscope can be modified for delayed fluorescence imaging (see Chapter 4) by the addition of two rapid shutters: one in the illuminating light path and one just below the eyepiece or camera (Tanke, 1989; Marriott *et al.*, 1991). The first shutter is used to produce repeating short bursts of excitation of fluorescence. The second is opened and closed out of phase with the first, and allows only photons emitted by slowly decaying fluorescence to reach observer or

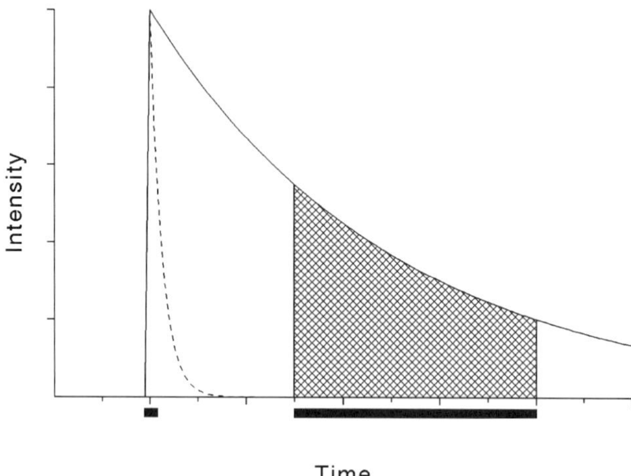

Figure 3.11 Delayed fluorescence microscopy: after a short pulse of excitation (short bar) a shutter in front of the camera is opened (long bar). The light detected during this interval (hatched area) is delayed fluorescence (solid curve) from special fluorochromes or phosphors, whereas short-lifetime autofluorescence (dashed curve) is rejected.

camera. Early experiments used spinning discs as shutters, but synchronization and vibration problems caused these to be replaced by liquid crystal shutters, which have no moving parts, but do relinquish sensitivity, since they have (at most!) 50% transmission. Using special phosphorescent or delayed fluorescent probes, the system is capable of reducing background fluorescence very effectively, since most autofluorescence has short lifetimes (Figure 3.11).

3.7 MICROSCOPE AUTOMATION

The goal of image analysed microscopy is automation of the microscopic measurements. Adding a camera and image processing system only automates the eye and brain of the observer. Complete automation also requires automation of the hands: the computer must have full control over all microscope settings. Many research microscopes either have automation built in, or can be upgraded by adding motorized stage movement, autofocus, or illumination or filter wheel control. Such additional features do come at a cost, so several low-budget solutions have been developed by researchers in a number of laboratories (e.g. exposure and illumination control, Wilkinson *et al.*, 1993; Langendijk *et al.*, 1995).

Whether factory made or 'home grown', automated control of the microscope can improve reproducibility of the results, increase the feasible maximum number and size of samples that can be processed, and reduce tedium (see Price and Gough, 1994; Bloem *et al.*, 1995). However, full automation does require good quality slides, or sophisticated software that can reliably discriminate objects of interest from contamination. Here, as in many other cases, there is a trade-off between the time (and money) spent achieving full automation, and the actual cost of a small amount of user interaction during each measurement.

3.7.1 Stage control

One feature that can readily be automated is the motion of the microscope stage. By linking stepper motors to the control wheels of the stage, x, y and z motion may be controlled fully by a computer. All leading manufacturers of microscopes support the addition of automated stage motion.

3.7.2 Autofocus

Autofocus systems may use a number of criteria to determine the best focus position, based on either image contrast, image sharpness, or a combination of both (Groen *et al.*, 1985; Price and Gough, 1994). Simple criteria may be the sum of squares of a gradient, or Laplacian filtered image (sharpness criteria) or the grey-level variance or standard deviation (contrast criteria). Combined sharpness criteria are the grey-level variance of a sharpened image, or autocorrelation function based criteria. Price and Gough give a comparison of 11 different criteria. The best focus was obtained using the variance of Laplacian filtered images, or the variance of the sharpened image.

Whatever the focus criterion, the basic procedure is the same. The first step is to take images at different focus (z) positions, and the criterion $C(z)$ is calculated for each z. After this, the peak of $C(z)$ must be determined. This can be done by simply taking the z value at which $C(z)$ reaches a maximum, or by computing a weighted average (Price and Gough, 1994):

$$z_{\text{opt}} = \frac{\sum_{z} zC(z)}{\sum_{z} C(z)} \tag{3.19}$$

This latter measure can improve precision because it uses all data, rather than just the maximum. A problem exists when using autofocus on fluorescence images. During the acquisition of the images for autofocus, the fluorescent dyes used may fade (photobleaching; see Chapter 4). Besides, living cells may be killed or damaged either by the exciting (UV) light or by toxic reaction products of the degraded fluorochromes. Due to these considerations it is probably better to focus under, for example, phase-contrast illumination, and to acquire the fluorescence image after switching the illumination mode. One word of warning is called for, since Price and Gough (1994) reported that the position of focus for phase-contrast and (blue) DAPI fluorescence can be significantly different, possibly due to the distribution of the stain within the cell, or simply axial chromatic aberration. This difference was systematic, and could be corrected in their case.

Research microscopes of all leading manufacturers can be equipped with autofocus.

3.7.3 Illumination and filter control

Many high-end research microscopes also allow addition of full computer control over the full illumination pathway. This is ideal for any application that requires multiple images in different illumination modes (see Chapters 10 and 11). When changing the settings manually, the operator may inadvertently jar the stage, causing misalignment of the resulting images. By eliminating this source of error, computer control ensures optimal alignment of images taken using different illumination modes.

3.7.4 Registration of microscope settings

A useful addition, not only for computer analysed microscopy, but also for routine photomicrography, is the possibility of automatically logging the microscope settings. Not only does this provide accurate records automatically (e.g. for inclusion in laboratory information management systems), but it is also possible for each user to store preferred settings for different experiments or staining techniques. The settings can be recalled with the same ease with which they were stored, greatly reducing the probability of errors.

3.8 SCANNING PROBE MICROSCOPY

When a higher resolution than can be obtained by light microscopy is needed, the traditional approach has been the use of electron microscopes. Although extremely high resolutions may be achieved, the fixation required for these techniques may cause severe distortion of the specimens (Vardi and Grover, 1992), and precludes the study of living cells at high resolution. The recent development of scanning probe microscopy (SPM), in which a probe scans the surface of a specimen under a wide range of conditions, allows imaging and manipulation of biological specimens at extremely high (molecular level) resolution under physiological conditions (Fritz *et al.*, 1994, 1995). In the following sections a number of different types of scanning probe microscopes and (potential) applications in microbiology are discussed.

3.8.1 Atomic force microscopy

The atomic force microscope (AFM) is one of the simplest forms of SPM. While scanning the surface of the specimen, the attractive or repulsive force between the tip of the probe is measured. This information may be used in different ways to create an image. One approach keeps the force on the tip constant by varying the z position of the tip. These 'equi-force' z positions are then used to create an 'altitude map' of the specimen, which may be visualized in various ways (Figure 3.12). Alternatively, the probe may be held at a constant z position during the scan, and variations in the force are transformed into grey levels instead. To prevent damage to fragile biological specimens, the so-called tapping-mode has been developed (Hansma *et al.*, 1994). In this case, the probe is lifted off the surface of the specimen, moved to a new (x, y) position, lowered until the force on the probe reaches a predefined value, lifted again, and so on (Figure 3.12). This prevents damage caused by scraping the probe over the surface.

AFM has been used to image many biomedical specimens, from strands of DNA immobilized on mica (e.g. Hansma *et al.*, 1993), to whole, living cells (e.g. Braunstein and Spudich, 1995). The structure of bacterial membranes has been investigated using AFM (Firtel and Beveridge, 1995; Müller *et al.*, 1995), including the binding

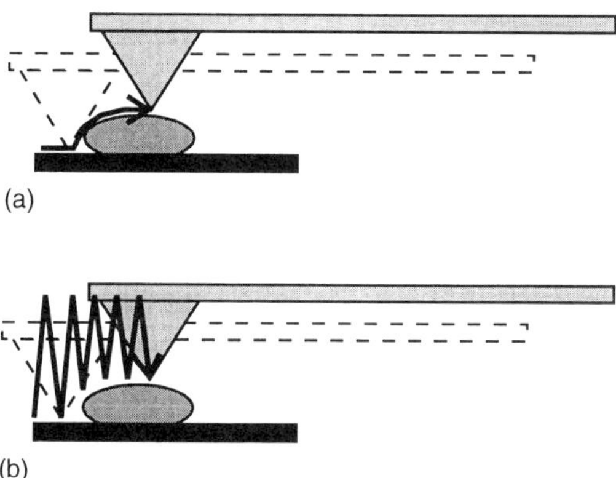

Figure 3.12 Standard and tapping mode AFM. In (a) standard mode AFM the probe tip scans the surface of the specimen along a curve at which the tip experiences a constant force from the specimen. In (b) tapping mode AFM the probe is repeatedly lifted from and lowered onto the specimen to a point at which the tip experiences a given force. The shape of the specimen is derived from the track followed by the probe.

of antibodies to such membranes (Müller *et al.*, 1996). Applications on whole bacteria are fewer (Firtel and Beveridge, 1995), although the possibility has been demonstrated by images of accidentally contaminating bacteria in other specimens (Figure 3.13). AFM may be used in combination with immunogold labelling, allowing accurate location of epitopes under physiological conditions.

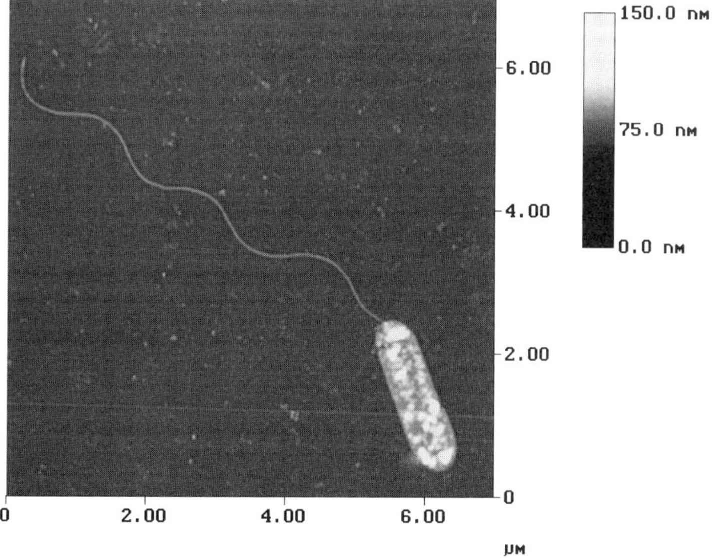

Figure 3.13 Image of an unknown bacterium found on AFM specimen (image made using Nanoscope® (Digital Instruments, Santa Barbara, CA), courtesy of J.S. Siino and E.M. Bradbury, Department of Biological Chemistry, University of California at Davis).

3.8.2 Near field scanning optical microscopy

Near field scanning optical microscopy (NSOM) is capable of optical imaging without being limited by diffraction, and therefore by the wavelength of light (Lewis, 1991; Morson *et al.*, 1995). The principle is based on tunnelling of photons in much the same way as in the case of TIRF (Section 3.6.1). In this case, light tunnels through the tip of the probe, which is far smaller than the wavelength of light. An exponentially decaying evanescent wave around the tip illuminates only a minute region on the surface of the specimen. The resulting fluorescence is measured and the probe is moved to the next position. The probe tip can be used as an atomic force microscope as well, obtaining structural and fluorescence data simultaneously. Like TIRF, NSOM is limited to surface fluorescence, but unlike TIRF it has a resolution in the x and y directions far beyond conventional microscopes, and is not restricted to flat surfaces.

A powerful demonstration of the capabilities of NSOM has been given by Ha *et al.* (1996), who were able to measure interactions between individual pairs of fluorochromes, using multicolour NSOM.

3.8.3 Magnetic force microscopes

In magnetic force microscopy, the tip of the probe is magnetic, allowing magnetic field strengths to be determined within the specimen. Originally used in research into, for example, magnetic media for data storage, an application into measurement of magnetic fields of magnetosomes of magnetotactic bacteria has been reported (Dahlberg *et al.*, 1995).

3.8.4 Simultaneous classical and scanning probe microscopy

A number of manufacturers have reduced the size of SPM instruments so they can be fitted onto normal (e.g. Surface Imaging Systems, Germany; DME, Denmark) or inverted (Digital Instruments, USA) optical microscopes. Others have integrated SPMs into a confocal laser scanning microscope (Schabert *et al.*, 1994).

3.9 FUTURE PROSPECTS

The quality and range of instruments available has increased over the years. No residual aberrations can be seen in visual work in current models of high-end research microscopes. However, for the most exacting, quantitative work, it is essential to measure the small errors (particularly chromatic and geometric errors) that are still present, whatever the claims of the manufacturers. Very often these claims are based on work with large, eukaryotic cells, not the small prokaryotes most microbiologists work with. A shift of 0.5 μm caused by residual optical wedge in filters may be insignificant when imaging cells several tens of micrometres across, but is unacceptable in the case of bacteria, which may be only 0.2–0.5 μm wide. Misidentification of cells in images taken through different filters becomes inevitable. Digital image processing techniques may be used to correct such residual errors.

In 1989, Tanke wrote a review article with the title 'Does light microscopy have a future?' (Tanke, 1989). Then as now, the answer to this question is a resounding 'yes'. Light microscopy is the method *par excellence* for the study of micro-organisms under (near) physiological conditions. Alternatives, such as SPM, will complement, not replace, traditional optical methods. In fact, the light microscope has gained in importance, rather than diminished, because we can now quantify what we see with far more ease and precision, thanks to the possibilities offered by modern detectors and the power of modern computers.

REFERENCES

Agard, D.A., Hiraoka, Y., Shaw, P. and Sedat, J.W. (1989) Fluorescence microscopy in three dimensions. In *Fluorescence Microscopy of Living Cells in Culture* (Taylor, D.L. and Wang Y.-L., eds), Part B, pp. 353–77. Academic Press: San Diego, CA.

Axelrod, (1989) Total internal reflection microscopy. In *Fluorescence Microscopy of Living Cells in Culture* (Taylor, D.L. and Wang, Y.-L., eds), Part B, pp. 236–70. Academic Press: San Diego, CA.

Bloem, J., Veninga, M. and Shepherd, J. (1995) Fully automatic determination of soil bacterium numbers, cell volumes and frequencies of dividing cells by confocal laser scanning microscopy and image analysis. *Appl. Envir. Microbiol.* **63**: 926–36.

Brakenhoff, G.J., Blom, P. and Barends, P. (1979) Confocal scanning light microscopy with high aperture immersion lenses. *J. Microsc.* **117**: 219–32.

Brakenhoff, G.J., Van Spronsen, E.A., Van der Voort, H.T.M. and Nanninga, N. (1989) Three-dimensional confocal microscopy. In *Fluorescence Microscopy of Living Cells in Culture* (Taylor, D.L. and Wang Y.-L., eds), Part B, pp. 379–98, Academic Press: San Diego, CA.

Braunstein, D. and Spudich, A. (1995) Structure and activation dynamics of RBL-2H3 cells observed with scanning force microscopy. *Biophys. J.* **66**: 1717–25.

Bruins, S., De Jong, M.C.J.M., Heeres, K., Wilkinson, M.H.F., Jonkman, M.F. and Van der Meer, J.B. (1994) Fluorescence overlay antigen mapping of the epidermal basement membrane zone: I. geometric errors. *J. Histochem. Cytochem.* **42**: 555–60.

Dahlberg, D.E., Proksch, R.B., Moskowitz, B.M., Bazylinski, D.A. and Frankel, R.B. (1995) Microbes, magnetism and microscopy. *J. Magn. Magn. Mater.* **140–44**: 1459–61.

Denk, W., Strickler, J.H. and Webb, W.W. (1990) Two-photon laser scanning fluorescence microscopy. *Science* **248**: 73–6.

Firtel, M. and Beveridge, T.J. (1995) Scanning probe microscopy in microbiology. *Micron* **26**: 347–62.

Fritz, M., Radmacher, M. and Gaub, H.E. (1994) Granula motion and membrane spreading during activation of human platelets imaged by atomic force microscopy. *Biophys. J.* **66**: 1328–34.

Fritz, M., Radmacher, M., Cleveland, J.P., Allersma, M.W., Stewart, R.J., Gieselmann, R., Janmey, P., Schmidt, C.F. and Hansma, P.K. (1995) Imaging globular and filamentous proteins in physiological buffer solutions with tapping mode atomic force microscopy. *Langmuir* **11**: 3529–35.

Groen, F.C.A., Young, I.T. and Ligthart, G. (1985) A comparison of different focus functions for use in autofocus algorithms. *Cytometry* **6**: 81–91.

Ha, T., Enderle, T., Ogletree, D.F., Chemla, D.S., Selvin, P.R. and Weiss, S. (1996) Probing the interaction between two single molecules: fluorescence resonance energy transfer between a single donor and a single acceptor. *Proc. Natl. Acad. Sci. USA* **93**: 6264–8.

Hansma, H.G., Bezanilla, M., Zenhausern, F., Adrian, M. and Sinsheimer, R.L. (1993) Atomic force microscopy of DNA in aqueous solutions. *Nucl. Acids Res.* **21**: 505–12.

Hansma, P.K., Cleveland, J.P., Radmacher, M., Walters, D.A., Hillner, P.A., Bezanilla, M., Fritz, M., Vie, D., Hansma, H.G., Prater, C.B., Massie, J., Fukunaga, L., Gurley, J. and Elings, V. (1994) Tapping mode atomic force microscopy in liquids. *Appl. Phys. Lett.* **64**: 1738–40.

Herman, B. and LeMasters, J.J. (1992) *Optical Microscopy: Emerging Methods and Applications*. Academic Press: London.

Hoffman, R. (1977) The modulation contrast microscope: principles and performance. *J. Microsc.* **110**: 205–22.

Holmes, T.J., Bhattacharya, S., Cooper, J.A., Hanzel, D., Krishnamurti, V., Lin, W.C., Roysam, B., Szarowski, D.H. and Turner, J.N. (1995) Light microscopic images reconstructed by maximum likelihood deconvolution. In *Handbook of Biological Confocal Microscopy*, 2nd edn (Pawley, J.B., ed.), pp. 389–402. Plenum Press: New York.

Jericevic, Z., Wiese, B., Bryan, J. and Smith, L.C. (1989) Validation of an imaging system: evaluate and validate a microscope imaging system for quantitative studies. In *Methods in Cell Biology*, vol. 30, *Fluorescence Microscopy of Living Cells in Culture* (Taylor, L.S. and Wang, Y.-L. eds), Part B, pp. 47–83. Academic Press: San Diego, CA.

Keevil, C.W. and Walker, J.T. (1992) Nomarski DIC microscopy and image analysis of biofilms. *Binary Comput. Microbiol.* **4**: 93–5.

Langendijk, P.S., Schut, F., Jansen, G.J., Raangs, G.C., Kamphuis, G., Wilkinson, M.H.F. and Welling, G.W. (1995) Quantitative fluorescence *in situ* hybridization of *Bifidobacterium* spp. with genus-specific 16S rRNA-targeted probe and its application in fecal samples. *Appl. Environ. Microbiol.* **61**: 3069–75.

Laurent, M., Johannin, G., Gilbert, N., Lucas, L., Cassio, D., Petit, P.X. and Fleury, A. (1994) Power and limits of laser scanning confocal microscopy. *Biol. Cell* **80**: 229–40.

Lewis, A. (1991) The optical near field and cell biology. *Semin. Cell Biol.* **2**: 187–92.

Longhurst, R.S. (1973) *Geometrical and Physical Optics*, 3rd edn. Longman: London.

Lowy, R.J. (1995) Evaluation of triple-band filters for quantitative epifluorescence microscopy. *J. Microsc.* **178**: 240–50.

Magde, D., Elson, E.L. and Webb, W.W. (1974) Fluorescence correlation spectroscopy II: an experimental realization. *Biopolymers* **13**: 29–61.

Marriott, G., Clegg, R.M., Arndt-Jovin, D.J. and Jovin, T.M. (1991) Time resolved imaging microscopy. Phosphorescence and delayed fluorescence imaging. *Biophys. J.* **60**: 1374–87.

Morson, E., Merrit, G., Smith, S., Langmore, J.P. and Kopelman, R. (1995) Implementation of an NSOM system for fluorescence microscopy. *Ultramicroscopy* **57**: 257–62.

Müller, D.J., Schabert, F.A., Büldt, G. and Engel, A. (1995) Imaging purple membranes in aqueous solutions at sub-nanometer resolution by atomic force microscopy. *Biophys. J.* **68**: 1681–6.

Müller, D.J., Schoenenberger, C.A., Büldt, G. and Engel, A. (1995) Immuno-atomic force microscopy of purple membrane. *Biophys. J.* **70**: 1796–801.

Pawley, J.B. (1995) *Handbook of Biological Confocal Microscopy*, 2nd edn. Plenum Press: New York.

Petersen, N.O., Johnsen, D.C. and Schlesinger, M.J. (1986) Scanning fluorescence correlation spectroscopy II: application to virus glycoprotein aggregation. *Biophys. J.* **49**: 817–20.

Petersen, N.O., Hodellius, P.L., Wiseman, P.W., Seger, O. and Magnusson, K.E. (1993) Quantitation of membrane receptor distributions by image correlation spectroscopy: concept and application. *Biophys. J.* **65**: 1135–46.

Ploem, J.S. (1967) The use of a vertical illuminator with interchangeable dichroic mirrors for fluorescence microscopy with incident light. *Z. Wiss. Mikrosk.* **68**: 129–42.

Price, J.H. and Gough, D.A. (1994) Comparison of phase-contrast and fluorescence digital autofocus for scanning microscopy. *Cytometry* **16**: 283–97.

Russ, J.C. (1990) *Computer Assisted Microscopy*. Plenum Press: New York.

Schabert, F., Knapp, H., Karrasch, S., Haring, R. and Engel, A. (1994) Confocal scanning laser-scanning probe hybrid microscope for biological applications. *Ultramicroscopy* **53**: 147–57.

Summers, R.G., Stricker, S.A. and Cameron, R.A. (1993) Applications of confocal microscopy to studies of sea urchin embryogenesis. *Methods Cell Biol.* **38**: 265–87.

Tanke, J. (1989) Does light microscopy have a future? *J. Microsc.* **155**: 405–18.

Taylor, L.S. and Wang, Y.-L. (eds) (1989) *Fluorescence Microscopy of Living Cells in Culture*, Part B. Academic Press: San Diego, CA.

Van Kempen, G.M.P., Van Vliet, L.J., Verveer, P. and Van der Voort, H.T.M. (1997) A quantitative comparison of image restoration techniques for confocal microscopy. *J. Microsc.* **185**: 354–65.

Vardi, E. and Grover, N.B. (1992) Shape changes in *Escherichia coli* during agar filtration. *Cytometry* **14**: 173–8.

Waggoner, A., DeBassio, R., Conrad, P., Bright, G.R., Ernst, L., Ryan, K., Nederlof, M. and Taylor D. (1989) Multiple spectral parameter imaging. In *Methods in Cell Biology*, vol. 30, *Fluorescence Microscopy of Living Cells in Culture* (Taylor, L.S. and Wang, Y.-L., eds), Part B, pp. 449–78. Academic Press: San Diego, CA.

Wang, Y.-L. and Taylor, L.S. (eds) (1989) *Fluorescence Microscopy of Living Cells in Culture*, Part A. Academic Press: San Diego, CA.

White, N.S., Errington, R.J., Fricker, M.D. and Wood, J.L. (1996) Multidimensional fluorescence microscopy: optical distortions in quantitative imaging of biological specimens. In *Fluorescence Microscopy and Fluorescent Probes* (Slavik, J., ed.), pp. 47–56. Plenum Press: New York.

Wilkinson, M.H.F., Jansen, G.J. and Van der Waaij, D. (1993) Very low level fluorescence detection and imaging using a long exposure charge coupled device system. In *Biotechnology Applications of Microinjection, Microscopic Imaging, and Fluorescence* (Bach, P., Reynolds, C.H., Clarck, J.M., Mottley, J. and Poole, P., eds), pp. 221–30. Plenum Press, New York.

Willis, B., Turner, J.N., Collins, D.N., Roysam, B. and Holmes, T.J. (1993) Developments in three-dimensional stereo brightfield microscopy. *Microsc. Res. Technol.* **24**: 437–51.

Wilson, T. and Sheppard, C. (1984) *Theory and Practice of Scanning Optical Microscopy*. Academic Press: San Diego, CA.

Wolf, D.E. (1989) Designing, building, and using a fluorescence recovery after photobleaching instrument. In *Methods in Cell Biology*, vol. 30, *Fluorescence Microscopy of Living Cells in Culture* (Taylor, L.S. and Wang, Y.-L., eds), Part B, pp. 271–306. Academic Press: San Diego, CA.

Young, I.T. (1989) Image fidelity: characterizing the imaging transfer function. In *Methods in Cell Biology*, vol. 30, *Fluorescence Microscopy of Living Cells in Culture* (Taylor, L.S. and Wang, Y.-L., eds), Part B, pp. 1–45. Academic Press: San Diego, CA.

4

Single and Multiple Spectral Parameter Fluorescence Microscopy

Jan Slavik

Czech Academy of Sciences, Prague, Czech Republic

4.1 INTRODUCTION

4.1.1 Fluorescence and its parameters

Fluorescence is one of the techniques of choice in biology and medicine. Fluorescence microscopy is predicted to be the fastest growing market of all the optical analytical instruments. The progress can be illustrated using the example of the intracellular ionic measurement. The classical pH measurement of the past used cell suspension in a cuvette and yielded only one average value for the whole suspension. With fluorescent probes and imaging techniques, time and space resolved maps of various ions inside the cells and in their immediate neighbourhood can be generated.

The great advantage of fluorescence techniques in biology and medicine is the possibility to investigate samples *in vivo* or *in vitro* under natural biological conditions. The measurement is non-destructive and almost non-invasive, due to the low concentrations of non-toxic dyes (fluorescence can work with 10^{-11} to 10^{-12} M concentrations of the dye, while absorption spectroscopy requires 10^{-8} M and nuclear magnetic resonance (NMR) spectrometry at least 10^{-5} M). Fluorescence yields more data than absorption spectrometry (two spectra – excitation and emission spectra – and fluorescence lifetime). The fact that a substance must both absorb and emit light can be advantageous in its identification. In the course of fluorescence measurement, the following observables are determined: the intensity of fluorescence, emission spectrum, excitation spectrum, absorption spectrum, quantum yield, polarization and lifetime. The parameters characterizing the fluorescence are generally independent, but there are some relations between them, which may be used as a control check for the correctness of the measurement. The detailed definition of fluorescence parameters and the relations among them can be found elsewhere in the luminescence literature (e.g. Parker, 1968; Lakowicz, 1983; Slavik, 1994) (Figure 4.1).

Digital Image Analysis of Microbes: Imaging, Morphometry, Fluorometry and Motility Techniques and Applications. Edited by M.H.F. Wilkinson and F. Schut.

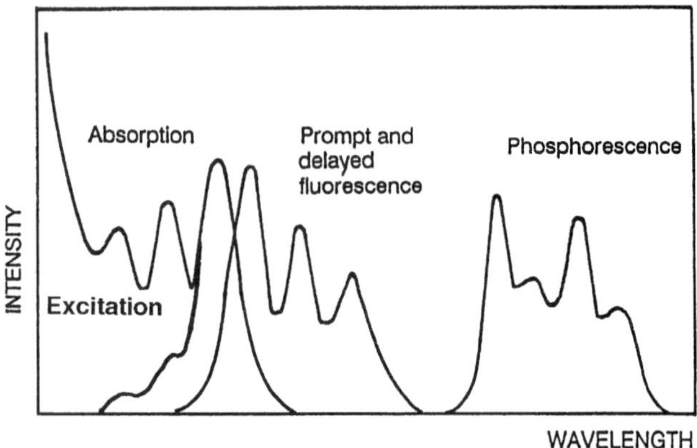

Figure 4.1 The relative positions of absorption, excitation, fluorescence (prompt as well as delayed) and phosphorescence.

4.1.2 Absorption and excitation spectra

In order to be emitted the light quantum must be absorbed. The intensity of fluorescence is directly proportional to the intensity of absorption of light and the quantum yield. The absorption of light follows the well-known Lambert–Beer law.

The absorption spectrum reflects the dependence of the degree of light absorption by the sample on the wavelength (or frequency or energy) of the light. The excitation spectrum may be viewed as a specially distorted or transformed absorption spectrum with respect to the 'efficiency' (i.e. quantum yield) of fluorescence. It should resemble the absorption spectrum, but is not identical with it. The excitation spectrum is obtained as the fluorescence intensity is measured at a fixed emission wavelength, while the excitation wavelength is scanned. In large, complex molecules the shape of the excitation spectra is stable and does not depend on the emission wavelength at which it is monitored. Thus the excitation spectrum can be used for the characterization of a dye.

4.1.3 Emission spectrum

The basic parameter of fluorescence is the intensity expressed as a number of emitted quanta in the given wavelength range per unit time. For instance, for the detector window set at 450 nm with slit width 2 nm, it corresponds to the number of photons emitted per second in the wavelength range from 449 to 451 nm. Since the absolute measurement of light intensity is very complicated, fluorescence intensity is usually expressed in relative units (see also Section 10.6 on calibration).

The emitted light, correctly called luminescence, comprises fluorescence, delayed fluorescence and phosphorescence. Fluorescence is usually by far the strongest emission. The fluorescence spectrum is generally close to the absorption spectrum, which it may even partially overlap (Stokes' shift). The absorption and fluorescence spectrum shapes may be nearly mirror images of each other. The phosphorescence spectrum appears generally at lower temperatures only, and is similar in shape to the fluorescence spectrum, but it is red-shifted as a whole (the triplet energy level involved in phosphorescence is lower than the corresponding singlet energy level involved in fluorescence). Delayed fluorescence appears in some compounds and it is spectrally identical with the (prompt) fluorescence, but it appears later in time after irradiation. Mutually interacting molecules give dimer, excimer and exciplex emission spectra, red-shifted with respect to the fluorescence of monomers, and their shape is entirely different due to molecular interactions in the ground and excited states. Interactions with solvent or aggregation of dye molecules also cause further red shifts.

A single, pure compound should exhibit only one fluorescence spectrum. If the shape of the emission spectrum changes when the wavelength of the exciting light is varied, the presence of more than one fluorescent compound or impurities should be suspected.

Quantum yield describes the 'efficiency' of fluorescence defined as the ratio of the number of quanta emitted per number of quanta absorbed. It ranges from 1 ('100% efficiency') to 0 ('zero efficiency' = non-fluorescent). Usually it is calculated or measured relatively by comparison with a standard compound of known quantum yield. Fluorescence light is polarized and the degree of polarization reflects the mobility of fluorophores. The fluorescence lifetime parameter characterizes the decay of fluorescence intensity when the excitation is terminated or, more correctly, as a response to an extremely short excitation pulse.

Time-resolved fluorescence (TRF) means that the fluorescence signal can be resolved with respect to the time interval following the excitation. This approach distinguishes the time profile of the signal on a time scale comparable to the lifetime of the chromophore (typically with nanosecond resolution). The temporal changes of polarization, emission or excitation spectra on this time scale include an enormous amount of information on the movement of molecules, and thus represent a significant improvement over the usual 'steady-state' fluorescence. On the other hand, the equipment required is sophisticated and expensive.

4.2 CLASSICAL AND CONFOCAL FLUORESCENCE MICROSCOPY

When selecting a microscope we can choose between the 'classical' upright model and the inverted model. The classical microscope observes the sample from the top (the cover slip is above the sample); the inverse microscope observes the sample from the bottom. For biological and medical experiments, the inverted microscope is more convenient because it allows free access to cells from above. This makes it possible to add various biologically active substances at any time during the measurement. Cells can be perfused, or mechanically handled (micro-electrode impalement), and the temperature of the medium can be directly measured with an inserted thermocouple etc. The optical resolution and stability of the image are slightly impaired. Excitation light (called epifluorescence) can come through the objective lens or through the condenser. Fluorescence often requires the use of special lenses because the normal lens may not transmit ultraviolet (UV) or far red excitation light and may exhibit autofluorescence.

The confocal microscope (CLSM: confocal laser scanning microscope) is an improved version of the classical optical microscope. In the confocal microscope, the sample is not illuminated as a whole, but only one point after another. The microscope uses point illumination of the sample and a point collection of the signal. The word 'confocal' describes the optical coincidence of the illuminated and detected points. In contrast to the classical microscope, the out-of-focus information is optically rejected; this effect is particularly prominent in turbid samples. The point illumination allows 'slicing up' of the sample optically and can be used as a basis for a subsequent three-dimensional computer reconstruction of the image of the sample. In a typical confocal microscope, a laser beam is gradually deflected, and so it scans the sample and the image is built up point by point in a way analogous to that in a scanning electron microscope. The principles of confocal microscopy permit easy ratio imaging, multiple labelling and lifetime imaging (see also Chapters 16 and 17).

4.2.1 Limitations of optical microscopes: sensitivity, temporal resolution, spatial definition

The higher sensitivity of fluorescence techniques in comparison with absorption techniques is due to the fact that the absorption signal is related to the 100% incident light intensity while fluorescence signals are detected against the zero background – 'darkness' – and therefore the separation of the dye-related signal is easier. The intensity of fluorescence of a given substance is proportional to the product of concentration of the dye, intensity of the exciting light, absorption (extinction) coefficient and fluorescence quantum yield. Generally, concentrations as low as 10^{-11} M can be detected without problems, while in absorption spectrometry 10^{-8} M is a usual limit.

It is usually the background fluorescence that limits the measurements. Most of the experiments are thus 'blank-limited' rather than 'instrument-limited'. The sensitivity of detectors such as a photomultiplier or a cooled charge-coupled device (CCD) camera can indeed reach the theoretical limits, so they are capable of counting individual photons (Chapter 2). The 'blank' background value arises from scattered light, luminescent impurities in the medium and reagents, and very often from the background luminescence of the sample. Luminescence of the microscopic glass, cover glass, lens, filters, immersion oils etc. can be insidious, especially after UV excitation.

The temporal resolution limits imposed by the dye are far below those imposed by the detection technique, i.e. by the time required for the complete detection and storage of one image. Grabbing of one image typically takes about 40 ms (television frequency), but there are grabbers faster than 1 ms. The scanning of one picture in CLSM takes approximately the same time interval. The change of an excitation or emission filter in a filter wheel requires typically 20–50 ms.

The spatial resolution of an optical instrument is limited by the laws of optical diffraction. In a classical fluorescence microscope the theoretical limit is about 150–200 nm; in a confocal microscope perhaps 100–50 nm. The resolution of a microscope can be enhanced by applying special mathematical deconvolution procedures, which may also remove the fluorescence flare and suppress the degradation of contrast by out-of-focus blurring (Section 17.2.4d). In any case, all brightly fluorescent objects smaller than the resolution limit will be visible as luminous points.

There are limits for imaging. Imaging techniques using high magnification microscopes equipped with highly sensitive cameras are capable of registering individual photons. They are also reaching the theoretical limits – there may be simply too few dye molecules in the sample. Attempts to maximize the temporal and the spatial resolution always bring a trade-off between sensitivity and time resolution. A dye concentration of 1 mM corresponds to 4 molecules per $0.2 \times 0.2 \times 0.2$ μm, and there is a limit to the number of excitation–emission cycles that one dye molecule can survive before it is photodegraded, i.e. bleached (for fluorescein it is about 10^5 times). With respect to the sensitivity of intensified or cooled CCD cameras and the quantum yield of the dye, the above concentration limits the number of images that can be captured to several hundreds or thousands, then the dye is lost. For longer measurements, either a lower resolution or a higher dye concentration must be employed. In conclusion, consider the following calculation. Suppose that a xenon lamp provides blue–green light that represents 1.3×10^{14} photons per second per cm^2. If the sample is 10 mm thick and the fluorophore has an extinction coefficient of 50 000, a dye concentration of 10 μM would absorb about 0.1% of the excitation light. Provided that the quantum yield is 0.5 and 10% of the signal reaches the detector, the image would be formed by approximately 2×10^5 photons. For 512×512 pixels it is less than about 1/10 photon per pixel per 1/30 s. If we require 50 counts per pixel for a reasonable picture we need to accumulate 500 frames, i.e. 16 s. If we use quin2, which survives only two or three excitation–emission cycles, this would limit the number of frames to 2000–3000, i.e. just six pictures could be taken before the dye was completely degraded. With a fluorescein-labelled sample it would correspond to about 10^5 pictures.

4.2.2 Shading correction

Shading correction (see Section 2.4.7) is a correction for the variation of the intensity response function of the photodetector according to the spatial position of the image. It can include multiplicative as well as additive corrections. Shading is important for image intensifiers and tube cameras where different regions of the camera target have different light response functions. Usually, the sensitivity in the central part of the viewing field is greater than at the edge, which means that evenly illuminated areas appear to be brighter in the centre than in the corners (this is a useful test for the necessity of shading correction). The shading correction is easy to apply in computer processing of pictures, where it can be implemented in the grabbing procedure. In CCD cameras the intensity response function is almost identical for all pixels. Ratio imaging techniques implement the shading already in the image processing procedures.

4.2.3 Geometric distortion

All imaging systems introduce a certain degree of geometric distortion (Section 3.5.4). Some distortion is introduced by the objective lenses, relay lenses and image intensifiers. Some cameras, such as intensified vidicons, exhibit a pincushion distortion; other cameras, such as CCD cameras, are virtually distortion-free.

Corrective geometric operations, including spatial warping, rotation and translation, can be introduced easily in digital image processing (Section 3.5.6).

4.2.4 Fluorescence recovery after photobleaching

Fluorescence recovery after photobleaching (FRAP) is a special technique used for the measurement of lateral diffusibility of molecules in membranes. The technique is based on the continuous monitoring of fluorescence intensity from a selected small spot in the membrane. In this spot, the intensity of the excitation light is temporarily enhanced about 10 000-fold. Consequently, most of the dye within the spot photobleaches. When the excitation level is reduced back to the monitoring intensity, the intensity of fluorescence is significantly reduced. If there is no freedom of motion of molecules in and out of the spot, the fluorescence intensity will remain at this level indefinitely, while if the molecules are completely free to diffuse in and out of the spot, the fluorescence intensity will recover to the prebleach level. The reality lies between these two extremes. The diffusion coefficient is obtained by fitting the recovery data to appropriate diffusion equations. The technique provides two measures of diffusibility: the fraction of the molecules free to diffuse and the diffusion coefficient of that fraction. The technique requires a specially modified fluorescence microscope, a fast detector and a good computer program calculating the values of the diffusion coefficient directly.

4.3 INSTRUMENTATION FOR SPECTRAL MEASUREMENT

4.3.1 Light sources

For the measurement of spectral properties of fluorescence it is necessary to know the spectral properties of both the light source and the detector, then by inserting optical filters and/or monochromators one can monitor the excitation and emission spectra of the fluorescence of the sample. An ideal excitation light source should deliver any selected wavelength at a reasonable intensity. In practice, an arc lamp or laser, or exceptionally a classical incandescent light bulb or light-emitting diode (LED), is generally used.

Arc lamps are usually filled with xenon or mercury or a mixture of both elements. A low-pressure mercury lamp gives sharp lines with almost no continuous background, while high-pressure mercury lamps provide some weak background. Xenon lamps have a lower light intensity in the UV range than mercury lamps, but they are preferred due to their almost spectrally smooth output (with the exception of a few small peaks around 450 nm). The fluctuation of the light output (flickering) which appears after longer operation time must be electronically corrected or the lamp must be changed regularly. Xenon–mercury lamps have higher intensities in the UV region than pure xenon lamps, but their spectrum is dominated by intense mercury lines. Hydrogen and deuterium lamps are widely used in absorption UV spectroscopy but are rarely employed in fluorescence. Their UV spectrum is fairly constant spectrally, but the intensity is too low. The light output of halogen lamps is low in the visible range and almost negligible in the UV range.

Lasers represent a very efficient source of light, with an effective intensity about one million times higher than that of an arc lamp. However, the light is emitted only at distinct wavelengths. Semiconductor, excimer or ion (gas) lasers usually emit only at one single wavelength, or a set of several fixed wavelengths. Dye lasers make it possible to vary the output wavelength in a limited range of approximately 40–70 nm for each dye. The dyes can be changed according to the required output wavelength range, and there are commercially available dyes for virtually any wavelength in the range 300–1000 nm. However, the prices of most of the above mentioned lasers are exorbitant, and thus in the field of microscopy an argon ion laser with 488 nm and 514 nm lines is by far the most common laser source. Lasers for UV excitation and experimental set-ups with (deep red to infrared; IR) two-photon excitation (Section 3.4.2) are still rare.

4.3.2 Light detectors and image intensifiers

Light detectors for fluorescence spectroscopy are represented mostly by photomultipliers and CCD cameras. Their properties have been described in detail by Van Vliet *et al.* in Chapter 2. The most important points are summarized briefly below.

The key element is the photocathode, which determines both the radiometric and the spectral sensitivity of the photomultiplier. The efficiency of the photocathode is very high, reaching 20–50% in its maximum wavelength range.

With respect to the electronic circuitry, both cameras and photomultipliers can operate either in an analog or in a photon-counting mode. The analog signal is proportional to the photon flux, and pulses corresponding to individual photons are averaged. In the photon-counting mode the photomultiplier operates as a detector responding to individual photons. A cooled CCD camera in a slow-scan mode operates in the photon counting mode.

Image intensifiers can be placed between the imaging lens and the sensor (camera). They can be removable or incorporated into the camera body. The gain, the level of illumination of the resulting image and its contrast (gamma) can be varied. Typically, the brightness of the image on the output side is 10^2 to 10^5 higher than that on the projected image on the input side. There are two types of image intensifiers: microchannel plates and conventional image intensifiers. The microchannel plate intensifier is a matrix composed of several million glass microcapillaries. Each microcapillary corresponds to one image pixel. Functionally, a microchannel plate can be described as an array of photomultipliers without separate dynodes. The amplification effect of the capillaries arises from the special coating of their inner side by a material of low conductivity and a high secondary electron emission. When a voltage ot about 1000 V across the microcapillaries is applied each capillary works as a channel electron multiplier with a gain of about 10^4. When several microchannel plate intensifiers are connected in series, a useful gain of up to 10^{10} can be obtained. The inner diameter of a capillary is around 10–20 μm and in a bundle they form a disc thinner than 1 mm. A conventional image intensifier is a vacuum tube that uses photoelectrons to form a replica brighter than the original image. The amplified replica is focused on the photocathode on the phosphorescent screen. The vacuum tube contains a photocathode and an anode. The photocathode emits electrons in proportion to the number of incident photons. As in the photomultiplier, emitted electrons are accelerated by the potential difference between the two electrodes. Using magnetic or electrostatic lenses, they are made to form a secondary phosphorescent image on a phosphorescent screen. Each accelerated electron generates about a thousand photons on the screen, which shows a bright replica of the original image. Taking into account the quantum efficiency of the photocathode, this means that the amplified image is about 200 times brighter than the original image. The resulting image can be viewed by the camera sensor, or used as a template image for an additional image intensifying stage. The image intensifiers can be connected together in a several cascade steps until a satisfactory light level at the output is reached (see also the appendix to Chapter 2).

4.3.3 Monochromators and filters for ratio imaging

Ratio imaging requires either two detectors or the possibility of changing the filters in a rapid and easy way. Filters are mounted on mechanical sliders, wheels or rotating cubes. Their movement is usually driven by a stepper motor, which will move between several positions and stop while images are captured through a particular filter. The time required for a filter change is around 10–100 ms. Monochromators can be used for ratio measurement in conjunction with a rotating dichroic mirror. In CLSM, the fluorescence emission can be detected simultaneously at several wavelengths depending on the number of photodetectors. The fluorescence beam is split according to the wavelength into two or more beams, and each beam is directed to another detector. The resulting images are stored separately and later undergo further processing independently of each other. Various mathematical operations (ratioing) may then be performed. The light delivered by the light source and the light detected by the detector can be spectrally modified by insertion of an optical filter or (rarely in imaging) by a monochromator.

4.3.3a Monochromators

Light can be distinguished with respect to its wavelength by insertion of a prism or a grating in the path of the light beam. Prisms use the dependence of the refractive index of glass on the wavelength. Generally, the longer the wavelength, the more the light beam is bent upon its passage through a prism or a combination of prisms. Gratings use the wavelength-dependent diffraction properties of the light striking a regularly etched surface (grooves). The advantage of monochromators in comparison to filters is the possibility of continuous tuning of the wavelength; their disadvantage is in lower efficiency and the inevitable addition of a certain proportion of unwanted stray light. This stray light is composed of all wavelengths and may cause severe trouble when interfering with the weak fluorescence of a turbid sample.

4.3.3b Glass filters

In microspectrofluorometry and time-resolved spectroscopy filters are preferred because they lose much less light than monochromators. In polarization measurements filters are preferred to monochromators because they do not alter the polarization of the transmitted light.

The simplest types of filters are made of tinted glass, and their absorption properties are determined by the dye present in the glass. It is easy to make a high quality cutoff filter, which blocks all wavelengths above a limiting wavelength edge. Such a filter may be called a long-pass filter. Making a short-pass filter, on the other hand, is more difficult, and good filters transmitting only in the UV or blue region are relatively rare. By combination of different dyes, uniformly grey filters can be prepared. Thermal filters are a special type of these filters which block infrared radiation, absorb it (the filter warms up) and let only visible light through. Tinted glass filters have a transmission curve independent of the angle of incident light and look like coloured glass (Figure 4.2).

Interference filters are prepared from transparent glass plates on which a thin layer, or several layers, of a dielectric compound is condensed. The selective

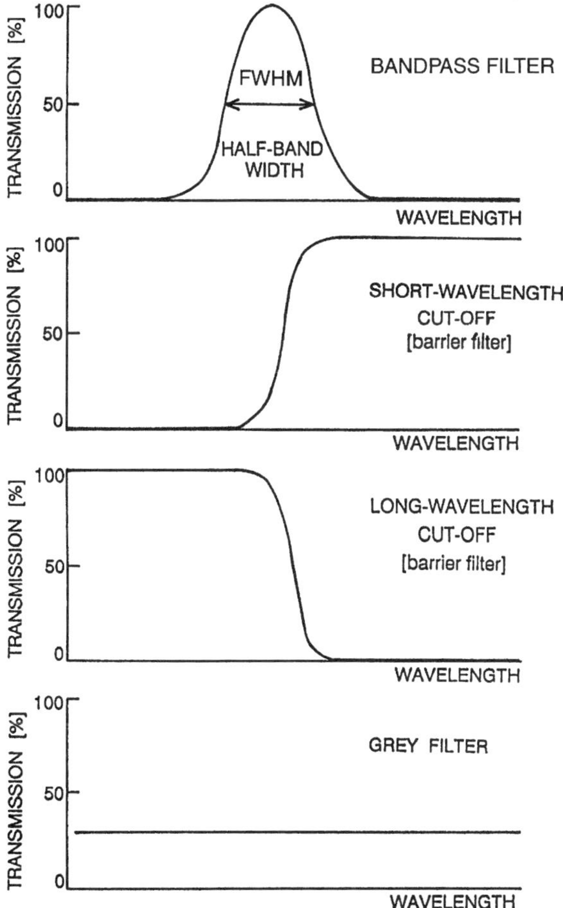

Figure 4.2 Basic types of transmission curves of optical filters.

transmission (or reflection) of these filters is caused by the interference of light determined by the thickness and index of refraction of the layers. Any kind of transmission curve can be prepared by application of several layers of dielectric compounds of appropriate thickness and index of refraction. In this way, broad or very narrow bandpass filters of desired half-width and maximum wavelength can be prepared easily, as well as short-pass filters, long-pass filters and filters with variable transmission wavelength (wedge filters).

Interference filters can be recognized by the mirror-like face of the filter. Their transmission curves change slightly with the incident angle (collimated light is thus preferred). A special kind of interference filter is a semitransparent mirror. According to its construction, a semitransparent mirror lets infrared radiation pass through and reflects only the visible portion of the light, or lets the visible light go through and reflects heat only.

To avoid possible confusion, it should be noted that the apparent colour of the filter is given by the colour that is transmitted. Thus a red filter transmits red light; an ultraviolet filter transmits only UV light and appears black. For more confusion, filters that block ultraviolet light are often also called UV filters (Figure 4.3).

Last but not least, one should not forget that most optical components, solvents and chemicals look clear, non-fluorescent and transparent. This, however, may lead to the illusion that they are also transparent in the UV and IR regions as well as in the visible range. In reality, limited transparency is found in many cuvettes, solvents and microscope lenses, and many coloured glass filters exhibit a strong autofluorescence in UV light and thus cannot be used for fluorescence measurement.

4.3.3c Microscope filter cubes

For routine work it is convenient to have a complete filter set optimized for each dye. There are commercially available filter cube sets for epifluorescence, which comprise an appropriate excitation filter, semitransparent mirror and emission filter. The filter cube ensures that the excitation light of the desired wavelength passes through the excitation filter and is then reflected on the sample but cannot be seen in the eyepiece or detected by the camera. The fluorescence, on the other hand, can pass through the cube and reach the eyepiece, camera or photodetector. Filter cubes are available to fit the most frequently used dyes (such as fura-2, fluorescein, rhodamine, Texas Red). If you use a special dye with special requirements for excitation and emission wavelengths you will need a custom-made filter, which may be expensive.

4.3.4 Detection for ratio imaging

4.3.4a Spectral response of the detector

The spectral response of a vidicon camera, photomultiplier or intensifier is given by the photocathode (in CCDs it is the semiconductor material). In intensified cameras,

Figure 4.3 (a) The schematics of a raw fluorescence spectrum, and the methods of removing the impurities in the case of (b) excitation spectra measurement and (c) emission spectra measurement.

FLUORESCENCE INTENSITY [relative units]
1
0
SCATTERED EXCITATION AND STRAY LIGHT
RAMAN PEAK
FLUORESCENCE BACKGROUND
fluorescence of dye
2nd order excitation
DYE COMPONENT (ESTIMATE)
STRAY LIGHT COMPONENT (ESTIMATE)
300
400
500
600
EMISSION WAVELENGTH [nm]

(a)

FLUORESCENCE INTENSITY [relative units]
BARRIER FILTER (in the excitation beam)
BANDPASS FILTER (in the emission light-path)
excitation spectrum
1
0
100
0
TRANSMITTANCE [%]
300
400
500
600
WAVELENGTH [nm]

(b)

FLUORESCENCE INTENSITY [relative units]
BANDPASS FILTER (in the excitation beam)
BARRIER FILTER (in the emission light path)
stray-light components are cut off by band-pass filter
dye fluorescence
1
0
100
0
TRANSMITTANCE [%]
300
400
500
600
WAVELENGTH [nm]

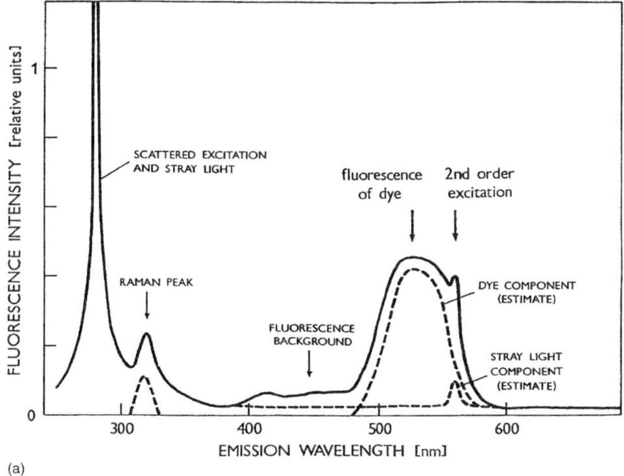

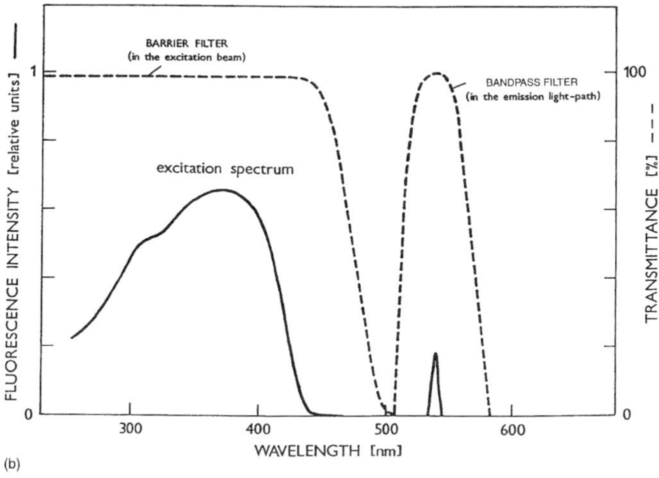

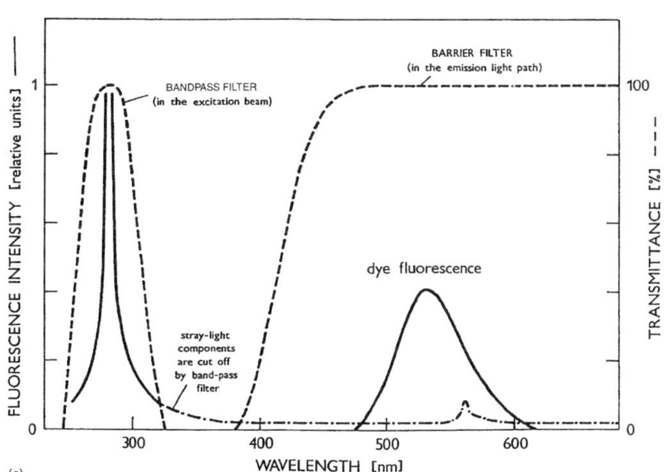

(c)

the spectral response of the intensifier determines the overall spectral sensitivity, and the spectral characteristics of the output screen of the intensifier must match the spectral sensitivity of the camera.

4.3.4b Radiometric response of the detector

The intensity response of a vidicon camera has a typical 'S' shape similar to that of a photomultiplier or to a sensitometric curve of a film, but its central part is almost linear. The value of the slope in the linear central part is called gamma, and in many cases the camera may be adjusted to different gamma values (usually 0.45, 0.7 or 1) or continuously varied (in some CCD cameras in an interactive way). Using photographic terminology, this corresponds to the 'contrast' of the film. The usable dynamic range of the camera is given by the difference between the minimum and the maximum detectable signal. The possible nonlinearity of the sensitometric curve can eventually be corrected for (rectified) by computer image processing using the so-called look-up tables (LUT; see Section 5.2.4b). The maximum signal is limited by the ability of the scanning beam to read the value at the target (pixel), while the minimal signal is usually determined by the noise. The signal-to-noise ratio can be improved by averaging several consecutive images (also known as Kalman filtering; see Section 16.4.1). As most of the noise appears due to random processes, the improvement is proportional to the square root of the number of averaged pictures. The practical limit is averaging of several hundreds of images, depending on the type of algorithm used for averaging. It is interesting to recall that human vision also uses averaging over about 0.2 s, which corresponds in video frequency to averaging five to six video frames. Averaging of a smaller number of video images results in apparent degradation of the image over what a viewer can see on the video monitor. This is of particular importance if a tape video recorder is used for storing pictures.

Ratio imaging requires that the signal from each pixel is registered in all (two or more channels) in an identical fashion. This means that the 'automatic gain controls' marked, for example, AVC in the photomultiplier control or AGC in the CCD camera must be turned off and manually overridden. The 'darkness' (no fluorescence signal) must correspond to the zero level of the fluorescence background signal and the gain ('contrast') must be set to the same value in all detected channels, preferably to the value $\gamma = 1$. Both of these values, i.e. the background signal not originating from the dye and the gain (it may be called contrast or slope), are constant in all calibration and experimental procedures (Figure 4.4).

4.3.5 Incubation chambers

The use of cell incubation chambers requires an inverted fluorescence microscope. Objective lenses with a long working distance are preferred; immersion objectives with a short working distance may accommodate only chambers with very thin bottom plates. Simple chambers can be designed on the basis of a Petri dish with an opening in its base covered by a cover slip glued from the bottom side. Adhering cells can be cultivated either directly in a microscope chamber or grown separately on tiny glass pieces cut from the cover slip slides, kept in the incubator and then transferred into the microscope chamber only immediately prior to measurement.

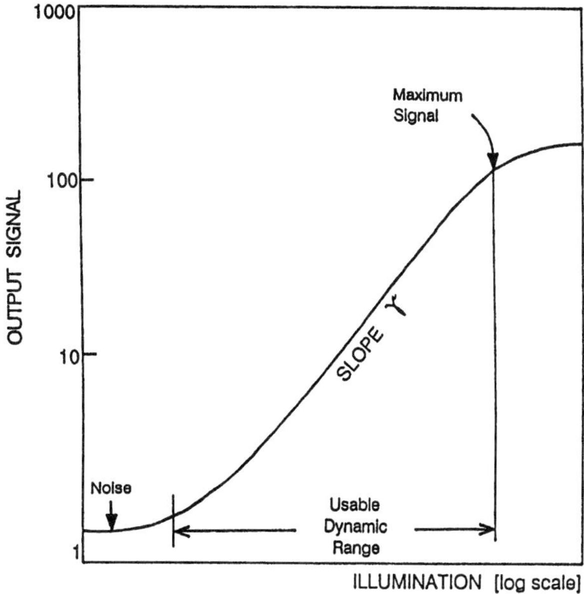

Figure 4.4 The usable range of the response of a detector in ratio measurements.

The perfusion flow system is based on some form of continuous supply of a fresh warm buffer and continuous removal of excess buffer. The tops of the incubation chambers may be closed or open, or the medium covered with a thin layer of silicone oil if access is required. Heating of the chamber to temperatures around 37 °C may be achieved in several ways. The whole microscope can be placed in a warmed room or a temperature-controlled box or it may be internally heated. The chamber may be warmed by a water bath, warm air or infrared light. The temperature is monitored by a thermistor or a thermocouple located as close as possible to the viewed cells. Many (expensive) commercial designs are available.

4.4 SPECTRAL AND LIFETIME FLUORESCENCE IMAGING

Fluorescence lifetime imaging has suddenly emerged as a useful and promising high technology fluorescence method. Often combined with spectral imaging, it is finding more and more use in fluorescence microscopy. In fact, it may be understood as ratioing in time.

Fluorescence lifetime is a parameter characterizing the rate of decay of fluorescence intensity after a very short excitation. The emission of a photon is a statistical phenomenon. Molecules of the dye emit fluorescence independently of each other; some molecules will emit photons sooner, some later. In the simplest case, following a short excitation pulse, the decay is exponential with a constant τ called the fluorescence lifetime:

$$I = I_0 \cdot e^{-t/\tau} \tag{4.1}$$

Statistically, 37% of molecules (1/e) will emit the photon between time zero and time $t = \tau$. The remaining 63% will emit photons after that period. Using sophisticated equipment, photons that were emitted immediately after excitation can be distinguished from those emitted later. In this way the decay of fluorescence intensity or, better, the time-resolved spectra (i.e. changes of spectra with time) or time-resolved polarization of fluorescence (i.e. changes of anisotropy with time) can be measured. Time-resolved fluorescence spectroscopy allows subtle solvating effects to be revealed, by monitoring the movement of the fluorescent probe in the nanosecond time range, and distinguishes the signal from molecules that existed for only a very short time in the excited state from that of molecules that were in the excited state for longer. The 'early' photons are often quite different from the 'late' emitted photons, and these differences make it possible to draw conclusions about the behaviour of the molecule in the excited state, particularly its interactions with neighbouring molecules (e.g. membrane fluidity).

4.4.1 Principles of the lifetime measurement

The measurement of the fluorescence lifetime can now be implemented relatively easily into fluorescence microscopy, both in the classical wide-field and in the confocal laser scanning microscopy. There are basically two different principles on which the measurements can be based: pulse or modulation techniques. Both of these techniques are described in detail by Lakowicz (1983).

4.4.1a Pulse excitation

In the first approach, extremely brief (usually sub-nanosecond) excitation pulses are used, usually obtained from a pulse laser. Gating of the image intensifier or the photomultiplier is a very elegant way of measuring fluorescence lifetime (see appendix to Chapter 2). The high voltage is varied so that the image intensifier of the photomultiplier is 'on' only during a short time interval after the application of the exciting pulse. In the meantime it is blind. 'Gated' in this way, it detects the incoming light only during a short 'time window'. This window may be 1 ns or narrower, and it can be moved relative to the 'time zero' of the excitation pulse. Ideally, the whole fluorescence decay curve is recorded step by step by gradually moving this window in the course of the experiment. In reality, using imaging with biological samples, two fixed time windows are employed, and the ratio of both signals calculated, assuming exponential fluorescence decay. Such an approach is highly advantageous for lifetime imaging, as instead of calculating the whole decay curve from each pixel, only a ratio of two values from each pixel is taken into calculations. In this way the calculation of the lifetime is much faster and simpler. Otherwise, it is clear that this technique gives a very weak signal requiring long accumulation times, and that the processing of such a large amount of data from a series of microscopic images is very demanding on hardware and computer time.

4.4.1b Modulation excitation

Modulation techniques use another approach. They are based on a continuous excitation light source sinusoidally modulated at a high frequency. They employ the time lag between the absorption and emission as it appears for the sinusoidally modulated excitation light. The emission is delayed in phase and demodulated with respect to the excitation (i.e. the amplitude of the modulation is decreased in the emitted fluorescence). Both the phase delay θ and the demodulation factor m can be used for the calculation of the fluorescence lifetime.

The theory of modulation techniques is briefly as follows. Consider a dye with exponentially decaying fluorescence of lifetime τ. When excited by the excitation light I modulated by frequency ω it gives the fluorescence signal J. One measures either the phase shift θ ($\tan\theta = \omega\tau$) between the excitation light I and fluorescence signal J, or the decrease in the degree of demodulation m between the excitation I and fluorescence signal J. The demodulation factor m characterizes the decrease of modulation of the fluorescence signal J relative to the excitation signal I, $m = \cos\theta = (1 + \omega\tau)$. Generally, the calculation of the lifetime τ is easy and simple with single exponential decay curves, but difficult with non-exponential fluorescence decays. Then the decay curve is approximated as a sum of two or three exponentials and the measurement of θ or m is performed at several modulation frequencies. A special feature of the modulation technique is its capability to suppress or enhance the signal of a given decay time from a heterogeneously decaying sample (the so-called phase-sensitive detection of fluorescence; PSDF).

4.4.2 Spectral and time-resolved fluorescence spectra

The time-resolved emission spectra (TRES) measurement consists of measuring temporal changes of the shape of the emission spectra on a nanosecond time scale. Spectral lifetime imaging is very promising and may be based on any of the three techniques named above.

4.5 SPECTRAL AND POLARIZATION IMAGING

As follows from the theory, there are two conditions that must be fulfilled when absorption is to take place. Not only must the light be of the right wavelength (the energy of the absorbed photon must fill the gap between the energy of the molecule in the ground state and the excited state), but the absorbed light must also have the right orientation of the plane of polarization relative to the molecule. The transition moment has a defined orientation in the molecule. Accordingly, the probability of absorption depends on the polarization of light as well as on the wavelength of the incoming light. The issue is closely related to the fluidity measurement of biological membranes, when fluorescent probes are located in the biological membrane and their rotational freedom of movement inside the membrane may be restricted (mapping of membrane fluidity). The measurement of polarization of emitted light requires additionally a polarizer (usually a dichroic filter).

4.6 THE SIMPLIFIED THEORY OF FLUORESCENCE

The correct application of fluorescence techniques requires a certain degree of understanding of the instrumentation and skill in the interpretation of results. It seems almost indispensable to consult both the experimental protocol and the results with an experienced spectroscopist when beginning experimental work. The greatest danger is the possibility of interference by a foreign signal, such as autofluorescence of the background or light scatter in biological samples, and mistaking it for a signal arising from the fluorescent dyes. Understanding the instrumentation helps not only in finding and separating the right fluorescence signal from the detected signal, but also gives a feeling for the limitations and inherent imperfections of the fluorescence microscopic set-up, and for the ways in which the fluorescence signal may be distorted in the measurement process.

Light covers a small part of the electromagnetic spectrum, namely wavelengths between approximately 10 nm and 1000 mm. The so-called visible light comprises wavelengths from 400 to 700 nm. It is perceived by the human eye and distinguished on the basis of colour as violet, blue, green, yellow, orange and red. At the short-wavelength side of this visible spectrum is ultraviolet (UV) light (200–400 nm), which may sometimes be visible as bluish-grey light. At the long-wavelength side is infrared (IR) light, seen as dark red or simply perceived as heat. White light contains all visible wavelengths more or less equally represented; polychromatic light contains more wavelengths and monochromatic light comprises waves of only a single wavelength.

4.6.1 Quantum mechanical description of light

The theoretical description of light reflects the duality of the quantum mechanical description of the world (see Section 2.2). Light can be regarded as a wave or as a particle. In the wave description it is represented by a composition of two waves, one electric and one magnetic, coupled with each other. The waves oscillate perpendicularly to each other and perpendicularly to the direction of the propagation of light. In the corpuscular description, light is considered as a stream of particles — photons – which move in the direction of propagation of the light beam. Both the wave description and the corpuscular description are equivalent and complementary to each other. Some phenomena are easier to explain using the corpuscular nature of light, while others can be explained better by considering light as a succession of waves. The electrical part of the electromagnetic wave plays a much more important role than the magnetic part, and is responsible for the phenomenon of optical absorption and fluorescence.

4.6.1a Intensity of light

The magnitude of the electric wave cannot be measured directly. The measurable parameter is the intensity of light defined as the mean value of the square of its amplitude over a time period. Following the terms of the corpuscular theory, the light beam is a stream of photons with energy $E = h\nu = hc/\lambda$ and the intensity of light is taken as the number of photons transmitted per unit time interval (e.g. per second). As most light detectors (photomultipliers, video cameras, photographic film or the human eye) respond to the number of quanta the most common way of expressing light intensity is the number of quanta per second.

4.6.1b Wavelength

Light can be described either by its frequency ν (frequency of the oscillation of the electromagnetic wave) or more often by its wavelength λ. They are related to each other by the velocity of light c.

4.6.1c Polarization

The phenomenon of polarization can best be explained using the wave description. A real light beam is a superposition of a large number of various sinusoidal electromagnetic waves. If all waves of the beam oscillate in the same plane (their electric vector oscillates in the same plane) then the light is called fully polarized. If the distribution of these planes of oscillation is quite random, then the light is called unpolarized. For example, the light emanating from a light bulb or arc lamp is unpolarized (type of 'black body' radiation), while a laser beam may (but need not) be polarized. The passage of an unpolarized light beam through a polarizer (represented, for example, by a Polaroid sheet) converts unpolarized light into polarized. The fluorescence emission is unpolarized or partly polarized, and the degree of polarization is related to the order and mobility of the fluorescence molecules. The main application field of polarization imaging is the monitoring of plasma membrane fluidity employing fluorescent dyes incorporated into the membrane double layer.

4.6.1d Coherence

In luminescence spectroscopy, coherence plays only a marginal role. The emission of a fluorescence photon is a random process that occurs separately in the individual molecules, thus the resulting fluorescence light is not coherent.

4.6.2 Quantum mechanical description of molecules

Following the duality of the quantum mechanics theory, molecules can be described either as a bundle of waves or as a group of particles. The interaction with light may then be understood either as induced oscillations of electrons and nuclei due to the force of the oscillating electric field of the light or as consequences of collisions of electrons and nucleons with photons.

Molecules are allowed to exist only in certain so-called quantum mechanically (q-m) permitted states that possess definite values of internal energy. A molecule can change its state from one such q-m permitted state to another only if the energy difference between the two states is supplied from the outside (e.g. by the absorption of light or by chemical reaction) or if it is liberated (lost) to the outside (e.g. by light emission). In the usual schematics the energies of states in which the molecules may exist are shown as lines (levels). The y-axis (ordinate) corresponds to the value of energy, taking the ground state energy of the molecule as the zero reference level.

In terms of electrons, such transition between two molecular states can be described approximately as a transition of one electron from one molecular orbital to another. Molecular orbitals can accommodate two electrons. In the ground state of the molecule, all of the molecular orbitals with the lowest energy are already occupied; those with higher energy are empty. This situation changes when the molecule absorbs a light quantum, when one of the electrons is promoted from the molecular orbital with lower energy into another orbital with higher energy. The molecule is then described as excited.

4.6.3 Quantum mechanical description of absorption and emission of light

The quantum theory requires that absorption or emission of the light quantum may take place only if two conditions are fulfilled, i.e. the 'condition of energy' ($h\nu = E_1 - E_2$) and the 'condition of polarization' of light.

Chronologically, the absorption of a light quantum takes about 10^{-15} s, then the molecule 'relaxes' within 10^{-12} s into the excited state with somewhat lower energy, and eventually returns to the ground state, typically after 10^{-8} s. The return to the ground state may be accompanied by the emission of a photon. This is called a luminescence photon, and – depending on the route taken – further sub-classified as fluorescence, phosphorescence or delayed-fluorescence photon. The most common case of luminescence is fluorescence. There is also a non-radiative process of return from the lowest excited state to the ground state, when no photon is emitted, which is called internal conversion.

The real situation of a dye molecule in solution is more complicated due to interactions with the surrounding molecules, relaxation of the solvent molecules, transfer of the excitation energy from one molecule to another, a variety of photochemical reactions quenching the fluorescence, and so on.

4.6.4 Specificity of large organic molecules

The description of fluorescence in terms of electron transitions based on molecular orbitals introduced by Kasha is very convenient because it allows one to speak about σ, π, n and l electrons (and their promotions, i.e. transitions) in a similar way as is conventional in the chemical description of molecules. Electrons responsible for absorption and fluorescence in the visible range are usually π electrons. A visible light quantum has enough energy to facilitate the π to π^* transition (this abbreviation denotes a shift of a π electron from its ground state to a π excited state) but not sufficient for a change of state of the σ electrons. Quite often, n and l electrons are involved in the visible absorption and fluorescence processes.

Large molecules with a large number of electrons and nucleons have many vibrational and rotational modes available for a ready redistribution of energy and consequently their fluorescence properties are distinctly different from those of small molecules composed only of several atoms. Their spectrum, fluorescence lifetime and quantum yield are independent of the excitation wavelength. With a few exceptions they exhibit only one broad main absorption band and one broad fluorescence band. Comparison of the shapes of the absorption spectrum (the lowest absorption band) and the fluorescence spectrum (or the phosphorescence spectrum) shows a nearly mirror-like symmetry.

Virtually all dyes used as fluorescent probes comply with the above description. The only exceptions are lanthanide or uranyl labels.

4.7 GENERAL RULES RELATING CHEMICAL STRUCTURE AND FLUORESCENCE PROPERTIES

Using the classical chemical description formalism, the transitions ascribed to the σ excited states give fluorescence with photons of energy corresponding to far UV light, while the transitions associated with the π electron systems or with n or l electrons correspond to the UV and visible region. The fluorescence takes place from the lowest transition. The σ to σ^* electron transition has the lowest energy only in saturated compounds, where the absorption maximum lies below 200 nm and the fluorescence is in the UV range. The π to π^* transitions correspond to the rise of one of the electrons from the electron pair occupying a π orbital to a π^* orbital and back. The corresponding absorption is strong, with a molar extinction coefficient ε usually between 10 000 and 100 000, and the fluorescence quantum yield is high (up to 1). The maxima of absorption and emission shift to longer wavelengths with increasing length of the conjugated double bonds, as the energy difference between the ground state π and the lowest excited orbital π^* decreases. Molecules containing oxygen, nitrogen or sulphur atoms may have a lone electron pair. Then the transition from the n orbital to an upper π^* orbital is generally the transition of lowest energy and fluorescence arises from this transition. Such transitions have a low absorption ($\varepsilon < 2000$), a low quantum yield and a long lifetime.

4.7.1 Fluorophores

The process of absorption and emission is indeed localized to a small region within the molecule, so it is appropriate to speak about chromophoric or fluorophoric groups rather than about whole fluorescent molecules. The word *chromophore* relates to absorption, while the words *fluorophore* and *fluorochrome* relate to fluorescence. Nevertheless, in everyday use all expressions are equivalent. Molecules that absorb and fluoresce in the visible region appear coloured (note that the colour perceived by our eyes is complementary to that which is absorbed).

Many compounds absorb in UV light only and thus appear colourless. With prolongation of conjugate bond chains and increasing complexity of aromatic cycles the absorption peak shifts toward longer wavelengths – to blue, blue–green, green, yellow, red and even to near IR. Generally, the corresponding colours of the dye as seen in solution then change from yellowish to blue–green. The colour of a substance in a powdery or crystalline state may be different from that in solution. For instance, fluorescein is an orange powder but it forms a yellow solution with a green fluorescence, and UV and blue absorption bands. The value of pK may change upon excitation. Usually the pK value decreases after excitation of phenols and aromatic amines and increases in aromatic carboxylic acids and aromatic ketones. Virtually all organic molecules are capable of fluorescence but the phenomenon is often limited to extremely low temperatures (liquid nitrogen or helium) and UV excitation wavelengths. There are only very few compounds that exhibit an intense visible fluorescence when illuminated with visible or near UV light at room temperature.

4.7.2 General rules

It is very difficult to relate the chemical structure of a dye to its fluorescence properties. However, some general rules relating molecular structure and fluorescence can be formulated.

4.7.2a Wavelength (spectral) rules

(a) A π electron system must be involved if excitation occurs at wavelengths above 220 nm.
(b) The absorption and fluorescence spectra shift to longer wavelengths with more extensive conjugation.
(c) A 'cross-link' of the conjugated chain shifts the absorption and fluorescence spectra to a shorter wavelength.
(d) Substitution of side groups usually results in a red shift. In some cases, new types of excited states are introduced and the fluorescence transition type changes (cf. the effect of substitutions on fluorescein emission).

4.7.2b Quantum yield (fluorescence intensity) rules

(a) Highly absorbing compounds should also be highly fluorescent.
(b) The introduction of heavy atoms into the fluorescent molecule reduces its fluorescence. This quenching effect is similar to the effect of heavy atom ions in the close molecular neighbourhood of the dye. Two of the best quenchers are iodine and bromine. For instance, the quantum yields of fluorescein, eosin (tetrabromofluorescein) and erythrosin (tetraiodofluorescein) in water are 0.85, 0.22 and 0.02, respectively. This effect of bromination has been used, for example, to prepare a non-fluorescent modification of the calcium ionophore A-23187.
(c) An increase in molecular rigidity often means an enhancement of fluorescence, as can be seen in the example of fluorescein and phenolphthalein or rhodamine B and malachite green. The coplanarity of the rings allowing interaction of the separate electron systems may also play a significant role. The quantum yield of antracene is also a sensitive probe of environmental rigidity.
(d) Systems displaying a π^* to n transition have a low fluorescence intensity.
(e) Charge transfer states can give rise to fluorescence.
(f) *Ortho-para* substituents do not interfere with the fluorescence ability but *meta*-directed groups (except for CN) reduce fluorescence.
(g) The quantum yield decreases with increasing temperature.
(h) The excited fluorophore may form a permanent or a transient fluorescent complex with another fluorophore of the same kind or another kind. The

permanent complex is called a dimer (photodimer); the transient complex is called an excimer (when formed with the same type of fluorophore) or an exciplex (when formed with another type of fluorophore). The transient complexes disaggregate after the release of the fluorescence photon. The fluorescence of aggregates is red-shifted with respect to the fluorescence of the monomeric molecule.

4.7.2c Photobleaching

Molecules in the excited state are more prone to chemical reaction. In terms of fluorescence, this often means an irreversible destruction of the dye, with the final product being non-fluorescent. In this way all fluorescent dyes will eventually fade after a certain number of excitation–emission cycles. The number of excitation–emission cycles is different for each dye; it represents, for example, 3–4 cycles for quin2, but several thousand for fluorescein and even more for some other dyes. The usual mechanism of this 'passing away' is oxidation, which may slowed down by addition of antioxidants. In the positive sense photobleaching can be used for the measurement of mobility of membrane proteins (FRAP; see above).

4.8 AUTOFLUORESCENCE OF CELLS (FLUORESCENCE OF CELLULAR COMPONENTS)

When working with fluorescent dyes in biology, the fluorescence of the dye must be easily distinguishable from the background fluorescence of biological samples. This may sometimes be difficult, because all biological material exhibits some fluorescence, which may interfere with the emission from fluorescent probes and labels. It is actually the background fluorescence of biological samples that limits the sensitivity of fluorescence measurement and determines the minimal usable concentration of the dye. The autofluorescence of biological samples is higher when UV excitation is used, hence fluorescent probes with visible excitation are preferred. A simple subtraction of the signal after and before addition of the dye may not be the right solution of the problem because the mere addition of the probe may slightly quench the autofluorescence, and the simple subtraction could reduce it too much. It has been estimated, for example, that autofluorescence of a single 3T3 fibroblast cell is equivalent to about 34 000 fluorescein molecules (excitation wavelength 488 nm). One should be aware that not only cells but also media may be highly fluorescent (e.g. phenol red, collagenase impurities) and the measuring device itself may add its share (cell chamber, optical filters, objective lens). In fixed cells, autofluorescence can be quenched by washing in PBS with 0.1% sodium borohydride for 30 min. The fluorescence of cells, known from the literature as autofluorescence, background fluorescence or intrinsic fluorescence, may be characterized with respect to the emission wavelength in the following way.

Proteins exhibit strong fluorescence in the range 330–350 nm after excitation at about 280 nm. The fluorescence arises mainly from aromatic amino acid residues of tryptophan, tyrosine and phenylalanine. Phenylalanine fluoresces with a maximum at 282 nm (when incorporated in a protein usually at 283 and 295 nm), tyrosine at

303 nm and tryptophan at 340 nm (when incorporated in a protein at 333 nm). Their quantum yields are 0.04, 0.2 and 0.2, respectively. Fluorescence of proteins reflects their conformational changes as well as the binding of ligands. Changes are reflected also in the fluorescence lifetime, which may range from 1 to 7 ns, but in denatured proteins it is rather uniform, with values close to 2.5 ns.

Nucleic acids also exhibit intense fluorescence in the 300–350 nm range with absorption peaking at 250–300 nm. It is mainly the mitochondria, nucleus and the contractile apparatus that exhibit the most intense fluorescence in this wavelength region. The character of the UV emission from mitochondria is determined by the ratio of oxidized and reduced forms of respiratory enzymes and in this way it reflects their functional state.

In mammalian cells, the fluorescence in this wavelength range is dominated by metabolites, mainly by reduced pyridine nucleotides (NADH, NADPH; absorption maximum 340 nm, emission 460 nm) and oxidized forms of flavoproteins (FAD, FMN; absorption maximum 450 nm, emission 515 nm). The fluorescence of vitamins, especially of vitamin A (with a characteristic fast bleaching) and riboflavin is important. In plant cells, lignins dominate, with an intense green fluorescence.

Red fluorescence originates mostly from porphyrins. A strong red fluorescence is exhibited by cytochromes, peroxidase, haemoglobin, myoglobin and chlorophyll, i.e. haem-containing proteins.

4.9 CONCLUDING REMARKS

Due to the increase of the usable spectral range (new UV objective lenses, IR sensitive detectors) the space for the application of fluorescent probes has become larger. This allows monitoring fluorescence spectra of several fluorescent probes simultaneously. In terms of ratio imaging it is possible to measure concentrations of several ions, e.g. pH and calcium, simultaneously with another parameter like membrane potential or cytoskeleton changes. Ideally, you can just sit down in a chair, sip coffee, look at the computer screen and wait for new scientific discoveries.

REFERENCES

Lakowicz, J.R. (1983) *Principles of Fluorescence Spectroscopy*. Plenum Press: New York.
Parker, C.A. (1968) *Photoluminescence of Solutions*. Elsevier: Amsterdam.
Slavik, J. (1994) *Fluorescent Probes in Molecular and Cellular Biology*. CRC Press: Boca Raton, FL.

FURTHER READING

Pawley, J. (1989) *The Handbook of Biological Confocal Microscopy*. IMR Press: Oxford.
Ploem, J.S., and Tanke, H.J. (1987) *Introduction to Fluorescence Microscopy*. Oxford University Press: New York.
Slavik, J. (ed.) (1996) *Fluorescence Microscopy and Fluorescent Probes*. Plenum Press: New York.
Slavik, J. (ed.) (1998) *Fluorescence Microscopy and Fluorescent Probes*, Vol. 2. Plenum Press: New York.

5

Some Practical Considerations for the Development of Image Analysis Applications in Microbiology: Ensuring Software and Data Durability

Michael H.F. Wilkinson

University of Groningen, Groningen, The Netherlands

5.1 INTRODUCTION

When the development of an image analysis application in microbiology is started, there are a great number of important decisions to take. Which camera, microscope, hardware, software and data storage method should be used? However, before any of these questions can be answered, the key decision is about which software approach will be adopted. This is not a matter of which programming language or data structures are to be used, but whether the software will be designed and written in-house, or whether third party software will be used, either an off-the-shelf package, or tailor-made. Writing your own software gives great flexibility, but it is also a great burden, and brings with it a number of risks. The alternatives also have their pros and cons. To appreciate these problems, it is necessary to give a brief discussion of the many different computer architectures used in image analysis, before exploring software development methodologies. The discussion is aimed at non-computer-scientists, explaining backgrounds, jargon and the consequences of choosing particular hardware or software.

The main aim of the chapter is to help prevent system and data obsolescence. Far too often computer systems lie idle after researchers with experience with that system have left, or after the manufacturer or supplier has gone out of business, or that particular product line has been discontinued. It is of course impossible to guarantee that these things will never happen, but several measures can be taken to reduce the probability that the complete system becomes obsolete, and the entire investment is ultimately wasted. Software, which codifies the image processing ideas underlying your system, and data, which codify the results, should at least be usable on other computers, even if the original hardware is no longer available. The main aim of such measures is to make the software and stored data as independent of specific vendors as possible. Catch-phrases that are frequently heard in this

Digital Image Analysis of Microbes: Imaging, Morphometry, Fluorometry and Motility Techniques and Applications. Edited by M.H.F. Wilkinson and F. Schut.

context are portability, machine or platform independence, 'open systems', 'high-level languages', object-oriented programming (OOP), documentation and file format standards. This chapter discusses the meaning and importance of these phrases.

It may not contain very much that is new for computer scientists. The ideas presented here can be found in many computer science textbooks (e.g. Wirth, 1976; Wilson and Clark, 1988; Friedman, 1991), and can also be found in Allen *et al.* (1985), who describe the decisions underlying the design of GIPSY (Groningen Image Processing System), an astronomical image processing system at the Kapteyn Laboratory, University of Groningen. This system has been ported to at least four different hardware platforms and operating systems successfully, exemplifying the success of their design decisions.

5.2 HARDWARE ARCHITECTURES FOR IMAGE ANALYSIS

5.2.1 The basic single processor machine

John Von Neumann is credited with the basic design for the programmable computer. A Von Neumann machine consists of a central processing unit (CPU) and random access memory (RAM). The key idea in Von Neumann's design is the fact that both program instructions and the data on which the program acts are stored in the same memory (though not of course at the same location). This design allows great flexibility. Loading a program reduces to reading it into memory from some permanent storage device (punched cards, tape, magnetic discs etc.). By contrast, earlier computers were often programmed by setting switches, or connecting cables differently, effectively rewiring the computer. Imagine loading a word processor by rewiring your PC! Another feature of the Von Neumann machine is that the processor can access one memory location at a time, and performs instructions serially. This limits the speed of computations.

Modern CPU designs are primarily aimed at increasing speed. Unlike earlier CPUs, more than one instruction may be carried out simultaneously by tricks called 'microparallellism' or 'pipeline processing'. A simple example of the latter is 'instruction prefetching', common to many processors: while the arithmetic part of the processor is performing one instruction, the next instruction can be fetched from memory by another part of the processor.

5.2.2 The multiple processor machines

Multiple processor machines offer a relatively simple means to increase processing power, by division of labour over more than one processor. Although PCs and workstations are usually considered to be single-processor machines, most do have multiple processors on board: one general purpose CPU and several supporting processors with specialized tasks. The best known examples are the 'graphics accelerators' found on modern display drivers. Some disc controllers have their own CPU. These control writing and caching of data while the main processor performs other tasks. A number of image processing boards for personal computers also have specialized processors to perform image processing tasks, also to relieve the CPU.

5.2.3 Massively parallel machines

Massively parallel computers carry the idea of multiple processors to the extreme. These architectures allow hundreds or thousands of instructions to be carried out simultaneously. Not all problems in computer science benefit particularly from massive parallelism, since certain tasks are serial by nature; one computation must be performed after another. However, many image processing tasks can benefit greatly from parallel architectures. Take the addition or subtraction of two images of 512 × 512 points. If we had a machine with 512 processors, each could process just one row of each image, speeding up the calculation by a factor of 512. Many supercomputers have such architectures, but they are usually well beyond the budget of groups working in the field of image analysis of microbes.

5.2.4 Auxiliary hardware for image processing

5.2.4a Image capture and storage expansion boards

The most common piece of auxiliary hardware in PC-based image processing systems is the frame grabber. Its purpose is to digitize images from video cameras and store them in memory (see Chapters 2 and 13). Many of these boards do not use the RAM on the main board of the computer, but store the images in their own RAM. Many also have a video output port, allowing connection to a separate monitor to display the image captured, or the results of image processing operations. Some slow-scan cameras come equipped with a special expansion board which performs the same tasks as a video frame grabber, namely capture, storage and display of images from the camera. Other slow-scan cameras connect either to a serial port or to the external connector of a SCSI (Small Computer Systems Interface) hard disc controller. SCSI interfaces are parallel, and far faster than standard serial interfaces. In either case, no image capture board is needed.

Older (or modern low-cost) frame grabbers only accept video input in a limited number of formats (RS-170, CCIR, PAL, NTSC etc.), and are restricted by these formats in resolution (typically 512 × 512, 512 × 768 or 480 × 640). These boards accept only analog input, and have analog-to-digital converters (ADCs) with 8 bit (256 grey level) resolution. More advanced types can be programmed to accept non-standard video cameras, up to some maximum frame rate and resolution. The ADCs are usually 8 bit, sometimes more. Some also accept digital video input at higher grey-level resolution.

There is an obvious advantage of storing the images in separate memory on the capture boards: more memory is available. However, reading from and writing to memory on expansion boards is often slower than main-board RAM. This slows image processing operations considerably if the CPU has most of the processing burden, since before the CPU can perform any calculation it must retrieve the image from the extension board. Results must be written to the extension board memory too (see Sections 13.4.3 and 13.4.6). Furthermore, the organization of memory on the capture boards into images of fixed size can be restrictive. Main-board RAM can be organized in any imaginable way. Besides, the maximum amount of memory on the frame grabber can be restricted too, and expansion of it often requires expensive special-purpose memory chips (e.g. dual-ported video RAM or VRAM).

Some image capture boards perform more than just the basic tasks. They have an array of special hardware that has been optimized to perform specific image processing tasks at high speed, rather than being 'jacks-of-all-trades' like the CPU. In these cases, significant speed gains can be obtained with the images in the memory of the capture board. The CPU need no longer access the entire image; it only has to send commands to the image processing hardware on the board. Boards equipped with such hardware are of course more expensive than those without. The increasing speed of CPUs in recent years has reduced the need for such hardware, except for critical (real-time) applications. In the following sections some types of extra hardware found on image capture boards are discussed. In each case, typical features are discussed. Figure 5.1 shows a block diagram of a fictitious advanced frame grabber board showing a series of advanced features. The possibilities of each of these units on individual boards will vary considerably. Jähne (1995) gives a detailed discussion of the architecture of different frame grabbers and the hardware extension available on some of them.

5.2.4b Look-up tables, video-keying and graphics overlays

Most, if not all, frame grabbers support at least one or more look-up tables (LUTs). A LUT is simply a memory bank: the output value is the contents of the memory address indicated by the input value (Figure 5.2). LUTs may serve three distinct

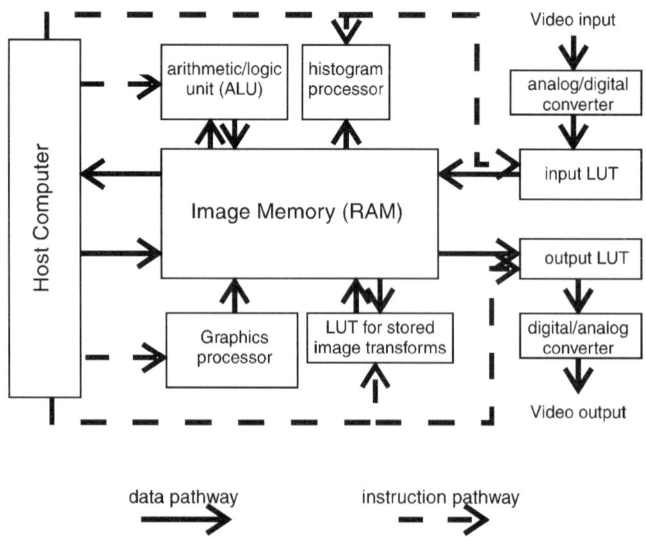

Figure 5.1 Schematic diagram of frame grabber, showing data and instruction pathways. For details on different units see text.

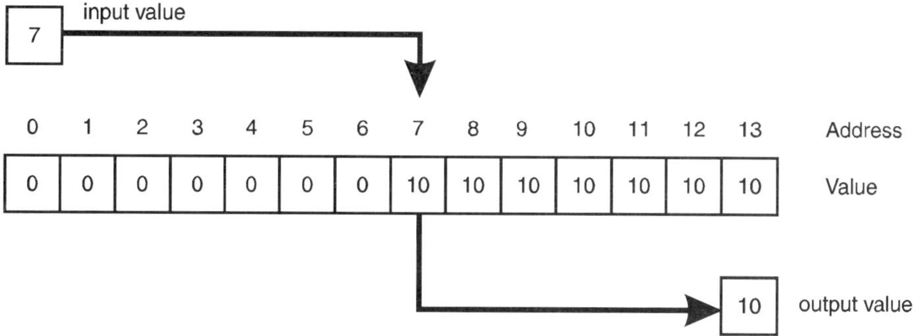

Figure 5.2 A look-up table (LUT). The input value is used as an address to find the corresponding output value. The LUT shown thresholds the input signal at 7: any input value below 7 returns a zero output; 7 and higher return an output of 10. Colour LUTs return three output values (for red, green and blue) for each input value.

functions:

(a) Input LUTs change the grey levels of the incoming image by translating the output of the ADC (or digital input) to new values. This may be used, for example, to invert the incoming image, change its contrast or threshold it globally for real-time segmentation (Section 13.4.3).
(b) Output LUTs change the appearance of the image on the screen without changing the grey levels of the image. These are equivalent to *palette registers* or *colour palettes* in normal video graphics boards (e.g. VGA). They may be used for 'pseudo-colour' display: translating grey levels into a range of colours can increase the visibility of detail (see Section 12.3 and Plate 2). This is especially useful with images with a far larger grey-level range (e.g. 12 bits per pixel) than the human eye or display monitor can register.
(c) LUT operations on stored images: all point operations on single images (Sections 1.4.1 and 6.7.1) can be performed very rapidly using special LUT hardware. The same functionality as input LUTs is provided, only in this case the input data come from a previously stored image, rather than from the data stream of the ADC or digital input port.

Many boards also allow so-called *video-keying*, which allows graphics to be overlaid on a live video image from the camera. When video-keying is enabled, the live video image is shown on the output monitor, except for those pixels that have particular values in a stored image (usually those that have a least significant bit (LSB) of 1). For these pixels, their grey level (or its colour palette equivalent) in the stored image is sent to the output. In this way, text, histograms, mouse cursors etc. may be displayed on top of the live video image.

Another way of achieving this is by using *graphics overlays*. These differ from video-keying operations in that they may be shown over any other image, either stored or live. This makes the technique more versatile than video-keying. The graphics or text to be displayed as an overlay is usually stored as a separate image with a reduced grey level range (typically 4 or 16). When graphics overlaying is enabled, any pixel in the overlay that has a non-zero grey level is displayed on top of the current image. A board may have multiple overlay images, allowing rapid switching between different messages or graphs.

5.2.4c Histogram processor

A histogram processor can compute image histograms rapidly. This is very useful for contrast stretching, histogram equalization and histogram-based thresholding techniques (Section 6.7.3). The speed-up may be considerable: computing a histogram using a MATROX MVP-AT processor with its histogram processor takes 67 ms. Computing the histogram by transferring the image data to the CPU costs 1–2 s on 80486-based PCs, limited mainly by the low access rate of the frame grabber's RAM. Some histogram processors can also compute grey level profiles of images, integrated intensities of regions in the image and so on.

5.2.4d Arithmetic logic units and neighbourhood processors

Arithmetic logic units (ALUs) provided on some boards can perform arithmetic and logical point operations between images at high speed (approximately 33 ms on the MVP-AT). These operations may include addition, subtraction, division (ratioing), multiplication and bitwise logical operations such as AND, OR and XOR (exclusive or). Here, too, gains in speed can be substantial, especially if the access rate to image memory is low.

Neighbourhood processors accelerate certain local operations, such as convolution with $n \times m$ kernels (up to some maximum size), median filters and operations from mathematical morphology (erosions, dilations etc.).

5.2.4e Graphics accelerators

Graphics accelerators perform 'primitive' graphics operations such as drawing lines, rectangles, circles, ellipses, characters etc. From a pure image processing point of view these may not be very interesting, except when producing comments or instructions to the user on the image processor display.

5.2.4f Processing windows and regions of interest

Many boards allow restriction of any operation to a rectangular area of the image, called the processing window. Other boards allow restriction of any operation to areas of any desired shape: regions of interest (ROI). When image processing operations are carried out by the hardware on the board, such hardware facilities for ROI or processing window definition are a must. If the CPU performs these tasks, software control is always possible, e.g. using schemes as in Sections 6.9 and 11.4.1.

5.2.4g Digital signal processors

Unlike the specialized hardware in the previous sections, digital signal processors (DSPs) are more-or-less general purpose processors, which have usually been optimized in some way for signal and image processing tasks, i.e. they can perform practically any task the CPU can, but they outperform it in these latter fields. Also, unlike many of the specialized processors, which usually process individual commands as they are received from the CPU, DSPs are fully programmable, and can work very much in parallel with the CPU. This means that they are far more flexible than the previous kinds of hardware. A single DSP can replace any or all of the specialized hardware in the previous sections.

5.2.4h Flexibility versus speed trade-off

When selecting hardware, there is a trade-off between flexibility and speed. Very often, specialized hardware outperforms a machine equipped with only a CPU, but only at specific tasks. As software demands change, such hardware may become obsolete quickly. In the past, CPUs in PCs were simply not fast enough, and memory limits were too restrictive for image processing, which meant that some auxiliary hardware was required. Now, with much faster CPUs and much larger memory spaces, there is less need for such hardware. Jähne (1995) illustrates this trend by certain new PCI-bus frame grabbers, which consist of just a little video circuitry and an ADC. Images are transferred directly to main-board RAM, where all further processing takes place. Thus, in many cases it makes more sense to invest in RAM, hard disc space and high speed CPUs. Only for time-critical applications may specialized hardware still be necessary (Chapter 13).

5.3 SOFTWARE CONSIDERATIONS

5.3.1 Languages, assemblers, compilers, interpreters

Programming languages are usually classified into different levels. At the lowest level is *machine code*, a series of numbers that correspond on a one-to-one basis with the instructions and data, in a format specific to a particular processor. They are usually represented in binary or hexadecimal (base sixteen) notation. Since it is hard to read, and therefore edit, such strings of numbers, *assembly language* was developed. Each instruction is represented by a mnemonic string, e.g. 'add' for addition, 'sub' for subtraction, etc. Before a program written in assembly language can be run on a computer, it must be converted into machine code, which is done by an *assembler*, a program that converts assembly language text into machine code.

Although far easier to read than machine code, assembly language suffers from the same drawback: it is suitable only for one type of machine, or at most a small family of compatible processors (e.g. Intel 80X86, Motorola 680X0 family). A program written in one type of code must be rewritten completely for a different machine. To solve this kind of problem, intermediate and high level *programming languages* were developed. Examples are C (intermediate), FORTRAN, Pascal, Modula 2 and 3 (all high level). These languages allow programmers to specify the required actions of a program at a higher level of abstraction. For example, if a is to be set to the value of the product of some quantities b and c in Pascal the programmer would write:

```
a:=b*c
```

(provided the meanings of *a*, *b* and *c* had been defined beforehand), whereas in assembly language it would be necessary to specify instructions to first load the value of *b* from the proper memory location into one of the CPU registers, then load the value of *c* into another, compute the product, and store the result in the memory location of *a*. Of course, the Pascal statement must be translated into machine code before it can be used, and this is done by a *compiler*, which is to a programming language what an assembler is to assembly language, or by an *interpreter*. Both take a file containing the programming language instructions (the *source code*) and convert it into machine code (*object code*) for one particular machine. Compilers take the whole source file and convert it into machine code, which may then be run as a program. Interpreters convert one statement, execute the machine code immediately, then turn to the next statement, etc. until the program terminates. Because Pascal and other programming languages do not specify instructions in terms of atomic machine actions, but in more abstract terms, the source code can be used on any machine that has a compiler for that particular language. Therefore such programs are said to be *portable*.

Apart from complete programs, compilers are capable of creating *object modules*, or ready-to-use chunks of object code, which can be combined with other chunks to form complete programs. This allows a subroutine for, say, a median filter to be compiled only once, and used many times in other programs. Object modules may be stored in separate files (object files), or may be combined in *libraries*, which usually contain a number of related object modules, which are generally used together. The process of combining multiple modules into a program is called *linking*, and the program that performs this task is called a *linker*. The linker must be given the names of files containing the main program module, and any auxiliary object modules and libraries.

5.3.2 Software toolkits for image analysis

Building a complete image processing system from scratch, using only program editor, compiler, assembler and linker, would invariably involve 'reinventing the wheel' a great number of times. Rather than spending most of the time optimizing standard algorithms such as median filters, edge detectors, morphological operators etc., one would like to focus on the interesting issues: which combination of image processing steps leads to the desired result. There are many software tools that allow researchers to do just this.

The simplest of these tools consist of ready-made libraries of image processing routines. The programmer must still write the main program by hand, but can simply refer to library routines whenever necessary. Furthermore, the programmer can add to this functionality by combining a series of standard functions into new object modules, or even add completely new functions, based on low-level image processing operations (e.g. read pixel, write pixel). Some of these libraries are supplied with image processing hardware such as frame grabbers, and are specifically intended for one particular model or series. Others are suitable for a range of hardware configurations. Most of these libraries for PCs are intended for use with the C or C++ programming language.

At the other end of the spectrum is a range of highly sophisticated image processing packages (e.g. IBAS, Kontron, Germany; Quantimet, Leica, Cambridge, UK; Visilog, Noesis, Canada), which often provide the same kind of functionality as image processing libraries, but in a much more user-friendly package. The user can grab images, activate image processing routines, and perform measurements by pointing and clicking on toolbars or menus interactively. Automation is achieved by writing *macros*, series of instructions, usually written in a proprietary macro language, that can be recalled and run using some form of an interpreter. This process is similar to building a program using standard library modules, but is a lot easier. If your image processing tasks require a simple, short sequence of standard procedures, such macros are an ideal solution. Rewriting them for other brands of image processing systems requires little effort, provided the macros are short. However, it is usually impossible, or unwise, to write completely new image processing routines using macro languages – impossible if the language does not allow access to individual pixels; unwise if the language is not portable across different software platforms, i.e. if it is not a standard programming language. Besides, scanning through all pixels in an image using interpreted, not compiled, code can be terribly slow. To alleviate these restrictions, most vendors can supply compilers and libraries for standard programming languages with their systems. Using these, programmers can add any new image processing function to the system.

In between the extremes of bare library and integrated point-and-click image processing packages, there is a range of CASE tools (Computer Aided Software Engineering), which take much of the burden off the programmer (e.g. SCIL-image, TNO, The Netherlands; AVS, Advanced Visual Systems, USA). Unlike integrated image processing systems, many of these tools are aimed at making stand-alone programs, which only need the correct hardware, not any auxiliary software (such as an interpreter), to run properly. There are also specialized CASE tools for creating good graphical user interfaces for stand-alone programs (e.g. CASE-W, Caseworks, USA; XVT, XVT Software, USA). These let programmers simply 'paint' the screen layout they desire, and automatically generate source code from this specification. In many cases this code is portable across a wide variety of platforms (Macintosh OS, MS-Windows, X-Windows etc.). Integrated image processing systems always supply a user interface.

5.3.3 Ensuring durability

Ensuring durability of software in a scientific environment is difficult, since much of the software written in a university setting is invariably written by (PhD) students or other workers with temporary contracts. These problems are not insurmountable, if certain 'good programming practices' are implemented as early in the project as possible.

A further problem for software durability stems from the diversity of hardware and software platforms discussed in the previous sections. The challenge facing software designers is not so much to write a program that can run properly on the platform used at any given moment, but to design a program that can be adapted easily to any type of hardware capable of performing the task. Here, too, good design can ensure, or at least improve, durability.

5.3.3a Structured, modular and object-oriented programming techniques

Structured programming is any form of programming that separates the problem to be solved into small, manageable units, and relies on so-called closed control structures, such as if–then–else, while–do, repeat–until constructs offered by most modern programming languages, to keep the code readable. Early programs frequently consisted of a single 'monolithic' block of code, with a liberal sprinkling of 'GOTO' statements and corresponding 'labels'. This caused the sequence of execution of instructions to be interrupted frequently by jumps to different locations in the code. It can be very hard to determine what the resulting 'logical spaghetti' actually does, let alone change it to perform a slightly different task. Figure 5.3 shows two different pieces of code, one using well-chosen names for variables and closed control structures, the other violating as many concepts of good programming practice as could be managed in such a short piece of code. Both pieces of code actually do the same thing. A classic on structured programming is the book by Wirth (1976).

Breaking up the problem into smaller units is achieved by dividing it up into separate functions and procedures (or subroutines), each of which has just one simple task to perform, possibly by calling other functions and procedures. Each procedure or function should ideally be short enough to fit on a single page (or even single screen).

Modular programming improves upon structured programming by allowing different parts of a program to be stored in different files. Each module consist of two parts: (i) the *interface*, which specifies what each procedure and function looks like to the outside world, and (ii) the *implementation* part, which contains the actual code. Some languages split the implementation part into the implementation itself, and a section used for initialization, specifying which actions must be taken when the program starts. Any feature of the module that is not specified in the interface is hidden from the outside world and cannot be accessed by the program. Changing the implementation while keeping the interface the same should therefore not affect the program in any way. This means that more than one implementation of each module may exist, e.g. for different image processing hardware. Besides, more efficient algorithms may be implemented in the module without the need to recompile all of the source code in the program. Once the module has been recompiled, the programs need only be linked.

Object-oriented programming (OOP) takes the idea of hiding implementation details from other modules one step further. In modular programming, the interface will frequently contain precise descriptions of new data types, e.g. defining an image as a two-dimensional array of 8-bit bytes. Modules using this definition use this information, for example to reserve sufficient memory for an image. If you wish to increase the range of grey levels in images, you might want to change the definition of an image to a two-dimensional array of 16-bit integers. Since the

```
PROGRAM SimpleStatistics (INPUT,OUTPUT);
(* Computes mean and variance of data in input. Input data must
   be a single collumn of floating point numbers                    *)

VAR Item,
    Sum,  SumSquare,
    Mean, Variance   : REAL;

    NumData          : INTEGER;

BEGIN
  Sum:=0;  SumSquare:=0;  NumData:=0;  (* Initialization *)
  WHILE NOT EOF DO                     (* Read to End of File *)
    BEGIN
      READLN(Item);
      Sum:=Sum+Item;
      SumSquare:=SumSquare+Item*Item;
      NumData:=NumData+1;
    END;
  IF NumData>1 THEN                  (* Check for sufficient data *)
    BEGIN
      Mean:=Sum/NumData;
      Variance:=(SumSquare-NumData*Mean*Mean)
                /(Numdata-1);
      WRITELN ( 'Mean     : ', Mean );
      WRITELN ( 'Variance : ', Variance );
    END
  ELSE
    WRITELN('Insufficient data');
END.
```

```
PROGRAM Stat (INPUT,OUTPUT);
LABEL 11, 12, 13, 14;
VAR x,s,s2:REAL; n:INTEGER;
BEGIN
s:=0;s2:=0; n:=0;
11:IF EOF THEN GOTO 12;
READLN(x); s:=s+x; s2:=s2+x*x; n:=n+1; GOTO 11;
12:IF n<=1 THEN GOTO 13;
s:=s/n; s2:=(s2-n*s*s)/(n-1);
WRITELN ( 'Mean     : ', s );
WRITELN ( 'Variance : ', s2 );
GOTO 14;
13:WRITELN('Insufficient data');
14:
END.
```

Figure 5.3 Two Pascal programs performing the same task. The top uses comments, good layout, illustrative names of variables and closed control structures to provide clear, readable source code. The bottom uses no comments or layout, short variable names and GOTO statements, and reuses variables, resulting in practically unreadable code.

definition is in the interface, *all* modules using this interface must be recompiled. In OOP, data types are not defined in terms of what they contain, but in terms of what things can be done with them. An OOP approach to the definition of an image would be to specify that the *class* (not data type) 'image' exists, and that variables of that class can be created, destroyed, captured from a camera, added, subtracted, convolved, stored on disc and retrieved etc., by calling appropriate procedures and functions. Only the implementation section actually knows what the structure of the image is. This means that only this module need be compiled to perform the change outlined above. Again, there is an increase in code reusability.

5.3.3b Layered approach

In any large software system, sooner or later there will be a problem that can only be solved in a platform-dependent (e.g. hardware or operating system) way. In the context of image analysis, there is the example of access to the image processing hardware, which cannot be addressed in a machine-independent way. Such parts of the program must be rewritten when the software is ported to other platforms, e.g. when a new frame grabber is introduced. To ensure that only a limited part of the system needs to be changed in such an event, a layered software design is called for. This layering is achieved by putting all hardware-dependent tasks into a limited set of modules. In the GRID system we use the following classification of modules (Meijer, 1991; see Figure 5.4):

(a) Level 1: Application programs. These may only contain machine-independent code, using only external declarations found in interfaces from levels 2 and 3.
(b) Level 2: Machine-independent support routines. Both the implementation and interface are portable to other platforms; these may therefore only use declarations from interfaces at levels 2 and 3. Examples are image filtering or addition using pixel access functions from level 3.

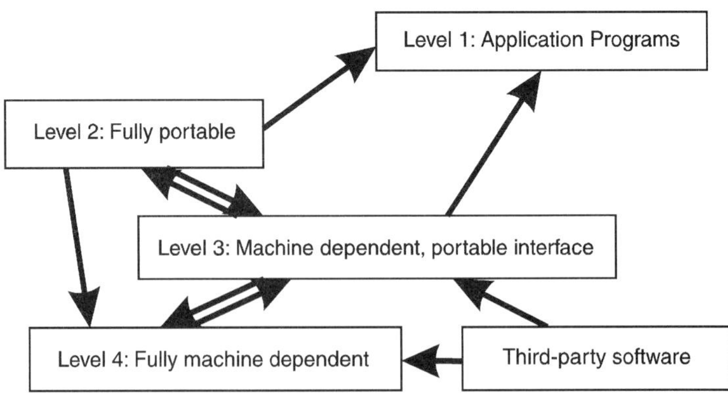

Figure 5.4 Software organization in the GRID system. Any module or application program may use any support routine from object modules from either level 2 or level 3. However, only routines in levels 3 and 4 may access level 4 and third party software. If either the machine or the third party software is replaced, only levels 3 and 4 need be rewritten.

(c) Level 3: Machine-*dependent* routines with machine-*independent* interface. These may access interface declarations from levels 2, 3 and 4 internally. One example is access to the image stored in frame grabber memory: READPIXEL(X,Y) can be implemented on any machine, so the interface is machine-independent. However, the *way* the pixel is read may be very different on different machines.
(d) Level 4: Fully machine-dependent routines. These may provide access to specific registers on image capture boards. Any routine accessing such a routine is automatically hardware-dependent.

One further note is required to explain what we mean by machine-dependent in this context. The GRID system currently works on two types of image processing boards: the older PIP-1024 and the MVP-AT (both MATROX Ltd, Dorval, Quebec). MATROX provides a software library to perform tasks such as image capture, access to (rows of) pixels etc. In theory, these should be level 3, yet we have classified them as level 4 ('third-party software' in Figure 5.4). In effect, we view these libraries as the 'machines' we work on. The reason for this is that the names of the corresponding procedures in the two libraries are different, along with the orders in which parameters must be given and the allowed ranges of certain parameters. Therefore, we have made (level 3) interface modules, linking these libraries to the rest of our system. By doing this we were able to port the system from the PIP-1024 to the MVP-AT in just three weeks, because only the interface library needed to be rewritten. There is some overhead involved with an interface library, which does sacrifice some speed, but this is not detectable in most cases. A possible exception is time-critical work such as used in tracking swimming microbes (Chapter 13).

5.3.3c Choice of language and toolkits

The choice of programming language and toolkits is almost a matter of religion among programmers. Endorsing one almost immediately alienates 50% of all programmers, who will instantly consider you a heretic. No programming language is ideal from all points of view, and it is probably not surprising that many practical systems have actually been written using a mixture of at least two languages (often one higher level language and assembler). A more thorough comparison of programming languages can be found in the textbooks by Wilson and Clark (1988) and Friedman (1991).

Considering portability, the C, C++ and FORTRAN 77 languages are most widely available. C and C++ offer considerably more flexibility and opportunity for structured programming than FORTRAN 77. C and C++ are certainly not ideal, in particular regarding readability of the code. Both languages 'overload' operators, leading to the unfortunate situation that one symbol may have different meanings depending on the context. For example, &A is the address of A, A&B performs a *bitwise* AND operation (a bit in the result of A&B is 1 if the corresponding bits in A and B are 1), and A&&B is a *logical* AND operation (the result is FALSE unless A and B are TRUE). This does not increase the readability of the program code. In fact, there is a notorious kind of 'party game' played by C programmers called 'guess what this line of code does'. Pascal-like languages like Modula (2 and 3) are far superior in this regard. Nevertheless, C and C++ are simply the most commonly used programming languages for this kind of work. Using proper programming style and documentation standards, readability can be ensured.

It is beyond the scope of this chapter to compare the plethora of graphical user interface design tools. Criteria that should be considered are (i) which platforms are supported and (ii) what level of functionality is offered (from simple library with manual up to screen painter with code generator).

5.3.3d Essential features of image processing packages

Rather than giving a comparison of the capabilities of a selection of all the image processing packages available (which would be out of date by the time of the next version), I would like to list the features I would like to see in a package, again from the point of view of portability and flexibility:

(a) Full programmability in a commonly used programming language (preferably C/C++).
(b) Interfacing capability to a wide range of computer platforms and user interfaces (DOS, Macintosh, MS-Windows, UNIX etc.).
(c) Support of multiple resolutions both in physical dimensions (x, y and z) and in grey levels: the 8 bit per pixel maximum seen in many older tools is no longer sufficient, given the increasing resolution of cameras. Ideally, both integer numbers (8, 16 or even 32 bit) and floating point numbers (single and double precision) should be available for processing.
(d) Interfacing capability to a wide range of image capture (video cameras, slow-scan CCDs etc.) and processing hardware.
(e) Interfacing capability to a wide range of microscope automation hardware.
(f) Support for commonly used image file formats.

It is difficult to put an exact ranking on the relative importance of these demands. The important thing to remember is that what might seem luxury now (e.g. 16 bits resolution per pixel) may be essential tomorrow. A software system can and therefore should be flexible enough to cater for future needs.

5.3.3e Programming and documentation standards

When it is decided to develop image analysis software, one of the first things to do is to decide on software and documentation standards, and then to adhere to them strictly. Even if programmers think their style of programming is 'self-documenting', most would have a hard time interpreting their own code a few months or years later. Programming and documentation standards should ensure durability, and clarity of the source code.

Many researchers tend to view writing documentation of code as boring and a distraction from their real work. However, I (and many others, no doubt) have frequently found that writing documentation *before* writing a line of code resulted in much better code, both in readability and efficiency, because writing the documentation forced me to think about the problem thoroughly. Thus, writing documentation should be viewed as part of writing code, not a boring extra chore. In the GRID system (Meijer, 1991; Wilkinson *et al.*, 1994), the documentation standards specified that the following items must be reported, using standard headings:

(a) *Problem definition*: a concise definition of the problem to be solved by the program or module.
(b) *Functional specification*: an exact specification of what functionality is offered by the program or module. Each of the tasks to be performed by each procedure (subroutine) or function is clearly defined.
(c) *User's manual*: clear definition of what questions are asked of the user, which messages are put on the screen etc.
(d) *Programmer's manual*: specifies the precise data type, function and procedure declarations, including requirements of input, memory use, data output etc.
(e) *Error conditions*: a list of error codes returned by each subroutine, plus a concise explanation. Our convention is that an error code of zero signifies 'no error', positive values are 'exceptions', which the program should be able to handle, and negative values flag unrecoverable errors.
(f) *Internals*: a specification of which internal representation of the data, which algorithms etc. are used.
(g) *Externals*: list of which data types, variables, procedures and functions from other modules are used.

Despite our best efforts and intentions, our documentation is not as complete as we would like it to be.

Apart from documentation standards, there should also be programming standards. These should include naming conventions for variables, data types, functions and procedures, forbidden control structures (such as GOTO statements), program layout etc.

In a scientific environment however, there is also a need for 'quick-and-dirty' programs for experiments. To facilitate this, without compromising the durability of the system, the GIPSY standards (Allen *et al.*, 1985) allow researchers to run their own programs (at their own risk) within the system, but without the programs becoming part of the system. The program remains in the researcher's own directory, which is not specified in the PATH environment variable. Other users can also access these programs, by specifying the directory of the researcher before the command. If many users find the program handy, it can be included in the system, but only after the documentation has been provided and the program has been altered to meet the programming standards.

5.4 DATA DURABILITY CONSIDERATIONS

5.4.1 Image file storage

An important consideration when developing an image processing system is choosing a file format for storing your images. For convenience, files are very often stored in a machine-dependent way, e.g. if the images of interest are always

512 × 512 pixels, at one byte per pixel, simply store consecutive pixels as bytes in a binary file. This approach works well as long as the hardware does not change, and the images are never imported into other software packages, e.g. for purposes of publication. In practice, both reasons for abandoning hardware-dependent storage emerge sooner or later. It is almost always better to use some standard format to store your images.

The next problem is to decide which of the plethora of image file formats to choose. The key issues on which this decision should be based are portability, flexibility, storage space requirements and extensibility. Portability requires the format to be readable on most modern hardware and software platforms. Flexibility requires the format to be able to store different image types (binary black and white, grey scale, true colour, palette colour), at as many different resolutions (in grey level, colour or X–Y space) as possible.

Storage space requirements depend on image compression techniques available within the file format. There are two classes of compression techniques: lossy and non-lossy. The latter allow exact reconstruction of the original image, whereas lossy methods allow reconstruction of an approximation of the original image. For display purposes, such lossy methods are usually adequate, but for exact measurement non-lossy methods must be used. Table 5.1 contains a comparison of a number of popular image file formats.

At the moment, the most portable file formats are probably GIF (Graphics Interchange Format), TIFF (Tag Image File Format), JPEG (Joint Photographic Experts Group), and the PBM/PGM/PPM group of formats. Of these formats, only TIFF supports more than 8 bits per sample, and any number of samples per pixel. Furthermore, TIFF can store images of any size, provided they fit within a file of 4 GB, which is a limit posed by many file systems anyway. Besides, TIFF can be extended to include extra data concerning calibration, physical scale etc. It has a number of non-lossy compression algorithms to limit the amount of disc space used.

Table 5.1 A comparison of file formats.

	TIFF	GIF	TGA	BMP	PCX	JPEG
Bits per pixel	$1..n \times 32$	1..8	16, 24, 32	4, 8, 24	1, 2, 4, 8, 24	6–24
Multiple images	+	+	[−]	[−]	[−]	[−]
Palette	+	+	[−]	+	+	[−]
RGB	+	[−]	+	+	+	+
Grey scale	+	+	[−]	[−]	+	+
Compression:						
Uncompressed	+	[−]	+	+	+	[−]
Runlength	+	[−]	+	[−]	+	[−]
Packbits	+	[−]	[−]	[−]	[−]	[−]
LZW	+	+	[−]	[−]	[−]	[−]
Lossy method	+	[−]	[−]	[−]	[−]	+
Expandable	+	+	[−]	[−]	[−]	[−]

TIFF, Tag Image File Format; GIF, Graphics Interchange Format (CompuServe); TGA, TrueVision Targa; BMP, Windows 3.X Bitmap; PCX, Paintbrush files; JPEG, Joint Photographic Experts Group; LZW, Lempel-Ziv & Welch compression method (non-lossy, general purpose).

Lossy compression (JPEG type) has become available in version 6.0 of the format. Multiple images may be included in a single file. By contrast JPEG only supports lossy compression, and GIF supports only palette colour or grey scale images. The flexibility of TIFF comes at the price of somewhat increased complexity. However, source code in C to read and write TIFF images is widely available.

All of the above formats have been designed for two-dimensional images. TIFF can be (or may already have been by the time of publication) extended to include three dimensions. All considerations for choosing two-dimensional file format hold when choosing a format for three-dimensional storage, but it is hard to say which formats will prove 'winners' at this point in time.

Those wishing to implement import or export functions for almost any image file format should consult the book by Murray and VanRyper (1996), which includes a CD-ROM with source code.

5.4.2 Numerical data formats

The numerical data produced by image analysis are handled more easily than the images themselves. Many software tools allow export of data in special spreadsheet formats or special-purpose 'portable' formats such as DIF (data interchange format). Since these formats are widely accepted, it is perfectly safe to use them. However, if you are writing in a programming language lacking output facilities in such complex formats, the best thing to do is to export the data in a simple ASCII text format, such as comma separated variables or fixed column width tables. These can be read by almost any package, so it is probably not worth the effort to write spreadsheet input and output routines from scratch.

5.4.3 Physical storage

The final problem in data durability is the physical storage format. In the past, various tape formats have been used for mass storage of data. These have a number of problems: (i) magnetic media degrade slowly, so the data have to be read and rewritten from time to time, (ii) tapes are sequential devices, unlike discs, so retrieving a file may be very time consuming, and (iii) especially in the PC market, tape formats have come and gone. Floppy discs are impractical due to the sheer bulk of data, and not as safe as magnetic tape (as a rule). DAT tapes are fine for monthly system backups, but lack the long-term storage capacity needed for permanent records to be kept. They share the serial access problem of all tape-based media.

By contrast, optical and magneto-optical (MO) discs have a long life span and very fast access rate. The main problem is that both storage and retrieval require a special drive, which is not available on most systems. At the moment, the ideal solution seems to be the recordable compact disc (CD-R). It is durable, affordable and has a high access rate. Although writing to the disc requires special hardware, retrieval is possible on most computers equipped with CD-ROM (some older models are reported to have problems reading CD-R discs).

Finally, although optical media are more durable than their magnetic counterparts, they do not have infinite life span. It pays to make multiple copies of important data.

5.5 TEAM DURABILITY

It may have struck the reader that much of this chapter is motivated by a single fear: that the programmer should leave the team. To a certain extent this is indeed the case, since departing programmers can take vital information with them from the team. This problem is neither restricted to image analysis in microbiology, nor to departing programmers. When working on any multidisciplinary research project, it is important to retain a certain critical mass of expertise in all disciplines involved. What is seen in many successful long-term research programmes in biomedical image analysis is that the team itself is made up of people from two different organizations: usually one biomedical university department and one department of computer science, often from the same university, or a software company (e.g. Chapter 11). This makes the team itself more durable, i.e. any person leaving can be replaced by other team members, since a large pool of experts in both disciplines is available.

5.6 CONCLUDING REMARKS

It should be clear that there is no absolute answer to the choice between writing software oneself or buying packages. The main problem of the former is that it requires on-site availability of programming skills, besides knowledge of image analysis. Another problem often overlooked is that writing software also means writing manuals, both for users and for (future) programmers. Data compatibility with standard packages is also a problem that must be addressed, chiefly by using recognized image file formats. Finally, the programs must be designed to be portable, i.e. it should be possible to adapt the software to new computers easily. All of these problems can be solved, but require thought and planning, besides manpower.

By contrast, buying an off-the-shelf image analysis package, many of which are available, has the advantage that it comes with the manuals, on-line support etc., but it might lack one or two features that happen to be essential for your particular problem. Lack of flexibility was the main problem of earlier off-the-shelf packages. Furthermore, it is sometimes forgotten that knowledge of image analysis is still required on-site, to make the correct decision on which of the options to use, and in what order, to achieve the desired result. Buying an image analysis package is more like buying a programmer's toolkit than buying a word processor, in that it can only function properly in the hands of highly skilled people.

It is often countered that off-the-shelf packages are sufficiently advanced to do any job you wish by stringing a number of commands together in macros. Besides, many suppliers of image processing packages are happy to assist in the development of such macros. Indeed, modern packages support an impressive array of functions, and powerful macro languages, which allow skilled hands to tackle many problems. However, image processing and computer vision are areas of active research, and many new algorithms cannot (or not efficiently) be programmed in terms of (necessarily older) algorithms supplied with your package. A package that does not allow programming in a high level language all but precludes collaborations with, for example, computer science departments.

Modern image analysis packages can offer both a plethora of built-in functionality and the opportunity to integrate software written by the scientists themselves. This combines the ease of use and user support of off-the-shelf packages, with the power and flexibility of writing your own software. It relieves much of the tedium of writing basic image processing procedures and functions, and should increase productivity considerably. However, care must be taken that the program code written by the scientists does not become too dependent on the software package used. Creating a portable interface between home-grown and third party software can ensure proper independence, at a negligible cost in computing time. Again, computing time is cheap compared to programmer time.

The development of digital image processing applications in any field requires a large investment in hardware, software and staff. Ultimately, the latter two, and especially the last, are the most precious, since hardware is often obsolete by the time you buy it. At the outset, this is often not appreciated because (as a computer science saying of unknown origin goes) hardware is sexy, software is not. This means that it is easier to get $30 000 funding for more megahertz, megabytes and megapixels than it is to get $3000 for a graphical user interface (GUI) toolkit contained on 10 floppy discs, even when you point out that it would cost the in-house programmer more than a year to come up with anything half as sophisticated or portable, as could be achieved with the toolkit in a month.

In summary, after choosing which microscope and camera type are needed, most people would first choose the computer hardware, and see which software supports it only afterwards. Based on the preceding discussion, the reverse is probably better: decide which software system offers the flexibility, portability and power required, and then choose which of all the hardware platforms the package supports you wish to have (or can afford).

REFERENCES

Allen, R.J., Ekers, R.D. and Terlouw, J.P. (1985) The Groningen Image Processing System. In *Proceedings of the International Workshop on Data Analysis in Astronomy* (Di Gesu, V. and Scarci, L., eds), pp. 271–93. Plenum Press: New York.

Friedman, L.W. (1991) *Comparative Programming Languages.* Prentice Hall: Englewood Cliffs, NJ.

Jähne, B. (1995) *Digital Image Processing: Concepts, Algorithms and Scientific Applications,* 3rd edn. Springer-Verlag: Berlin.

Meijer, B.C. (1991) Medeyes: an extensible language for image analysis. *Binary Comp. Microbiol.* **3**: 57–60.

Murray, J.D. and VanRyper, W. (1996) *Encyclopedia of Graphics File Formats.* O'Reilly and Associates, Inc.: Sebastopol, CA.

Wilkinson, M.H.F., Jansen, G.J. and Van der Waaij, D. (1994) Computer processing of microscopic images of bacteria: morphometry and fluorimetry. *Trends Microbiol.* 2: 485–9.

Wilson, L.B. and Clarck, R.G. (1988) *Comparative Programming Languages.* Addison-Wesley: Reading, MA.

Wirth, N. (1976) *Algorithms + Data Structures = Programs.* Prentice Hall: Englewood Cliffs, NJ.

6

Automated and Manual Segmentation Techniques in Image Analysis of Microbes

Michael H.F. Wilkinson

University of Groningen, Groningen, The Netherlands

6.1 INTRODUCTION

The goal of image segmentation techniques is to classify pixels meaningfully, allowing the identification of semantically different regions. The simplest form of segmentation, which is frequently encountered in microscopic images in microbiology, identifies pixels as being part of an object (foreground) or of the background. More complicated forms might assign pixels to numerous different classes (categories of cell, different organelles, types of tissue etc.). Thus, segmentation of images is a crucial step in obtaining image understanding. The quality and interpretation of all measurements on different parts of an image depend critically on the ability of the segmentation method to assign each pixel to the appropriate class. To measure, for example, the volume of a bacterium in an image, we must of course first be able to determine which pixels actually belong to the object. The same applies to practically any other type of measurement.

The human observer is particularly adept at making the type of decision required of segmentation algorithms. Even without any training in microbiology, most observers would identify the 'entities' seen in a typical microscopic image of microbes. Figure 6.1A shows a fairly typical image of untreated faecal material, which contains large clumps of debris, apart from a large number of bacteria. Despite the confusing background, a human observer could count the number of bacteria with a fair degree of accuracy. It would be tedious, but also fairly easy. By contrast, designing a computer program to perform the same task is far from trivial.

Fortunately, there is a host of publications on the topic of image segmentation in the computer science literature. It should be stressed that none of the techniques is generally applicable. It is often seen that algorithms that perform well on particular categories of images fail dismally on others. The different approaches to segmentation may be divided into a number of categories, which are discussed in Section 6.2. In this section a number of rules of thumb on the applicability of different techniques and degree of automation required in different applications are given. Section 6.3 deals with the validation of segmentation results, a topic not often

Digital Image Analysis of Microbes: Imaging, Morphometry, Fluorometry and Motility Techniques and Applications. Edited by M.H.F. Wilkinson and F. Schut.

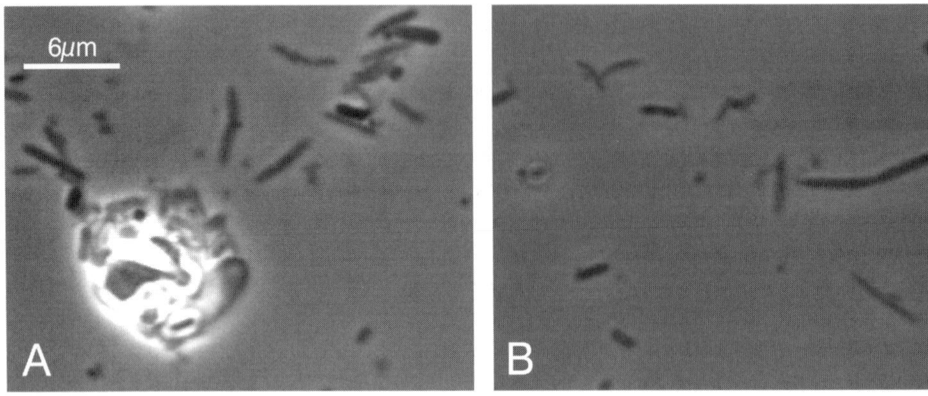

Figure 6.1 Phase contrast image of two slides of rat faeces: (A) smear of untreated faecal suspension showing large clumps of debris apart from many bacteria and (B) smear of faecal suspension treated according to Meijer *et al.* (1990; see also Chapter 9) showing a large reduction in the amount of debris.

dealt with in detail in the image processing literature (Lee *et al.*, 1990; Pal and Pal, 1993).

Many segmentation methods have been applied to different types of microscopic images successfully. One of the great advantages of working with microscopic images from the computer science point of view is that the image content can often be controlled very well compared to the images encountered in more general computer vision problems. Indeed, it is useful to distinguish two types of image encountered in image analysed microscopy: (i) 'easy' images, in which objects stand out from a background which itself shows little detail, and (ii) 'difficult' images, in which there is a confusing background, which shows detail of the same order of magnitude as the objects. Examples of easy and difficult images are given in Figure 6.1. The usual approach for successful segmentation of a difficult image is to try to convert it to an easy image by some means. Section 6.4 describes a number of computational approaches to this kind of problem. However, before trying to improve the images by digital image analysis, it is much better to improve the slides! Very often, better sampling, staining, fixation and illumination (e.g. the use of narrowband excitation to reduce background fluorescence) methods can transform difficult images into easy ones with far better results than can ever be obtained digitally. Thus, one should first optimize the manufacture of slides, and only afterwards tackle any remaining problem with uneven background.

Sections 6.5, 6.6 and 6.7 deal with the three most important categories of image segmentation methods in some detail. The emphasis is on image thresholding (Section 6.7), which is one of the simplest, very successful and probably best researched of the techniques used in image analysed microscopy. Section 6.8 describes methods to separate touching objects. Section 6.9 deals with several representation and storage schemes for the segmentation results. In Section 6.10 conclusions are given and future prospects discussed.

6.2 CATEGORIES OF SEGMENTATION TECHNIQUES

All segmentation techniques are concerned with finding boundaries of some kind. The most obvious approach is to detect boundaries in the image itself, e.g. by detecting edges between regions of different brightness, or with a different texture. This approach is probably close to the way our own visual system detects objects in an image (Hubel and Wiesel, 1977). Most of the information in the image is contained in the edges, which is why we have no difficulty in recognizing cartoon drawings, which are essentially just collections of edges. Numerous algorithms use edge detection as a basis for image segmentation.

A different approach tries do identify regions directly, rather than the boundaries between them. Within this category two subclasses may be distinguished. The first works in *image space*, dividing the image into connected regions in which all pixels share certain properties. The other approach works in *data space*, using the grey level (or colour) histogram of the image, classifying pixels purely on the grounds of their numerical values, rather than including connectivity. Although this method may produce slightly more jagged edges in the spatial domain, and some form of boundary smoothing is often needed (Section 1.6.1c), it is by far the simplest approach to image segmentation. Because of its simplicity, it is one of the fastest, most popular segmentation techniques used in image analysed microscopy.

A totally different classification of segmentation techniques concerns the degree of user interaction in the process, or, conversely, the degree of automation. Certain techniques are purely manual: the user sets a brightness threshold or draws the outline with a mouse, trackball or other pointing device. At the other extreme are fully automatic algorithms, which require no user supervision at all. In between are methods that require some user interaction. Many of these latter algorithms do segment the image automatically into objects and background, but require the user to edit out objects that are not considered relevant. Fully manual methods have obvious advantages: they come standard with most image processing packages, and can segment any image an operator can interpret correctly. The obvious drawback is tedium. Any form of user interaction, whether in fully manual or semi-automated techniques, can introduce user biases. On the other hand, there are many situations in which no other option exists, e.g. when the contaminating objects are of similar size and shape to the objects of interest. In such cases, semi-automated techniques, which only allow the user to reject objects, but do determine the shape of each object automatically, are probably an optimum choice. If the contamination is sufficiently different in size or shape from the objects of interest, fully automated methods should be considered. Both semi-automated and fully automated methods are generally less widely applicable, each method being specialized for a specific category of images.

Apart from these considerations, the choice of manual versus (semi-)automatic segmentation depends largely on the number of images and image segments to be processed. It does not make sense to spend more time developing and testing an automatic algorithm than the time needed to separate the regions of interest (ROIs) manually. Similarly, if a semi-automated method performs well, one should weigh the development time for a fully automatic algorithm against the time spent actually editing the images. For short runs, e.g. for pilot studies, it is probably best to use a manual method, unless a good semi-automatic algorithm is already available. Fully automatic algorithms should only be developed for large numbers of slides, and only when slide preparation has been optimized.

6.3 VALIDATION

One of the key problems when evaluating methods is finding an objective means to quantify errors. There are two reasons for this: (i) it is not easy to find an objective 'best' segmentation and (ii) it may not be straightforward to quantify differences in segmentation results. The first problem is by far the more important, and in the next subsections a number of 'gold standards' for use within microscopy are discussed. Once a suitable comparison has been found, validation can focus on two separate issues: (i) accuracy of the delineation of the ROIs in the image and (ii) accuracy of measurements derived from such shapes. Most validation schemes used within this field are of type (ii) (e.g. Sieracki *et al.*, 1989a). Although (i) implies (ii), the reverse is not necessarily true. This difference can be demonstrated quite easily. Consider the two segmented shapes of the same object in Figure 6.2. If shape A is the 'gold

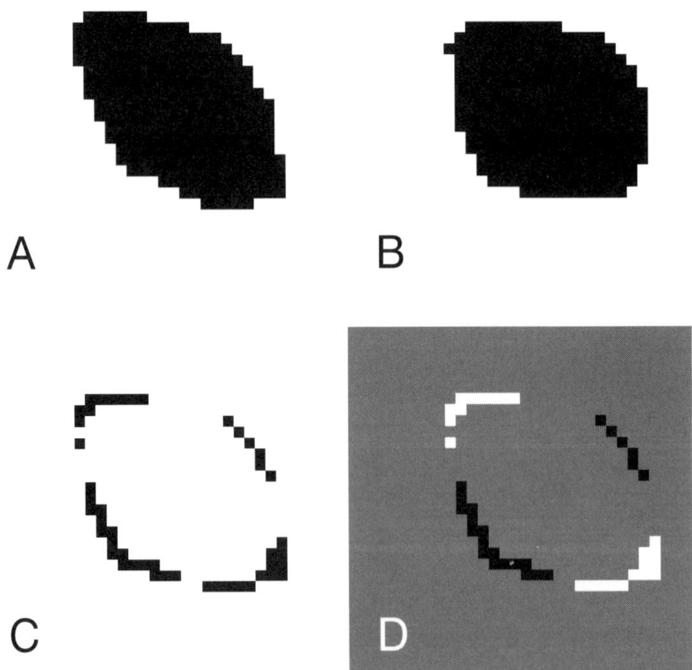

Figure 6.2 An illustration of segmentation errors: (A) 'correct' segmentation, (B) same object derived from segmentation method to be validated, (C) result of XOR operation between the two images: 46 pixels have been misclassified. (D) Subtraction of (A) from (B) shows 23 pixels misclassified as object in black, and 23 misclassified as background in white. While retaining the surface area, the new method shortens and widens this object.

standard', the error in shape B may be measured as the number of pixels in image B that have been classified differently from the corresponding pixels in image A. Image C contains all such pixels, obtained by performing a pixel-by-pixel exclusive or (XOR) operation on A and B. The relative error can be computed by dividing the number of foreground pixels in C by the number of foreground pixels in A, or the total number of pixels in the image. In this example the relative error is 18.4%. However, measuring surface area by counting the total number of foreground pixels yields the same result in A and B. The relative error in the measurement of surface area is therefore zero.

A more general approach for quantification of segmentation errors has been introduced by Levine and Nazif (1982). Let the reference segmentation divide an image of total area A, into N regions $\{R_1, R_2, ..., R_N\}$, with areas $\{A_1, A_2, ..., A_N\}$, and the test segmentation find M regions $\{T_1, T_2, ..., T_M\}$, with areas $\{\alpha_1, \alpha_2, ..., \alpha_M\}$. For each region T_j, find the region R_k, so that the area of the intersection $(T_j \cap R_k)$ is maximal. The *undermerging* error (E_{um}) is then defined as:

$$E_{\mathrm{um}} = \sum_{j=1}^{M} \frac{\{A_k - (T_j \cap R_k)\}(T_j \cap R_k)}{A_k} \tag{6.1}$$

The *overmerging* error is defined as:

$$E_{\mathrm{om}} = \sum_{j=1}^{M} \{\alpha_j - (T_j \cap R_k)\} \tag{6.2}$$

A normalized total error is then:

$$E_{\mathrm{tot}} = \frac{\sqrt{E_{\mathrm{om}}^2 + E_{\mathrm{um}}^2}}{A} \tag{6.3}$$

This measure is zero only when the two segmentations are identical. The method can be adapted to any kind of segmentation. The method does not say much about the *kind* of error; it just gives a total magnitude.

One other important criterion is the sensitivity to noise, or the detection limit of a method. When trying to find a faint object in a noisy background, we want to feel confident that the detected object is real, i.e. is not a random fluctuation in the noise. It is often possible to compute a theoretical detection limit of a segmentation method based on a model of the noise, for a given probability of getting spurious responses. Noise models are usually Gaussian of variable standard deviation, although Poisson may also be used.

6.3.1 Comparison with visual methods

The human observer is particularly good at discerning shapes of objects, even amidst a confusing background. An obvious step, which should always be performed, is a visual comparison of the result of segmentation with the original grey scale image. If a method yields shapes that do not seem to correspond to a human observer's idea of the shape of the object, the segmentation method is almost certainly wrong, and should be considered suspect.

The common-sense approach of checking your segmentation algorithm visually can be improved by quantitative versions of this approach: comparison of automated and manual methods. Often, manual segmentation is known to work well (Viles and Sieracki, 1992), but is too laborious for routine use. In that case, it is enough to compare the results of the manual method with the automatic methods to validate your results. Comparison, using either the strict criterion of the total error or comparison of measurements, can then be used to assess the accuracy of the automated method (Viles and Sieracki, 1992; Wilkinson, 1996). Although straightforward enough, one minor pitfall must be avoided: manual segmentation should be done first, so that the operator does not see the shape derived by the automatic method. Otherwise, the operator might be biased towards selecting this shape manually. Preferably, multiple manual segmentations, made by different operators, should be compared to the automated method, to remove any bias a single user might have.

6.3.2 The use of objects of known size and shape

Any segmentation method should also be validated using objects of known size and shape. The most common method in microscopy uses fluorescently labelled latex beads of known diameter (Sieracki *et al.*, 1989a; Vogt *et al.*, 1991; Viles and Sieracki, 1992). These allow accurate calibration of any segmentation technique, including manual methods. This type of validation is discussed in detail in the next chapter.

One of the problems of the use of these latex beads, which is not always appreciated, is the fact that they come in just one shape: spherical. Certain segmentation methods may produce different systematic errors for different shapes. Some methods could be particularly sensitive to this, since they are optimized for particular shapes. To circumvent this problem, computer generated images of objects could be used. By modelling the imaging process of your acquisition system (microscope, camera, digitization etc.) for a simulated object of any desired shape, it is possible to generate an artificial set of images that contain objects of known dimensions. The segmentation result can then be compared to the appropriate projection of the artificial object's shape. Sensitivity to noise, image contrast and background fluctuations can then be measured as a function of object size and shape in a tightly controlled way. One caveat remains: the synthetic images must model the imaging process in the real acquisition system closely, or the results will not bear a close enough resemblance to the real performance of the segmentation algorithm being tested.

6.3.3 Comparison with high definition measurements

One of the chief problems when studying bacteria is their small size relative to the wavelength of visible light. This means that blurring by the point-spread function of the optics forms a fundamental limit to the accuracy of segmentation. If we could obtain a high resolution image of each object by other than optical means, segmenting the image with any reasonable algorithm – even the algorithm to be validated – should result in an image of much higher accuracy than could ever be achieved optically. Comparison of the segmented shapes of individual objects in the optical image with the segmented shapes obtained from the high definition image could be used to validate an algorithm.

What remains is to obtain such a high definition image of the objects on the light microscope stage. Fixation and preparation methods for light and electron microscopy are so different that the latter is probably not a good source of high definition images (Vardi and Grover, 1990), especially if we wish to establish a one-to-one relationship between objects in the high and low definition images. By contrast, atomic force microscopy (AFM) can be carried out simultaneously with optical acquisition under physiological conditions (Radermacher *et al.*, 1992). This precludes the type of shrinkage and other deformation problems prevalent in electron microscopy (Vardi and Grover, 1990). The resolution of AFM should be quite enough for the calibration and validation purposes envisaged here. To the best of my knowledge, no such calibration has been performed to date.

6.4 PRE-PROCESSING TECHNIQUES: SIMPLIFICATION OF THE IMAGE BEFORE SEGMENTATION

6.4.1 Noise reduction by smoothing: pre- and post-processing approaches

High frequency noise in the image, caused by any of the sources discussed in Chapter 2, can cause significant problems for segmentation techniques. Straight edges can become jagged; solid objects may seem to have holes in them (Figure 6.3). Worse, certain methods to compute ideal boundaries may become biased by the presence of noise (Kittler *et al.*, 1985), causing the sizes of objects to be either overestimated or underestimated. Pre-processing the image with any of the noise reduction methods in Section 1.4.2a may reduce these problems, but always at the cost of some detail. As a rule, median filters have less tendency to blur edges,

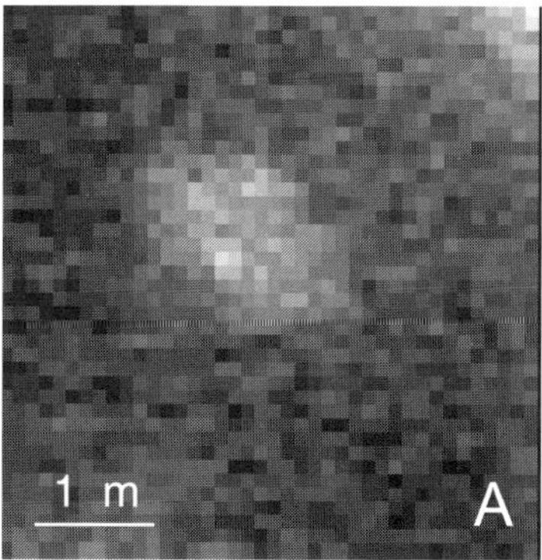

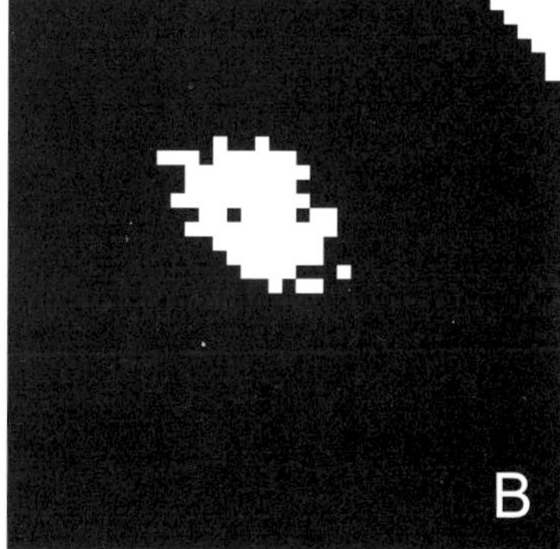

Figure 6.3 The effect of noise on segmentation: (A) original noisy fluorescence image of bacterium and (B) thresholded result shows jagged edges and holes.

causing less distortion, yet they tend to remove small point-shaped objects more readily than do averaging filters, regardless of the brightness (Figure 6.4).

Selection of kernel size is another issue. Small kernel filters do not reduce noise as well as large kernel filters, yet they preserve more detail. Whenever smoothing of some kind is used, there is this trade-off between noise reduction and preservation of detail. A number of authors use 3×3 average, Gaussian or median filters to pre-process the images, although larger sizes have been reported (Bloem *et al.*, 1995).

Recently, a number of more advanced smoothing algorithms have been developed, including parametric relaxation (Li and Lee, 1995) and non-isotropic diffusion techniques (Perona and Malik, 1990), which show a great deal of promise. These algorithms all strive to reduce noise without blurring edges or corners. Carried far enough, these methods can actually segment the image completely, by smoothing each region to a constant value. Relaxation techniques were in fact proposed as segmentation, rather than smoothing, techniques. Although their performance is impressive in images of fairly large objects, none has been tested on biomedical images, and especially microscopic images of bacteria or other small cells. Furthermore, the computational burden and memory requirements may be too great for many image processing systems found in biomedical settings.

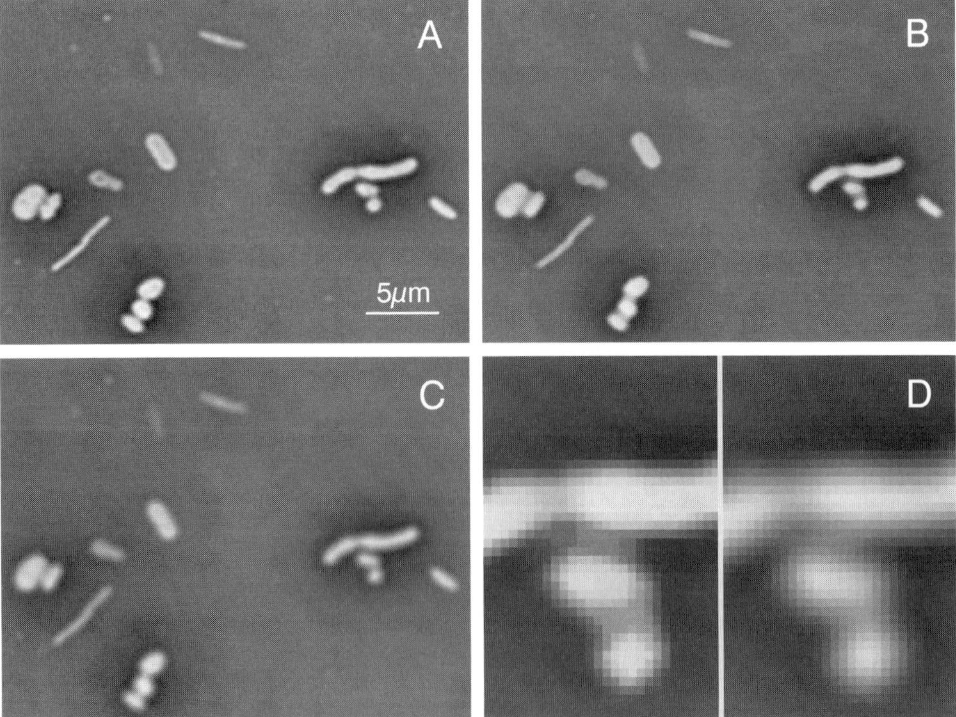

Figure 6.4 Noise removal before segmentation: (A) original image, (B) median filtered using 5×5 pixel kernel and (C) averaged using 5×5 kernel (equal weights). (D) Enlargements of parts of (B) and (C). Median filtering (left in (D)) reduces edges less than averaging.

Besides pre-processing, some types of noise may be reduced by post-processing of the segmented image. Binary opening and closing (Section 1.6.1c) may be used to smooth irregular boundaries.

6.4.2 Background elimination by highpass filtering

The background in many images does not contain high spatial frequency components, i.e. it is relatively smooth compared to the size of the objects of interest. Various methods of background removal rely on this fact. Most of these methods try to construct a smoothed edition of the original image, which no longer contains the objects, but resembles the background in the original image closely. Subtraction of this lowpass filtered edition of the image results in a highpass filtered version of the original image. Alternatively, a highpass filter of some kind could be used directly (Viles and Sieracki, 1992). Ideally, the result should contain only objects on a smooth background.

A common approach uses grey-scale closing (Section 1.6.2b). If the objects are brighter than the background, they can be removed by passing the image through a minimum filter with a kernel size that must be at least half the diameter of the largest object. If necessary, repeated use of the minimum filter removes all traces of objects. The result of such an operation is shown in Figure 6.5B. This image contains only background information, yet it is too dark on average. Therefore, the same number of calls to a maximum filter with the same kernel size is used to increase the brightness (Figure 6.5C), the result is subtracted from the original and a constant background is added (Figure 6.5D). For dark objects on a light background, simply reverse the order in which minimum and maximum filters are used.

6.4.3 Optical/digital methods for background removal

Meijer *et al.* (1991) use a slightly different approach to background removal, although it is also based on highpass filtering. In this case, an out-of-focus image is acquired of the field of view of interest. This out-of-focus image is an optically blurred, and therefore lowpass filtered, version of the original. Subtraction from the original, with the addition of a constant background, yields results comparable to the use of digital lowpass filters. The main problem lies in the difficulty of reproducing these results accurately, since the defocusing is done manually. Only when digital control over all stage motions is available can this method become fully reproducible. Nonetheless, the method is particularly useful when working with small computers with limited speed. If sufficient computing power is available there is no need to employ this method.

A variation of this method may be suitable for removal of autofluorescence background in fluorescence microscopy, using in-focus, rather than out-of-focus, images. The method, introduced by Szollosi *et al.* (1995), uses multiple colour images of the same field of view using different sets of excitation and emission filters. One set is centred on the excitation and emission bands of the fluorochrome in use, and the two others detect emission on either side of the central band. A weighted average of the images of fluorescence at either side of the wavelength band of interest is then taken to estimate an in-focus image of autofluorescence in

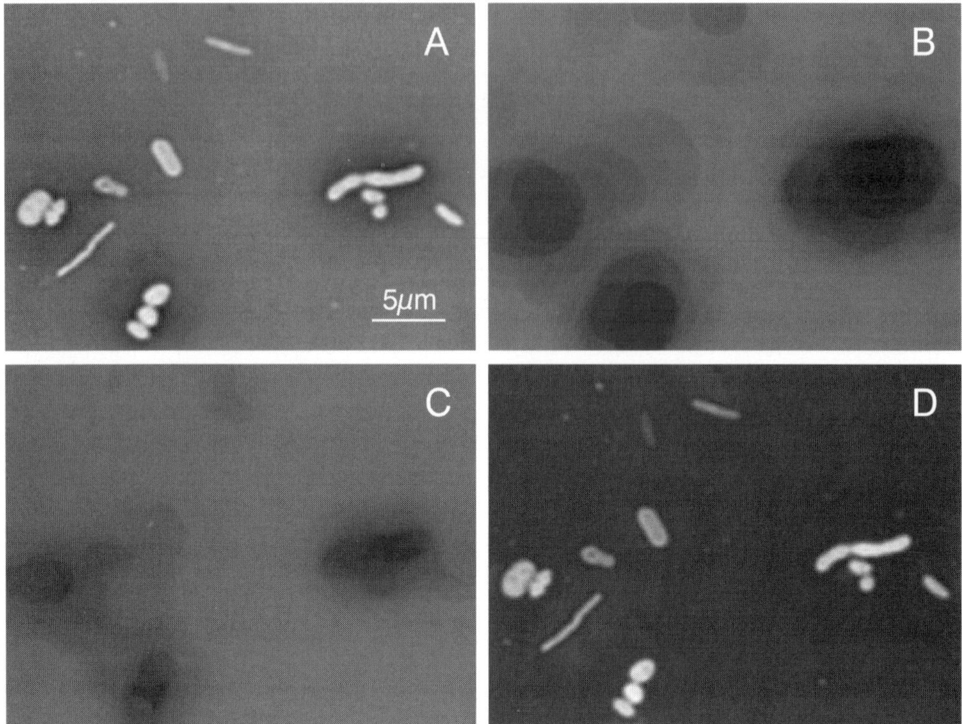

Figure 6.5 Background removal by grey scale morphology: (A) original image, (B) grey scale dilation of (A) using circular kernel of 15 pixel radius and (C) grey scale erosion of (B) using same kernel size. (D) Subtracting (C) from (A) yields smooth background.

the central band. This estimate can then be subtracted from the original image, resulting in a much reduced background. Care must be taken to avoid or correct for image shifts and scale changes due to optical imperfections when using multiple wavelength bands (Bruins *et al.*, 1994).

A final variant of this methodology uses two images in the same wavelength band, thus avoiding image shifts and scale changes completely. The method relies on the presence of photobleaching (fading) of the fluorochrome after prolonged illumination, and on the fact that most (but not all) autofluorescence shows little photobleaching compared to most fluorochromes. After focusing the image under phase-contrast illumination, the shutter of the excitation light source is opened, and an exposure is taken immediately. After this, the shutter of the excitation light source is kept open for a certain time span to photobleach the fluorochrome used. Then, a second fluorescence image is acquired at the same exposure time as the first, and the two images are subtracted, yielding a much reduced background in the final image.

One word of warning is in order when using any of these methods in fluorescence microscopy. Both images will contain photon noise (Section 2.4.1a); subtracting the two will increase this by some 40% on average. Using a weighted average of a couple of images is less damaging, but will nonetheless increase photon noise as well. As so often, bias is only reduced at the expense of increased variance. Brightfield and phase-contrast images have a better signal to noise ratio as a rule, so it is much safer to use optical background removal there.

6.4.4 Remapping texture to intensity

In the previous discussion we have assumed that the objects are distinguishable from the background by differences in brightness. Although this is the case for many types of stain, it is not true in a significant number of cases, e.g. in slides of tissue sections. Here, ROI may show differences in texture when compared to the background, instead of grey scale or colour. In such cases, a number of strategies to remap texture information to grey scale exist. Once this has been done, segmentation techniques for grey level difference may be used to segment the image.

The simplest method computes the variance of grey levels in the surroundings of each pixel. This allows separation of smooth and 'grainy' parts of an image. By computing variances for different kernel sizes, it is possible to obtain information on the mean 'grain size' in various parts of the image. However, no information about the shape and orientation of the grains is obtained. Furthermore, variance filters also respond to boundaries between regions of different mean brightness. More advanced operators include those of Haralick *et al.* (1973), the local Hurst operator (Russ, 1990), and Gabor filters (Fogel and Sagi, 1989; Jain and Farrokhnia, 1991).

It goes without saying that any form of remapping texture to grey scale information implies a loss of resolution, since texture can only be defined in terms of multiple pixels. Therefore, these methods are largely irrelevant to the problem of segmenting bacteria, which show little internal detail. When working with, for example, phase contrast, interference contrast or Hoffman modulation contrast images of larger, unstained, eukaryotic cells, texture may well be the best method to distinguish the cells from the background.

6.5 BOUNDARY AND EDGE BASED METHODS

6.5.1 Types of edge detectors: first and second derivatives

Before we can speak of edge detectors we need a clear definition of an edge. Van Vliet *et al.* (1989) give the following definition of an edge:

> A simply connected contour, one pixel thick, at the centre of the slope between two adjacent regions with a considerable difference in grey level.

Quite apart from the qualitative nature of the phrase 'considerable difference in grey level', this definition contains three important criteria for 'good' edge detectors:

(a) The resulting contour must be connected (no holes or interruptions).
(b) The edge must be as thin as possible.
(c) The edge must be well localized (centred on the 'true' position of the edge).

The first two criteria are reasonably self-explanatory, but the definition of the centre of the slope needs sharper definition.

If objects have step edges (Figure 6.6), and these are blurred with the point-spread function of the optics, we can reconstruct the exact location of the edges by determining where the first derivative has a maximum and the second a zero crossing, at least in theory. In practice, close proximity of several edges, noise, and spatial and grey level discretization disturb this idealized picture. Nonetheless, the two-dimensional equivalents of first and second derivatives, and especially colocalization of zero crossings in the second and high values in the first derivative, are the chief means of detecting edges in images (Canny, 1986; Van Vliet *et al.*, 1989).

Apart from linear operators (Chapter 1), nonlinear versions have been proposed (Lee *et al.*, 1986; Van Vliet *et al.*, 1989). The Lee edge strength measure first blurs the images with some smoothing filter of width W. Consequently, the minimum and maximum grey levels in the neighbourhood N (also of width W) of each pixel are computed. The edge strength is then given as:

$$ES(x, y) = \min\left\{ \max_{(x',y') \in N} p(x', y') - p(x, y),\ p(x, y) - \min_{(x',y') \in N} p(x', y') \right\} \tag{6.4}$$

A nonlinear Laplacian (*NLLAP*), discussed by Van Vliet *et al.* (1989), is given as:

$$NLLAP(x, y) = 2p(x, y) - \max_{(x',y') \in N} p(x', y') - \min_{(x',y') \in N} p(x', y') \tag{6.5}$$

Strictly speaking, this is -1 times the *NLLAP* filter, to keep its sign equal to the usual definitions of the 3×3 Laplacian (Section 1.4.2b). A combination of the two filters was used by Van Vliet *et al.* in an effective way to find edges that meet all criteria. The process has a number of stages: (i) the image is smoothed using a Gaussian of width σ, (ii) the smoothed image is passed through the *NLLAP* filter of size W, (iii) the smoothed image is also passed through the *ES* filter, also of width W, (iv) in the *NLLAP* filtered image zero crossings are detected: pixels at zero crossings are assigned a level of 1; others 0, (v) the image containing the zero crossings is

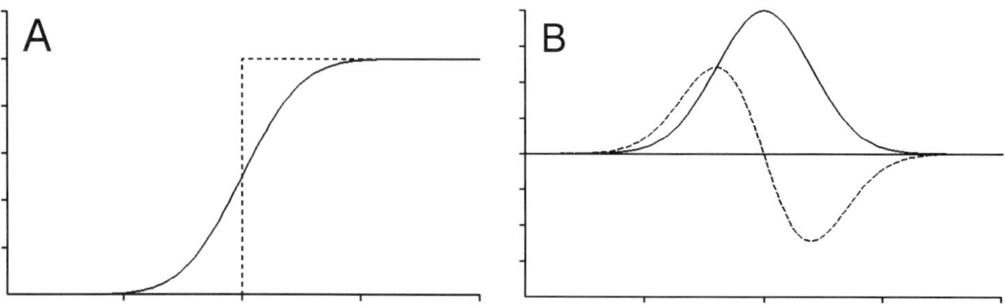

Figure 6.6 The principle of edge detection: (A) step edge in image (dashed) convolved with Gaussian point spread function (solid); (B) gradient of blurred edge (solid) shows maximum; second derivative shows zero crossing.

multiplied with the edge strength image and (vi) this final image is thresholded to reject edges with low gradients.

The zero crossing detector they use is very simple: the *NLLAP* image is divided into positive, negative and zero pixels by thresholding (Section 6.7). Two Euclidean distance maps (Section 1.6.1e) are then computed, one containing the distance of each zero pixel from the nearest positive pixel, and one containing the distance to the nearest negative pixel. By comparing these distances all zero pixels are assigned a negative or positive value: if the distance to a negative pixel is smaller, the zero pixel becomes negative; otherwise it becomes positive. In the resulting image, all positive pixels that are 8-connected (Section 1.2.4) to a negative pixel are then considered to be edge pixels. They can be found by subtraction of a 4-connected eroded version of the image (Section 1.6.1a). Such an edge has the advantage of *always* being a single pixel wide, and *never* having holes. The optimal smoothing width σ and neighbourhood width W depend on the signal-to-noise ratio (SNR). Below SNR = 3, Van Vliet *et al.* propose $\sigma = 3.0$ and $W = 7$, for $3 < \text{SNR} < 10$, $\sigma = 2.0$ and $W = 5$.

Figure 6.7 shows the steps for a fluorescence image of bacteria. The last stage, thresholding the resulting edges, is the most problematic one, since the elimination of pixels below an arbitrary threshold may cause the result to violate the closure of boundaries requirement (Figure 6.7E). Canny (1986) discusses two possible solutions: (i) accepting or rejecting complete contours on the basis of the average

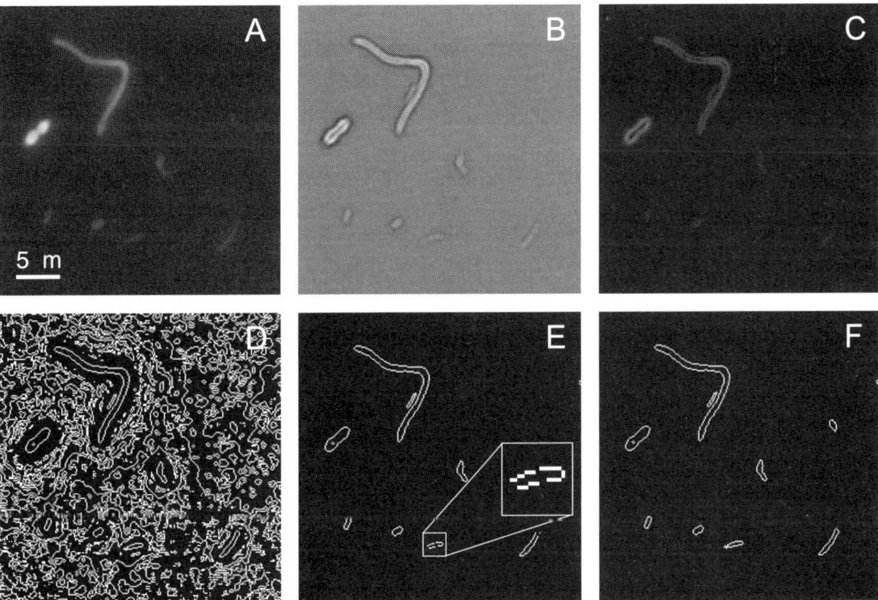

Figure 6.7 Edge detection according to Van Vliet *et al.* (1989): (A) original image, (B) after smoothing and NLLAP filter, (C) after smoothing and ES filter, (D) edges from zero crossings in (B). (E) After thresholding the product of (C) and (D) spurious edges are gone, but objects are lost, and connectedness is no longer guaranteed. (F) Using the mean edge strength instead to accept or reject entire contours retains connectedness.

edge strength and (ii) using a form of hysteresis that does not necessarily result in closed contours. In the case of images of bacteria, using the average edge strength to accept or reject seems to work well (Figure 6.7F). Setting thresholds for acceptance of edges is usually done heuristically, since little theory has been developed on this topic. Venkatesh and Rosin (1995) were probably the first to discuss thresholding of edge detector output thoroughly.

Van Vliet's algorithm has not been applied to images of bacteria before, but Viles and Sieracki (1992) developed a method inspired by it, which is discussed in Section 6.7.5. A more complete discussion of edge detection methods is given by Pal and Pal (1993).

One method, which I call 'optical edge detection', should be mentioned here. Pons *et al.* (1993) made use of the fact that it is precisely the edges of cells which show up in phase-contrast images of yeast cells. Simple thresholding of such images using any of the methods in Section 6.7 results in an image of the cell edges. However, such edges are neither guaranteed to have single pixel thickness nor to be continuous. Pons *et al.* use the following approach: gaps in the boundaries are deleted by binary closing (Section 1.6.1c); after this, hole filling (Section 6.5.4) is used to fill the boundaries. Although the method might lack the mathematical rigour of, for example, the approach of Canny (1986) it is highly effective within its context.

6.5.2 Manual outline tracing

Manual tracing out of the outline of an ROI, with, for example, a mouse or trackball, is the simplest contour-based method. The user simply points and clicks on successive points along the outline of each object, and the computer stores them in a list. A disadvantage of this approach noted by Russ (1992) is that human operators tend to draw the outline just outside the actual border of the cell. The small size of bacteria means that this bias results in a larger relative error than in the case of large cells.

Within microbiology, manual outline tracing has been used for segmenting biofilm images (Chapter 15). In this case there are just a few large objects (often one) in the image, so the method becomes feasible.

6.5.3 Object model based techniques

Object model based techniques all assume that there is some prior knowledge of the shapes of the objects to be searched for. The parameters describing the model can be adjusted in some way to move the hypothetical edge through the image until an optimum fit is obtained. This type of approach usually generates smooth edges, since some penalty is assigned to jagged edges in the optimization process. One example is the 'snakes' approach (Kass *et al.*, 1987). This approach might be thought of as fitting a flexible, thin spring to the edges in the image, in such a way as to minimize both the distance between the edges and the spring, and the tension on the spring. Since sharp curves need high tension, the border becomes smooth automatically. Snakes pose no other limits on the shape of the object. Another approach uses a parameterization of the object boundary (Cootes *et al.*, 1995). Principal component analysis of the parameters in a learning set of shapes of interest yields an optimal set of parameters describing the shapes in this set. The edges in the image are then fit to these parameters iteratively. Once an optimum fit has been found, the algorithm can decide whether the object found is an object of interest based on the parameter values. This algorithm explicitly searches for a certain type of object, ignoring others. Because of the large computational burden, none of these methods has been used for the segmentation of images of bacteria. Some of these methods are discussed by Cootes *et al.* (1995).

6.5.4 Boundary fill techniques

Since many applications need information on all of the pixels within an object, it may not be sufficient to label only the edge pixels. Three different approaches are used to derive this information from the edge pixels. Hole filling is a simple iterative procedure that is guaranteed to find all possible light regions completely surrounded by darker areas (or the reverse). It is simple to implement, but it also fills real holes in objects (Figure 6.8). Alternatively, region growing (Section 6.6.1) may be used, which uses the presence of edge pixels as stop criteria, rather than region criteria. This has two problems: (i) one must be able to determine which side of an edge is inside the object and (ii) the entire object might not be filled by a single application of region growing (see Figure 6.8). The first problem may be solved by determining grey levels, or colours, at both sides of the edge. The second is usually solved by repeating the process as many times as needed to fill the entire object.

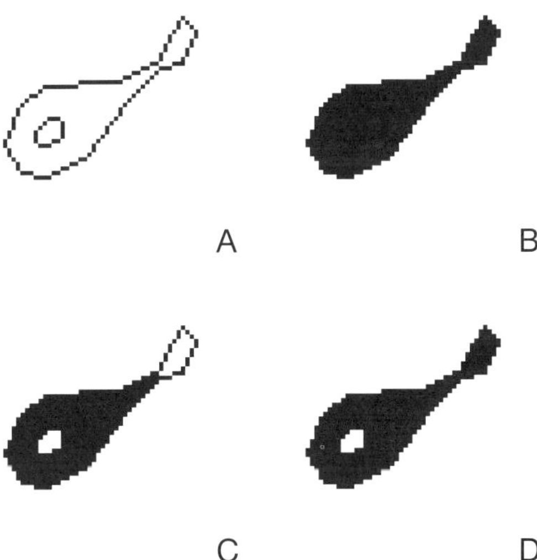

Figure 6.8 Filling edges: (A) edges of object with hole and narrow 'tail', (B) hole filling fills all enclosed regions, and also the hole in the object, (C) a single region fill misses part of the object, since part of the tail is not 4-connected to the rest of the inside of the object, (D) multiple region fills or a single polygon fill (after transformation of edge pixels to polygon format) give a correct fill.

A third approach to boundary filling, which does not suffer from either of the above drawbacks, is polygon filling. The boundary may always be represented as a polygon consisting of a series of line segments. Given this representation, one of a number of scan-line polygon fill algorithms may be used (Burger and Gillies, 1989) to fill the entire region. However, if a polygon representation is used, there is no particular need to fill the region in the image, unless for display purposes.

6.6 REGION BASED IMAGE SPACE METHODS

6.6.1 Region growing

Region growing attempts to find connected regions that are smooth in some sense, e.g. grey level, colour or texture do not vary much over the region. Region growing starts at a pixel specified either by a user or by some automated algorithm. All neighbouring pixels are compared to this 'seed' pixel, and some criterion of similarity is used to decide which of these pixels belong to the same category as the seed. These pixels are added to the region, and all pixels neighbouring this updated region are then inspected in their turn. This process is repeated until none of the pixels adjacent to the region is sufficiently similar. The process is a series of conditional dilations, the condition being the similarity measure. The entire process must be repeated for each connected region in the image. Many variants of region growing exist, which differ chiefly in the similarity measure used. Two main categories exist: (i) those using region criteria and (ii) those using boundary criteria. The former compare the grey level or colour of candidate pixels to the regional mean or median, or the seed pixel's grey level or colour. The latter use the presence of a boundary at the candidate pixel. The presence of a boundary is detected using one of many possible edge detectors (Sections 1.4.2b, 1.7.3a and 6.5.1).

The main problem of region growing is that it is difficult to supply the seed pixels automatically. Besides, different seed pixels may produce different region shapes, even when using the same similarity measure. It is also difficult to ensure that all ROIs have been detected.

6.6.2 Split-and-merge

Split-and-merge algorithms use the reverse approach to region growing. They consider arbitrary regions of the image, and check whether they are uniform enough to be considered as belonging to a single class. If not, the region is split in some arbitrary way. Once the image has been split into smooth regions, neighbouring regions are merged if they are similar enough. If the latter stage is not performed, one would be left with regions that do not conform to the shapes of the objects in the image, since the splitting is along arbitrary lines.

Many split-and-merge algorithms use a so-called quadtree to split the image. At the highest level is the entire image. If it is not uniform enough, it is split into four quadrants. Each quadrant is then inspected in its turn, and split into four again when necessary. This process continues until the finest level of subdivision is reached – single pixels, which are always uniform. Those quadrants that have not been subdivided any further (the 'leaves' of the tree) are then compared to each of their neighbouring leaves. If they are sufficiently similar, they are merged. In the final result, each connected region is represented by a list of leaves of similar properties. Split-and-merge algorithms have not been used for segmentation of images of bacteria, so little can be said about their performance in this field.

6.7 THRESHOLDING AND OTHER DATA SPACE METHODS

6.7.1 The thresholding principle

As has been mentioned before, threshold selection is one of the simplest, most common forms of image segmentation. In its basic form, threshold selection is done on a pixel by pixel basis, by comparing the grey level of each pixel with some predetermined value T. If the grey level is below the threshold, the pixel is assigned to category 1; if it is higher to category 2. What happens at equality is a matter of taste; some algorithms assign it to category 1, others to 2. The actual interpretation of each of the categories (foreground or background) depends on the type of image. In the case of microscopic images of single-celled organisms, this depends largely on the type of stain used. In most applications within this field, just two categories are present (cell and background), but more categories, and therefore multiple thresholds, may be defined. Thresholds are most often defined globally, i.e. the same value of each of the thresholds applies to each pixel. This simplifies the computations greatly, through the use of look-up tables (LUT; see Sections 1.4.1 and 5.2.4b). Once the thresholds have been selected, it can be decided into which category each of the N grey levels belongs. Each of the N entries in a grey scale LUT is then filled with numbers corresponding to each category:

```
for (i=0;i<N;i++)
   LUT[i]=Category(i);
```

in which `Category` is a function that returns to which category a pixel with grey level i belongs. After this, special hardware, if available, may be used to remap the grey levels of all pixels in the source image to the category levels in the destination image. If no hardware is present, applying the simple assignment:

```
p[x][y]=LUT[p[x][y]];
```

to each of the pixels in the image suffices to remap the pixels. This approach is much faster than doing (multiple) comparisons on each pixel.

The speed gain of global thresholding does come at a price. If the objects are not all equally bright, or if the background shows residual variations at large spatial scales, a global threshold will always fail for certain parts of the image (Sieracki *et al.*, 1989a; Viles and Sieracki, 1992; Wilkinson, 1996). Unless a considerable amount of computing time is to be spent correcting for these defects (e.g. Bloem *et al.*, 1995), local threshold techniques, which compute different thresholds for each region of the image, are called for in such situations. In some applications, e.g. motion analysis (Chapter 13), there are such strict constraints on the time spent segmenting the image that global thresholding is the best that can be done in the time available.

Manual threshold setting is almost exclusively global, unless one goes to the lengths of dividing each image into sub-images, each containing a single cell, and setting a separate threshold for each. Obtaining samples of some hundreds or thousands of cells, using this kind of technique, is not for the faint-hearted.

Apart from manual thresholding, in which the user selects an 'optimum' value for the threshold, there is a host of literature on the topic of automatic threshold selection methods (Lee *et al.*, 1990; Pal and Pal, 1993; Sahoo *et al.*, 1988; Parker, 1997). Some work purely from the grey level histogram, or statistics derived from it, whereas others include edge information as well. Several of the most important methods are discussed in the following sections. The discussion focuses on two-category segmentation. It will also be noted which techniques can be extended readily to multiple thresholds.

Whatever the method, thresholding noisy images will produce errors, since there will be overlap between the distributions of background and foreground pixels. In practice, a signal-to-noise ratio of at least 3 (preferably >5) is needed for reliable thresholding.

6.7.2 Using a fixed arbitrary threshold after edge enhancement

A number of workers in the field use a fixed arbitrary threshold, which is applied to a pre-processed image, in which the background has been reduced to (practically) zero, and the edges of objects have been enhanced. These methods have been applied to both bacteria (Bjørnsen, 1986; Bloem *et al.*, 1995) and fungi (Cox and Thomas, 1992; see also Chapter 14). By 'arbitrary threshold' I mean that the threshold level is determined not from the image itself (as in the next sections) but from a 'learning set' of images acquired previously. In the case of Bloem *et al.*, it was determined visually from this learning set that a threshold of 25 selected all objects safely. This threshold was applied to all other images in the experiment. The following paragraphs discuss the algorithm of Bloem *et al.*, since it is described in far more detail than that of other authors. However, both methods are very similar in their basic approach, and share the same strengths and weaknesses.

Figure 6.9A–F show the most important steps. First the image is smoothed using a 5×5 pixel averaging filter (Figure 6.9A) and subsequent grey scale opening and closing. After this, the background is flattened using the method in Section 6.4.2 (Figure 6.9B), and the image is sharpened with a nonlinear edge enhancement filter, which computes the minimum and maximum pixel values within a neighbourhood of each pixel (Figure 6.9C). If the pixel's grey level is closer to the maximum than to the minimum, it is assigned the maximum; otherwise it is assigned the minimum. This sharpens the edges to such an extent that the shapes of objects become largely independent of the chosen threshold (Figure 6.9D–F). After these pre-processing stages, the global threshold is used to segment the image.

It is not clear that this level of pre-processing is necessary: similar results might be obtained using local thresholding, on a less thoroughly pre-processed image, using less computer time. It is also an unresolved issue whether nonlinear edge enhancement alters the shapes of objects in a systematic way. Nonetheless, the

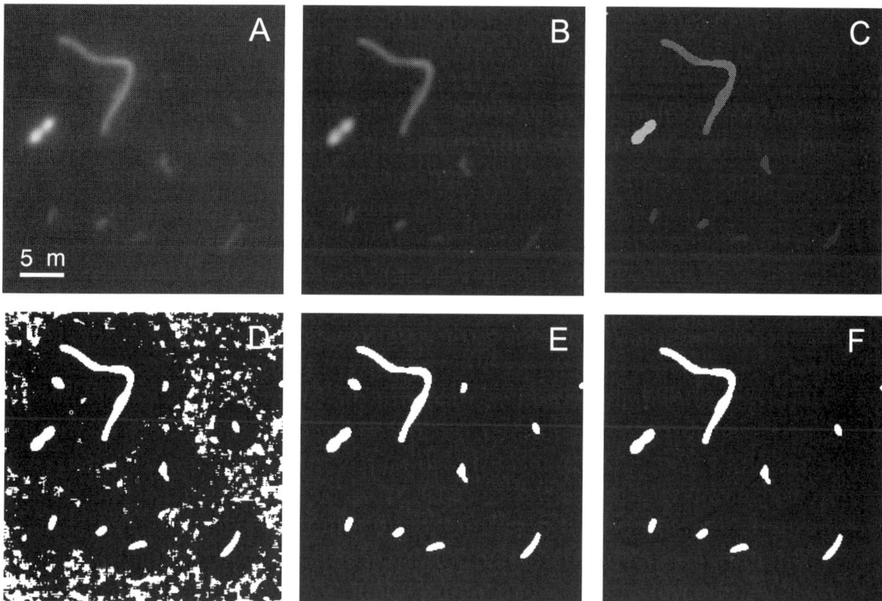

Figure 6.9 Background removal and nonlinear edge enhancement as in Bloem *et al.* (1995): (A) 5×5 average filtered image, (B) background removal with grey scale morphology, (C) nonlinear edge enhancement, (D–F) global thresholding at grey levels of 1, 4 and 6, respectively. Only in a very small range of settings are all objects detected. The shape of each individual object does stay the same over a wide range of settings.

method is a very good demonstration of what can be done to turn difficult images into easy ones.

The aim of Bloem and co-workers was to eliminate manual editing of the segmented images completely. Within the context of the experiment described, they have succeeded in this aim, but this has as much to do with the slide manufacture as the image processing involved. Unwanted detail is eliminated on morphological grounds: it is either too small (noise filtering) or too large (top hat filter) to fall into the category of interest. Should the slides contain background detail of similar sizes to the objects of interest, this kind of morphological method fails.

One word of warning concerning the use of a predetermined threshold is in order. To use a fixed threshold with impunity there must be *a priori* knowledge that all objects in all slides are brighter than the threshold. Changes in staining procedures, illumination intensity and detector sensitivity cause problems for such systems, since they do not adapt the threshold to the image statistics. The narrow range of grey levels that detect all objects in Figure 6.9A is evident in Figure 6.9D–F. Evidently the staining procedure used in this slide is not suitable for this type of algorithm.

6.7.3 Threshold selection from histograms

There are various ways in which the grey level histogram may be used to determine an 'optimum' threshold. One of the simplest approaches is demonstrated in

Figure 6.10. If the histogram shows two well-separated modes, the lowest point in the 'valley' between the two modes is an excellent threshold. However, if the modes overlap, or if they have very different sizes, it is often impossible to obtain a good threshold using this technique. Various other techniques have been proposed to obtain the threshold from histograms. A number of them have been reviewed by Lee *et al.* (1990).

One group of techniques focuses on modifying the histogram in such a way as to increase the depth of the value or the sharpness of the peaks. Weszka and Rosenfeld (1979) discuss the use of gradient information for histogram-based threshold selection. The basic idea is that the 'valley floor' in the histogram is composed mainly of points that lie on object/background boundaries, and therefore have a high gradient value. Either by using only points with a gradient below a certain threshold, or by assigning low weights to high gradient pixels, it is possible to deepen the valley of the histogram. Alternatively, it is possible to use only the pixels with high gradient value, and use the mode, median or mean of the resulting histogram. Given that high gradient points should most often be edge points, this statistic should yield a good threshold. A related method is discussed in Section 6.7.4.

Histogram modification by the use of quadtrees has also been proposed (Wu *et al.*, 1982). First, the image is split up into a hierarchy of quadrants in the same way as in the split-and-merge algorithm. In this algorithm, quadrants are split if the standard deviation of the grey levels is larger than one third of the standard deviation of the entire image. Within each quadrant each pixel is given the mean grey level of the quadrant. The histogram of this modified image shows sharper peaks than the original. A refinement to this approach uses only quadrants larger than one pixel. The single pixel quadrant should occur most often near feature boundaries, and therefore contribute little to the peaks. In a similar vein to the gradient approach, the mean grey level of single-pixel quadrants was thought to yield a good threshold, but this approach does not perform well (Wu *et al.*, 1982; Lee *et al.*, 1990).

One good general-purpose technique has been proposed by Otsu (1979). It is based on discriminant analysis, and works by maximizing the between-categories

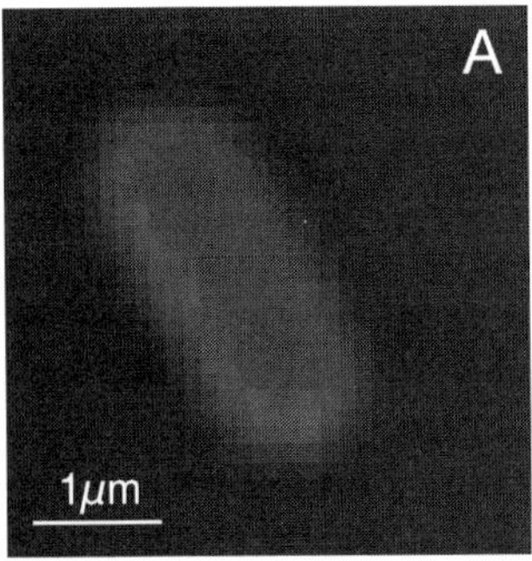

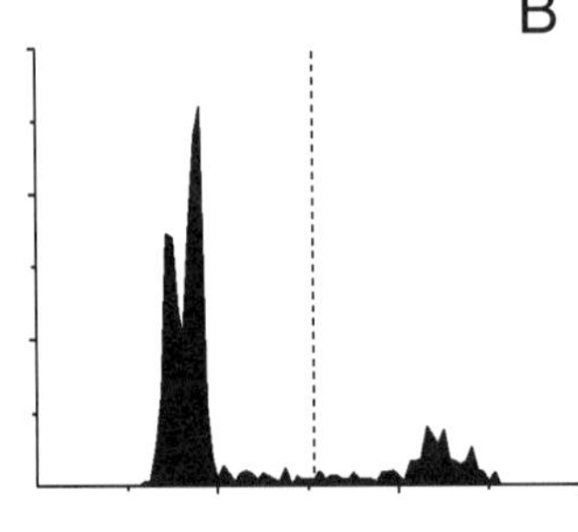

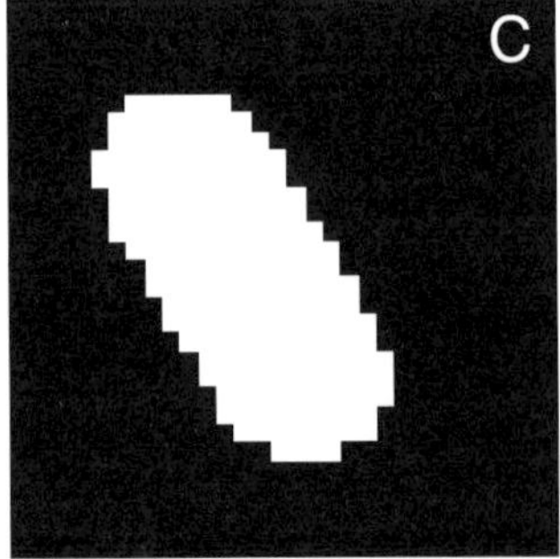

Figure 6.10 Using the bimodality of the image histogram for thresholding: (A) original image; (B) grey level histogram showing two modes and proposed threshold grey level (dashed line); (C) thresholded result.

variance (σ_B^2) over all possible thresholds (T). This statistic can be computed quite simply:

$$\sigma_{\mathrm{B}}^2 = \frac{[\mu_{\mathrm{tot}}\omega(T) - \mu(T)]^2}{\omega(T)[1 - \omega(T)]} \tag{6.6}$$

with

$$\omega(T) = \sum_{i=0}^{T} p_i \quad \text{and} \quad \mu(T) = \frac{1}{\omega(T)} \sum_{i=0}^{T} ip_i$$

In these equations p_i is the probability of a pixel having grey level i, and μ_{tot} is the mean grey level of the entire image. Therefore, $\omega(T)$ is the probability of a pixel having a grey level of T or lower, and $\mu(T)$ is the mean grey level of those pixels. By computing $\sigma_B^2(T)$ as a function of T and selecting the maximum value, an optimum threshold is obtained.

Another category is based on information theory, and is usually referred to as entropic, or maximum entropy threshold, algorithms. The first was introduced by Pun (1981), and a number of modifications have been published (Kapur *et al.*, 1985, Beghdadi *et al.*, 1995). The methods seek to maximize the information content of the segmented image, which is measured as the *a posteriori* entropy of the images (a discussion of entropy as a measure of information is given in Section 9.4.3). As a rule, entropic thresholding algorithms tend to fail on images that show large areas of noisy background with a few small objects. Figure 6.11 shows a fluorescence image of bacteria thresholded globally with the methods of Otsu (1979) and Kapur *et al.* (1985).

Most histogram methods can be extended to multiple thresholds quite readily, e.g. if multiple modes are obvious in the histogram. As a rule, histogram-based techniques cannot be readily extended to local thresholding. One problem is determining the number of modes. Further problems are caused by the fact that histograms become very sparsely populated, and therefore noisy, when small regions are used. Furthermore, high grey scale resolution (14, 16 or 18 bit per pixel) images, obtained with slow-scan charge-coupled device (CCD) cameras, have histograms with very many entries, again leading to sparsely populated histograms. For 8 bit per pixel images, Chow and Kaneko (1972) partially circumvent these problems by using overlapping regions (which allows the use of larger size), and by smoothing the histogram of the sub-images.

6.7.4 Robust Automatic Threshold Selection (RATS)

The RATS algorithm was first proposed by Kittler *et al.* (1985) as a robust and simple method for computing thresholds fully automatically by using edge information in the image as well as grey level information. Several modifications of the original method have been proposed (Kahmoun and Astruc, 1994; Wilkinson, submitted), and it has been used to segment images of bacteria

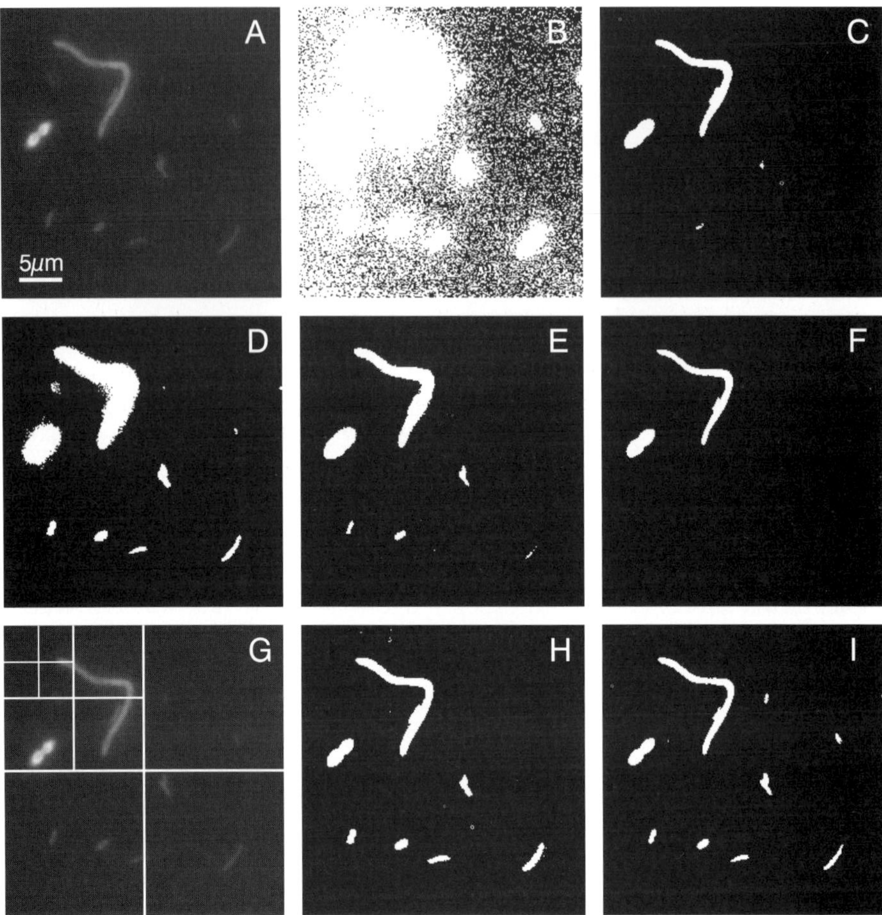

Figure 6.11 The performance of a number of thresholding methods on the same fluorescence image of bacteria: (A) original image, (B–F) global thresholding: (B) entropic thresholding (Kapur *et al.*, 1985), (C) Otsu method (1979), (D) RATS without noise bias correction, (E) RATS, rejecting gradients below 5σ, (F) RATS, square Sobel method, rejecting gradients below 3σ (Wilkinson, submitted). (G) Quadtree recursively subdivides an image into a hierarchy of rectangles, (H) and (I) local thresholding with 64 sub-images in four-level quadtree: (H) Otsu's method using local variance to select valid thresholds and (I) RATS, using gradients as in (F), sum of gradients selects valid local threshold.

(Meijer *et al.*, 1990; Wilkinson 1996) and cellular nuclei (Meijer *et al.*, 1997). The method computes the optimum threshold from simple image statistics. In the absence of noise, for an image A, the optimum threshold is:

$$T = \frac{\sum_A e(x, y)p(x, y)}{\sum_A e(x, y)} \tag{6.7}$$

with

$$e(x, y) = \max\{|p(x+1, y) - p(x-1, y)|, |p(x, y+1) - p(x, y-1)|\}$$

The equality in equation (6.7) was originally proven for $e(x,y)$ alone, but the proof can easily be extended to many other edge detectors (Wilkinson, submitted). The threshold is thus computed as an average of the grey levels in the image, weighted by the local gradient. This is similar to the idea expounded by Weszka and Rosenfeld (1979), who used the mean grey level of high gradient pixels as the threshold. Kittler *et al.* (1985) have demonstrated that this statistic yields optimum thresholds in noisy images when the background and objects each occupy 50% of the image area. If, as is often the case with images of bacteria, the background occupies far more than 50%, and there is noise in the background, the surface area of the objects will be overestimated, since the low-level gradients in the background will bias the thresholds computed from equation (6.7). A number of modifications to RATS can improve this. First of all, it is possible to use edge detectors with a low response to noise and a high response to edges. One method (Kahmoun and Astruc, 1994) uses higher powers of the gradient:

$$T = \frac{\sum_A g^m(x, y)p(x, y)}{\sum_A g^m(x, y)} \tag{6.8}$$

where $g(x, y)$ is the gradient and m an integer number. These were tested for m ranging from 1 to 10. Although the bias is reduced, the variance of T increases progressively. One other option is the use of Sobel kernels (Section 1.4.2b) to compute the derivatives in x and y (Δ_x and Δ_y) Ideally, the gradient should be:

$$g(x, y) = \sqrt{\Delta_x^2 + \Delta_y^2} \tag{6.9}$$

rather than simply taking the maximum value, as in Kittler's $e(x,y)$. However, taking a square root is computationally inefficient, so equation (6.9) is rarely used. Yet if we use Kahmoun and Astruc's idea of higher order filters, and use $m = 2$, we can avoid taking a square root:

$$g^2(x, y) = \Delta_x^2 + \Delta_y^2 \tag{6.10}$$

This turns out to be a very efficient edge detector in this context (Wilkinson, submitted). Apart from showing good performance in noisy images, it also deals with edge curvature better than many others, and is computationally efficient.

Although the bias is reduced to some extent by more efficient gradient operators, they are not sufficient. Kittler *et al.* (1984) exclude pixels with gradients below some threshold $\lambda\sigma$ (or $\lambda^m\sigma^m$ for higher order filters as in equation (6.8)), with σ the standard deviation of the noise, and λ a parameter with which to tune the sensitivity of the algorithm. Using Sobel filters, the bias becomes negligible for

$\lambda \approx 3$; for the original edge detector $\lambda \approx 5$ yields good results (Wilkinson, submitted). The disadvantage is that objects that do not differ by $\lambda\sigma$, or more, from the background cannot be detected. In practice, a signal-to-noise ratio of at least 3 is needed for reliable thresholding anyway, so the Sobel filter method does not relinquish any real sensitivity.

Quite unlike many histogram-based methods, RATS is very suitable for local thresholding (Figures 6.11 and 6.12). Following Kittler *et al.* (1984, 1985), the image is subdivided into a quadtree hierarchy, with the finest subdivisions having 16×16 or 32×32 pixels. The numerator and denominator of equation (6.8) are computed for all quadrants. However, the threshold T is only considered valid if the denominator (the sum of weights) sufficiently exceeds the value expected for a flat background containing only noise. If the gradient operator itself has been thresholded, each area with a denominator at least three times the gradient threshold can be assumed to yield a valid T. For every other area its parent is inspected: if the parent meets the weight criterion, its optimum threshold is assigned to its child; otherwise its parent is inspected, right up to the global threshold if necessary. The thresholds are then assigned to the centre point of each of the finest image subdivisions, and some form of interpolation is used to assign thresholds to individual pixels. Kittler *et al.* (1985) proposed a polynomial fit to the centroid values as interpolation, but simple bilinear interpolation can perform well (Wilkinson, 1996).

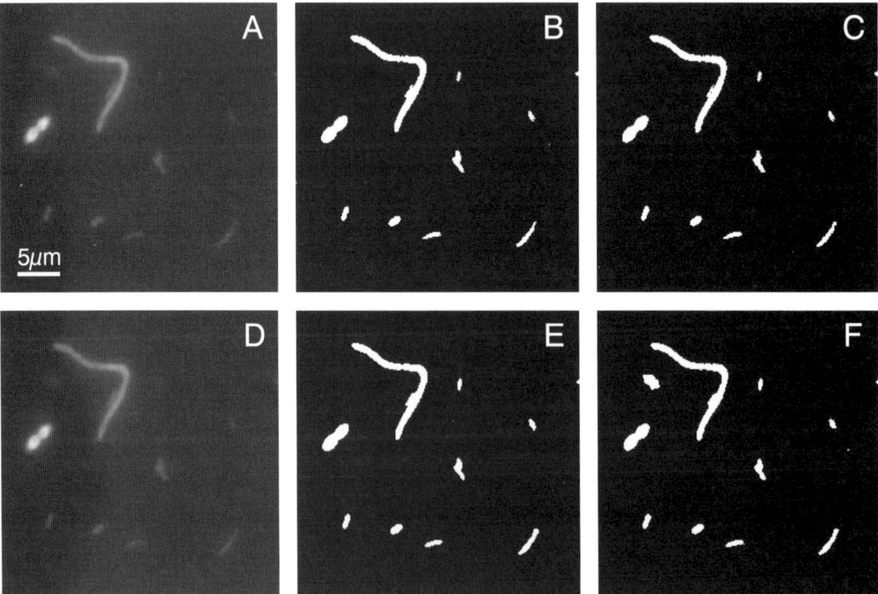

Figure 6.12 Increasing the sensitivity of local thresholding using RATS by noise filtering and increasing the number of sub-images: (A) original image, (B) RATS result using 64 sub-images, (C) RATS result using 256 sub-images, (D) 3×3 median filtered image, (E) and (F) results for (D) using same method as (B) and (C).

The chief disadvantage is that RATS is not easily applicable to multiple thresholds, as it is specifically designed to work with just two categories. Furthermore, if two objects of greatly differing brightness are close together, the method may fail to detect the fainter one, unless very fine subdivisions are used. A related method, proposed by Yanowitz and Bruckstein (1989), could perform slightly better on this count. In their algorithm, they first compute the edge locations using the Canny edge detector. They then compute threshold levels for the entire image by interpolation of the grey levels of these edge pixels by assuming that the threshold levels must fit a potential surface. The method has been applied to bacteria by Dubuisson *et al.* (1994). One of the drawbacks of the method is a much higher computational burden (3500 iterations, according to Dubuisson *et al.*) than RATS (two scans of the image).

6.7.5 Marr–Hildreth and edge strength method

Viles and Sieracki (1992) observed that fluorescent images of bacteria showed little internal detail, but showed great inter-individual variations in intensity. Within each bacterium there is a smooth transition of intensity from centre to edge, with a negative second derivative throughout the cell. In theory, filtering the image with the Laplacian second derivative operator and thresholding the resulting image at zero grey level should then segment the image properly. In practice, a smoothed version of the Laplacian, the Marr–Hildreth filter, must be used, because of noise sensitivity. The Marr–Hildreth filter can be implemented by first applying a Laplacian, followed by a Gaussian smoothing filter of standard deviation σ. By varying σ, the filter can be tuned to detail of certain sizes. A large σ results in less sensitivity to noise, at the cost of sensitivity to small objects. However, even the Marr–Hildreth filtered image shows spurious responses in the background when thresholded at zero, so a second pass is made through this image, to determine which connected regions contain at least one pixel with sufficient edge strength. The edge strength is defined as the maximum grey level difference between a pixel and each of its 4-connected neighbours. Empirically, for the cooled slow-scan CCD camera used by Viles and Sieracki, it was found that an edge strength of 10 eliminated the background fluctuations well. Although intended for slow-scan CCDs, its performance on an image captured with a long exposure, video rate system can be quite impressive (Figure 6.13). As the authors note, internal detail, as is seen in larger cells, causes the method to break down. On good quality images of small objects, with large inter-individual differences in brightness, it performs well.

A number of measures can be used to improve the performance of the method for 8 bit per pixel video camera images as in Figure 6.14. First, a gradient preserving pre-filter, such as a median filter, may be used. Second, σ may be increased to reduce noise responses. Finally, the edge strength may be computed more robustly using, for example, a Sobel filter or the Lee edge strength (Section 6.5.1). For small objects and small σ, it is best to compute the edge strength in the Marr–Hildreth filtered image, as in the original implementation. For larger σ, and larger objects, edge detection in the original rather than filtered image is to be preferred.

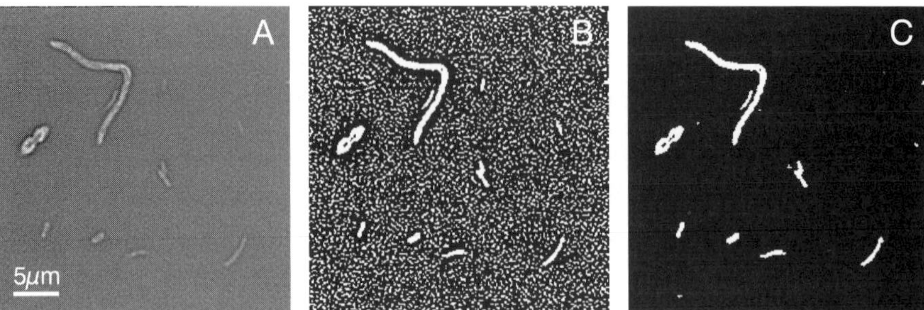

Figure 6.13 The Marr–Hildreth/edge strength algorithm (after Viles and Sieracki, 1992): (A) Marr–Hildreth filtered image, using Gaussian filter of $\sigma = 1.0$ and 3×3 kernel size; 127 has been added to each grey level to ensure that the results stay within the range 0–255; (B) image (A) segmented at 127: all objects except one are detected, along with much spurious detail; (C) all objects with maximum edge strength in the Marr–Hildreth filtered image of less than 10 are removed: some objects are lost, and most, but not all, background removed.

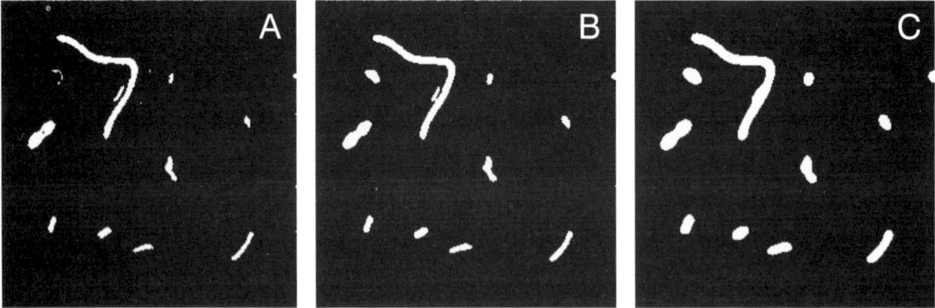

Figure 6.14 Reducing noise sensitivity in the MHES algorithm by using 5×5 median pre-filter, using Sobel edge strength in median filtered, instead of Marr–Hildreth images, and increasing σ: (A) results for $\sigma = 1.0$; (B) $\sigma = 1.7$; (C) $\sigma = 3.0$. In all cases the edge strength threshold was 3. Note the increasing size of the objects at high σ.

6.7.6 Multivariate thresholding and clustering

Multivariate thresholding techniques are needed when using multiband images. The best known examples are (RGB) colour images, but the output of texture analysis may also be multiband, each band being the response of a single filter. In true colour images, the data are often transformed to a more suitable system of colour coordinates, such as hue, saturation and brightness (HSB). This is because such systems are a better reflection of our own colour perception, and manage to group similar colours together in a more intuitive way (e.g. Gong and Sakauchi, 1995).

Whatever the nature of the image data, the multidimensional measurement space must be divided into regions corresponding to the different classes of pixels in the image. In the simplest but also the crudest case, these regions are delineated by a series of thresholds, one or more for each band. This means that the regions in measurement space are rectangles (for two bands) or blocks (for three or more bands). Very often no adequate set of thresholds can be found, because the shapes of the clusters of points in data space do not fit into such shapes. A better approach to this problem is cluster analysis (in analogy to Otsu's method for grey scale images). Several methods to do this are reviewed by Pal and Pal (1993). Strictly speaking, these methods are not thresholding techniques, yet they too decide a pixel's category based on the content of the pixel, rather than the context. The texture segmentation approach of Jain and Farrokhnia (1991) demonstrates how the context of a pixel can be included in multiband image segmentation by clustering, simply by adding the x and y coordinates to the data to be clustered.

Since colour segmentation has not been used on images of bacteria to date, little can be said about the performance of different methods.

6.7.7 Finding connected components in thresholded images

Once each pixel has been classified as foreground or background, a second stage is usually needed to find connected components in this image. One procedure to do this is by region growing, since the similarity of pixels belonging to a specific class is optimal in thresholded images. In contrast to normal grey scale images, region growing does not need any user interaction in this case. Assuming we are not interested in the background, the thresholded image is scanned from top to bottom and left to right to find the first foreground pixel. This pixel serves as the seed for the region growing procedure. During region growing, all members of the connected component are labelled with a different grey level, to mark them as 'found'. After the connected component has been stored and all member pixels marked, the image is scanned for the next foreground pixel, starting just after the seed position of the last object detected. This process is repeated until no foreground pixels remain. The shapes of the regions are stored either in empty bitmap images or in one of the representation schemes outlined in Section 6.9.

Alternatively, contour tracing may be done, also fully automatically, to detect all boundary pixels in the thresholded image. The procedure is similar to the one outlined above, except that the contours are traced, instead of the internal pixels. The natural storage format for connected components when contour tracing is used is discussed in Section 6.9.2.

Instead of storing all connected components in different images, it is possible to give each connected component a unique grey level or label. A fast method of uniquely labelling all connected regions in an image can be found in Rosenfeld and Kak (1982). Initially, the image contains pixels of value zero (background) and the grey level maximum ($N-1=$ foreground: 255 for 8 bits per pixel, 4095 for 12 etc.). Furthermore, an equivalence table of $N-1$ integers is initialized, so that each ith entry has value i. The image is scanned from top to bottom and left to right, and each foreground pixel is compared to its top and left-hand neighbours. If these neighbours are both zero, the pixel is given a new label (one for the first such pixel, two for the next etc.). If one neighbour is zero but the other is non-zero, the pixel is given the latter value. If both neighbours are non-zero, but different in value, the pixel is given the lower value of the two, and the entry in the equivalence table of the higher value is given the lower. During the second scan, each non-zero pixel is given the equivalent value of its grey level. This means that if the entry in the table equals the grey level, the pixel is unchanged; otherwise, the value in the table is used in the same manner: if its entry is equal to its value, it becomes the new grey level; otherwise, the entry value is used etc. After this scan each connected region has been given a unique grey level. Counting of all objects above a certain area reduces to making a histogram and counting the number of entries of grey levels larger than zero, which have a pixel count larger than the area threshold.

The disadvantage of this method is that it does not always work. If more than $N-1$ objects exist it must obviously fail, but in reality it fails at much lower numbers of objects. If we limit the type of object to circles with radius R, $(N-1)/R$ is the actual number of objects that may be present before the algorithm fails. In images of a few hundred objects with radii of some 10 pixels, this requires at least 12–16 bits per pixel. Large amounts of single-pixel debris can also cause problems with this algorithm. Furthermore, the method does not store the objects in an easy format for further processing: to find all pixels of a particular object, all pixels of the image must be scanned, or one of the above approaches must be used. New approaches of connected region labelling alleviate some of the problems of this earlier implementation, and fast hardware has been developed to perform these algorithms (Fong, 1984; Nicol, 1995), yet the number of regions must still be smaller than $N-1$.

6.8 SEPARATION OF TOUCHING OBJECTS

One of the most difficult problems when working on images of cells is that of touching objects. Even from a computer scientist's point of view, there are numerous problems when trying to separate a single connected region into its constituent objects (Russ, 1992). In microbiology, the problems may be compounded further by the fact that it is very often hard to tell when to classify a compound object as multiple, separate entities, rather than a single dividing cell (see Chapters 7, 8 and 9). One important case is that of various types of cocci. On the one hand, single cells may be considered separate, autonomous entities. On the other hand, the shape of the clusters they form yields vital information on the species. If we would like to detect streptococci by image analysis, it would not do to break up all strings of circular objects.

The need for separation not only depends on the nature of the objects, but also on the pre-processing and segmentation algorithms used. If an image is smoothed a great deal, thin 'valleys' between objects might disappear. Conversely, the second derivative method of Viles and Sieracki (1992) is very sensitive to such valleys, and should separate many touching objects automatically. Boundary- and edge-based segmentation are also usually better at separating objects than threshold-based methods.

A number of post-processing algorithms to separate aggregates of cells into single objects have been developed to date. Some approaches work solely on the basis of the binary image of the object (Russ, 1992; Dubuisson *et al.*, 1994): one such approach is the watershed method discussed in Section 1.7.1b. Dubuisson *et al.* detect indentation points by measuring the local curvature. This is computed as the difference between the slopes of the vectors from each boundary point to points k pixels to the right and left on the boundary. The angle is 180° for a straight edge, larger for convex parts of the boundary and smaller at concave parts. Angles lower than a threshold T are considered as sharp indentations. The same authors found that $k = 4$ and $T = 150°$ yielded good results. Once the list of indentation points has been made, the object is split using the following rules:

(a) If an indentation point has three adjacent boundary pixels, the removal of the indentation point will split the object into two parts. After this, the indentation point is removed from the list.
(b) If the indentation point has two adjacent boundary pixels and another indentation point is close to it, split the object by removing all pixels on a line between these two points, including the points themselves. Both points are deleted from the list.
(c) If the indentation point has two adjacent boundary pixels and no other indentation point is near to it, split the object along a line perpendicular to the boundary at that point, and remove the indentation point from the list.

Dubuisson *et al.* also describe methods to join two objects that have been split by the segmentation algorithm, by linking elongated objects which lie close enough together and have approximately the same orientation.

Others methods include grey level information as well (Bloem *et al.*, 1995; Fernàndez *et al.*, 1995). The approach used by Bloem *et al.* first finds local maxima of the grey levels within each object (or clump). A local maximum is defined as a region completely surrounded by pixels of a lower grey level. Once these areas of local maxima have been determined, the SKIZ ('skeleton by influence zones') of the resulting image (Section 1.6.1f) or skeleton of the background is determined. The lines formed by the SKIZ are then used to divide the clumps into their constituents. A problem noted by Bloem *et al.* is that faint objects connected to larger, brighter ones need not show local maxima, and are therefore not segmented.

Fernàndez *et al.* (1995) first find the most concave points on the outline of the object. Then they try to link these points into pairs, demanding that: (i) the points are not too far apart and (ii) the sum of gradients on the line connecting the two does not exceed a threshold. The first criterion ensures that minor dents in large objects are not linked; the second that the 'path' between them is relatively level. When a pair meets both criteria, the line is dilated, and a watershed algorithm is used to find a path along the bottom of the valley. Although watershed is used for the ultimate segmentation, the over-segmentation (or undermerging) error discussed in Section 1.7.1b is avoided. This method has been used on microscopic images of plant cells.

6.9 STORAGE OF IMAGE SEGMENTS

In many cases, it is desirable to have an efficient internal representation of the shapes of the regions of interest detected by any of the segmentation techniques described previously. The term 'efficient' has a twofold meaning in this context:

(i) requiring little memory space and (ii) allowing easy extraction of the required information. The simple storage in bitmap form (as an image of an individual object) is not always the most efficient in both of these senses simultaneously.

6.9.1 Scan-line based storage of region shapes

A popular method for storage of binary shapes is based on a scan-line or runlength representation, for use in both computer graphics (Burger and Gillies, 1989) and image processing (Young *et al.*, 1981). To represent a binary image of an object, we only need to store the lines (or rows) in the image which actually contain some object pixels. Empty lines contain no information, and may therefore be scrapped. This idea can be extended further, by storing only the points on each scan-line where transitions from background to object and vice versa occur. These points occupy only a small fraction of the entire scan-line, yet provide a complete description of the object. Therefore, any manipulation or measurement to be performed on the object can be defined in terms of these boundary points. Since the number of these points is very small compared to the number of pixels in a bitmap needed to represent the object, this can speed up processing the object.

There is one small price to pay for this advantage: unlike the total number of pixels in an image, the number of boundary points of an object is not known beforehand, so they must be stored in a dynamic structure like a linked list. One such representation is shown in Figure 6.15. It is an adaptation of the scheme of Young *et al.* (1981) developed by Meijer *et al.* (1990), and uses a single linked list, in which each node contains both the x and y coordinates, in such a way that they can

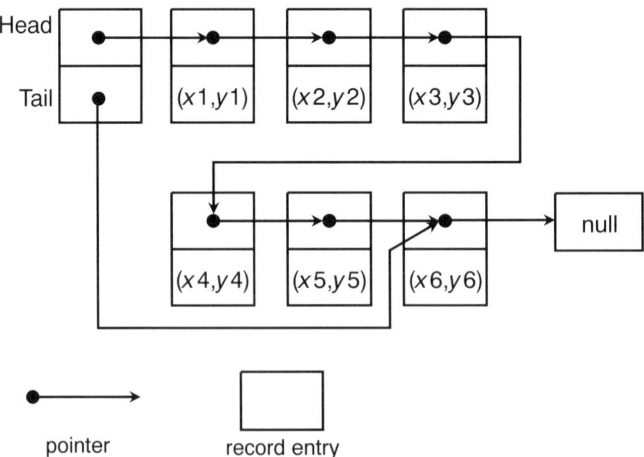

Figure 6.15 Linked list structure used by Meijer *et al.* (1990) to store and process binary images of bacteria. Each node (record in computer memory) of the list contains the coordinates of a point in the image where the grey level changes from background to foreground or vice versa, and a pointer to the next node (= its memory address). The (x, y) pairs can also be accessed as a single key, by which the list is sorted. A header structure points to the first and last nodes of the list; the 'next' pointer of the last element points to the 'null' address, to flag the end of the list.

be read as a single key by which to sort the linked list:

$$key = 65536 \cdot y + x$$

This allows the program to access each node either as a single 4-byte key for sorting, or as two 2-byte integers, representing x and y coordinates. One might think of this scheme as a transform of the image, from a raster of X by Y pixels, to one of 1 by XY pixels. This single row is scanned for transitions.

To convert a bitmap into a scan-line representation, the image is scanned row by row, from top to bottom and from left to right. Each pixel's state (foreground or background) is compared with the last stored state (that of the previous pixel). If they are different, a new element, containing the current x and y positions, is added to the list. In the case of the first pixel, there is no last stored state. Therefore, an arbitrary choice must be made, as to whether the last state at the beginning of the scan is foreground or background. The most sensible choice is to assume that the last state is background at the beginning of each scan, since the number of background pixels is much larger than the number of foreground pixels.

The advantage of scan-line storage is that any object shape, even one with holes in it or consisting of more than one connected region, can be stored in the same data structure. Furthermore, the representation is translated to and from bitmap representation easily. This makes it very easy to perform measurements on the grey levels of the corresponding area in, for example, the fluorescent image of the same field of view. Such applications are less straightforward with some other schemes.

6.9.2 Contour encoding of region shapes

Contour coding is most compatible with the output of outline traces produced manually or by automatic edge detection methods. The data structure most suitable for storage of contours is the linked ring: a linked list in which the tail element's 'next' pointer is set to the head element's address. Each element need only store the *direction* in which the next boundary point lies. The number of boundary points stored is always larger than or equal to the number of points stored in scan-line storage, but since there are only four or eight possible directions in which the next point can lie, only two or three bits of storage per element are needed. In practice, both methods have similar memory requirements.

A far more important consideration is the type of information to be derived from the ROI, and the nature of the ROI shape. If the ROI has holes in it, more than one linked ring must be used to define its contour. This complicates the data structure somewhat, since it must now also reflect the hierarchy of which contour lies within which.

Many morphological operators can be implemented efficiently using only the contours (Verwer and Van Vliet, 1988). The use of contour code is a very elegant solution for the Fourier descriptor method of shape characterization (Gonzales and Wintz, 1987).

6.9.3 Data compression for disk storage

Most representations that are efficient internally (in memory) are also highly efficient for disc storage of images. Each internal representation is closely related to a specific data compression algorithm used in various image file formats. Scan-line storage is closely related to runlength code (RLC), which is used in numerous popular file formats, such as TIFF. Similarly, contour encoding may also be stored on disc efficiently (Freeman, 1961), though it is not a common choice in image file formats.

Storing the internal structure in a related file format is usually faster than storage in the same format directly from bitmap. Besides, when read from file again, the data may be translated to the required internal representation rather than to a bitmap, with further speed gains.

The great utility of such schemes can be seen from a simple data compression scheme for storage of binary images of full fields of view containing roughly 100 objects, and of individual objects (bacteria) in each field of view (Wilkinson, 1994). Images containing just a single object could be compressed in just under 40 bytes on average, instead of 24 kilobytes, which represents uncompressed storage of the bitmaps. The average compression ratio was more than 600 times.

A detailed description of file formats and compression techniques is beyond the scope of this chapter, but can be found in many textbooks (Gonzalez and Wintz, 1987; Lindley, 1991; Murray and VanRyper, 1996).

6.10 DISCUSSION

6.10.1 The current state of affairs

Segmentation is a very difficult field, in part because there is no particular agreement on quality criteria for segmentation results. A method that is optimal in one sense, e.g. minimizing fragmentation of the shape, may perform poorly in another, e.g. percentage of error (Sahoo *et al.*, 1988; Lee *et al.*, 1990). Thus there is no single optimum algorithm: it all depends on the application. Quite apart from the quality of the results, other factors play a role in deciding which algorithm to choose, e.g. the computer time and memory size needed. The evaluation of these factors is also highly application-dependent. If a system is fully automated, as in the case of Bloem *et al.* (1995), it does not matter that it needs several minutes to process each image, since the operator need not be present. By contrast, if even a small amount of editing is necessary to eliminate occasional debris, it is essential that the algorithm can present its results in under 10 seconds, which is an approximate 'limit of user irritation'. Real-time applications, e.g. for motility measurement, require computing times of 30 ms (one video frame) or less (see Chapter 13).

A great deal of work has been done on the segmentation of bacteria or other cells against a relatively smooth background (Bjørnsen 1986; Sieracki *et al.*, 1989a; Viles and Sieracki 1992; Wilkinson, 1996). Methods such as RATS can deliver accurate results at the expense of very little computing time, for objects of a wide variety of sizes and shapes. Application-orientated techniques, such as developed by Viles and Sieracki, may be restricted in their use to certain sizes and shapes of objects, but may outperform more general-purpose algorithms on their 'home turf'. Only a head to head test will tell which algorithm is best for a particular application.

By contrast, the work of Bloem *et al.* (1995) is unique in tackling images in which the background is far from smooth. Although the use of fixed thresholds should perhaps be replaced by automatic or adaptive methods, the pre-processing stages of the algorithm could be useful in all cases when debris cannot be eliminated from the slides.

6.10.2 Extending the methods to three-dimensional images

The literature on segmentation of three-dimensional images, such as those obtained with confocal microscopes (Chapters 16 and 17), is far less extensive than that on two-dimensional segmentation. Nonetheless, there is quite a range of techniques available, usually derived from two-dimensional applications. All pure data-space techniques (histogram-based thresholding, colour clustering etc.) can be applied to three-dimensional images without any modification, because they do not use any spatial information. Thresholding schemes such as RATS and the Marr–Hildreth/edge strength method also lend themselves to three dimensions, by replacing two-dimensional first derivative, Laplacian and Gaussian filters by their three-dimensional counterparts. In three dimensions, local thresholding with RATS requires an octtree, rather than quadtree hierarchy (each 'parent' having eight instead of four 'children'), but this too is a simple adaptation. Within the context of RATS, it can even be shown that squared gradients provide the same optimization for curvature bias in the three-dimensional case as in two dimensions (Wilkinson, submitted).

Contour tracing is harder to adapt to three dimensions, since it is a surface that must now be traced, not a simple contour. Nonetheless, three-dimensional boundary tracing algorithms have been developed (e.g. Gordon and Udupa, 1989; Udupa and Ajjanagadde, 1990). Likewise, region growing is also somewhat more complicated in three dimensions than in two dimensions. By contrast, split-and-merge is comparatively simple in three dimensions: as in the case of RATS, the quadtree structure must be replaced by an octtree.

There is quite an extensive literature on connected component labelling and object representation in three dimensions (Udupa, 1982; Udupa and Ajjanagadde; 1990, Thurfjell *et al.*, 1992, 1995). The latter authors use a representation of the image in which only boundary voxels are stored, in a two-dimensional (y, z) array of pointers to lists of boundary pixels. Each entry in these lists contains the x position of the voxel, a 6-bit field indicating which neighbouring pixels are background, and a label field, which contains the label of the connected region to which the voxel belongs. Apart from considerably reducing memory space needed, the scheme allows rapid erosion, dilation and connected component labelling.

Biomedical applications involving three-dimensional imaging usually centre on computed X-ray tomography (CT), nuclear magnetic resonance (NMR) imaging and positron emission tomography (PET). By contrast, three-dimensional segmentation techniques have not been used much (if at all) within the field of microbiology. Given the increasing popularity of confocal microscopy and other three-dimensional imaging techniques, three-dimensional segmentation will certainly become more important in this field too.

6.10.3 Future prospects

Apart from the more or less classical image segmentation techniques discussed in this chapter, several newer techniques are emerging. All techniques discussed are 'crisp' or 'hard' segmentation: each pixel is assigned to a particular class, without any 'grey area' in between. In fuzzy segmentation, pixels are assigned probabilities of belonging to a specific class. It is thought that such techniques may reduce noise sensitivity and allow inference from incomplete data. Neural networks have also been applied to the problem of image segmentation, and appear to be very robust in noisy situations. Both classes of technique are included in the review by Pal and Pal (1993). They also discuss relaxation labelling at some length, which was only mentioned briefly in Section 6.4.1.

The drawback of many of these new techniques is that they require considerable computing power and memory, and have often been designed for massively parallel computers. Besides, using these techniques for the comparatively simple task of segmenting microscopic images of bacteria might be considered swatting a fly with a sledgehammer. The classical techniques can often cope well enough. Nonetheless, when dealing with more complex images, e.g. of flocs or other three-dimensional communities, these more complicated methods may be useful, as computing power becomes ever cheaper.

Snakes and similar deformable models could be used too, especially to provide sub-pixel resolution estimates of the shapes. Such an approach is currently being developed at Michigan State University (A.K. Jain and L. Hong, personal communication). The simplicity of the three-dimensional shapes of bacteria (Sieracki *et al.*, 1989b) should be a great bonus in this kind of work. Furthermore, the success of simple techniques can be used to reduce the computational burden, by providing a good first estimate for the more advanced iterative procedures.

Quite apart from using advanced algorithms, there are many permutations and combinations of existing algorithms which should be explored. All of the techniques already in use for segmentation of bacteria have their strengths and weaknesses, and judicious combination of steps in pre-processing, segmentation, storage schemes and separation of touching objects from different authors may lead to new systems, which show more of the strengths and fewer weaknesses.

ACKNOWLEDGEMENTS

This work was supported by the generous financial support of the Institute for Microbiology and Biochemistry, Herborn-Dill, Germany, and the International Study Group for New Antimicrobial Strategies (ISGNAS).

REFERENCES

Beghdadi, A., Le Negrate, A. and Viaris de Lesegno, P. (1995) Entropic thresholding using a block source model. *Graph. Models Image Proc.* **57**: 197–205.
Bjørnsen, P.K. (1986) Automatic determination of bacterioplankton biomass by image analysis. *Appl. Environ. Microbiol.* **51**: 1199–1204.
Bloem, J., Veninga, M. and Shepherd, J. (1995) Fully automatic determination of soil bacterium numbers, cell volumes, and frequencies of dividing cells by confocal laser scanning microscopy and image analysis. *Appl. Environ. Microbiol.* **63**: 926–36.
Bruins, S., De Jong, M.C.J.M., Heeres, K., Wilkinson, M.H.F., Jonkman, M.F. and Van der Meer, J.B. (1994) Fluorescence overlay antigen mapping of the epidermal basement membrane zone: I. geometric errors. *J. Histochem. Cytochem.* **42**: 555–60.
Burger, P. and Gillies, D. (1989) *Interactive Computer Graphics*. Addison-Wesley: Reading, MA.
Canny, J. (1986) A computational approach to edge detection. *IEEE Trans. Pattern Anal. Mach. Intell.* **6**: 679–98.
Chow, C.K. and Kaneko, T. (1972) Automatic boundary detection of the left ventricle from cineangiograms. *Comput. Biomed. Res.* **5**: 388–410.
Cootes, T.F., Taylor, C.J., Cooper, D.H. and Graham, J. (1995) Active shape models – their training and applications. *Comput. Vis. Image Understanding* **61**: 38–59.
Cox, P.W. and Thomas, C.R. (1992) Classification of fungal pellets by automated image analysis. *Biotech. Bioengng* **39**: 945–52.
Dubuisson, M.P., Jain, A.K. and Jain, M.K. (1994) Segmentation and classification of bacterial culture images. *J. Microbiol. Meth.* **19**: 279–95.
Fdez-Valdivia, J., García, J.A., Perez de la Blanca, N., Garrido, A. and Fuertes, J.M. (1995) A new methodology to solve the problem of characterizing 2-D biomedical shapes. *Comput. Meth. Prog. Biomed.* **46**: 187–205.
Fernàndez, G., Kunt, M. and Zrÿd, J.P. (1995) A new plant cell image segmentation algorithm. In *Image Analysis and Processing, 8th International Conference ICIAP '95 Proceedings* (Braccini, C., DeFloriani, L. and Vernazza, G., eds), pp. 229–34. Springer-Verlag, Berlin.
Fogel, I. and Sagi, D. (1989) Gabor filters as texture discriminator. *Biol. Cybernet.* **61**:103–13.
Fong, A.C. (1984) A scheme for reusing label locations in real time connected component labelling of images. In *Proceedings of the 7th International Conference on Pattern Recognition*, Montreal, pp. 243–5. IEEE Press: Washington, DC.
Freeman, H. (1961) On the encoding of arbitrary geometric configurations. *IRE Trans. Electron. Comp.* **EC-10**: 260–8.
Gong, Y. and Sakauchi, M. (1995) Detection of regions matching specified chromatic features. *Comput. Vis. Image Understanding* **61**: 263–9.
Gonzales, R.C. and Wintz, P. (1987) *Digital Image Processing*, 2nd edn. Addison-Wesley: Reading, MA.
Gordon, D. and Udupa, J.K. (1989) Fast surface tracking in three-dimensional binary images. *Comput. Vis. Graph Image Proc.* **45**: 196–214.
Haralick, R.M., Shanmugam, K. and Dinstein, I. (1973) Textural features for image classification. *IEEE Trans. Systems Man Cybernet.* **3**: 610–21.
Hubel, D.H. and Wiesel, T.N. (1977) Functional architecture of macaque monkey visual cortex. *Proc. R. Soc. London* **B198**: 1–59.
Jain, A.K. and Farrokhnia, F. (1991) Unsupervised texture segmentation using Gabor filters. *Pattern Recogn.* **24**: 1167–87.
Kahmoun, F. and Astruc, J.P. (1994) Optimum edge detection for object-background picture. *CVGIP: Graph. Models Image Proc.* **56**: 25–8.
Kapur, J.N., Sahoo, P.K. and Wong, A.K.C. (1985) A new method for gray-level picture thresholding using the entropy of the histogram. *Comput. Vis. Graph Image Proc.* **29**: 273–85.
Kass, M., Witkin, A. and Terzopoulos, D. (1987) Snakes: active contour models. In *Proceedings of the 1st International Conference on Computer Vision*, pp. 259–68. IEEE Computer Society Press: Washington, DC.
Kittler, J., Illingworth, J., Föglein, J. and Paler, K. (1984) An automatic thresholding algorithm and its performance. In *Proceedings of the 7th International Conference on Pattern Recognition*, Montreal, pp. 287–9. IEEE Press: Washington, DC.
Kittler, J., Illingworth, J. and Föglein, J. (1985) Threshold selection based on a simple image statistic. *Comput. Vis. Graph. Image Proc.* **30**: 125–47.
Lee, J.S., Haralick, R.M. and Shapiro, L.S. (1986) Morphologic edge detection. In *Proceedings of the 8th International Conference on Pattern Recognition*, Paris, pp. 369–73. IEEE Press: New York.
Lee, S.U., Chung, S.Y. and Park, R.H. (1990) A comparative performance study of several global thresholding techniques for segmentation. *Comput. Vis. Graph. Image Proc.* **52**: 171–90.
Levine, M.D. and Nazif, A.M. (1982) An experimental rule based system for testing low level segmentation strategy. In *Multicomputers and Image Processing: Algorithms and Programs* (Preston, K. and Uhr, L., eds), pp. 149–60. Academic Press: New York.

Li, C.H. and Lee, C.K. (1995) Image smoothing using parametric relaxation. *Graph. Models Image Proc.* **57**: 161–74.

Lindley, C.A. (1991) *Practical Image Processing in C.* Wiley: New York.

Meijer, B.C., Kootstra, G.J. and Wilkinson, M.H.F. (1990) A theoretical and practical investigation into the characterization of bacterial species by image analysis. *Binary Comput. Microbiol.* **2**: 21–31.

Meijer, B.C., Kootstra, G.J., Oenema, D.G. and Wilkinson, M.H.F. (1991). Effects of ceftriaxone on faecal flora: analysis by micromorphometry. *Epidemiol. Infect.* **106**: 513–21.

Meijer, C., de Vries, E.G.E., Dam, W.A., Wilkinson, M.H.F., Hollema, H.A., Hoekstra, H.J. and Mulder, N.H. (1997) Sensitive immunocytochemical detection and quantitation of cisplatin-induced DNA-platination with double fluorescence video microscopy. *Br. Cancer J.* **76**: 290–8.

Murray, J.D. and VanRyper, W. (1996) *Encyclopedia of Graphics File Formats.* O'Reilly and Associates, Inc.: Sebastopol, CA.

Nicol, C. (1995) A systolic approach for real time connected component labelling. *Comput. Vis. Image Understanding* **61**: 17–31.

Otsu, N. (1979) A threshold selection method from gray-level histogram. *IEEE Trans. Systems Man Cybernet.* **9**: 62–6.

Pal, N.R. and Pal, S.K. (1993) A review on image segmentation techniques. *Pattern Recogn.* **26**: 1277–94.

Parker, J.R. (1997) *Algorithms for Image Processing and Computer Vision*, pp. 116–49. Wiley: New York.

Perona, P. and Malik, J. (1990) Scale space and edge detection using anisotropic diffusion. *IEEE Trans. Pattern Anal. Mach. Intell.* **12**: 629–39.

Pons, M.N., Vivier, H., Remý, J.F. and Dodds, J.A. (1993) Morphological characterization of yeast by image analysis. *Biotechnol. Bioengng* **42**:1352–9.

Pun, T. (1981) Entropic thresholding, a new approach. *Comput. Graph. Image Proc.* **16**: 210–39.

Radermacher, M., Tillman, R.W., Fritz, M. and Gaub, H.E. (1992) From molecules to cells: imaging soft samples with the atomic force microscope. *Science* **257**: 1900–5.

Rosenfeld, A. and Kak, A.C. (1982) *Digital Picture Processing*, vol. 2, 2nd edn., pp. 241–3. Academic Press: Orlando, FL.

Russ, J.C. (1990) Processing images with a local Hurst operator to reveal textural differences. *J. Comp. Assist. Microsc.* **2**: 249–57.

Russ, J.C. (1992) *The Image Processing Handbook*, pp. 255–77. CRC Press, Boca Raton, FL.

Sahoo, P.K., Soltani, S., Wong, A.K.C. and Chen, Y.C. (1988) A survey of thresholding techniques. *Comput. Vis. Graph. Image Proc.* **41**: 233–60.

Sieracki, M.E., Reichenbach, S.E. and Webb, K.L. (1989a) Evaluation of automated threshold selection methods for accurately sizing of microscopic fluorescent cells by image analysis. *Appl. Environ. Microbiol.* **55**: 2762–72.

Sieracki, M.E., Viles, C.L. and Webb, K.L. (1989b). Algorithm to estimate cell biovolume using image analyzed cell microscopy. *Cytometry* **10**: 551–7.

Szollosi, J., Lockett, S.J., Balazs, M. and Waldman, F.M. (1995) Autofluorescence correction for fluorescence in situ hybridization. *Cytometry* **20**: 356–61.

Thurfjell, L., Bengtson, E. and Nordin, B. (1992) A new three-dimensional connected component labelling algorithm with simultaneous object feature extraction capability. *CVGIP: Graph. Models Image Proc.* **54**: 357–64.

Thurfjell, L., Bengtson, E. and Nordin, B. (1995) A boundary approach for fast neighbourhood operations on three-dimensional binary data. *Graph. Models Image Proc.* **57**: 13–19.

Udupa, J.K. (1982) Interactive segmentation and boundary surface formation for 3D-digital images. *Comput. Graph. Image Proc.* **18**: 213–35.

Udupa, J.K. and Ajjanagadde, V. (1990) Boundary and object labelling in three dimensional images. *Comput. Vis. Graph Image Proc.* **51**: 355–69.

Van Vliet, L.J., Young, I.T. and Beckers, G.L. (1989) A nonlinear Laplace operator as edge detector in noisy images. *Comput. Vis. Graph. Image Proc.* **45**: 167–95.

Vardi, E. and Grover, N.B. (1990) Shape changes in *Escherichia coli* B/r *A* during agar filtration. *Cytometry* **14**: 173–8.

Venkatesh, S. and Rosin, P.L. (1995) Dynamic threshold determination by local and global edge evaluation. *Graph. Models Image Proc.* **57**: 146–60.

Verwer, B.J.H. and Van Vliet, L.J. (1988) A contour processing method for fast binary neighborhood operations. *Pattern Recogn. Lett.* **7**: 27–36.

Viles, C.L. and Sieracki, M.E. (1992) Measurement of marine picoplankton size by using a cooled, charge-coupled device camera with image analyzed fluorescence microscopy. *Appl. Environ. Microbiol.* **58**: 584–92.

Vogt, R.F., Cross, G.D., Philips, D.L., Henderson, L.O. and Hannon, W.H. (1991) Interlaboratory study of cellular fluorescence intensity measurement with fluorescein-labeled microbead standards. *Cytometry* **12**: 525–36.

Weszka, J.S. and Rosenfeld, A. (1979) Histogram modification for threshold selection. *IEEE Trans. Systems Man Cybernet.* **9**: 38–52.

Wilkinson, M.H.F. (1994) A simple data compression scheme for binary images of bacteria compared to commonly used image data compression schemes. *Comput. Meth. Prog. Biomed.* **42**: 255–62.

Wilkinson, M.H.F. (1996) Rapid automatic segmentation of fluorescent and phase-contrast images of bacteria. In *Fluorescence Microscopy and Fluorescent Probes* (Slavik, J., ed.), pp. 261–6. Plenum Press: New York.

Wilkinson, M.H.F. (submitted) Optimizing edge detectors for robust automatic threshold selection: coping with edge curvature and noise. *Graph. Models Image Proc.*

Wu, A.Y., Hong, T.H. and Rosenfeld, A. (1982) Threshold selection using quadtrees. *IEEE Trans. Pattern Anal. Mach. Intell.* **4**: 90–4.

Yanowitz, S.D. and Bruckstein, A.M. (1989) A new method for image segmentation. *Comput. Vis. Graph. Image Proc.* **46**: 82–95.

Young, I., Peverini, R., Verbeek, P. and Van Otterloo, P. (1981) A new implementation for the Minkowski operators. *Comput. Graph. Image Proc.* **17**: 189–210.

Part II

Single-Celled Organisms

7

Enumeration and Sizing of Micro-organisms using Digital Image Analysis

Michael E. Sieracki[1] and Charles L. Viles[2]
[1] *Bigelow Laboratory for Ocean Sciences, Maine, USA*
[2] *University of North Carolina, Chapel Hill, North Carolina, USA*

7.1 INTRODUCTION

Many aspects of the study of micro-organisms depend upon accurate measurement of the numbers and sizes of cells. In microbial ecology, estimates of biomass and size distributions are critical to many of our theories of how populations and communities interact and function. A fundamental goal of ecology is to explain observed distributions of organisms in nature. This goal obviously depends upon accurate data on the abundance and location of these organisms.

Other applications that depend upon accurate measurements of cells include fluorescent physiological indicators of cells in nature, phylogenetic stains that label species or group-specific nucleotide sequences, environmental monitoring, and, of course, a wide array of biomedical and clinical applications. Some of these topics are discussed in detail in other chapters of this book. The accuracy and utility of all of these methods depend upon good cell discrimination and measurement.

Traditional methods based on visual cell counting by microscopy are tedious, and the sample analysis rate is low and limited by technician fatigue. Observations are often qualitative. Measuring cell sizes using visual microscopic methods reduces the rate at which samples can be analysed, and also introduces operator error and subjectivity. For these reasons, it is unsurprising that automated methods using image analysis are being more widely used for micro-organism studies, particularly in the light of technological improvements in cameras, computers and image analysis software.

Other methods of determining the biomass of micro-organisms in nature include bulk measures, such as chlorophyll as a proxy for phototrophic biomass, or lipids (Findlay and Dobbs, 1993) or ATP content (Karl, 1993) for total biomass. These bulk methods do not allow discrimination of cell types and the resulting measurements can be confounded by dead cells and detritus. For many research questions, single-cell measurements are necessary. They can provide discrimination of cell types and allow quantification of intra-population variation. Since many properties and

Digital Image Analysis of Microbes: Imaging, Morphometry, Fluorometry and Motility Techniques and Applications. Edited by M.H.F. Wilkinson and F. Schut.

processes of natural assemblages are not normally distributed, this ability to measure variation within populations and describe the distribution curves is critically important to understanding the functioning and trends of natural systems. In addition to automated microscopy, single-cell analysis is also accomplished by flow cytometry and Coulter counting technology for cells. These methods are best for cells already in suspension, such as natural plankton samples, body fluid samples and cells grown in liquid culture media.

Many types of microscope samples are readily adaptable to automated image analysis techniques. The focus of our work has been with fluorescence microscope images of natural planktonic assemblages. These provide images of considerable contrast since the fluorochrome-stained cells become light sources and can be placed on dark, non-fluorescing background substrates.

Accurate, automated detection, counting and sizing of cells depends critically on sample preparation (see appendix), system optics and image acquisition. These factors combine to determine the overall quality of a digital image. High quality image acquisition should not be compromised with the idea that image processing techniques will be used afterwards to correct a poor quality image. The previous chapter dealt with the issues of image segmentation – that is, the automated, or semi-automated, discrimination of objects of interest from background in an image. The choice of segmentation method is critical to accurate detection and measurement of cells of interest, so the two topics are closely related.

This chapter reviews the state of knowledge in the detection and sizing of micro-organisms by automated imaging cytometry. The methods are relatively straightforward and have now been applied to a variety of systems, cell types and research applications. A digital image is essentially a scene that has been sampled discretely. A class of measurement errors is associated with image sampling (i.e. digitization) and these errors will be discussed. Important measurements like area, shape, intensity and biovolume are discussed next. We then consider validation and verification methods for confirming the sizing capabilities of imaging systems. There are some problematic conditions that will be considered as special cases for cell counting. Digital imaging cytometry will be compared to flow cytometry with focus on the relative strengths and weaknesses of these two technologies. Our experience has been primarily with natural marine plankton communities. The protocol for preparing samples for image analysis of picoplankton is given in the appendix to this chapter.

7.2 IMAGE SAMPLING

Image sampling is perhaps one of the least appreciated aspects of image analysis, particularly among application scientists. We must first realize that the act of measurement itself causes a loss of information. Measurement is intrusive and generally affects the object being measured. For example, in fluorescence microscopy, we generally fix and stain samples and place them on a microscope slide. In most systems, we then form a two-dimensional image from this three-dimensional scene. It is the two-dimensional image that we measure.

In digital systems, the two-dimensional image is a discrete representation of two of the three real dimensions in which the object resides. This process of forming a discrete representation from an inherently continuous scene is called *sampling*. A convenient way to think about sampling is to imagine a two-dimensional *sampling grid* that overlays a continuous scene. Each element of the grid is a pixel. If an object fills more than half the pixel, we turn the pixel on; otherwise, the pixel is off. When we are done, we have an image of the scene where objects are blocky approximations of real life (Figure 7.1). Very large objects are represented accurately because they are composed of many pixels – the sampling grid is fine-grained relative to the objects. On the other hand, very small objects may often be inaccurately represented or missed entirely because they are composed of very few pixels – the sampling grid is coarse-grained relative to these objects.

The rest of this section is devoted to driving home this very intuitive idea that high sampling rates yield better size estimates and that low sampling rates can often give unpredictable results. Our goal is not to provide a solid mathematical foundation for understanding the sampling phenomenon – there are other texts and papers that already do so (Hamming, 1983; Park *et al.*, 1984; Pratt, 1991). Rather, we wish to give enough intuition to the application scientist to understand the sampling process and its ramifications on measurement.

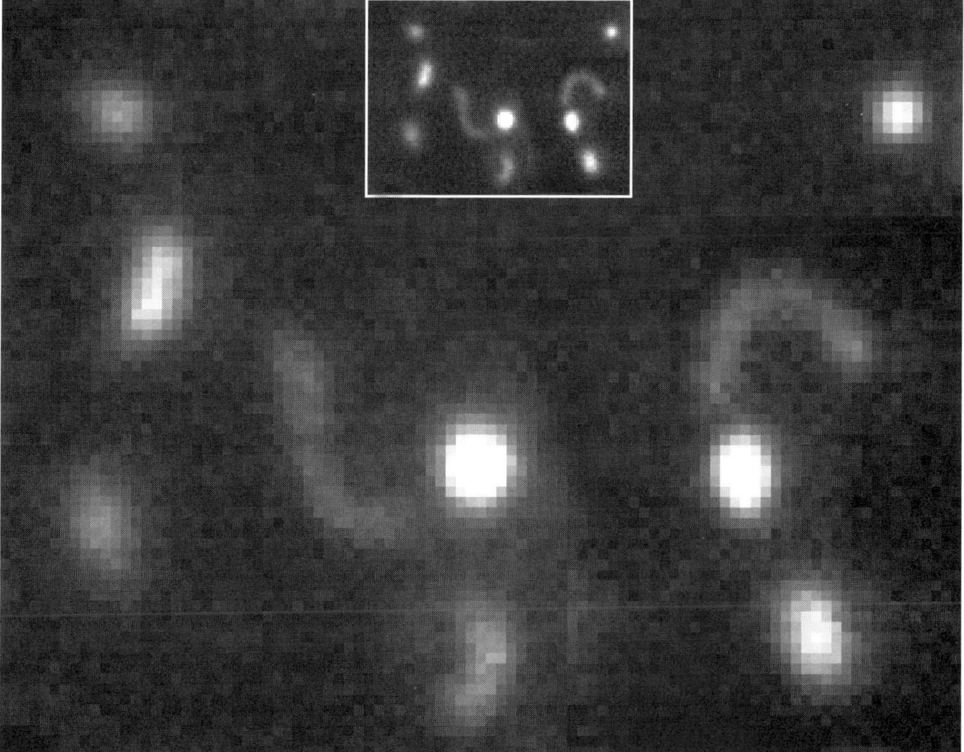

Figure 7.1 Oceanic bacteria, stained with DAPI, imaged by a cooled CCD camera. The image was enlarged by pixel replication to show its digital nature. The original image is shown in the inset.

7.2.1 Sampling density

Sampling density, or the number of samples per unit length, has a dramatic effect on measurement precision. In most cases, we expect that the sampling grid is laid randomly across the scene of interest. If we image the scene multiple times, the orientation of the grid with respect to the scene is different each time. In probabilistic terms this sample-scene phase is the juxtaposition of two uniformly distributed random variables, one describing the horizontal dimension and the other the vertical dimension. In a measurement environment, we expect different measurements of the same object each time we image the scene. Practically speaking, if we measure a cell at the margin of a microscope field, and then re-adjust the microscope stage so that the cell is centred in the field, then we will probably get a different size estimate.

To illustrate this idea, consider a simple simulation, the general form of which was described previously by Young (1988). Suppose we take a circle of unit diameter and randomly overlay a sampling grid that has, for example, three samples per unit (Figure 7.2). We count each pixel in the grid whose centre is contained in the circle as part of the circle and find six pixels (Figure 7.2a). Pixels whose centre is outside the circle are not counted. Now we adjust the sample-scene phase at random and again count the number of included pixels, this time finding seven (Figure 7.2b). In our simulation, we repeated this procedure 1000 times and found a mean

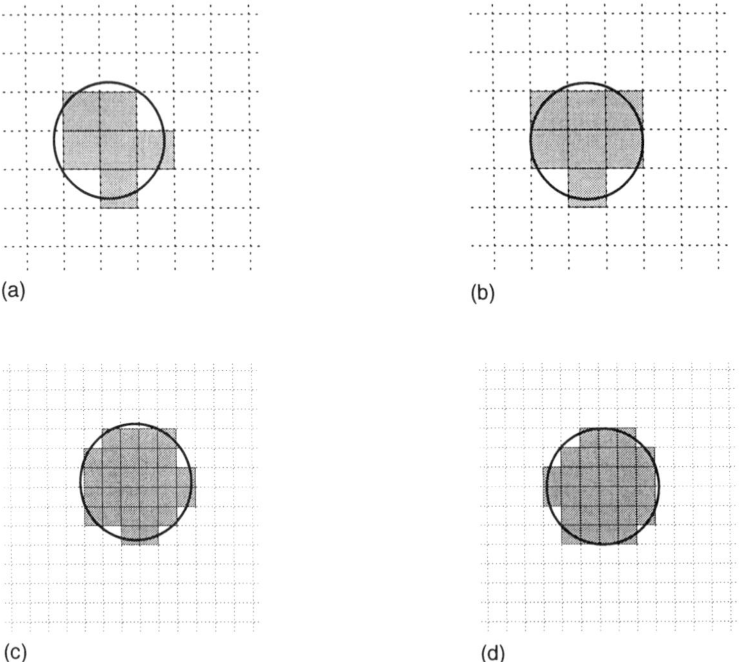

Figure 7.2 A simple illustration of the effect of sampling density on measurement. Two trials at low density (top) yield estimates of (a) 6 and (b) 7 pixels (17% difference). Two trials at bottom yield estimates of (c) 28 and (d) 29 pixels (3% difference).

estimated size of 7.09 pixels, a standard deviation of 0.75 pixels, and a coefficient of variation (CV) of 10.5%. Analytically, we expect the measured size to be 2.25π, or about 7.07, pixels. So if we measure the circle enough times, we will get very close to its actual area. But, if we measure the circle only once, we can get as few as 4 or as many as 9 pixels (these numbers are from the 1000 run data).

As we might expect, if we increase the sampling density, we still have the same problem, but it is not nearly as pronounced. In our simulation, if we double the sampling density to six samples per unit, we get minimum and maximum estimates of 26 and 32 pixels respectively. Most importantly, the CV is reduced to 4.1%. The results of the simulation of measured sizes at various sampling densities are summarized in Table 7.1.

Young (1988) summarizes these findings in an empirically derived formula that gives the expected error due to sampling given a particular sampling density S:

$$\text{Error }(\%) = 58.5 \cdot S^{-1.6} \tag{7.1}$$

There is *always* error associated with sampling an image. The magnitude of that error is directly related to the sampling density.

7.2.2 Other effects of insufficient sampling

7.2.2a Aliasing

If you have ever watched a wagon wheel turn backwards in an old Hollywood western then you are familiar with a phenomenon called aliasing. Imagine marking a particular spot on the wagon wheel with a red dot. The camera filming the scene shoots one frame every sixteenth of a second. As the wagon wheel starts rolling, the red dot spins forwards, as we would expect. When the wheel reaches 4/3 revolutions per second (rps) the camera repeatedly catches the dot at 1 o'clock, 2 o'clock, 3 o'clock etc. in successive frames and in that order. The dot appears to be moving forward. As the wheel moves up to 4 rps, the camera catches the dot at 3 o'clock, 6 o'clock, 9 o'clock and 12 o'clock. To the eye, there appear to be four stationary dots on the wheel. If we speed the wheel up to 16 rps, the dot will appear to be stationary because the wheel makes a full revolution for every frame shot. If we slow the wheel down just a little, the dot appears to move backwards because in

Table 7.1 Effect of changing sampling density on variation in measured cell size. Results from simulation, 1000 runs at each density.

Density (pixels/unit)	Actual (pixels)	Mean (pixels)	Minimum (pixels)	Maximum (pixels)	Standard deviation	CV (%)
1	0.79	0.79	0	1	0.41	52.0
3	7.07	7.09	4	9	0.75	10.5
6	28.27	28.28	26	32	1.15	4.1
9	63.62	63.45	60	69	1.46	2.3
12	113.10	113.05	108	116	1.72	1.5

successive frames the dot does not quite make it all the way around to the position it was in during the previous frame. If we speed the wheel up so that it rotates slightly faster than 16 rps, the dot appears to move forward slowly, just as it would if the wheel were actually rotating at a very slow rate. Because of sampling, high rotational frequencies are aliased into low rotational frequencies. Aliasing is caused by insufficient temporal sampling (by the movie camera) of a rapidly changing scene (the wagon wheel).

Aliasing occurs with spatial frequencies as well. Consider the high frequency sinusoidal-shaped curve depicted in Figure 7.3. With a little imagination, this might represent cross-sections of uniformly placed cells on a microscope slide. If the cross-section (the sine curve) is sampled at low frequency, then the high frequency nature of the original curve has been completely lost. A reconstruction using the sample points can only yield a low frequency result. At the sample points, there is no way to tell high frequency from low frequency.

Sampling theory tells us that sampling must occur with at least twice the frequency of the smallest feature that needs to be detected in order to avoid aliasing. In light microscopy, this should be considered a starting point for determining the appropriate sampling density for a particular application. Because light microscopy is bandlimited, somewhat higher sampling frequencies may be needed in order to ensure detection of desired features (Harms and Aus, 1984; Young, 1988). In our work with bacteria using fluorescence microscopy, there is very little internal structure that can be discerned, thus we have been mainly concerned with sampling at sufficient intensity to detect the entire cell.

A more concrete example of aliasing is easily imaginable. If two cells are very close together, as depicted in the cross-section of Figure 7.4, then aliasing can cause the edge between the two cells to be missed entirely when the sampling frequency is too low. The result is detection of one larger cell rather than two smaller cells. Only

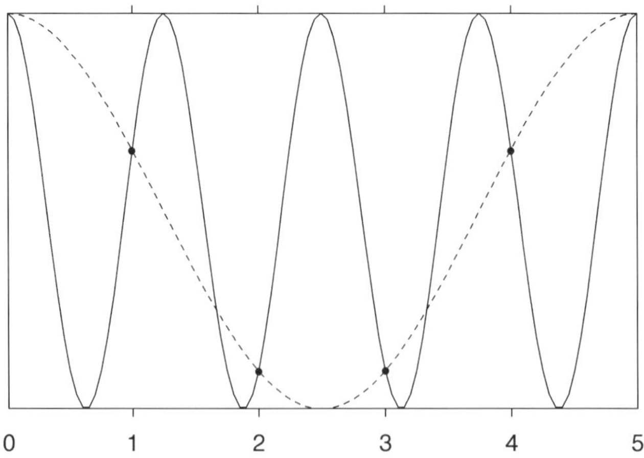

Figure 7.3 In this classic illustration of aliasing, two sine curves, one of high frequency and one of low frequency, are sampled at the same density. The sampling is at too low a frequency to capture the high frequency curve. With information only at the sample points (dots), the two curves are indistinguishable.

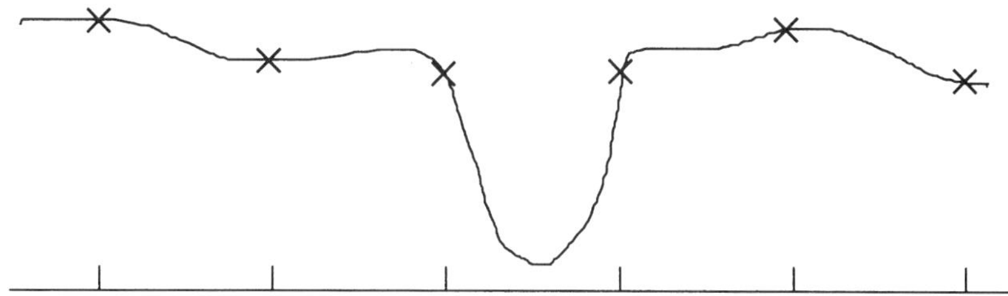

Figure 7.4 Features that are close to the same spatial frequency as the sampling frequency are lost through aliasing. In this cross-section, the boundary between adjacent cells has been lost. The resulting image will show a single cell.

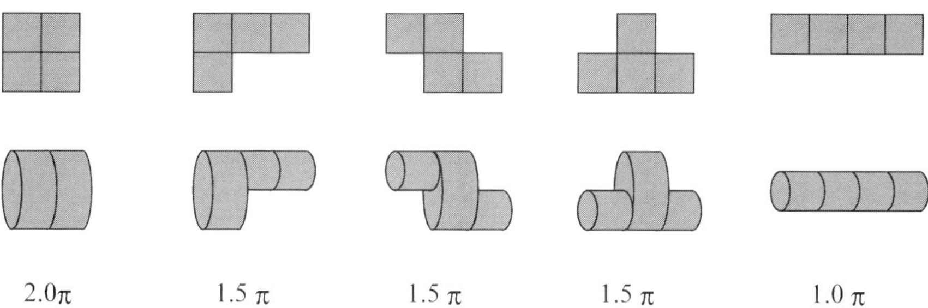

Figure 7.5 At very low sampling densities, cells belong to one of just a very few size classes. Here a 4-connected four-pixel cell can only belong to one of three volume size classes, no matter how the pixels are arranged.

when the sampling rate is increased can the border between the two cells be detected.

7.2.2b Discrete size classes

It is a fact of life that we must sometimes sample at less than desirable densities. In these cases, we must keep in mind how the now very discrete nature of the image affects derived measurements. For example, consider a four-pixel cell as depicted in Figure 7.5. There is only a very small number of ways in which four pixels can be arranged in a 4-connected image, thus there can only be a very small number of possibilities for size measurements derived from the image. For our biovolume algorithm (described later in this chapter), the five possible arrangements for a four-pixel 4-connected blob (discounting rotations and symmetry) yield only three distinct volume measurements. So in this case it would be incorrect (and misleading!) to show size classes with 0.25π or finer increments.

7.3 CELL ENUMERATION

The accurate counting of cells in digital images is a matter of cell detection, and detection of an object in a digital image is closely related to segmentation – the separation of the objects of interest from the rest of the image (background). Therefore, accurate counting (and sizing, for that matter) depends crucially on using an optimal segmentation method. These methods were reviewed in the previous chapter.

The simplest approach to counting particles in an image is the use of a global threshold. Most image analysis software systems now include simple particle counting algorithms based on single or multiple global thresholds. Global thresholds are applied to entire images, or across multiple images collected under constant optical conditions. Sometimes the cell counting is done on a binary image produced by the thresholding. Cell count, size and shape can be determined in a binary image, but the intensity information is lost. A binary image produced by a segmentation method can be used as a 'cookie cutter' or mask to extract only the pixels belonging to cells of interest from the original image. This is done by simply multiplying the binary and the original image. The result has a background of zero while the cells retain their original grey scale information. Often 'filters' can be set that limit the count to objects meeting specific size, shape or brightness criteria.

Common counting algorithms are based on counting the objects (i.e. groups of contiguous pixels) above or below a specified brightness threshold, or within a range of values. In a typical counting algorithm, the image array is scanned pixel-by-pixel in a left to right, top to bottom order. When a pixel above (for fluorescence images) or below (for transmitted light images) is encountered that was not previously seen, it is marked as detected. Then either the 4- or 8-connected pixels adjacent to the 'seed' pixel are examined. This is often done in a recursive way until all of the pixels making up the cell are counted and marked. The algorithm then returns to the seed pixel and continues through the image.

Regardless of the method used to segment the image, additional data can be obtained as each pixel within the cell of interest is examined. Simply counting the pixels will yield the cell area. Calculating the sum and the sum of squares of the pixel brightness values can be used to calculate the mean and variance of the pixel brightnesses making up each cell. The (x, y) positions of the pixels can be used to calculate cell shape factors such as perimeter, length, width and orientation. In colour images, the brightness values in red, green and blue colour space can be measured.

7.4 PIXEL CALIBRATION

In order to convert spatial pixel measurements into absolute units a conversion factor needs to be determined. The most straightforward way to do this is with a stage micrometer. Stage micrometers are commonly available that have gradation marks every 10 or 100 μm. For each optical configuration of the microscope imaging system (especially for each objective lens used) one or several images of the micrometer scales should be taken. Then count the number of pixels between successive lines. If your camera is aligned with the microscope you can simply read off the x positions in the image. Since the actual edges of the micrometer gradations are not sharp, especially at high magnifications, it is best to take counts between multiple (successive) gradations and average them.

7.5 SHAPE MEASUREMENT

Morphometric parameters can be calculated from the previous measures such as the ratio of length to width and circularity *C*:

$$C = \frac{P}{2\sqrt{\pi A}} \tag{7.2}$$

Circularity is the ratio of the observed perimeter *P* to the perimeter of a circle with the same area *A* as that measured. In this formulation a circle has a circularity of 1 and other shapes have *C* larger than 1.

Another type of shape factor derived from a two-dimensional image of a cell is the moment of inertia (Horn, 1986). This is essentially a calculation of the variance of all of the pixel positions in the *x* direction, *y* direction or a combination of the two. For example, the moment of inertia in the *x* direction is calculated as follows:

$$m_x = \sum_{i_{\min}}^{i_{\max}} x_i^2 - x_c \sum_{i_{\min}}^{i_{\max}} x_i \tag{7.3}$$

where *x* is the *x* position of each pixel making up the cell, x_c is the centre *x* position (i.e. the mean of the *x* values), and the summation is over only those pixels that are part of the cell.

If this is calculated after the cells have been rotated so that their longest axis is horizontal, then the *x* moment relates to the cell length and the *y* moment is related to the cell width. The moments, however, contain the spatial information from all of the pixels while length and width information is derived only from the most extreme pixels. The moments can also be weighted by the pixel intensity values (Section 1.8.2b).

For problems such as automated classification of larger plankton cells with relatively complex shapes with taxonomic meaning, it may be necessary to use frequency domain analysis to discriminate cell types accurately. Culverhouse *et al.* (1996) have successfully used two-dimensional fast Fourier transformation to yield classification parameters used as inputs to a neural network for this purpose. Detailed description of these frequency domain methods is beyond the scope of this chapter but can be found elsewhere (e.g. Bracewell, 1986).

7.6 INTENSITY

Intensity (brightness) statistics can include the minimum and maximum pixel brightness, the mean and variance of the brightness values, and the sum of the brightness values of all of the pixels making up a cell. In colour images analogous data can be collected for colour parameters in either RGB space or other colour representations. Chapters 10 and 11 deal with intensity measurement in detail.

7.7 BIOVOLUME ESTIMATION

Cell biovolume is of interest because biovolume measurements in conjunction with cell counts and appropriate N or C per unit volume conversion factors can give better estimates of cell biomass than other methods. Accurate estimates of cell biomass are required for many applications, including ecological modelling and analysis of nutrient fluxes. In image analysis, there are at least three methods to obtain biovolume. Two require access to a two-dimensional image, and the third requires a three-dimensional image as might be obtained using digital optical section microscopy (Chapter 3) or confocal laser scanning microscopy (CLSM, Chapters 3, 16 and 17).

7.7.1 Volume from shape formulas

Of the two-dimensional methods, the easiest to implement is one where volume is calculated from simple length and width measurements and the application of simple geometrical formulas. The particular formula used depends on the cell population being examined. Typical formulas include a prolate spheroid:

$$\frac{\pi}{6} W^2 L \tag{7.4}$$

and a cylinder with hemispherical ends:

$$\frac{\pi}{4}(L - W)W^2 + \frac{\pi}{6} W^3 \tag{7.5}$$

where L is length and W width.

There are a variety of methods by which the length and width of cells can be calculated. In an interactive mode the user can select points representing the ends of the cells along the longest axis for the length and points representing a perpendicular line for the width. Many commercially available software packages will calculate the distances between two points selected in this way. Automatic measurement of length and width requires that the cell orientation relative to the sampling grid (pixels) be taken into account. In the method for biovolume (described in Section 7.7.2), we show how to calculate the orientation of the axis of elongation. Once this is known the perimeter points can be rotated so that the axis is horizontal, then the length and width are simply the difference between the maximum and minimum x and y values, respectively.

The use of simple geometrical formulas for volume is essentially the same as the method used in visual cell measurement (e.g. Edler, 1979). The main problem with using formulas is that they do not take the natural shape variations in real cell populations into account. To the extent that a cell is not shaped like a circle, ellipse etc., the measurement is inaccurate.

7.7.2 Volume from discrete integration

A second two-dimensional technique rectifies this problem in the following way. Instead of using simple formulas, the algorithm estimates volume by stepping along the long axis of a cell and measuring the pixel thick 'discs' that comprise a three-dimensional projection of the cell. The width of each disc is determined by measuring parallel to the minor axis of the cell for each pixel step along the long axis.

The first step in the solid of revolution technique is to find all perimeter points of the cell. We assume that the image has already been segmented (see Chapter 6) and that cells are represented as 4-connected objects above a threshold. The perimeter points are generated by starting at a seed point on the cell's border and running an edge-following algorithm (e.g. Ballard and Brown, 1982; see also Section 13.4.5), adding each perimeter point to an array as the point is encountered.

We wish to rotate the cell's edge to a standard orientation to do the volume measurement. To do so, we need the cell's centroid, and an angle that represents the orientation of the cell's axis of elongation (Horn, 1986). The centroid, denoted (x_c, y_c), is easily obtained from the individual pixels that comprise the cell by summing the x and y coordinates, respectively, and dividing each sum by the number of pixels in the cell. The angle of the axis of elongation θ can be obtained from the cell's second moments using the following equation:

$$\tan 2\theta = \frac{b}{a - c} \tag{7.6a}$$

where a, b and c are the second moments defined as:

$$a = \sum (x')^2,\ b = \sum x'y',\ c = \sum (y')^2 \tag{7.6b}$$

The summation is over all pixels in the cell, and $x' = x - x_c$ and $y' = y - y_c$. Once the perimeter and angle of the axis of elongation have been obtained, we can rotate each point in the perimeter array to a standard position that represents the cell lying horizontally, its axis of elongation now parallel to the x axis. The rotation equations depend upon the coordinate system in use. For coordinate systems with the origin at the top left and x and y positive going to the right and down, respectively, then the transformations are:

$$x_{\text{rotated}} = x_c + (x - x_c)\cos(-\theta) + (y - y_c)\sin(-\theta) \tag{7.7a}$$

and

$$y_{\text{rotated}} = y_c - (x - x_c)\sin(-\theta) + (y - y_c)\cos(-\theta) \tag{7.7b}$$

This new perimeter array may have 'holes' caused by the rotation. A simple way to remedy this is to traverse the array looking at adjacent elements. If the difference in the x coordinates is large enough to create a hole, then a new perimeter point can be added to fill the hole. Since the cell is now parallel to the x axis, the discs that correspond to the digital solid of revolution can be found by searching the rotated array for the points with the minimum and maximum y extent for each x coordinate that appears in the perimeter array. Further details of how this can be accomplished are given in Sieracki *et al.* (1989b).

The above technique amounts to a digital solid of revolution around the cell's long axis (Figure 7.6). The attraction to this algorithm is that it automatically accounts for the natural undulations and shape variability that are to be expected in cell populations. Verification of the algorithm shows that it performs as well as other algorithms based on simple formulas for cells of simple shapes, and is drastically better for more complex shapes (see Table 7.2).

7.7.3 Volume using three-dimensional techniques

Both two-dimensional algorithms make an important assumption – the object being imaged has simple geometric symmetry in the third dimension. Clearly not all cell populations will obey this assumption. To the extent they do not, volume estimates are biased. In systems that can form three-dimensional images (Verity *et al.*, 1996) like those using digital optical section microscopy, the above drawback is not present. These systems determine volume by simply counting the number of 'voxels' (three-dimensional pixels) that make up the image. After simple conversions to account for voxel asymmetry, a volume estimate is obtained. The main drawback to this method is the high degree of operator intervention required to form the three-dimensional image. As a result, sample throughput is very low. The issue of thresholding the three-dimensional image to get the three-dimensional object still remains.

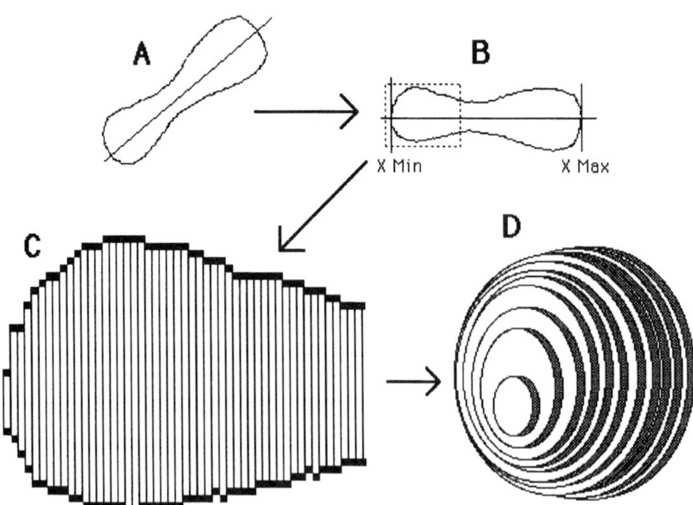

Figure 7.6 Steps taken in our algorithm to estimate biovolume from a two-dimensional image. The digital perimeter of a cell (A) is rotated so that its axis of elongation is horizontal (B). The one-pixel wide slices of the cell (C) are considered to be thin cylinders (D) and their volumes are calculated and summed. Reprinted from Sieracki and Webb (1991), by permission of Springer-Verlag.

Table 7.2 Comparison of three methods that use a single two-dimensional image to obtain biovolume. All three methods do well on the simple shapes represented by straight wire, rice and pasta grains. However, the integration technique works much better than either the prolate spheroid or cylinder method on non-uniform shaped objects (bent wire). All volumes are in cubic millimetres. Data taken from Sieracki *et al.* (1989b).

Object	Sample size	Direct measure	Integration	Prolate spheroid	Cylinder
Straight wire	30	73.6	68.6	64.8	86.9
Pasta	30	29.1	27.3	31.3	40.7
Rice	30	12.7	12.3	12.8	17.1
Bent wire	10	100.3	98.7	158.5	208.5

7.8 VALIDATION OF CELL COUNTING AND SIZING

Cell counts by standard visual microscopy methods yield CVs on the order of 10–20% (Kirchman *et al.*, 1982). The accuracy of the count depends on the uniformity of the cell dispersion on the microscope slide, the variation of cell types being enumerated, the amount of foreign material in the sample and the number of cells counted. Automated methods can increase the number of cells counted per sample, thereby increasing the precision of the cell count. The other factors will still be present, however. Typically automated counting by image analysis is validated by comparing repeated counts with traditional visual methods (Sieracki *et al.*, 1985; Bjørnsen, 1986; Bloem *et al.*, 1995). More accurate and precise methods such as Coulter counters or flow cytometry might be better ways of validating the counts obtained by image analysis. The sample type, however, must be appropriate for the instruments used. Fluorescent microbeads, for example, are easy to count by many methods. Similarly, pure cultures are usually counted more easily than natural samples with mixed assemblages. For the researcher in a particular field it is most convincing to use real samples, rather than a surrogate that works well with various methods.

Validation of cell sizing is also important with an automated system. The most common approach is to use fluorescing latex microspheres, which are produced with very high precision in size, especially for spheres less than 10 μm in diameter (Bjørnsen, 1986; Sieracki *et al.*, 1989a). The problem of accurately sizing fluorescing cells from a mixed assemblage is largely a problem of accurate segmentation. Such cells usually have a wide variation in fluorescence intensity and size. Fluorescence and size are correlated, so a method is necessary that can accurately size cells that are small and dim or large and bright, as well as the occasional small, bright and large, dim cells. Since we found fluorescent microspheres to be considerably brighter than most cells, we used a set of microspheres with a range of fluorescence intensities, and a neutral density filter in the emission path of the microscope so that their images simulated the images of the cells we intended to measure (Sieracki *et al.*, 1989a).

Standard fluorescent microspheres, however, are different in many ways from cells in their optical properties. As mentioned, their fluorescence is usually much brighter than that of stained cells. They also have different light scattering properties that are important when using them with flow cytometry (described below). Also, unlike cells, they are nearly perfect spheres. Microspheres can be very useful for standardization purposes and for monitoring system performance. The fact that a system can measure microspheres very accurately and precisely, however, does not guarantee that it will do well with a particular cell type or cells from particular samples. As previously discussed for validation of cell counting, it is best to demonstrate that a system can measure the particular cell types of interest. The difficulty here is that the methods for cell size measurement are not very good, and all have considerable sources of error. The standard methods include: (i) visual microscopy using an ocular micrometer to measure cell dimensions, (ii) Coulter sizing by electrical impedance (Kachel, 1990) and (iii) flow cytometry sizing by forward-angle light scatter (Salzman *et al.*, 1990). It is probably best to choose a validation method that has been used traditionally in your particular application. As one example, we used two visual methods: measuring cells by ocular micrometer with the microscope only, and using an interactive threshold on the display of a digital image to obtain the 'best' outline of each cell (Sieracki *et al.*, 1989a).

Verification of volume calculating algorithms is very important and can require some ingenuity. Direct verification is best but cannot always be done. For example, to verify the two-dimensional solid of revolution technique (Sieracki *et al.*, 1989b), we used objects like rice grains and bent pieces of solder. This allowed us to determine volume directly through water displacement and compare these estimates to those obtained through image analysis. If volume measurements are only going to be used as an intermediate step towards calculating, for example, biomass, then biased volume estimates might be usable if the bias is accounted for in the volume to mass conversion. For example, if carbon biomass is measured directly for a particular cell type and volume estimation technique, then the associated mass/volume ratio (i.e. carbon density) might be used for subsequent measurements of that cell type without the more expensive direct biomass measurement (Fagerbakke *et al.*, 1996).

7.9 PROBLEMATIC CONDITIONS

Some of the problematic conditions that can confound measurements include artefacts related to the imaging system, as well as those related to sample preparation and the natural variation of cell populations.

7.9.1 Asymmetric pixels

In some imaging systems, particularly older video-based systems, sampling is performed at higher densities along the vertical axis than the horizontal axis. Video cameras producing NTSC colour or RS-170 monochrome generate frames with a 5 : 4 or 4 : 3 horizontal/vertical aspect ratio. Digitizing this stream without adjustment yields asymmetric pixels, which are longer horizontally than vertically. When we say the pixels are asymmetric, we mean that they represent an area in the object plane that has unequal horizontal and vertical dimensions. Charge-coupled device

(CCD) based camera systems that sample the detectors directly do not normally have this problem.

Imaging systems with asymmetric pixels must be calibrated to yield accurate results. Using a single pixel-to-length conversion factor in these situations is a grave mistake in that simple length estimates can be off by as much as 25% based solely on the orientation of the object (see Figure 7.7). Such errors are compounded in multidimensional calculations like biovolume.

It is relatively easy to check and correct for asymmetric pixels. By measuring a calibrated slide both vertically and horizontally and counting the number of pixels along a known length, an accurate ratio can be calculated. This ratio must then be used in all calculations that involve the size of the cell, including area, length, width, volume and orientation. Operations like rotations that use the object's spatial information are also affected.

For example, if a calibration indicates that the vertical to horizontal ratio is 0.8, then area measurements must be corrected as in the following pseudo-code:

```
YCONV=0.8;
actual_area=pixel_count*YCONV*horizontal_pixels_per_mcm;
```

An alternative method to correct for asymmetric pixels is to remap the image to a coordinate system with symmetric pixels. If this is done before measurement, then corrections for subsequent measurements are unnecessary.

7.9.2 Dividing and overlapping cells

Most slide preparations yield cells that are randomly positioned and orientated. This can cause cells to be overlapping or touching in the two-dimensional image. Cells can also appear to be overlapping or touching if they are in the process of dividing or have recently divided. In pure cultures, as well as in natural assemblages, the frequency of dividing bacterial cells has been used as a quantitative indicator of cell growth rate (Hagström *et al.*, 1979), so it can be useful to discriminate dividing from single cells in a sample. The frequency of

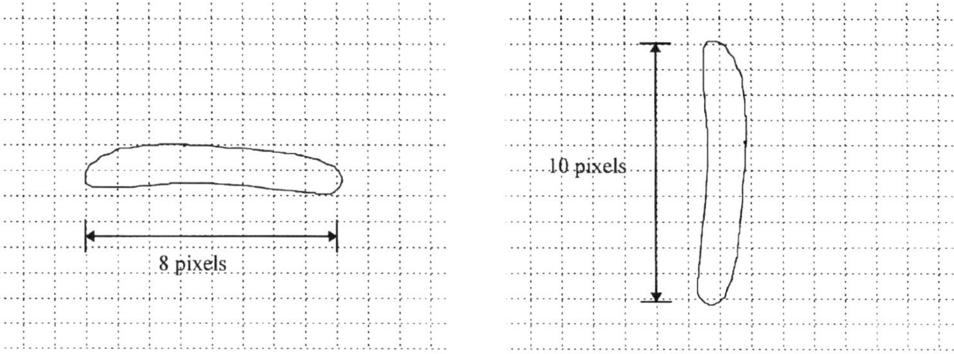

Figure 7.7 With asymmetric pixels, measurements can be different depending upon the orientation of the object. The asymmetry must be corrected for in software.

overlapping cells can be reduced by diluting a dense sample before preparing slides.

One method (Møller *et al.*, 1995) to discriminate dividing cells automatically is based on the conditions that: (i) there is a constriction in cell outline and/or (ii) there is a valley in the grey level values. After segmentation and rotation of each object (equations 7.6 and 7.7), so that the longest axis is parallel to the x axis, all the pixels of a cell are summed in the y direction. The summed ys are examined over the range of xs. Dividing cells are those that are oblong and have a minimum in the plot of summed ys as a function of x (Figure 7.8). Møller *et al.* (1995) have implemented this approach in a software system called Cellstat.

Another approach was reported by Bloem *et al.* (1995), who described a fully automated system for counting and measuring bacteria from soil, a difficult type of sample to work with. Dividing cells are determined in segmented cells as those with multiple local maxima within them. A simple, non-automatic, approximate way to handle double or dividing cells is to label the data file manually post-analysis (e.g. Chapter 11). Then the cell count can be incremented and the cell size/biomass halved for each one encountered.

7.9.3 Cells crossing image edges

When samples are analysed for cell counts and sizing, random microscope fields of the slide are typically chosen for digitization. Since the cells are randomly located on the slide, this results in occasional cells that are intersected by the image edge and are only partially in the image. Counting and sizing such partial cells will obviously bias both the size measurements and the cell counts. There are several

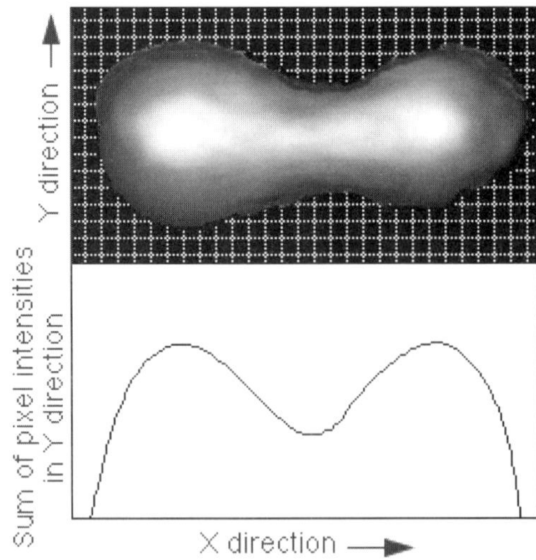

Figure 7.8 Automated determinations of dividing cells are made by summing the pixel intensities in the y direction, and then looking for a minimum in the sum plotted over the x direction. The presence of a distinct minimum in an oblong cell indicates that the cell is dividing.

approaches to ameliorating this potential source of error. One way is to count only cells that intersect two of the four image edges, such as the top and right. Since larger cells have a higher likelihood of being intersected by the image edges, however, this method has a bias against the larger cells. And the size of cells not completely in the image is not known.

Another approach is to consider the image measurement area to be only those pixels inside a specified margin around the image (Figure 7.9). The margin should be at least the length of the expected cell size. Gunderson (1974) uses this approach, then counts and measures only cells that intersect two edges of the margin (e.g. the top and right edges). Using mathematical morphology representations (see Section 6.9) this method can be computationally efficient. An alternative method is to calculate the centroid of each cell and then count only those cells whose centroids fall within the margins. This method, however, requires that the entire outlines of the cells that cross the margins are within the digitized image so that the centroid can be calculated. This method is useful when the centre of mass is calculated anyway for other cell measurements.

7.9.4 Detecting rare events

There are several applications that require the detection of a very few 'positive' cells in a sample that is either very sparse of cells or contains many other cell types. An example of the first case is the detection of the pathogens *Giardia* and *Cryptosporidia* in drinking water (Nieminski *et al.*, 1995). Detection of rare cancer cells among large numbers of normal cells is an example of the latter application. Mesker *et al.* (1994) describe a system for the automated detection of particular cancer cells in blood. The system uses an automatic microscope stage and employs multiple imaging

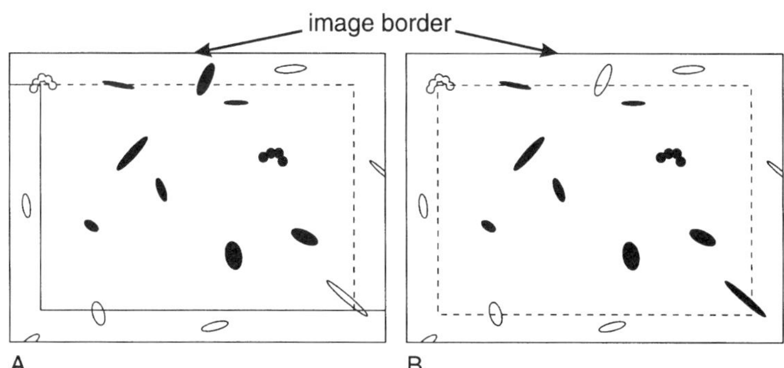

Figure 7.9 Methods for avoiding errors associated with cells crossing the edges of images. Only the black objects are counted. In method A (Gunderson) the image is split into three regions: (i) the 'don't care' region (top/right), (ii) the count frame (middle) and (iii) the 'reject' region (bottom/left). Objects that intersect the count frame (centre area), *and* do not intersect the 'reject' region are counted. In method B only objects that have a centre of mass inside the count frame are counted. In both methods, if the images were spaced regularly over the slide, so that count frames were adjacent and cover the whole surface area, all objects would be counted *exactly* once. Hence there is no counting bias. Sizing bias is only zero if the margin between count frame and image border is larger than or equal to the largest cell diameter.

cameras for different magnifications (2× objective and 1/2× objective) of the same field of view. For a well-defined cell system the instrument is able to find one positive cell in 10^6 negatives. The authors note that preparation of the immunocytochemically stained cells is critical. Imaging has the advantage over flow cytometry in this application, in that it allows visual verification of potentially positive cells and it allows the rare cells to be relocated after analysis since the position on the slide is recorded.

7.10 COMPARISON OF FLOW AND IMAGING CYTOMETRY

Another technology that is used to measure parameters about individual cells is flow cytometry. This approach uses complex instrumentation, including fluid flow control, laser optics, photodetectors, signal processing and computers to measure cells. A sample of fluid containing the cells in suspension is hydrodynamically focused into a thin stream so that the cells pass in single file. The sample stream intersects an interrogation region where light from the laser is scattered by the passing cells and excites fluorescence. The scattered light and emitted fluorescence are measured by light detectors focused on the interrogation region. These signals are then digitized, stored and displayed in real time on a computer display. In general, forward-angle light scatter is related to cell size. Light scattered at right angles to the incident beam is influenced by the particle refractive index and granularity, in addition to size. Fluorescence emission can be measured for a number of spectral bands using dichroic mirrors and bandpass filters. There are several good sources for further information on flow cytometric principles at both the introductory (Givan, 1992) and more advanced (Shapiro, 1988; Melamed *et al.*, 1990) levels.

Flow cytometers are available at most large universities and research hospitals so investigators may have the choice of using either flow or imaging cytometry for their particular application. There are some considerations in making this choice. Table 7.3 summarizes the similarities and differences between these two approaches to cytometry, and some of the major points are discussed here. If measurements are to be made on living cells, such as the application of fluorescent indicators of physiological states of the cells, then it is important to consider the amount of manipulation disturbance of the cells in the measurement process. If cells are growing in suspension, then flow cytometry might be advantageous. If the cells are growing on a solid substrate, they can be observed under a microscope with minimal disturbance. Another important consideration is the detail of information needed about each cell. Flow cytometry can measure only a limited number of optical parameters – typically forward and side angle scatter and three or four fluorescence bands (occasionally more). Many flow cytometers use a wide-angle forward-light scatter detection system, which is equivalent to forward scatter *per se*, but does not necessarily relate directly to size. For applications requiring more detailed cell information such as morphology, intracellular localization of fluorescence or measurements on cell organelles, imaging cytometry is more appropriate. Another important distinction is that flow cytometry can measure cells at a much higher rate than is usually attainable using imaging cytometry. This high

Table 7.3 Comparison of the features and measurements of flow and imaging cytometry.

Feature or measurement	Flow cytometry	Imaging cytometry
Sample preparation	Live samples or simple fixation and cryostorage	Fixed samples, fluorochrome staining and slide preparation
Physical cell handling	Cells are moving rapidly in a flow stream	Cells are fixed on a filter; individual cells can be revisited
Fluorescence excitation	By specific laser lines or arc lamp with filter sets	Excitation by arc lamp and filter sets, multiple excitations easy, laser for confocal systems
Analysis rate	10^4–10^6 cells per minute	1–100 cells per minute
Data per cell	3–6 data points per cell	8–1000 pixels per cell yield 10–20 cell parameters
Cell count precision	High	Medium
Cell count accuracy	Determined by errors in measuring volume analysed	Accuracy dependent primarily on simple volumetric measurement
Cell size	Indirectly determined from light scatter (or Coulter volume)	Size measured from two-dimensional image
Cell shape	Indirectly measured from scatter (nonlinear relationship)	Shape factors measured from two-dimensional image
Intracellular resolution	Intracellular arrangements not resolvable	Organelles (nucleus, chloroplasts, vacuoles) can be resolved
Images of cells	Imaging in flow possible	Visual verification of cell/particle detection
Sorting	Yes	No

rate, allowing more cells from a sample to be counted, makes a significantly higher cell count precision practical (Monger and Landry, 1993). Cell sizing is generally more accurate using imaging cytometry than by flow cytometry since size is measured directly from the two-dimensional image.

For our own work on natural populations of planktonic cells in the oceans we use imaging cytometry as a validation method of the flow cytometric analyses (Sieracki *et al.*, 1995). Imaging is used to verify cell identifications and to calibrate the forward scatter signal to biovolume. Imaging cytometry is also more practical for analysis of rarer cells, such as protozoa and larger phytoplankton, which exist in nature at concentrations below a few hundred per millilitre. This is because samples can easily be concentrated on a membrane filter for image analysis.

7.11 FUTURE DIRECTIONS

The vast improvements in digital technology appear to be continuing, with more powerful computers and better imaging systems becoming less costly and more widely available. As more scientists gain access to the hardware and software of image analysis it will be used in more varied applications. Several approaches warrant consideration in a discussion of future directions. The merging of the techniques of imaging and flow cytometry into a single instrument has already begun in the field of plankton ecology (Hüller *et al.*, 1994; Sieracki and Sieracki, 1996). This will be especially important for larger cells and multicellular organisms, where images can be extremely useful for identification. Automated classification of organisms from digital images is another application where advances can be made in the near future. The reduction of image information to key parameters for recognition and discrimination is necessary. An example application is the automated detection of algal species with the potential to produce toxins that can enter the human food supply. Culverhouse *et al.* (1996) show that images of over twenty dinoflagellate species (some toxic) can be discriminated accurately using a neural net classifier with two-dimensional Fourier components of the images as inputs.

The application of fluorescent 'phylogenetic stains', oligonucleotides that label species- or group-specific gene sequences, will allow identification of specific cell types from natural mixed assemblages (DeLong *et al.*, 1989; Britschgi and Giovannoni, 1991). Fluorescent probes for ionic states, membrane gradients and other physiological indicators of living cells are also rapidly being developed in the biomedical field (e.g. Wang and Taylor, 1989; Waggoner, 1990; see also Chapters 4 and 12) and will find application in microbiology. These techniques will require image analysis of the fluorescent markers since the fluorescence is often very faint and only cooled CCD imagers can detect them accurately.

ACKNOWLEDGEMENT

This work was partly funded by research grants (to M.E.S.) OCE-9102154 and OCE-9423535 from the US National Science Foundation.

REFERENCES

Ballard, D.H. and Brown, C.M. (1982) *Computer Vision*. Prentice Hall: Englewood Cliffs, NJ.

Bjørnsen, P.K. (1986) Automatic determination of bacterioplankton biomass by image analysis. *Appl. Environ. Microbiol.* **51**: 1199–204.

Bloem, J., Veninga, M. and Shepherd, J. (1995) Fully automatic determination of soil bacterium numbers, cell volumes, and frequencies of dividing cells by confocal laser scanning microscopy and image analysis. *Appl. Environ. Microbiol.* **61**: 926–36.

Bracewell, R.N. (1986) *The Fourier Transform and its Applications*, 2nd edn, revised. McGraw-Hill: New York.

Britschgi, T.B. and Giovanonni, S.J. (1991) Phylogenetic analysis of a natural marine bacterioplankton population by rRNA gene cloning and sequencing. *Appl. Environ. Microbiol.* **57**: 1707–13.

Culverhouse, P.F., Simpson, R.G., Ellis, R., Lindley, J.A. , Williams, R., Parisini, T., Reguera, B., Bravo, I., Zoppoli, R., Earnshaw, G., McCall, H. and Smith, G. (1996) Automatic classification of field-collected dinoflagellates by artificial neural network. *Marine Ecol.-Prog. Ser.* **139**: 281–7.

DeLong, E.F., Wickham, G.S. and Pace, N.R. (1989) Phylogenetic stains: ribosomal RNA-based probes for the identification of single cells. *Science* **243**: 1360–3.

Edler, L. (1979) Recommendations on methods for marine biological studies in the Baltic Sea: phytoplankton and chlorophyll. *The Baltic Marine Biologist*, no. 5.

Fagerbakke, K.M., Heldal, M. and Norland, S. (1996) Content of carbon, nitrogen, oxygen, sulfur and phosphorus in native aquatic and cultured bacteria. *Aquatic Microbial Ecol.* **10**: 15–27.

Findlay, R.H. and Dobbs, F.C. (1993) Quantititive description of microbial communities using lipid analysis. In *Handbook of Methods in Aquatic Microbial Ecology* (Kemp, P.F., Sherr, B.F., Sherr, E.B. and Cole, J.J., eds), pp. 347–58. Lewis Publishers: Boca Raton, FL.

Givan, A.L. (1992) *Flow Cytometry: First Principles.* Wiley-Liss: New York.

Gunderson, H.J.G. (1974) Simple counting rules for eliminating edge effects in planar sampling. In *Quantitative Analysis of Microstructure in Biology, and Medicine Materials Development* (Exner, H. ed.). Riederer: Stuttgart.

Hagström, Å., Larsson, V., Horstedt, P. and Normark, S. (1979) Frequency of dividing cells, a new approach to the determination of bacterial growth rates in aquatic environments. *Appl. Environ. Microbiol.* **37**: 805–12.

Hamming, R.W. (1983) *Digital Filters*, 2nd edn. Prentice Hall: Englewood Cliffs, NJ.

Harms, H. and Aus, H.M. (1984) Estimation of sampling errors in a high-resolution TV microscope image-processing system. *Cytometry* **5**: 228–35.

Horn, B.K.P. (1986) *Robot Vision.* MIT Press: Boston, MA.

Hüller, R., Glossner, E., Schaub, S., Weingärtner, J. and Kachel, V. (1994) The macro flow planktometer: a new device for volume and fluorescence analysis of macroplankton including triggered video imaging in flow. *Cytometry* **17**: 109–18.

Kachel, V. (1990) Electronic resistance pulse sizing: Coulter sizing. In *Flow Cytometry and Cell Sorting* (Melamed, M.R., Lindmo, T. and Mendelsohn, M.L., eds), pp. 45–80. Wiley-Liss: New York.

Karl, D.M. (1993) Total microbial biomass estimation derived from the measurement of particulate adenosine-5′-triphosphate. In *Handbook of Methods in Aquatic Microbial Ecology* (Kemp, P.F., Sherr, B.F., Sherr, E.B. and Cole, J.J., eds), pp. 483–94. Lewis Publishers: Boca Raton, FL.

Kirchman, D., Sigda, J., Kapuscinski, R. and Mitchell, R. (1982) Statistical analysis of the direct count method for enumerating bacteria. *Appl. Environ. Microbiol.* **44**: 376–82.

Melamed, M.R., Lindmo, T. and Mendelsohn, M.L. (1990) *Flow Cytometry and Cell Sorting.* Wiley-Liss: New York.

Mesker, W.E., van der Burg, M.J.M., Oud, P.S., Knepfle, C.F.H.M., Ouweker-v. Vlezen, M.C.M., Schipper, N.W. and Tanke, H.J. (1994) Detection of immunocytochemically stained rare events using image analysis. *Cytometry* **17**: 209–15.

Møller, S., Kristensen, C.S., Poulsen, L.K., Carstensen, J.M. and Molin, S. (1995) Bacterial growth on surfaces: automated image analysis for quantification of growth rate-related parameters. *Appl. Environ. Microbiol.* **61**: 741–8.

Monger, B.C. and Landry, M. (1993) Flow cytometric analysis of marine bacteria with Hoechst 33342. *Appl. Environ. Microbiol.* **59**: 905–11.

Nieminski, E.C., Schaefer, F.W. and Ongerth, J.E. (1995) Comparison of two methods for detection of *Giardia* cysts and *Cryptosporidium* oocysts in water. *Appl. Environ. Microbiol.* **61**: 1714–19.

Park, S.K., Schowengerdt, R. and Kaczynski, M. (1984) Modulation-transfer-function analysis for sampled image systems. *Appl. Opt.* **23**: 2572–82.

Pratt, W.K. (1991) *Digital Image Processing*, 2nd edn. Wiley: New York.

Salzman, G.C., Singham, S.B., Johnston, R.G. and Bohren, C.F. (1990) Light scattering and cytometry. In *Flow Cytometry and Cell Sorting* (Melamed, M.R., Lindmo, T. and Mendelsohn, M.L., eds), pp. 81–108. Wiley-Liss: New York.

Shapiro, H.M. (1988) *Practical Flow Cytometry*, 2nd edn. Alan R. Liss: New York.

Sieracki, C.K. and Sieracki, M.E. (1996) A high throughput volume particle in-flow imaging system. *Proc. SPIE* (*Ocean Optics XIII*, Ackleson, S.G. and Frouin, R., eds), **2963**: 886–91.

Sieracki, M.E. and Webb, K.L. (1991) Applications of image analyzed fluorescence microscopy for characterizing planktonic protist communities. In *Protozoa and their Role in Marine Processes* (Reid, P.C., Turley, C.M. and Burkill, P.H., eds), pp. 77–100. Springer-Verlag: Berlin.

Sieracki, M.E., Johnson, P.W. and Sieburth, J.McN. (1985) The detection, enumeration and sizing of aquatic bacteria by image analyzed epifluorescence microscopy. *Appl. Environ. Microbiol.* **49**: 799–810.

Sieracki, M.E., Reichenbach, S.E. and Webb, K.L. (1989a) Evaluation of automated threshold selection methods for accurately sizing microscopic fluorescent cells by image analysis. *Appl. Environ. Microbiol.* **55**: 2762–72.

Sieracki, M.E., Webb, K.L. and Viles, C.L. (1989b) Algorithm to estimate cell biovolume using image analyzed microscopy. *Cytometry* **10**: 551–7.
Sieracki, M.E., Haugen, E. and Cucci, T.L. (1995) Overestimation of heterotrophic bacteria in the Sargasso Sea: direct evidence by flow and imaging cytometry. *Deep-Sea Res.* **42**: 1399–1409.
Verity, P.G., Beatty, T.M. and Williams, S.C. (1996) Visualization and quantification of plankton and detritus using digital confocal microscopy. *Aquatic Microb. Ecol.* **10**: 55–67.
Waggoner, A.S. (1990) Fluorescent probes for cytometry. In *Flow Cytometry and Cell Sorting* (Melamed, M.R., Lindmo, T. and Mendelsohn, M.L., eds), pp. 209–26. Wiley-Liss: New York.
Wang, Y. and Taylor, D.L. (1989) *Fluorescence Microscopy of Living Cells in Culture. Part A. Fluorescent Analogs, Labeling Cells, and Basic Microscopy.* Academic Press: New York.
Young, I.T. (1988) Sampling density and quantitative microscopy. *Anal. Quant. Cytol. Histol.* **10**: 269–75.

APPENDIX: PROTOCOL FOR PICOPLANKTON BIOMASS ANALYSIS

The following method for slide preparation for picoplankton analysis is based on the method of Porter and Feig (1980). There are several sources for alternative methods, including Kemp *et al.* (1993) and a review by Kepner and Pratt (1994). Most of these are variations and modifications of the basic idea.

Protocol A.1 Picoplankton biomass analysis.

Reagents:
- Glutaraldehyde 50% (v/v)
- 4′,6-Diamidino-2-phenylindole (DAPI), working solution 250 μg/ml

Supplies:
- 0.2 μm pore size polycarbonate filters, 25 mm, black
- 25 mm cellulose acetate backing filters
- Glass microscope slides
- No. 1 cover slips
- Immersion oil
- Slide boxes

Equipment:
- Vacuum pump
- Filter holder set, including fritted glass base, funnel and clamp
- Filter manifold or side-arm flask
- Pipetters and tips
- Forceps
- Timer

Procedure

1 Preserve and fix 100 ml of sample by adding 0.6 ml of 50% stock glutaraldehyde to yield 0.3% final concentration. Place samples in refrigerator for at least one hour of fixation before processing.
2 Mount filters in filter holders with a pre-wetted cellulose acetate backing filter under the polycarbonate filter.

3 Measure the appropriate volume of sample (see Notes below) into the filter funnel.
4 Add 20 μl of the DAPI working solution per millilitre of sample.
5 Allow staining for at least four minutes.
6 Apply vacuum and filter just to dryness.
7 Carefully lift polycarbonate filter off with forceps.
8 Breathe on slide to moisten it slightly.
9 Place filter on slide so that it is flat.
10 Place one small drop of immersion oil directly onto the filter and place cover slip on top.
11 Place slide flat in the dark for several minutes so that the oil spreads over the filter.
12 Place slide vertically in slide box and place in freezer immediately.
13 Blank slides should be prepared periodically to test for bacterial contamination of fixative, stains, pipettes or glassware. In this case substitute 0.2 μm filtered seawater or distilled water for the sample and follow all steps as above.

Sample analysis

1 Thaw slide at room temperature.
2 Scan under epifluorescence illumination at intermediate magnification (e.g. 400×) to ensure sample is evenly distributed over the filter area.
3 Switch to 1000× magnification and randomly choose microscope fields (or sub-areas of the field) for counting.
4 Count at least seven fields and 600 cells to yield a precision of ±10% at the 95% confidence level.
5 To calculate back to cells per millilitre use:

$$\text{Cells/ml} = \frac{F}{CF \times V}$$

where F is the mean cell count per field, CF is a conversion factor equal to the ratio of the area of the field of view to the area of the entire filter and V is the original sample volume filtered (see Note 9 below).

Notes

1 Fixed samples can be stored cold and dark for up to 24 h before processing.
2 Samples and slides should be kept out of bright light during handling and processing.
3 Sample volumes: optimal volumes of sample filtered depend on cell concentrations. Relatively large numbers of cells per microscope field facilitate image analysis, but too many cells overlapping each other (or detritus) will greatly slow analysis. So a balance between these is necessary and can only be determined by looking at representative slides while preparing them. Optimal volumes generally range from 1 ml (estuarine, coastal waters) to 5 ml or higher in oligotrophic waters.

4 Keep filter vacuum low: less than 5–10 cmHg.
5 Excess water left on the filter will crystallize and damage cells during freezing and thawing.
6 Backing filters can be reused a few times by flipping them over between uses.
7 Common mistakes are (i) placing filter on slide upside down, (ii) using two cover slips and (iii) too much immersion oil.
8 Slides can be stored for up to a year (or more?) without losing fluorescence.
9 When calculating cells per millilitre remember to account for the volume of fixative added to the sample. In the protocol above the fixative adds 0.6% to the sample volume.

Turley and Hughes (1992) have shown significant bacterial cell losses in samples fixed with aldehyde and stored in liquid at room temperature. In open ocean samples heterotrophic bacteria counts can be confounded by the presence of abundant *Prochlorococcus* populations that are not visually distinguishable (Sieracki *et al.*, 1995). Zweifel and Hagström (1995) report that 30–90% of fluorescing bacteria-sized objects in samples prepared this way are actually 'ghost cells' composed of lipid membranes but containing no DNA (DAPI becomes non-specific at seawater salinities). There continues to be debate over how many of the cells seen in nature are active, dormant, living etc.

References

Kemp, P.F., Sherr, B.F., Sherr, E.B. and Cole, J.J. (1993) *Handbook of Methods in Aquatic Microbial Ecology*. Lewis: Boca Raton, FL.

Kepner, R.L. and Pratt, J.R. (1994) Use of fluorochromes for direct enumeration of total bacteria in environmental samples: past and present. *Microbiol. Rev.* **58**: 603–15.

Porter, K.G. and Feig, Y.S. (1980) The use of DAPI for identifying and counting aquatic microflora. *Limnol. Oceanogr.* **25**: 943–8.

Sieracki, M.E., Johnson, P.W. and Sieburth, J.McN. (1985) The detection, enumeration and sizing of aquatic bacteria by image analyzed epifluorescence microscopy. *Appl. Environ. Microbiol.* **49**: 799–810.

Sieracki, M.E., Haugen, E. and Cucci, T.L. (1995) Overestimation of heterotrophic bacteria in the Sargasso Sea: direct evidence by flow and imaging cytometry. *Deep-Sea Res.* **42**: 1399–1409.

Turley, C.M. and Hughes, D.J. (1992) Effects of storage on direct estimates of bacterial numbers of preserved seawater samples. *Deep-Sea Res.* **39**: 375–94.

Zweifel, U.L. and Hagström Å. (1995) Total counts of marine bacteria include a large fraction of non-nucleoid-containing bacteria (ghosts). *Appl. Environ. Microbiol.* **61**: 2180–5.

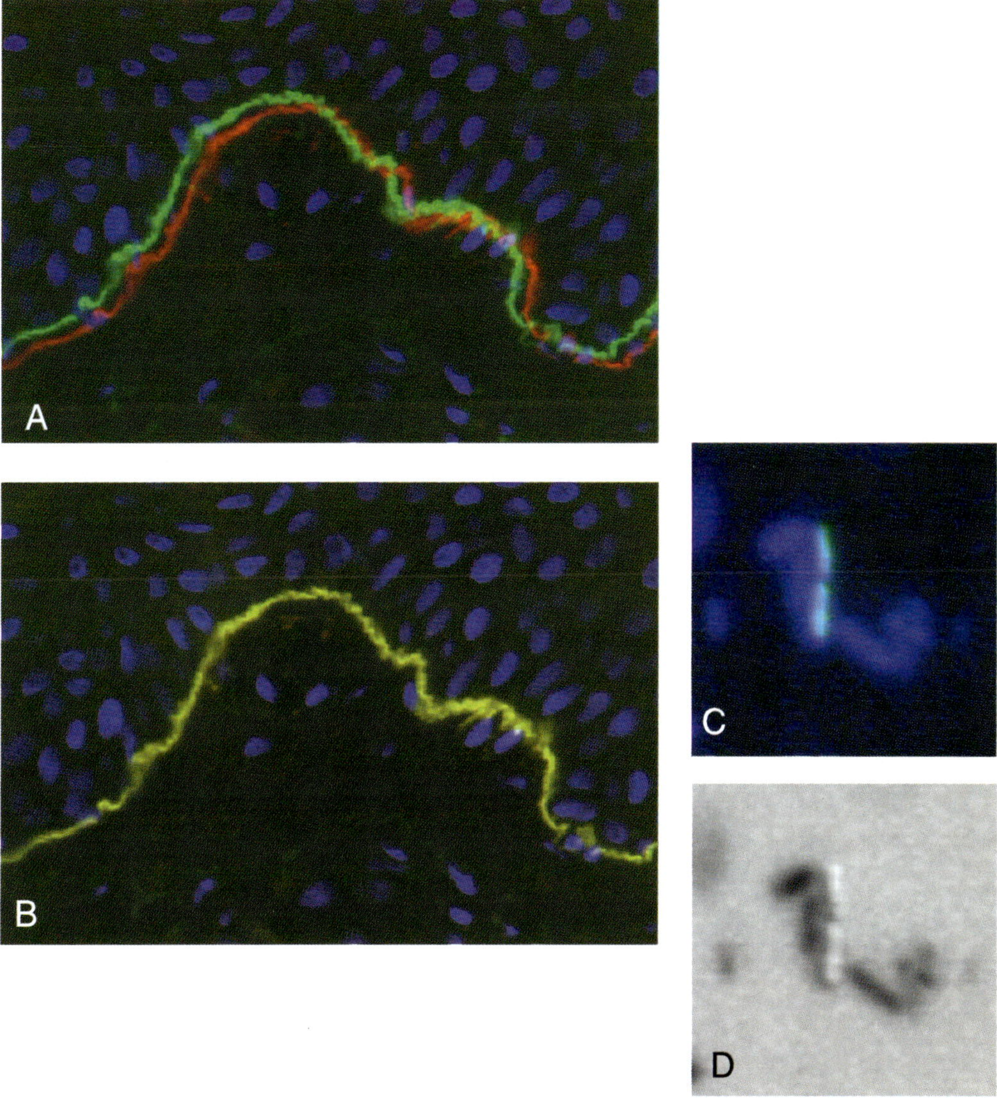

Plate 1 Image shifts in multiple colour fluorescence microscopy. (A) Uncorrected image of triple stained cryosection of human skin tissue, using 4′6-diamidino-2-phenylindole (DAPI, blue) for the nuclei, and fluoresceine isothiocyanate (FITC, green) and lissamine rhodamine sulfonyl chloride (LRSC, red) to double-stain the epidermal basement membrane antigen collagen Type VII. Prismatic errors in the filters cause considerable shift between the colour components, which were recorded separately on a monochrome CCD system, mounted on a Leitz Orthoplan microscope (c. 1975). Scale changes due to lateral chromatic aberration are also present, yet not visible. (B) Same image corrected by image processing, using experimentally predetermined magnitude of errors (images courtesy of M.C.J.M. de Jong and S. Bruins, Department of Dermatology, University Hospital, Groningen, The Netherlands). (C) Double-stained image of bacteria, using DAPI to stain the DNA and an FITC-labelled probe targeted against 16S rRNA of *Bifidobacterium* spp. (Langendijk *et al.*, 1995), recorded with the same camera type mounted on an Olympus BX60 (c. 1995). Newer optics show far smaller prismatic errors. (D) Difference of green and blue colour planes of image in (C): note the relief-like image of the double-stained bacterium in (C), indicating a residual image shift, which should be corrected for critical work (images courtesy of F. Schut, Microscreen BV, Groningen, The Netherlands).

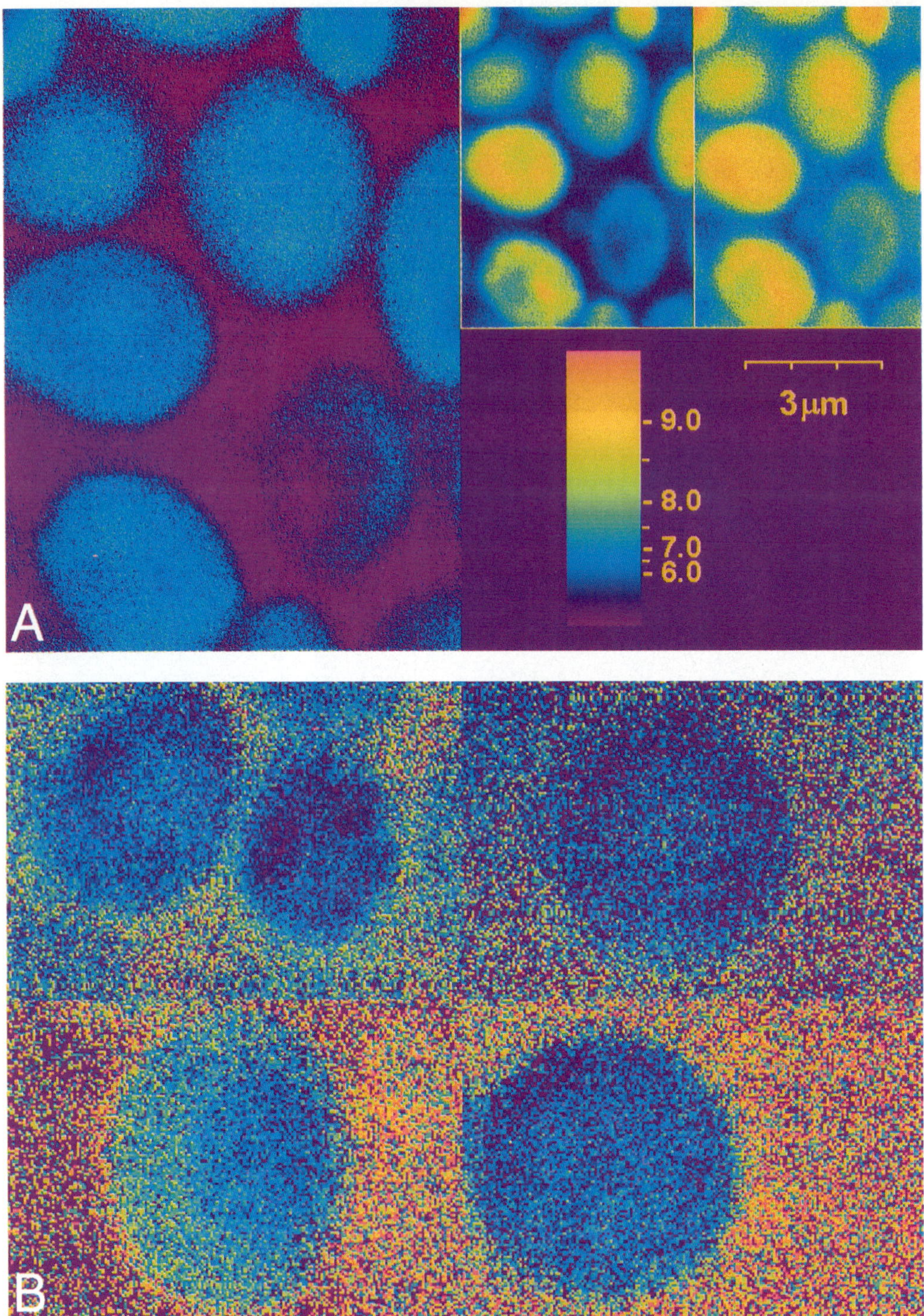

Plate 2 (A) Intracellular pH map of a group of yeast cells filled with carboxy-SNARF. The small images in the upper right corner are original fluorescence pictures; the larger is a ratio image with pseudocolors corresponding to the pH scale. (B) Enlargements of pH maps of yeast cells stained as in (A). The position of the acidic vacuoles can be seen quite clearly and the intravacuolar pH value can be estimated.

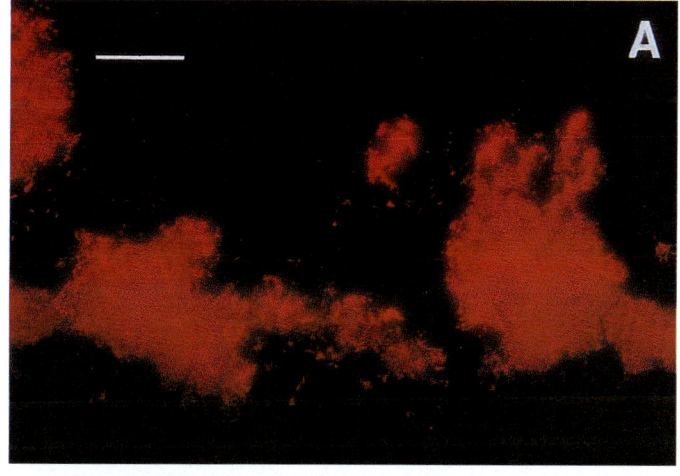

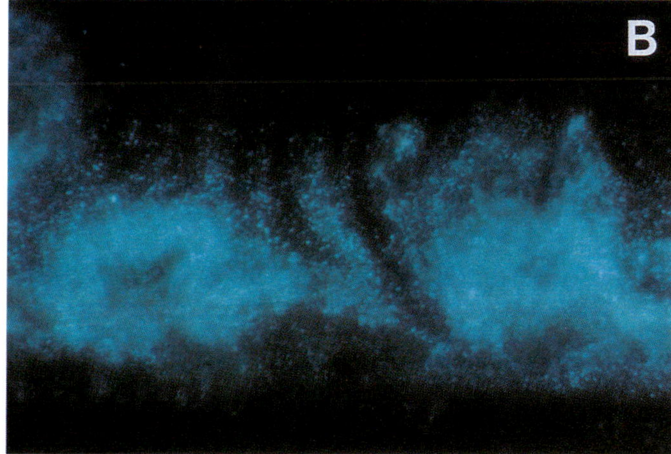

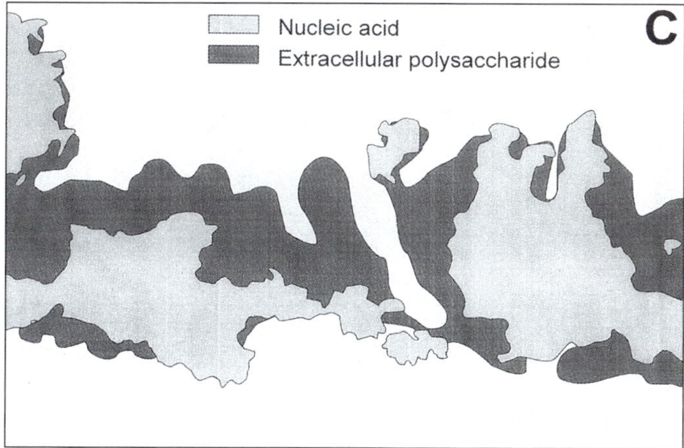

Plate 3 Cellular nucleic acid (A) and extracellular polysaccharide (B) staining of a frozen cross-section of a two-species (*K. pneumoniae, P. aeruginosa*) biofilm. Panels A and B were captured using Olympus G and U filter cubic units, respectively. The exposure time for panels (A) and (B) was 0.5 s. Panel (C) was created by superimposing segmented image (A) and (B) using Corel Draw 5.0 software. The scale bar is 100 μm. Reproduced by permission of Elsevier Science Ltd.

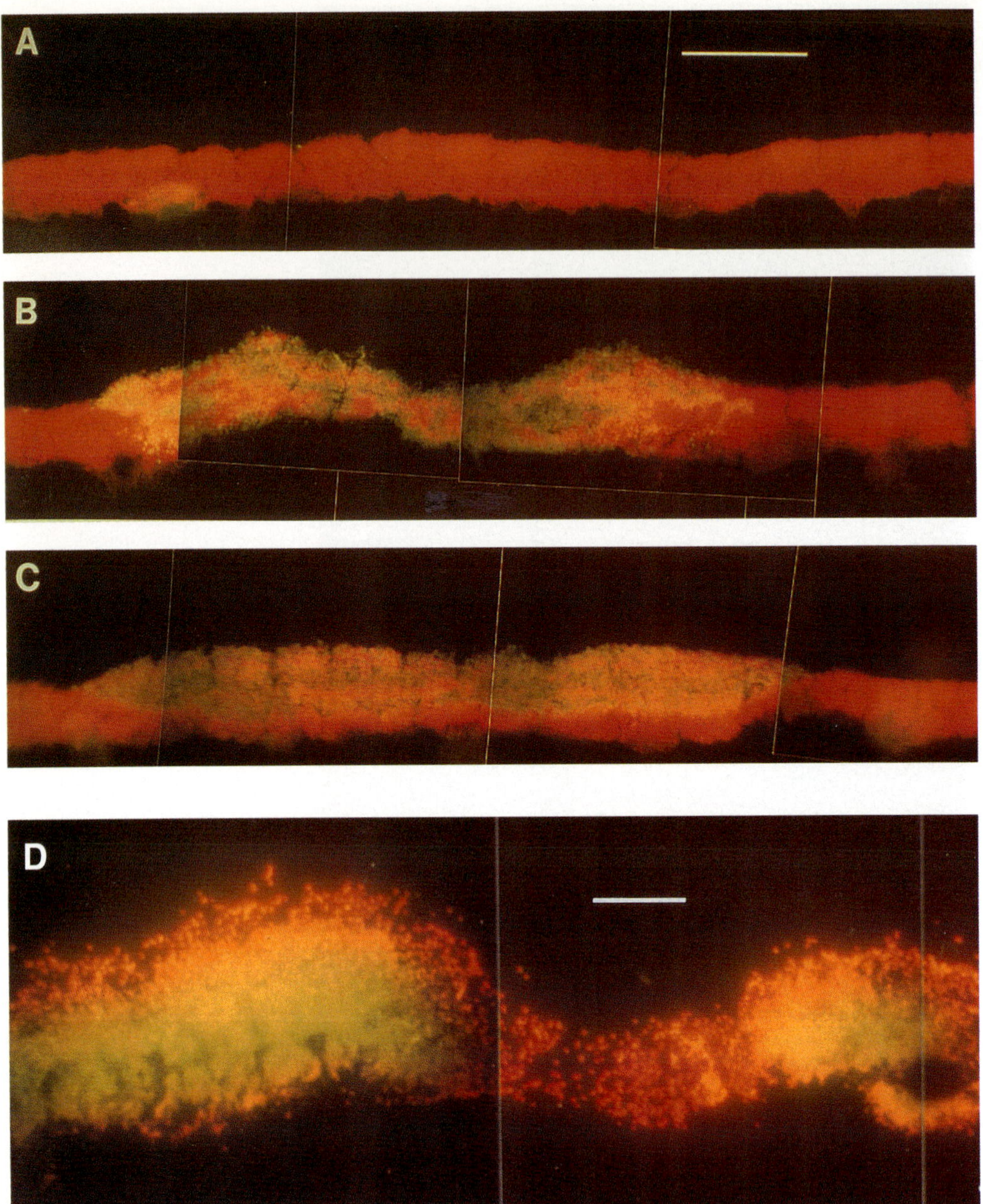

Plate 4 (A–C) Photomicrographic montage showing distribution of *K. pneumoniae* in biofilm cross-sections. Green indicates fluorescein fluorescence and the presence of *K. pneumoniae*; red indicates propidium iodide, which stains nucleic acids of either bacterial species. Regions lacking green fluorescence indicate the presence of *P. aeruginosa* alone. Where fluorescence from propidium iodide and fluorescein overlap the colour appears yellow; this indicates the presence of *K. pneumoniae* alone or a mixture of *K. pneumoniae* and *P. aeruginosa*. The images were captured using a Olympus B filter cubic unit at contiguous locations at an exposure time of 4 s. The scale bar is 50 μm. Reproduced by permission of Springer-Verlag, New York Inc. (D) Photomicrograph of portions of acridine orange stained cross-sections of *K. pneumoniae* biofilm grown in continuous culture. The bulk fluid was at the top and the substratum at the bottom. The images were captured using an Olympus B filter cubic unit at contiguous spots and the exposure time was 2.5 s. The scale bar is 100 μm. Reproduced by permission of the American Institute of Chemical Engineers.

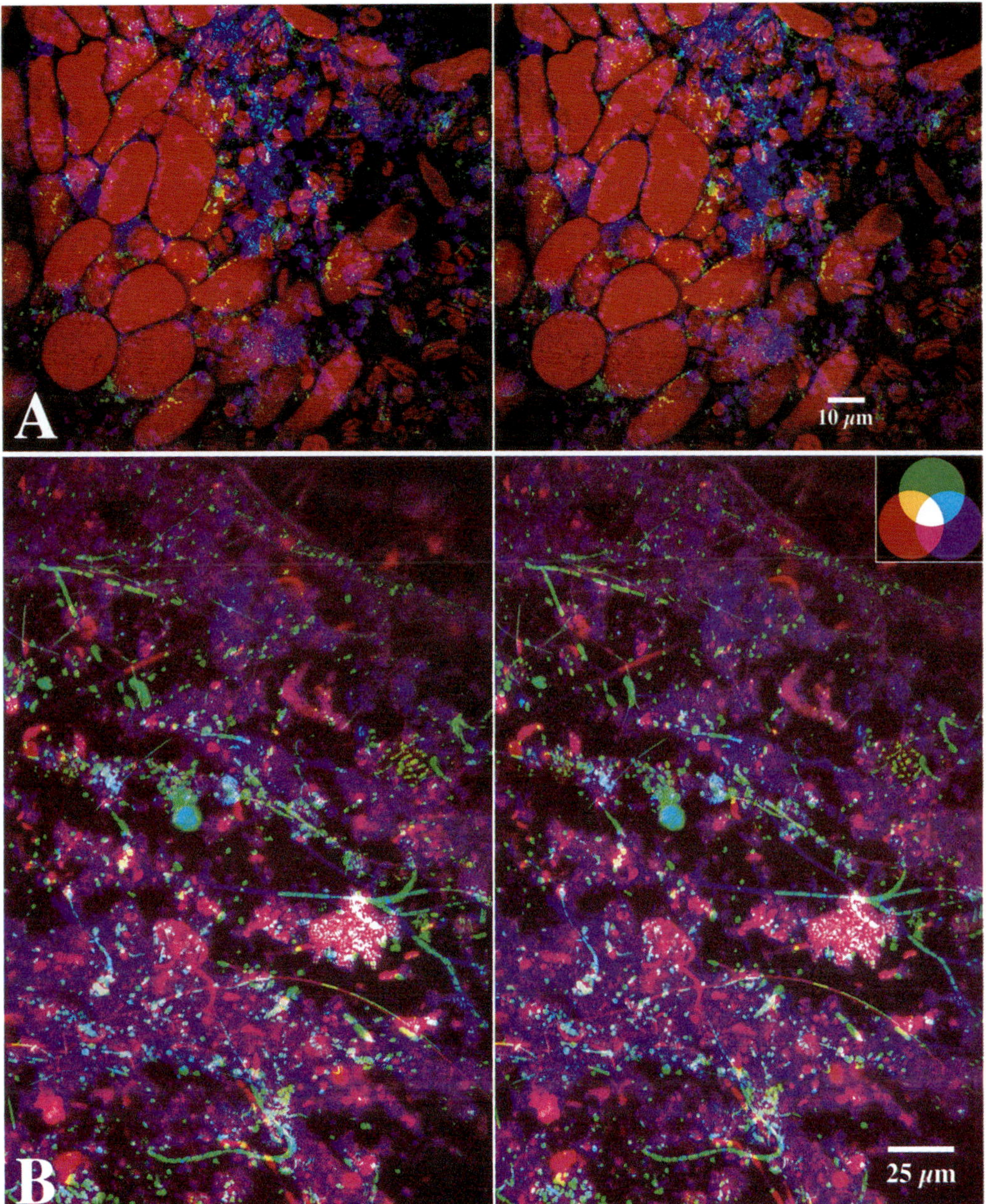

Plate 5 (A) A three-colour image showing the distribution of bacterial cells stained with SYTO™ 9 (green channel), exopolymer stained with *Triticum vulgaris* lectin TRITC (blue channel) and autofluorescence of algal cells (red channel) in a complex aquatic biofilm community. The images were compiled into two stacks, one offset by 5° and aligned side by side to create a three-colour stereo pair. (B) The results of combined staining of a river biofilm community developed in an annular flow reactor, with SYTO™ 9: *Ulex europeaus* lectin FITC (green channel), *Triticum vulgarus* lectin TRITC (red channel) and *Arachis hypogea* lectin (blue channel). A colour wheel is provided to aid in the interpretation of regions where possible combinations of the probes occur.

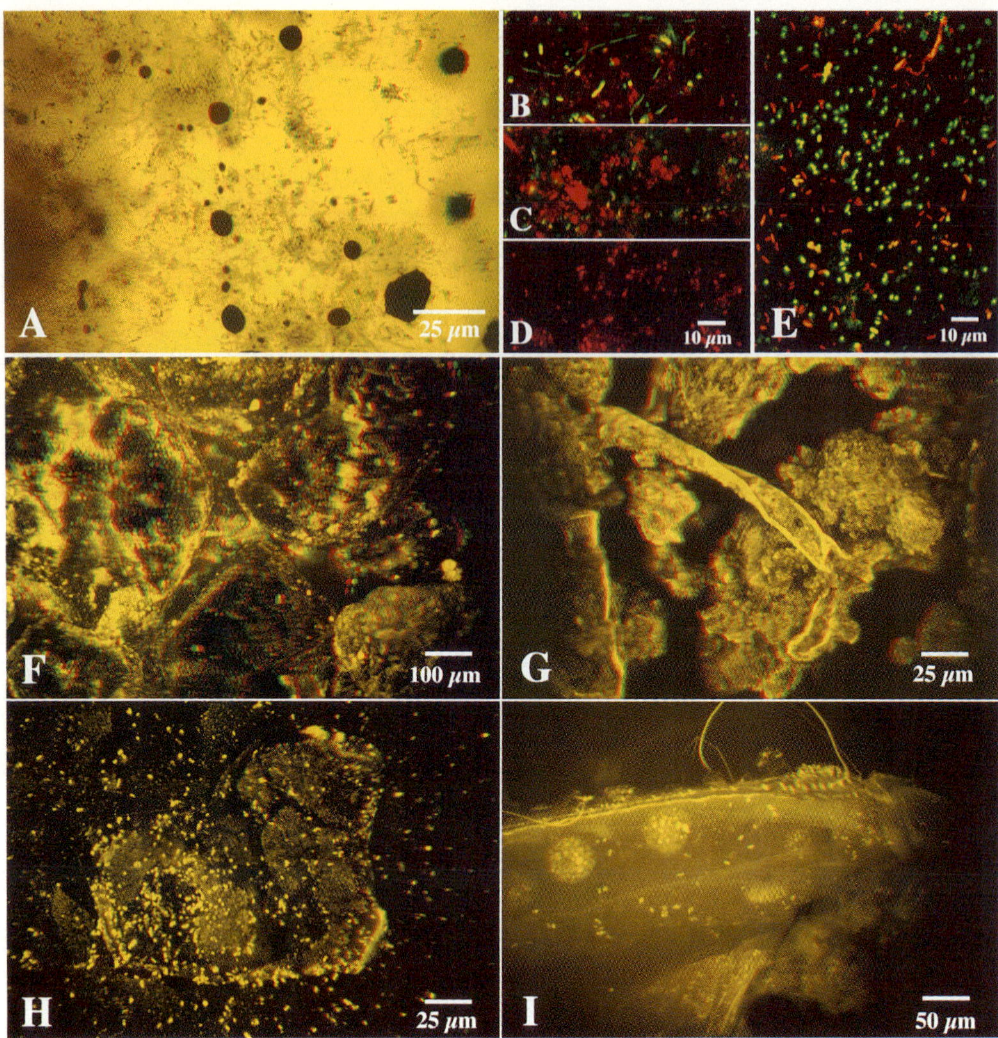

Plate 6 (A) Three-dimensional stereo projection of the *xy* series shown in Figure 16.6. (B–D) Application of live/dead (green/red respectively) staining showing the effect of chlorine treatment: (B) control, (C) 0.5 mg/l and (D) 4.0 mg/l. (E) Dual-channel image showing the presence of Gram-negative (green) and Gram-positive (red) bacteria in a river biofilm. (F) Three-dimensional projection of autofluorescence in an aquifer sand pack. (G) Three-dimensional projection of SYTO™ 15 stained rhizosphere soil. (H) Three-dimensional projection of a *Thiobacillus ferrooxidans* biofilm on a weathered pyrrhotite grain surrounded by pyrite. (I) Three-dimensional image of a mayfly infected with apostomes and covered with a bacterial biofilm.

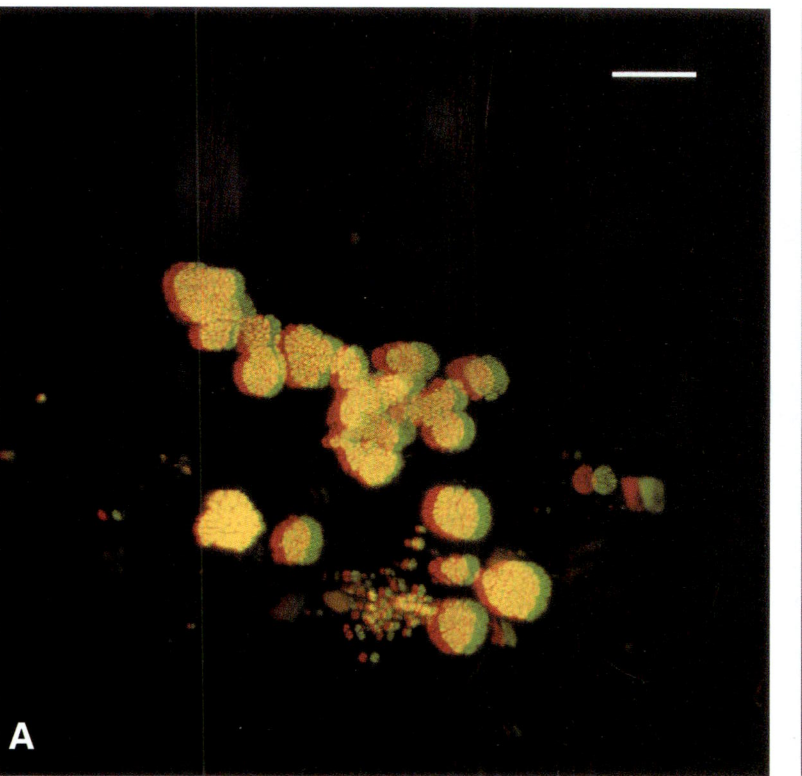

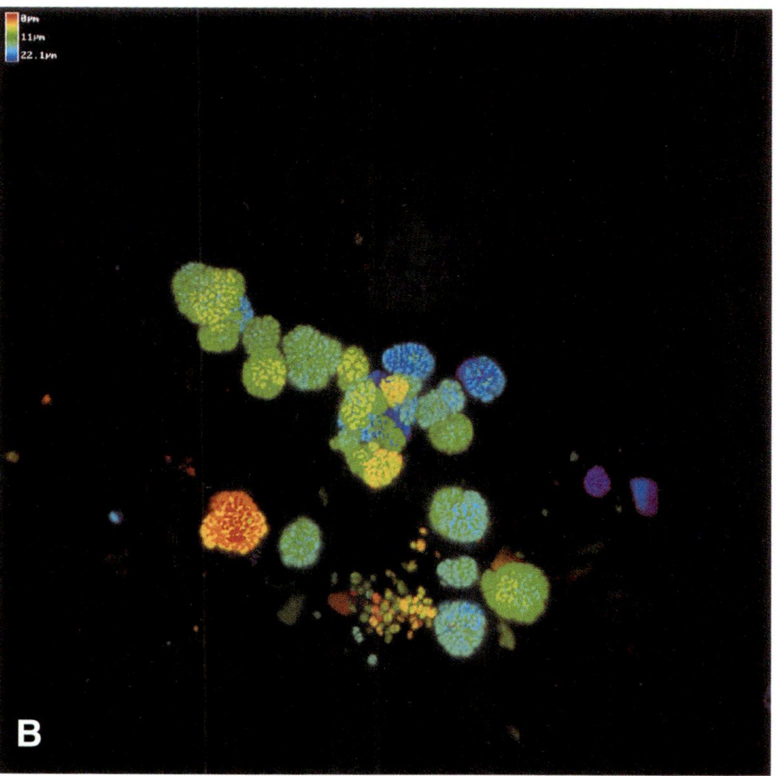

Plate 7 Three-dimensional visualization of specifically stained bacteria in activated sludge. (A) A three-dimensional red–green anaglyph projection constructed from a set of 22 serial optical sections taken at 1.0 µm intervals showing ammonia-oxidizing bacteria identified with Cy3-labelled probe NEU (Wagner *et al.*, 1995) in a 22 µm thick activated sludge floc. The three-dimensional reconstruction should be viewed through red–green glasses. (B) Corresponding coloured depth profile. The colours and the corresponding depths are indicated. The scale bar is 10 µm and applies to both photomicrographs.

8

Morphometry of Yeast

Marie-Noëlle Pons and Hervé Vivier

Laboratoire des Sciences du Génie Chimique, Nancy, France

8.1 INTRODUCTION

Yeasts are single-celled non-motile eukaryotes belonging to the fungi, which reproduce either by sexual or asexual proliferation cycles. They are divided into two large groups, depending upon their ability to produce different types of sexual reproductive organs:

(a) ascomycetes produce ascospores inside the cell,
(b) basidiomycetes produce external spores (basidiospores).

The biological cycles of yeasts have some similarities with the cycles of filamentous species, and yeasts are capable of producing chains and hyphae. There are two methods of asexual reproduction:

(a) by budding,
(b) by formation of intracellular transversal walls.

Dimorphism and even polymorphism are not unusual, and depend on external conditions such as temperature, pH and sugar concentration.

Free cells, with single cells or association of two cells, are the most active from the metabolic point of view. The cell wall, of thickness between 150 and 230 nm, is an exoskeleton which gives the cell its rigidity and shape. Free cells have a characteristic morphology: spherical, ovoid, cylindrical, apiculated, bottle-shaped, pyramidal (Figure 8.1). They are 2.5–10.5 μm wide and 4.5–21 μm long (Bonaly, 1991). These dimensions depend on the culturing conditions (Scherr and Weaver, 1953), such as temperature, pressure and gravity (Walther *et al.*, 1996), on the age of the micro-organism or on possible mutations (Shimosaka *et al.*, 1991; Chun and Goebl, 1996). For example, cells tend to become elongated in a nitrogen-limited medium (Brown and Hough, 1965).

The *pseudo-mycelium* is formed when, after budding, daughter cells remain associated with each other. Chains of a few cells or even branched systems can be observed. In some cases, the formation of pseudo-mycelium is favoured by anaerobic conditions (Figure 8.2).

Digital Image Analysis of Microbes: Imaging, Morphometry, Fluorometry and Motility Techniques and Applications. Edited by M.H.F. Wilkinson and F. Schut.

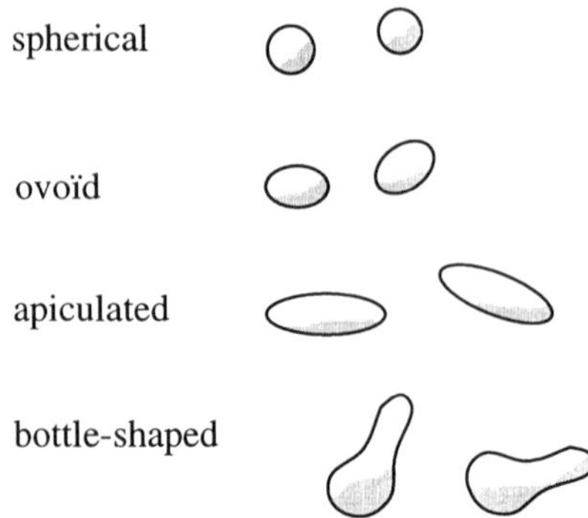

Figure 8.1 Schematic typical morphologies of free yeast cells.

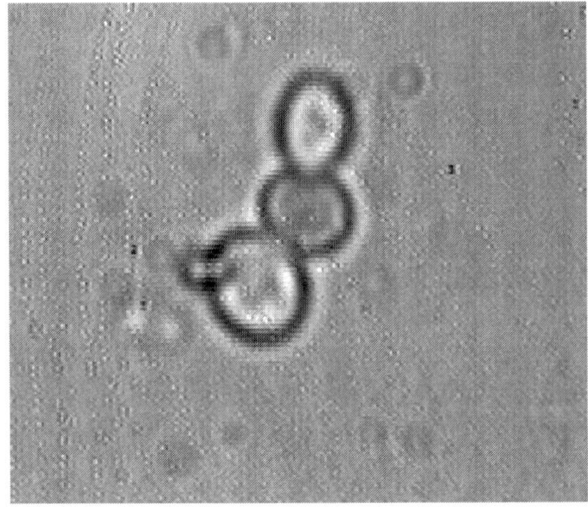

Figure 8.2 Formation of pseudo-mycelium by *Saccharomyces cerevisiae* (late stage of anaerobiosis).

Finally, a true *mycelium* separated by septa can be observed, resulting from an important elongation of cells. In some yeast cells, such as *Candida albicans*, mycelium can be obtained by apical growth of a germ tube, such as can be seen classically in *Penicillium* spp.

The budding process leaves very clear scars on the surface of the cell. The location and size of these scars are typical of the yeast type (Figure 8.3). When the cell is roughly spherical (e.g. *Saccharomyces cerevisiae*) budding generally takes place anywhere on the surface. Bud scars can cover the birth scar partially and even completely, but never cover other bud scars. In apiculated yeasts (*Kloechera*, *Hanseniaspora* and *Saccharomycodes* spp.), the budding is bipolar: the birth scar is covered by successive budding scars. Conversely, *Pitysporum* strains are of the monopolar type. Finally, *Sterigmatomyces* mother cells produce fine filaments (sterigmates), at the extremities of which daughter cells develop. Separation takes place later by the formation of a septum in the central region of the sterigmate.

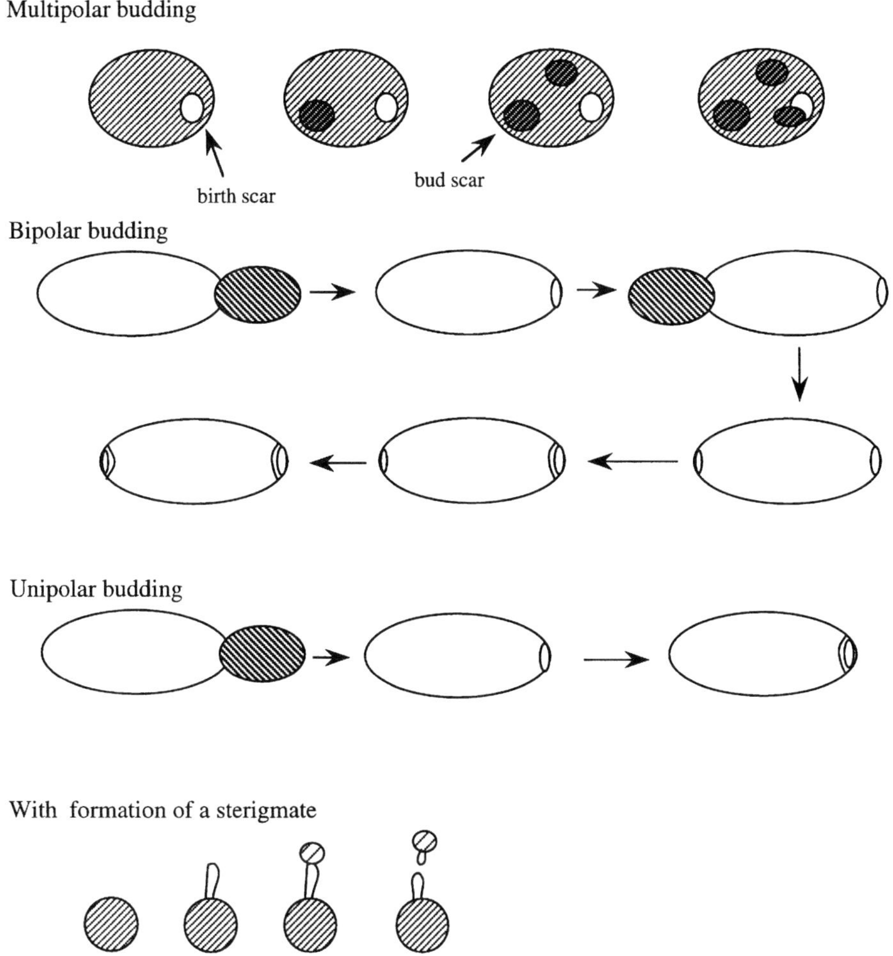

Figure 8.3 Budding sites.

8.2 MICROSCOPIC VISUALIZATION OF YEAST CELLS

Yeast cells can be observed with a light microscope at 500× magnification at least, and at higher magnification (1000× to 1500× with phase contrast and oil immersion lens) for a more detailed analysis. Figure 8.4 presents monochrome images of *Saccharomyces cerevisiae* cells observed with a Leitz Dialux microscope equipped with a zoom lens and a tube video camera connected to an image grabbing board. For complete automation of the image capture, options for automatic slide motion, auto-focus and brightness control are now available.

Staining of yeast cells can enhance details. A mixture of methylene blue and basic fuchsin stains the nuclei blue and the cytoplasm magenta. Methylene blue in citrate buffer is a classical viability test for yeast single cells: the dye permeates through the

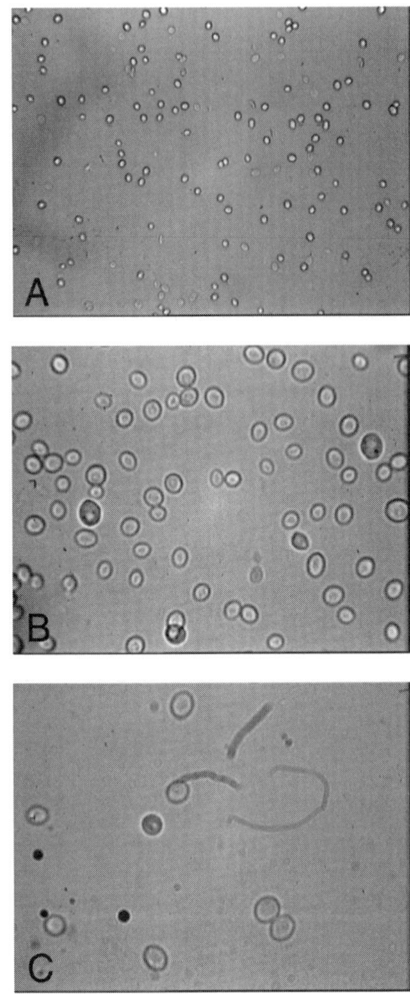

Figure 8.4 Images of *Saccharomyces cerevisiae* at magnifications of (A) 400×, (B) 1000× and (C) 1500×.

membrane of 'dead' cells, while live cells remain unstained. In Figure 8.5A two single unstained, i.e. live, cells can be seen on the left-hand side, while the single dark cell in the lower right-hand corner is a 'dead' cell. One of the elements of the doublet, in the middle of the image, has started to pick up the stain, indicating the beginning of cell degeneration. In fact, this test characterizes dying cells and dead cells which are lysed, with cytoplasm and other intracellular components being released in the medium (Figure 8.5B). Acridine Orange is also used for differentiating between live and dead yeast cells. It is a well-known metachromatic stain, which fluoresces orange when bound to single-stranded and green when bound to double-stranded nucleic acids. The colour, which ranges from orange to yellow and green, can give an indication of the relative proportions of RNA (mostly single-stranded) and DNA (mostly double-stranded). However, the analysis of colour is not straightforward

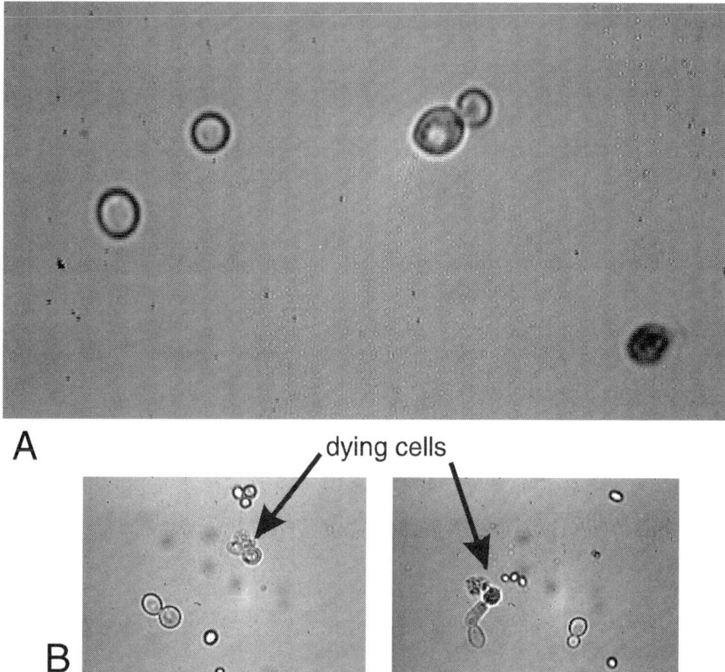

Figure 8.5 Methylene blue viability test on *Saccharomyces cerevisiae*. (A) See text for details; (B) dead cells with cytoplasm released in medium after membrane disruption.

because the differences can be very subtle. For automated analysis, colour image processing is required, whereas on a grey level image only intensity is taken into account. Another difficulty is the occurrence of a halo around cells, which should be minimized by careful selection of the dye concentration.

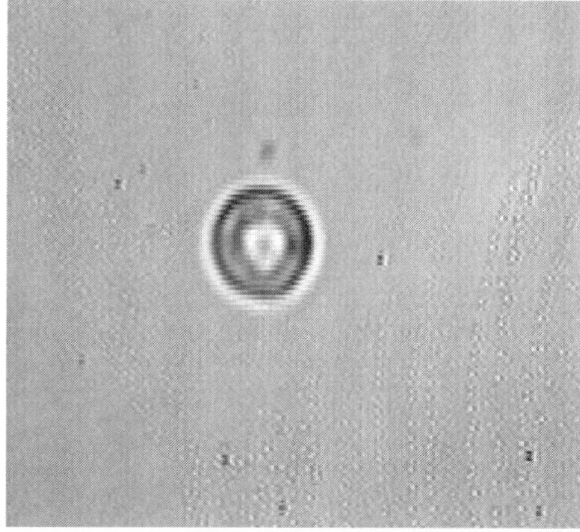

Figure 8.6 Large vacuole in a *Saccharomyces cerevisiae* cell. The white halo outside the cell is due to phase contrast.

Mithramycin and DAPI (4′,6-diamidino-2-phenylindole), which bind to DNA, are used to analyse the cell cycle by staining of nuclei and mitochondria. Vacuoles (Figure 8.6), whose number and size vary largely as a function of the age of the cells, are easily stained with neutral red. Calcofluor staining is used to visualize the bud scars (Baroni *et al.*, 1989; Huls *et al.*, 1992; Walther *et al.*, 1996).

The cell wall of single yeast cells is not as rigid as the walls of filaments, and the addition of staining solutions can modify the size of the cell by changing the osmotic pressure (Berner and Gervais, 1994; Srinorakutara *et al.*, 1996). This effect can be seen in Figure 8.7. *Saccharomyces cerevisiae* cells from the same sample have been examined without staining and with methylene blue staining. Dead cells appear to

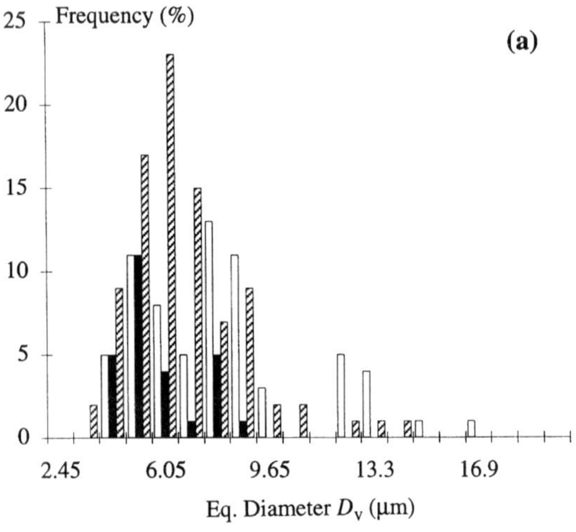

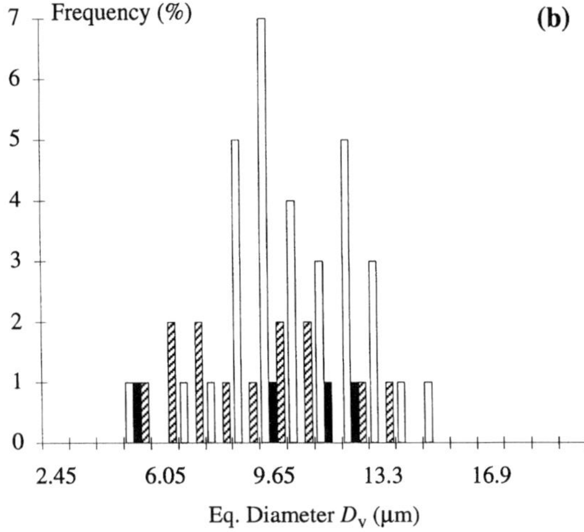

Figure 8.7 Effect of methylene blue staining on size distribution of *Saccharomyces cerevisiae*: without staining (hatched area); with staining: live (white area) and dead (black area) cells. (a) Single cells and (b) doublets.

be smaller than live cells, with the equivalent diameter D_{eq} following the rule:

$$D_{eq,\,dead} \leqslant D_{eq,\,no\ stain} \leqslant D_{eq,\,live} \tag{8.1}$$

Scanning electron microscopy reveals details on the surface of cells, such as bud scars. Careful preparation is necessary because the sample must be fixed and dehydrated before examination. Glutaraldehyde and osmium tetroxide are generally used (Watson and Arthur, 1977; Bonaly, 1991). Dehydration causes shrinkage of the cells and may modify the morphology (Huls *et al.*, 1992).

8.3 MORPHOMETRY OF *SACCHAROMYCES CEREVISIAE*

Saccharomyces cerevisiae is the yeast cell that has received the most attention because of its industrial significance (production of baker's yeast and production of alcohol in distilleries and alcoholic beverages such as wine, beer and cider). Besides, the small size of its genome has made it a good candidate for genetic engineering applications.

8.3.1 Current state of affairs

Division of *Saccharomyces cerevisiae* is asymmetrical: the parent or mother cell develops a bud or daughter. Immediately after division, the daughter cell is usually smaller than its parent cell. The daughter cell needs to attain a certain size (G_1 phase) in order to initiate the DNA division cycle and be able to grow a bud (S phase) (Figure 8.8). During the G_1 phase, storage substances such as glycogen or trehalose are synthesized. The duration of the division cycle depends not only on the yeast species but also on culture conditions (nutrients, aeration, temperature, pH). *S. cerevisiae* daughter cells have a longer division cycle than parent cells. This has been shown by Lord and Wheals (1981), who studied the kinetics of *S. cerevisiae* proliferation by time-lapse cinephotomicrography. Special growth media with a high refractive index were selected to reveal intracellular details under a phase-contrast microscope. One of these media contains (per litre solution) 300 g polyvinylpyrrolidone, 70 g purified agar, 30 g peptone, 15 g yeast extract, 30 g carbon source and 50 µg adenine (Robinow, 1975; Lord and Wheals, 1980). Note that such media are only required for studies of cells growing in a chamber on a

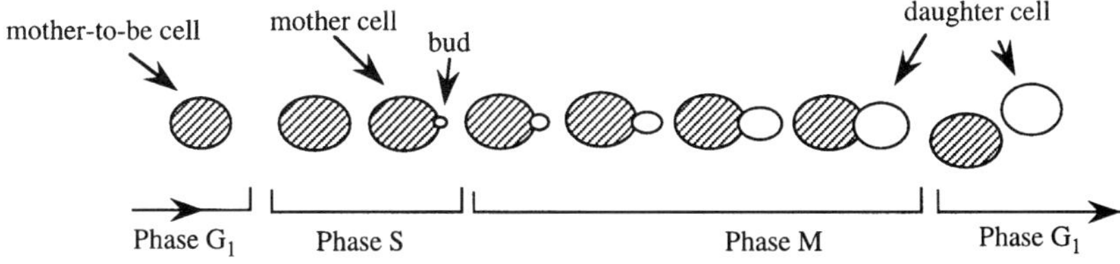

Figure 8.8 Time sequence of budding in a *Saccharomyces cerevisiae* cell.

microscope stage. The cell cycle was registered on 16 mm films, which were later projected onto a screen for manual analysis (Lord and Wheals, 1980; Wheals, 1982).

For more process-orientated applications, it is generally required to examine a large number of cells. This can be automated using image analysis. Costello and Monk (1985) were probably the pioneers in the application of image analysis to yeast cell sizing, but they did not make any assumption about the cell shape. Berner and Gervais (1994) assumed that yeast cells are spherical in their study on the response of cells to osmotic shifts. They observed the cells under direct light with a magnification of about 1000×. The treatment is relatively simple: after thresholding, an erosion–dilation step is sufficient to remove all particles of debris, which are much smaller than the cells. Every object remaining in the field is considered to be a cell and is individually analysed to give its projected area A. The equivalent diameter is then calculated according to:

$$D_{eq} = 2\sqrt{\frac{A}{\pi}} \tag{8.2}$$

This is also the approach adopted by Yamashita *et al.* (1993) for their on-line measurement of cell size distribution and concentration of yeast. A classical method of circle detection, the Hough transform, is applied to the binary image obtained by automated threshold of the edge-enhanced image (Figure 8.9). Good agreement with visual detection was reached. However, Hough transforms are expensive methods in terms of time and memory. Furthermore, the method does not distinguish between budding and non-budding cells.

S. cerevisiae cells are more generally considered to be prolate ellipsoids (Baroni *et al.*, 1989; Pons *et al.*, 1993). The simplest method is to assume that the projection of the non-budding cell onto the image is an ellipse (Figure 8.10). The maximal Feret diameter F_{max} gives the major axis length L, and the minimal Feret diameter F_{min}, or the Feret diameter in the direction perpendicular to the direction of F_{max}, gives the minor axis length B. Different algorithms have been proposed to determine the Feret diameters (Russ, 1995, pp. 514–5) and could be implemented by researchers, but this is now a standard procedure in many image analysis software packages. The shape factor F_H is computed as (Hirano, 1990):

$$F_H = \frac{\pi(L + B)}{2P} \tag{8.3}$$

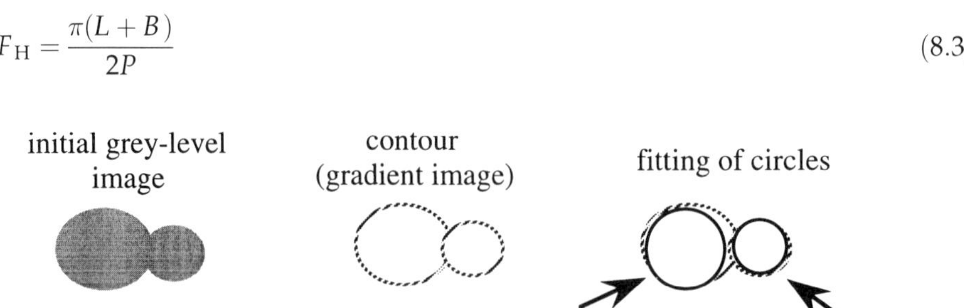

Figure 8.9 Finding yeast cells by fitting of circles according to Yamashita *et al.* (1993).

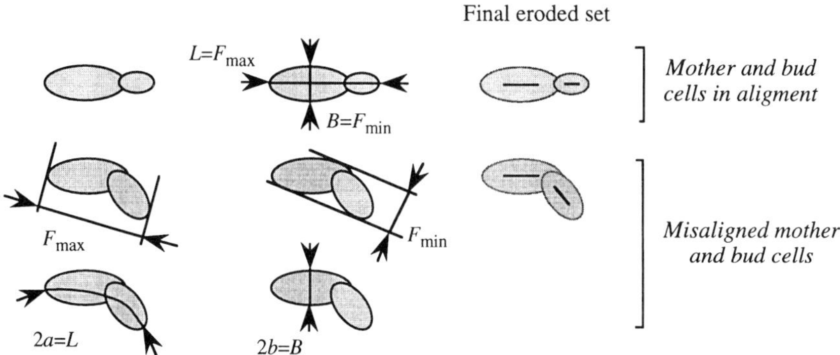

Figure 8.10 Basic size measurements on a *Saccharomyces cerevisiae* cell according to the alignment between mother and daughter cells.

where P is the perimeter. F_H is equal to 1 for a circle and close to 1 for an ellipse, i.e. a non-budding cell. When the shape is irregular, i.e. as a bud develops, F_H decreases ($0.7 \leqslant F_H < 1$). F_H is equal to 0.7 for two adjacent circles. The budding rate, as defined by Hirano (1990), is the ratio of the number of budding cells (i.e. exhibiting F_H lower than a preset value) to the number of non-budding cells.

The assumption of a prolate ellipsoid is not completely true when the cell is budding. Huls *et al.* (1992) determined the contour of the cell and its axis of symmetry on epifluorescence images. The volume of the cell is calculated next. However, the volume integration technique is sophisticated and could not be fully automated (Figure 8.11): using a light pen the operator has to select the valid objects, i.e. the isolated cells, within an image in order to exclude aggregates or touching single cells. After image enhancement by a median filter to help the contour detection, the constriction corresponding to the bud neck is determined from the bending profile (Figure 8.11c). When the bud is too small to be detected, the operator can interact with the computer. The mother and daughter cells are separated and an ellipse is fitted to each part (Figure 8.11f).

Although *S. cerevisiae per se* are relatively easy to examine at a magnification of 1000× with most imaging systems, the quantification of internal details such as vacuoles is more difficult. It has been attempted by Zalewski and Buchholz (1996) in order to provide an indication of the culture vitality without using any staining technique: vacuoles can be detected due to their high grey level with respect to the membrane and the rest of the cell (Figure 8.6). Finally, some yeast-based bioprocesses are using highly flocculent strains: image analysis provides a gentle means to determine the floc size, as flocs are rather fragile structures and should be manipulated with great care (Vicente *et al.*, 1996).

8.3.2 Size and shape quantification

For a bioprocess application a compromise should be reached between the degree of automation of the image analysis procedure, the number of cells examined, the

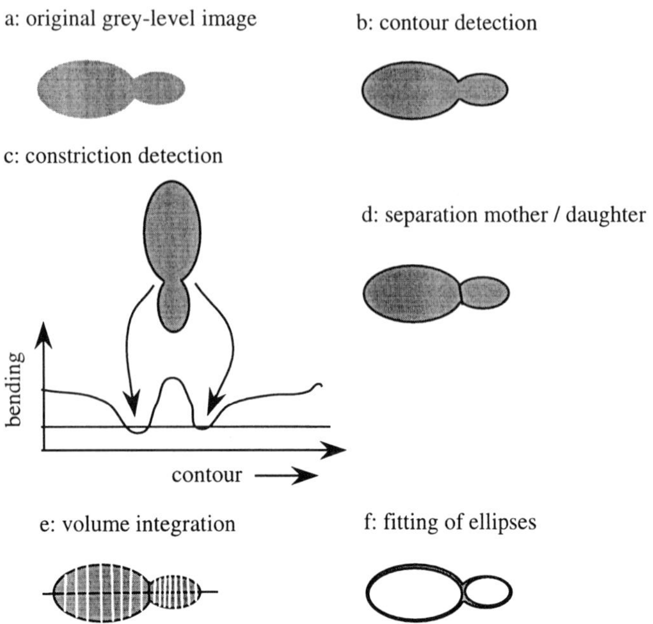

Figure 8.11 Calculation of cell volume according to Huls *et al.* (1992).

resolution (i.e. the combination of the microscope magnification, the camera and the grabbing board characteristics) and the precision of the computed parameters. Using an imaging system with a resolution of 0.04 μm^2 per pixel, an *S. cerevisiae* cell, of approximately 10 μm diameter will have an apparent surface area of about 2000 pixels, which is fully sufficient for sizing and morphological characterization. But an *Escherichia coli* bacterial cell, which is a 1–3 μm long rod, will have an apparent area of about 50 pixels: this makes sizing and, especially, morphological characterization of these micro-organisms much more difficult.

Figure 8.12 presents typical images of single yeast cells, doublets and triplets. A doublet is constituted of a mother cell and a non-detached daughter cell. A triplet is constituted of a mother cell with two daughter cells. Larger clusters may be found (Figure 8.2), depending on strain (Zalewski and Buchholz, 1996) or culture conditions. The purpose of the image analysis is to compute the number of cells (single, doublets, triplets), their size (equivalent diameter D_{eq}) and their shape.

8.3.2a Descriptors of elongation

Different macroscopic shape descriptors can be used (Figure 8.10). For example, circularity (or compacity):

$$C = \frac{P^2}{4\pi A} \tag{8.4a}$$

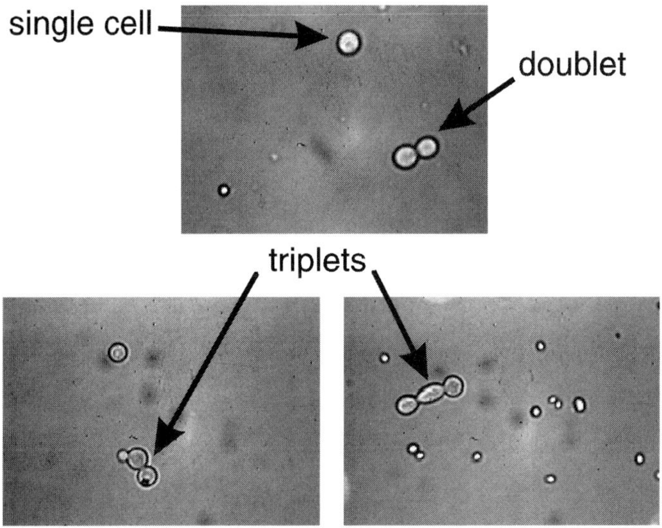

Figure 8.12 *Saccharomyces cerevisiae*: single cells, doublets and triplets.

C is equal to 1 for a disc and increases with elongation, which can be expressed as:

$$\frac{F_{\max}}{F_{\min}} \tag{8.4b}$$

or

$$\frac{a}{b} \tag{8.4c}$$

The latter is the ratio of the major half-axis to the half-minor axis of the ellipse equivalent to the cell projection, in which a is given by the geodesic diameter of the cell. The geodesic diameter is obtained by a two-step procedure. In the first step, a geodesic propagation of the labelled or binary image (Figure 8.13A), containing all of the silhouettes of the cells, is performed. The resulting image is a grey level image (Figure 8.13B), where each grey level has a value equal to its propagation function, $t_X(x)$. The propagation function is defined for an object X by the relation:

$$t_X(x) = \text{Sup}\{d_X(x, y); \quad y \in X\} \tag{8.5}$$

where x and y represent pixels belonging to the object X, and $d_X(x, y)$ is the shortest distance between x and y under the condition that the entire path between the points is included in X. In Figure 8.13B, the propagation function is coded so that the darkest levels correspond to the largest values of $t_X(x)$. The second step consists of finding the maximum value of $t_X(x)$ within each object. This can be done by propagation of the maxima within each object X, which assigns the maximum value

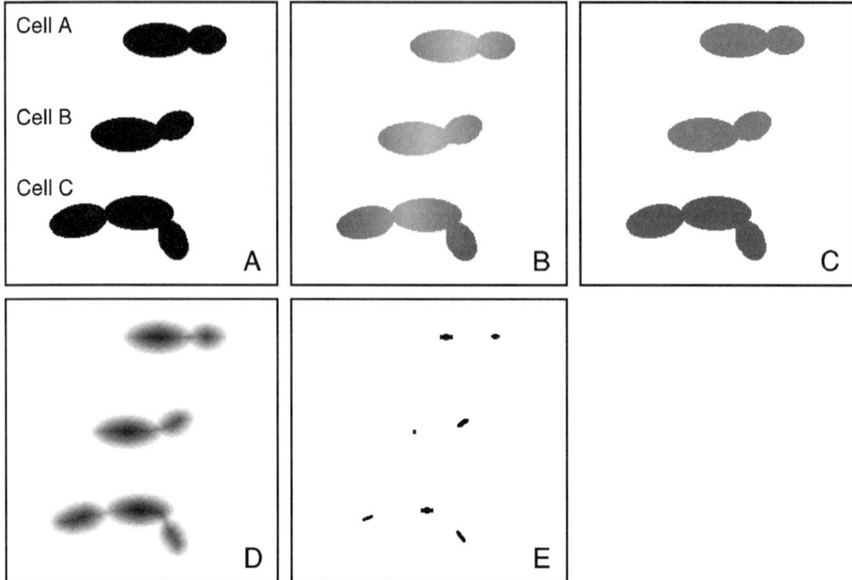

Figure 8.13 Measurement of length on typical *Saccharomyces cerevisiae*: (A) binary image with cell silhouettes, (B) geodesic propagation, (C) propagation of the maxima of the geodesic distance, (D) Euclidean distance map and (E) ultimate eroded set.

to each point of the object (Figure 8.13C). The parameter a is different from $F_{max}/2$ when the mother cell and the bud are not in perfect alignment (Figure 8.2). Table 8.1 shows a comparison of the values of F_{max} and $2a$ obtained for the three model cells of Figure 8.13.

Parameter b is obtained by computing the distance from every pixel of the cell, or cell assembly to the nearest contour point, i.e. computing the Euclidean distance map (EDM) of each object (see Section 1.6.1e; Russ, 1995, pp. 463–79). The EDM construction is available in some commercial software, but efficient algorithms like the one described by Danielsson (1980) can be implemented easily. Here, the procedure is applied to objects that are large enough to avoid the appearance of any spurious maxima (Figure 8.13D). The darkest zones correspond to the largest values of the distance function. Again a compromise should be established here: although a/b should theoretically be more precise than the elongation factor given by

Table 8.1 Comparison of length measurements on yeast cells shown in Figure 7.13.

Cell	Geodesic diameter ($2a$) (a.u.)	F_{max} (a.u.)	F_{min} (a.u.)	$2b$ (a.u.)
A	98	98	32	31
B	97	99	35	31
C	132	134	53	31

a.u., arbitrary units.

F_{max}/F_{min}, it has to be recognized that:

(a) Computation of a and b is more time consuming than computation of F_{max} and F_{min}, as the former requires two separate steps.
(b) The number of skewed cells is generally low, so that the error made by using the elongation given by F_{max}/F_{min} is not large.

8.3.2b Volume estimates

The volume of the equivalent ellipsoid is:

$$V = \frac{4\pi ab^2}{3} \tag{8.6a}$$

or

$$V = \frac{4\pi \frac{F_{max}}{2}\left(\frac{F_{min}}{2}\right)^2}{3} \tag{8.6b}$$

from which an equivalent diameter to the ellipsoid volume, D_v, can be computed:

$$D_v = \sqrt[3]{\frac{6V}{\pi}} \tag{8.7}$$

D_v could be used for comparison with volume-based distributions obtained using other instruments, e.g. a Coulter counter. The cell volume computation described by Huls *et al.* (1992) was not used here: the cost of implementing such a procedure for routine measurements was considered to be too high with respect to the potential increase in accuracy. A more detailed discussion of volume estimation can be found in Section 7.7.

8.3.2c Classification as singlets, doublets etc.

The classification into categories of single cells, doublets and larger aggregates is based on the number of elements of the final eroded set (Figure 8.10): aggregates with a final eroded set of two (three) elements are considered to be doublets (triplets). In this case, the final eroded set has been obtained by a series of erosions, but the local EDM maxima can also be used to represent this set. The latter method is usually faster. No attempt was made to separate, by image treatment, mother cells from daughter cells, in order to compute their size separately. It was considered that the resolution of the image was not sufficient to provide reliable information of that kind.

8.3.2d Image capture and segmentation

The size parameters and shape descriptors are determined on the binary image (Figure 8.14B) obtained form the grey level image (Figure 8.14A) captured on the phase-contrast microscope (Pons *et al.*, 1993): the images are focused in such a way

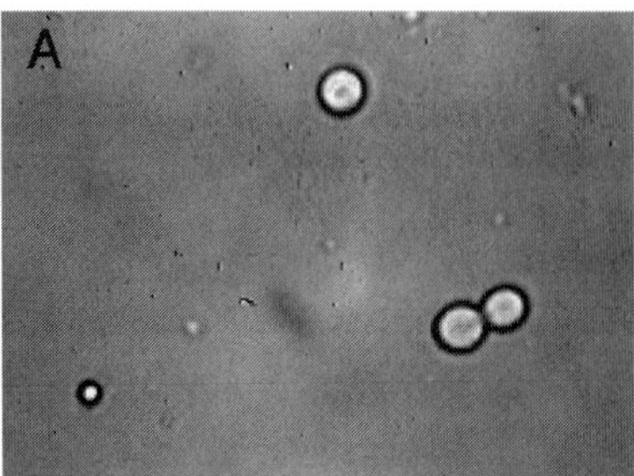

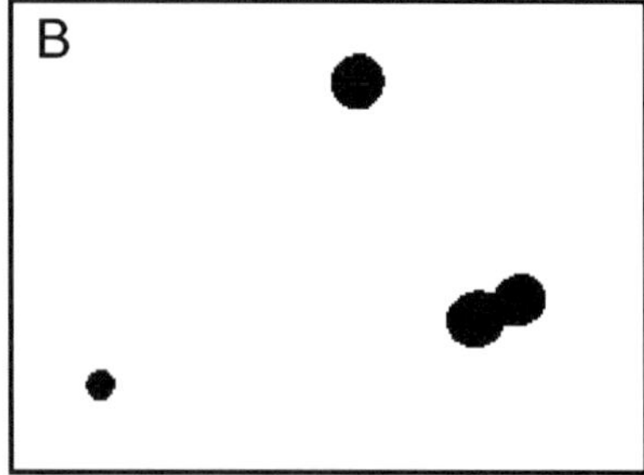

Figure 8.14 (A) Grey-level image and (B) corresponding binary image.

that the halo around the cell membrane appears dark and thin. In the absence of automated brightness control, the upper threshold is selected interactively by the operator to select the cell membrane as object. The lower threshold is set to 0. The binary image contains the cell membranes and occasional debris from the growth medium. The rest of the image treatment is fully automated and run under the interpreter mode of Visilog 4.1.4 (Noésis, Orsay, France) on a workstation or a PC. To ensure that the cell boundary is closed, a two-iteration dilation is applied, followed by a hole-fill procedure and a two-iteration erosion. Open cell boundaries may arise when very small buds are out of the focus plane (Figure 8.2). Border killing is applied to remove any object in contact with the image frame. Small debris is removed from the binary image by a three-iteration erosion followed by a reconstruction procedure.

The number of images to be captured should be a compromise between the accuracy of the results, the total treatment time (including image capture) and the storage volume of images on disc. In the applications described below, typically 50 images are captured for each sample, which corresponds to a total of about 200–300 objects. Dilution may be adjusted in order to limit the number of touching objects. No effect of the number of images captured on the distribution of the number of sub-elements (Table 8.2) or on the size distribution of single cells (Figure 8.15) or doublets was found at a significance level of 5%.

8.3.3 Applications

Figure 8.16 presents the changes in size and morphological parameters observed during the anaerobic fermentation of grape juice by a baker's yeast (Fermipan,

Table 8.2 Effect of the number of images on object sub-element distribution.

Number of sub-elements	50 fields (%)	100 fields (%)	50 fields at random among 100 (%)
1	63	67	71
2	31	27	23
3	3	4	5
4	2	1	1

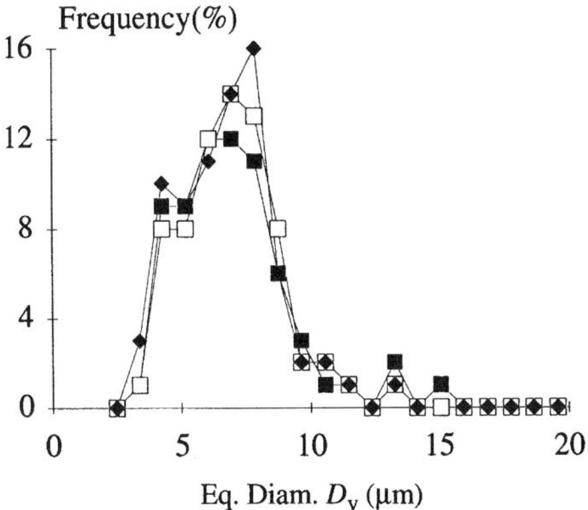

Figure 8.15 Effect of the number of examined images on size distribution of single cells: (■) 50 images; (□) 100 images; (◆) 50 images at random among 100.

Société des Produits du Maïs, Clamart, France) with an initial cell concentration of 2.75 g/l (Pons *et al.*, 1993). Figure 16a and b show the general fermentation kinetics during this experiment: consumption of substrate (glucose, fructose), production of metabolites (glycerol and volatile substances such as ethanol, acetaldehyde, fusel alcohols) and biomass concentration as measured by optical density.

The percentage of single cells (Figure 8.16c) decreases shortly after inoculation. Simultaneously the percentage of doublets increases, which indicates biomass growth. After 10 h, the percentage of single cells increases again. There is no net increase of the mean equivalent diameter during the culture. A very slight shift of the size distributions for single cells and doublets can be seen before 10 h (Figure 8.17), which is in agreement with size distributions obtained with a Coulter counter.

The average diameter for all types of cells is almost identical and the doublets have only slightly larger diameters than single cells: 8.5 μm for doublets and 6.5 μm for single cells. If we model a doublet by two juxtaposed spheres of equal diameter D, i.e. assuming that the mother cell has the same diameter as the daughter cell, the equivalent diameter given by image analysis, based on the projected area, is:

$$D_{eq} = 2\sqrt{\frac{2}{\pi}\frac{\pi D^2}{4}} = \sqrt{2}D = 1.41D \tag{8.8a}$$

In terms of volume, the volume equivalent diameter of a doublet is:

$$D_v = D\sqrt[3]{2} = 1.26D \tag{8.8a}$$

Experimentally:

$$\frac{D_v}{D} = \frac{8.5}{6.5} = 1.31 \tag{8.8c}$$

which is in good agreement with the theoretical value.

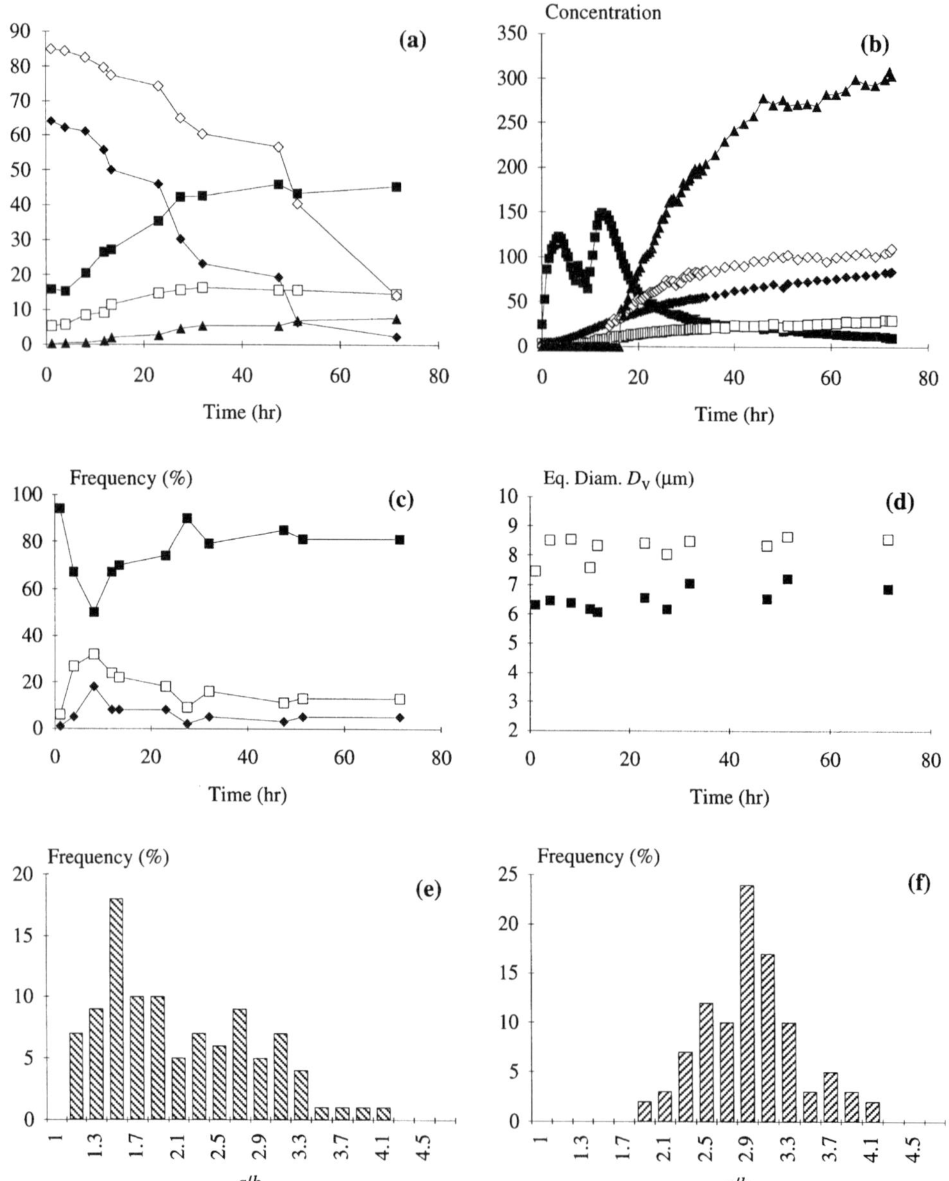

Figure 8.16 Anaerobic conditions. (a) Off-line analysis: (■) optical density (×100), (□) cell number ($\times 10^{-4}$), (◆) glucose (g/l), (◇) fructose (g/l), (▲) glycerol (g/l). (b) On-line analysis of volatile components: (■) acetaldehyde (mg/l), (□) ethyl acetate (mg/l), (◆) ethanol (g/l), (◇) isobutanol (mg/l), (▲) isoamyl alcohol (mg/l). (c) Distribution of (■) single cells, (□) doublets and (◆) clusters of more than two cells. (d) Equivalent diameter D_v) of (■) single cells and (□) doublets. (e, f) Distribution of elongation for (e) single cells and (f) doublets at $t = 6$ h.

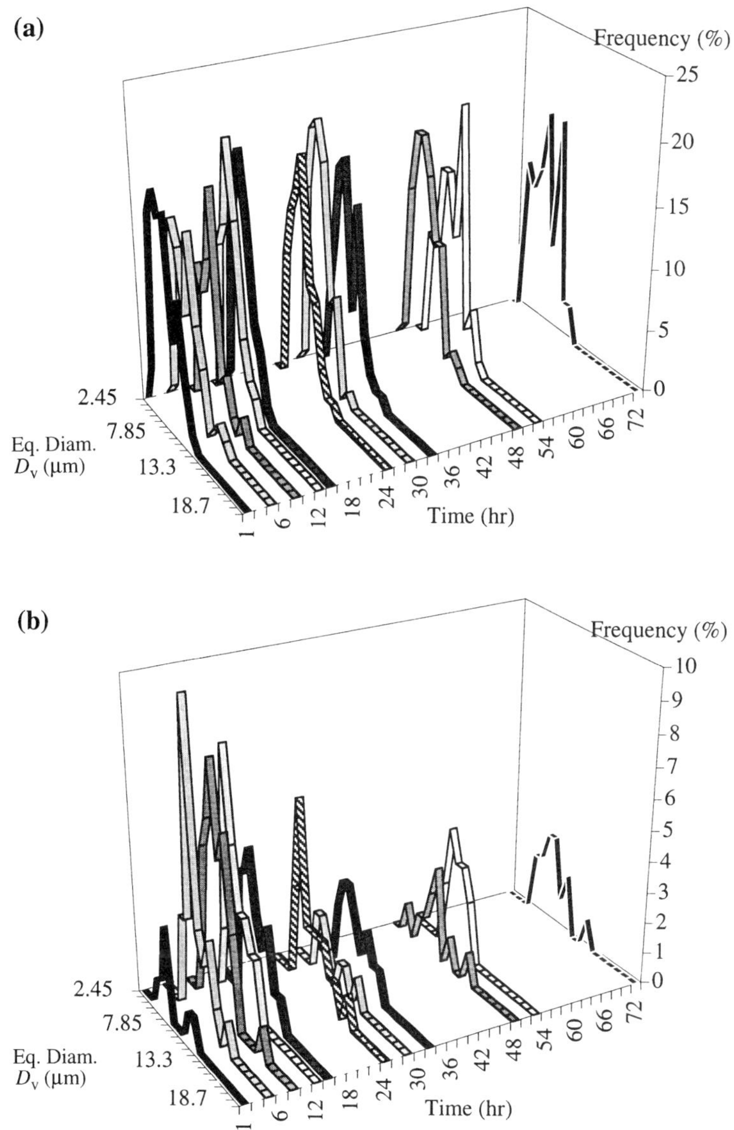

Figure 8.17 Equivalent diameter D_v distributions versus time for (a) single cells and (b) doublets.

The elongation descriptor a/b has an average value of 1.5 for single cells and 3 for doublets. In fact, the term 'single cell' encompasses non-budding cells, like the one shown in Figure 8.12, but also budding mother cells, whose bud is not large enough to create a second sub-element (S phase). A similar analysis can be made for doublets.

Figure 8.18 compares the morphological kinetics observed for three strains of *S. cerevisiae* during anaerobic fermentation of grape juice. Strain A is a baker's yeast,

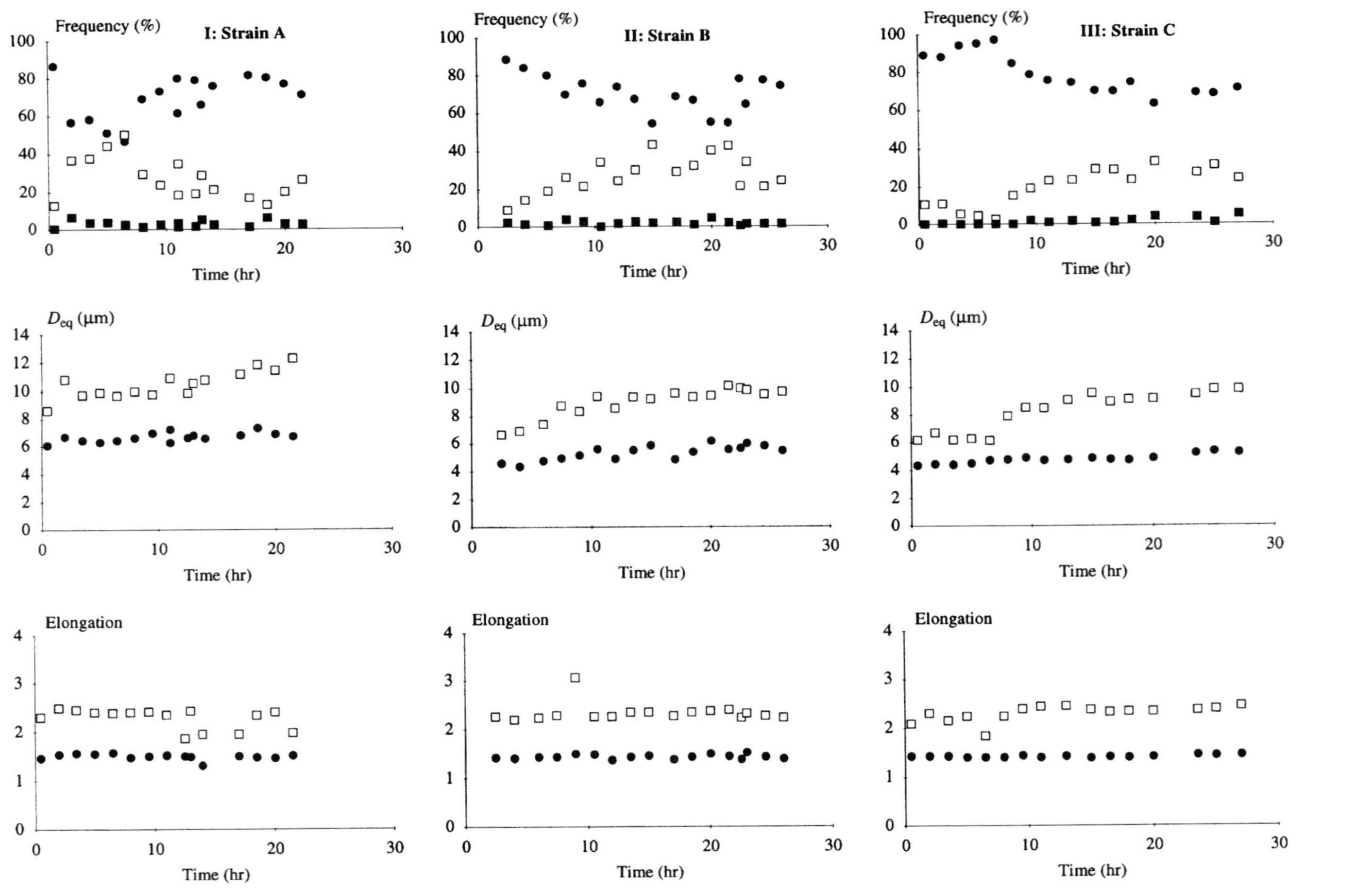

Figure 8.18 Comparison of *Saccharomyces cerevisiae* strains: I: baker's yeast (strain A) , II: wine yeast B, III: wine yeast C; (●) single cells, (□) doublets and (■) clusters.

(Fermipan™), strains B and C are wine yeasts (Littorale Oenologie, Béziers, France). Strain C is a killer-type yeast. The fermentation conditions are identical and the experiments are run with an initial yeast concentration 5.5 g/l.

In all experiments a decrease in the number of single cells and a simultaneous increase in the number of doublets are initially observed. However, the minimum of the proportion of single cells is obtained after 6 h for strain A, and around 20 h for strains B and C. Cells of strain A are larger than those of strains B and C. The ratio D_{eq}(doublets)/D_{eq}(single) is constant for strain A during the course of the fermentation, but increases for the other two strains, for which the proportions of doublets increase significantly: there is indication of a long lag time observed for strains B and C.

The average elongation is given by F_{max}/F_{min} in this experiment, and is similar for all single cells. This value is slightly higher for doublets of strain A than for the other doublets. It should, however, be mentioned that average values give incomplete information about the shape distribution: closer examination of the elongation distribution reveals that some cells of strain C (about 5%) exhibit a very large elongation ratio (up to 5) toward the end of the fermentation, due to their irregular shapes and their tendency to form a pseudo-mycelium (Figure 8.19).

Figure 8.20 presents another application of the characterization of *S. cerevisiae* morphology, this time under aerobic batch growth conditions (Pons *et al.*, 1997). A gradual decrease of the proportion of doublets and increase of the proportion of single cells are observed as the substrate (glucose) is depleted.

8.3.4 Off-line versus on-line

Applications shown so far have been run off-line, which means that samples have been taken manually, appropriately diluted by the operator, deposited on a glass slide and examined. The ultimate goal would be to provide on-line monitoring of the morphological characteristics.

The first approach is to use a side-loop, by pumping the broth through a flow chamber. Microglass channels have been proposed, either with a normal microscope (Lichtfield *et al.*, 1992; Yamashita *et al.*, 1993) or with an inverted

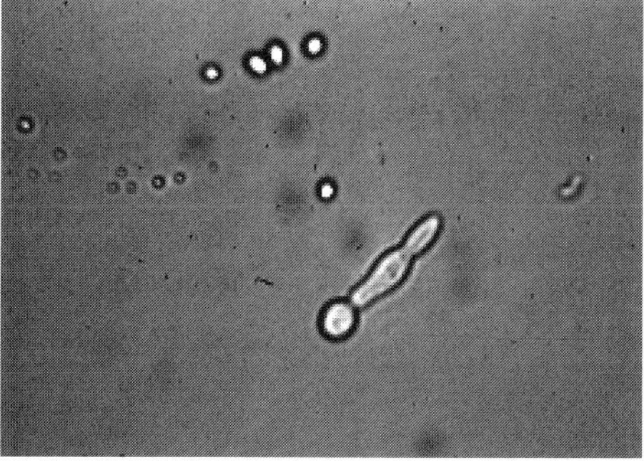

Figure 8.19 Very elongated *Saccharomyces cerevisiae* (strain C) at the end of the run under anaerobic conditions.

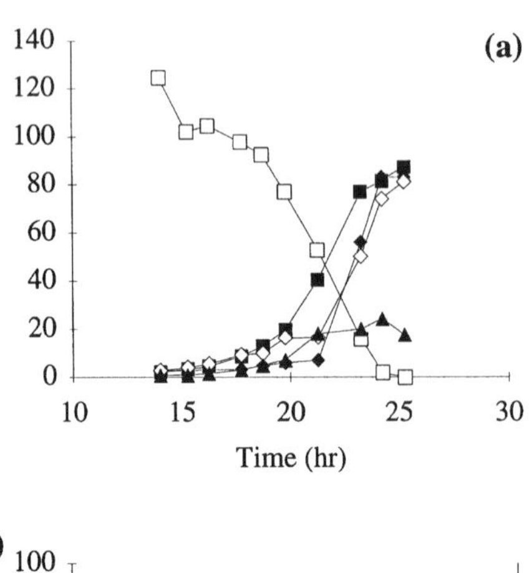

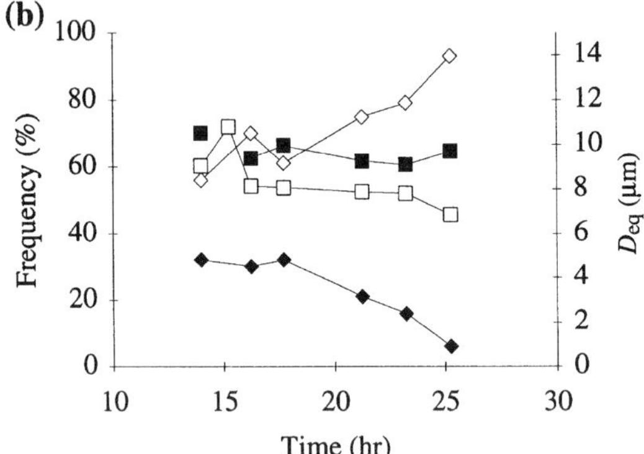

Figure 8.20 Aerobic growth. (a) (■) optical density (×10), (□) glucose (g/l, ×10), (◆) acetaldehyde (mg/l), (◇) acetate (mg/l), (▲) ethanol (g/l, ×10). (b) (◇) Percentage and (□) D_{eq} of single cells; (◆) percentage and (■) D_{eq} of doublets.

microscope (Zalewski and Buchholz, 1996). It is often necessary to stop the flow to be able to focus the cells correctly (Yamashita *et al.*, 1993; Zalewski and Buchholz, 1996). Dilution can be automated (Zalewski and Buchholz, 1996).

Some investigators are reluctant to use side-loops, foreseeing problems of asepsis, clogging of tubing, and so on. The ultimate solution would be to have an *in situ* microscope like the one developed by Suhr *et al.* (1995). A front-end head is inserted directly in one of the fermenter ports and mounted on an epifluorescence microscope. Still images generated using pulsed illumination are captured by a silicon-intensified target camera. Although for the on-line applications using a side-loop the image treatment is more or less the same as for the previously described off-line analysis, it is completely different here: the on-line images are highly blurred and noisy (Figure 8.21), and require sophisticated image treatment based on a depth-from-focus algorithm to form the binary image (Figure 8.21b), from which information on size and shape can be extracted. With available computing systems the data output rate is still too slow for real-time control applications – one image containing 5–10 cells per minute, when about 200 cells are necessary to obtain good statistics. Using this *in situ* microscope, physiology

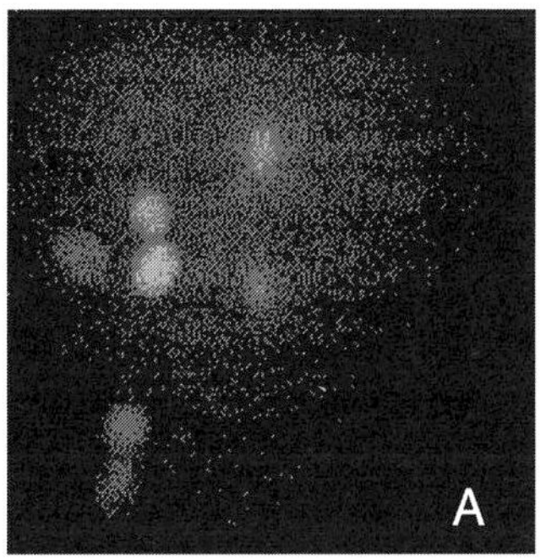

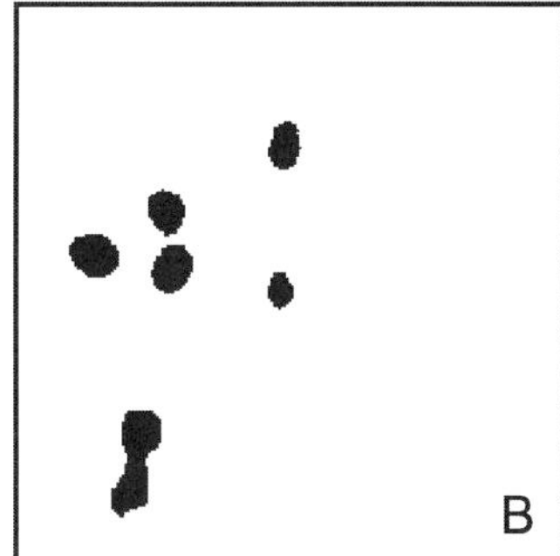

Figure 8.21 *In situ* images of *Saccharomyces cerevisiae*. (A) Grey level image and (B) corresponding binary image. Reproduced by permission of T. Scheper, Institut für Technische Chemie, Universität Hannover.

assessment can be combined with sizing, since NADH (auto)fluorescence is modulated by metabolic changes.

8.4 MORPHOMETRY OF POLYMORPHIC STRAINS

8.4.1 *Kluyveromyces marxianus*

As stated in the introduction, polymorphism is not uncommon in yeasts. *Kluyveromyces marxianus* var. *marxianus* is used for the fermentation of cheese whey permeate to ethanol and can produce various enzymes of industrial interest: the NRRLy2415 strain is known for its dimorphism, and the observed morphologies can range from simple ovoid yeast cells to branched mycelial forms (Figure 8.22). A semi-automated image analysis procedure has been developed by O'Shea and Walsh (1996a, b) to classify the morphologies into six classes (Table 8.3). As in any image analysis problem, a compromise has to be made concerning the number of fields to be analysed and the number of cells per fields, i.e. the magnification chosen for the visualization system. In any case a minimum number of cells, 200–300, should be analysed. The option selected by O'Shea and Walsh was to minimize the number of fields (6), which did not make the morphology characterization task

Table 8.3 Morphologies displayed by *Kluyveromyces marxianus* var. *marxianus* NRRLy2415 (O'Shea and Walsh, 1996a, b).

Class	Name	Description
1	Yeast	Spherical or ellipsoidal single cell
2	Elongated yeast	Elongated ellipsoidal single cells
3	Filament	Rod-like cell with no visible constriction
4	Double yeast	Budding yeast or elongated cell, with a visible constriction
5	Double filament	Joined filaments
6	Mycelium	Three or more cells joined together

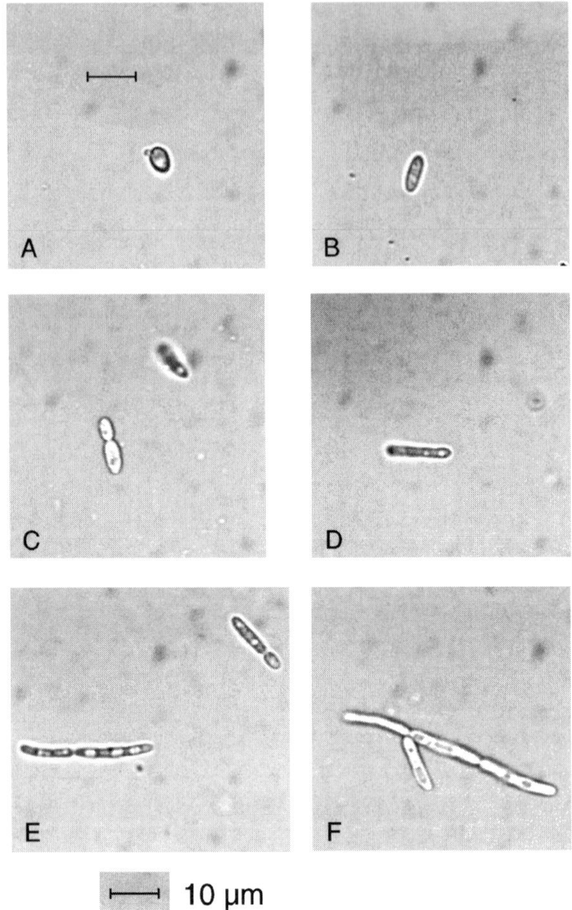

Figure 8.22 Morphological forms displayed by *Kluyveromyces marxianus* var. *marxianus* NRRLy2415. (A) yeast, (B) elongated yeast, (C) double yeast, (D) filament, (E) elongated filament and (F) branched mycelium. Reproduced by permission of P. Walsh.

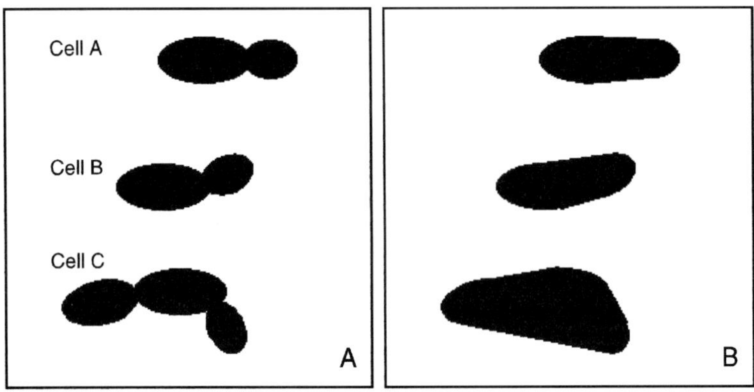

Figure 8.23 Definition of the convex hull: (A) silhouettes and (B) convex hulls.

easy, due to the small size of the objects in pixels. Part of the classification is based on the comparison of the cell perimeter P with its convex hull perimeter P_c. The convex hull is the smallest convex polygon into which the cell fits (Figure 8.23). The complete algorithm is summarized in Figure 8.24.

Less than 6% error with respect to visual classification was obtained. The volume of individual and double cells was estimated and typical morphological properties of mycelial cells were measured. However, O'Shea and Walsh state that it is not easy

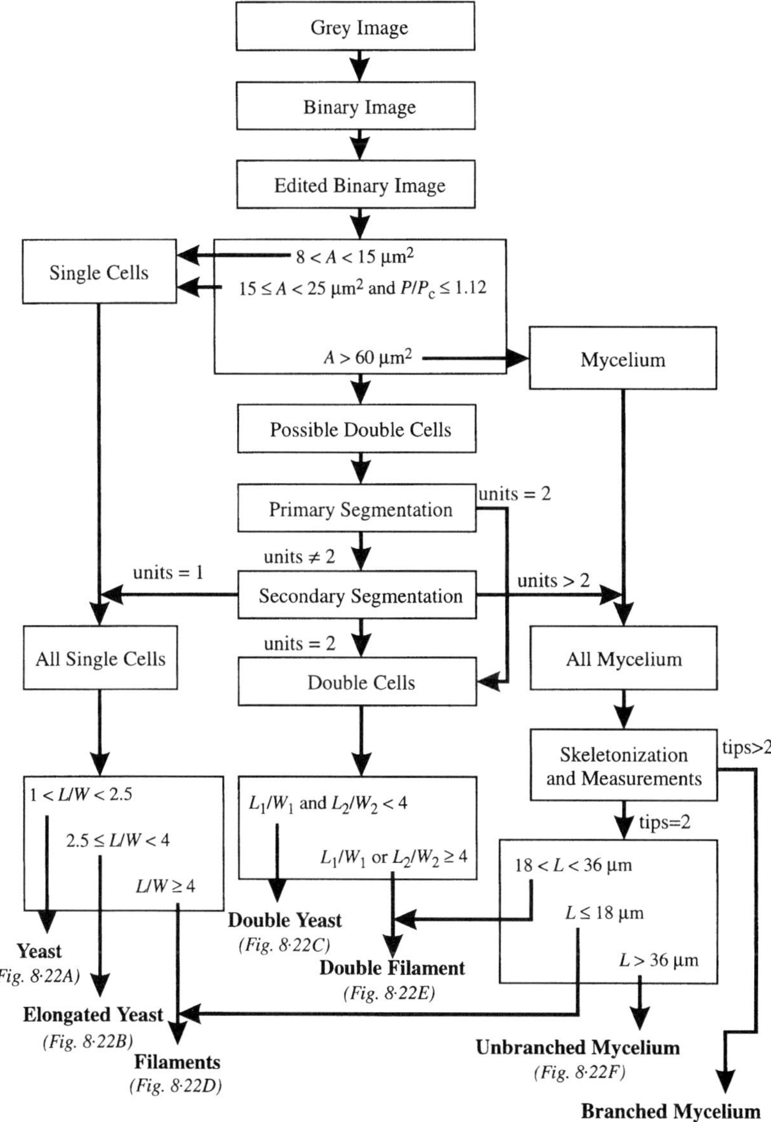

Figure 8.24 Algorithm for morphological classification of *Kluyveromyces marxianus* var. *marxianus* NRRLy2415 (O'Shea and Walsh, 1996a, b). A, projected area; P, cell perimeter; P_c, convex hull perimeter; L, length; W, breadth.

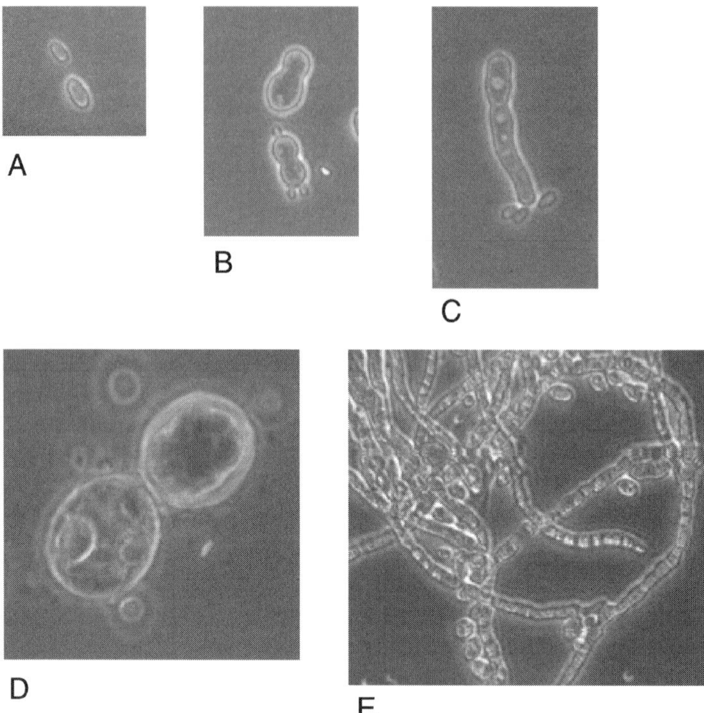

Figure 8.25 Morphological forms displayed by *Aureobasidium pullulans*: (A) yeast-like cells, (B) young blastospores, (C) swollen blastospores, (D) chlamydospores and (E) mycelium. Reproduced by permission of Elsevier Science Ltd from Y. Shabtai, *Comp. Chem. Engng* **20** (1996) S321–6.

to provide a classification into distinct morphological classes: the micro-organism undergoes a continuous change.

8.4.2 *Aureobasidium pullulans*

A similar comment is made by Guterman and Shabtai (1996), who attempted to characterize the complex morphology of *Aureobasidium pullulans*, a basidiomycete producing pullulan. They distinguish five morphological states (Figure 8.25): yeast cells, young blastospores, swollen blastospores, chlamydospores and mycelia. Human experts may sometimes disagree on the classification. Here, classical image analysis is combined with neural networks and fuzzy logic technologies in order to match the classification results made by experts.

8.5 CONCLUSIONS

Because of the large interest in *Saccharomyces cerevisiae*, applications of automated morphometry to yeast cells have largely been devoted to this species: size and budding characteristics are generally addressed. There is a trend to extend this technique to other yeast species, especially the ones prone to polymorphism. In that field there is a clear relationship with the work being done on classical filamentous species (see Chapter 14.). One can go even further than simple morphometry: using dyes, cell differentiation can be monitored (Vanhoutte *et al.*, 1995) and this could improve our understanding of the cell physiology.

It should also be mentioned that some of the measurements described here for *S. cerevisiae* can be applied to other types of cells: mammalian cells, growing in suspension, are good candidates for such morphometric analysis. Because of their relatively large size they are easily observed by optical microscopy. Such extension cannot be made without caution for very small cells such as *E. coli*, for which the imaging is difficult and large errors can occur.

At present, it is possible to foresee the routine use of automated morphometry in industrial laboratories for off-line control of cell growth and cell quality. Unfortunately, the attempts to obtain such information *in situ* or at least on-line have had limited success to date. It is also true that automated image analysis, even when using dedicated computer systems, remains slow compared with Coulter counting or flow cytometry (see Section 7.10). However, it provides information on morphology that is not given by any other technique. It can also be added that the widespread availability of user-friendly, powerful and low-cost computer systems able to deal with images easily will facilitate the development of such techniques in industry, where quality control is more and more important.

REFERENCES

Baroni, M.D., Martegani, E., Monti, P. and Alberghina, L. (1989) Cell size modulation by CDC25 and RAS2 genes in *Saccharomyces cerevisiae*. *Molec. Cell. Biol.* **9**: 2715–23.

Berner, J.L. and Gervais, P. (1994) A new visualization chamber to study the transient volumetric response of yeast cells submitted to osmotic shifts. *Biotechnol. Bioengng* **43**: 165–70.

Bonaly, R. (1991) Morphologie et reproduction asexuée des levures. In *Biotechnologie des Levures* (Larpent, J.P., ed.), pp. 3–22. Masson: Paris.

Brown, C.M. and Hough, Y.S. (1965) Elongation of yeast cells in continuous culture. *Nature* **206**: 676–8.

Chun, K.T. and Goebl, M.G. (1996) The identification of transposon-tagged mutations in essential genes that affect cell morphology in *Saccharomyces cerevisiae*. *Genetics* **142**: 39–50.

Costello, J.P. and Monk, P.R. (1985) Image analysis method for the rapid counting of *Saccharomyces cerevisiae* cells. *Appl. Environ. Microbiol.* **49**: 863–6.

Danielsson, P.E. (1980) Euclidian distance mapping. *Comp. Graphics Image Proc.* **14**: 227–48.

Guterman, H. and Shabtai, Y. (1996) A self tuning vision system for monitoring biotechnological processes. I. Application to production of pullulan by *Aureobasidium pullulans*. *Biotechnol. Bioengng* **51**: 501–10.

Hirano, T. (1990) Method for measuring the budding index of yeast. *ASBC J.* **48**: 79–81.

Huls, P.G., Nanninga, N., van Spronsen, E.A., Valkenburg, J.A.C., Visscher, N.O.E. and Woldringh, C.L. (1992) A computer-aided measuring system for the characterization of yeast populations combining 2D-image analysis, electronic particle counter and flow cytometry. *Biotechnol. Bioengng* **39**: 343–50.

Lichtfield, J.B., Reid, J.F. and Richburg, B.A. (1992) Machine vision microscopy for on-line sampling, analysis and control. In *Modeling and Control of Biotechnical Processes*, pp. 275–8. IFAC: Colorado.

Lord, P.G. and Wheals, A.E. (1980) Asymmetrical division of *Saccharomyces cerevisiae*. *J. Bacteriol.* **142**: 808–18.

Lord, P.G. and Wheals, A.E. (1981) Variability in individual cell cycles of *Saccharomyces cerevisiae*. *J. Cell Sci.* **50**: 361–76.

O'Shea, D.G. and Walsh, P.K. (1996a) Morphological characterization of the dimorphic yeast *Kluyveromyces marxianus* var. *marxianus* NRRLy2415 by semi-automated image analysis. *Biotechnol. Bioengng* **51**: 679–90.

O'Shea, D.G. and Walsh, P.K. (1996b) Morphological characteristics of the yeast *Kluyveromyces marxianus* var. *marxianus* NRRLy2415 immobilized in calcium alginate gel beads. *IChemE Res. Event, Eur. Conf. Young Res. Chem. Eng. 2nd* **2**: 1061–3.

Pons, M.N., Vivier, H., Rémy, J.F. and Dodds, J.A. (1993) Morphological characterization of yeast by image analysis. *Biotechnol. Bioengng* **42**: 1352–9.

Pons, M.N., Litzén, A., Kresbach, G.M., Ehrat, M. and Vivier, H. (1997) Study of *Saccharomyces cerevisiae* yeast cells by field flow fractionation and image analysis. *Sep. Sci. Technol.* **32**: 1477–92.

Robinow, C.F. (1975) The preparation of yeasts for light microscopy. In *Methods in Cell Biology* (Prescott, D.M., ed.), vol. 11, pp. 1–22. Academic Press: New York.

Russ, J.C. (1995) *The Image Processing Handbook*, 2nd edn. CRC Press: Boca Raton, FL.

Scherr, G.H. and Weaver, R.H. (1953) The dimorphism phenomenon in yeasts. *Bacteriol. Rev.* **17**: 51–92.

Shimosaka, M., Masui, S., Tagawa, Y. and Okazaki, M. (1991) Cell aggregation and elongated cell morphology caused by a mutation conferring increased sensitivity to indole-3-acetic acid in the yeast *Saccharomyces cerevisiae*. *J. Ferment. Bioengng* **72**: 485–7.

Srinorakutara, T., Zhang, Z. and Thomas, C.R. (1996) Osmolarity effect on yeast cell strength and cell size. *IChemE Res. Event, Eur. Conf. Young Res. Chem. Eng. 2nd* **1**: 151–3.

Suhr, H., Wehnert, G., Schneider, K., Bittner, C., Scholz, T., Geissler, P., Jähne, B. and Scheper, T. (1995) In situ microscopy for on-line characterization of cell-populations in bioreactors, including cell-concentration measurements by depth from focus. *Biotechnol. Bioengng* **47**: 106–16.

Vanhoutte, B., Pons, M.N., Thomas, C.R., Louvel, L. and Vivier, H. (1995) Characterization of *Penicillium chrysogenum* physiology in submerged cultures by color and monochrome image analysis. *Biotechnol. Bioengng* **48**: 1–11.

Vicente, A., Meinders, J.M. and Teixeira, J.A. (1996) Sizing and counting of *Saccharomyces cerevisiae* floc populations by image analysis, using an automatically calculated threshold. *Biotechnol. Bioengng* **51**: 673–8.

Walther, I., Bechler, B., Müller, O., Hunzinger, E. and Cogoli, A. (1996) Cultivation of *Saccharomyces cerevisiae* in a bioreactor in microgravity. *J. Biotechnol.* **47**: 113–27.

Watson, K. and Arthur, M. (1977) Cell surface topography of *Candida* and *Leucosporidium* yeasts as revealed by scanning electron microscopy. *J. Bacteriol.* **130**: 312–7.

Wheals, A.E. (1982) Size control model of *Saccharomyces cerevisiae* cell proliferation. *Molec. Cell. Biol.* **2**: 361–8.

Yamashita, Y., Kuwashima, M., Nonaka, T. and Suzuki, M. (1993) On-line measurement of cell size distribution and concentration of yeast by image processing. *J. Chem. Engng Jpn* **26**: 615–9.

Zalewski, K. and Buchholz, R. (1996) Morphological analysis of yeast cells using an automated image processing system. *J. Biotechnol.* **48**: 43–9.

9

Optimized Population Statistics Derived from Morphometry

Bart C. Meijer[1] and Michael H.F. Wilkinson[2]

[1]*Regional Laboratory for Public Health, Groningen, The Netherlands*

[2]*University of Groningen, Groningen, The Netherlands*

9.1 INTRODUCTION

In this chapter we concern ourselves with shape as an indicator of biodiversity in a microbial system. After an introduction to the gut microflora, we discuss the morphometric measurements used to quantify the shape of a single bacterium. In the same section, the rationale is presented for logarithmic transformation and principal components transformation on the raw measurements. An experimental interlude follows on the effect of grid anisotropy on the principal component scores.

After this we are in a position to describe quantitatively not only individual bacteria, but mixtures of several species of bacteria. In the central section about morphometric distributions, graphical techniques are presented first, followed by a discussion of measures for location and variety.

Returning briefly to experimental matters, we discuss the effects of freezing and slide preparation parameters on our measurements. As an application the effect of ceftriaxone administration on gut microflora is described in the last section.

9.1.1 The gut microflora

In the gut of humans and other mammals, complex microbial ecosystems exist that share important characteristics. In those microfloras, bacteria of hundreds of species coexist (Moore and Holdeman, 1974). Nevertheless, the intact mammalian gut microflora is strongly selective: bacteria of non-indigenous species will fail to colonize the gut, unless they are administered in large quantity (Van der Waaij and Berghuis, 1974). This phenomenon has become known as colonization resistance (Van der Waaij *et al.*, 1971). A healthy flora will keep out most pathogens. Because of this, gut microflora balance and its disturbance have become important topics in human and veterinary medicine.

Gut microflora is a difficult subject of study, from both practical and theoretical points of view. On the practical side, the sheer number of species to isolate, the

Digital Image Analysis of Microbes: Imaging, Morphometry, Fluorometry and Motility Techniques and Applications. Edited by M.H.F. Wilkinson and F. Schut.

fastidious growth requirements of many of them, and the fact that many species are obligate anaerobes and grow very slowly hamper efficient analysis by classical microbiological means.

Theoretically, if 400 species are present in a micro-ecological system, the number of possible pairwise interactions is already very large (of the order of 80 000), and there are indications that even more complex relations operate between bacteria and host, or among bacteria (Freter, 1988). Possible mechanisms for such interactions are environmental factors such as redox potential, pH, bile salts concentration and the composition of the mucus layer; bacteriocines secreted by some bacteria to inhibit others; and last but not least, immunologic factors. The immune system is immensely complex in its own right (Roitt *et al.*, 1985), and the gut wall as a whole is the most important site of interaction between the immune system and the environment.

A detailed mechanistic picture of the interaction among micro-organisms in the gut would require adequate observation and description, candidate mechanisms for bringing about the states described, proof that such mechanisms operate *in vivo*, and an understanding of how these states are maintained in equilibrium. The formidable complexity of the system, the many unknown parameters and the difficulty of anaerobic culturing preclude a direct approach to the problem. Therefore we attempt to describe the system at a more global level.

Useful methods include digital image analysis, applied to slide preparations of bacteria in faeces. This chapter describes how the morphometrical distribution of bacteria on such slides is summarized in a few parameters. The link with exact size measurement and, for instance, biomass determination, is left to Chapter 7. Here we concern ourselves only with the distribution of shapes.

Topics covered include slide preparation, measurements on single bacteria, logarithmic transform, principal components analysis, precision of measurement, and methods to summarize important properties of the resulting distribution in general parameters. One such parameter, morphometric entropy, is theoretically interesting and applicable to other ecosystems as well. It is therefore discussed in more detail.

9.2 MORPHOMETRIC PARAMETERS OF SINGLE BACTERIA

Bacterial forms are three-dimensional. By putting the bacteria on a slide and recording only the projection of its shape on the microscopic object plane, information is lost. Chapter 7 describes methods to reconstruct volume from two-dimensional shape. In this chapter we confine ourselves to analysis of bacterial shape in two dimensions. Likewise, the grey-scale image of the bacterium does not concern us here: our analysis starts after bacteria and background have been separated by the methods described in Chapter 6.

Intuitively one expects bacteria to have only a few morphological characteristics. This is borne out by measurement. Bacterial forms differ from one another in size, roundness and regularity of contour. Surface area (A), perimeter (P), moment of inertia (I) and surface area of the convex hull (H) together provide sufficient information.

Originally, we measured and recorded more variables. They were the erosion width, the skeleton length and the projections of the bacterial shape on the x and

y axes. Erosion width and skeleton length were obtained by mathematical morphology methods (Serra, 1982). Selection of variables was done using principal component analysis. Modified scatter plots were inspected to determine whether additional principal components, derived from the eight variables, carried meaningful morphological information. Both methods are detailed below. This selection process left us with area, perimeter, moment of inertia and convex hull area.

The perimeter is measured by counting the number of pixels in the edge of the figure. This edge can again be found using the methods of mathematical morphology, which are detailed in Section 1.6. Surface area of the bacterium and its convex hull result from counting the pixels in the entire figure of the bacterium and in its convex hull, respectively.

For the moment of inertia, we imagine that the bacterial figure is rotated around an axis through its centre of gravity and perpendicular to the object plane, and that all pixels have the same 'mass', set equal to 1. We get:

$$I = \sum_i (x_i - \bar{x})^2 + (y_i - \bar{y})^2 \tag{9.1}$$

where the x_i are the x coordinates of all pixels belonging to the object, and $\bar{x}$ is their mean; likewise for y_i and $\bar{y}$.

9.2.1 Units of measurement

The distance between adjacent pixels corresponds to a distance, d, in the object plane. This distance is measured in micrometres (μm). It is easiest to measure in pixels (e.g. count pixels in the bacterial figure for A) and afterwards to divide the result by d (for P), d^2 (for A and H), or d^4 (for I).

9.2.2 Linearizing and eliminating theoretical dependencies

The raw data from the measurements have undesirable properties from a statistical point of view. Firstly, the variables have skewed distributions, as illustrated in Figure 9.1 for the moment of inertia of a slide from a *Klebsiella pneumoniae* culture. Secondly, all variables depend on the size of the bacterium, and so on one another. Finally, those relationships are not linear (Figure 9.2).

These difficulties can be overcome by converting all variables except the surface area to make them size-independent and then taking natural logarithms. Thus, we define form factors F_1 and F_2 and concavity factor C:

$$F_1 = \frac{P^2}{A} \tag{9.2}$$

$$F_2 = \frac{I}{A^2} \tag{9.3}$$

$$C = \frac{H}{A} \tag{9.4}$$

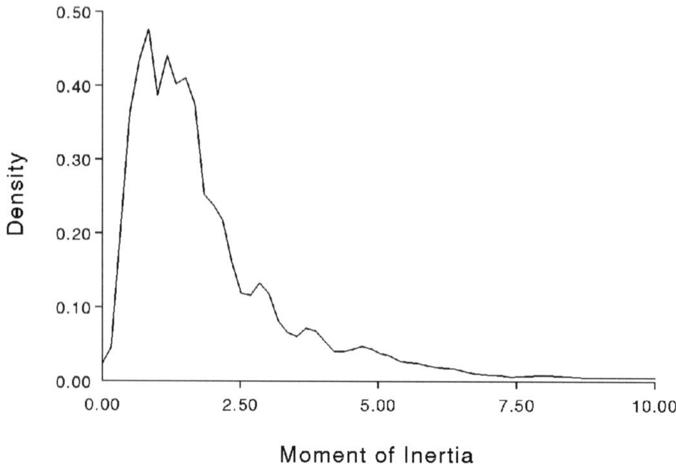

Figure 9.1 Distribution of the moment of inertia for *K. pneumoniae.*

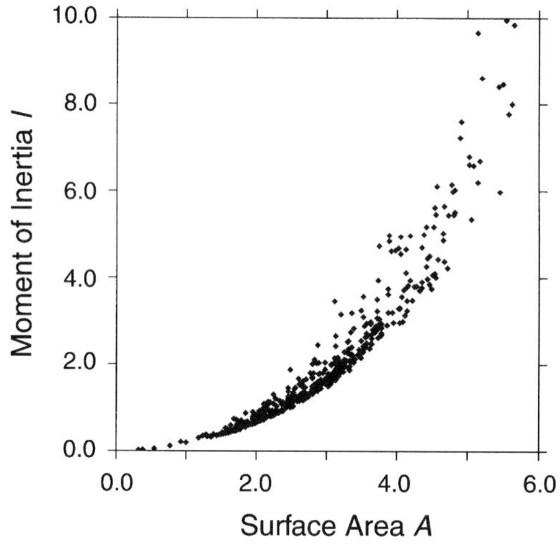

Figure 9.2 Scatter plot of surface area (μm^2) versus moment of inertia μm^4) for *K. pneumoniae.*

and by taking logarithms we get

$$a = \log A \tag{9.5}$$

$$f_1 = 2 \log P - \log A \tag{9.6}$$

$$f_2 = \log I - 2 \log A \tag{9.7}$$

$$c = \log H - \log A \tag{9.8}$$

The first variable, a, measures size; f_1 indicates the irregularity of the contour; f_2 reflects elongation of overall form; and large values for c will occur if the bacterium is curved or otherwise concave. Figure 9.3 shows the resulting improvement in distribution characteristics, independence of size and linearity.

9.2.3 Principal components analysis

We are not done yet. If we were interested in the distribution of a single variable, say x, we might compute its average m and variance s^2, and compute further statistics from the standardized variable:

$$z = \frac{x - m}{s} \tag{9.9}$$

This standardized version has an average of 0 and a variance of 1.

If our aim were to compare various collections of observations $\{x_i\}$, obtained over time, we would compute m and s from the first sample and use the same values every time, so that differences in position and width of the distributions would not go undetected.

Transformation to principal components can be viewed as the multivariate analog of standardization. In this case, the original variables are transformed to principal components, which again have average 0 and variance 1, and are linearly independent of one another. Another, more graphical, way to view principal component transformation is first translating the axes of the coordinate system in such a way that the origin is at the centre of the data. After this the axes are rotated to be parallel to the axes of the (ellipsoidal) data distribution. Finally, the scale along each axis is changed so that the variance of the data along each axis is unity.

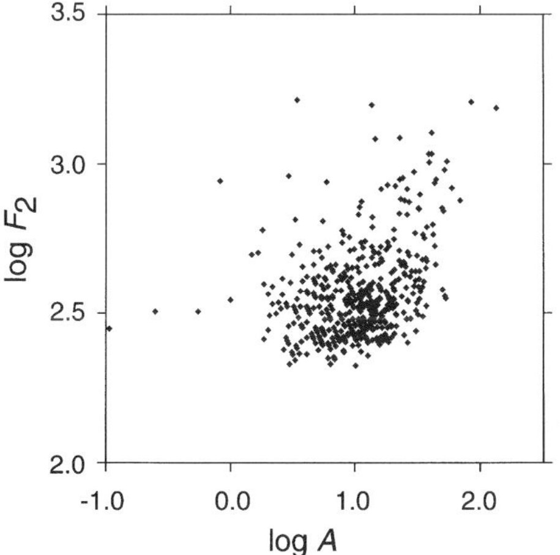

Figure 9.3 Scatter plot: log surface area versus log second form factor for *K. pneumoniae*.

Familiarity with multivariate statistics is probably necessary to read the following, more formal, paragraphs. Readers may wish to consult a textbook, such as the one by Mardia *et al.* (1979). Following their approach we will focus on the reasons why transformation to principal components is useful in our case.

Let $\mathbf{X}$ be the standardized data matrix: its elements x_{ij} contain the values of observation of variable j (there are p variables, so $j = 1..p$) on bacterium i (the sample consists of m bacteria, so $i = 1..m$), standardized so that, over all bacteria, the variables have a mean of 0 and a variance of 1. The sample correlation matrix $\mathbf{C}$ will then be:

$$\mathbf{C} = \frac{1}{m-1}\mathbf{X}'\mathbf{X} \tag{9.10}$$

and because $\mathbf{C}$ is symmetrical and positive, it can be written as:

$$\mathbf{C} = \mathbf{G}\mathbf{L}\mathbf{G}' \tag{9.11}$$

In this equation, $\mathbf{L}$ is a $p \times p$ matrix with the eigenvalues of $\mathbf{C}$ as its diagonal elements, and zero everywhere else; $\mathbf{G}$ is the matrix of corresponding eigenvectors. $\mathbf{L}$ (and $\mathbf{G}$) are sorted in descending order: $\lambda_{11} \geqslant \lambda_{22} \geqslant \cdots$. If it exists, (i.e. if all eigenvalues are non-zero) the inverse of $\mathbf{L}$ is found by substituting the diagonal elements by their reciprocals: $\mathbf{L}^{-1} = \{\lambda_{ii}^{-1}\}$. It even has a meaning to take the square root of $\mathbf{L}$, and the inverse of that: $\mathbf{L}^{-1/2} = \{\lambda_{ii}^{-1/2}\}$. As expected, $\mathbf{L}^{-1}\mathbf{L} = \mathbf{I}$, and $\mathbf{L}^{-1/2}\mathbf{L}\mathbf{L}^{-1/2} = \mathbf{I}$.

Because $\mathbf{G}$ is orthogonal, the inner products of column i with column j are 1, if $i = j$, and 0 otherwise. For this reason, $\mathbf{G}'\mathbf{G} = \mathbf{G}\mathbf{G}' = \mathbf{I}$: the transpose and the inverse of $\mathbf{G}$ are the same. Now, we wish to transform the data matrix $\mathbf{X}$ to some $\mathbf{Y} = \mathbf{X}\mathbf{A}$, so that the covariance of $\mathbf{Y}$ will be the identity matrix:

$$\mathbf{C}_{\mathbf{y}} = \mathbf{A}'\mathbf{X}'\mathbf{X}\mathbf{A} = \mathbf{A}'\mathbf{C}_{\mathbf{x}}\mathbf{A} = \mathbf{I} \tag{9.12}$$

Let us try $\mathbf{A} = \mathbf{G}\mathbf{L}^{-1/2}$. We get:

$$\mathbf{A}' = \mathbf{L}^{-1/2}\mathbf{G}' \tag{9.13}$$

and

$$\mathbf{C}_{\mathbf{y}} = \mathbf{A}'\mathbf{G}'\mathbf{L}\mathbf{G}\mathbf{A} = \mathbf{L}^{-1/2}\mathbf{G}'\mathbf{G}\mathbf{L}\mathbf{G}'\mathbf{G}\mathbf{L}^{-1/2} \tag{9.14}$$

This looks more daunting than it is. We get:

$$\mathbf{L}^{-1/2}\mathbf{G}'\mathbf{G}\mathbf{L}\mathbf{G}'\mathbf{G}\mathbf{L}^{-1/2} = \mathbf{L}^{-1/2}\mathbf{L}\mathbf{L}^{-1/2} = \mathbf{I} \tag{9.15}$$

which proves that our transformation is the desired one.

This is all very nice, but what are the advantages of such an elaborate transformation? Firstly, the original data are transformed so that they will scatter as spherically as possible (Figure 9.4). This carries the mundane advantage that

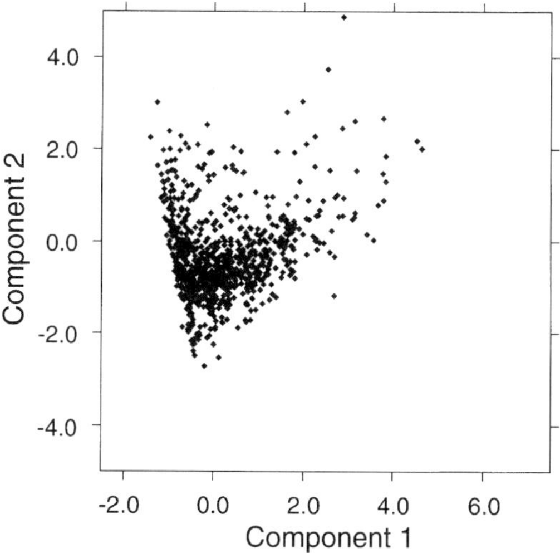

Figure 9.4 Scatter plot of morphometric distribution of faecal bacteria.

graph paper will be used economically when we make scatter plots, but it is more important for the easy determination of the entropy and density distribution of a sample (as described below). Secondly, components associated with low eigenvalues are often dropped entirely. The resulting approximation of the original data is optimal in a least-squares sense. Thus, the information in the data can be stored as economically as possible. (Incidentally, this is the basis of the Hotelling technique of image compression.)

In practice the transformation is done in two stages. First, the original variables are standardized, as above. Next, a linear transformation is applied, yielding the component scores. Accordingly, one needs means and standard deviations for the first step and component score coefficients (i.e. the **A** matrix) for the second. By using those statistics and coefficients we can again compute standard component scores to compare sets of observations. As in the standardization case, values of statistics and coefficients are computed once and for all from the first set of samples (the learning set) measured in a project of this kind.

9.3 EXPERIMENTAL TESTING OF THE PRECISION

The four measurement variables have now been transformed to four principal components. The first of these explains most of the variance, followed by the second, and so on. It is to be expected that one or more of the components associated with the smallest eigenvalues will not contain much useful information. Those components need not be considered in further analysis. An experiment clarifies these ideas and tells us which components to keep and which to discard.

9.3.1 Slide preparation

Obviously, a technique is required that provides optimal separation between bacteria and other matter. The author and his colleagues have obtained good results by simple means, as follows.

A solution of 2.5 g of Tween-80 per litre of distilled water was prepared – the suspension fluid – then 0.5 g of faeces was thoroughly mixed with 4.5 ml of suspension fluid, using a Vortex mixer and glass beads. To separate bacteria from debris, the suspension was centrifuged at low acceleration for 5 minutes, so that the bacteria remained in the supernatant fluid, which was decanted and centrifuged again, at $9000 \times g$, for 10 minutes. The bacteria, now in the pellet, were suspended again in 4.5 ml of suspension fluid. The acceleration for the first centrifugation step was determined experimentally (see below).

From the final suspension slides were made, either unstained, to be viewed using phase contrast microscopy, or stained with nigrosine: a nigrosine solution of 100 g/l was mixed in equal quantities with the bacterial suspension, and from the resulting mixture 5 µl was spread on an object glass, using a second object glass at a 45° angle. After drying, a cover slip was sealed on to prevent scratching.

9.3.2 Image analysis

An Olympus BH2 microscope was used to view the slides, with a 100× objective lens (numerical aperture 1.30) and a Sony CH 1400 video camera. Total magnification was such that the distance between adjacent pixels in the digital image corresponded to 0.126 μm in the object.

In this study, image analysis proceeded as in earlier work: after obtaining the image, an out-of-focus image was subtracted from it to eliminate slow variations in background intensity. The y axis was mapped to three quarters of its original length to correct for pixel aspect ratio. Averaging with a 3×3 kernel was performed. Then, the Robust Automatic Threshold Selection (RATS) algorithm described by Kittler *et al.* (1985; see Section 6.7.4) was used to separate objects and background. The objects in the resulting binary image were screened visually by the operator and artefacts were deleted. Remaining objects, including diplococci, streptococci, etc. were analysed as if they were single bacteria. From each portion of faeces, at least 1000 bacteria were analysed whenever possible. Four measures were taken of each bacterium: surface area (A), perimeter (P), moment of inertia (I) and convex hull area (H). The raw data were transformed logarithmically and converted to principal components, all as described above.

9.3.3 Results: the effect of grid anisotropy

In the experiment, fields of view were recorded twice, the second time with the camera rotated through a random angle between 0° and 90°. Analysis of variance was performed on the paired measurements on 92 bacteria. The results are shown in Table 9.1.

Clearly, differences in bacterial shape are the chief determinants of principal components 1, 2 and 3, whereas the rotation, which acts as a random disturbance,

Table 9.1 Analysis of variance for anisotropy (92 bacteria were measured twice).

Effect of		SS	MS[1]	F	P
Bacteria	PC 1	115.570	1.270	46.335	0.000
(df = 91)	PC 2	171.262	1.882	18.459	0.000
	PC 3	91.910	1.010	12.007	0.000
	PC 4	33.397	0.367	0.394	1.000
Rotation	PC 1	2.482	0.027		
(df = 92)	PC 2	9.384	0.102		
	PC 3	7.728	0.084		
	PC 4	85.652	0.931		

[1] SS, sum of square deviations from mean; MS, mean square deviation from mean: variance.

explains most of the variance in component 4. We must obviously keep the first three components, and discard the fourth.

9.4 MORPHOMETRIC DISTRIBUTIONS

9.4.1 Graphical display

In principal component space every bacterium measured yields a measurement point, or vector, which consists of the three principal components obtained for it. By measuring a large number of bacteria (around 1000) we get a cloud of points, which we consider as a sample drawn from some distribution.

When examining large, multivariate data sets, it is very valuable to be able to visualize the distribution of the data in the multidimensional data space, or at least in lower dimensional projections of the data space. In practice, this usually means projection onto two-dimensional data plane, or occasionally a three-dimensional data space. In the following discussion we focus on two-dimensional representation, although the theory can be extrapolated to more than one dimension in a number of cases.

9.4.1a Morphograms: modified scatter plots

In general statistics it is customary to show bivariate sample data graphically in a scatter plot. Scatter plots are useful in our case, too. Figure 9.4 is a scatter plot made from data obtained from human faecal flora. In Figures 9.5 and 9.6 a modification (dubbed *morphogram*) is shown. The plotted points are replaced by drawings of the bacteria themselves, drawn with their centres of gravity on the spot corresponding to their component scores.

Conceptually this is straightforward enough, yet there are a number of pitfalls to avoid. First of all, not all objects should be printed, since the touching and overlapping objects would clutter the graph completely. To prevent this, the plot area is represented by a 512×512 bitmapped image, which is at first empty. After each object is translated to its position in the graph (and corresponding bitmap), the original shape is dilated at least once. If this dilated shape overlaps any object pixels in the bitmap,

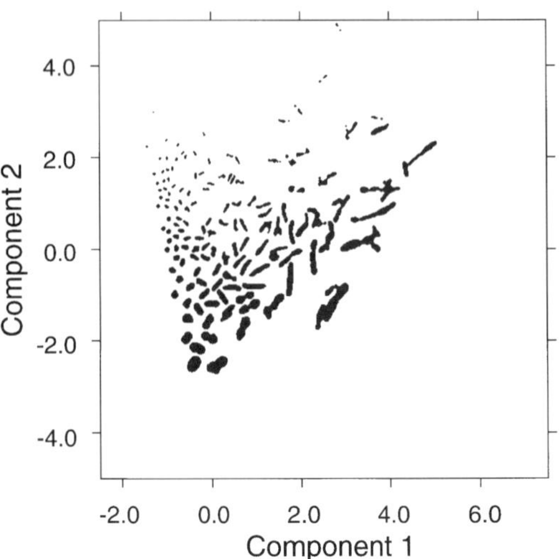

Figure 9.5 Modified scatter plot (morphogram) of components 1 and 2. The objects are drawn on the spots corresponding to their component scores. Reproduced from *Epidemiol. Infect.* 1991, **107**, 383–91, by permission of Cambridge University Press.

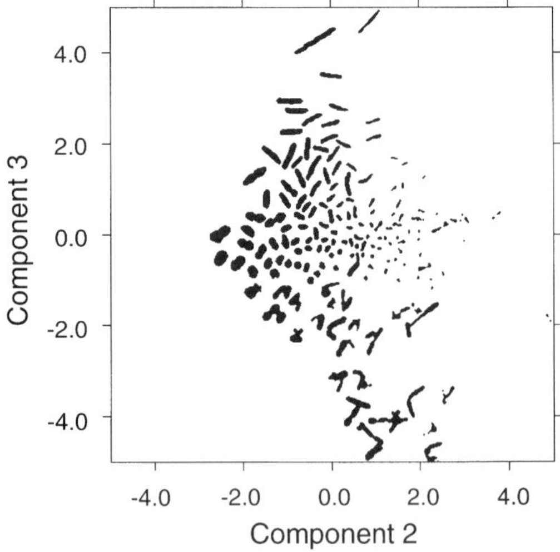

Figure 9.6 Morphogram of components 2 and 3. The objects are drawn on the spots corresponding to their component scores. Reproduced from *Epidemiol. Infect.* 1991, **107**, 383–91, by permission of Cambridge University Press.

the object is not plotted, since it would touch an object that has already been plotted. If no overlap is detected, the original object is added to the bitmap image, and scaled to fit the resolution of the graph (which can have an arbitrary size).

This brings us to the second problem. The images of the individual bacteria are stored in a scan-line format, as discussed in Section 6.8. This format is not easily scaled to an appropriate size in pixels for graphs of arbitrary resolutions (in dots per inch). Therefore, each bacterial shape that must be added to the graph must first be converted to a polygon, which is an easily scalable boundary representation. For further implementation details see Wilkinson and Meijer (1995). Note that the *pointset* representation discussed in Section 11.4.1, which stores both a scan-line and a boundary representation of each object, should be highly suitable for generating morphograms.

9.4.1b Density estimation

In the univariate case, the most commonly used estimates of the population density distribution are histograms. However, in the multivariate case, kernel density estimation is much to be preferred over multivariate histograms. Although the latter are trivial to compute, the number of bins (and the amount of memory required) rises exponentially with the number of dimensions, for a given resolution per axis. This also means that the number of data points needed to obtain reasonably accurate density estimates rises very rapidly indeed. An alternative to histograms is the scatter plot, which requires smaller numbers of data, but may easily be confusing, especially when the numbers of points are large. Morphograms suffer from similar drawbacks. A thorough discussion on different representations can be found in Silverman (1986). Kernel estimation can be compared to making scatter plots. Instead of plotting a symbol at the correct location in the graph, each data point is represented by a small probability distribution, or kernel function (K) centred on the data point. These kernels are summed, and the normalized result yields an estimate of the density distribution.

A number of statistical packages support kernel smoothing, notably XploRe (Härdle, 1995). The following treatment of this method is intended for the mathematically inclined, and those interested in implementing this method in their software. Less mathematically inclined readers might wish to compare Figures 9.4, 9.5, 9.7 and 9.8 which show a scatter plot, morphogram, and kernel estimates of the same distribution. For convenience of notation, we represent positions in data space as vectors ($\mathbf{x}$). The N data points are given as $\mathbf{x}_i$. After normalization, the sum of all these probability distributions yields the density estimate $\hat{p}(\mathbf{x})$ at any point:

$$\hat{p}(\mathbf{x}) = h^{-d}N^{-1}\sum_{i=1}^{N} K\left(\frac{\mathbf{x}_i - \mathbf{x}}{h}\right) \tag{9.16}$$

in which h is the smoothing or window width, and d is the number of dimensions. Many different kernels have been proposed (Silverman, 1986). Given that the kernels should be probability density functions with zero mean and unit variance, it has been shown by Epanechnikov (1969), that given an optimum window width h, the smallest mean integrated square error (MISE) in $\hat{p}(\mathbf{x})$ is obtained using:

$$K_e(\mathbf{u}) = \begin{cases} \dfrac{d+2}{2c_d}(1 - \mathbf{u}\cdot\mathbf{u}) & \text{if } \mathbf{u}\cdot\mathbf{u} < 1 \\ 0 & \text{otherwise} \end{cases} \tag{9.17}$$

in which c_d is the volume of the unit sphere in d-dimensional space. This kernel is known as the Epanechnikov kernel. Note that the form of K_e in equation (9.17) does not have unit variance. This is compensated by multiplying the desired window width by the square root of the variance of K_e. One 'optimum' choice h_{opt} for the window width h is given by:

$$h_{\text{opt}} = \left\{ \frac{8(d+4)(2\sqrt{\pi})^2}{c_d} \right\}^{1/(d+4)} N^{-1/(d+4)} \sigma \tag{9.18}$$

in which σ is the square root of the average of the variances of the data in each of the dimensions. When statistical inferences are made from the density distributions, it is probably best to use robust estimators for these variances. In the context of graphical display, or exploratory data analysis, Silverman refers to h_{opt} as in equation (9.18) as the automatic rather than optimal window width, since it is mainly used to give the user a good starting point. Values slightly smaller than h_{opt} are usually better, since it is easier to smooth an under-smoothed distribution in one's mind than it is to sharpen an over-smoothed graph.

In the simplest approach, the same, fixed width, circular (spherical in three dimensions) kernel is used for each data point. One adaptation, which becomes necessary when the data have radically different standard deviations in the different dimensions, is to use elliptical kernels (in two dimensions), in which the major and minor axes are proportional to the standard deviations (this is equivalent to rescaling the data). For the jth dimension we then have an automatic smoothing width of:

$$h_{\text{opt},\,j} = \left\{ \frac{8(d+4)(2\sqrt{\pi})^d}{c_d} \right\}^{1/(d+4)} N^{-1/(d+4)} \sigma_j \tag{9.19}$$

with σ_j the standard deviation in the jth dimension. The density estimate becomes:

$$\hat{p}(\mathbf{x}) = N^{-1} \det \mathbf{H} \sum_{i=1}^{N} K(\mathbf{H}(\mathbf{x}_i - \mathbf{x})) \tag{9.20a}$$

with the matrix $\mathbf{H}$ defined as:

$$\mathbf{H} = \begin{pmatrix} h_{11}^{-1} & 0 & \dots & 0 \\ 0 & h_{22}^{-1} & & \vdots \\ \vdots & & \ddots & 0 \\ 0 & \dots & 0 & h_{dd}^{-1} \end{pmatrix} \tag{9.20b}$$

and therefore:

$$\det \mathbf{H} = \prod_{j=1}^{d} h_{jj}^{-1} \tag{9.20c}$$

with h_{jj} the window width in the jth dimension. Using $h_{jj} = h_{\text{opt},j}$ we have the automatic choice for $\mathbf{H}$. When using principal components, this adaptation is not necessary.

Better results may be obtained if the smoothing is adapted to the local density: little smoothing where the data are abundant, more smoothing when the data are sparse. This can be achieved through the adaptive kernel approach (Breiman *et al.*, 1977), which varies the width of the kernel according to the value of a 'pilot' density estimate $\hat{p}_{\text{pilot}}(\mathbf{x}_i)$, e.g. computed, using a fixed kernel method. The adaptive estimate becomes:

$$\hat{p}(\mathbf{x}) = h^{-d}N^{-1}\sum_{i=1}^{N}\lambda_i^{-d}K\left(\frac{\mathbf{x}_i - \mathbf{x}}{h\lambda_i}\right) \tag{9.21a}$$

with

$$\lambda_i = \left\{\frac{\hat{p}_{\text{pilot}}(\mathbf{x}_i)}{g}\right\}^{-\alpha} \tag{9.21b}$$

in which g is the geometric mean of the pilot densities, λ_i is the local bandwidth parameter, and α the sensitivity parameter. Breiman *et al.* (1977) suggest that $\alpha = 1/d$, others put it at $1/5$ (Parzen, 1962) or $1/2$ (Abrahamson, 1982). In the two-dimensional case, the choices of Breiman *et al.* and Abrahamson are equivalent, so we have used $\alpha = 1/2$ in our work (Wilkinson and Meijer, 1995). Adaptive kernel estimation can be combined with the anisotropic smoothing of equations (9.20a,b):

$$\hat{p}(\mathbf{x}) = N^{-1}\det\mathbf{H}\sum_{i=1}^{N}\lambda_i^{-d}K\left(\frac{\mathbf{H}(\mathbf{x}_i - \mathbf{x})}{\lambda_i}\right) \tag{9.22}$$

In our implementation, the pilot estimate is computed according to equations (9.19) and (9.20) at a grid of, for example, 32×32 points, rather than at the data points $\mathbf{x}$ themselves. To obtain an estimate at the data point, bilinear interpolation from the four nearest grid points is performed (see Section 3.5.6). For large N, this method is considerably faster than direct computation at each data point. After computing the pilot estimate, the program computes the final estimate according to equation (9.22) at the same grid of points, using user-defined values of the window widths h_{jj}.

9.4.1c Related techniques: nonparametric regression methods

Very often, scientists wish to determine how one parameter depends on one or more others. The usual approach is to use some parametric model, and to fit this to the data in some way, by minimizing some penalty function (usually the sum of squared errors). In some cases there may be *a priori* knowledge that the data should fit a particular model, but in many cases no such knowledge is available. A common approach in these cases is to use a (multivariate) linear parametric model. Mathematically speaking, one might look at this as fitting the data to a first order Taylor expansion of the 'true' mathematical form of the relationship. Provided the range of the independent data is small enough, nonlinear behaviour is expected to be small. If nonlinearities do show up, second, or even third and higher order, terms might be included (quadratic, cubic, quartic etc.). Alternatively, one might approximate the data by some arbitrary, smooth curve. Collectively such methods are known as nonparametric regression analysis. These approaches have the advantage that no attempt is made to fit the data in a linear (quadratic, ...) 'straitjacket'. Of course, these methods do not allow estimation of parameters. There is an excellent textbook on these methods by Härdle (1995).

The method that has received most attention in the literature is kernel smoothing, which is very similar to kernel density estimation. One of the simplest forms of this method is the Naradaya–Watson estimator:

$$\hat{y}(\mathbf{x}) = \frac{h^{-d}N^{-1}\sum_{i=1}^{N} K\left(\frac{\mathbf{x}_i - \mathbf{x}}{h}\right) y_i}{h^{-d}N^{-1}\sum_{i=1}^{N} K\left(\frac{\mathbf{x}_i - \mathbf{x}}{h}\right)} = \frac{\sum_{i=1}^{N} K\left(\frac{\mathbf{x}_i - \mathbf{x}}{h}\right) y_i}{\sum_{i=1}^{N} K\left(\frac{\mathbf{x}_i - \mathbf{x}}{h}\right)} \tag{9.23}$$

This states that the predicted value $\hat{y}(\mathbf{x})$ is just the weighted average of the data values y_i. The weights are given by the values of the kernel functions at $\mathbf{x}$. As with kernel density estimation, smoothing may be adaptive, preserving detail where possible.

Other methods include spline interpolation, and k-nearest neighbour (k-NN) methods (for more details see Härdle, 1995).

9.4.1d Displaying the results

Displaying a two-dimensional map of any function of two variables is relatively straightforward. The two most popular methods are the surface plot and the contour plot (Figures 9.7 and 9.8). Algorithms to draw them are abundant in the literature (Burger and Gillies, 1990), and many scientific graphics programs support them as well. One useful addition to standard contour plots, which show relative abundances but no morphological information, is the addition of a morphogram in the same graph (Figure 9.8). To do this, a morphogram is computed and plotted in the usual way, except that the bacteria are shaded rather than given a solid fill pattern. The contour plot is printed as an overlay over the morphogram.

Useful tools for graphical display of scientific data include AVS (Upson *et al.*, 1989), Khoros and the public domain packages GNUplot and XploRe. All of these allow programmers to extend the graphics options with their own code. Application-specific modifications such as the morphogram can be added without the need to write a graphics package from scratch. The graphs in Figures 9.4–9.9 were made using DATAPLOT (Wilkinson and Meijer, 1995).

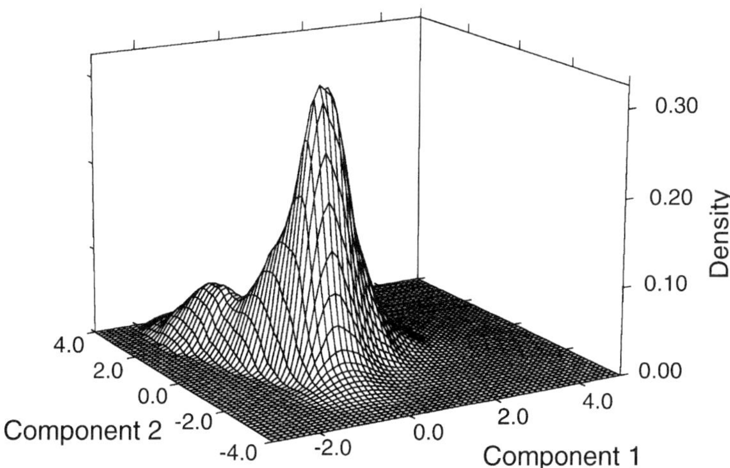

Figure 9.7 Surface plot of adaptive kernel density estimate of the same distribution as in Figures 9.4 and 9.5.

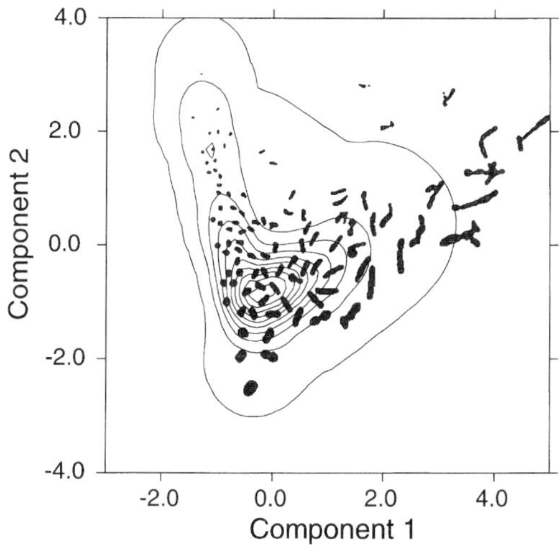

Figure 9.8 Modified scatter plot with contour plot of the adaptive kernel density estimate of gut flora before treatment with an antibiotic. Reproduced from *Epidemiol. Infect.* (1991), **106**: 513–21, by permission of Cambridge University Press.

9.4.1e Insights derived from graphical display

In Figure 9.5, we find the round shapes immediately right of a rather sharp left border. This explains why no bacteria are plotted left of this border: they would need to be rounder than round.

These morphograms and contour plots provide us with a reason to rotate the first two components somewhat: if we rotate both axes through about 8° in a positive direction, we get a first component depending nearly exclusively on roundness, or elongation, and a second component depending on size. This is borne out by the fact that the component loading of surface area on the first component can be made zero. In the practical work described below, this rotation was not applied, however.

Figures 9.8 and 9.9 show modified scatter plots combined with contour plots of the density estimates from the same subject before and after treatment with an antibiotic (ceftriaxone) that affects gut microflora. It is immediately apparent which shapes were eliminated.

9.4.2 Measures for location and variety

In variables assumed to be normally distributed, averages are used to indicate the location of their distribution, and covariance as a measure of variety. From the scatter plots our principal components appear to have a far from normal distribution. In our search for quantities that inform us about location and variety we have to include other candidates too.

9.4.2a Location

Most obvious candidates are averages and medians, computed for the three principal components. Especially for the first component, the presence or absence of few very long rods can make a difference that over-represents the importance of those shapes. In general, averages are less stable than medians. On the other hand, the vector of three averages (the centroid) is not moved relative to the data when components are rotated, whereas the 'centroid' composed of medians does vary with component rotation. For most purposes, component medians are the best measure of location.

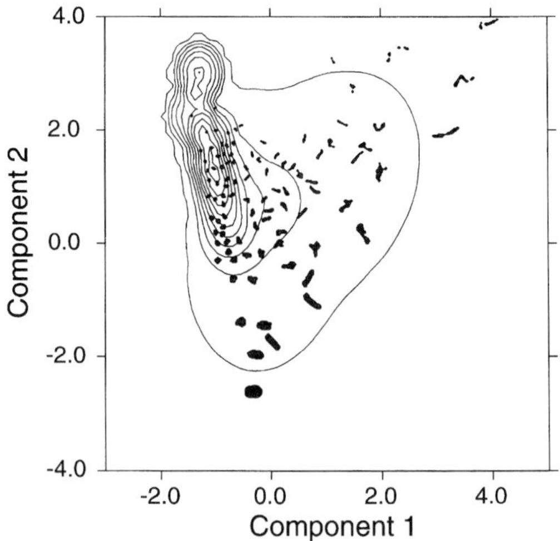

Figure 9.9 Modified scatter plot with contour plot of the adaptive kernel density estimate of gut microflora immediately after treatment with an antibiotic. Reproduced from *Epidemiol. Infect.* (1991), **106**: 513–21, by permission of Cambridge University Press.

9.4.2b Variety

In everyday statistics, the covariance matrix, or its determinant, is the obvious way of quantifying variety in a multivariate distribution. In our case it proves rather useless. It is too sensitive to irrelevant outliers such as very long rods, and not sensitive enough to more subtle changes in flora composition. A more general approach to the notion of variety is needed.

9.4.3 Entropy

9.4.3a Historical background

The usefulness of entropy was discovered almost accidentally. In the first stages of development it was assumed that bacterial shapes could be classified in separate groups (morphotypes), such as 'small cocci', 'large rods' etc. Accordingly, an attempt was made at cluster analysis by the method of *k* means: the probability density in principal component space was written as a weighted sum of multivariate normal distributions, as follows:

$$f(x) = \sum_i w_i f_N(\mathrm{x}; \mu_i, \Sigma_i) \tag{9.24}$$

$f_N(\mathrm{x}; \mu_i, \Sigma_i)$ being the density function of cluster i. Faecal floras from healthy and untreated subjects could be fitted reasonably well by this method when eight clusters were used.

To test for goodness of fit, a likelihood ratio method was used. An estimate of log-likelihood, based on the data themselves using the same methods as used for the entropy, was compared to the expected log-likelihood of the fitted distribution:

$$l = \int f(x) \log f(x) \, \mathrm{d}x \tag{9.25}$$

computed directly by Gauss–Hermite integration (e.g. Press *et al.*, 1992). This method was next applied to floras obtained after treatment with antibiotics. Following the idea of constant morphotypes, the cluster distribution parameters μ_i and Σ_i were kept fixed, and only the weights w_i were allowed to vary. Both the log-likelihood obtained from the data and the expected value for the fitted distribution were computed along with estimates and error bounds for the weights. It turned out that the weights reacted erratically to the changes brought about by the treatment, whereas both log-likelihood measures increased consistently with antibiotic treatment. Furthermore, likelihood ratio testing showed that our weighted-cluster method did not consistently yield good fits. We discarded clustering and kept log-likelihood, or entropy. Besides being useful, the concept of entropy is theoretically interesting for its links to systems theory (Shannon and Weaver, 1949).

9.4.3b Definition

In the case of a random variable X which assumes values x_i with probabilities $p_i(i = 1, ..., I)$, we compute entropy by:

$$S = -\sum_{i=1}^{I} p_i \log(p_i) \tag{9.26}$$

It is easily verified that entropy is maximal if all the p_i are equal, and that it equals zero if only one of the p_i equals 1, and the others are zero. Obviously, the distribution with all p_i equal has highest variety, and the one with only one different from zero has no variety at all. Alternatively, entropy tells us how much information we need to find the value of a variable X, if we only know its distribution. Note that the particular values x_i do not enter into the definition: it does not matter how far apart they are. Ostensibly, entropy is a rather abstract measure of variety.

9.4.4 Estimation of entropy

In practice we do not know the probabilities p_i. We estimate them by performing, say, n drawings X_j from the distribution and counting the numbers n_i of times the values x_i are returned:

$$\hat{p}_i = \frac{n_i}{n} \tag{9.27}$$

If we write the estimated value as the sum of the true value and an error term:

$$\hat{p}_i = p_i + \varepsilon_i \tag{9.28}$$

we find that the expectation value of the error term is zero, and its variance is:

$$s^2_{\varepsilon_i} = s^2_{\hat{p}_i} = \frac{np_i(p_i - 1)}{n^2} = \frac{p_i q_i}{n} \tag{9.29}$$

9.4.4a Bias and variance of the entropy estimator

We now approximate $\hat{p} \log \hat{p}$ by a second-order Taylor expansion:

$$\hat{p} \log \hat{p} \approx p \log p + \varepsilon(1 + \log p) + \frac{\varepsilon^2}{2p} + \cdots \tag{9.30}$$

The expectation of the first term is $p \log p$, the second term has zero expectation, and, from the binomial distribution, that of the third is

$$E\left(\frac{\varepsilon^2}{2p}\right) = \frac{1}{2p} E(\varepsilon^2) = \frac{npq}{2pn^2} = \frac{q}{2n} \tag{9.31}$$

For the entropy we get:

$$-\sum_{i=1}^{I} \hat{p}_i \log \hat{p}_i \approx -\sum_{i=1}^{I} p_i \log p_i - \sum_{i=1}^{I} \frac{q_i}{2n} \tag{9.32}$$

and because:

$$\sum_i q_i = I - 1 \tag{9.33}$$

we find that the entropy estimator is biased:

$$E\left(-\sum_i \hat{p}_i \log \hat{p}_i\right) \approx -\sum_i p_i \log p_i - \frac{I-1}{2n} \tag{9.34}$$

To obtain an approximation for the variance we use Gaussian error analysis:

$$\sigma^2(\hat{p}_i \log \hat{p}_i) \approx \sigma^2(\hat{p}_i)\left\{\left.\frac{\mathrm{d}(p \log p)}{\mathrm{d}p}\right|_{p=p_i}\right\}^2 = \frac{p_i q_i}{n}(\log p_i + 1)^2 \tag{9.35}$$

and sum over i.

9.4.4b *Entropy of continuous distributions; discretization*

For a continuous probability density $f(x)$, entropy is an integral:

$$S = -\int_{-\infty}^{\infty} f(x) \log(f(x)) \, \mathrm{d}x \tag{9.36}$$

For practical purposes, we write the integral as a sum:

$$S = -\sum_i \int_{x_i}^{x_i + \Delta x} f(x) \log(f(x)) \, \mathrm{d}x \tag{9.37}$$

and approximate that sum by assuming that f is nearly constant in the interval from x to $x + \Delta x$ so that:

$$\int_{x_i}^{x_i + \Delta x} f(x) \, \mathrm{d}x = p_i \approx f(x_i)\Delta x \tag{9.38}$$

For the approximation we then get:

$$-\int f(x) \log(f(x)) \, \mathrm{d}x \approx -\sum_i p_i \log\left(\frac{p_i}{\Delta x}\right) = -\sum_i p_i \log(p_i) + \log(\Delta x) \tag{9.39}$$

because the p_i sum to 1. We recognize the equation for entropy in the discrete case, which we can estimate using equations (9.24) and (9.25). In the case of an M-dimensional multivariate density $f(\mathbf{x})$, Δx is replaced by $\Delta \mathbf{x} = \Delta x_1 \Delta x_2 \ldots \Delta x_M$.

9.4.4c. Entropy and systems theory

For a wide class of dynamic systems it can be shown that in equilibrium system entropy reaches a maximum. Every disturbance of this equilibrium will at first cause a decrease in system entropy. If we were able to measure system entropy directly, that would be a sensitive indicator of almost any disturbance. Unfortunately, many more quantities would enter into the definition of system entropy for a multispecies microflora than we could enumerate, let alone measure.

Morphometric entropy, a quantity we can measure, is not system entropy. Compared to the latter, it is incomplete: only morphologic variables enter into it. Because morphometric entropy depends on – arbitrary – morphologic properties of the populations that constitute the bacterial ecosystem, it is also distorted. Results from systems theory can nevertheless be used informally to understand and predict changes in morphometric entropy.

9.4.5 Effect of centrifugation, freezing and slide preparation on measurement

We now have the means to determine important properties of a collection of shapes: its location in component space and the amount of variety. It would be useful if these quantities varied because of interesting influences, such as the use of antibiotics by the person whose gut flora was being examined. But before we can confidently ascribe variations to such scientifically worthy causes we should first be sure that the experiment itself does not cause them. We present an experiment that determines the influences of freezing and thawing faeces before examination, and centrifugation time and acceleration.

9.4.5a Freezing

Fresh faeces were collected from three healthy volunteers. Using the optimal centrifugation procedure, nigrosine-stained slides were prepared twice: once within one hour of defecation, and once after the faeces had been stored at −20 °C for 7 days. The slides were image-analysed and morphometrical distributions were computed. The relative importance of the effect of freezing was assessed by analysis of variance on means and medians of the three principal components, and entropy.

9.4.5b Centrifugation

Nigrosine-stained slides were made from one sample of faeces. 0.5 g of faeces was suspended in 4.5 ml Tween-80, 0.25%, using a Vortex mixer. To separate bacteria from debris, the suspension was centrifuged for 5 minutes at low acceleration so that the bacteria remained in the supernatant fluid, which was decanted and centrifuged again, for 15 minutes at 9000 × g. The bacteria, now in the pellet, were suspended again in 4.5 ml Tween-80, 0.25%. From this suspension a slide was made as described above. This was repeated for several accelerations and durations of the first centrifugation step.

The slides were subjected to image analysis, yielding a morphometrical distribution for each. From each distribution the seven parameters mentioned above were computed.

9.4.5c *Results*

Table 9.2 presents the results of centrifugation experiment. We observe no significant change between accelerations of 10 and $30 \times g$ and conclude that $30 \times g$ is the highest of the accelerations tested that had no significant influence on any of the seven sample properties.

The results of analysis of variance for storage at −20 °C are shown in Table 9.3. Freezing obviously has no significant influence on means, medians or entropy.

Table 9.2 Distribution statistics for different centrifugation accelerations.

	Acceleration (g)				
	10	30	100	300	1000
Mean 1	−0.004(0.028)	0.025(0.026)*	−0.084(0.025)*	−0.022(0.027)*	−0.357(0.019)
Mean 2	−0.038(0.026)*	−0.126(0.024)	−0.075(0.025)*	−0.012(0.024)*	0.150(0.019)
Mean 3	−0.002(0.025)	−0.014(0.025)	0.040(0.023)*	−0.077(0.025)*	−0.250(0.015)
Median1	−0.278(0.029)	−0.267(0.025)	−0.334(0.028)	−0.271(0.031)*	−0.690(0.014)
Median2	−0.154(0.031)	−0.227(0.029)	−0.202(0.036)*	−0.094(0.032)*	0.101(0.027)
Median3	−0.093(0.017)	−0.104(0.021)	−0.093(0.023)	−0.065(0.025)*	−0.365(0.007)
Entropy	4.315(0.10)	4.348(0.10)	4.236(0.10)	4.239(0.10)*	3.809(0.07)

* The difference is significant (pairwise t test; $p < 0.05$). With the medians, a normal approximation is used.

Table 9.3 Analysis of variance for storage at −20 °C.

Effect of	On variable	Sum of squares	Mean square	F	P
Freezing	Mean 1	0.000	0.000	0.000	0.998
(df = 1)	Mean 2	0.002	0.002	1.182	0.338
	Mean 3	0.015	0.015	0.233	0.655
	Median 1	0.000	0.000	0.002	0.963
	Median 2	0.005	0.005	1.937	0.236
	Entropy	0.020	0.020	0.204	0.675
Residual	Mean 1	0.268	0.067		
(df = 4)	Mean 2	0.008	0.002		
	Mean 3	0.257	0.064		
	Median 1	0.309	0.077		
	Median 2	0.011	0.003		
	Median 3	0.368	0.092		
	Entropy	0.394	0.098		

9.5 APPLICATION: EFFECT OF CEFTRIAXONE ON FAECAL FLORA

9.5.1 Introduction

To put the preceding material in perspective, excerpts follow from a paper, 'Effects of ceftriaxone on faecal flora: analysis by micromorphometry' (Meijer *et al.*, 1991), reproduced by permission of Cambridge University Press. Material treated above is omitted or shortened.

In order to elucidate the effect of ceftriaxone therapy on the morphology of gut microflora, eleven human volunteers were treated with ceftriaxone, 1 g daily, given intramuscularly in one dose. Treatment continued for five days. Faecal microflora was analysed by the means described above, before, during and after the treatment period. Ceftriaxone is a relatively new cephalosporin, known for its broad antimicrobial spectrum and long serum half-life. It is mainly excreted by the liver, and therefore may reach high levels in the gut lumen.

9.5.2 Materials and methods

Subjects. Eleven healthy human volunteers (mean age: 34; range: 22–42; eight male, three female) were treated once daily with 1 g of ceftriaxone, injected intramuscularly in combination with lidocaine, for five days. Informed consent was obtained from all volunteers; the study was approved by the Commission for Medical Ethics of the Academic Hospital of Groningen. Faeces were collected from 7 days before to the 22nd day after the beginning of treatment. It was not possible to obtain faeces every day from each volunteer: 218 specimens were analysed in total. There were at least two pre-treatment specimens from each volunteer. Days were numbered as follows: the first dose was given on day 0, previous days being numbered −1, −2 etc. On day 0, faeces were collected before the first dose was given. The last dose was given on day 4.

Slide preparation and image analysis. Microscopic slides were prepared from the faecal samples, using the nigrosine smear technique described above. Image analysis and subsequent data processing have also described above. Entropy was calculated by binning the data in a three-dimensional histogram with 4096 entries, using equations (9.26) and (9.27).

Ceftriaxone concentrations in faeces. De Vries-Hospers *et al.* (1991) determined faecal ceftriaxone concentrations every day. Data from these measurements were used with the authors' permission.

9.5.3 Statistical analysis

For each specimen, medians were computed of the three component scores. This yielded three numbers for each available combination of subject and day. They were used separately in statistical testing. Overall differences between days were tested nonparametrically, according to Friedman (Lehmann, 1975). Before applying the Friedman test, missing data were filled in: if no measurements were available for a subject on a certain day, we substituted the arithmetic mean of all days for that subject. This method reduces the sensitivity of the test, but it is conservative.

Next, pairs of subsequent days were tested for significant difference, using a Wilcoxon symmetry test (Lehmann, 1975). No special treatment of missing data was necessary here: subjects with one or both days of the pair missing were not included in the test.

9.5.4 Graphic representation

Modified scatter plots, as detailed above, were made of selected specimens. The four morphometric variables – the three medians and entropy – were plotted against time for two subjects chosen to represent extremes of possible response to treatment. In order to facilitate comparison, each variable was standardized as follows: for each subject a pre-treatment average was computed and subtracted from the values for that subject. Next, a pre-treatment pooled standard deviation was computed and the values for all subjects were divided by that standard deviation.

9.5.5 Results

For nearly all specimens, the number of bacteria analysed was between 500 and 1000. Due to killing of bacteria by the ceftriaxone, in some cases this number was not achieved in twenty fields of view. These cases were subject 9 (162 bacteria on day 6), subject 10 (384, 56, 74 and 679 bacteria on days 3–6, respectively) and subject 11 (281 bacteria on day 2).

For all three component medians and entropy, the overall differences between days proved significant (Friedman test; $P < 0.1\%$). The Wilcoxon tests between subsequent days showed significant differences for all medians and entropy between days 1 and 2 ($P < 5\%$). Due to missing data, only 174 of 216 possible pairs (80.6%) were complete. Between days 1 and 2, the comparison was applied to 9 of the 11 subjects.

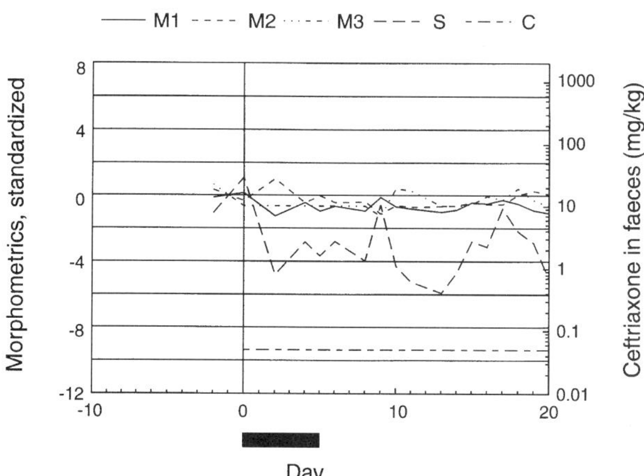

Figure 9.10 Standardized medians 1, 2 and 3 (M1, M2, M3), standardized morphometric entropy (S) and ceftriaxone concentration (C) as functions of time for subject 2. The black bar indicates the treatment period. Reproduced from *Epidemiol. Infect.* (1991), **106**: 513–21, by permission of Cambridge University Press.

There were 40 faecal specimens from the pre-treatment period. Friedman and Wilcoxon tests on this period yielded no statistical significance.

Figure 9.8 shows a modified scatter plot from the pre-treatment period (subject 3, day −6). In Figure 9.9, taken on day 5 from the same subject, the change is notable. Figures 9.10 and 9.11 show the medians of the three principal components, the entropy and the faecal ceftriaxone concentration as functions of time for subjects 2 and 3, respectively. The ceftriaxone concentration is plotted against a logarithmic axis. The medians and the entropy were standardized as described above and

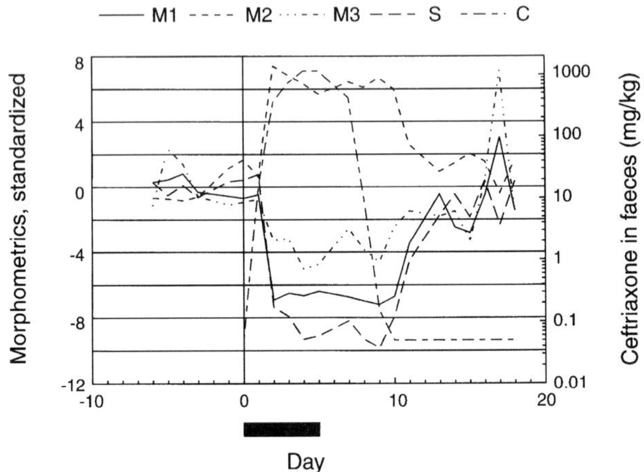

Figure 9.11 Standardized medians 1, 2 and 3 (M1, M2, M3), standardized morphometric entropy (S) and ceftriaxone concentration (C) as functions of time for subject 3. The black bar indicates the treatment period. Reproduced from *Epidemiol. Infect.* (1991), **106**: 513–21, by permission of Cambridge University Press.

Table 9.4 Summary of patterns found in morphometrical parameters and faecal ceftriaxone concentration. Reproduced from *Epidemiol. Infect.* (1991), **106**: 513–21, by permission of Cambridge University Press.

Subject	Ceftriazone detected in faeces	Median 1	Median 2	Median 3	Entropy
1	no	d	i	d	d
2	no	–	–	–	d
3	yes	d	i	d	d
4	yes	d	i	d	d
5	yes	–	i	–	d
6	yes	d	i	–	d
7	no	d	i	d	d
8	no	–	–	–	d
9	no	–	i	–	d
10	yes	–	i	–	d
11	yes	–	–	–	d

i, Increased significantly ($p < 0.05$) during or after treatment; d, decreased; –, no significant change.

plotted against a linear axis. A significance level of 5% is attained when the absolute value of the standardized variable exceeds 2.04 (Student distribution, 29 degrees of freedom, two-tailed; for practical purposes: 2.0).

In Table 9.4 the behaviours of the median of component 1, the entropy, and the ceftriaxone concentration as functions of time are summarized. Median 1 shows significant change in 5 of 11 subjects (45%), median 2 in 8 of 11 (73%), median 3 in 4 of 11 (36%) and the entropy in all 11 subjects.

9.6 DISCUSSION

9.6.1 The ceftriaxone experiment

The results show that digital micromorphometry is sensitive to the effects of ceftriaxone on faecal microflora. Moreover, the influence on morphometric parameters is consistent when various subjects are compared. Generally, the following effects are seen during treatment. Principal component 1 is decreased, component 2 is increased and component 3 is decreased a little. Morphologically, the changes correspond to a relative preponderance of coccoid, small, convex bacteria. The morphometrical diversity as measured by our estimates for entropy is decreased. For those subjects in whom the faecal level of ceftriaxone is above the detection limit, the duration and severity of these effects correlate well with duration and height of ceftriaxone level in the faeces. Interestingly, a consistent effect on the morphometric entropy is also found in those subjects in whom faecal ceftriaxone remained below the detection limit. In this experiment, entropy was the most sensitive morphometric indicator of disturbance in gut microflora.

Digital micromorphometry is fast compared to culturing anaerobic bacteria. It can be used to examine specimens from patients regularly and so monitor their gut flora.

9.6.2 Application to other ecosystems and other types of data

The methods described in this chapter were developed for use in gut microbiology. In view of their general nature, it is to be expected that they are applicable to other microbial ecosystems as well. In particular, morphometric entropy, which is most sensitive to system disturbance, may prove useful.

The conceptual framework provided in this chapter can also be extended to any number of measured parameters. Although the morphology of bacteria may not provide more than three (maybe four) useful principal components, similar analysis of combined multiple spectral band fluorescence and morphometry data could well yield more components.

REFERENCES

Abrahamson, I.S. (1982) On bandwidth variation in kernel estimates – a square root law. *Ann. Statist.* **10**: 1217–23.

Breiman, L., Meisel, W. and Purcell, E. (1977) Variable kernel estimates of multivariate densities. *Technometrics* **19**: 135–44.

Burger, L. and Gillies, D. (1990) *Interactive Computer Graphics*. Addison-Wesley: Reading, MA.

De Vries-Hospers, H.G., Tonk, R.H. and Van der Waaij (1991) Effect of intramuscular ceftriaxone on aerobic oral and faecal flora of 11 healthy human volunteers. *Scand. J. Infect. Dis.* **23**: 625–33.

Epanechnikov, V.A. (1969) Nonparametric estimation of a multidimensional probability density. *Theor. Probab. Appl.* **14**: 153–8.

Freter, R. (1988) Mechanisms of bacterial colonization of the mucosal surfaces of the gut. In *Virulence Mechanisms of Bacterial Pathogens* (Roth, J.A., ed.), pp. 45–60. American Society of Microbiology: Washington, DC.

Hřdle, W. (1995) *Applied Nonparametric Regression Analysis*. Cambridge University Press: Cambridge.

Kittler, J., Illingworth, J. and Föglein, J. (1985) Threshold selection based on a simple image statistic. *Comp. Vis. Graph. Image Proc.* **30**: 125–47.

Lehmann, E.L. (1975) *Nonparametrics: Statistics Based on Ranks*. Holden-Day: San Francisco, CA.

Mardia, K.V., Kent, T. and Bibby, J.M. (1979) *Multivariate Analysis*. Academic Press: London.

Meijer, B.C., Kootstra, G.J., Geerstma, D.G. and Wilkinson, M.H.F. (1991) Effects of ceftriaxone on faecal flora: analysis by micromorphometry. *Epidemiol. Infect.* **106**: 513–21.

Moore, W.E.C. and Holdeman, L.V. (1974) Human faecal flora: the normal flora of 20 Japanese Hawaiians. *Appl. Environ. Microbiol.* **27**: 961–79.

Parzen, E. (1962) On estimation of a probability density function and mode. *Ann. Math. Statist.* **33**: 1065–76.

Press, W.H., Teukolsky, S.A., Vetterling, W.A. and Flannery, B.P. (1992) *Numerical Recipes in C*, pp. 153–4. Cambridge University Press: Cambridge.

Roitt, I.M., Brostoff, J. and Male, D. (1985) *Immunology*. Churchill Livingstone: Edinburgh.

Serra, J. (1982) *Image Analysis and Mathematical Morphology*. Academic Press: London.

Shannon, C.E. and Weaver, W. (1949) *The Mathematical Theory of Communication*. University of Illinois Press: Urbana, IL.

Silverman, B.W. (1986) *Density Estimation for Statistics and Data Analysis*. Chapman & Hall: London.

Upson, C., Faulhaber Jr, T., Kamins, D., Laidlaw, D., Schlegel, D., Vroom, J., Gurwitz, R. and Van Dam, A. (1989) The Application Visualization System: a computational environment for scientific visualization. *IEEE Comp. Graph. Applicat.* **9**(4): 30–42.

Van der Waaij, D. and Berghuis, J.M. (1974) Determination of the colonization resistance of the digestive tract of individual mice. *J. Hyg. Camb.* **72**: 379–87.

Van der Waaij, D., Berghuis-de Vries, J.M. and Lekkerkerk, J.E.C. (1971) Colonization resistance of the digestive tract in conventional and antibiotic-treated mice. *J. Hyg. Camb.* **69**: 405–11.

Wilkinson, M.H.F. and Meijer, B.C. (1995) DATAPLOT: a graphical display package for bacterial morphometry and fluorimetry data. *Comp. Meth. Prog. Biomedicine* **47**: 35–49.

10

Quantitating Single-Colour Fluorescence: Immunofluorescence and Fluorescence *In Situ* Hybridization

Frits Schut[1], Michael H.F. Wilkinson[2], Gijsbert J. Jansen[2] and Dirk van der Waaij[2]

[1] *Microscreen BV, Centre for Microbial Detection and Identification Technology, Groningen, The Netherlands*

[2] *University of Groningen, Groningen, The Netherlands*

10.1 INTRODUCTION

The use of fluorescence measurement in microbiology is not as recent as we are often led to think. One of the most common and yet most cumbersome of microbiological routines, that of identifying individual bacterial species, was challenged by the use of fluorescence early this century when bacteriologists illuminated isolated bacterial colonies with ultraviolet (UV) light (see references in Radley and Grant, 1933). In those days, the primary fluorescence (or autofluorescence) of certain compounds naturally present in living cells was evaluated for taxonomic purposes. Initial results with various types of 'paratyphoid bacteria' and 'dysenteric bacteria' were promising, but it was soon realized that autofluorescence properties of individual species and strains did not harbour information that was taxonomically sufficiently discriminatory. The use of fluorescence as an instrument to investigate the keeping properties of foodstuffs proved more useful. It proved possible to determine the freshness of fish and meat by observing the change in the fluorescence of the surfaces of these products as a result of the presence and activity of bacteria. The discovery of highly fluorescent species of bacteria from fish and shrimps (such as *Microspira phosphoreum*, now *Photobacterium phosphoreum*) originates largely from this time. Fluorescence microscopy, mainly for the detection of *Mycobacterium tuberculosis*, was already widely used by microbiologists in the late 1930s and early 1940s. Since then, applications in fluorescence microscopy have steadily gained in importance. Of these techniques, immunofluorescence (IF) and fluorescence *in situ* hybridization (FISH) are presently the best known.

Digital Image Analysis of Microbes: Imaging, Morphometry, Fluorometry and Motility Techniques and Applications. Edited by M.H.F. Wilkinson and F. Schut.

It has been noted in Chapter 4, and by many other authors, that the plethora of fluorescent staining techniques currently available gives the microscopist an incredibly versatile set of tools for the study of living or fixed cells, whether single cells, communities of microbes, or tissue. However, the full power of these techniques can only be realized if the results can be quantified in some way. To illustrate this, Taylor and Heimer (1975) quote William Thomson (later Lord Kelvin), who stated in 1891:

> When you can measure what you are speaking about and express it in numbers you know something about it; but when you cannot measure it, when you cannot express it in numbers, your knowledge is of a meagre and unsatisfactory kind: it may be the beginning of knowledge, but you have scarcely, in your thoughts, advanced to the stage of *science*, whatever the matter may be.

It is therefore not surprising that a great deal of effort has been put into quantifying the results of staining techniques. Three distinctly different forms of quantification may be recognized: (i) quantifying the fluorescence intensity (per cell, μm^2, pixel), (ii) quantifying the absolute amount, or concentration of the stain, and (iii) determining the number of sites (pixels, cells, organelles) that have been stained (are 'positive'). Types (i) and (ii) may be expressed either in SI units, or more commonly as a fraction of the fluorescence or concentration of some reference, often dubbed arbitrary units. Quantification of the actual numbers of molecules per cell, or μm^3, is much to be preferred over quantification of fluorescence strength, since the latter is very much dependent on the experimental situation. Similarly, absolute quantification is much to be preferred to quantification relative to some arbitrary fluorescence standard, since results may be compared much more readily between laboratories. However, absolute, molecular level quantification is by no means easy. A number of approaches have been put forward, but all of these suffer from problems, so most laboratories simply work with some fluorescence reference, which will allow at least testing of the reproducibility of the results within the laboratory. Provided the reference can be made available to other laboratories, it should be possible to compare results between them (Vogt *et al.*, 1991).

Type (iii), determining the number of positively stained sites, is dimensionless by nature, and is generally speaking the easier form of quantification, though numerous pitfalls exist. The chief problem is which cells are to be considered 'positive' and which not. The problem reduces to signal classification, rather that quantification.

In this chapter we discuss different approaches to signal quantification in single-colour fluorescence measurement, and illustrate these with two different techniques: (i) serum immunoglobulin titre measurement by quantitative indirect immunofluorescence and (ii) quantification of FISH results. Slide preparation, fluorescence intensity calibration, image acquisition and subsequent analysis, and statistical analysis methods of the results are presented.

10.2 IMMUNOFLUORESCENCE

Since the technique was originally described by Coons and Kaplan in 1950, immunofluorescence microscopy has been widely used in a multitude of scientific disciplines. The possibility to examine the morphology of fluorescent objects under the microscope and to localize antigens in microscopical preparations of cells or tissues rendered the technique particularly useful in qualitative analysis of complex antigenic substrates. The technique combines the specificity of the antigen–antibody bond with the sensitivity of fluorescence detection.

10.2.1 Basic principle

Immunoglobulins (Ig) are glycoproteins that can be found in the body fluids, in mucous surfaces and specifically in the serum of mammals, where they can make up 25% of the proteins present. An Ig is called an antibody when the antigen for which the Ig is specific is known. After immunization of a mouse or rabbit with a specific (bacterial) antigen, immunoglobulins are collected from the globulin portion of the serum by ammonium sulphate precipitation. Specific IgG or IgA fractions are isolated by chromatography and, if necessary, treated with pepsin to split the $F(ab)_2$ fragments from the Fc fragment. The IgGs, IgAs or F(ab)s isolated in this way exhibit mixed specificity since over 10^6 species of immunoglobulins can be present in serum. After labelling with a fluorochrome and a final gel filtration to remove unbound dye, an appropriate dilution of the antibodies is used in the immunofluorescence assays.

Briefly, the experimental procedure in immunofluorescence facilitates the binding of a primary antibody labelled with a fluorochrome to an antigen (mostly an extracellular structure like a flagellum, pilus or lipopolysaccharide) immobilized on a microscope slide. Alternatively, a labelled secondary antibody may be attached to an unlabelled primary antibody. The secondary antibody is commonly referred to as the conjugate. Over the years, immunofluorescence has evolved to give rise to the three basic types of immunofluorescence currently in use:

(a) direct immunofluorescence (IF)
(b) indirect immunofluorescence (IIF)
(c) sandwich immunofluorescence (SIF)

In IF assays the conjugate is bound directly to the fixed antigen. This method requires large amounts of antigen on the cell surface since the ratio of the number of detectable antigens to conjugates is at most 1. In IIF assays the primary antibody is tagged with a conjugate. Because several molecules of conjugate (typically 2–5) may bind to one primary antibody, some signal amplification is achieved (Figure 10.1). SIF assays are employed to detect the presence of antibody receptors in which a primary antigen is immobilized on the surface of the slide and a secondary antibody specific for the antigen to be tested is used for final detection.

10.2.2 Limitations of polyclonal antibody IF

In the early days, antigen-specific antibodies were collected from the mixed globulin fraction of the (anti-)serum of a mammal after immunization with the antigen. The quality of the obtained antiserum is in this case pivotal to the success of the IF method, which generally yields ambiguous results due to cross-reactivity

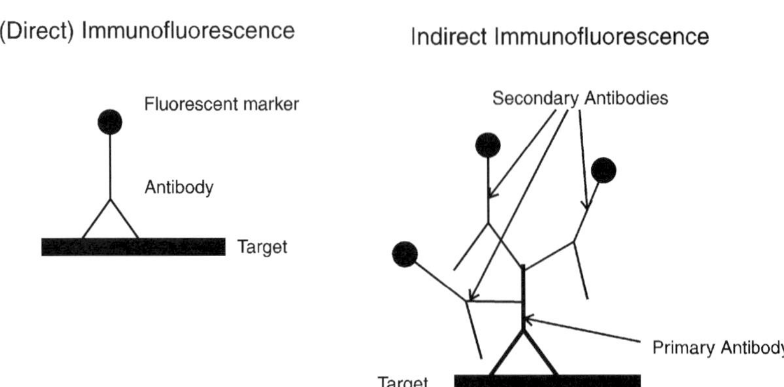

Figure 10.1 The signal amplification of indirect immunofluorescence (IIF) compared to direct immunofluorescence (IF). In IF the fluorescent marker is attached directly to the primary antibody, whereas in IIF a secondary antibody is used, several of which may bind to the primary.

between the secondary antibody and native antibodies in the serum. Although the specificity of the antigen–antibody bond is high, cross-reactivity with related antigens can also occur. This is called undesired specific staining (USS). Furthermore, the Fc region of an antibody can weakly bind to other proteins in the sample such as those of non-related bacterial species. This process results in non-specific staining (NSS). Increasing the sensitivity of the method, e.g. by increasing the number of tagged fluorochromes per antibody, will concomitantly increase the NSS. USS can be diminished by absorption of undesired staining antibodies to the cross-reacting antigens and the subsequent removal of these from the sample by centrifugation. NSS can be reduced by removing the Fc fragment. The resulting F(ab) fragments exhibit increased specificity. Furthermore, reduction of NSS is obtained by using competitor antibodies with a different specificity and a different fluorescent label in large excess over the test antibodies so that they first occupy the non-specific binding sites. In general, the major limitation of the polyclonal antibody method is the mixed specificity of the preparations of immunoglobulins when obtained via the method described above.

10.2.3 Improvements of IF by monoclonals

A major breakthrough in immunofluorescence was achieved when the first monoclonal antibodies (MoAbs) were successfully obtained by fusing nuclei of antibody-producing mouse spleen cells to nuclei of mouse myeloma cells and re-cloning the antibody-producing hybridomas. Mass production of the monospecific MoAbs and labelling of these with different types of fluorochromes resulted in a vast array of ultra-pure and commercially available conjugates. The importance of this technique cannot be overstated. Practically all pathogenic bacteria are suitable for IF labelling with monoclonal antibodies, and clinical applications have been numerous. There is, however, one important limitation to immunological methods: the antigen (bacterium) must be obtained in pure culture, and should preferably be alive. Furthermore, when composition analyses of complex mixed microbial populations are to be conducted with immunofluorescence methods, the specificity of antibodies is so large that vast numbers of antibodies are required to 'cover' the entire population. Group-specific antibodies can hardly be developed, let alone designed. In these instances, FISH can be a major improvement. Finally, low fluorescence signals in IF methods are often still positive due to low antigen titres. In FISH methods, however, low fluorescence signals are indicative of the absence of hybridization and are therefore negative. This makes the evaluation of FISH results somewhat easier than IF acquired results.

10.2.4 Serological staining

The aim of serological techniques is to determine the concentration (or titre) of specific antibodies in whole serum. These antibodies may be present in the serum due to contact between the immune system and a novel antigen. In the assay, in which the antigen is coated to a slide and the natural antibodies in the serum are used as primary antibodies, a titre (reciprocal of the highest (two-fold serial) dilution that is still positive) to the antigen is determined. Labelled mouse anti-human IgA or IgGs are used as secondary antibodies; the method is thus indirect. Acquisition of the absolute intensity of the resulting signal is important because the ratio of positive to negative signals determines the titre (see Section 10.6.3). This can be done efficiently using image analysis (Apperloo-Renkema *et al.*, 1991; Jansen *et al.*, 1993a,b).

10.2.5 Sample preparation for IIF

For both IF and FISH the procedures for sample preparation are virtually identical. Fixation with formalin increases the primary fluorescence of proteins via denaturation. Although this is not a major problem when the fluorescence signal from secondary antibodies is strong, autofluorescence can be a major problem when weak signals are expected. Long fixation times (> 24 h) should therefore be avoided. While formaldehyde fixation is a standard procedure in FISH applications, it is not in IF. In IF applications, where proteinaceous epitopes often serve as antigen, 1% paraformaldehyde will not seriously affect tissue and membrane bound antigens in lymphocytes, but 2–4% formaldehyde or 0.1% gluteraldehyde may cause a rapid loss of antigenicity (Smit *et al.*, 1974).

Protocol 10.1 Sample preparation for IIF.

Cell preparation for image analysis of faecal bacteria

1 Suspend 0.5 g of faeces in 4.5 ml of a 0.5% solution of Tween 20 in sterile water.
2 Suspension of aggregates is facilitated by vortexing with glass beads (diameter 3 mm).
3 Centrifuge the suspension for 10 min at a very low gravity (e.g. $100 \times g$) and dilute the supernatant (cells) 25 times in a 0.5% Tween 20 solution.

Indirect immunofluorescence

1 Degrease microscope slides in 100% acetone for 10 min and air dry.
2 Pipette 10 μl of unfixed cells to the slide.
3 After drying at 40 °C for 15 min, the cells are fixed to the slide by submersion in acetone for 10 min.
4 Incubate the cell smears for 45 min at room temperature with 20 μl diluted serum (1:20 in phosphate buffered saline (PBS), pH 7.2).
5 Wash the slides three times in PBS.
6 Incubate with 20 μl of a diluted solution (1:100 in PBS) of FITC-conjugated goat anti-human F(ab′)2 IgA (Kallestad, Texas, USA) for 1 h at room temperature in a moist chamber;
7 Wash the slides three times in PBS and mount in an appropriate mounting medium (e.g. Vectashield, Vector Laboratories Inc., Burlingame, CA, USA).

10.3 FLUORESCENCE *IN SITU* HYBRIDIZATION

10.3.1 Principle of FISH

One strength of the IF method, i.e. the fact that it is based on the presence of a very specific antigen on the surface of the bacterium, also marks its limitation. Expression of these cell-surface structures is a phenotypic process, and is therefore inherently dependent on the growth conditions of the organism. Although the expression of the nucleic acid targets used in FISH is also determined by the growth rate, the uniformity in the quantity of the target is much better than in IF. FISH is a DNA-probe based technology. Like all DNA-probe methods, the principle is based on the tendency of DNA or RNA to bind specifically to its inverted complement (hybridize). A strand of 5′-AATGGC-3′ will specifically hybridize to a strand of 3′-TTACCG-5′ when brought together under non-denaturing conditions (i.e. high salt or low temperature). In contemporary FISH procedures in microbiology the probe consists of a fluorescently labelled deoxyoligonucleotide of between 15 and 25 nucleotides in length. Its target for hybridization is formed by the 16S or 23S ribosomal RNA molecules inside the microbial cell (Figure 10.2). Unlike proteinaceous targets, the stability of nucleic acid targets is very high. The performance of FISH is therefore relatively independent of growth conditions.

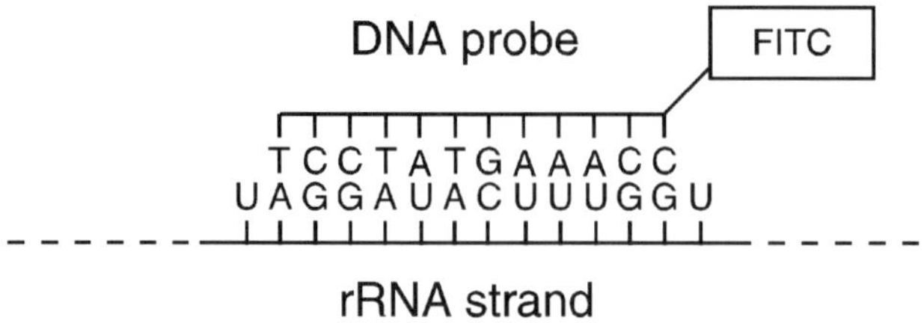

Figure 10.2 The principle of rRNA targeted FISH. The fluorescently labelled DNA strand attaches to the complementary code on the single-stranded rRNA.

10.3.2 Criteria for 16S rRNA probe design

Each living cell contains ribosomes necessary for protein synthesis. Several tens of thousands of ribosomes can be present in actively growing cells. Because of differences in the preservation of mutations within the ribosomal RNA, rRNA molecules harbour both hypervariably and conserved regions as well as regions of intermediary variation. Therefore, the bacterial ribosome is uniquely suited as a phylogenetic (evolutionary) or taxonomic marker (Olsen *et al.*, 1986; Pace *et al.*, 1986). Since synthetic deoxyoligonucleotide probes can be designed to hybridize specifically to a certain sequence of nucleotides in the rRNA molecules, the specificity of the oligonucleotide probes can theoretically be adjusted to fit any taxonomic level. The 16S rRNA sequence of some 5000 bacterial species is presently known (Larsen *et al.*, 1993) and most bacterial groups can therefore be included in searches for taxon-specific 16S rRNA sequences.

Species and group-specific hybridization probes can now be developed on a rational basis, the essence of which lies in the identification of a unique sequence of approximately 20 nucleotides within the sequence of the target organism(s). Shorter stretches (10 bases) will lack specificity. Long sequences (50 bases) have the disadvantage that the discriminatory mismatches do not result in a significant decrease of the dissociation temperature of the non-target/probe complex. The dissociation temperature (T_{d}) of a perfect hybrid can be estimated from the relationship (Stahl and Amann, 1991):

$$T_{\mathrm{d}} = 81.5 + 16.6 \log M + 0.41[\%(G + C)] - 820/n \qquad (10.1)$$

where M is the concentration of monovalent cations (in moles), $\%(G + C)$ the number of strong bonds in the hybrid, and n the length of the oligonucleotide.

Mismatches are best situated 2–5 bases from the 3′ or 5′ ends, as this will result in a relatively long non-hybridizing overhang. Mismatches in the mid-position can, on the other hand, be compensated for by perfect hybridization on both ends, which will result in a low number of effective mismatches. G–T base pairings should be regarded as a perfect match.

10.3.3 Validation of *in situ* probes

The 16S rRNA hybridization technique for *in situ* detection of specific bacterial taxa was first described by Giovannoni *et al.* (1988). Since then, a massive amount of 16S rRNA sequence data have accumulated in the EMBL and GenBank databases. The alignment efforts of the ribosomal database project (RDP) have contributed greatly to the functionality of these data. Presently, species and group-specific hybridization probes can be designed and tested against the entire database by using the CHECK-PROBE command in the RDP services. Apart from database testing, the specificity of probes requires empirical testing against a large number of reference RNAs. This can best be done using so-called phyloblots – a set of isolated rRNAs, PCR-amplified rDNAs or genomic DNAs from a representative selection of phylogenetically distinct micro-organisms immobilized on a membrane in an orderly arrangement to produce a sort of 'reference card'. Simultaneous, if not prior, to tests on the specificity of the probe, it should be established whether the probe is functional in whole cells of the target organism, i.e. whether the target region is available for hybridization, or whether there is any interference with ribosomal secondary or tertiary structure. All too often, a seemingly appropriate probe does not perform at all in whole cells (e.g. Langendijk *et al.*, 1995).

10.3.4 Current limitations in the application of 16S rRNA probes

In the free-water phase of aquatic environments, populations of bacteria range from 10^5 to 10^7. This relatively low number is mainly due to the marginal availability of growth substrates. In certain pockets of activity, such as biofilms or aggregates of organic material, cell densities can reach much higher levels. It is also in these nutrient-rich regions that bacterial cells reach their highest metabolic activity. The detection and identification of the low-density/low-activity bacterial populations in free-water phases confronts marine and freshwater microbiologists with grave problems. The 'great plate-count anomaly' (Staley and Konopka, 1985), or the phenomenon that only about 0.1–1% of the bacteria present in natural samples can form colonies on agar media, is believed to be the result of bacteria with low activity but with a high metabolic potential that refuse to grow on the nutrient-rich media presented to them (Schut *et al.*, 1997). The lack of suitable isolation procedures for aquatic bacteria has now in part been overcome by the ability to identify the bacteria *in situ* with DNA probes. These probes are directed towards the naturally amplified ribosomal RNA targets, of which some 20 000 are present in each growing cell. The low activity of aquatic bacteria, resulting in low ribosomal contents, limits the applicability of rRNA probes in these environments. Recently, cyanine dyes (e.g. Cy3 or Cy5) have been used successfully in whole-cell fluorescence *in situ* hybridization protocols on various groups of bacteria in an alpine lake (Alfrieder *et al.*, 1996). Such methodologies are fundamental in understanding the ecology of the microflora inhabiting these oligotrophic environments.

10.3.5 Sample preparation for FISH

In this example, we demonstrate the use of quantitative FISH for the optimization of hybridization conditions. This can best be done by using pure cultures. To this end, growing cultures are harvested at mid-log, i.e. at a cell-density below the maximum attainable at the medium of choice. Most individual cells in exponentially growing cultures have similar cellular composition and contain large numbers of ribosomes. Therefore, cells from mid-log cultures exhibit the most homogenous distribution of fluorescence signals over the population. This allows for easier optimization of the various parameters in the protocol. Harvesting of the cells typically occurs by centrifugation at between 3000 and 10 000 × *g*.

Fixation of washed cells in 4% formaldehyde or paraformaldehyde is necessary in most instances, although it is claimed that this will increase the autofluorescence. Gluteraldehyde at low concentrations (e.g. 0.5%) can also be used. Although killing of cells is instantaneous with formaldehyde, true fixation is not. Usually, overnight fixation is recommended. But since speed is one of the advantages of FISH over traditional methods, overnight fixation is never to be preferred. We found that a

4-hour fixation in fresh paraformaldehyde at 4 °C is a suitable compromise. Prolonged fixation and storage of samples in paraformaldehyde should be advised against although prolonged storage of samples for some period prior to analysis is necessary in many cases. Storage of cells in 50% ethanol is reported to be a suitable option for long-term storage in the refrigerator. We have found that cells start to disintegrate in such solutions within 2 weeks. Freezing of the samples at −20 °C can lead to cell rupture due to slow growth of ice crystals. Formaldehyde fixation, if necessary in combination with freezing at −20 °C, can be a very useful compromise. The best way to store cells for prolonged periods of time before the actual analysis is to go through the first steps of the FISH protocol (see below) up to the stage where the cells are dehydrated on glass slides. The immediate preparation and ribosomal staining of cells is, however, definitely preferable.

Protocol 10.2 Sample preparation for FISH.

1 Fix the sample for 4 hours in 4% fresh paraformaldehyde, pH 7 (use pH paper, not a pH meter!).
2 Wash the cells in PBS and spot a few millilitres on the surface of a microscope slide.
3 Dry the slide for 15 min and dehydrate the cells by submerging the slide in a graded ethanol series (50%, 70%, 98%, 2–3 min each).

For long-term storage, store the slides in a desiccator over dry silica. In this way, cells can be stored for years without significant loss of FISH fluorescence signals.

The dehydrated cells on the glass slide can also be used directly for FISH (see steps 4–6).

4 Apply a small amount of hybridization mixture to the cells and cover the suspension with a cover glass. The probe (added mostly in a volume 1/10 of the hybridization buffer and at a final concentration of 5 ng/μl) can be added immediately; pre-hybridization is not necessary for FISH procedures.
5 Incubate the cells for 1–16 h in a buffer-saturated chamber, e.g. a Petri dish, at a temperature of 5–10 °C below the dissociation temperature of the probe–target complex.
6 Upon hybridization, wash the cells by submersion of the slide in hybridization buffer for 5 min at a stringent temperature (usually also 5–10 °C below the dissociation temperature of the probe–target complex).

In contrast to the larger 'genomic probes', hybridization rates of short oligonucleotides like 16S rRNA probes are not elevated by the addition of dextran sulphate in the hybridization mixture. Formamide may be used to improve specificity of the probe by increasing the stringency of hybridization. It is also believed that formamide may 'loosen' the tertiary structure of the ribosomal RNA, thereby improving target accessibility. In analogy, reducing hybridization temperatures (e.g. to 37 °C) may increase the association rate between probe and target, as a consequence of which the hybridization time can be shortened, but this may also reduce the accessibility of the target. Unless there are obvious other reasons, it is most convenient to use similar temperatures for the hybridization and washing steps in FISH protocols.

Although the protocol described above can be used for pure cultures, other protocols should be used for quantitative analysis (cell counting) of natural mixed populations of bacteria. During the various washing and dehydration steps of the immobilized cells, bias may occur due to the fact that different cells will detach from the glass surface at different rates. To reduce these effects, the surface of the slides may be coated with poly-L-lysine beforehand, to improve adhesion. For quantitative analysis, however, membrane filtration techniques (0.2 μm pore-size polycarbonate) in combination with epifluorescence microscopy or flow cytometric methods perform better in this sense (see Chapter 7, appendix). Alternative methods used to ensure proper adhesion are discussed in Section 11.3.

10.3.6 The dissociation temperature of probe–target complexes

During the design of an oligonucleotide probe, one may encounter targets with a minimum number (e.g. one) of mismatches. In such instances, specific hybridization conditions have to be determined. Phyloblots or comparative southern hybridization of the probe to target and non-target RNAs will not create the conditions encountered during *in situ* probing. Although the dissociation temperature for oligonucleotides can readily be estimated from equation (10.1), the best way to determine the optimal hybridization and washing temperature in FISH is by empirical testing. When keeping the hybridization temperature low (e.g. 37 °C) and increasing the washing temperature, determination of the fluorescence signal from individual cells by image analysis at different temperatures results in a curve from which the dissociation temperature can readily be calculated. As the

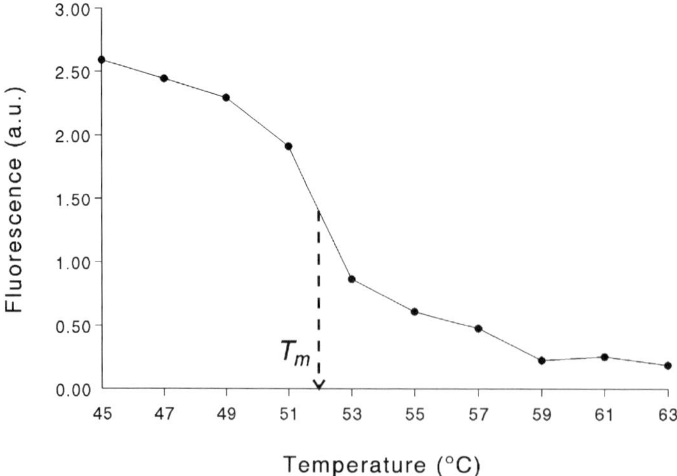

Figure 10.3 Determining the optimum hybridization temperature empirically, by measuring fluorescence strength as a function of hybridization temperature. The point in the curve with the steepest slope yields the empirical melting temperature.

washing temperature determines the stringency in such experiments it should be prolonged to 20 min. In Figure 10.3, the dissociation temperature was determined for a probe with a calculated T_d of 65 °C. The empirically determined melting temperature of this probe was determined at 52 °C. The highest specificity of this probe with maintenance of signal is thus achieved at 50 °C. Hybridization should therefore occur at this temperature.

10.4 SIGNAL OPTIMIZATION

10.4.1 Anti-fading

Tremendous advances have been made in the field of signal prolongation and amplification. The initial conjugates were all labelled with fluorochromes, mostly derivatives of fluorescein or rhodamine. FITC (Riggs *et al.*, 1958), the most stable form of the fluorescein derivatives, reacts with the amino groups of lysine residues under weakly alkaline conditions. Further fluorochromes that found wide use were lissamine rhodamine (RB 200) (Chadwick *et al.*, 1958) and tetramethyl rhodamine isothiocyanate (TRITC) (Hiramoto *et al.*, 1958). Since the sulphonyl chloride of RB 200 needed to be bench-top synthesized in order to reduce fluorescence heterogeneity, TRITC has found wider application. Although these substances fluoresce brightly when excited with the appropriate wavelength, they are not very photostable. Indeed, prolonged excitation may completely eliminate the signal derived from these fluorochromes. Several new photostable fluorochromes have been introduced, such as those of the Oregon Green™ series, SYTOX™ and YOYO™.

Besides the use of more photostable dyes, anti-fading reagents can be added to the mounting medium. Many of these reagents have been proposed, ranging from simple salts such as sodium iodide and sodium azide (Böck *et al.*, 1985), to anti-fluorescein antibodies (Abuknesha *et al.*, 1992). Some of these substances are highly toxic (e.g. sodium azide, and the very popular *p*-phenylenediamine or PPD), and should be handled with care. To minimize health hazards, it may be advisable to use ready-to-go commercial anti-fading mounting media (Longin *et al.*, 1993), e.g. ProLong, VectaShield and CitiFluor.

A number of media (both commercial and 'home made') have been compared by Böck *et al.* (1985). Many of these anti-fading reagents suffer from problems, such as quenching of the initial fluorescence intensity, toxicity and increasing the background fluorescence. PPD in particular shows a strong, wide-spectrum background fluorescence, which overlaps with the fluorescence spectra of most fluorescein- and rhodamine-based dyes (Böck *et al.*, 1985). In our own work, we have compared a commercial PPD-based medium and the addition of sodium iodide (2.5% w/v) to a simple glycerol/PBS mounting medium. Although PPD provides more anti-fading protection under wideband excitation, using narrowband excitation (Olympus filterset MNIB-A for FITC) no difference in protection was observed. However, the fluorescence intensity of unstained bacteria (i.e. the autofluorescence) is up to three times higher with PPD (Figure 10.4). Since autofluorescence is the limiting factor in most fluorescence microscopy (Tanke,

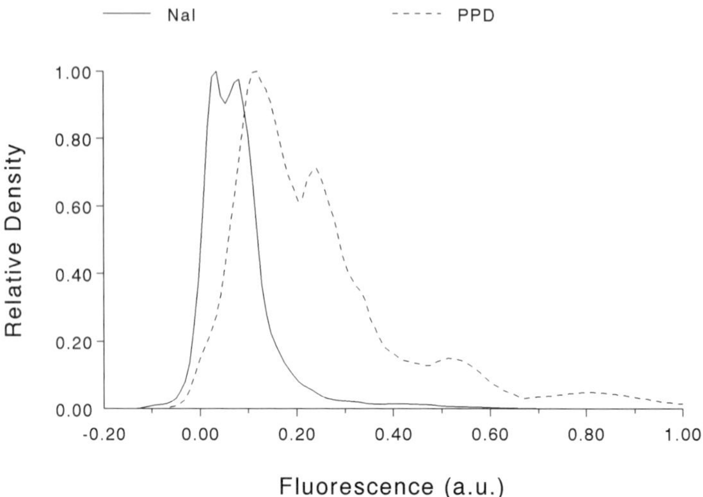

Figure 10.4 Distribution of autofluorescence of faecal bacteria using two different anti-fading reagents. The x axis gives the mean difference in intensity inside and outside each cell (see Section 10.5.3); the y axis gives the probability density relative to the mode of the distribution. Autofluorescence is increased some threefold by use of PPD containing anti-fading reagent.

1989), the NaI medium is much to be preferred in this setting. An added bonus is its low cost and low toxicity.

10.4.2 Signal amplification

Signal amplification has been obtained by alternative methods for conjugate detection. Two well-known examples are (i) the immuno silver gold method and (ii) methods that make use of enzyme-linked conjugates. In the immuno silver gold method, a conjugate labelled with colloidal gold is used to bind to the primary antibody, or to an epitope. Subsequently, ionic silver is added to the immune complex. Due to electrostatic attraction, the silver ions precipitate on the gold to form a large complex that can be detected readily under the light microscope.

The second group of methods makes use of enzyme-linked conjugates. After the conjugate is bound to the primary antibody or the epitope, a substrate is added. The enzyme converts the substrate into a solid, coloured product that can be observed easily by normal light microscopy. Well-known enzymes that are regularly employed in this way are alkaline phosphatase, β-galactosidase and horseradish-peroxidase. Depending on experimental conditions, it may be convenient to add the conjugate and the enzyme separately. Under those circumstances it is advisable to label the conjugate with streptavidin and the enzyme with biotin. The bond between streptavidin and biotin is one of the strongest intermolecular forces known to date and is, therefore, very suitable to link enzyme and conjugate. This latter method also adds flexibility to the procedure since various enzymes may be tested simultaneously. Very impressive signal amplification may be reached when polymeric streptavidin is added. This results in a large complex of biotinylated enzyme interlinked with polymeric streptavidin molecules. This technology makes it possible to bind 20–30 molecules of the enzyme per epitope.

Novel methods of signal amplification in FISH methods make use of enzyme-mediated precipitation of fluorescent radicals (such as the tyramide–fluorescein complex in combination with peroxidase, e.g. Lebaron *et al.*, 1997; Schönhuber *et al.*, 1997) or branched DNA (bDNA) in order to obtain 10–100 times amplification of fluorescent signals relative to conventional monolabelled methods.

Once sufficient signal amplification has been realized, the microscopic image needs to be evaluated. Quantitative analysis of a microscope image is preferred over qualitative evaluation because of the inherent subjectivity of the latter method. In the early 1970s, Taylor and Heimer (1975) described a method for quantifying the light emitted by immunofluorescence of FITC-labelled objects in a microscope field using a photomultiplier tube. Although this method yields reproducible results, it has never become generally employed. Nowadays, flow cytometry and image analysis are commonly used in the analysis of fluorescently-labelled bacteria. In the next section the image processing steps required for quantitative analysis are discussed.

10.5 IMAGE PROCESSING STEPS FOR QUANTITATIVE FLUORESCENCE MEASUREMENT

10.5.1 Image acquisition and sample preparation

In most cases, we need to measure the fluorescence of all cells in a sample, not just the fluorescent ones. If this is the case, it is necessary to acquire two images per field of view: (i) one that contains all objects, and (ii) one that contains the fluorescence information (not necessarily acquired in that order). The best way to acquire image (i) is usually dictated by the properties of the stain used in image (ii). The following considerations for image (i) are important:

(a) Switching from image (i) to (ii) should not cause image shifts or scale changes.
(b) Image (i) should be easy to segment.
(c) The imaging method for (i) should show the objects completely: dyes that stain objects only partially should be avoided. As a minimum requirement, all parts of the cell stained in image (ii) should be visible in image (i).
(d) Any stain used for image (i) must not interfere chemically or optically with the stain used in image (ii).
(e) The illumination for image (i) should not cause fading of image (ii).

The first consideration implies that switching between the images should preferably be achieved by changing the illumination light path, not the emission light path, in classical microscopes. In confocal laser scanning microscopes, the situation is different, since spatial information is contained mostly in the illumination, rather than the emission, light path (Chapter 3). If image (ii) is acquired using classical epifluorescence microscopy, two approaches may be recognized:

(a) Image (i) is a fluorescence image. This requires a double passband fluorescence filter. If the images are acquired sequentially rather than simultaneously, switching between bandpass filters in the excitation pathway switches between the two images. Otherwise a colour camera is needed.
(b) Image (i) is acquired using transmitted illumination. This leaves us with a choice of illumination and staining options.

If possible, the switch should be computer controlled, to prevent any shaking of the microscope. If image shifts and scale changes are unavoidable it is possible to align them, either manually (Section 11.4.2g) or by image processing (Section 3.5.6; Plate 1).

The second consideration reduces the applicability of Nomarski differential interference contrast (DIC) and Hoffman modulation contrast (Sections 3.3.3 and 3.3.4.), since these produce images of objects that differ in texture, rather than brightness. Texture segmentation methods exist, but they all reduce the effective resolution (Section 6.4.4). Any image acquired at low photon counts is also unfavourable, because of low signal-to-noise ratio (SNR) (Section 2.4.1a).

The third consideration eliminates many DNA stains (e.g. Hoechst 33342), since they only stain the nucleoid region, and even RNA stains (16S rRNA targeted FISH) and DNA/RNA stains (acridine orange) since these do not stain the cellular envelope, and produce systematically smaller images of cells than, for example, phase contrast (e.g. Wilkinson, 1996).

The fourth precludes the use of any stain that causes quenching of fluorescence, shows autofluorescence in the same wavelength band as (ii), interferes with binding of probes to target, etc.

The final consideration is less severe than it may sound. If image (ii) is acquired first, no detrimental effect on the measurements will result, provided the illumination used for image (i) is limited to the field of view. Otherwise, there may be fading in parts of the next field of view measured. As a rule, it is best to acquire images at long wavelengths first, since the lower energy photons needed for the excitation of long wavelength emissions are less likely to do damage than high energy photons.

Avoiding fading in image (ii) is also an important consideration when selecting fields and focusing. If this is done using wavelengths and intensities that cause fading of the fluorescence in image (ii), results will depend on the time spent selecting and focusing, and will therefore not be reproducible. Fading in image (i) is not a problem as long as all objects can be distinguished readily from the background.

These considerations have led to the choice of phase contrast for image (i) by a number of groups independently (Apperloo-Renkema *et al.*, 1991; Jansen *et al.* 1993a,b; Nwoguh *et al.*, 1995). In the following discussion, we assume that image (i) is acquired in phase-contrast mode. However, none of the further steps in the measurement process depends on this fact. The only requirement is having two pixel-aligned images of the same field of view, one containing shape information, one fluorescence information. Figure 10.5A,B shows such an image pair.

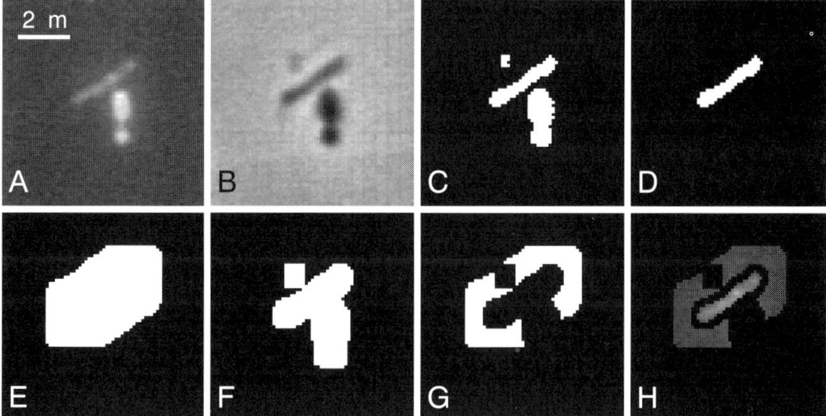

Figure 10.5 Image processing steps in quantitative fluorescence measurement: (A) detail of fluorescence image of faecal bacteria stained with FITC labelled EUB338 probe; (B) phase-contrast image of the same area; (C) segmented phase contrast image; (D) isolated object; (E) object dilated with square 17×17 kernel; (F) segmented image dilated by 5×5 kernel; (G) difference (F)−(G) yields neighbourhood of object excluding any other object, and a 'neutral zone' around all objects; (H) binary masks (D) and (G) overlaid on (A).

10.5.2 Pre-processing and segmentation

10.5.2a Shape image

Image (i) can be pre-processed and segmented using any of the methods in Chapter 6. We use background subtraction according to Meijer *et al.* (1990) (see also Section 6.4.3), 3×3 median filtering, and Robust Automatic Threshold Selection (RATS; Section 6.7.4). Figure 10.5C shows the result of segmenting 10.5B. Each object is defined as a single connected component, which can be measured individually (Figure 10.5D). No attempt at separating touching cells is made by us, unlike certain other groups (Bloem *et al.*, 1995; Møller *et al.*, 1995; Nwoguh *et al.*, 1995). These methods are discussed in Sections 1.7.1b, 6.8, 7.9.2 and Protocol 11.3.

10.5.2b Fluorescence image

The pre-processing of image (ii) may consist of full image calibration (Section 2.4.7), which removes dark current and variations in camera gain. The resulting image must be stored in floating point numbers to prevent round-off errors. Alternatively, it is possible to correct only for dark current, and do the gain correction later, which allows the use of integer (8, 16 or more bits per pixel) mathematics during most of the processing (Wilkinson, 1994). This can be an advantage in certain PC-based image processing systems.

Note that no smoothing of image (ii) should be done unless the result is stored in floating point numbers, again to avoid round-off error. In any case, smoothing may cause a decrease in the fluorescence within the object by blurring the boundaries. Likewise, using certain morphological filters, such as the 'delineation' filter used by

Bloem *et al.* (1995; see also Section 6.7.2) can increase the mean fluorescence within each object, by assigning the maximum signal within the object to the whole object. In our own application, only dark current is subtracted from image (ii).

There are several other pre-processing techniques, which are aimed at removing the blur caused by the finite size of the point-spread function of the optics (Section 2.2.1.). Van Kempen *et al.* (1997) have shown that such image restoration methods can improve the accuracy of measurements of fluorescence of adjacent objects, by reduction of the glare surrounding objects. These methods are generally rather computationally intensive, yet given the increasing processing power of PCs, they will almost certainly become more important in the near future.

10.5.3 Analysis of individual objects

To assess what fluorescence may be attributed to the object, it is necessary to measure the fluorescence in image (ii) within the object, and to determine which part of this is attributable to the background fluorescence of the slide. This latter estimate may be obtained in two ways: (i) globally, e.g. by using the mode of the grey-level histogram as estimate, and (ii) locally, by determining the mean fluorescence intensity in the neighbourhood of the object. The first is the easier: simply subtract the mode of the histogram from all pixels (e.g. by using a look-up table), and determine the fluorescence within each object afterwards.

If the background is very smooth, this method is fine, but if there is any variation, a local estimate is better. Obtaining this local estimate requires some thought. If we measure the fluorescence just outside the object, it may be contaminated by glare, caused by blurring of the image by the optics, out-of-focus information, and scattering in the mounting medium. Therefore, a narrow 'neutral' zone just outside the object should not be used. Besides, any other object in the immediate vicinity of the cell under consideration should also be removed from the estimate of the background.

This leads to the simple algorithm shown in Figure 10.5E–H. First, the object is dilated with a large, square structuring element (in our case 17×17 pixels is optimal). Then, the segmented image containing *all* objects is dilated with a 5×5 square kernel. Figure 10.5G shows the result of the pointwise logical operation 'E AND NOT F': any pixel that is white (considered 'TRUE') in Figure 10.5E, AND is NOT white in Figure 10.5F, is white in Figure 10.5G. The effect of this is that we now have a 'region of interest' (ROI) containing those pixels that are nearer than some upper limit to the object, but not closer than two pixels from it, or any other object. Figure 10.5H shows the two ROIs (object and neighbourhood) overlaid on Figure 10.5A. Any difference in mean brightness between the two ROIs is considered to be attributable to the presence of the object.

In its original application to immunofluorescence, the algorithm was modified slightly, to account for the fact that the fluorescence stems from a region slightly larger than the phase-contrast image, since the fluorescence stems from the surface of the cell, not the cytoplasm (Figure 10.6). To account for this, the object ROI was dilated once with a 3×3 structuring element, increasing the size of the object by one layer of pixels. We have later found that this also accounts for the glare seen around cells stained using FISH (Langendijk *et al.*, 1995). A similar observation is made in the next chapter (Section 11.4.1).

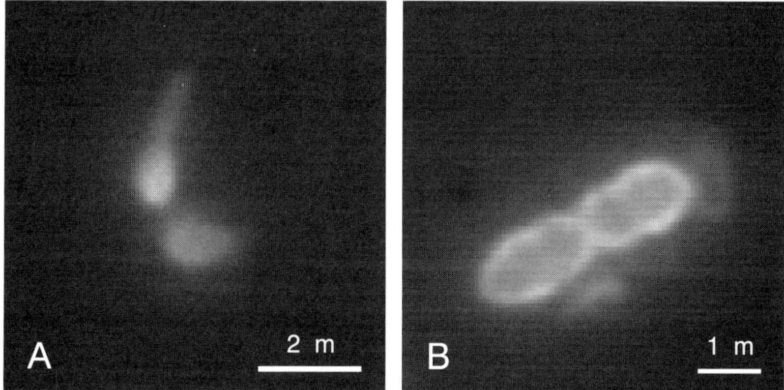

Figure 10.6 Difference in fluorescence distribution between FISH and IF: (A) FISH image of *Bifidobacterium longum*: fluorescence stems from the cytoplasm; (B) IF image of *Klebsiella* sp.: fluorescence stems from the surface of the cell, yielding a more annular fluorescence pattern.

If image (ii) was fully calibrated beforehand, we now have a measure of surface fluorescence intensity. In our application we perform a further post-processing correction for differences in gain between pixels, since this is not done in the pre-processing stage. During analysis, a shading image obtained from a solid fluorescence reference is stored in memory as well. The calibrated mean surface fluorescence brightness f_{ROI} of any ROI is then given by:

$$f_{\mathrm{ROI}} = \frac{\sum\limits_{\mathrm{ROI}} I(x, y)}{\sum\limits_{\mathrm{ROI}} S(x, y)} \tag{10.2}$$

where $I(x, y)$ is the intensity in image (ii) at point (x, y) and $S(x, y)$ is the brightness in the corresponding pixel in the shading image. Summations are carried out over all pixels in the ROI. This calibration is carried out for both ROIs, and the difference is the mean surface fluorescence of the object.

Converting mean surface fluorescence to integrated fluorescence is a matter of multiplication with the number of pixels in the object (projected surface area in pixels). Conversion to volume fluorescence intensity requires division of integrated intensity by the estimated volume of the object, obtained by any of the means in Section 7.7.

10.5.4 Statistical analysis of populations

10.5.4a Population measures of level of fluorescence

The simplest population measures of fluorescence are the sample mean and standard deviation, or median and inter-quartile range. The latter measures are more robust against outliers, which occur quite frequently in fluorescence distributions. Also, because of the non-negativity of fluorescence, distributions of faintly fluorescent objects are inherently skewed. As with morphological measures of diversity (Section 9.4.3), the entropy of the fluorescence distribution could be used instead of the standard deviation or inter-quartile range, but this has not been attempted to date, in part because there is no particular model to use for the interpretation of such an entropy.

The most complete description of the population is of course the density distribution, obtained, for example, using a kernel estimate such as in Figure 10.4 (see also Section 9.4.1b). Its main use is in visualization of the results. It can also be used in analysis, e.g. for determining the mode(s) of the distribution, and for cluster analysis.

It is also possible to investigate the distribution of fluorescence as a function of morphology. In the study of faecal flora, a number of attempts to do this have been made by Apperloo-Renkema *et al.* (1992) and Jansen *et al.* (1993b). They subdivided the morphometrical space defined by the first two principal components of the morphometrical distribution (Section 9.2.3) into 100 squares. Within each of these squares, the median and first and third quartiles of the fluorescence distributions were computed (if possible: not all bins contained data). The patterns of fluorescence as a function of shape could be compared by a variety of techniques (Jansen *et al.*, 1993b). Although intriguing inter-individual differences and temporal shifts were seen, the lack of a theoretical framework to explain these phenomena has hampered their interpretation.

Whatever population measure is used, the outcome of the measurements obtained with the relevant stain must be compared to the outcome of a negative control: simple subtraction in the case of measures of location, variance and entropy. Correction of inter-quartile ranges is less straightforward. Besides computing the magnitude of the difference, it is important to know whether differences are significant. If the distributions are sufficiently Gaussian to use mean and standard deviation, Student's t- and F-tests can be used to compute the significance of the differences in location and variance, respectively. In the case of fluorescence distributions, it is usually better to use distribution-free tests, which can be found in textbooks (e.g. Sprent, 1993).

10.5.4b Determining numbers of positive objects

This is essentially a problem of classification very reminiscent of the problem of thresholding from histograms (Section 6.7.3). In both cases data points must be classified into one of two classes: foreground and background in the case of thresholding, positive and negative in the case of cell classification. However, there are two important differences: (i) unlike pixel classification, there is no contextual information (such as gradients) that can aid threshold selection, and (ii) in many cases it is sufficient to know how many cells are positive, not which ones.

The simplest way to determine the number of positive cells is by applying a manual threshold (or gate, in flow cytometry parlance). This is an efficient method only if the data are clearly bimodal and there is a distinct inter-modal region containing few cells. In such cases, most methods will give good results, and the problem may be considered trivial. Good staining protocols and proper selection of bandpass filters can indeed produce such well-defined clusters. As so often, optimal imaging leads to simple analysis.

The type of data used can also have a profound impact on the sharpness of the clusters. If we assume that the fluorochrome and autofluorescence are evenly distributed throughout each cell, the integrated fluorescence signal will simply be the volume fluorescence intensity multiplied by the cell volume. This might be a good model for 16S-rRNA targeted FISH, for example. Let us assume we have spherical cells with radii drawn from some distribution with mean R and standard deviation σ_R, and which have volume fluorescence intensities drawn from a distribution with mean F and standard deviation σ_R. The coefficients of variance of the distributions of volume (I_{vol}), projected surface (I_{surf}) and integrated fluorescence (I_{int}) will be, respectively:

$$\frac{\sigma_{\text{vol}}}{I_{\text{vol}}} = \frac{\sigma_F}{F} \tag{10.3a}$$

$$\frac{\sigma_{\text{surf}}}{I_{\text{surf}}} = \sqrt{\left(\frac{\sigma_F}{I_{\text{surf}}}\frac{\mathrm{d}I_{\text{surf}}}{\mathrm{d}F}\right)^2 + \left(\frac{\sigma_R}{I_{\text{surf}}}\frac{\mathrm{d}I_{\text{surf}}}{\mathrm{d}R}\right)^2} = \sqrt{\left(\frac{\sigma_F}{F}\right)^2 + \left(\frac{\sigma_R}{R}\right)^2} \tag{10.3b}$$

$$\frac{\sigma_{\text{int}}}{I_{\text{int}}} = \sqrt{\left(\frac{\sigma_F}{I_{\text{int}}}\frac{\mathrm{d}I_{\text{int}}}{\mathrm{d}F}\right)^2 + \left(\frac{\sigma_R}{I_{\text{int}}}\frac{\mathrm{d}I_{\text{int}}}{\mathrm{d}R}\right)^2} = \sqrt{\left(\frac{\sigma_F}{F}\right)^2 + 9\left(\frac{\sigma_R}{R}\right)^2} \tag{10.3c}$$

Thus, if the radii vary a great deal, it is best to use the volume fluorescence intensity, rather than the integrated fluorescence intensity (see Figure 10.7). If the coefficient of variance of the radii is far smaller than that of the staining intensity, e.g. when

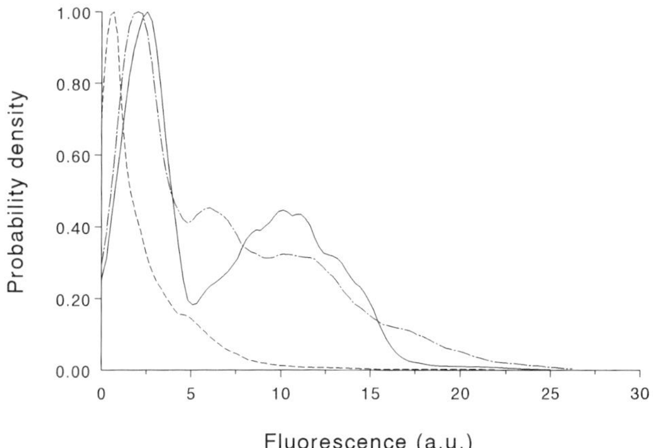

Figure 10.7 Different measures of fluorescence: density plots as in Figure 10.4 of simulated data of 200 fluorescent spheres with mean radius 10, s.d. = 3.1, and volume fluorescence intensities with means 2 (s.d. = 1.4) and 10 (s.d. = 3.1) for negative and positive objects, respectively. Solid curve, I_{vol}; dot–dash, I_{surf}; dashed, I_{int}. The three graphs have been scaled in x and y to fit a single scale. I_{vol} is most clearly bimodal, whereas I_{int} shows no significant bimodality. Separating positive from negative cells is therefore more reliable using I_{vol}.

studying lymphocytes, all measures perform equally. In mixed populations of bacteria this need not be the case.

In the case of IF it is the surface that is fluorescently labelled, so projected surface fluorescence intensity is the best of the three, by similar reasoning. Better still would be the true surface fluorescence intensity. To compute this parameter, we need an estimate of the true surface area of the cell, which could be computed in a similar way to the volume estimates in Chapter 7. To date, this parameter has not been used in analysis of populations of bacteria.

Langendijk *et al.* (1995) estimated the number of positive cells by determining which percentage of cells in the mixed population lie above the *N*th percentile of the negative control distribution (Figure 10.8). Usually *N* was chosen as 95 or 99. Though simple, it was pointed out that the method has a number of disadvantages. First, the percentage of 'positives' in the negative control is $100 - N$. In general, if the true number of positives is low, this method overestimates their number. Second, if the number of positives is high, the method may underestimate their number by classifying part of the lower tail of the distribution as negative, especially when the negative control contains a few brightly fluorescent outliers.

Within the context of the experiment, the method was nonetheless satisfactory. Its aim was to compare three different probes for the genus *Bifidobacterium* in an objective fashion, not to determine the fraction of Bifidobacteria within a mixed population. A ranking of probes using this method is certainly possible.

Sladek and Jacobberger (1993) discuss the problem of mixture analysis within the context of flow cytometry, but their conclusions are equally applicable in image analysis. One of their methods, called match region subtraction, assumes that the negative distribution is known, and that a second sample contains a mixture of a certain fraction α of this known distribution and an unknown 'positive' distribution.

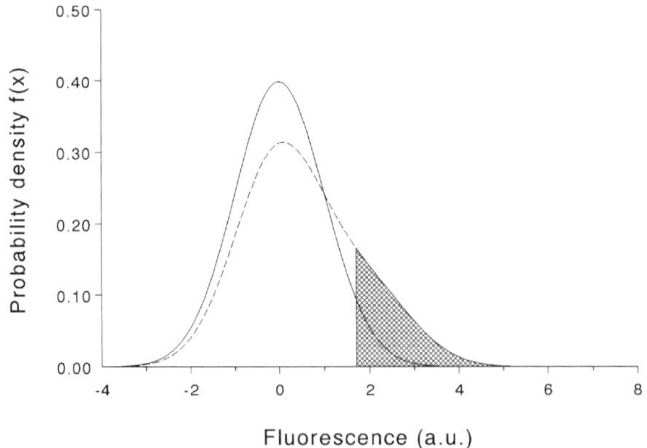

Figure 10.8 Method of Langendijk *et al.* (1995) for the determination of the number of positive objects in a mixture, where the distribution of the negative population (solid line) can be measured. In this example we have two unity variance Gaussian distributions with means of 0 (negative) and 2 (positive); 25% of the mixture (dashed line) consists of positive cells; all cells above the 95th or 99th percentile of the negative control distribution are considered positive (hatched area), yielding estimates of 25.1% (95th percentile) and 18.3% (99th percentile).

The goal of the analysis is to determine the fraction $(1 - \alpha)$ of this latter component. To do this they assume that there is an interval (the match region) in which the distribution of the mixture contains only members of the negative control distribution (Figure 10.9). Within this interval they fit the negative distribution to the mixture, using:

$$f_{\mathrm{mix}} = \alpha f_{\mathrm{neg}}(x) \tag{10.4}$$

where f_{mix} is the distribution function of the mixture and f_{neg} the distribution of the negative control. This method was used by Van der Waaij *et al.* (1994) on flow cytometry data of intestinal bacteria.

When the data are binned (as is normally the case in flow cytometry) the histogram serves as an estimate of the density distribution. If the data are real valued (as they are in our image analysis system), density estimates, as discussed in Silverman (1986; see also Section 9.3) could be used. Alternatively, the cumulative distribution could be used, which is estimated by:

$$P(x) \equiv \int_{-\infty}^{x} f(x')\,\mathrm{d}x' \approx \frac{n_{\mathrm{data} \leqslant x}}{N_{\mathrm{data}}} = \hat{P}(x) \tag{10.5}$$

where $n_{\mathrm{data} \leqslant x}$ is the number of data points smaller than or equal to x and N_{data} is the total number of data. Minimizing the mean square error in the fit in equation (10.4) over a match interval $[a, b]$ leads to:

$$\alpha = \frac{\hat{P}_{\mathrm{mix}}(b) - \hat{P}_{\mathrm{mix}}(a)}{\hat{P}_{\mathrm{neg}}(b) - \hat{P}_{\mathrm{neg}}(\mathrm{a})} \tag{10.6}$$

A slightly cruder method, which is a correction of the method of Langendijk *et al.* (1995) discussed above, uses the same assumptions as the method of Sladek and

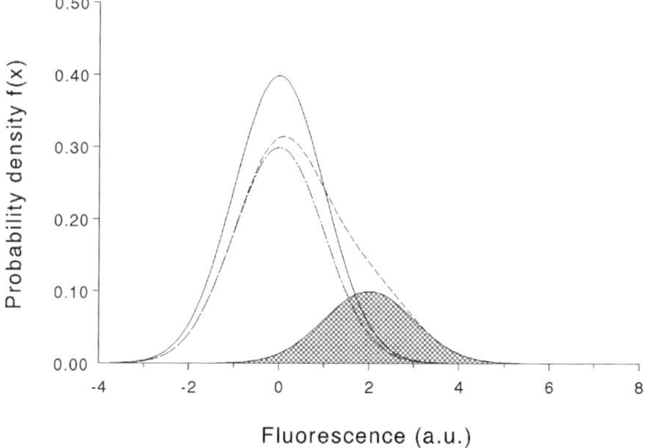

Figure 10.9 Match region method (Sladek and Jacobberger, 1993) for the same distribution as in Figure 10.3: solid line, negative distribution; dashed, mixture; dot–dash, fit of the negative distribution to the mixture over a match region (bar) yields an estimate of 25.0% of positive cells (hatched area).

Jacobberger (1993), but does not perform a fit over the interval (see Figure 10.10). Instead, it uses the ratio of the cumulative probabilities at some threshold T as an estimate of α:

$$\alpha = \frac{\hat{P}_{\text{mix}}(T)}{\hat{P}_{\text{neg}}(T)} \tag{10.7}$$

effectively using a match interval of $[-\infty, T]$. The bias of these methods is zero only when there is no overlap between the two distributions in the interval. Thus the bias is reduced by choosing a small match region; however, the variance decreases with increasing size of the match region since more data points are involved in the estimate. This leads to the main problem of these methods: selecting the proper interval. Usually, this is done manually, but automatic methods could be developed. We are currently working on the development of such methods. Sladek and Jacobberger (1993) also discuss the situation when both distributions are known. Estimating α then boils down to finding the minimum of the integrated square error (ISE):

$$\text{ISE} = \int_{-\infty}^{\infty} \{f_{\text{mix}}(x) - \alpha f_{\text{neg}}(x) - (1-\alpha) f_{\text{pos}}(x)\}^2 \, \mathrm{d}x \tag{10.8}$$

as a function of α. In practice, this is rarely possible. Only when the component populations can be isolated by some other means can mixtures be analysed in this way. When analysing mixtures of different types of white blood cells, this may be feasible. In the case of bacteria this is complicated by the fact that the expression of the target of the probes (either FISH or IF) depends strongly on the environmental conditions. This makes comparison of the distributions found in pure cultures with those from some other environment difficult.

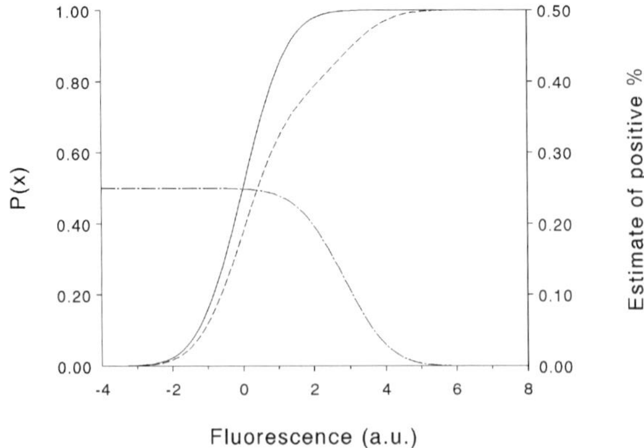

Figure 10.10 Simplified match region method using cumulative probability distribution functions. Solid line, negative distribution; dashed, mixture; dot–dash, ratio of the two cumulative probability distribution functions yields a consistent estimate of 25.0% for all $x < 0$.

10.6 CALIBRATION AND PERFORMANCE VALIDATION SCHEMES

Many authors (Sisken, 1989, and references therein) recognize two forms of calibration in quantitative fluorescence microscopy: (i) system calibration, which is used to evaluate and correct for drifts and imperfections in the performance of the imaging system, and (ii) analytical calibration, which attempts to calibrate the fluorescence data obtained under given experimental conditions into physical units that are of relevance to the (biomedical) problem under investigation.

10.6.1 System calibration

For a fluorescence imaging system to be calibrated, all differences in gain (or relative sensitivity) and baseline (offset) between pixels must be corrected, using the shading correction scheme in Section 2.4.7, equation (2.3). However, this scheme only ensures that the gain of all pixels is constant over the entire image, not that the gain is constant *in time*. To achieve this latter calibration, we need some fluorescence reference slide, which has the following characteristics:

(a) Its fluorescence properties are uniform over the entire field of view of the imaging system.
(b) Its spectral properties (emission and excitation) closely match those of the fluorochrome(s) used.
(c) Its fluorescence properties do not change in time.
(d) Its thickness is approximately the same as that of the samples used.
(e) The optical pathway (and in particular focusing) when recording the reference can accurately reproduce that of normal samples.

Of the different types of fluorescence standards available (Sisken, 1989; Vogt *et al.* 1991; Vrieling *et al.*, 1993) the most obvious class that can meet all requirements is solid fluorescence references. These may consist of a thin piece of fluorescent glass (e.g. uranyl glass) or polymer (Starna polymer), or a thin layer of agar with a known amount of fluorochrome added. Glass has the advantage of a practically unlimited shelf life, whereas the agar method can replicate the fluorescence properties of fluorochromes in the actual experiment more accurately. Once a correct reference has been selected, reference images can be made, and shading correction can be done using equation (2.3). To reduce photon and read-out noise, it is best to take the average of a number of reference images, e.g. using a Kalman running average filter (Section 16.4.1, equation (16.1)).

10.6.2 Calibration in physical units

10.6.2a Photometric calibration

As was explained in Section 2.4.3, it is fairly straightforward to calibrate a photon noise limited detector system in terms of the numbers of *detected* photons per pixel, simply by using the properties of Poisson noise. It is slightly more difficult, but considerably more useful, to find out how many photons were actually emitted by the sample under study per unit of surface area. For this reason, Taylor and Heimer (1975) proposed *photometric* calibration of fluorescence in SI units. Their calibration scheme allows expression of the amount of light emitted per unit of surface area to be expressed in physical units, in this case candela per square metre (cd/m^2). They achieve this by comparing the fluorescence emission of a piece of uranyl glass to a standard light source, at a given setting of the optical system. Since the intensity of the standard light source is known, the uranyl emission can be calculated in terms of candela per square metre, and therefore all subsequent measurements calibrated using the uranyl glass, provided the settings of the illumination system can be reproduced accurately.

Although the results are expressed in SI units, comparison between laboratories using this type of scheme is not at all straightforward. The chief problem is that the amount of light emitted by a given number of fluorophores will only be the same in two experiments if the excitation conditions (spectrum, pH, exposure time etc.) are identical. Differences in spectral sensitivities of the detectors, passbands of emission filters and even ageing of the light source influence the outcome of photometric measurements.

10.6.2b Fluorometric calibration

From the previous discussion it should be clear that calibration in terms of fluorescence strength is needed, rather than in photometric terms. In flow cytometry this is usually achieved by using fluorescent microbeads of known fluorescence intensity. These beads may be obtained from a number of manufacturers, usually in suspension, sometimes ready mounted on slides. Usually the fluorescence strength of the beads is expressed in molecular equivalent of soluble fluorescein (MESF). It must be stressed that this does not mean that beads of, for example, 100 000 MESF fluorescence intensity contain 100 000 molecules of fluorescein. The fluorescence properties of bound and soluble forms of fluorophores may differ considerably, as in the case of FITC, which has a 50% lower quantum efficiency when bound to protein (see also Chapter 4).

Keeping these caveats in mind, it is probably safer to consider bead standards as a means to calibrate the instrument in fluorometric terms than in molecular terms. Furthermore, they are an excellent means to cross-calibrate flow cytometers and imaging systems, allowing ready comparison of results between these two techniques. The same caveats apply to standard solutions and agar with known concentrations of fluorochromes. Comparison is only feasible if the dyes in solution (or agar) experience the same or very similar environmental conditions as in the real sample.

System calibration using solid or liquid fluorescence references constitutes a different kind of fluorometric calibration from beads or surrogate cells. The latter provide calibration of the complete image processing procedure, not just the imaging part of the system (illumination system, objective lenses, camera and digitizer).

10.6.2c Molecular calibration

Rather than determining a derived property, such as the number of photons emitted, molecular calibration attempts to calibrate the measurements in terms of numbers of fluorescent molecules, antibodies etc. The only way to do this is to observe some property besides the fluorescence, which depends on the number of molecules directly. If the molecules are motile, fluorescence correlation spectroscopy could be used. Another way would be to use near-field scanning optical microscopy (NSOM), which allows direct observation of individual fluorophores.

10.6.3 Calibration in terms of earlier measurements or traditional, non-physical units

In many situations it is desirable to be able to compare data obtained by image analysis with data acquired, for example, by visual means. One example can be found in an application of indirect immunofluorescence (IIF; Section 10.2.5) to serological staining (Apperloo-Renkema *et al.*, 1991). Traditionally, the serum antibody concentration is expressed as a titre: the (log of the) number of times a serum must be diluted to get the minimum assay response still considered 'positive' by the observer. In practice (Apperloo-Renkema *et al.*, 1991), a slide with a series of 12 wells was made, each containing a sample of the bacteria. Serum was added to 11 of the wells in a dilution sequence (e.g. $16\times, 32\times, 64\times, \ldots, 16\,384\times$). The last well was a negative control, containing only PBS. After the full IIF staining procedure (Protocol 10.1), the observer must judge which is the highest dilution (lowest serum concentration) with a significantly brighter fluorescence than the negative control. By contrast, computerized IIF evaluation assumes that the fluorescence signal is a linear function of serum antibody concentration (i.e. there is no saturation of the number of available binding sites). If this is the case, the (mean or median) signal S_b from each of the serum dilutions should be:

$$S_\mathrm{b} = b + F[\mathrm{Ab}]C_\mathrm{n} \tag{10.9}$$

where b is the background fluorescence of the cells (a combination of non-specific and undesired specific binding of the fluorescent conjugate, and autofluorescence), [Ab] is the antibody concentration in the undiluted serum, F is the fluorescence per mol/l of antibody, and C_n is the serum concentration (one over the dilution factor), which is given by:

$$C_n = \begin{cases} 0 & \text{if } n = N_\mathrm{wells} \\ \dfrac{1}{D_1 2^n} & \text{if } n = 1, 2, \ldots, N_\mathrm{wells} - 1 \end{cases} \tag{10.10}$$

where D_1 is the first serum dilution (16 in the example), N_wells is the number of wells (12) and n is the well number, which runs from 1 to N_wells inclusive. Since D_1 and n are known for each well, it is possible to estimate F[Ab] from a linear fit of S_n versus C_n. The slope gives a measure of serum antibody concentration, but in unfamiliar units. If we assume that there is a constant visual detection limit D_vis at which fluorescence is significantly different from the background b, the dilution D

at which fluorescence is first detected is given by:

$$D_{\text{vis}} = \frac{F}{D}[\text{Ab}] \Leftrightarrow D = \frac{F}{D_{\text{vis}}}[\text{Ab}] \tag{10.11}$$

In other words, the slope $F[\text{Ab}]$ is a linear measure of the titre. The conversion factor can be computed by performing a linear fit of the visually determined titre as a function of the digitally measured slope $F[\text{Ab}]$. Apperloo-Renkema *et al.* (1991) performed this calibration for a single observer (the first author). This allowed them to compare results obtained by digital image analysis with observations made earlier by visual evaluation.

10.6.4 Validation

Besides calibration of the signal, the performance of the system must be validated in terms of SNR. Ideally, fluorescence measurement should be photon-noise limited (Section 2.4.1a). However, there are many other sources of noise outside the detector system that can contribute significantly to the total variance in the measurements.

One of these is 'molecular' noise, which is caused by the fact that the signal of interest derives from a discrete number of fluorescent markers (which may consist of any number of fluorochrome molecules). The noise in this signal has two components: (i) photon noise and (ii) noise due to the random nature of the label distribution. If the number of photons per fluorescence marker is much larger than 1, the theoretical maximum SNR is limited by the number of labels, rather than the number of photons. In small areas such as pixels, blurring by the optics (which smooths out the molecular component of the noise) must be taken into account, and further blurring can be expected if the labels have not been immobilized. In that case we approach the situation found in fluorescence correlation spectroscopy (FCS; Section 3.6.2), where photon noise dominates, both because the exposure time per sample is short and the molecules are motile.

In practice, physical limits can often not be attained, due to variations in slide manufacturing, sampling etc. In many medical settings, it is useful to distinguish three sources of variance: (i) intra-assay error σ^2_{system}, caused by errors in the complete measurement system, including slide manufacture, (ii) inter-assay variance σ^2_{health}, caused by normal, non-pathological or non-treatment-related variations, and (iii) pathological or treatment-related variations, σ^2_{patho}. Variance of type (ii) may be split into inter-individual differences and temporal fluctuations. Assuming these three sources of variance are independent and distributed normally, the total variance in a given data set is simply:

$$\sigma^2_{\text{total}} = \sigma^2_{\text{system}} + \sigma^2_{\text{health}} + \sigma^2_{\text{pathol}} \tag{10.12}$$

Thus, we can assume the presence of a pathological or treatment-related response, at the 5% level of confidence if:

$$\sigma^2_{\text{total}} > 4(\sigma^2_{\text{system}} + \sigma^2_{\text{health}}) \tag{10.13a}$$

or:

$$\sigma^2_{\text{pathol}} > 3(\sigma^2_{\text{system}} + \sigma^2_{\text{health}}) \tag{10.13b}$$

Hence, if our system is accurate enough to detect non-pathological variance (i.e. $\sigma^2_{\text{health}} \gg \sigma^2_{\text{system}}$) no real improvements can be made by better cameras etc. (Jansen *et al.*, 1993a), since we are limited by σ^2_{health}.

The intra-assay variance can be determined by repeating the same measurement for a number of times and computing the distribution (e.g. by any of the methods in Chapter 9). After checking that the distribution is indeed reasonably Gaussian (e.g. by applying the Lilliefors test; Lilliefors, 1967), the variance of the data may be used as an indicator of the intra-assay error. In a similar way we may determine the non-pathological variance by determining the total variance in the absence of treatment or disease ($\sigma^2_{\text{pathol}} = 0$). The difference between total and intra-assay variance gives σ^2_{health}.

10.7 DISCUSSION

10.7.1 The importance of applications to the intestinal microflora

Through evolution inter-micro-organism defences have gradually developed and adapted to changing compositions of ecosystems. In animals that developed an intestinal tract a gradually increasingly complex defence system has evolved. The host defence system developed to supplement *and* to control the composition of the ecosystem of the gastrointestinal tract. In higher animals the host's defence apparatus functions through the concerted action of many different cell types, which together form the 'immune system'. The original type of inter-bacterial defence mechanism, however, still exists in the gastrointestinal tract and provides the first barrier to foreign micro-organisms. This barrier is called colonization resistance (CR) and is predominantly formed by the intestinal microflora.

It appears that the CR is stronger and more efficient, the more complex the intestinal flora is. Because the stability of the composition also plays a role, a detailed study of the composition of the intestinal microflora in individuals with a high CR as well as in those with a low CR is mandatory. This is only possible using a technique with which rapid and reproducible analysis of the microflora can be accomplished at low cost. With quantitative morphological analysis (Chapter 9) and FISH techniques, a sufficient level of understanding of the mechanism(s) of the CR may be acquired, to make adequate modulation of the composition of a patient's intestinal microflora possible, from a low CR level to a high CR level.

Constant interactions of components of the gut microflora with cells of the host defence apparatus maintain a condition that efficiently controls micro-organisms which, albeit on a small scale, continuously penetrate the mucosal layer. This process also includes the potentially pathogenic microbes. It is considered important, therefore, to learn much more about these immuno-physiologic interactions, which may reach as far as the bone marrow and the thymus. This can be accomplished by studying consecutive samples of washed faecal flora suspended in serum of the same individual and incubated with fluorescent anti-isotype sera. Such information may become of great practical value clinically.

To extend our knowledge about these important host–microflora interactions high priorities should be given to further development and improvement of techniques that make a rapid quantitative and qualitative analysis of the intestinal (faecal) flora possible. Only with specific and reproducible information about the interactions of the microflora with the host can this complex defence system be studied efficiently. Image analysis techniques, and especially FISH and IF quantification, are among the foremost techniques capable of providing these data.

10.7.2 Future developments

Although in this chapter we limited our discussion of the fluorescence measurements to IF and FISH, the algorithms used can be applied to any fluorescent probe. The algorithm could be extended to multiple spectral band measurements. In that case, correction of image shifts and scale changes (and possibly changes in depth of focus) should be added as pre-processing stages for the fluorescence images. Furthermore, multiple fluorescence references are required: one for each spectral band.

The crucial advantage of image analysed fluorescence microscopy over flow cytometry is the simultaneous accessibility of morphological information – both the overall cell shape and the probe distribution. Future developments that could use this advantage may determine the distribution of DNA within bacteria, e.g. to detect whether multiple nucleoid regions are present, which could be used to determine the division rate of cells. If probes against higher taxonomic groups are used, the morphological diversity (as measured by the entropy, see Section 9.4.3.) could yield valuable clues about the species diversity within each cluster. Besides, determining the volume fluorescence intensity, rather than integrated intensity as in flow cytometry, could be essential in separating weak positive signals from the background.

Although in previous years much attention has been paid to the taxonomic identification of the micro-organisms visible under the microscope, future developments will focus increasingly on the determination of the *in situ* activity of micro-organisms. New developments are therefore expected in the field of messenger RNA (mRNA) detection and quantification. Specific mRNA species, 'freshly' transcribed from the DNA template and immediately covered by ribosomes, may in many instances be localized in restricted areas of the cell due to the spatial organization of the transcription and translation process in polysomes. Since the areas that produce fluorescence are in such cases extremely small, flow cytometric detection and quantification is not likely to be the most promising method. Expression of genes may also vary spatially in consortia of micro-organisms (Chapters 15–18), and it is here that fluorometry by image analysis will be indispensable.

REFERENCES

Abuknesha, R.A., Al-Mazeedi, H.M. and Price, R.G. (1992) Reduction of the rate of fluorescence decay of FITC- and carboxyfluorescein-stained cells by anti-FITC antibodies. *Histochem. J.* **24**: 73–7.

Alfrieder, A., Pernthaler, J., Amann, R., Sattler, B., Glöckner, F.O., Wille, A. and Psenner, R. (1996) Community analysis of the bacterial assemblages in the winter cover and pelagic layers of a high mountain lake by in situ hybridization. *Appl. Environ. Microbiol.* **62**: 2138–44.

Apperloo-Renkema, H.Z., Wilkinson, M.H.F., Oenema, D.G. and Van der Waaij, D. (1991) Objective quantitation of serum antibody titres against *Enterobacteriaceae* using indirect immunofluorescence, read by video camera and image processing system. *Med. Microbiol. Immunol.* **180**: 93–100.

Apperloo-Renkema, H.Z., Wilkinson, M.H.F. and Van der Waaij, D. (1992) Circulating antibodies against faecal bacteria assessed by immunomorphometry: combining quantitative immunofluorescence and image analysis. *Epidemiol. Infect.* **109**: 497–506.

Bloem, J., Veninga, M. and Shepherd, J. (1995) Fully automatic determination of soil bacterium numbers, cell volumes, and frequencies of dividing cells by confocal laser scanning microscopy and image analysis. *Appl. Environ. Microbiol.* **63**: 926–36.

Böck, G., Hilchenbach, M., Schauenstein, K. and Wick, G. (1985) Photometric analysis of antifading reagents for immunofluorescence with laser and conventional illumination sources. *J. Histochem. Cytochem.* **33**: 699–705.

Chadwick, C.S., McEntegart, M.G. and Nairn, R.C. (1958) Fluorescent protein tracers: a trial of new fluorochromes and the development of an alternative to fluorescein. *Immunology* **1**: 315–27.

Coons, A.H. and Kaplan, M.H. (1950) Localization of antigen in tissue cells. II. Improvements in a method for the detection of antigens by means of fluorescent antibody. *J. Exp. Med.* **91**: 1–14.

Giovannoni, S.J., DeLong, E.F., Olsen, G.J. and Pace, N.R. (1988) Phylogenetic group-specific probes for identification of single microbial cells. *J. Bacteriol.* **170**: 720–6.

Hiramoto, R.N., Engel, K. and Pressman, D. (1958) Tetramethyl rhodamine as an immunochemical fluorescent label in the study of chronic thyroiditis. *Proc. Soc. Exp. Biol. Med.* **97**: 611–14.

Jansen, G.J., Wilkinson, M.H.F., Deddens, B. and Van der Waaij, D. (1993a) Statistical evaluation of an improved quantitative immunofluorescence method of measuring serum antibody levels directed against intestinal bacteria. *J. Microbiol. Meth.* **17**: 137–44.

Jansen, G.J., Wilkinson, M.H.F., Deddens, B. and Van der Waaij, D. (1993b) Characterization of human faecal flora by means of an improved fluoro-morphometrical method. *Epidemiol. Infect.* **111**: 265–72.

Langendijk, P.S., Schut, F., Jansen, G.J., Raangs, G.C., Kamphuis, G., Wilkinson, M.H.F. and Welling, G.W. (1995) Quantitative fluorescence *in situ* hybridization of *Bifidobacterium* spp. with genus-specific 16S rRNA-targeted probe and its application in fecal samples. *Appl. Environ. Microbiol.* **61**: 3069–75.

Larsen, N., Olsen, G.J., Maidak, B.J., McCaugey, M.J., Overbeek, R., Macke, T.J., Marsh, T.L. and Woese, C.R. (1993) The ribosomal database project. *Nucl. Acids Res.* **21**: 3021–3.

Lebaron, P., Catala, P., Fajon, C., Joux, F., Baudart, J. and Bernard, L. (1997) A new, whole-cell hybridization technique for detection of bacteria involving biotinylated oligonucleotide probe targeting rRNA and tyramide signal amplification. *Appl. Environ. Microbiol.* **63**: 3274–8.

Lilliefors, H.W. (1967) On the Kolmogorov–Smirnov test for normality with mean and variance unknown. *J. Am. Stat. Soc.* **64**: 399–402.

Longin, A., Souchier, C., Ffrench, M. and Byron, P. (1993) Comparison of anti-fading agents used in fluorescence microscopy: Image analysis and laser confocal microscopy study. *J. Histochem. Cytochem.* **41**: 1833–40.

Meijer, B.C., Kootstra, G.J. and Wilkinson, M.H.F. (1990) A theoretical and practical investigation into the characterization of bacterial species by image analysis. *Binary Comput. Microbiol.* **2**: 21–31.

Møller, S., Kristensen, C.S., Poulsen, L.K., Carstensen, J.M. and Molin, S. (1995) Bacterial growth on surfaces: automated image analysis for quantification of growth related parameters. *Appl. Environ. Microbiol.* **61**: 741–8.

Nwoguh, C.E., Harwood, C.R. and Barer, M.R. (1995) Detection of induced beta-galactosidase activity in individual non-culturable cells of pathogenic bacteria by quantitative cytological assay. *Molec. Microbiol.* **17**: 545–54.

Olsen, G.J., Lane, D.J., Giovannoni, S.J., Pace, N.R. and Stahl, D.A. (1986) Microbial ecology and evolution: a robisomal RNA approach. *Ann. Rev. Microbiol.* **40**: 337–65.

Pace, N.R., Stahl, D.A., Lane, D.J. and Olsen, G.J. (1986) The analysis of microbial populations by ribosomal RNA sequences. *Adv. Microb. Ecol.* **9**: 1–55.

Radley, J.A. and Grant, J. (1933) *Fluorescence Analysis in Ultra-violet Light*. Chapman & Hall: London.

Riggs, J.L., Seiwald, R.J., Burckhalter, J.H., Downs, C.M. and Metcalf, T.G. (1958) Isothiocyanate compounds as fluorescent labeling agents for immune serum. *Am. J. Pathol.* **34**: 1081–97.

Sani, G., Citti, U. and Caramazza, G. (1964) *Fluorescence Microscopy in the Cytodiagnosis of Cancer*. Charles C. Thomas: Springfield, IL.

Schör, W., Fuchs, B., Juretschko, S. and Amann, R. (1997) Improved sensititvity of whole-cell hybridization by the combination of horseradish peroxidase-labeled oligonucleotides and tyramide signal amplification. *Environ. Microbiol.* **63**: 3268–73.

Schut, F., Prins, R.A. and Gottschal, J.C. (1997) Oligotrophy and pelagic marine bacteria: facts and fiction. *Aquat. Microb. Ecol.* **12**: 177–202.

Silverman, B.W. (1986) *Density Estimation for Statistics and Data Analysis.* Chapman & Hall: London.

Sisken, J.E. (1989) Fluorescent standards. In *Methods in Cell Biology,* vol. 30, *Fluorescence Microscopy of Living Cells in Culture* (Taylor, L.S. and Wang, Y.L. eds), Part B, pp. 113–25. Academic Press: San Diego, CA.

Sladek, T.L. and Jacobberger, J.W. (1993) Flow cytometric evaluation of retroviral expression vectors: comparison of methods for analysis of immunofluorescence histograms derived from cells expressing low antigen levels. *Cytometry* **14**: 23–31.

Smit, J.W., Meijer, C.J.L.M., Decary, F. and Felkamp-Vroom, T.M. (1974) Paraformaldehyde fixation in immunofluorescence and immunoelectron microscopy. *J. Immunol. Meth.* **6**: 93–8.

Sprent, P. (1993) *Applied Nonparametric Statistics.* Chapman & Hall: London.

Stahl, D.A. and Amann, R. (1991) Development and application of nucleic acid probes. In *Nucleic Acid Techniques in Bacterial Systematics* (Stackebrandt, E. and Goodfellow, M., eds), pp. 205–48. Wiley: Chichester.

Staley, J.T. and Konopka, A. (1985) Measurement of *in situ* activities of nonphotosynthetic microorganisms in aquatic and terrestrial habitats. *Ann. Rev. Microbiol.* **39**: 321–46.

Tanke, H.J. (1989) Does light microscopy have a future? *J. Microsc.* **155**: 405–18.

Taylor, C.E.D. and Heimer, G.V. (1975) Quantitive immunofluorescence studies. *Ann. N.Y. Acad. Sci.* **254**: 151–6.

Van der Waaij, L.A., Mesander, G., Limburg, P.C. and Van der Waaij, D. (1994) Direct flow cytometry of anaerobic bacteria in human feces. *Cytometry* **16**: 270–9.

Van Kempen, G.M.P., Van Vliet, L.J., Verveer, P. and Van der Voort, H.T.M. (1997) A quantitative comparison of image restoration techniques for confocal microscopy. *J. Microsc.* **185**: 354–65.

Vogt, R.F., Cross, G.D., Philips, D.L., Henderson, L.O. and Hannon, W.H. (1991) Interlaboratory study of cellular fluorescence intensity measurement with fluorescein-labeled microbead standards. *Cytometry* **12**: 525–36.

Vrieling, E.G., Draaijer, A., Van Zeijl, W.J.M., Peperzak, L., Gieskes, W.W.C. and Veenhuis, M. (1993) Effect of labelling intensity, estimated by real time confocal laser scanning microscopy, on flow cytometric appearance and identification of immunochemically labelled marine dinoflagellates. *J. Phycol.* **29**: 180–8.

Wilkinson, M.H.F. (1994) Shading correction and calibration in bacterial fluorescence measurement by image processing system. *Comput. Meth. Programs Biomed.* **44**: 61–7.

Wilkinson, M.H.F. (1996) Rapid automatic segmentation of fluorescent and phase-contrast images of bacteria. In *Fluorescence Microscopy and Fluorescent Probes* (Slavik, J., ed.), pp. 261–6. Plenum Press: New York.

11

Determining Biochemical and Physiological Phenotypes of Bacteria by Cytological Assay

Andrew S. Whiteley[1], Ravinder Grewal[1], Andrew Hunt[2] and Michael R. Barer[1]

[1] *University of Newcastle, Newcastle upon Tyne, UK*

[2]*Adaptrix, Gateshead, UK*

11.1 INTRODUCTION

This chapter is concerned with chromogenic and fluorogenic assays of bacterial activity applied to determine the properties of single cells, non-growing cells and heterogeneous cell populations. Both the technical approach and the subsequent image processing are described. A central feature of our image analysis software (codenamed 'Bugs') is that, after segmentation, the representation of each cell and the data associated with it can be stored and accessed individually. The advantages of this object-based approach will be evident in the sections describing our software techniques.

11.1.1 Microbiology at the single cell level

The biochemistry and physiology of bacteria have traditionally been studied at the population level by bulk analysis. While considerable understanding has stemmed from such studies, there are two important disadvantages with this approach. Firstly, a monoculture comprising a substantial biomass is generally required, which may take time to produce, and secondly, propagation *in vitro* stimulates the organism to adapt its phenotype to suit the culture conditions rather than retain features relevant to the environment from which it was isolated.

Analysis of bacteria at the single cell (cytological) level has the potential to overcome most of these problems, particularly for environmental studies. Bulk analysis of environmental samples inevitably represents the sum of the properties of the numerous species present. In order to gain a thorough picture of the ecology of microbes, we need to be able to subdivide activities among the components of the community. The established approach to this problem is to isolate organisms

Digital Image Analysis of Microbes: Imaging, Morphometry, Fluorometry and Motility Techniques and Applications. Edited by M.H.F. Wilkinson and F. Schut.

selectively from the environment and analyse monocultures of the species recovered. However, apart from the time and effort required, there is always considerable uncertainty concerning the degree to which the isolated organisms are representative of the natural community. Moreover, the phenotypes observed are simply those that *can* be expressed *in vitro*, and it may be impossible to recreate the conditions required for the organisms to display properties relevant to the environments under investigation. This makes physiological studies on isolated monocultures difficult to relate back to the natural ecology of the organisms concerned. A further drawback of growth-dependent studies on bacteria concerns our capacity to culture all of the organisms present. Some current estimates suggest that less than 20% of the species in natural environments have been isolated and cultured in the laboratory (Wayne *et al.*, 1987), though others are less optimistic (Stackebrandt and Rainey, 1996).

By contrast, cytological analysis of bacterial phenotypes is culture-independent, rapid and requires only sufficient biomass to be visualized by microscopy; it therefore enables each of the problems outlined to be surmounted. The value of the cytological approach is not confined to environmental studies. The conventional, bulk assay-based approach to bacterial physiology overlooks the possibility of heterogeneity within bacterial populations. Bulk measurements are generally divided by the number of cells present to estimate cellular activity. The central assumption is that the assayed activity is distributed evenly between all of the organisms in the population. However, the authors have not obtained cytological results which would support this in practice, indeed we have consistently observed heterogeneities in even the most defined of bacterial preparations (e.g. of mid-exponential phase culture) after simple cell staining procedures. We should really ask what advantage there might be for a microbe to maintain homogeneous populations. It makes more sense for there to be some (possibly programmed) diversity in the phenotypes present in a single population – this should increase the chances of survival in a changing environment, and the genes responsible would therefore confer a selective advantage. The cellular basis for this proposed heterogeneity or diversity can only be demonstrated by cytological assays.

One area of bacteriology that has provoked much recent interest in cytological assays has been the determination of viability and, in particular, the so-called 'viable but non-culturable' (VBNC) state (Oliver, 1993). While some authors appear to accept cellular activities such as substrate responsiveness in the nalidixic acid induced cell elongation assay (Kogure *et al.*, 1979) as a working definition of viability, we consider that, in operational terms, proof of extended replication (in most cases, culturability) provides the only proper unambiguous demonstration of viability in microbiology, and that no single cytological assay can be considered an adequate demonstration of cell viability or non-viability in these terms (Barer *et al.*, 1993). However, we also agree with McFeters and colleagues (1995) that 'specific measures of physiological or metabolic activity provide more descriptive information concerning the potential physiological activity of bacteria and their dynamics within natural or engineered microbial communities than the ability of cells to grow and form colonies on a specific medium'. The VBNC hypothesis originated as an explanation of phenomena related to normally readily culturable bacteria such as *Vibrio cholerae* and *Campylobacter jejuni*. In the context of VBNC cells of these species, it is hypothesized that the organisms can make a transition to and from a non-culturable state. Although important research has been done in the past 15 years in this area, the VBNC characteristics of these and other bacteria remain difficult to interpret or define in exact physiological terms. In these types of experiments cytological assays are often used to determine viability. However, as far as we are aware, no established cytological assay has the capacity to predict whether an individual non-culturable cell has the ultimate potential to be cultured.

In adopting this view we have no desire to denigrate those cytological assays that purport to demonstrate viability; we merely wish to point out their limitations. Many assays have been well-standardized against specific mechanisms of cell death such as heat treatment or antibiotic effects and are satisfactory in those specific circumstances (e.g. Gant *et al.*, 1993; Bovill *et al.*, 1994; Diaper and Edwards, 1994; Deere *et al.*, 1995; Jepras *et al.*, 1995; Mason *et al.*, 1995). However, the validity of these assays cannot simply be extrapolated to situations where different or undefined cidal mechanisms are operating without proper research to demonstrate their applicability. Lack of activity cannot be considered universally to indicate death (spores are inert) and, conversely, active cells do not necessarily have potential for further replication (e.g. lethally irradiated cells and minicells). Cytological assays report on the physiological and biochemical status of cells, and retained activity in non-culturable cells should be seen as significant in its own right irrespective of its relationship, if any, to viability.

The cytological approach described here is advanced as an alternative means of studying micro-organisms, not as a solution to the viability problem. Our approach is centred on light microscopy as the means of data acquisition. However, multiple parameter flow cytometry is an important alternative technique, which can also be employed to address many of the issues raised above (Davey and Kell, 1996). A flow cytometer can analyse signals from up to 1000 cells per second, and this enables low frequency sub-populations to be detected. The technique is predominantly dependent on rapid fluorescence analysis (this restricts the range of cytological methods applicable and their sensitivity). Although sorting is possible, each cell can, in practical terms, only be analysed once by flow cytometry. Cell suspensions must be free from clumps and other particulate matter, and substantial cell numbers ($>10^4$/ml) are required. Microscopy, on the other hand, can utilize the full range of cytological methods and sequential analyses can be performed on the same cell. Further, the direct observation involved in microscopy allows the source of signals to be confirmed as bacteria and the recognition of spatial interrelationships and heterogeneity.

In contrasting the properties of flow cytometry and microscopy, we seek simply to clarify their suitability for different applications, not to promote one approach over the other. Nonetheless, here we concentrate on cytological analyses based on light microscopic, cytochemical techniques.

11.2 'MICROBIAL CYTOCHEMISTRY'

Microbial cytochemistry refers to chemical, biochemical and physiological methods of analysis that, when applied to micro-organisms, give a cell-localized result which can be detected by microscopy. Providing information that is relatively specific in chemical terms can be obtained, any microscopic method of illumination and signal detection may be included. This chapter is concerned with techniques that use brightfield, phase-contrast and epifluorescence illumination systems for light microscopy (LM). In contrast to most electron-based illumination systems, which require fixation, all of these provide opportunities for rapid analysis of living material. Cytochemistry is related to histochemistry and shares many common procedures. However, the requirements of tissue preparation and sectioning for histochemistry alter certain features of the reactions so that they are not simply interchangeable. Indeed, where fixation is required for LM bacterial cytochemistry, such as in fluorescence *in situ* hybridization (FISH), the process needs to be optimized for application to bacterial preparations.

Many textbooks contain sections on the staining of bacteria. These are primarily aimed at demonstrating the organisms and their subdivision into classical cell wall types, i.e. via Gram staining. Reviews of the more functional aspects of bacteria that can be demonstrated by chromophore or fluorochrome methods can be found in specific literature (Vendrely, 1955; McFeters *et al*, 1995).

The particular constraints faced by microbial cytochemistry are illustrated in Figure 11.1. Although not required for some applications, attachment or immobilization is essential for work that requires more than one image to be processed for an individual cell. Attachment can also improve the sensitivity of fluorescence techniques, since it allows image acquisition over longer time periods. We have used deposition of cell suspensions by centrifugation onto aminopropyltriethoxysilane (APS) treated glass (Barer, 1991) in most of our work. Suspensions are placed in a chamber attached to a cover slip or slide (Barer, 1991; Walker *et al.*, 1992) and deposition achieved in a conventional centrifuge or cytospin apparatus. It was initially envisaged that reactions would be carried out

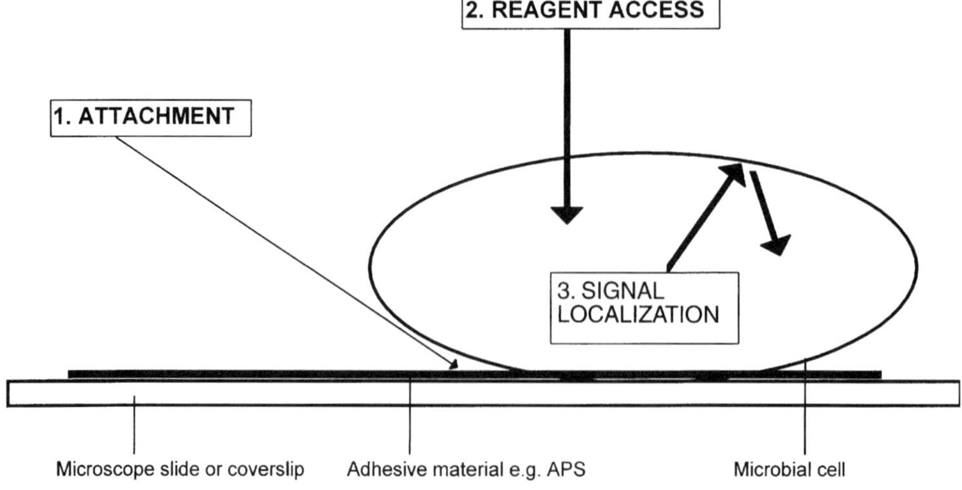

Figure 11.1 Technical problems faced by microbial cytochemistry.

Table 11.1 Some methods in microbial cytochemistry.

Property demonstrated	Signal molecule	Detection method	Comment
β-Galactosidase (*lacZ*)	Indigo compound (indoxyl substrate)	Bf	Slow and insensitive – X-gal can be used as the substrate (Barer, 1991)
	Formazan (indoxyl substrate)	Bf/Fl*	Rapid, indoxyl hydrolysis product reduces TS, need to inhibit endogenous TS reduction.
	Fluorescein (fluorescein digalactoside substrate)	Fl	Rapid, potential to detect a single enzyme molecule, fluorescein leaks out (mount in silicone oil) (Lewis *et al.*, 1994; Walker *et al.*, 1994; Nwoguh *et al.*, 1995)
Cytochrome oxidase	Indamine (diaminobenzidine substrate)	Bf	Simple, rapid and well localized but hazardous reagents (Barer and Marsh, 1992)
Production of reducing equivalents/dehydrogenase enzymes	Formazans (endogenous or exogenous oxidizable substrates)	BF/Fl*	Careful selection of TS required (Thom *et al.*, 1993), INT most widely applicable, exogenous substrates enhance formazan production over endogenous baseline (Gribbon and Barer, 1995; Whitely *et al.*, 1996)
Energization/charge across cytoplasmic membrane	Rhodamine 123 (R123)	Fl	Unreliable with Gram-negative organisms. Low R123 concentration avoids uncoupler insensitive accumulation (Kaprelyants and Kell, 1992)
Dye exclusion	Propidium iodide	Fl	Combines well with R123. Binds to RNA and DNA
Ribosomal RNA	Fluorescently labelled oligonucleotides	Fl	Determinative and physiological information. Balance between permeabilizing and cell integrity difficult
Lipid domains	Alkylated fluorescein (C_8–C_{18}), dodecyl bodipy, NBD-stearate, Nile red	Fl	Hydrophobicity and amphipathic properties predict localization (Horobin and Rashid, 1990; Rashid and Horobin, 1990). All except Nile red delivered by liposomes

Bf, brightfield; Fl, fluorescence; *, fluorescence possible with cyanoditolyl tetrazolium chloride.

on immobilized cells because this facilitates reagent exchange and washing steps. However, the tendency for high background signals to develop in some reactions and processing aspects of certain procedures make it advisable to base the decision to immobilize before or after the reaction on direct experience. Alternatives to centrifugation and APS treatment include gelatine subbing, and embedding of cells within deeper matrices (Willaert and Baron, 1996). Consideration must always be given to the impact of the immobilization procedure on the cytochemical result. If applied before the reaction, attachment should not interfere with the property to be demonstrated or with reagent access; if applied after the reaction, it should not affect signal localization or intensity.

Reagent access is perhaps the greatest challenge facing microbial cytochemistry. For any intracellular activity, reagents must pass through the cell envelope, and any procedure applied to permeabilize this structure must not cause loss of the target activity or delocalization of the reaction signal. There are no universal answers to this problem, and although models aimed at predicting the localizing properties of cytochemical reagents are available (Horobin and Rashid, 1990; Rashid and Horobin, 1990), which can be helpful in making initial reagent selections, a good deal of empirical development is often necessary. Essentially, a balance has to be made between dealing with the hydrophobic and hydrophilic barriers presented by microbial cell envelopes.

Finally, once the cytochemical reaction is complete, the reaction signal (a dye or probe bound to a specific target molecule or a chemical or enzyme reaction product) must, at the very least, remain localized in the cell in which the reaction took place. There is no problem if the signal molecule or complex has a high affinity for molecules or molecular complexes (e.g. ribosomes) retained within the cell. However, the affinities involved are often marginal, and reaction products may be permeant to what is left of the cell envelope. Where localization is a problem, two principal approaches can be adopted. The first is to acquire images of the result as soon as possible (even during the reaction), before the signal delocalizes, and the second is to mount the cell preparation in a medium that restricts signal diffusion. In the latter case, one general principle is to use a hydrophobic mountant (e.g. silicone oil) if the signal is water soluble, and a water-based mountant if the signal is non-polar.

These principles underpin both general and specific aspects of the cytochemical procedures established in this laboratory. Several of these are summarized in Table 11.1, and selected protocols are described below, with reference to specific examples.

11.3 EXAMPLES OF CYTOCHEMICAL PROTOCOLS AND THEIR ANALYSIS

Two experimental approaches are described here: (i) the simultaneous measurement of specific metabolic activity followed by on-stage FISH, in a single field of view (up to 200 cells), and (ii) the monitoring of reporter gene activity in more substantial populations (2–10 000 cells). Attachment to APS-coated cover slips is necessary in both cases.

Protocol 11.1 3-Aminopropyltriethoxysilane coating of glass cover slips and attachment of bacteria.

APS reacts with free hydroxyl groups on the glass surface. The resultant covalently-coated surface carries a positive charge at physiological pH and is more hydrophobic (less wetable) than the original glass.

1 Cover slips (22 mm × 66 mm rectangular or 19 mm diameter circular cover slips, thickness no. 1, BDH Chemicals, Lutterworth, Leicestershire, UK) are used for most of our applications but slides can also be used. These are briefly cleaned by immersion in acetone and wiped clean with tissue. More stringent cleaning protocols are often used in histochemical procedures (Van Prooijen-Knegt *et al.*, 1982) where APS coating is a common practice for *in situ* hybridization studies.
2 Working in a chemical fume hood, lay the cover slips flat in a suitable container (glass dish or similar) and immerse the slides in a freshly prepared 2% (v/v) solution of APS (Sigma Chemicals, Poole, Dorset, UK) in acetone and leave at room temperature for between 16 and 24 hours. We routinely store the undiluted APS stock solution at 4 °C for a maximum of 1 month, as we find that activity can fall off rapidly with longer storage periods.
3 Remove the APS/acetone coating solution and immerse the slides in acetone for 5 minutes to remove excess APS.
4 Remove the acetone and individually wash the cover slips by sequentially immersing them in two distilled water washes. Note that care must be taken to identify the upper (coated) surface.
5 Drain the cover slips and gently blot with tissue paper. Leave at 37 °C for 1 hour to dry.
6 Mark the uncoated surface and store the cover slips for up to 1 month in a clean dry Petri dish.
7 Bacterial suspensions are deposited onto the coated slides by centrifugation. Cell densities in the range 10^5–10^9/ml, combined with a sample volume of 30–50 μl can give satisfactory results, but 10^8/ml is generally optimal. Cover slips are assembled into a mounting chamber – essentially a moulded silicone superstructure, which is held onto the cover slip surface to give a watertight seal. We use a Bellco microslide chamber (unfortunately, no longer available) for work with large cover slips (up to ten cell monolayers per cover slip) and our own design (Walker *et al.*, 1992) for small cover slips. The latter design, which gives up to four monolayers, is particularly useful for work with pathogenic organisms requiring biological containment. Standard centrifugation conditions are 1000 × *g* for 2 min.
8 The supernatant is removed and the chamber assembly dismantled, taking care not to interfere with the deposited monolayers. Further processing depends on the procedure. If cytochemical reactions were done prior to immobilization then the cover slips are dried and mounted (cell deposits face down) in an appropriate mountant. If the reactions are to be done with immobilized bacteria, the chamber may or may not be dismantled, depending on the procedure and the organism. For example the β-galactosidase reactions (see below) are done in assembled chambers while the INT and FISH reactions described below are done after disassembly.

Three further protocols are now described to illustrate particular techniques and analysis procedures established in this laboratory.

Protocol 11.2a Sequential metabolic and 16S rRNA analysis: kinetic determination of *p*-iodonitrotetrazolium violet (INT) reduction.

1 Microscope and camera set-up. We routinely use a DIAPHOT 300 inverted microscope (Nikon UK Limited, Surrey, UK) equipped with standard epifluorescence attachments and a 100 W xenon light source. The inverted microscope provides a stable platform for experiments and there is no need to change the field of view or light path from the objective to the detector to obtain phase contrast, brightfield or fluorescence images. In practice, we do not leave the fluorescence filtersets in the optical path for brightfield data acquisition, although this could be done if image shifts between filter positions were a significant problem. Images are acquired with a Peltier-cooled, integrating charged-coupled device (CCD) camera ('Coolview', Photonic Science, East Sussex, UK) linked to a DT2867LC frame grabber (Data Translation, Berkshire, UK) in a 50 MHz 486 PC (Elonex) and stored at 8-bit resolution in Foster-Findlay (FFA) format as 768 × 512 square pixel *.imf files (Foster Findlay and Associates, Newcastle upon Tyne, UK). Reaction temperatures are maintained by use of a heated stage (model MS100, Linkam Instruments, with a 16 mm aperture Nikon heated stage insert), which is thermally isolated from the metal microscope stage by inserting insulating material (e.g. polystyrene or cardboard) between the conventional and heated stages. The heated stage must be pre-calibrated against the temperature controller display to ensure accuracy. Reaction temperature is confirmed by means of a thermocouple placed into test solutions on a dummy assembly (see below).

2 Reaction procedure: immobilize 50 µl of a 1×10^8 cell/ml bacterial suspension on a large APS-coated cover slip. The cells are maintained under a 50 µl droplet of the appropriate medium – usually a non-nutrient salt solution (nine salts solution (NSS; Nystrom *et al.* , 1986) in this example of work on a marine organism). APS coating causes droplets to maintain a tight meniscus. All solutions and the heated stage should be pre-warmed to the reaction temperature (in this case 30 °C).

3 Form a reaction chamber by placing a glass slide which has a 5 mm diameter hole drilled through it on top of the cover slip such that the droplet covering the cells resides within the hole. When an extended procedure is intended, a ring may be drawn round the cell deposit with a 'Pap' pen (Daido Sangyo Co. Ltd, Japan) to improve the seal between the cover slip and the drilled slide. If no sequential reactions (e.g. INT followed by FISH) are intended then INT reactions are generally done under an open 100 µl droplet (see below) without assembling a reaction chamber. Evaporation does not appear to pose significant problems with reaction times under 15 minutes.

4 Place the cover slip/slide assembly on the inverted microscope and view by phase contrast (×100 oil immersion). When a satisfactory field of view is obtained the slide assembly is carefully taped in place and the stage locked with a G-clamp.

5 Acquire and save an image (single video rate field image) of the cells under phase-contrast illumination. This will be used to produce the mask (binary) image, identifying all objects, for subsequent analysis.
6 Add 50 μl of pre-warmed 2 × INT solution (and substrate if required) to the droplet (total volume now 100 μl) to initiate the reaction and switch to brightfield optics. In practice, we add 50 μl of 10 mg/ml INT solution supplemented with 100 mM substrate to give a final reaction concentration of 5 mg/ml INT (10 mM) and 50 mM substrate. All reagents (INT and substrates) can be purchased from Sigma Chemicals, Poole, Dorset, UK. The INT solution is effectively saturated at 10 mg/ml and is difficult to solubilize in water without brief heating to 50 °C and sonication. Lower INT concentrations may be adequate for some applications, but reduction rates begin to fall at concentrations below 4 mM.
7 Acquire and store brightfield images of the proceeding reaction at appropriate time intervals (e.g. 1 minute). The time 0 (initial brightfield image) serves as a shade correction (Section 2.4.7) image for all subsequent images in the series. A de-focused brightfield image should also be acquired in case substantial cell movement invalidates the initial shade correction image.

Protocol 11.2b Sequential metabolic and 16S rRNA analysis: on-stage FISH.

On-stage hybridizations can be performed on previously unreacted cell suspensions, or after INT reactions such as those described in Protocol 11.2a. The latter approach allows co-determination of specific genotypic (16S rRNA) content and phenotypic traits (INT reduction) in the same cells. The approach is required if any sequential observations (e.g. kinetic or multiple properties) are to be made on individual cells, but is unnecessary if cell labelling from phenotypic cytochemical procedures does not quench the FISH signal, and can be retained through the hybridization procedure, as is the case with the cytochrome oxidase reaction product (Whiteley *et al.*, 1996). Reduced INT and several other formazans, including CTC-formazan, are extracted by the ethanol dehydration step. Formazans that resist extraction are either less effective cytochemical reagents than INT for bacterial application (e.g. nitroblue tetrazolium) or not commercially available (e.g. 2,3,5-tri-*p*-nitrophenyl-2*H*-tetrazolium chloride) (Thom *et al.*, 1993; Whiteley, unpublished observations). FISH cannot be performed *before* any measurement that requires active (or intact) cells, because of the required permeabilization.

1 Fix the cell monolayer with 50 μl freshly made 4% paraformaldehyde for 1 minute. This can be done directly after INT or other reactions have been completed.
2 Remove the paraformaldehyde and serially dehydrate the cells by adding and removing at 1 minute intervals 50, 80 and 96% ethanol. Care must be taken in the last ethanol dehydration (96%) to avoid complete evaporation (this disturbs the cell locations).
3 Add 20 ng fluorescently labelled probe in 0.9 M NaCl, 20 mM Tris-HCl and 0.1% SDS hybridization solution and set the stage temperature to the required stringency (e.g. 48 °C for EUB338).

4 Place a cover slip over the chamber opening (see Protocol 11.2a) and hybridize for the required time. In practice we have found 15 minutes to be sufficient for Gram-negative organisms if an integrating CCD camera is available. As yet, we have not applied the procedure to Gram-positive organisms.
5 Remove the probing solution and wash the cells twice at hybridization temperature by adding 100 μl of hybridization solution for 5 minutes. Appropriate hybridization temperatures, times and washing procedures should be established against known positive and negative control organisms.
6 Remove the final wash solution and cover the monolayer with silicone oil.
7 Acquire phase-contrast (mask) image and an epifluorescence (data) image with the appropriate filterset. The former may not be required if the mask image from a preceding INT reaction is still valid. With the unshielded xenon epifluorescence illumination image, integration is generally required over 100 European standard video rate frames (4 seconds) to obtain bright but unsaturated fluorescence images.

Protocol 11.3 Analysis of chromophore and fluorochrome signals with paired mask and data images.

Image acquisition

A standard microscope set-up is described in Protocol 11.2a, and image acquisition outlined in Protocols 11.2a and 11.2b. Where quantitative fluorescence studies are intended on preparations that fade on full-strength illumination (e.g. β-galactosidase assays), images are acquired with neutral density filters in the excitation path, and extended integration times on the Coolview camera. The appropriate combination of illumination and image acquisition time is ascertained by determining conditions under which two successive images give the same cellular intensity values. The software will accept images in most standard 8-bit formats.

1 Pre-processing: Bugs automatically initiates an analysis session by requesting mask (usually phase-contrast) and data image pairs together with their respective background shade correction images (if required). Shade correction (subtraction followed by scaling in the C_Images library, Foster Findlay Associates) is generally applied to brightfield data images. The images are loaded as a pair, zoomed up two-fold (the rationale for this is explained in Section 11.4) and displayed side by side as vertically tiled windows. The images can be viewed in scroll or stretch mode. In scroll mode they are linked such that moving the field of view in one image automatically moves the other image to the corresponding field. The mask image (usually phase-contrast) is processed to facilitate preparation of the mask binary image (a binary overlay which describes the objects to be analysed). At a minimum this involves one or two average filter passes with a 3×3 kernel (Section 1.4.2a) to smooth the effects of zooming up. In addition, contrast stretching may also be performed.
2 Segmentation: this is a two-stage process involving global image processing to select objects for consideration followed by single-object based review and editing. An initial binary mask is created by simple thresholding of the mask

image and separation of some touching cells by splitting (erosion followed by thickening). Additional rounds of thickening may be required in situations where object size may increase during the experiment (e.g. due to INT–formazan deposits). Clearly, a separate unthickened binary image is required if measurement of cell dimensions is desired in these circumstances. Objects for further analysis are selected by size (area) exclusion criteria.

3 Alignment: the initial binary is overlayed on the mask and data images and the objects unfilled. If necessary, global alignment (shifting the binary with respect to the data image) can then be performed such that the mask correctly overlays the cells in data image.

4 Object-based segmentation and analysis of individual cells: in the next stage, the accepted objects are reviewed individually through the 'Cell Browser' dialogue (Figure 11.2). This displays the relevant portion of the mask and data images with the unfilled and aligned binary mask applied to both. Individual objects can now be accepted, rejected or edited to give acceptable cell overlays. Individual cell alignments can also be made to cover minor shifts of cells during the course of an experiment. Each object is numbered and measured after acceptance or rejection; this enables criteria for rejection and bias to be assessed if necessary. Cells that the experimenter may wish to display later can be added to a cumulative gallery, and individual cells can subsequently be cross-referenced to their individual measurement values. Galleries can also be created after measurement and analysis. The standard densitometric and fluorimetric

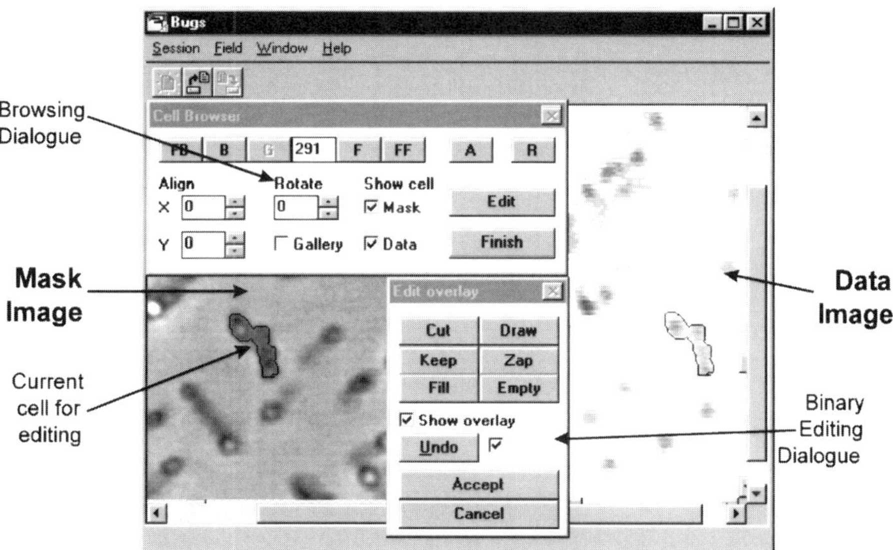

Figure 11.2 Screen shot of Bugs demonstrating the 'cell browser'. After detection the individual cells are stepped through and classified as accepted or rejected depending upon their suitability, and their mask edited when required. Mask editing can involve separating cells from each other or clumps and aligning binaries that have become offset (horizontal, vertical or rotational offsets). Once completed, the mask can be applied to subsequent data images and the mask for individual cells independently re-edited if required.

measurements are: (i) integrated optical weight (IOW), an estimate of the integrated optical density of all the deposits within a cell using:

$$\begin{aligned} \mathrm{IOW} &= \sum_{i=1}^{n} \{\log 255.5 - \log(g_i + 0.5)\} \\ &= n \log 255.5 - \sum_{i=1}^{n} \log(g_i + 0.5) \end{aligned} \tag{11.1}$$

where n is the number of pixels representing the cell and g_i is the grey level of the ith pixel in the series 1 to n, and (ii) integrated grey weight (IGW), which is the sum of all the pixel intensity values covered by the cell mask. Measurements (morphometric, densitometric and fluorimetric) are saved as comma-separated text files when the cell browsing stage is complete or copied to the Windows clipboard. When sequential data images are analysed under a single mask image, as in the kinetic determination of INT reduction (Protocol 11.2a), a new data image is loaded after the first has been analysed and the cell browsing stage is repeated using the binary image resulting from the first round of cell browsing. Only accepted cells are reviewed at this stage, and cells can only be excluded from further analysis at this stage, not added. In this way successive images are checked and analysed under the same mask and the sequential data set accumulated as a single file. Where a set of images is used to accumulate a substantial number of cells for population analysis, pairs of mask and data images acquired from a single cell preparation are processed (i.e. several fields of view) and the data added to the data file.

5 Analysis of distributions: frequency distributions are analysed using standard spreadsheet functions. Histograms of IOW or IGW, with and without normalization against cell profile area, are routinely plotted. Scatter plots of densitometric and fluorimetric data against cell profile area can be useful in identifying discrete sub-populations.

Results from two experiments are now presented to illustrate 'kinetic' (image series) analysis of densitometric data, and a more substantial population (image set) analysis of cellular reporter gene (*lacZ*) expression.

11.3.1 Substrate-enhanced tetrazolium reduction in stressed cells of *Vibrio vulnificus*

The experiment illustrated is part of our programme to establish a basis for applying cytochemical assays to environmental samples. The aim is to enable specific phenotypic traits of single cells to be determined, and to relate these to the identities of the cells as determined by DNA–rRNA *in situ* hybridization (ISH) reactions. We have reported our approach to co-localizing metabolic and determinative ISH signals within APS immobilized bacterial cells elsewhere (Whiteley *et al.*, 1996). Although not shown, the reactions discussed below were all compatible with subsequent ISH (Protocol 11.2b).

In the specific experiment presented, a culture of *Vibrio vulnificus* was nutrient and cold stressed (4 °C), in order to initiate transition of cells from a culturable to a non-culturable state (Oliver, 1993). The process is associated with a relatively constant total cell number, and with transition from bacilliary to coccoid morphology. Here, we demonstrate the use of our object-based analysis to monitor the kinetics of a substrate-enhanced tetrazolium reaction in a morphologically and metabolically heterogeneous bacterial population, which had previously been immobilized upon APS coated cover slips.

Figure 11.3 shows the results obtained when immobilized cells that had been maintained in a sterile, nutrient-free, artificial sea water medium at 4 °C for 2 days were reacted with INT in the presence of 50 mM galactose. Initial ($t = 0$) phase contrast (mask) and brightfield (shade correction) images were acquired before addition of the reaction mixture; subsequently, a series of brightfield (data) images of the same group of cells were taken at 1 minute intervals for kinetic analysis. The results illustrate the capacity of this approach to study and display the metabolic properties of individual cells.

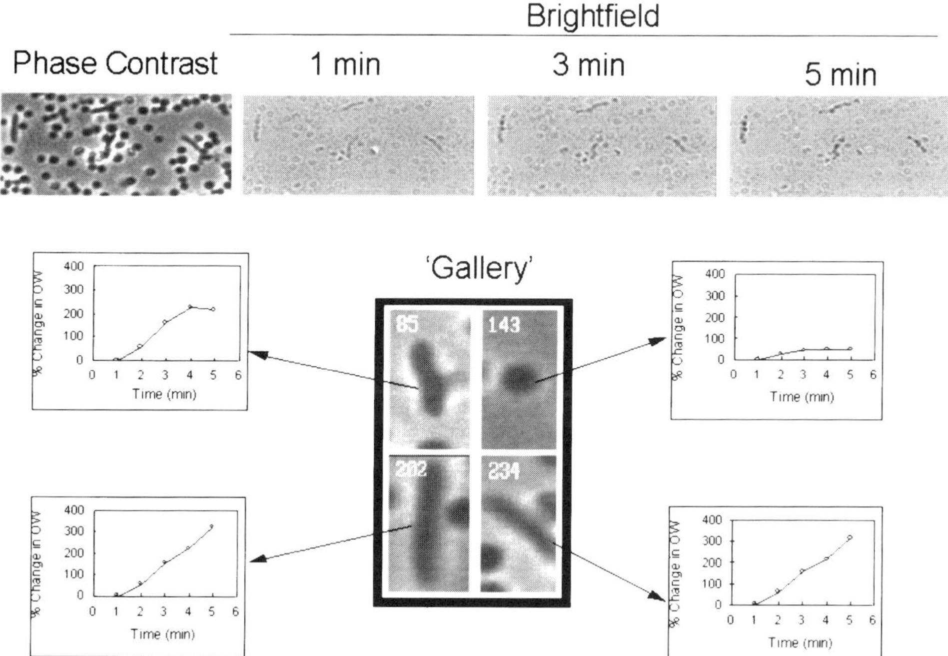

Figure 11.3 Time-series analysis of galactose-enhanced tetrazolium reduction expressed as percentage change in cell optical weight (OW) in *Vibrio vulnificus* during the transition between culturability and non-culturability. A single phase image was used to construct a binary overlay and this was applied to brightfield images of the same cells taken at 1 minute intervals in order to measure the kinetics of INT reduction. The initial phase image and three brightfield images (1, 3 and 5 min after INT and galactose addition) are shown. Four phase images of individual cells have been selected for display from the 'gallery' together with time-course plots of their associated INT reduction.

11.3.2 Cellular β-galactosidase reporter activity in carbon/nitrogen stressed cells of *Vibrio* sp. S14

This experiment was designed to compare the activity of a previously described Mu dI(*lac*) fusion to a carbon starvation induced gene in *Vibrio* sp. S14 (S141 M5, latterly J109; Östling *et al.*, 1995) detected by bulk *o*-nitrophenylgalactoside (ONPG) assay and cytological assay. The *lacZ* reporter fusion serves to indicate the level of expression of a gene whose function and phenotypic effects have yet to be determined. The results shown illustrate features of gene expression that can only be detected by cytological assay, and demonstrate application of our software to the cumulative analysis of a set (multiple pairs) of mask (phase-contrast) and data (fluorescence) images enabling assessment of substantial sample populations (2000–10 000 cells).

Carbon deprivation was achieved by preparing mid-exponential phase cells in MMM medium (Östling *et al.*, 1991), harvesting and washing twice in MMM medium minus glucose. Samples were taken at intervals for bulk (ONPG) and cytochemical assay up to 24 h. Individual cell β-galactosidase activity was detected by a cytological assay using the fluorogenic substrate, fluorescein di-β-D-galactopyranoside (FDG) in a single step assay (Nwoguh *et al.*, 1995). Key features of the FDG assay are permeabilization of cells by air drying and mounting in silicone oil to prevent reaction product (fluorescein) from diffusing out of the cells.

Images were acquired with the camera and microscope system described in Protocol 11.2a under the conditions outlined in Protocol 11.3, step 1. Analysis followed the image set pattern (Protocol 11.3, step 5). A key distinction between this analysis of a set of images, in contrast to the series of images discussed in the last section, is the number of cells assessed. The results shown in Figure 11.4 are based on analysis of around 2000 cells (approximately 10 mask and data image pairs) at each sample time. With our object-based cell-browser system such analyses can be achieved in under 2 hours. Previously, with general-purpose image processing and analysis packages that process images or regions of interest (ROIs) rather than objects, we had difficulty completing comparable tasks in less than a week!

Figure 11.4 shows the distributions of cell intensities obtained from samples taken at 0, 3, 17 and 24 h after initiating carbon starvation. Bulk β-galactosidase assay values increased steadily over a five-fold range between 0 and 10 h, and subsequently remained constant from 10 to 17 h. While the initial rise in cellular fluorescence at 3 h accords well with these results, the subsequent apparent lack of change between 3 and 17 h and the development of an intensely active population at 24 h were not reflected at the bulk level.

Although the lack of increase in cellular activity between 3 and 17 h was initially perplexing, it transpired that cell morphometry results for these samples showed a substantial fall in cell profile area between these times. Taking this and other results into account (the pattern is repeatable), the most likely explanation appears to be that at least one cell division has taken place, and levels of expression per cell are set at an early, low, level by 3 h. Net expression therefore increases per unit volume of the sample, mainly as a result of an increase in cell number.

In contrast, the development of the highly active population at 24 h appears to be attributable to increased expression per cell, within this sub-population.

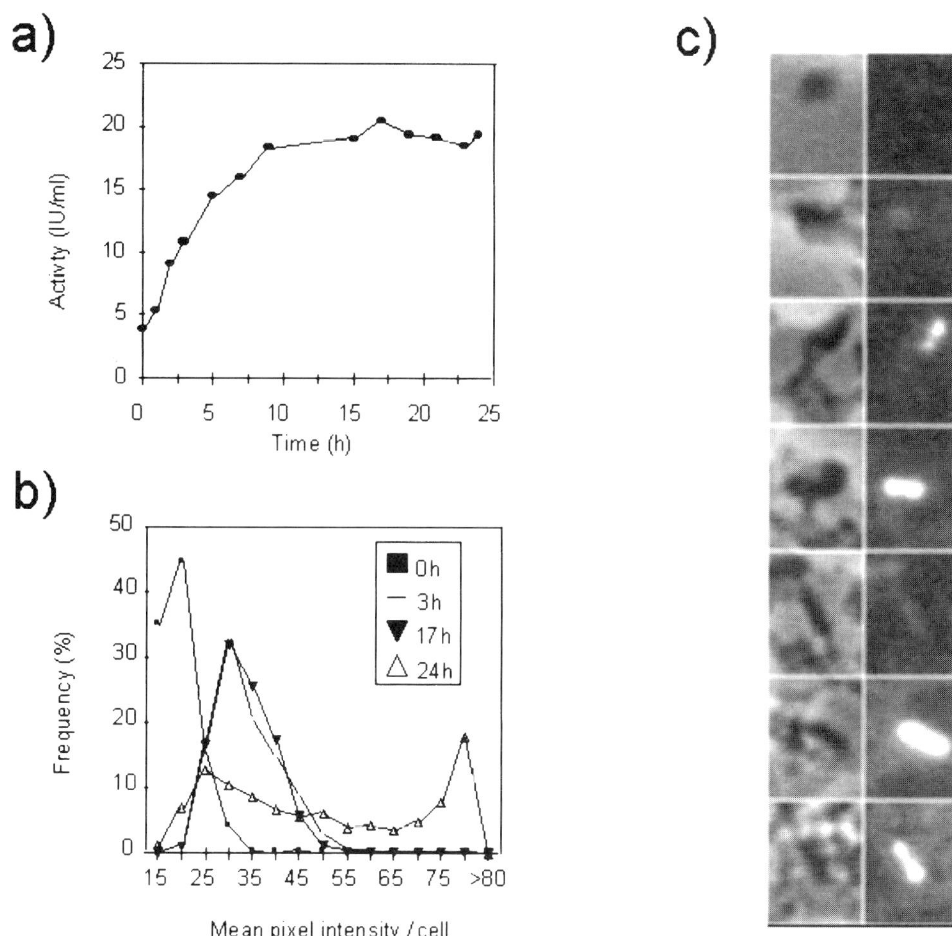

Figure 11.4 β-galactosidase activity in carbon-stressed cells of *Vibrio* sp. S14. Cell samples were obtained at intervals after initiating carbon stress and reacted with fluorescein digalactoside as described previously (Nwoguh *et al.*, 1995). Preparations were mounted in silicone oil and representative images acquired. (a) Bulk biochemical analysis of activity by ONPG assay. (b) Analysed cell intensity distributions (integrated grey weight/object area) at $t = 0$, 3, 17 and 24 h. (c) Selection of mask (phase) and data (fluorescence) pairs from the gallery function illustrating the range of cell associated activities at 24 h.

Demonstration of this sub-population suggests a previously unsuspected heterogeneous pattern of gene expression in carbon-starved cells, and illustrates the sort of novel information that cytological analysis can provide. With the development of exogenous substrate-independent reporter systems such as the green fluorescent protein (Dhandayuthapani *et al.*, 1995), considerable expansion in this type of assay may be anticipated. In the example shown, the software enabled rapid analysis of a substantial population and selection of representative images for display with the gallery function. These and other software issues are now considered in more detail.

11.4 THE SOFTWARE

Bugs is a Windows 3.1 Multiple Document Interface application developed specifically for microbial image analysis tasks such as those outlined above. The interface uses all the familiar Windows application features – menus, toolbars and dialogues. Image processing and analysis functions are provided by the C_Images library from Foster Findlay Associates, a hardware- and operating-system-independent image processing library.

11.4.1 Images, pointsets and objects

Using the C_Images library, images can be defined in host memory, on hard disc or on any supported framestore; the image handling is transparent to the programmer and this flexibility is exploited in the implementation of Bugs. However, the central feature of the program is its treatment of cells as individual objects. To enable viewing, editing and measurement of cells one at a time in different images, we need to describe and manage them separately from their underlying images.

In Bugs objects are captured and described using the C_Images implementation of 'pointsets', originally developed at the Woolfson Image Analysis Unit at the University of Manchester. A pointset specifies a set of connected points in the image and acts as a template, or stencil, when reading from and writing to an image. Using the C_Images library, it is only possible to access the image data via pointsets.

All pointsets have the following data in common:

(a) A specifier to indicate the type of pointset. There are five types of pointset: points, windows, arcs, boundaries and regions (see Figure 11.5).
(b) The origin of the first point in the pointset relative to the image origin. Adjusting the origin of the pointset allows it to be moved around over the underlying image.
(c) A workpoint to indicate the pixel being accessed by the pointset.
(d) The number of points from the workpoint to the end of the pointset row.

Each pointset builds on this by adding its own data structures:

(a) Point: the simplest pointset, representing a single point specified by the pointset origin.
(b) Window: a rectangle of points, specified by its origin and the width and height of the rectangle in pixels.
(c) Arc: a line of points linked by vectors. The first point is located by the origin; thereafter each point is described by the vector linking it to the previous one. The vector takes the values 0..8, 0..7 to represent the direction to take to the next pixel; 8 represents the end of the vector list. Each vector can be thought of as a compass bearing required to take you to the next pixel. Starting with 0 as west and working anti-clockwise around the pixel, 1 is south-west, 2 south, ... 7 north-west (see also Section 13.4.5, on the boundary tracing algorithm).
(d) Boundary: an arc that forms a loop. The last vector indicates a move to the origin (cf. contour representation, Section 6.9.2).

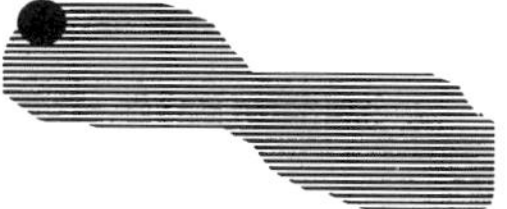

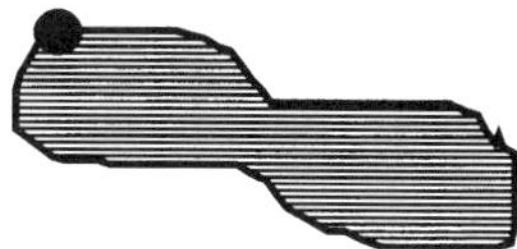

Figure 11.5 Diagram illustrating the terms 'pointset' and 'objects' as used in Bugs.

(e) Regions: areas of points described as a list of chords, representing a row of points. Each chord contains the length of the chord in pixels and the offset to the last pixel of the previous chord. The pointset origin specifies the first point of the first chord (cf. runlength representation, Section 6.9.1).

In the analysis, each object in an image (in this case each cell) is represented by a composite of a boundary and region pointset. Morphometry and intensity measurements are made from these objects.

Pointsets have four essential features for object-based analysis:

(a) They are independent of images. Although a pointset may be defined in an image, either by drawing or object detection, once described it can be used on any image. For example, a pointset drawn in a binary overlay can be drawn into another displayed image, or used to take measurements from an image in memory, or used as template for an image processing operation on a disc image. In Bugs, the original disc image is used for all intensitometric measurements. This has the advantage that the full 8-bit resolution is preserved (2-bit planes are used for overlays in the visible data image).
(b) They are templates for image processing operations. Processing can be performed inside an area, or list of areas, described by a pointset. This ability to process selectively has the further advantage of accelerating the analysis by

concentrating exclusively on the regions described by the pointsets – a particularly useful feature for bacteriological analysis, where the cells may occupy well under 10% of the image pixels.

(c) They can be easily manipulated on images. Library calls and simple manipulation of the pointset origin allow them to be moved, drawn, hidden and edited in the mask and data images.

(d) Since pointsets can easily be stored and retrieved from disc files, a cell database can be constructed and cells re-examined or the analysis extended at a later date, or for graphical display in morphograms (see Section 9.4.1a)

11.4.2 The analysis process

Integration of the various elements of Bugs, and an indication of areas where further development is intended, can be shown by following a typical analysis sequence. An image series analysis is described (one mask, many data images); image sets (multiple pairs of mask and data images acquired from one sample) may be manipulated in a similar way. Both patterns of analysis (series and sets) are outlined in a flow diagram (Figure 11.6). Series analyses are useful for sequential images acquired from a single-cell population, as in kinetic studies, or multi-parameter fluorescence labelling. Sets are used to expand the cell numbers analysed from a single sample.

11.4.2a Image capture

Image capture is from 8-bit monochrome standard video and cooled chip CCD cameras. In order to get the best possible images two issues need to be addressed in future versions of the software:

(a) Dynamic range: the current acquisition techniques could be enhanced to make full use of the available 8-bit data range for both phase-contrast and fluorescence images. When dealing with integrated capture it is quite difficult to ensure that the captured dynamic range is optimal. Further work on calculating the optimum integration for a given sample would provide more available data for object separation and measurement purposes. Alternatively, higher dynamic range (slow scan) cameras (10–16 bits per pixel) can be used.

(b) Flare: fluorescence images are subject to flare, which distorts the shape of the bacteria with respect to their mask image. Therefore de-blurring procedures, from within Bugs or via third-party packages, would enhance the quality of the images prior to analysis.

11.4.2b Creating a group

The software is designed to mimic the organization of our experiments. Each group of experiments is used to gather a series (or set) of analyses into a coherent whole. Each group has its own directory, each experiment its own subdirectory containing the original image files, cell database, data files and miscellaneous files. Bugs creates and manages the directory structure itself, hiding the file and directory

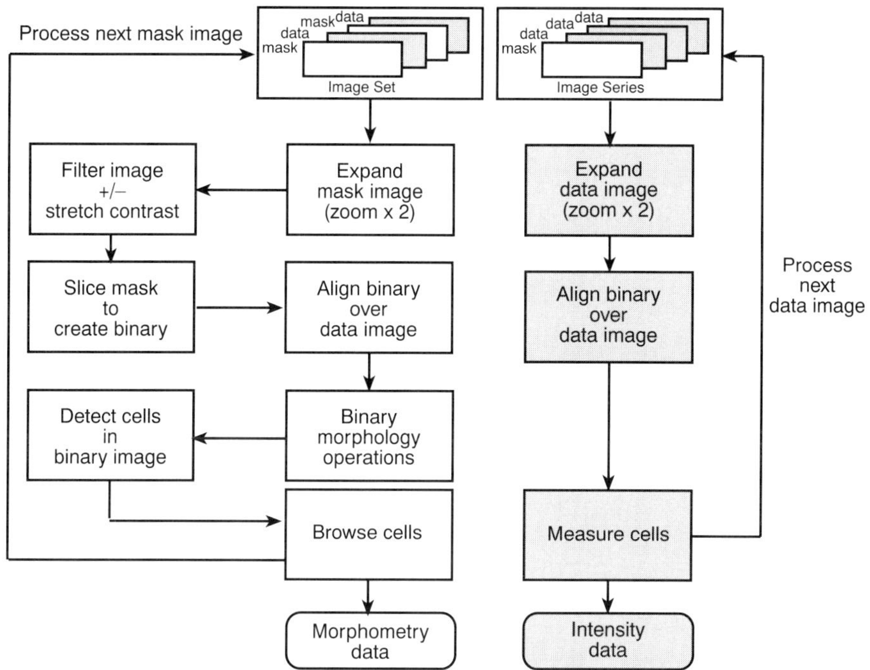

Figure 11.6 Flow diagram of the image processing steps within Bugs. Images are grouped as 'sets' (pairs of mask and data images) or 'series' (one mask image and multiple data images of the same cells). All binary processing and manipulation are carried out on the mask image (unshaded boxes) and simultaneously applied to the data image (shaded boxes) prior to measurements. Morphometry measurements are extracted from the mask image while intensitometric measurements are extracted from the data image. Once the measurements are complete, a loop is present to load either the next images in a 'set' (mask and data image pair) or the next image in a 'series' (e.g. next data image of a time series).

management from the user, who only needs to work in terms of groups and series or sets.

When creating the group or series the user enters the names of the group and series from user dialogues, which create the necessary directory structure and prompt the user to select the image files he or she wishes to analyse. To create a series, or add a new series to a group, the user must select a mask image and one or more data images associated with that mask. Before the images are opened and displayed, the user is given the opportunity to select a shade correction image that can be applied to the original.

11.4.2c Image expansion

At this stage, with all of our image capture systems, the original image has dimensions of around 768×512 pixels, each of which represents approximately 0.01 μm^2. Although this sampling ratio is sufficient to support the optical resolution (Section 2.2.1), the number of pixels representing a bacterium remains low (approximately 50–500). This gives rise to two problems after the binary image of the cells has been created from the mask:

(a) To separate cells and remove spurious pixels, the binary images are processed using dilations and erosions. Such operations on very small objects often markedly distort their shape or eliminate them altogether, thereby invalidating the measurements.
(b) When separating a binary object by editing, a line two (empty) pixels across must be drawn through the object to ensure that 8-connectivity (Section 1.2.4) is broken. On a very small object such a cut may remove a significant area of the object from measurement.

To overcome these effects, each mask or data image is geometrically expanded (zoomed up) by a factor of two, to give working images of 1536×1024 pixels. Each object now occupies four times as many pixels; binary morphology and editing operations have a proportionately less effect. All the object measurements are scaled back to compensate for the expansion. This processing route was chosen rather than changing the sampling ratio by 'empty' magnification (which increases the required exposure time and causes photobleaching problems), or increased CCD array size as a compromise between memory requirements, desirable population sizes for analysis and equipment cost.

The expanded mask and data images are presented side by side to the user for processing. Both are saved to temporary disc files, where they may be used to undo image operations, or supply uncorrupted data for measurement purposes. No further processing of the data image occurs prior to measurement. The original images are not affected in any way and can be used in other analyses or to regenerate the current analysis later.

At present Bugs uses a simple pixel replication algorithm for zooming up the images to working size, producing the characteristic 'blockiness' of adjoining pixels having the same brightness values. To overcome this, more sophisticated expansion algorithms (bilinear or bicubic interpolation) are required to expand the image smoothly. This would generate a 'more detailed' image for processing.

11.4.2d Pre-processing mask image

The goal of this step is to provide a good image for thresholding. A high quality mask image will allow an accurate thresholding to produce a binary image in which the cells are clearly separated. Two types of processing can be carried out, reducing expansion artefacts and noise from the image, and enhancing the contrast. In the current implementation the following smoothing functions are provided:

(a) Mean filter, to remove noise by averaging over a 3×3 kernel.
(b) Median filter, remove noise retaining sharp edges.
(c) Gaussian filter, weighted average towards the central value of a 3×3 kernel.

Contrast enhancement is available via manual manipulation of a contrast look-up table (LUT; see Sections 1.4.1a and 5.2.4b). Current mask image contrast enhancement is performed on a global basis, but one part of the image may be 'enhanced' at the expense of another. Local contrast enhancement routines would allow a more selective enhancement, leading to more clearly delineated objects (Russ, 1994).

11.4.2e Thresholding

The manual slicing method used in bugs is very simple to implement but, since it is a global operation, it is very difficult to create a binary image that accurately reflects the underlying image (Section 6.7), and the user has to perform further binary enhancement and/or editing before detecting the objects. For example it is not possible to threshold cells containing refractile inclusions (e.g. formazan deposits in phase-contrast images) in one pass, nor can it deal effectively with uncorrected uneven background illumination. The current solution adopted for the inclusions problem is to create a 'cumulative slice' routine to allow preliminary segmentations of the inclusions to be dilated and added to subsequent binary images. More sophisticated approaches to thresholding the mask image, such as Rapid Automatic Threshold Selection (RATS; Section 6.7.4), are required to reduce further the post-processing of the binary image.

11.4.2f Binary processing

The available binary operators include opening, closing, erosion, dilation, thickening and thinning (Section 1.6.1). In addition to the binary morphology operating on small objects, large objects in the binary image can be added or rejected using a binary editor.

It is inevitable that some of the objects are going to be touching in the binary image: current object separation techniques are based on global binary image processing, which can be a blunt instrument for separating objects of different sizes in the image. Object separation based on the distance transform and watershed filter (Russ, 1994; Section 1.7.1b) should greatly improve the object separation before detection.

11.4.2g Binary alignment

If the mask and data images are not in perfect registration, intensity measurements for each object will be inaccurate, because the origin of the object in the mask image will not correspond to the origin of the object in the data image (the origin of the detected object is set relative to the image in which it was detected). To compensate for the misalignment, the user can align the two images, creating an offset for each object when measured on that image.

The binary overlay from the mask is copied into the data binary overlay, where it can be moved across the underlying image. When the object binary and the underlying objects overlap, the (x, y) offsets of the data image with respect to the mask image are calculated and stored with the cells, as a coarse alignment for each cell. Performing this step here means that the alignment of the cells can be done once for all the cells, and a fine adjustment need only be made if a cell is offset from the overall alignment.

When analysing subsequent data images, if the user is confident that the cell positions relative to the image origin are stable, then there is no need to visit each of the cells in turn to accept the alignment. The user can perform the overall alignment, and measure all of the cells in the new data image with one command; the software will adjust the cell origins and perform the cell measurements without further user involvement.

11.4.2h Cell detection

When the binary image has been processed and aligned, the cells are detected, by scanning the binary image from the top left to bottom right, encoding each object as it is found (connected component labelling; Section 6.7.7). In order to reduce the number of objects to be examined in the classification phase of the analysis, each detected object is tested against selection criteria (maximum and minimum object area). The user can enter the selection criteria manually or select 'exemplary cells' with a mouse click. These objects are measured and the detected areas become the upper and lower limits. Below the lower area limit objects are deemed to be too small to be of interest, and are discarded. Above the upper area limit cells are retained but marked as rejected (if this object represents a cluster of cells it may be edited later when browsing). Each object that survives selection is stored sequentially in a disc file in the cell database for this analysis.

11.4.2i Cell browsing and classification

After detecting a collection of potential cells the user must examine each of the objects in turn, to select which of them qualify as cells and will be measured. Since each of the objects has been stored in the cell database, it is possible to recall each object and review it in both images using the cell browser (Figure 11.2). In practice, we have found this a particularly useful feature, which accelerates processing and results in operators becoming familiar with their results in an interactive and informative manner. The browser allows the user to move back and forth through the cells in the database. Because the images are too large to be viewed simultaneously they are scrolled in synchrony to bring the displayed object, drawn as an outline, into view. The user can view each cell in the context of the mask and data images before classifying it. Each object may be classified in the following way:

(a) Pending: cells awaiting classification. At the start of the classification all of the detected objects falling within the acceptance criteria are classified as pending.
(b) Accepted: the user has classified this object as a valid cell and it is measured in the current, and subsequent data images.
(c) Rejected: the user has classified this as not being a cell – it may be debris or a cluster of cells that can not be separated easily. To enable analysis of such 'failed' objects, they are measured but the data is clearly marked as rejected.
(d) Parent: when an object is divided into two or more daughter cells by an editing operation, the original cell is tagged by the software as a parent, to exclude it

from measurement (because it occupies the same space as its daughters, it would produce duplicated data). Once a cell has been marked as a parent it cannot be re-edited or classified.

(e) Out of bounds: if an object is very close to the edge of an image, adjusting the alignment of the cell may move part of the object outside the boundary of the image. Since an object cannot be measured outside of its image, the cell is marked as out of bounds when this happens, whatever its previous status.

Well-separated objects can be classified as either accepted or rejected by the user. Classification moves on to the next object. If necessary, the user can go back and change the classification of such cells.

If objects represent touching cells the user can edit them, to split adjoining cells or to remove attached debris from a cell. When an object has been satisfactorily separated, the remaining object(s) can be accepted as child cells of the original. When a cell is split in this way it is automatically classified as a parent and excluded from further editing, classification and measurement.

Although the mask overlay was aligned on the data image before the cells were detected, it is possible that some individual cells will be offset from the overall alignment. The browser allows each cell to be individually aligned. The user moves the outline of the cell on the data image until it directly overlaps the underlying data image. The new cell offset is stored with the cell and applied when the object is measured in the data image.

11.4.2j Cell measurements

As the user accepts or rejects each cell, it is measured using the data image stored on disc. The morphometric measurements, taken directly from the object, do not depend on the underlying data image. However, intensity measurements must be made for each cell in each of the data images. To obtain the intensity measurements for the object in the current data image, the object is aligned with that image before the pixel values represented by the object can be measured. When the user finishes the browsing classification step, the morphometric and intensity data are saved to data files.

If the user is processing a series of images, processing subsequent data images is simplified by the work that has already been done. The user selects the next data image, which is loaded, shade corrected and expanded as before. Since the objects have been detected, the user only needs to align the new image and measure the objects. If aligning the whole image seems to be satisfactory, repeating the browsing task would be tedious and error prone. The user can select a 'measure all' function, which will iterate through the cell database measuring each cell against the new data image without further interaction from the user. If necessary, the user can re-browse the cells to check and adjust individual alignments if that seems appropriate.

An important feature of the cell classification and measurement features is that *both* rejected and accepted cells are measured. This provides opportunities to look at the criteria used to reject objects, and to determine whether any unacceptable systematic bias has resulted from this process.

11.4.2k Gallery

While browsing, the user may notice cells that he or she would like to extract from the image for further analysis or publication purposes. For this purpose, the gallery function allows the user to create a collage of cells of interest into a new image. Cells of interest are tagged for the gallery during browsing. When browsing has been completed and a gallery requested, a rectangular area around each of the tagged cells is copied from its original image into a new gallery image. Gallery images can consist of cells from the mask image, data image or both images arranged in a variety of formats (examples are shown in Figures 11.3 and 11.4).

11.4.2l Data export

Supplying the accumulated data for further analysis is done using comma-separated text files, or by copying the data to the clipboard. In the future, direct communication between Bugs and third-party spreadsheets, statistics and graphics programs will be established. The user should be able to step seamlessly from measurement to analysis of the data. A logical, and appealing, extension of this procedure would be to create a two-way connection. Data points of interest selected by the user would be linked to the corresponding cells and images, and could be used to allow the user to revisit those cells.

11.4.2m Image versus object processing

Bugs is currently biased towards handling cells rather than the images that contain them. Dealing with cells as objects has a number of advantages:

(a) Fine control: the user can manipulate individual entities from very large populations.
(b) Cells represented in several images – in a time sequence, for example – can be extracted with ease into a composite 'gallery' image for comparison or publication.
(c) Over a series of images, cells may be offset from their original positions, the ability to align (or even rotate) cells as individuals, in addition to aligning the images as a whole, providing greater accuracy in the measurements.
(d) Each cell, or cluster of cells, can be edited manually in the case of difficult preparations, which do not lend themselves to fully automated object separation techniques.

The downside of the purely object-based approach is of course the need for user intervention in the analysis, particularly in the editing phase. This approach was chosen for three reasons. First, trying to perform this kind of analysis with a general-purpose image analysis package and its rigid adherence to image-based processing was frustrating and time consuming. Naturally, its replacement was built with a view to overcoming the weaknesses of the earlier approach. Second, the ability to examine cells individually, particularly when browsing, allows the operator to view large numbers of cells and get a 'feel' for the data. Third, Bugs is still in an experimental development stage; whole image processing techniques have not been forgotten – rather, they have been put in second place to the cell handling. With the experience gained from extended use, we want to improve the image processing leading up to segmentation, to provide better cell separation and cut down on the amount of manual intervention.

11.5 CONCLUSIONS

The approach described here provides a starting point for investigation of the physiology and biochemistry of micro-organisms at the single-cell level. With careful attention to appropriate controls, the methods are applicable to any sample material that can be prepared for microscopy. Because cell growth is not required, the methods can be used to investigate organisms that are not growing, those that cannot (as yet) be propagated *in vitro*, including cells in the putative VBNC state, and cells whose metabolic properties may be specific to positional relationships that exist in natural environments. Although we may gain more information about 'VBNC' cells by cytochemical investigation, for the reasons outlined earlier we do not think that the outstanding controversies will be resolved in this way, unless a cytological demonstration which is inextricably linked to cellular replicative potential can be developed.

In the future, more extensive rRNA sequence data will undoubtedly become available, and this will enable application of co-localization techniques to a wider range of organisms. Further developments in this area could include development of more cytochemical reagents that are not extracted by the FISH process (obviating the need for on-stage hybridization) and improvements in the sensitivity of FISH demonstration so that it can be applied to other genetic elements including DNA and mRNA.

Wider application of the cytochemical approach will facilitate development of a clearer view of what micro-organisms are actually doing in natural environments. Such environments may include ecological niches that are of particular significance to human activities, such as agriculture and sewage disposal. More intimate associations between micro-organisms and mankind leading to colonization and infection, where virtually nothing is known about the metabolic activity of organisms *in situ*, are also potentially amenable to cytochemical investigation.

The laboratory and software techniques presented here are a staging post towards the long-term objective of establishing a baseline set of analyses which can be applied to characterize microbes at the single-cell level. All areas described are under active development to overcome the biological, optical, electronic and computational problems faced. Nonetheless, even at its present level, the system does enable collection of substantial data sets, which appear to fall into coherent patterns after analysis. The final arbiter of their value will lie in the extent to which the results of such analyses can extend our knowledge and understanding of micro-organisms.

ACKNOWLEDGEMENTS

The authors gratefully acknowledge the financial support of the BBSRC, NERC and Wellcome Trust in relation to different aspects of the work presented. Jorgen Östling and Staffan Kjelleberg kindly provided the J109 strain of *Vibrio* sp. S14.

REFERENCES

Barer, M.R. (1991) New possibilities for bacterial cytochemistry – light microscopic demonstration of beta-galactosidase in unfixed immobilized bacteria. *Histochem. J.* **23**: 529–33.

Barer, M.R. and Marsh, P.J. (1992) Rapid cytochemical demonstration of cytochrome-oxidase activity in pathogenic bacteria. *J. Clin. Pathol.* **45**: 487–9.

Barer, M.R., Gribbon, L.T., Harwood, C.R. and Nwoguh, C.E. (1993) The viable but non-culturable hypothesis and medical microbiology. *Rev. Med. Microbiol.* **4**: 183–91.

Bovill, R.A., Shallcross, J.A. and Mackey, B.M. (1994) Comparison of the fluorescent redox dye 5-cyano-2,3-ditolyltetrazolium chloride with p-iodonitrotetrazolium violet to detect metabolic-activity in heat-stressed *Listeria monocytogenes* cells. *J. Appl. Bacteriol.* **77**: 353–8.

Davey, H.M. and Kell, D.B. (1996) Flow cytometry and cell sorting of heterogeneous microbial populations: the importance of single-cell analysis. *Microbiol. Rev.* **60**: 641–96.

Deere, D., Porter, J., Edwards, C. and Pickup, R. (1995) Evaluation of the suitability of bis-(1,3-dibutylbarbituric acid) trimethine oxonol, (diba-c-4,(3)(-)), for the flow cytometric assessment of bacterial viability. *FEMS Microbiol. Lett.* **130**: 165–9.

Dhandayuthapani, S., Via, L.E., Thomas, C.A., Horowitz, P.M., Deretic, D. and Deretic, V. (1995) Green fluorescent protein as a marker for gene-expression and cell biology of mycobacterial interactions with macrophages. *Molec. Microbiol.* **17**: 901–12.

Diaper, J.P. and Edwards, C. (1994) The use of fluorogenic esters to detect viable bacteria by flow-cytometry. *J. Appl. Bacteriol.* **77**: 221–8.

Gant, V.A., Warnes, G., Phillips, I. and Savidge, G.F. (1993) The application of flow-cytometry to the study of bacterial responses to antibiotics. *J. Med. Microbiol.* **39**: 147–54.

Gribbon, L.T. and Barer, M.R. (1995) Oxidative-metabolism in nonculturable *Helicobacter pylori* and *Vibrio vulnificus* cells studied by substrate-enhanced tetrazolium reduction and digital image-processing. *Appl. Environ. Microbiol.* **61**: 3379–84.

Horobin, R.W. and Rashid, F. (1990) Interactions of molecular probes with living cells and tissues. 1. Some general mechanistic proposals, making use of a simplistic Chinese box model. *Histochemistry* **94**: 205–9.

Jepras, R.I., Carter, J., Pearson, S.C., Paul, F.E. and Wilkinson, M.J. (1995) Development of a robust flow cytometric assay for determining numbers of viable bacteria. *Appl. Environ. Microbiol.* **61**: 2696–701.

Kaprelyants, A.S. and Kell, D.B. (1992) Rapid assessment of bacterial viability and vitality by rhodamine 123 and flow-cytometry. *J. Appl. Bacteriol.* **72**: 410–22.

Kogure, K., Simidu, U. and Taga, N. (1979) A tentative direct microscopic method of counting living bacteria. *Can. J. Microbiol.* **25**: 415–20.

Lewis, P.J., Nwoguh, C.E., Barer, M.R., Harwood, C.R. and Errington, J. (1994) Use of digitized video microscopy with a fluorogenic enzyme-substrate to demonstrate cell-specific and compartment-specific gene-expression in *Salmonella enteritidis* and *Bacillus subtilis*. *Molec. Microbiol.* **13**: 655–62.

Mason, D.J., Lopezamoros, R., Allman, R., Stark, J.M. and Lloyd, D. (1995) The ability of membrane-potential dyes and Calcafluor White to distinguish between viable and nonviable bacteria. *J. Appl. Bacteriol.* **78**: 309–15.

McFeters, G.A., Yu, F., Pyle, B.H. and Stewart, P.S. (1995) Physiological assessment of bacteria using fluorochromes. *J. Microbiol. Meth.* **21**: 1–13.

Nystrom, T., Marden, P. and Kjelleberg, S. (1986) Relative changes in incorporation of leucine and methionine during starvation survival of two bacteria isolated from marine waters. *FEMS Microbiol. Ecol.* **38**: 285–92.

Nwoguh, C.E., Harwood, C.R. and Barer, M.R. (1995) Detection of induced beta-galactosidase activity in individual non-culturable cells of pathogenic bacteria by quantitative cytological assay. *Molec. Microbiol.* **17**: 545–54.

Oliver, J.D. (1993) Formation of viable but nonculturable cells. In *Starvation in Bacteria* (Kjelleberg, S., ed.), pp. 239–72. Plenum Press: New York.

Östling, J., Goodman, A. and Kjelleberg, S. (1991) Behavior of incp-1 plasmids and a minimum transposon in a marine vibrio sp – isolation of starvation inducible lac operon fusions. *FEMS Microbiol. Ecol.* **86**: 83–94.

Östling, J.O., Flardh, K. and Kjelleberg, S. (1995) Isolation of a carbon starvation regulatory mutant in a marine *Vibrio* strain. *J. Bacteriol.* **177**: 6978–82.

Rashid, F. and Horobin, R.W. (1990) Interaction of molecular probes with living cells and tissues. 2. A structure–activity analysis of mitochondrial staining by cationic probes, and a discussion of the synergistic nature of image-based and biochemical approaches. *Histochemistry* **94**: 303–8.

Russ, J.C. (1994) *The Image Processing Handbook*, 2nd edn, pp. 674 ff. CRC Press: Boca Raton, FL.

Stackebrandt, E. and Rainey, F.A. (1995) Partial and complete 16S rDNA sequences, their use in generation of 16S rDNA phylogenetic trees and their implications in molecular ecological studies. In *Molecular Microbial Ecology Manual* (Akkermans, A.D.L., Van Elsas, J.D. and De Bruijn, F., eds), pp. 3.1.1:1–17. Kluwer: Dordrecht.

Thom, S.M., Horobin, R.W., Seidler, E. and Barer, M.R. (1993) Factors affecting the selection and use of tetrazolium salts as cytochemical indicators of microbial viability and activity. *J. Appl. Bacteriol.* **74**: 433–43.

Van Prooijen-Knegt, A.C., Raap, A.K., Van der Berg, M.J.M., Vrolijk, J. and Van der Ploeg, M. (1982) Spreading and staining of human metaphase chromosomes on aminoalkysilane-treated glass. *Histochem. J.* **14**: 333–44.

Vendrely, R. (1955) Histochemistry of bacteria. *Int. Rev. Cytol.* **4**: 115–42.

Walker, D.R., Nwoguh, C.E. and Barer, M.R. (1994) A microchamber system for the rapid cytochemical demonstration of beta-galactosidase and other properties in pathogenic microbes. *Lett. Appl. Microbiol.* **18**: 102–4.

Wayne, L.G., Brenner, D.J., Colwell, R.R., Grimont, P.A.D., Kandler, O., Krichevsky, M.I., Moore, L.H., Moore, W.E.C., Murray, R.G.E., Stackebrandt, E., Starr, M.P. and Truper, H.G. (1987) Report of the ad-hoc committee on reconciliation of approaches to bacterial systematics. *Int. J. System. Bacteriol.* **37**: 463–4.

Whiteley, A.S., O'Donnell, A.G., MacNaughton, S.J. and Barer, M.R. (1996) Cytochemical colocalization and quantitation of phenotypic and genotypic characteristics in individual bacterial-cells. *Appl. Environ. Microbiol.* **62**: 1873–9.

Willaert, R.G. and Baron, G.V. (1996) Gel entrapment and micro-encapsulation – methods, applications and engineering principles. *Rev. Chem. Engng* **12**: 5–205.

Wolf, P.W. and Oliver, J.D. (1992) Temperature effects on the viable but non-culturable state of *Vibrio vulnificus*. *FEMS Microbiol. Ecol.* **101**: 33–9.

12

Microspectrofluorometry: Measuring Ion Concentrations in Living Microbes

Jan Slavik

Czech Academy of Sciences, Prague, Czech Republic

12.1 INTRACELLULAR IONIC CONCENTRATIONS

The intracellular ionic composition is closely related to the cellular activity. The pH and concentrations of calcium, sodium and potassium play key roles in controlling cell metabolism. Calcium regulates many important cellular functions, and cells spend a great amount of their energy on maintaining the level of cytosolic free calcium. The cytoplasmic pH is known to regulate enzyme activities and the transmembrane electrochemical potential, which is generally accepted as a link between electron transport and ATP synthesis in bacteria, mitochondria and chloroplasts. Such data reveal the importance of knowledge of local values of ionic concentrations. Special emphasis should be given to the transmembrane gradients of ions, which are indispensable in the study of bioenergetics, membrane transport or signal transduction. The evaluation of an ion gradient obviously requires detailed knowledge of ion concentrations, and fluorescent probes have revealed an enormous amount of information on intracellular behaviour of ions. The most important findings may be divided into three groups. First, populations of cells are heterogeneous with respect to their average cytoplasmic ionic composition. It has been shown that cells in suspension show inter-individual differences in pH or pCa values (Figure 12.1). Second, in a variety of cell types the level of intracellular free calcium in a particular cell often suddenly increases and then returns to the original value. Such 'spikes' usually appear asynchronously every several tens or hundreds of seconds (Figure 12.2). Obviously, they can be revealed only by microscopic observation of individual cells, but not by studying whole populations of cells in cuvettes, where they disappear in the average value. Third, ratio imaging has shown that the changes in calcium concentrations and, to some extent, pH are local. The rise in calcium concentration travels through the cell cytoplasm like a wave. The famous classical example for calcium is the local intracellular increase of calcium level close to the place where sperm has entered the sea urchin egg followed by its propagation across the cell (Tsien and Poenie, 1986). The movement of such calcium 'waves' is still unclear. It has been explained on the basis of

Digital Image Analysis of Microbes: Imaging, Morphometry, Fluorometry and Motility Techniques and Applications. Edited by M.H.F. Wilkinson and F. Schut.

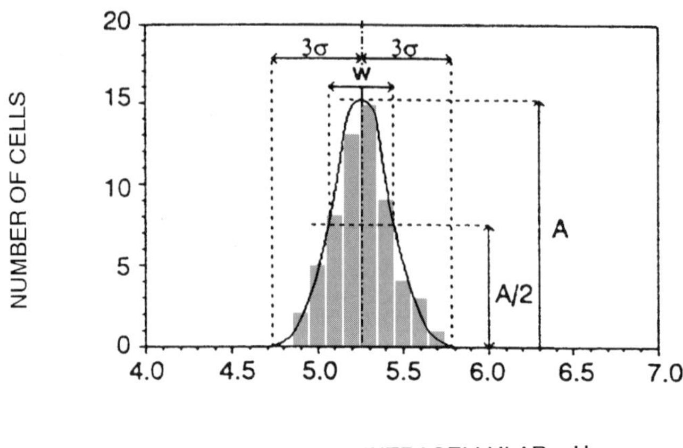

Figure 12.1 Distribution of intracellular pH of yeast cells. The distribution follows a Gaussian curve with mean pH value 5.25, $\sigma = 0.18$, $w = 0.42$, 60 cells evaluated (after Cimprich *et al.*, 1995).

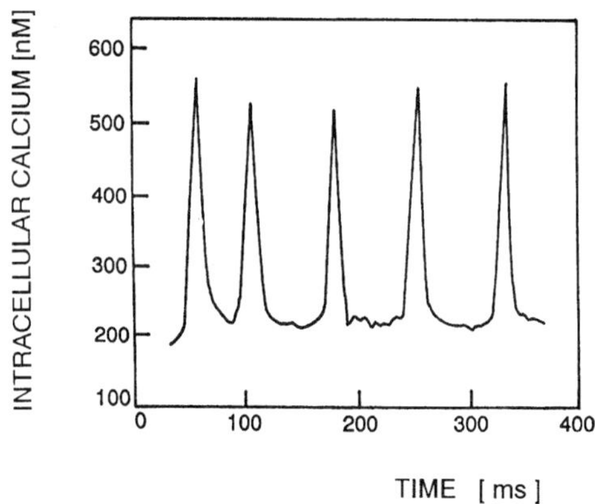

Figure 12.2 Calcium oscillations in a single hepatocyte loaded by fura-2 (redrawn after Rooney TA, Sass EJ and Thomas AP, *J. Biol. Chem.* **264**, 1713, 1989).

imperfect mixing due to the slow diffusion rate or as a serial saltatory movement along calcium pools. In the case of pH, the local heterogeneity appears in the perimembrane region and disappears when proton transport across the plasma membrane has ceased. The connection of confocal microscopy and digital image processing pushed the barriers further (Plate 2).

12.2 ION-SELECTIVE FLUORESCENT PROBES

The fluorescence emitted by a fluorescent molecule contains information not only on the fluorescent molecule itself, but also on fluorescence encoded information on neighbouring molecules. In this way, a dye molecule can be used as a tiny molecular 'reporter' informing about what is going on in its molecular neighbourhood. The encoding of the information is highly specific for each type of dye and it requires good knowledge of fluorescence mechanisms. Then it can be translated into terms of membrane fluidity, viscosity of the solvent, polarity of the solvent, distance between two chromophores, membrane potential and, especially, intracellular ion concentration. The ideal probe must not interfere with the event observed; according to Gregorius Weber, it must be a witness of but not an actor in the physiological drama (Figure 12.3).

Many dyes can be used as fluorescent probes for the measurement of local concentrations of various ions inside the cell and in intracellular organelles. There are fluorescent probes specific for intracellular pH, free calcium concentration, magnesium, sodium, potassium, chloride, zinc, iron and heavy metals. The discriminating selectivity and sensitivity of these fluorescent probes are very high. Although the dyes are far from perfect, due to the low performance or non-existence of competitive techniques, fluorescence analysis is the method of choice. It is no surprise that the demand for new fluorescent indicator dyes is high.

The basic principle of ion-sensitive dyes follows the old principle of the red–blue pH-driven transition of litmus paper. When the dye binds a particular ion, its excitation, emission or lifetime changes profoundly (see Chapter 4). Provided that both forms of the dye, the free form and the ion-bound form, can easily be distinguished spectroscopically, the concentration ratio of 'free' versus 'ion-bound' dye can be determined. This ratio depends on the concentration of the measured ion, and (using an appropriate standard calibration curve) its values can be converted into the actual local values of pH, pCa and the like. Because it is the fluorescence ratio that is determined, the techniques are usually called ratio imaging techniques. Such ion-sensitive fluorescent probes may be called *in situ* indicators in order to distinguish them from the distributive fluorescent dyes described later in this chapter. The distributive fluorescence dyes are weak acids or bases, the

Figure 12.3 Fluorescent dye as a molecular reporter.

undissociated forms of which are permeant and the dissociated form membrane-impermeant. The equilibrium intracellular concentration is reached after several minutes and the intracellular/extracellular concentration ratio is given by the ΔpH and p*K* of the dye. The intracellular and extracellular concentrations are also distinguishable by confocal laser scanning microscopy (CLSM).

To summarize, fluorescent probes for ions are compatible with all optical microscopic imaging techniques, including phase contrast and Nomarski differential interference contrast (DIC), and can be used with flow cytometry and cell sorting. Ratio imaging can be done in emission (emission ratio imaging), excitation (excitation ratio imaging) or lifetime (using the lifetime ratio as a parameter). The intracellular maps are displayed in pseudocolours on the computer screen, possibly in real time, with rates attaining the normal television rate of 25–30 images per second. Using several dyes simultaneously, it is possible to obtain maps of two or more ions at a time in a single cell, and also to combine them with, for example, cytoskeleton labelling or phase-contrast or Nomarski DIC images of the cell.

12.3 THEORETICAL BACKGROUND

The ion-sensitive dyes are assumed to bind a specific ion selectively, usually at a 1:1 stoichiometry. The binding is accompanied by a profound change in their fluorescent properties. The 'free' and 'bound' forms informally correspond to the red and blue forms of the litmus dye (Figure 12.4). The measurement consists of the determination of the concentration ratio *R* of the 'red' versus 'blue' forms, based on their fluorescence differences (Figure 12.5).

Let us assume that the fluorescent probe exists in two spectrally distinguishable forms. Then the measurement of the fluorescence intensity at two different wavelengths allows us to calculate their concentration ratio as follows. Let the first measurement be done at the excitation wavelength λ_1 and the emission wavelength λ_2, and the second measurement at excitation λ_3 and emission λ_4.

Then the two general equations determining the fluorescence intensities are as follows:

$$I(\lambda_1, \lambda_3) = KI_{\text{exc}}(\lambda_1)\{c_{\text{A}}\varepsilon(\lambda_1)\gamma_{\text{A}}(\lambda_3) + c_{\text{B}}\varepsilon(\lambda_1)\gamma_{\text{B}}(\lambda_3)\} \tag{12.1a}$$

$$I(\lambda_2, \lambda_4) = KI_{\text{exc}}(\lambda_2)\{c_{\text{A}}\varepsilon(\lambda_2)\gamma_{\text{A}}(\lambda_4) + c_{\text{B}}\varepsilon(\lambda_2)\gamma_{\text{B}}(\lambda_4)\} \tag{12.1b}$$

where $I(\lambda_i, \lambda_j)$ is the intensity of fluorescence at λ_j after excitation by λ_i, K characterizes the overall apparatus constant, c_{A} and c_{B} are the concentrations of each form, $I_{\text{exc}}(\lambda_i)$ is the intensity of the excitation light beam, $\varepsilon(\lambda_i)$ is the coefficient of extinction at λ_i, and $\gamma_{A,B}(\lambda_i)$ is the quantum yield of each form detected at wavelength λ_i.

The most usual choice is a double excitation (λ_1 differs from λ_3) or a double emission (same excitation wavelength but different emission wavelengths λ_2 and λ_4). Both approaches are called ratio measurement or the ratio technique and are very convenient in biological experiments, particularly in imaging, because the

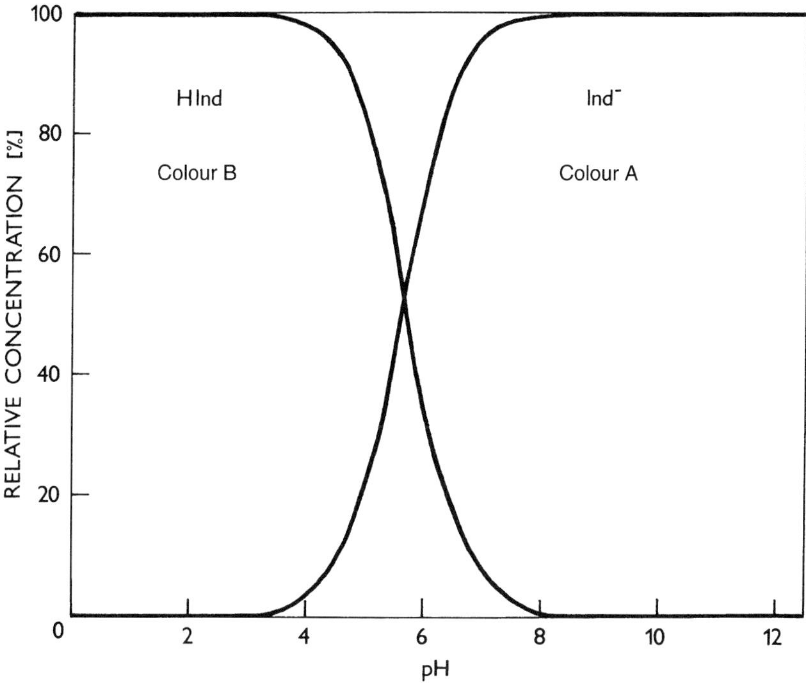

Figure 12.4 Schematics of a pH-dependent decrease of the concentration of one form of the pH indicator followed by the increase of the concentration of its other form.

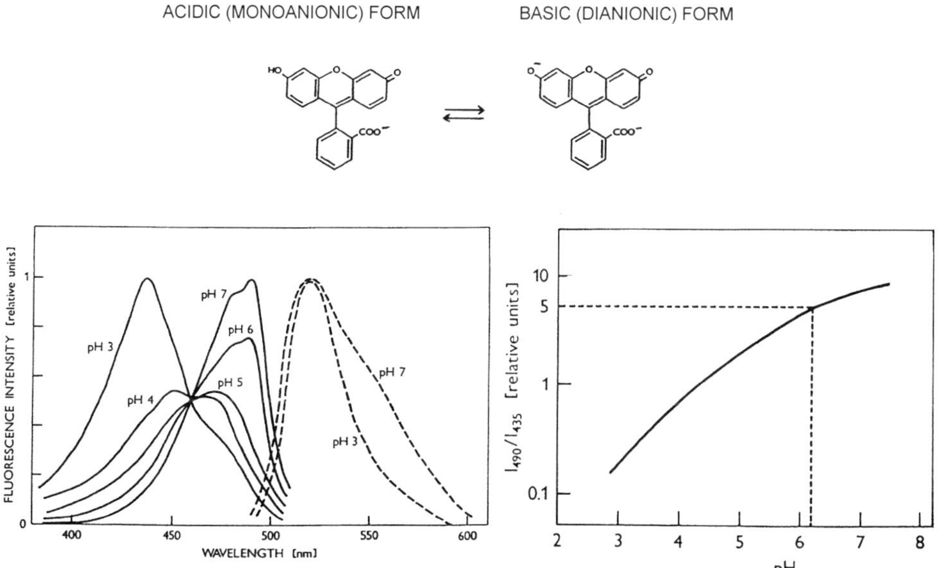

Figure 12.5 The pH response of fluorescein. In the left panel the solid lines represent the excitation spectrum and the dashed lines the emission spectrum.

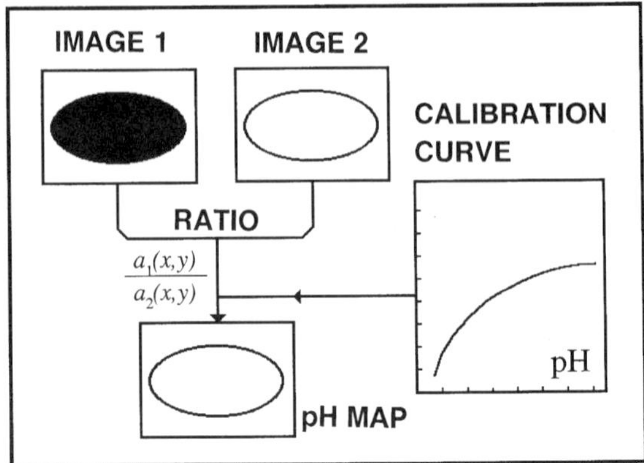

Figure 12.6 The basic principle of ratio imaging.

ratioing cancels out the effects of heterogeneities in camera sensitivity and unevenness of illumination (cf. shading correction, Section 2.4.7).

Both double emission and double excitation are easy to perform. Fura-2 may serve as an example of a dye for the excitation ratio measurement of calcium (340 nm versus 380 nm excitation, 510 nm emission) or carboxyfluorescein for pH, while indo-1 is an example of an indicator using two different emission wavelengths. Dyes like SNARF or SNAFL make it possible to choose either the ratio excitation or ratio emission (or use all four different wavelengths for excitation and emission – if there is a reason for it).

Based on these background principles, ratio imaging combined with confocal microscopy fulfils a large portion of the ultimate dream of intracellular measurement represented by a time-resolved, three-dimensional map of a particular ion concentration in a living cell. The technique uses two pictures of the same cell filled with fluorescent dye taken at two different excitation or emission wavelengths (depending on the type of dye). After subtracting the background fluorescence (if necessary) the images are ratioed (i.e. the value of fluorescence intensity at each point (pixel) in the first picture is divided by the value at the corresponding point in the second picture, one point after another, one line after another). The ratio of the two pictures computed in this way is then displayed. The computed values of the ratio R correspond to the local values of pH, pCa etc. Using the standard curve (for details see Sections 12.7 and 12.8) the local R values are converted to local pH or pCa values and displayed using pseudocolours usually based on the rainbow colour scale, or imitating the traditional colour scale of pH-indicating papers (Figure 12.6).

12.4 BASIC CONCEPT OF DISTRIBUTIVE pH DYES

For the intracellular pH assays, there is one more group of methods, namely distributive techniques. They are conceptually different because, unlike the ion measurement methods described above, they do not use *in situ* ion-sensitive indicators. They are also based on fluorescent dyes, mostly weak electrolytes, but it is the equilibrium distribution of a fluorescent weak acid or base between the extracellular and intracellular spaces that is measured. The equilibrium is usually established after several minutes, and the measurement is done under the assumption that the fluorescence of the intracellular dye is quenched.

The ideal distributive pH indicator should have a membrane-permeant undissociated form and a membrane-impermeant dissociated form. The ratio of intracellular to extracellular concentration of the dye (after equilibrium distribution has been reached) is determined by measuring fluorescence intensity or lifetime. CLSM is a useful technique, because it allows the determination of local dye concentrations, i.e. in the cytoplasm versus outside the cell. In principle, the determination of both concentrations can also be done using other techniques such as absorption, electron paramagnetic resonance (EPR), nuclear magnetic resonance (NMR) or radioactively labelled weak electrolytes.

The estimate of intracellular pH is based on the assumption that the uncharged form of the weak electrolyte is permeant while the charged form is membrane-impermeant. Consequently, weak acids are accumulated inside cells after addition to the cell suspension if the intracellular pH is lower than the extracellular pH and, by analogy, weak bases are accumulated if the intracellular pH is higher than the external one.

After the establishment of the steady-state equilibrium concentration of the indicator, the intracellular pH value is calculated from the ratio of analytical concentrations of the electrolyte in the internal and external fluids, the tabulated p*K* value and the extracellular pH value. Usually, the analytical concentration of the pH indicator is determined extracellularly and the counterpart, the intracellular concentration, is determined as the difference, using the total analytical amount of the indicator as the starting point. The theory of distribution measurements is described in detail by Kotyk and Slavik (1989).

12.5 LOADING OF CELLS WITH A RATIO ION-SENSITIVE DYE

12.5.1 Diffusion, membrane transport, endocytosis

The easiest way to load cells with an ion-sensitive dye is to incubate them in a solution of the dye and wait. With the exception of highly lipophilic or highly polar dyes, a certain level of intracellular concentration is reached after a longer incubation. The often slow equilibrating process may be accelerated by varying the external pH (so-called acid loading). This effect of acid loading is based on the assumptions that the uncharged forms of the dye will penetrate the membrane more easily, and if the dye is a weak acid that the proportion of uncharged form increases with pH. In the cytoplasm the dye dissociates back to the impermeant anionic state, and is trapped inside the cell. In plant cells acid loading has been used for calcium indicators fura-2, quin2 and indo-1, in bacterial cells for loading of free acid form of a pH-indicator BCECF using brief incubation at pH 2. Optimal incubation of barley protoplasts has been achieved at external pH 4.5, where the intracellular concentration of indo-1 reached 15 mM after 2 h loading. Some dyes may also be loaded into the cell using passive or active membrane transport processes via existing transport proteins (carboxyfluorescein, lucifer yellow and fura-2).

12.5.2 Use of membrane-permeable precursors

Acetylation offers the possibility of tricking the permeability barrier of the plasma membrane. Instead of a membrane-impermeant dye, its membrane-permeant precursor is used for the incubation of cells. The precursor (often non-absorbing and non-fluorescent) enters the cell, where it is enzymatically changed into a membrane-impermeant highly fluorescent probe and becomes trapped inside. In this way all cells are eventually filled with the fluorescent dye. This attractive idea, developed in our laboratory (Slavik, 1982), has become increasingly popular, especially with applications of calcium and pH-sensitive probes (Figure 12.7).

The precursors are fluorescent probes, the polar groups of which are 'protected' with ester groups, resulting in electrically neutral, lipophilic molecules that pass through the plasma membrane to the cytoplasm. The ester groups arise from acetate derivatives of aromatic alcohols (present in most fluorescent indicators) and acetoxymethyl (AM) esters of carboxylic acids (present in most ion indicators). Intracellular esterases cleave the esters, liberating the more polar, polyanionic indicators within the cell. The loading medium is usually prepared by a hundred- or thousand-fold dilution of the stock solution of the AM dye precursors into a (preferably mineral) medium. The stock solution of 1–10 μM AM dye in anhydrous dimethyl sulphoxide (DMSO) is prepared and stored desiccated at −20 °C. This is necessary to curtail the spontaneous hydrolysis that occurs in moist and alkaline (above pH 7!) environments. Because the melting of DMSO

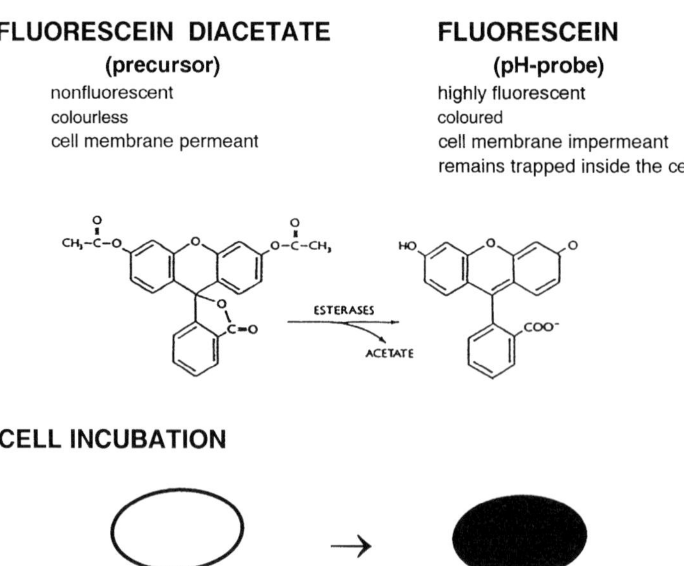

Figure 12.7 The principle of loading cells with a precursor of a fluorescent dye.

(18 °C) is sharply depressed by the water content, moisture pickup is easily seen as rapid melting of the solution upon removal from the refrigerator. Administration of serum, collagenases, amino acids or compounds with amino groups should be avoided until loading is complete, since they may already cleave the AM esters outside the cell, and thus prevent the loading. However, the addition of serum albumin (2%), plasma proteins or a non-ionic surfactant such as Pluronic (final concentration 0.025%) is sometimes reported to facilitate the loading, perhaps by helping to disperse the dye and to prevent formation of its aggregates, or by providing a 'carrier' function for the dye.

Typically, 2 μM BCECF/AM, 1–5 μM fura-2, 5–10 μM SBFI and 10–45 min incubation (in some cells 10–50 μM up to 3 h) are used. Final intracellular concentrations are estimated up to 10–50 μM for BCECF and 50–150 μM for fura-2. The loading procedure usually takes place at 37 °C, but quite often lower temperatures such as room temperature are preferred. In some cases even temperatures as low as 4 °C are used, to prevent sequestration of the dye. Washing of cells after loading is recommended, especially if the AM ester is already fluorescent (e.g. fura-2/AM).

The slow and incomplete intracellular hydrolysis of some AM dyes represents one of the major problems of loading. For instance, fura-2/AM has five acetoxymethyl groups that have to be cleaved. The incompletely cleaved fura-2/AM is also fluorescent, but it does not bind calcium. The non-hydrolysed intracellular dye is prone to be promoted further through the organellar membrane into organelles and may be sequestered there.

12.5.3 Disruptive ways of loading

Some techniques originally developed for transport of larger molecules such as proteins and nucleic acids into cells can be modified to introduce virtually any dye into cells. The strategies of temporarily breaking the plasma membrane and invading the cytoplasm can be divided into mechanical, chemical, electrical and 'vector' methods. Mechanical methods include glass microcapillary microinjection, osmotic shock, sonication, cell injury by vigorous agitation, scraping, scratching and sprinkling glass beads onto cells. In the chemical strategy cells are soaked in ATP or EGTA prior to loading. In the electrical strategy, cells are transiently permeabilized by application of a short high-voltage pulse. In the vector strategy, cells are loaded by fusing them with a liposome or red blood cell ghost filled with the selected fluorescent dye. Each of these strategies has its advantages and limitations. Some of the techniques are more convenient for cells in suspension; others are better for adherent cell culture cells.

12.6 FATE OF THE DYE INSIDE THE CELL

The fate of the dye in the cell is determined by its chemical structure. It defines whether the dye will stay in the plasma membrane or on the cell surface or whether it will proceed further into the cytoplasm and possibly inside some cellular compartments. These factors can actually be influenced by the choice of dye.

The interaction of a dye with cells can be described schematically at the cellular level (which cells are stained), at the subcellular level (which organelles or intracellular compartments are stained) and at the molecular level (which molecules of the cell interact with the dye).

In a simplified model the cell may be described as a set of compartments within compartments, where the membranes act as barriers that control the entry and the retention of probes on the basis of hydrophilicity–lipophilicity. Using this model it can be explained why extreme hydrophilic (i.e. lipophobic) probes will stay in the medium and will not enter the cells. On the other hand, highly hydrophobic (lipophilic) dyes will be trapped in the membrane lipid bilayer and stay there. They can proceed to the organelles only when the outer membrane is internalized or they end up in lipid droplets. Only moderately lipophilic probes will manage to go through the membrane and enter the cytoplasm. These dyes may then proceed further and pass through the organelle membranes and eventually be trapped there. This simple model must be supplemented by specific biological transit routes, such as carrier transport or endocytosis, which may transport the dye into the cell through the membrane. Furthermore, the uptake and the fate of the probe inside the cell are in reality not decided only on the basis of its hydrophilicity–lipophilicity. One must also consider other physicochemical properties of the dye, especially its electrical charge, its ionic or molecular size and its ability to bind non-specifically to membrane and cellular proteins.

Generally, the neutral form of a dye penetrates all membranes in a cell. If it is a weak acid or base it tends to establish local concentration equilibria according to local transmembrane ΔpH. As a consequence, the dye may accumulate within various subcellular compartments at significantly higher concentrations. For instance, fluorescein is assumed to be accumulated inside mitochondria whereas carboxyfluorescein is not. Both dyes, however, can cross the vacuolar membrane and are found in both the cytoplasm and the vacuoles. Consequently, both dyes reflect changes in cytoplasmic and vacuolar pH, while changes in intramitochondrial pH should be reflected by fluorescein and should not appear in the carboxyfluorescein signal. Charged forms of dyes may prefer some sites for purely electrostatic reasons. They are attracted by oppositely charged surfaces and often form aggregates with oppositely charged ions. The lipophilicity of the dye may emphasize the signal from the perimembrane region. Highly lipophilic dyes may end up in lipid droplets.

12.6.1 Compartmentalization and how to avoid it

Freely moving indicator molecules in the cytoplasm non-interacting with other intracellular components represent a dream of intracellular measurement with ion-concentration sensitive probes. It was taken for granted that these dyes disperse evenly in the cell interior, until confocal imaging of the calcium-indicating dye fura-2 showed that there may be some compartmentalization problems. Compartmentalization can be demonstrated by direct visualization of the dye by the microscope, on the basis of co-localization with an organelle-specific marker, or by application of digitonin or Triton in the presence of on external quencher such as manganese or iodide. In this way it has been shown that fura-2 is accumulated in lysosomes, in mitochondria, in the nucleus, in the sarcoplasmic reticulum and in secretory granules. Other dyes seem to be less prone to compartmentalization. Therefore, much effort has been devoted to seeking ways to avoid compartmentalization of fura-2. They are as follows. Loading conditions are optimized to achieve better cytosolic localization and full de-esterification. With fura-2/AM, optimal loading should be sought with a 1–5 μM concentration and 10 min to 2 h incubation at 22–37 °C. Concentrations higher than 5 μM produce only a small increase in fluorescence intensity but a significant enhancement of compartmentalization. Lower loading temperatures give more diffusely labelled cells. Therefore, if noncytosolic compartmentalization occurs, cells should be pre-chilled at 4 °C for 5–15 min before loading with fura-2 at room temperature or 37 °C (Roe *et al.*, 1990). The addition of probenecid or sulfinpyrazone also reduces the sequestration of fura-2.

Evidence exists that fura-2 microinjected in cells behaves differently from fura-2 in cells in which it has been loaded using incubation with AM fura-2. The reason is probably the existence of five acetoxymethyl ester groups on each molecule, the slow de-esterification, high fluorescence of the AM form and the low photostability of all forms appearing in the course of the de-esterification procedure. The sequestration process can be explained by considering three pathways: (i) diffusion of the original non-hydrolysed AM form of the dye from the cytoplasm across the organellar membrane into the organelle, (ii) active transport of the dye already hydrolysed in the cytoplasm from the cytoplasm into the organelles by a transport system for organic anions located in the organellar membrane, and (iii) endocytotic uptake of the AM form of the dye into organelles.

12.6.2 Binding of the dye inside the cell

Generally, there is always a certain amount of the dye bound somewhere in the cell. The choice of the optimal concentration is one of the keys to success. The recommended concentrations are in the range from 10 μM to 1 mM. If the concentration is too low, a large proportion of the dye molecules may be bound. If the concentration is too high, the presence of the dye may shift the measured cytoplasmic value of pH or pCa by its very presence, or as a side effect, by perturbation of cell metabolism. The proportion of bound dye can most easily be estimated by direct observation of cells under the fluorescence microscope. Indirectly, the proportion of bound dye can be estimated from the absorption and emission peaks (in bound dye they are usually red-shifted, e.g. 50% binding of fluorescein to BSA shifts the maximum of absorption from 492 nm to 512 nm). The degree of polarization of fluorescence of freely moving molecules in the medium is almost zero, while in the case of bound (i.e. fixed) molecules it may approach 0.4.

12.6.3 Leakage

It is difficult to bring the dye into the cell, but it may also be a problem to keep it there. Leakage of the dye from the cytoplasm into the extracellular medium is a common property of almost all cells and all dyes. The leakage can occur via numerous routes, which may be discussed in a parallel way to loading. The leakage depends on the cell type and the dye used. Generally, the use of dyes that have a higher charge results in lower leakage (but increases the danger of binding to membrane surfaces and decreases the stability of the p*K*). The non-specific leakage can be slowed by lowering the temperature (storing cells on ice). The leakage half-life of the dye may range from several weeks (carboxyfluorescein in liposomes) to less than a minute (fluorescein in *Bacillus acidocaldarius*).

Fortunately, imaging gives a good opportunity for spatial differentiation of the fluorescence signal from the external and internal sides of the cell. The fluorescence intensity of the leaked dye can also be reduced by the use of a quencher that does not penetrate inside the cell and stays in the outer medium. In the usual approach, $MnCl_2$ is added to the extracellular medium at a concentration of 100 μM to 1 mM. However, some $MnCl_2$ may also enter the cell after a while and quench the intracellular dye as well. For some dyes it is difficult to find an appropriate quencher; some dyes may require a specific quencher. Molecular Probes Inc. (Eugene, OR) offer membrane-impermeant antibodies, which are supposed to quench the dye by contact (e.g. anti-fluorescein). One 'unit' of anti-fluorescein is said to quench 96% of the fluorescence of fluorescein and 91% of that of carboxyfluorescein. Alternatively the concentration of the leaked dye may be lowered by repeated washing, continuous flow dialysis, continuous perfusion, rapid filtration or centrifugation, and possibly by enzymatic degradation of the extracellular dye.

12.6.4 Intracellular buffering as a side effect of loading

The effect of decayed AM groups is usually harmless. Their hydrolysis inside the cytoplasm produces two protons, one acetate and one formaldehyde. The direct effect of hydrolysis products can be tested by using the AM ester of a carboxylic acid. More important is the effect of intracellular buffering on the concentration of the measured ion due to the presence of the indicator. It is not critical for pH indicators, due to the millimolar intrinsic pH buffering, or for Na^+ or K^+ indicators because of millimolar Na^+ and K^+ levels. However, the intracellular calcium might be buffered by the indicator much more easily. The intracellular free Ca^{2+} level is about 100 nM, and the intracellular concentration of the fura-2 is about 1000 times higher (100 μM) with a p*K* of 200–300 nM. The intracellular concentration of quin2 after AM loading may even reach 0.1–5 mM. In these cases the kinetics of Ca^{2+} change, and transient Ca^{2+} responses can be altered by the very presence of the indicating dye. In spite of this, the calcium spikes and waves are easily observed. It seems that the buffering effects are compensated for by the intrinsic buffering of calcium in the cell.

The last – but not least important – point is the toxicity of the dye, and various side effects should be considered, such as that the calcium-binding dye may chelate heavy metals that are important in cell function.

12.7 THEORY OF THE CALIBRATION CURVE

The theory is exhaustively described by Slavik (1994). The simpler Ostwald approach is based on the assumption of a simple transition due to the binding of the ion. For pH dyes it yields a simple relation between pH, pK and the measured ratio R (after subtraction of fluorescence background):

$$\text{pH} = \text{p}K + \log R \tag{12.2}$$

or generally for any ion:

$$\text{pI} = \text{p}K + \log R \tag{12.3}$$

The Hantzsch theory is more general. It is based on the postulated existence of tautomeric forms of the indicator (characterized by pK_T) and it considers equilibria among at least three different forms of the indicator. The resulting equations are even more complicated. The pH value is given by R and a set of equilibrium constants (see Slavik (1994) for details). Most papers use empirically derived equations of the type of equation (12.3) where pK is replaced by an effective pK' value calculated to best fit the experimental points. The equations often include additional terms, such as limiting maximal or minimal intensity of fluorescence intensity in extreme points. This is the case of the widely used empirical Grynkiewicz equation derived for fura-2 calcium measurement:

$$[\text{Ca}^{2+}] = K_d \left(\frac{R - R_{min}}{R_{max} - R} \right) \frac{S_{f2}}{S_{b2}} \tag{12.4}$$

in which K_d is the *in vitro* dissociation constant, R_{min} is the fluorescence ratio at zero calcium concentration, R_{max} is the ratio at saturating calcium concentration, S_{f2} is the fluorescence signal at wavelength 2 (380 nm excitation for fura-2) at zero calcium concentration, and S_{b2} is the same at saturating concentration. At first sight, there is no apparent similarity between these equations and the above equation resulting from the Hantzsch theory. However, if rewritten and formally rearranged a striking similarity appears.

12.8 CALIBRATION PROCEDURES

The calibration of the ion-sensitive response of an ion-selective dye is difficult due to the complexity of its fluorescence response and the imperfections of the dye behaviour (Negulescu and Machen, 1989; Thomas and Delaville, 1991). Therefore, empirical calibrations are preferred, especially *in vitro* (based on measurements in a series of mineral buffers) and *in situ* (*in vivo*, inside the cell). Both approaches have their pros and cons: the resulting standard (calibration) curves of fluorescence response versus actual ionic concentration may sometimes differ substantially. Calibrations based on *ab initio* calculations using the tabulated pK value and the theoretical equations are rarely used.

In some dyes, the ion-dependent response is well reproducible, and the standard curve is stable and independent of the conditions under which it has been measured (fluorescein, quin2); in other dyes (fura-2) it represents a delicate and complicate process.

12.8.1 Calibration procedures *in vitro*

The *in vitro* procedures mostly use simple mineral buffers, rarely a cell lysate. The preparation of pH buffers is simple and is described elsewhere, while the preparation of calcium buffers requires more skill. The use of cell lysate as a general basis for further titration seems to imitate the cytoplasm properties better than a simple mineral buffer solution.

For *in vitro* pH calibration according to Negulescu and Machen (1989), a good approximation of cellular ionic content is 120 mM K^+, 10 mM Na^+, 1 mM Mg^{2+}, 30 mM Cl^-, 100 nM Ca^{2+} (or calcium-free), 100 mM gluconate (or cyclamate or some other impermeant ion), 10 mM glucose and 10 mM HEPES. This solution is then titrated to several different pH values. They used this buffer for the calibration of the pH-sensitive dye BCECF. Using the 439/490 nm excitation ratio they obtained a ratio of 2 at pH 5.5 and of 9–10 at pH 8.5.

For the *in vitro* calcium calibration procedure it is necessary to measure the fluorescence in calcium–EGTA buffers containing different amounts of free calcium and creating a standard curve relating the ratio values to calcium concentrations. More precise determination of free calcium concentration is done by potentiometric measurement of calibrating solutions. Molecular Probes offers two different sets of calibration buffers for calcium calibration. The *in vitro* calibration is convenient and justified when a 'simple and well-behaved' probe like quin2 is used, but it is apparently not reliable in the case of the capricious fura-2. The dependence of fura-2 fluorescence on viscosity, ionic strength, pH and polarity makes *in vitro* calibration extremely difficult, as a sufficiently precise simulation of the cytoplasmic milieu is almost impossible. The reliability of the calibration standard curve depends on the wavelengths selected for the ratio measurement. For instance, with fura-2 it is believed that 365 nm excitation is less sensitive to artefacts such as viscosity, and thus the 340/365 ratio is recommended instead of the usual 340/380 ratio (Roe *et al.*, 1990).

12.8.2 Calibration procedures *in vivo* (*in situ*)

The *in situ* procedures are based on the use of ionophores, together with the clamping of the intracellular concentration of the studied ion, by changing the external medium composition. In both procedures, the pI is changed step-wise, and the fluorescence values are recorded. The application of an ionophore allows alteration of the intracellular pI value, by changing the extracellular concentration of the ion, and waiting for the equilibration of the intracellular concentration with the extracellular to set in.

In the case of pH probes this is done with nigericin in the presence of high extracellular potassium (to eliminate the membrane potential effects). In the case of calcium and magnesium ionomycin or bromo-A23187 is used, while gramicidin D and tributyltin are applied simultaneously with nigericin for potassium or sodium calibrations, in order to clamp chloride concentration. The standard curve of the dye is then obtained by a simple titration by addition of small amounts of acid or base in the case of pH probes, and by addition of excess calcium ('high calcium concentration') or the removal of calcium by chelation with EGTA ('low calcium') in the case of calcium probes etc. The detailed description of these procedures is given below; several variants of calibration procedures for calcium were described and compared by Thomas and Delaville (1991).

12.9 EXAMPLES OF CONCRETE CALIBRATION PROTOCOLS

12.9.1 pH calibration

Nigericin is a K^+/H^+ exchange ionophore which allows H^+ to equilibrate freely across the plasma membrane. It is used to equilibrate the intracellular pH with the extracellular value by application of an approximately 10 mM concentration in the presence of a high potassium concentration (100–150 mM) (Thomas, 1979; Kitagawa, 1987; Olsnes, 1987; Eisner, 1989). The high potassium concentration is required to eliminate membrane potential. Thomas (1979) reports 1–10 mM, and Negulescu and Machen (1989) have used 10–50 mM nigericin to clamp the intracellular pH. They incubated cells with a high K^+ Ringer solution and subsequently titrated to several different pH values. The K^+ Ringer solution should ideally approximate the cellular ionic contents. Most crucial is the extracellular K^+ concentration. If it is higher than the intracellular concentration, the pH will be underestimated; if it is lower outside than inside, the pH will be overestimated. Using several pH values it is possible to construct a titration curve of fluorescence intensity ratio versus pH. Horowitz and Maxfield (1984) used 30 mM monensin to equilibrate the intracellular, intravesicular and extracellular pH.

12.9.2 Calcium calibration

The dye most widely used for *in situ* calibration is fura-2. The calibration procedure is usually simplified by application of the calibration formula from Grynkiewicz *et al.* (1985), and using only two extreme values for calcium concentration (namely 'zero' and 'infinite' concentrations), and the use of the original p*K* determined *in vitro* (!) which, for unknown reasons, has become the routinely used standard value under various experimental circumstances (though *in situ* it may be significantly different).

The two most widely used ionophores applied to the equilibration of the Ca^{2+} concentration between extracellular and intracellular environments are ionomycin and 4-bromo-A23187. The stoichiometry with calcium is 1:1 with ionomycin but 1:2 with 4-bromo-A23187. Both ionophores show a high selectivity for calcium over other divalent cations. However, they both transport Mn^{2+} and the rejection of Mg^{2+} is not high with 4-bromo-A23187. Ionomycin is more efficient in calcium transport than the other ionophores but its transport function is pH-dependent with a maximum at pH 9.5, but quite low below pH 7.0. In comparison, 4-bromo-A23187 is relatively unaffected by pH. Addition of BSA and FCS in order to facilitate loading is reported to reduce the potency of calcium ionophores (Lui and Herman, 1978; Deber, 1985, Negulescu and Machen, 1989; Thomas and Delaville, 1991).

Ionomycin and 4-bromo-A23187 are non-fluorescent in the ultraviolet (UV) region. For fura-2, indo-1 and quin2 it is necessary to use either ionomycin or the non-fluorescent brominated version of A-23187 as the fluorescence of non-brominated A-23187 is too high under UV excitation. However, the non-brominated A-23187 can be used with the longer wavelength indicators such as fluo-3, calcium green, calcium orange, calcium crimson and fura red. A-23187 is probably the most reliable ionophore, rapidly transporting both Ca^{2+} and Mn^{2+} in and out of cells. Instead of using ionophores, cells may be permeabilized with Triton X-100 to release the dye from the cytoplasm into the extracellular solution. The intracellular level of calcium can also be varied by adding membrane-permeant AM forms of the non-fluorescent chelators calcium BAPTA, EGTA and EDTA, which can be used in combination with calibrated calcium buffers.

The most commonly used procedure is the addition of ionophore at an extracellular Ca^{2+} concentration of 1–2 mM. Consequently, the dye is saturated with calcium (R_{max}), after which EGTA is added to the solution to chelate all the free calcium (R_{min}). In this way, measured ratios will be between the R_{max} and R_{min} values (see Grynkiewicz equation (12.4)). The R_{max} and R_{min} values should differ by a multiplicative factor of at least 20 for fura-2 and the 340/380 excitation ratio, in order to provide a check that the dye is properly and evenly distributed throughout the cytoplasm, and that it reacts adequately with the changes of free calcium concentration in the cell. If it is not so, there is probably a mistake in the experimental loading or calibration procedure. A number of detailed calibration protocols can be found in *Cellular Calcium* (special issue no. 11, 1990).

It has been reported in the literature that a single cell population can show a heterogeneity in fluorescence ratio at a given external Ca^{2+} concentration, simply because of poor dispersal of the ionophore. In such a case only some cells are being clamped at the selected level.

12.9.3 Magnesium calibration

The calibration procedure for magnesium follows the same lines and rules as those for calcium. Actually, the magnesium-sensitive dyes mag-fura and mag-indo are chemically very similar to the calcium indicators, and the calcium ionophore A-23187, and its brominated, non-fluorescent version 4-bromo-A-23187 can be used for *in vivo* calibration of magnesium. Molecular Probes offers a magnesium calibration buffer kit.

12.9.4 Sodium and potassium calibration

SBFI and PBFI require *in situ* calibration as the fluorescence intensity is viscosity-dependent, and the excitation ratio value is strongly affected by the ionic strength. Besides, the affinity for Na^+ *in vivo* is lower than *in vitro*. The *in vivo* calibration is most conveniently performed with gramicidin (Minta and Tsien, 1989; Negulescu *et al.*, 1990; Tsien, 1989). This antibiotic makes pores in intact cells which rapidly clamp intracellular sodium and potassium at equality with the extracellular levels. Therefore, the same ionophore and a similar calibration procedure can be used for both sodium and potassium dyes. Due to the more effective ionophore and larger ion fluxes, the calibration is considerably easier than that with fura-2. By variation of the extracellular concentrations of these ions their intracellular concentration can be altered. The optimal concentration seems to be 5–10 mM gramicidin D. In the sodium calibration process, two solutions of equal ionic strength are prepared. One is sodium-free and contains 130 mM potassium gluconate and 30 mM KCl. The other contains 130 mM sodium gluconate and 30 mM NaCl. Both solutions contain 10 mM HEPES, 1 mM $CaCl_2$ and 1 mM $MgSO_4$ and are titrated with NMG base to pH 7.1. By mixing these solutions, any sodium concentration between 0 and 150 mM can be prepared.

12.9.5 Chloride calibraton

The most widely used chloride dye, and thus the one most often considered in the literature, is SPQ. For *in situ* calibration two ionophores are used simultaneously, tributyltin, a Cl^-/OH^- exchanger, together with nigericin, a K^+/H^+ exchanger. Details on the calibration can be found in the literature (Chao, 1989; Verkman, 1989; Morris, 1982).

12.10 A LIST OF POPULAR ION-SENSITIVE DYES

There are several dyes that could be employed as ion-sensitive intracellular indicators. Many of them would probably do better than the present most popular indicators. However, the detailed characterization of a dye is rather time consuming and the application of an unknown dye risky, so the well-tried dyes and procedures are still preferred. This explains why, conservatively, only a few compounds are being used systematically. Virtually all of these are offered in the acylated form of the dye.

Instead of compiling a long list of literature, I refer the reader to the Molecular Probes catalogue (Haugland, 1996), where literally thousands of references can be found.

12.10.1 pH-sensitive dyes

12.10.1a Fluorescein and its derivatives

All fluorescein derivatives should principally follow the dissociation scheme of fluorescein, although with slightly different p*K* values. Similarly, their absorption, excitation and emission spectra should resemble those of fluorescein, with slightly shifted maxima. Their acylated forms are weakly absorbing or colourless and weakly fluorescent, similarly to the colourless and non-fluorescent fluorescein diacetate.

Fluorescein is the most well-known fluorescent dye. Its absorption and fluorescence properties have been thoroughly tested and are unambiguously pH-dependent (Figure 12.5). There are nine different forms of fluorescein, six of them existing in aqueous solution, with p*K* values of 2.2, 4.4 and 6.7. All of these six forms are absorbing and fluorescent, making the resulting pH dependence of excitation and emission fluorescence spectra rather complex. However, in the physiological pH range the monoanionic and dianionic forms predominate. The extraordinary stability of p*K* values, dependent on neither the composition of the medium nor the excitation (pK = 4.4 does not change with excitation at all, pK = 6.7 shifts to 6.9), makes this dye ideal for pH measurement, in contrast to other dyes, which exhibit a pK–pK^* shift typically of several pH units.

The emission of all six forms exhibits a maximum around 520 nm, making this wavelength ideal for monitoring of fluorescence intensity. The emission of cationic and dianionic forms decreases more rapidly on the red side than that of the neutral and monoanionic forms. The excitation spectrum shows consecutive contributions of excitation maxima at 435, 473 and 490 nm. The higher quantum yield of the cationic form relative to that of the neutral one makes it possible to extend the applicability of 490/435 excitation ratio to the measurement of pH values below pH 2.

The acylated form of fluorescein is colourless and non-fluorescent and the incubation of cells in its solution is used as a sensitive assay for cell viability. Viable cells develop intense fluorescence as they are gradually filled with intracellularly formed fluorescein. This phenomenon – termed 'fluorochromasia' by Rothman and Papermaster (1966) – and the above-mentioned pH dependence of fluorescence excitation led to the idea of the employment of intracellularly trapped fluorescein for the measurement of intracellular pH by fluorescence, as introduced by our laboratory in 1982 (Slavik, 1982).

12.10.1b Carboxyfluorescein and BCECF

The main aim of modifications of fluorescein was to increase its apparent pK to a value more convenient for pH measurement in mammalian cells. Electron-withdrawing substituents decrease the pK, while electron-donating groups raise it. In this way 4′,5′-dimethyl-5(or 6)-carboxyfluorescein (DMCF) and 4′,5′-dimethylfluorescein-dextran (DMFD) with a pK of 6.8–6.9 have been synthesized and with further substitution 2′,7′-bis(carboxyethyl)-5(or 6)-carboxyfluorescein appeared with a pK of almost 7.0.

The most popular derivatives of fluorescein for intracellular pH measurement are 5- (and 6)-carboxyfluorescein (CF) and 2′,7′-bis-(carboxyethyl)-5-(and 6)-carboxyfluorescein (BCECF).

BCECF formed by hydrolysis of BCECF-AM is assumed to yield diffuse fluorescence in the cytoplasm with insignificant compartmentalization into organelles, such as mitochondria or acidic organelles. Its leakage out of the cell is very slow, presumably due to its high electrical charge. The acidic form of BCECF bears a charge of −4 and the basic −5. However, in some cells a pK increase by 0.1–0.3 units with respect to buffers has been observed. This pK shift is probably caused by the sensitivity of pK to the ionic strength of the medium. Also, the spectra of intracellular BCECF are red-shifted by a few nanometres with respect to those in the buffer, indicating an interaction of BCECF with cell constituents (probably a binding to proteins or to positively charged surfaces).

The other fluorescein derivatives are 5-(and 6)-carboxy-4′,5′-dimethylfluorescein and chloromethylfluorescein (CMF). The latter compound shows good retention in cells and allows dual excitation, with pK = 6.5. The pH stability of acylated dimethylcarboxyfluorescein should be higher than that of carboxyfluorescein, as both compounds together with fluorescein diacetate decompose spontaneously at pH values above 6.5. Naphthofluorescein and its carboxy derivative show a pH-dependent red fluorescence with the relatively high p*K* of 7.6.

12.10.1c SNARFs and SNAFLs

SNARFs and SNAFLs represent a new series of pH indicators based on naphthofluorescein. They appear to be superior in some respects to the other pH indicators because they offer large spectral changes and may be employed both as dual-excitation as well as dual-emission probes. Their photophysical properties have been reviewed by Whitaker *et al.* (1991). Their p*K* of about 7.6 at room temperature and 7.3–7.4 at 37 °C makes them optimal for measuring pH changes in the physiological pH range from 6.3 to 8.6. They are also available in the AM version. SNARFs and SNAFLs can be used as dual-emission indicators with argon laser excitation lines at 488 or 514 nm, and thus are very convenient for confocal microscopy. Besides, their spectra are well separated from those of fura-2, which makes them suitable for simultaneous measurements of intracellular pH and pCa. In SNAFL the acidic form is more fluorescent, while in SNARF, the basic form is more fluorescent. The AM esters of SNARF and SNAFL derivatives are essentially colourless and non-fluorescent, or weakly fluorescent (until hydrolysed into the final product). The SNARF and SNAFL calceins are supposed to be retained better in cells than SNARF and SNAFL.

12.10.1d Azidofluorescein

Azidofluorescein diacetate has been used for photolabelling intracellular proteins and subsequent measurement of intracellular pH.

12.10.1e Pyranine and its derivatives

Pyranine (8-hydroxy-1,3,6-pyrene-trisulfonate) is an extremely pH-sensitive dye convenient for monitoring of intracellular pH (Figure 12.8). Both forms of pyranine have the same shape of fluorescence emission spectra with a maximum at 510 nm (so that fluorescein filters may be used). The pH-dependent excitation spectra show a 5000-fold change in the fluorescence intensity ratio after excitation at 400 nm (attributed to the trianionic form) and 450 nm (attributed to the tetra-anionic form) in the pH range from 4 to 10. The ratio is apparently unaffected by the usual quenchers (1 M Cl^-, Br^- or I^- and oxygen). The high electrical charge of both forms of pyranine makes its p*K* value exceptionally vulnerable to the ionic strength effects. The p*K* shift is reported to be from 7.2 (in a medium of low ionic strength) to 8.0 (in high ionic strength); *in situ* in fibroblasts it was 7.8. On the other hand, the membrane permeability of pyranine seems to be very low.

12.10.1f Pyranine derivative 1,3-dihydroxypyrene-6,8-disulphonic acid (DHPDS)

DHPDS retains the high pH-sensitivity of pyranine and its large usable pH range, but its p*K* is much more stable (actually, DHPDS has two p*K* values, 7.33 and 8.55). There is a large pH-induced shift in both excitation and emission, allowing ratio

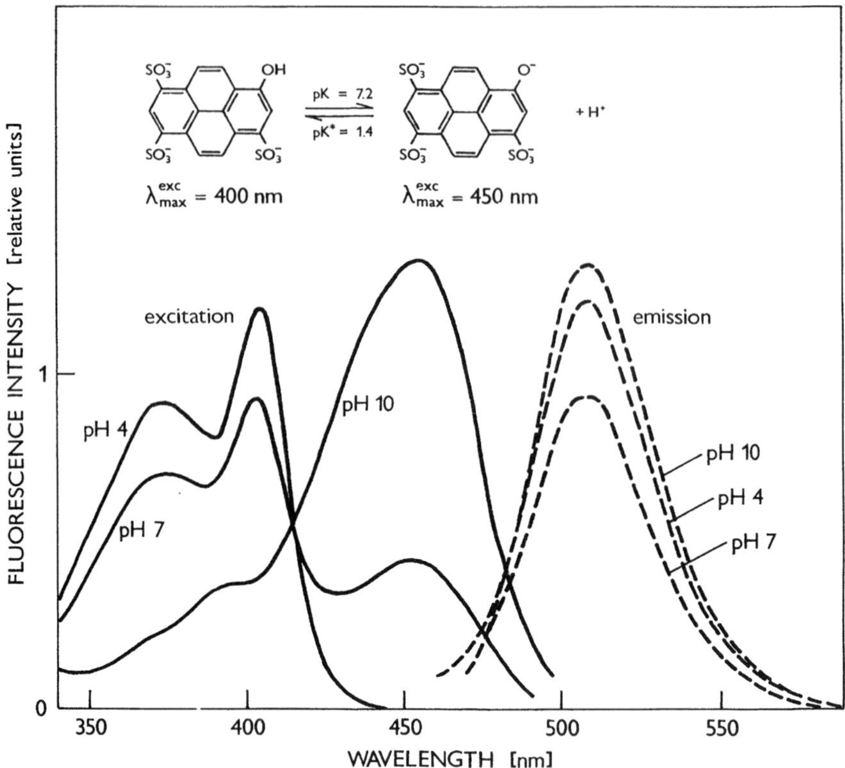

Figure 12.8 The pH-dependent excitation and emission spectra of pyranine.

imaging in either absorption or emission. The neutral form has maximum excitation at 405 nm and emission at 465 nm, the monoanionic form at 458 nm and 500 nm, respectively. The dianionic form exhibits no fluorescence. The p*K* of the monoanionic–neutral transition is approximately 2.3. No observable p*K* shift due to increased ionic strength has been observed.

Among others probes one should mention derivatives of umbelliferone (7-hydroxycoumarin), namely 4-methylumbelliferone (7-hydroxy-4-methylcoumarin), 4-chloromethyl-7-hydroxycoumarin (CMHC) and 6,7-dihydroxy-4-methylcoumarin. They are rather lipophilic and hence show a high rate of leakage, significantly higher than that of fluorescein and its derivatives. The high bleaching and the necessity of UV excitation together with the high leakage make them inconvenient for intracellular pH measurement. They are commercially available in AM versions.

12.10.2 Calcium-sensitive dyes

The most widely used fluorescent indicators for intracellular measurement of free calcium concentration are quin2, fura-2, indo-1 and fluo-3. Their structures share nearly identical binding sites modelled on the well-known Ca^{2+}-selective chelator

EGTA (ethylene glycol bis(β-aminoethyl ether)-*N,N'*-tetraacetic acid; Figure 12.9). The binding of Ca^{2+} diverts the nitrogen's lone pair of electrons away from the aromatic system, thus causing large spectral changes upon binding. The binding site binds Mg^{2+} by about five orders of magnitude more weakly than Ca^{2+}, because Mg^{2+} is too small to contact more than about half of the liganding groups simultaneously. Monovalent cations also cannot fit the binding pocket. At pH 7, EGTA is normally occupied by two protons, but the incorporation of the aromatic ring to make it fluorescent lowers the p*K* below 6.5, thus eliminating the pH-dependence of binding above pH 6.8.

The discovery of fura, quin and indo made the measurement of the free calcium level relatively easy to perform. The desire to measure and to map intracellular

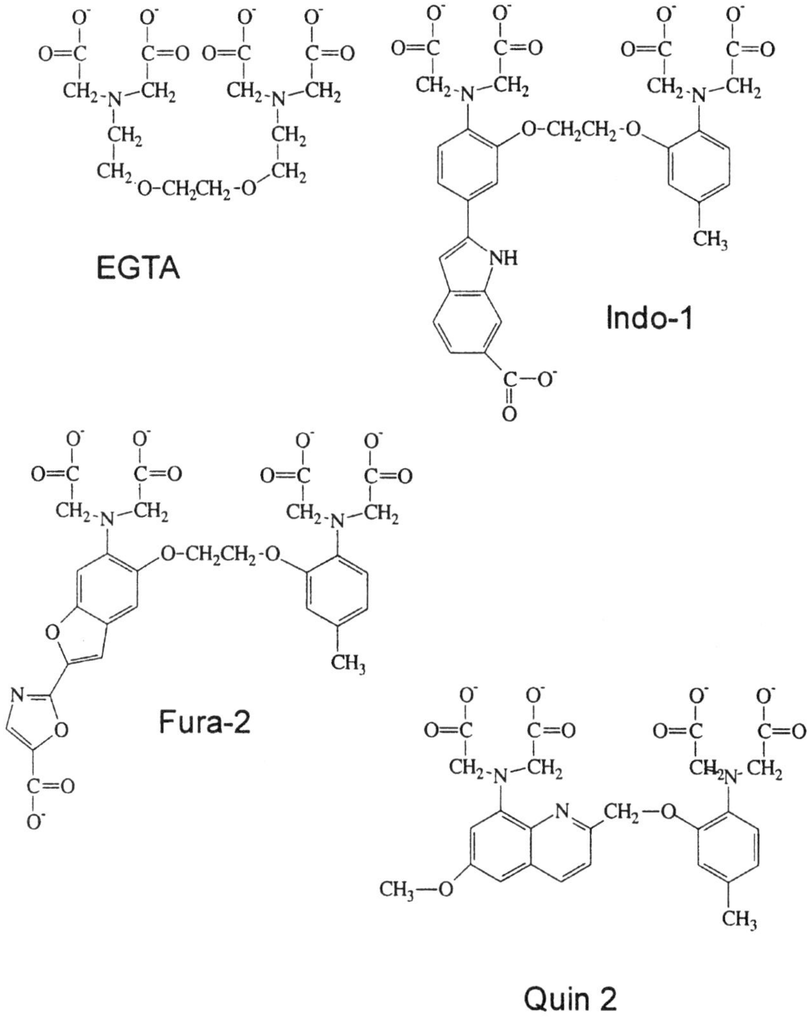

EGTA

Indo-1

Fura-2

Quin 2

Figure 12.9 Fluorescent indicators of calcium derived from EGTA.

calcium distribution has been driven by the understanding that many changes in cell function result from highly localized changes in Ca^{2+} activity.

12.10.2a Quin2

Quin2 was the first fluorescent indicator of calcium practically usable for intracellular measurement (Figure 12.10). It binds calcium at a 1:1 stoichiometry and an apparent p*K* of 114 nM in the presence of 1 mM magnesium. In the absence of magnesium it decreases to 60 nM. Furthermore, it is temperature-dependent, increasing about two-fold from room temperature to 37 °C. Its value also varies with ionic strength. The low p*K* value means strong binding, which consequently limits the applicability of quin2 for calcium measurement in the submicromolar concentration range. The selectivity of calcium over magnesium is about 10 000:1, worse than with fura-2 or indo-1. The binding of Ca^{2+} increases the intensity of fluorescence; binding of Mg^{2+} does not. Many ions, such as Mn^{2+}, Fe^{3+}, Zn^{2+}, Co^{2+} and Ni^{2+}, bind quin2 better than Ca^{2+} does and quench the quin2 fluorescence. In the free-acid form, quin2 forms a white or yellow powder; in the AM form it is a yellow gum or resin. The fluorescence emission increases about 6-fold upon calcium binding, when excited at 339 nm and emission monitored at 492 nm. Due to an excitation spectral shift, quin2 could be used in ratio mode of 339/365 or 340/395 for excitation and 492 nm for emission. Because of a low absorption coefficient ($\varepsilon = 6000\ cm^{-1}\ mol^{-1}$) and a rather low quantum yield, high intracellular concentrations of quin2 (at least 0.5 mM) are required to obtain a reasonably reliable signal. Often, millimolar to ten times millimolar concentrations of quin2 are used (which surely buffers the intracellular calcium). Photostability of quin2 is low:

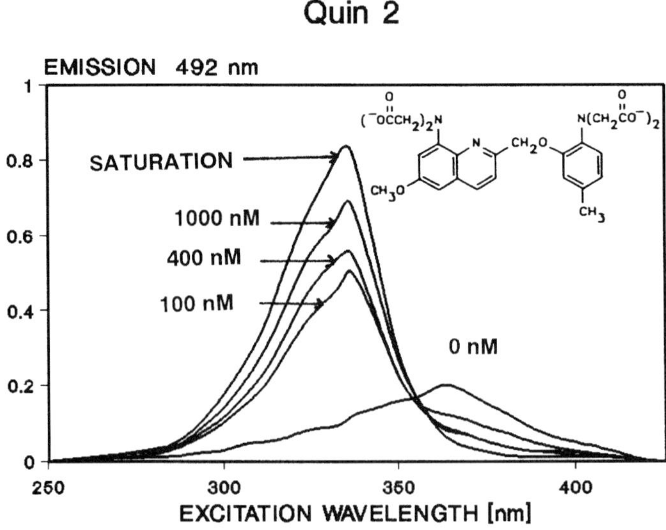

Figure 12.10 The calcium dependence of the excitation spectra of quin2 after excitation at 492 nm (redrawn after Thomas and Delaville (1991, p. 1)).

it bleaches after three or four excitation–emission cycles. The photoproducts of quin2 are also fluorescent. Summarizing, with the exception of some cases when loading with quin2 may be easier, there are no advantages of quin2 in comparison to fura-2.

The membrane-permeant quin2/AM has no calcium-binding properties. Inside the cell it is transformed into quin2 tetra-anion by four separate hydrolysis steps, one for each acetoxymethyl group. It is essential that this process is completed prior to the measurement. The incompletely hydrolysed molecules contribute to the fluorescence, yet they are crippled in the Ca^{2+} binding capability. The usual time for the incubation is about 1 h for full hydrolysis. Some cells are faster (e.g. hepatocytes), some slower, and some do not hydrolyse at all.

Recommended storage of the AM ester is in the desiccated form over silica gel at −20 °C since the AM ester can hydrolyse and/or oxidize and then the protonated free acid can gradually decarboxylate. The only derivative of quin2 known to be crystallizable and indefinitely stable at room temperature is the tetraethyl ester. The addition of 0.1–1 M KOH or NaOH (1:1) to the stock solution results in complete hydrolysis of AM, which can be monitored as a spectral shift of the emission peak from 430 to 490–500 nm. The properties of quin2 are reviewed by Tsien *et al.* (1984), with many inspiring ideas. For instance, as quin2 binds Mn^{2+} about 500 times better than calcium, Mn^{2+} concentrations greater than about 10 M will quench the fluorescence of quin2, even in the presence of millimolar calcium. Because Mn^{2+} has a very low membrane permeability in the absence of the ionophore, 50–100 mM external Mn^{2+} will immediately quench external quin2, while the intracellular dye is only slowly affected. To continue the experiment, Mn^{2+} can then be chelated with a slight excess of the diethylenetriaminepentaacetic acid (DTPA), a chelator whose Mn:Ca preference greatly exceeds that of quin2 or EGTA. This should restore the fluorescence intensity to the initial value. Sometimes the fluorescence restoration due to DTPA is larger than the quenching due to Mn^{2+}, because traces of heavy metals were already present in the medium.

All heavy metals, such as Mn^{2+}, Zn^{2+}, Fe^{2+} and Cu^{2+}, not only directly suppress quin2 fluorescence, but also competitively reduce Ca^{2+} binding. Testing for the dye quenching by endogenous heavy metals can easily be done with the membrane-permeant heavy-metal-specific chelator TPEN (N,N,N',N'-tetrakis(2-pyridylmethyl)ethylenediamine). TPEN has a much higher affinity for heavy metals but lower affinity for Ca^{2+} than quin2 does. Unlike DTPA, TPEN is almost uncharged at physiological pH and readily crosses membranes as the free base.

12.10.2b Indo-1

Indo-1 shows both excitation and emission shifts upon binding of calcium. The excitation shift is in the UV region (around 250 nm) and thus cannot be used in biological experiments. The emission shift allows emission ratio imaging and makes this dye invaluable for all cases when emission ratio is preferred (Figure 12.11). The fluorescence brightness, quenching, selectivity, kinetics etc. seem to be similar to fura-2. According to Grynkiewicz *et al.* (1985), the emission ratio 400/490 nm changes between $R_{max} = 2.60$ and $R_{min} = 0.121$, ($S_{f2}/S_{b2} = 2.01$; see equation (12.4)). The apparent dissociation constant is 250 nM.

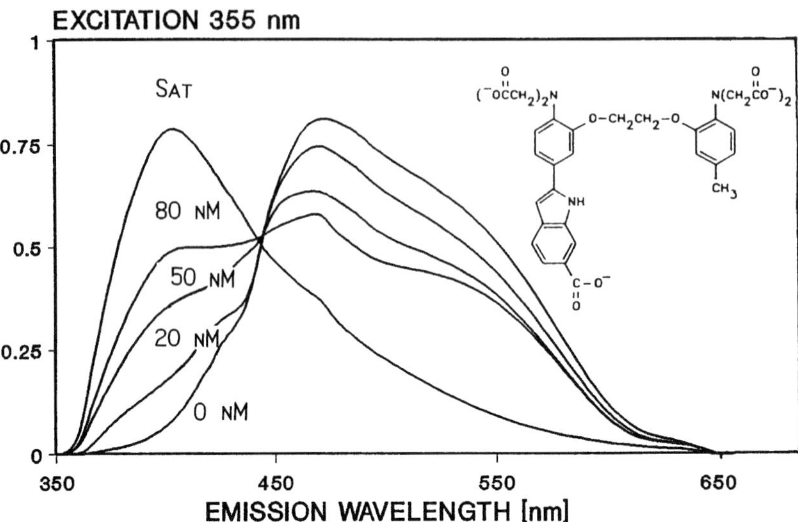

Figure 12.11 The calcium-dependent emission spectra of indo-1 after excitation at 355 nm (redrawn after Haugland (1996)).

The maximum of emission is 485 nm for the unbound form and 405 nm for the calcium-bound form. Ratio imaging can easily be done by excitation between 350 and 365 nm. The possibility of ratio imaging in emission makes indo-1 invaluable for flow cytometry and useful for confocal microscopy.

Major disadvantages of indo-1 compared to fura-2 are that it photobleaches several times faster than fura-2, and its emission spectrum in the blue and violet part of the spectrum coincides with that of cell pyridine nucleotides, so cell autofluorescence may become a problem.

12.10.2c Fura-2

Fura-2 is currently the most popular calcium indicator. Compared to quin2 and indo-1, it is significantly better. Its excitation wavelength is slightly longer, both the extinction coefficient and the quantum yield are higher, and photostability is better than that of indo-1 or quin2. The improved absorption and emission properties make fura-2 fluorescence about 30 times brighter and thus make it possible to use significantly lower concentrations. Intracellular concentrations of 20–30 μM are already sufficient to make all background and cell autofluorescence negligible. The fluorescence at 340 nm (excitation) increases about three-fold and at 380 nm it decreases about ten-fold upon calcium binding. This provides an excellent basis for ratio measurement and ratio imaging (Figure 12.12). The emission maximum remains unchanged around 505–520 nm in both forms of fura-2. The most popular

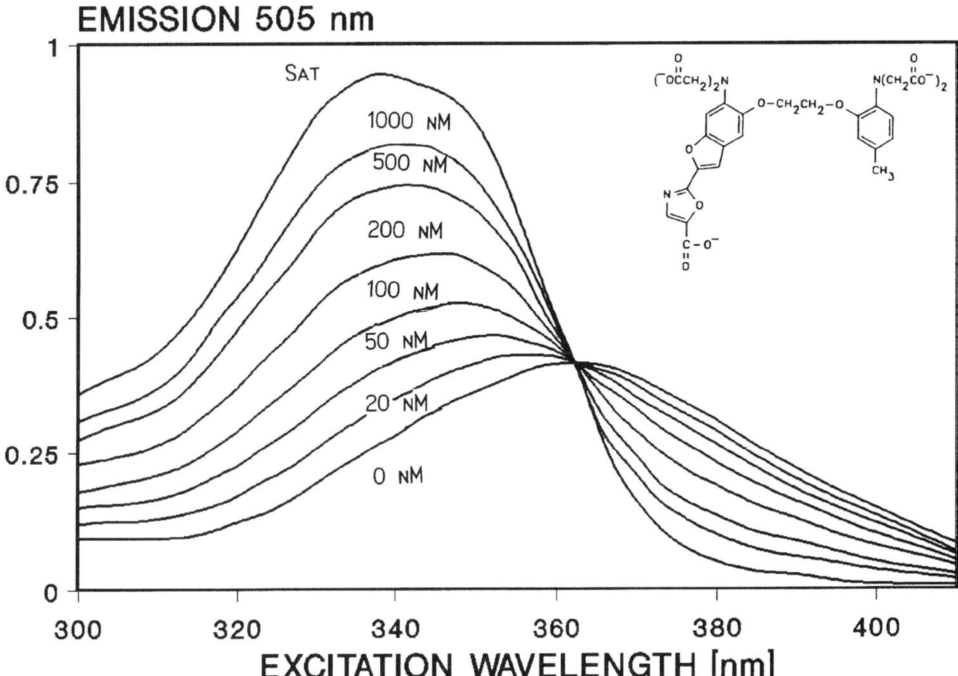

Figure 12.12 The dependence of the excitation spectra of fura-2 on calcium at 505 nm (redrawn after Tsien and Poenie (1986)).

wavelengths for ratioing are 340/380 for excitation, but other wavelength pairs (340/360 or 350/380) have also been used. Fura-2 shows about a 35-fold increase in the 340/380 ratio when the calcium concentration is varied from the minimal to the maximal value in buffer. The dissociation constant for calcium ranges from 135 nM (20 °C, no Mg^{2+}, 100 mM KCl) to 224 nM (37 °C, 1 mM Mg^{2+}, 120 mM K^+, 20 mM Na^+) to 774 nM (18 °C, no Mg^{2+}, 225 mM K^+, 25 mM Na^+); the most important factor seems to be the ionic strength rather than the magnesium concentration. The dissociation constant for Mg^{2+} is 6–10 mM, making magnesium interference negligible. Zn^{2+} resembles Ca^{2+} in fluorescence.

To some extent, fura-2 fluorescence also reflects environmental properties other than concentration of free calcium level. The polarity of the solvent has an effect on the excitation and emission spectra. As polarity decreases, excitation is red-shifted and quenching takes place. Simultaneously, the emission spectra undergo a blue shift. There is a viscosity effect. As viscosity increases, pK decreases, and the 340/380 ratio also decreases. The effect of viscosity is less if 340/365 is employed. The influence of viscosity is already significant at values one might normally encounter in the cell cytoplasm. In the cytoplasm the 340/380 ratio is 1.1- to 1.6-fold lower than in a buffer, probably due to viscosity. Changes in ionic strength can affect the pK of fura-2 over a 5- to 8-fold range. Williams and Fay (1989) claim a linear logarithmic relationship between pK and ionic strength. Furthermore, pH value has a significant effect on pK.

The calcium response of fura-2 is fast. The exponential time constant for the calcium response is $\tau = 1/(k_a[Ca^{2+}] + k_d)$ where the rate constant of association k_a is 6×10^8 mol^{-1} s^{-1} and the constant of dissociation k_d is 84–97 s^{-1} at 20 °C in 100 mM KCl. In the cytoplasm the rate constants seem to be 4–8 times lower than *in vitro*. Therefore, the composite relaxation time is about 5 ms at low calcium and about 1.5 ms at 1 mM calcium.

All dyes are vulnerable to quenching, and fura-2 is no exception. Addition of $MnCl_2$ (0.1 mM to 1 mM) to the extracellular medium bathing the cells will quench all extracellular fluorescence. However, after a while, some $MnCl_2$ may also enter the cell and quench fura-2 there. If heavy metals interfere, then the addition of TPEN (a chelator of heavy metals) will show them by quenching (one compares the intensity of fluorescence before and after the addition). There is a similarity in fura-2 and quin2 quenching.

Any fluorescent dye may undergo irreversible photobleaching after exposure to strong excitation light and thus lose its responsiveness to the environmental parameter. Fura-2 is certainly no exception. Its photobleaching results in the formation of a highly fluorescent intermediate, which binds calcium in the millimolar instead of the nanomolar range. Photobleaching can be reduced upon addition of an antioxidant in the incubation medium or by removing oxygen from the probe environment. This is difficult because cells do not survive these conditions for long. The only sensible solution to avoiding the problems of photodamage is to use low levels of excitation intensity and to keep exposure as short as possible.

12.10.2d Fluo-3

The main advantages of fluo-3 are the visible excitation (503–506 nm maximum), a 40- to 80-fold enhancement upon Ca^{2+} binding, and pK = 400 nM (weaker affinity) thus permitting measurement up to 5–10 mM Ca^{2+}. Its advantage could be in the possibility of simultaneous application of fluo-3 (visible excitation) and SBFI (UV excitation), thus not spectrally interfering with each other and allowing measurement of Na^+ and Ca^{2+} simultaneously. It is convenient for confocal microscopy because it can be excited by an argon laser, thus avoiding the necessity of expensive UV lasers. The AM version of fluo-3 AM is non-fluorescent in the presence of calcium until hydrolysed.

Since it is not based on a ratio, calibration is difficult. One of the solutions is to couple fluo-3 to another fluorophore used as a reference monitor of concentration. However, the effects of uneven bleaching of the two chromophores remain unsolved. The main problem of cellular applications is high leakage, which may be stopped by addition of probenecid.

With the increasing availability of lifetime imaging, fluo-3 seems to be an ideal candidate for confocal microscopy lifetime imaging.

12.10.2e Rhod-2

Rhod-2 may be used with rhodamine filters. It can also be excited with an argon laser, although with lesser efficiency than fluo-3. There is neither an excitation nor an emission shift upon calcium binding; only a 3.4-fold enhancement of fluorescence intensity.

12.10.2f Calcium green, calcium orange and calcium crimson

Calcium green, calcium orange and calcium crimson represent a new series of calcium indicators with long wavelength excitation – 505 nm, 550 nm and 590 nm, respectively. None of these dyes shows any spectral shifts. They differ in calcium affinity and in the extent of fluorescence enhancement upon calcium binding. Their AM forms are non-fluorescent.

12.10.2g Fura red

Fura red is another new fluorescent indicator for calcium. It allows dual excitation at 425–450 nm for the calcium-insensitive signal and at 480–500 nm for the calcium-sensitive signal. The emission maximum is about 660 nm. Unfortunately, the quantum yield is very low, which makes it necessary to use a higher concentration of the AM ester of the dye (possibly 10–25 mM). The fluorescence intensity decreases with calcium binding.

12.10.2h Magnesium dyes as indicators for calcium

The magnesium-sensitive indicators mag-fura-2, mag-fura-5, mag-indo-1 and magnesium green, though designed for magnesium, can also be used for measurement of high concentrations of calcium, when the classical calcium probes are already saturated. Such cases may arise when calcium spikes are monitored, as they may well exceed 1 mM free calcium concentrations. Employed as a calcium probe, mag-fura-2 has a pK equal to 50 mM, mag-fura-5 25 mM, and the pK of magnesium green or calcium green 5N is estimated to be in the range from 1 mM to 100 mM.

12.10.3 Magnesium-sensitive dyes

Raju (1989) developed three fluorescent indicators for magnesium by reducing the size of the binding site for calcium in calcium indicators. Consequently, the calcium affinity decreased and that for magnesium increased. The resulting compounds have Mg^{2+} dissociation constants in the millimolar range and their Ca^{2+} constant lower by one or two orders of magnitude. Therefore these compounds respond to magnesium concentration with maximal sensitivity in the 0.1–5 mM range, a range that is optimal for detecting physiological levels of this cation. The high dissociation constant for calcium causes them to be practically insensitive to intracellular calcium concentrations, since they are well above the usual physiological concentration of calcium (10 nM to 1 mM) and thus calcium fluctuations cannot interfere with magnesium measurements. The only exceptions are large calcium spikes, which may temporarily exceed 10 mM.

Due to the similarity in chemical structure to calcium indicators fura-2 and indo, it is convenient to derive the names of magnesium indicators from them. They are thus called mag-fura-2 (sometimes also furaptra), mag-fura-5 and mag-indo-1. Mag-fura-2 and mag-fura-5 show an excitation shift upon Mg^{2+} binding similar to fura-2, while mag-indo-1 shows both excitation and emission shifts similar to indo-1. Mag-fura-2 is reported to have a dissociation constant of 2.3 mM for magnesium and 10.3 mM for calcium, while mag-indo-1 has 3.8 mM for magnesium and 8.9 mM for calcium. The affinities are not temperature-dependent (in the temperature range 22–37 °C) or pH-dependent (in the pH range 5.5–7.4). The magnesium indicators have been used to monitor Mg^{2+} in beef heart mitochondria, murine pancreatic beta cells and cultured rat smooth muscle cells. The calibration is similar to that of calcium and it is described in detail in the previous paragraphs (ionophore 4-bromo-A23187). For *in vitro* calibration, Molecular Probes offers a magnesium calibration buffer kit.

12.10.4 Sodium- and potassium-sensitive dyes

SBFI represents at present the best indicator for sodium (Figure 12.13). It consists of a crown which is just large enough to fit an Na^+ ion but too small to accept K^+. Divalent cations are also rejected because there are no negative charges lining the cavity. The potassium indicator PBFI uses a larger size of the crown cavity, which results in potassium selectivity. The attached fluorophores are benzofurans similar to those used in calcium indicators quin2, indo-1 and fura-2, so that the spectral characteristics of SBFI are similar to fura-2. In contrast to calcium indicators, the sodium indicator SBFI and the potassium indicator PBFI consist of two identical benzofuran-based fluorophores because the organic synthesis was eased by preserving the symmetry around the crown ether ring.

The spectral responses of the sodium indicator SBFI and of the potassium indicator PBFI upon binding their respective ligands are similar to each other and similar to those of fura-2. The binding of the ion increases the overall fluorescence intensity about 2.5-fold, along with changes in the shape of the excitation spectra. Upon ion binding, the excitation wavelength maximum shifts slightly by 12 nm from 346 to 334 nm, but more important is that it changes the shape of the excitation spectra. Consequently, a large change in the 340/380 excitation ratio appears, which is very convenient because the same filterset as for fura-2 can be employed. However, the choice of wavelength ratio has been done arbitrarily, as this combination is widely used for fura-2. It is likely that other, perhaps better, choices of excitation wavelength pairs could be found. The emission spectrum has a maximum at 500–505 nm and does not change upon binding. Therefore, SBFI can be used for sodium ratio imaging. The extinction coefficients are greater than 40 000 cm^{-1} mol^{-1} at pH 7. Fluorescence is pH-insensitive at pH > 6.2, but is strongly affected by ionic strength and viscosity. The quantum yield is lower than of fura-2, so higher loading concentrations are required, and longer loading times, up to 4 h, may also be needed (with AM

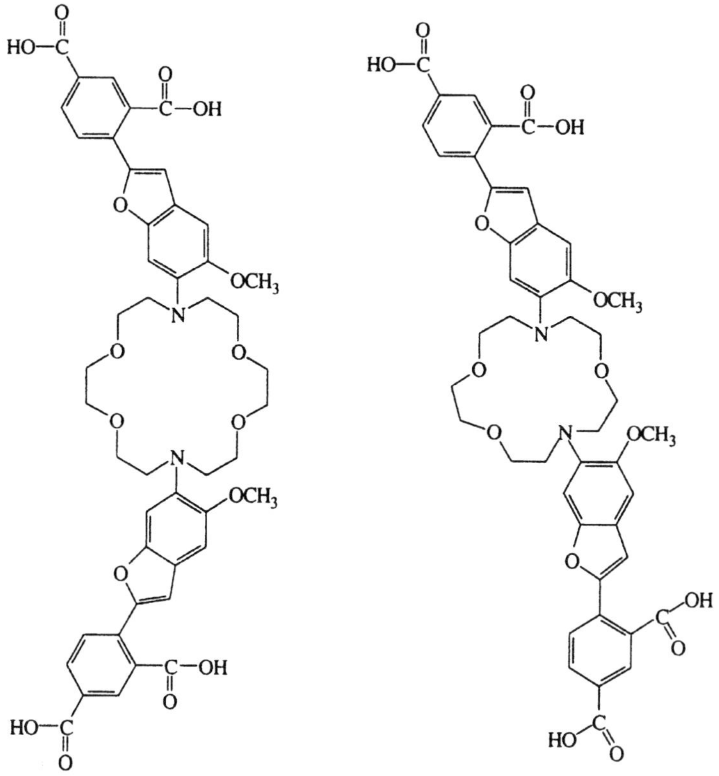

Figure 12.13 The sodium probe SBFI and the potassium probe PBFI.

forms, as both SBFI and PBFI are available both as free acid and as cell-permeant acetoxymethyl esters).

Although SBFI and PBFI are quite selective for Na^+ and K^+, there is a certain effect of K^+ on the Na^+ affinity of the Na^+ probe SBFI and, inversely, a certain effect of Na^+ on the K^+ affinity of the K^+ probe PBFI. The Na^+ probe SBFI is about 18-fold more selective for Na^+ than for K^+. The pK of SBFI for Na^+ is 9.4 mM in the absence and 17–19 mM in the presence of physiological concentrations of K^+, which is suitable for monitoring of Na^+ around the typical resting level 10–20 mM. The K^+ probe PBFI is only 1.5 fold more selective for K^+ than for Na^+, which is sufficient under normal physiological conditions.

Calibration of SBFI is done intracellularly using gramicidin D, that of PBFI using the potassium ionophore valinomycin. The effects of compartmentalization are reasonable.

There are many literature references on the application of SBFI, but almost none on PBFI. For instance, SBFI has been used for sodium measurement in fibroblasts and lymphocytes, in colonic crypts, rabbit gastric gland cells, murine pancreatic beta cells, human platelets, MDCK cells and guinea pig myocytes. On the other hand, there is only one paper on PBFI application so far, on liposome-reconstituted K^+/H^+ antiporters from beef heart and rat liver (Jezek *et al.*, 1990).

12.10.4a FCryp-2

In addition to SBFI, there is another new sodium indicator, FCryp-2, with reportedly excellent Na^+ affinity (p$K = 6$ mM) and K^+ rejection. When excited at 340 nm, the emission peak at 395 nm increases its intensity upon Na^+ binding, and the emission at wavelengths longer than 460 nm is decreased. Such an indicator could be convenient for emission ratio measurement and emission ratio imaging.

12.10.5 Chloride-sensitive dyes

It is known that the fluorescence of quinine and related compounds is quenched by Cl^- and the mechanism of this quenching is collision (dynamic). The theoretical calibration equation is thus (provided chloride is the only quencher of fluorescence):

$$\tau = \frac{\tau_0}{1 + [Cl^-]/K} \tag{12.5a}$$

which gives:

$$[Cl^-] = K(\tau/\tau_0 - 1) \tag{12.5b}$$

where τ and τ_0 are the lifetimes measured in the presence and absence of chloride ions, respectively. Other halides, such as bromide and iodide, as well as thiocyanate also quench the dye and may interfere, but they are usually absent from cells. Therefore, the appropriate *in situ* calibration and lifetime imaging allow direct determination of chloride concentration in millimoles per litre. The four most common probes used for the measurement of intracellular chloride are 6-methoxy-*N*-(sulphopropyl)quinolinium (SPQ), *N*-(sulphopropyl)acridinium (SPA), *N*-(6-methoxyquinolyl)acetic acid (MQAA) and 6-methoxy-*N*-ethylquinolinium (MEQ) (Figure 12.14). All of these dyes exhibit a concentration-dependent quenching by chloride and other halides, with no spectral shifts.

*12.10.5a 6-Methoxy-*N*-(sulphopropyl)quinolinium (SPQ)*

SPQ undergoes a 50% quenching in 10 mM chloride. The corresponding value of the quenching constant in the Stern–Volmer equation is 8.5 mM, a value that should provide good sensitivity to physiological alterations of Cl^-. The chloride quenching is less effective in cells ($K = 12$ M) than in aqueous solution ($K = 118$ M). In cytoplasm, the Stern–Volmer constant is apparently ten-fold greater (83 mM, after Krapf *et al.* (1988)) than in water. This drastic weakening of the $2Cl^-$-sensitivity of the dye has been ascribed to the combined effects of intracellular ions, viscosity and binding of the dye to cellular constituents. Consequently, calibration must be done *in situ*, in intact cells after each experiment by application of chloride/hydroxyl ionophores such as tributyltin together with potassium/proton ionophore nigericin, as described above. The excitation maximum is 344 nm, the extinction coefficient 3700 cm^{-1} mol^{-1} and the emission maximum 443 nm.

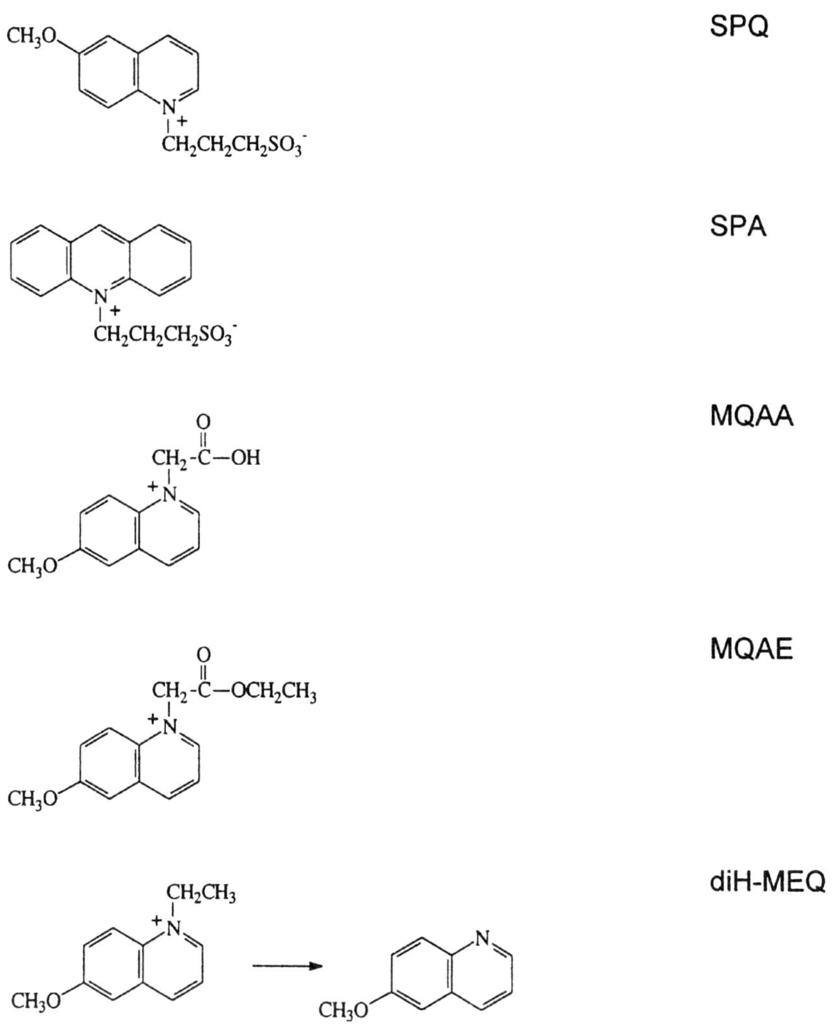

SPQ

SPA

MQAA

MQAE

diH-MEQ

Figure 12.14 Chloride probes.

SPQ is quite membrane-permeant, so there is no need for AM forms. Cells are simply loaded by soaking them in a high concentration of the dye. The rate of leakage is very high, with a half-life of 8–9 min.

12.10.5b N-(6-methoxyquinolyl)acetic acid (MQAA)

MQAA has excitation and emission wavelengths 10–15 nm longer than SPQ, thus facilitating the application of non-quartz optics. MQAA has a greater sensitivity to chloride ($K = 200$ M) than SPQ. In cells, the 50% decrease of fluorescence intensity requires 60 mM chloride. The rate of leakage of MQAA is less than 20% per hour at 37 °C, much slower than with SPQ. MQAA is also available in AM version (MQAE).

12.10.5c 6-Methoxy-N-ethylquinolinum (MEQ)

MEQ exhibits 50% quenching in 53 mM chloride solution. It appears upon intracellular oxidation of a lipophilic, chloride-insensitive compound diH-MEQ (6-methoxy-*N*-ethyl-1,2-dihydroquinoline) with excitation 316 nm and emission maximum 441 nm, $\varepsilon = 4300\ \text{cm}^{-1}\ \text{mol}^{-1}$). The oxidation occurs rapidly and results in trapping MEQ in the cytosol. The process is similar to the application of AM compounds. Since diH-MEQ is unstable upon storage, commercially available kits for preparation of diH-MEQ immediately prior to cell loading are recommended.

Calibration is done in the same way as with other chloride probes, namely *in situ*, using tributyltin (chloride/hydroxide ionophore) and nigericin (K^+/H^+ ionophore).

12.10.6 Other ion-sensitive dyes

12.10.6a Iron-sensitive dyes

There are also probes MA-DFO and NBD-DFO, which show a similar change in their fluorescence lifetime dependent on the concentration of Fe^{3+}.

12.10.6b Zinc-sensitive dyes

The fluorescent probe *N*-(6-methoxy-8-quinolyl)-*p*-toluenesulphonamide (TSQ) (Figure 12.15) is selective for Zn^{2+} even in the presence of physiological concentrations of calcium and magnesium ions. Its excitation maximum is 335 nm, emission 376 nm. It is not clear whether TSQ responds to the soluble or protein-bound zinc concentration in the cytoplasm.

12.10.6c Heavy metal-sensitive dyes

Quite recently new probes, APTA and APTRA, have been developed which can be used for Cd^{2+}, Hg^{2+}, Pb^{2+}, Ba^{2+} and La^{3+}.

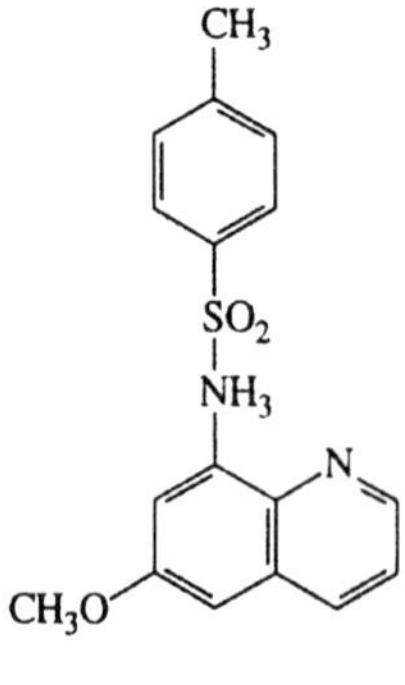

Figure 12.15 Zinc-sensitive fluorescent probe *N*-(6-methoxy-8-quinolyl)-*p*-toluenesulphonamide (TSQ).

12.11 CONCLUSION

There are a large number of fluorescent dyes that could be used as fluorescent probes of ionic composition. A new line of dyes with 488 nm excitation and dual emission have emerged on the market, with the increasing usage of confocal laser scanning microscopes and argon excitation line 488 nm. Please bear in mind that none of the ion-sensitive dyes on the market is perfect. The description of new dyes in catalogues may be alluring, but it is sometimes safer to use 'old faithful' probes with well-known ion-indicating properties and well-known side-effects.

REFERENCES

Chao, A.C. (1989) Fluorescence measurement of chloride transport in monolayer cultured cells. *Biophys. J.* **56**: 1071–5.

Cimprich, P., Slavik, J. and Kotyk, A. (1995) Distribution of individual cytoplasmic pH values in a population of the yeast *Saccharomyces cerevisae*. *FEMS Microbiol. Lett.* **130**: 245–52.

Deber, M.C. (1985) Bromo-A23187: a nonfluorescent calcium ionophore for use with fluorescent probes. *Anal. Biochem.* **146**: 349–54.

Eisner, D.A. (1989) A novel method for absolute calibration of intracellular pH indicators. *Pflugers Arch.* **413**: 553–8.

Grynkiewicz, C., Poenie, M. and Tsien, R.Y. (1985) A new generation of calcium indicators with greatly improved fluorescence properties. *J. Biol. Chem.* **260**: 3440–50.

Haugland, R.P. (1996) *Handbook of Fluorescent Probes and Research Chemicals*. Molecular Probes Inc.: Eugene OR.

Horowitz, M.A. and Maxfield, F.R. (1984) Legionelle pneumonia inhibits acidification of its phagosome in human monocytes. *J. Cell Biol.* **99**: 1936–44.

Jezek, P., Mahdi, F. and Garlid, K.D. (1990) Reconstitution of the beef heart and rat liver mitochondrial K^+/H^+ (Na^+/H^+) antiporter. *J. Biol. Chem.* **265**: 10522–8.

Kitagawa, S. (1987) Relationship of the effects of nigericin on the aggregation and cytoplasmic pH of bovine platelets in the presence of different cations. *Biochim. Biophys. Acta* **930**: 48–55.

Kotyk, A. and Slavik, J. (1989) *Intracellular pH and its Measurement*. CRC Press: Boca Raton, FL.

Krapf, R., Berry, C.A. and Verkman, A.S. (1988) Estimation of intracellular chloride activity in isolated perfused rabbit proximal convoluted tubules using a fluorescent indicator. *Biophys. J.* **53**: 955–60.

Lui, C. and Herman, T.E. (1978) Characterisation of ionomycine as a calcium ionophore. *J. Biol. Chem.* **253**: 5892–98.

Minta, A. and Tsien, R.Y. (1989) Fluorescence indicators for cytosolic sodium. *J. Biol. Chem.* **260**: 3440–9.

Morris, S.J. (1982) Calcium-promoted resonance energy transfer between fluorescently labeled proteins during aggregation of chromaffin granule membranes. *Biochim. Biophys. Acta* **693**: 425–32.

Negulescu, P.A. and Machen, T.E. (1989) Intracellular ion activities and membrane transport in parietal cells measured with fluorescent dyes. *Meth. Enzymol.* **172**: 38–75.

Negulescu, P.A., Harootunian, A.T., Tsien, R.Y. and Machen, T.E. (1990) Fluorescence measurement of cytosolic free sodium concentration, influx and efflux in gastric cells. *Cell Regul.* **1**: 259–65.

Olsnes, S. (1987) Effect of intracellular pH on the rate of chloride uptake and efflux in different mammalian cell lines. *Biochemistry* **26**: 2778–83.

Raju, B. (1989) A fluorescent indicator for measuring cytosolic free magnesium. *Am. J. Physiol.* **256**: C540–8.

Roe, M.W., Lemasters, J.J. and Herman, B. (1990) Assesment of fura-2 for measurement of cytosolic free calcium. *Cell Calcium* **11**: 63–70.

Rothman, D. and Papermaster, B.W. (1966) Membrane properties of living mammalian cells as studied by enzymatic hydrolysis of fluorogenic esters. *Proc. Natl. Acad. Sci. USA* **55**: 134–9.

Slavik, J. (1982) Intracellular pH of yeast cells measured with fluorescent probes. *FEBS Lett.* **140**: 22–26.

Slavik, J. (1994) *Fluorescent Probes in Cellular and Molecular Biology*. CRC Press: Boca Raton, FL.

Thomas, A.P. and Delaville, F. (1991) Fluorescent calcium indicators In *Cellular Calcium. A Practical Approach* (McCormack, J.G. and Cobbold, P.H., eds), pp.1–15. IRL Press: Oxford.

Thomas, J.A. (1979) Intracellular pH measurements in Ehrlich ascites tumor cells utilising spectroscopic probes generated *in situ*. *Biochemistry* **18**: 2210–18.

Tsien, R.Y. (1989) Fluorescent indicators of ion concentrations. *Meth. Cell Biol.* **30**: 127–56.

Tsien, R.Y. and Poenie, M. (1986) Fluorescence ratio imaging: a new window into intracellular ionic signalling. *TIBS* **11**: 450–5.
Tsien, R.Y., Pozzan, T. and Rink, T.J. (1984) Measuring and manipulating cytosotic free calcium with trapped indicators. *Trends Biochem. Sci* **9**: 263.
Verkman, A.S. (1989) Synthesis and characterisation of improved chloride-sensitive fluorescent indicators for biological applications. *Anal. Biochem.* **178**: 355–60.
Whitaker, J.E., Haugland, R.P. and Prendegast, F.G. (1991) Spectral and photophysical studies of benzoxanthene dyes: dual emission pH sensors. *Anal. Biochem.* **194**: 330–8.
Williams, D.A. and Fay, F.S. (1989) Intracellular calibration of the fluorescent calcium indicator fura-2. *Cell Calcium* **11**: 75.

FURTHER READING

Herman, B. and LeMasters, J.J. (1992) *Optical Microscopy: Emerging Methods and Applications*. Academic Press: New York.
Mason, W.T. (1993) *Fluorescent and Luminescent Probes for Biological Activity*. Academic Press: New York.
Nucitelli, R. (1994) *A Practical Guide to the Study of Calcium in Living Cells*, Methods in Cell Biology vol. 40. Academic Press: New York.
Slavik, J. (ed.) (1996) *Fluorescence Microscopy and Fluorescent Probes*. Plenum Press: New York.
Taylor, D.L. and Wang, Y.L. (eds) (1989) *Fluorescence Microscopy of Living Cells in Culture*, Part B, Methods in Cell Biology vol. 30. Academic Press: New York.
Wang, Y.L. and Taylor, D.L. (eds) (1989) *Fluorescence Microscopy of Living Cells in Culture*, Part A, Methods in Cell Biology vol. 29. Academic Press: New York.

13

Motility, Chemotaxis and Phototaxis Measurements by Image Analysis

Giovanni Cercignani[1], Sabina Lucia[2] and Donatella Petracchi[2]

[1] *Università di Pisa, Pisa, Italy*

[2] *Istituto di Biofisica del CNR, Pisa, Italy*

13.1 INTRODUCTION

The quantitative study of the motility of unicellular organisms is of great biological interest, since the possible survival of a microbial population is related to the motion activity of its individual cells. Automated analysis of motile cells behaviour is therefore a useful means to investigate how the local environment affects cells and their ability to survive. Many physical and chemical parameters in their immediate surroundings can influence the behaviour of single moving cells and several features of their motion can be measured to study their response to environmental stimuli. The aim is to gather insight into the mechanisms by which information on the outside world is sensed, transduced, integrated and used at the very elementary level of a unicellular swimming organism. In microbial sensory transduction, the chain of events leading from stimulus to motor response is made up of intracellular chemical reactions and current research in this field is aimed at a detailed investigation of its relevant molecular mechanisms. However, observation and quantitative analysis of motor responses are still the basic ways to study the output of living cell systems. They have been used to determine action spectra in response to light stimuli, to screen for mutants, to investigate the chemotactic effects of several substances, and in general for the study of any factor affecting cell motility or sensory transduction. For instance, a phenomenon of particular relevance that has kindled the attention of several researchers over the past few years is the influence of ultraviolet (UV) radiation on cell motility (Ekelund, 1990; Sgarbossa *et al.*, 1995; Gerber *et al.*, 1996).

Responses to stimuli of diverse nature have been observed and investigated. Most of these studies concern the chemotactic responses of bacteria and the phototactic responses of bacteria, flagellated algae or ciliated protozoa (for reviews, see Colombetti and Petracchi, 1989; Armitage, 1992; Petracchi *et al.*, 1994; Blair, 1995). Clearly, the biological meaning of the motor responses of unicellular microorganisms is to react to changes in the outside world in order to select the

Digital Image Analysis of Microbes: Imaging, Morphometry, Fluorometry and Motility Techniques and Applications. Edited by M.H.F. Wilkinson and F. Schut.

environment best suited to their survival. As a matter of fact, it is essential for photosynthetic organisms to find the best light conditions to carry out their light-dependent functions, while it is essential for any micro-organism to avoid excessive light intensities or potentially harmful wavelengths in the UV region, so that either photoaccumulation or photodispersal can be the net result of different light stimuli. As far as chemical stimuli are concerned, it is advantageous for the cell to search for higher or lower concentrations of particular substances, depending on their effects or usefulness in the cell's metabolism. Unicellular organisms have developed sensory mechanisms to perform this task. The ability of cells to detect and process stimuli allows them to accumulate in regions where the concentration of useful molecules (attractants) is higher and to avoid potentially harmful chemicals (repellents) by moving away from their source(s). A similar phenomenon occurs with light stimuli in phototactic micro-organisms, which either accumulate in or disperse from light traps – in analogy with chemotactic phenomena, one can speak of either attractant or repellent light.

When single cells of flagellated algae, like *Chlamydomonas reinhardtii*, *Haematococcus lacustris* etc., are observed, two distinct types of behaviour resulting in photoaccumulation can be described: (i) true phototaxis, meaning that an orientated motion occurs in the direction of the light source, and (ii) a phobic response. The latter is a step-up or step-down response, which consists of sudden changes in the direction of motion occurring when the light stimulus is turned on or off. Phobic responses are also present in some light-sensitive protozoa, as shown schematically in Figure 13.1. As far as prokaryotes are concerned, most studies have been performed on Bacteria like *Escherichia coli* or *Bacillus subtilis* and on Archaea

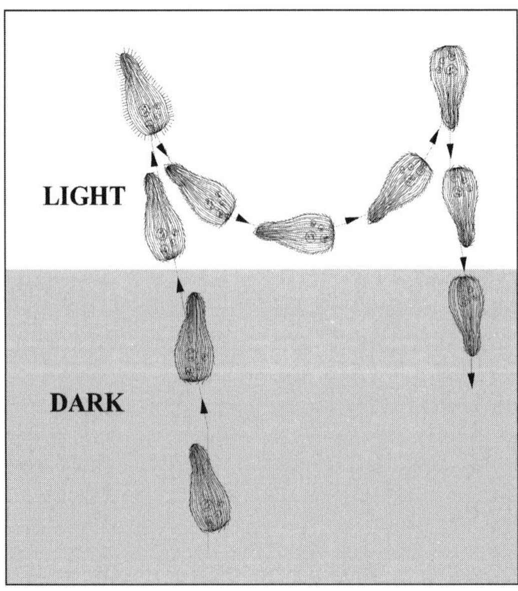

Figure 13.1 Schematic drawing of a step-up photoreaction in *Blepharisma japonicum*. Based on data from Ghetti (1991).

like *Halobacterium salinarum*. Here the phenomena of accumulation and dispersal arise solely from step-up and step-down reactions, respectively. For instance, the spontaneous swimming behaviour of *E. coli* consists of smooth swimming periods, called runs, interrupted by brief tumble periods eventually producing a reorientation in the cell motion. Similar behaviours can be described in *H. salinarum* or in *B. subtilis*, although the structural and molecular details of the phenomena can be quite different. Indeed, in *H. salinarum* there is no tumbling behaviour, since the mechanism producing a change in direction is different, but

Figure 13.2 Representative trajectories of *Halobacterium salinarum* under sequential light stimuli that elicit reversals. These paths were drawn by the program used by Lucia *et al.* (1996). Dots represent the position of reversals as identified by the program. False identifications are highlighted by thick arrows.

altogether the motion track of this micro-organism consists of more or less straight stretches separated by sudden inversions, called reversals. Typical trajectories of swimming *H. salinarum* cells in response to repellent stimuli can be seen in Figure 13.2. Indeed, chemotactic or phototactic stimuli affect the frequency of tumbles or of reversals. More precisely, it is the change in the level of these stimuli that increases or decreases the frequency of directional changes. It is easy to understand how the occurrence of this kind of response produces photoaccumulation or photodispersal. Consider, for instance, a spot of blue light impinging on a cell sample of *H. salinarum*. A cell entering the light spot experiences an increase of the blue light level and this elicits a reversal in motion, so that the cell moves away from the repellent light. The opposite occurs for cells leaving the illuminated region: they continue to swim away from the light region and show no change in the direction of motion for a time longer than usual, since the decrease in blue light is a stimulus that suppresses reversals. As a consequence, the blue spot becomes empty or almost empty of cells after some time. The opposite behaviour occurs with a red–orange (attractant) light spot.

13.2 THE MEASUREMENT OF PHOTORESPONSES AND THE POTENTIALITIES OF IMAGE ANALYSIS

Measurements of accumulation or dispersal offer a quite simple way to quantify behavioural responses in a sample. Within this context, starting in the mid-1960s (Lindes *et al.*, 1965), densitometric measurements have been employed by several groups to quantify this phenomenon. In practice, changes in the cell density of the sample are detected by the difference $I_1 - I_2$ in the transmitted light intensities collected at two distinct points of the sample container by separate photodetectors. The initial slope of this variable after the onset of the stimulus is used to reveal a response. This kind of apparatus made it possible to determine the action spectra for photosensitive micro-organisms for the first time, by evaluating the relative efficiency of light stimuli of different wavelengths. Obviously, in this method as well as in almost any other method used to investigate the behaviour of light-sensitive micro-organisms (recording and analysis of scattered light, video recording with subsequent analysis or on-line image analysis), it is mandatory to avoid the analysing light beam itself acting as a stimulus. This can be accomplished through the use of appropriate filters, e.g. using infrared illumination, which usually does not act as a stimulus and is readily sensed by a video camera.

However, variations in cell density are a late effect, and whenever the detection of early responses is essential, more accurate measurements are required to detect the different effects selectively. The most important point is that densitometry, by measuring a consequence of the motor responses and not motor responses themselves, does not allow measurement of the time course of photoresponses, which is very informative and sometimes critical for the interpretation and modelling of the sensor–effector system.

Different methods, mainly optical, have been developed to measure velocity distributions, flagellar beating and so on; they are sometimes (not always) more simple and easier to master than image analysis systems, but always yield partial results. On the other hand, methods based on image analysis appear to be very flexible tools, allowing the measurement of any motion parameter of interest, and can be adapted to many different schemes of experiments.

Before going into a full discussion of the methods for cell tracking using the image analysis approach, let us recall the most complete device used to study the sensory responses of a single micro-organism. The tracking microscope developed by Berg (1971) and Berg and Brown (1972) is a very sophisticated, computer-controlled system. It allows the observer to choose a moving target in a microscope field under darkfield illumination; a feedback circuit maximizes the light scattered by the target by moving (in three dimensions) the microscope stage, thus keeping the object stationary with respect to the microscope holder. The feedback signal allows the spatial coordinates of the chosen cell to be stored in computer memory for later analysis. At present, this system is the only one able to analyse cell motion in three dimensions. Using this system, Berg and Brown studied the chemotactic response of *E. coli* in terms of single-cell behaviour.

For slowly swimming bacteria, the detection of sudden direction changes can also be done by manual tracking of the cell in the microscope field. Despite its lack of sophistication, this method has been used in a series of works. It allows tracking of a single cell for a long while, which is of advantage when studying the temporal structure of a sequence of events, such as motion reversals (Schimz and Hildebrand, 1987; Krohs, 1994, 1995) or the behaviour elicited by periodic stimuli (Lucia *et al.*, 1992). The major drawback is the time required to achieve reliable statistics in the collected data.

Yet another method based on the observation of single cells has produced important results: in this case, cells tethered to the glass are observed. When a single flagellum or a flagellar bundle is tethered to the glass, the cell body rotates, acting as an amplifier of the flagellar motion. This simple method allows observation and study of the clockwise–counterclockwise alternation of the flagellar motion in *E. coli* and *H. salinarum* (Kobayashi *et al.*, 1977; Block *et al.*, 1983; Rudolph and Oesterhelt, 1996).

Image analysis methods certainly yield less complete information than the tracking microscope of Berg and Brown, which we briefly described above. They can only analyse the velocity components in the plane of focus. Notwithstanding this limitation, image analysis can give a wealth of information, since it allows reconstruction of the trajectories of micro-organisms as long as they are in focus and can thus yield measurements of various parameters, depending on the aim of the experiments. These are, for instance, the cell velocities, the curvatures of the trajectories, the time intervals between singular events like reversals, and the statistics of the angles taken after a reversal or tumble in bacterial motion, or in the phobic response of unicellular algae and protozoa. Image analysis can obtain population measures, since appropriate computer programs can follow the trajectories of many cells at a time. Any and every geometrical property of each trajectory can in principle be studied. Last but not least, methods based on image analysis, once they have been set up and implemented, can also be used by people not particularly skilled. The relevance of these methods is also attested by the regular publication of books on the specific topic (Noble and Levine, 1986; Häder, 1992). However, the potential flexibility of these methods depends on the structure of the software used in the analysis, as will become clear from the following discussion.

13.3 A SURVEY OF IMAGE ANALYSIS APPLIED TO STUDIES ON MICRO-ORGANISM MOTION

The first attempts at using analysis of trajectories to study sensory transduction in unicellular organisms date back to the end of the 1970s and the beginning of the 1980s. In those years it was unthinkable to carry out the analysis during the experiment. Indeed, early attempts at reconstructing the trajectories were done more or less by hand. For instance, the photobehaviour and action spectrum for the step-down photophobic response in *Euglena gracilis* were studied by using a video recorder to record the motion of several cells (typically 20 per field) and analysing the trajectories at a later time (Barghigiani *et al.*, 1979). The same approach was used two years later in the study of the phototaxis in *Eu. gracilis* and *Ochromonas danica* (Häder *et al.*, 1981). The analysis was done by replaying a videotape frame by frame, after which the swimming paths were traced on a transparent plastic sheet placed over the monitor. A photophobic step-down reaction was identified as an arrest in forward movement upon switching off of the light stimulus, followed by several rotations of 180° with respect to the previous direction of motion. In Figure 13.3 we show a reproduction of the paths traced most patiently by hand on the transparent plastic sheet. In an automated analysis system, the rotations would be identified automatically, and each parameter (percentage of cells responding to light, time lag of the response, duration of the rotatory behaviour and so on) could be measured, with the great advantage of doing this within the time interval of the experiment. To have the results displayed at the end of each experiment is really of great help. Not only does it save time, but it also enables the experimenter to adjust the parameters of the stimulation during the run of experiments, and to make controls, which would otherwise be impossible. The aim of getting results displayed at the time of the experiment does not imply necessarily that the analysis is in real time. As we will discuss below, if it is possible to perform the analysis in the dead time required between subsequent experiments, this is sufficient, since from the point of view of the experimenter this does not make much difference compared with real-time analysis.

The first endeavours in automated analysis of micro-organism motility go back to the mid-1980s. In 1982, an off-line system was developed at Hokkaido University in which the movement of freely swimming micro-organisms was recorded on a videotape by a video camera mounted on a darkfield microscope, and then automatically analysed (Takahashi and Kobatake, 1982). This system was used to perform measurements on the photobehaviour of *H. salinarum* and it was implemented on a PC (64 kbytes of memory) equipped with a video digitizer. On receiving the marks of the light stimuli, the computer triggered the video digitizer to convert the signal to digital information and then transferred the data to the main memory. During this routine, the data were converted to binary values (either 0 or 1) using a suitable threshold. An entire frame was divided into 256 × 256 pixels, each represented by a single bit. The sampling of a single picture was

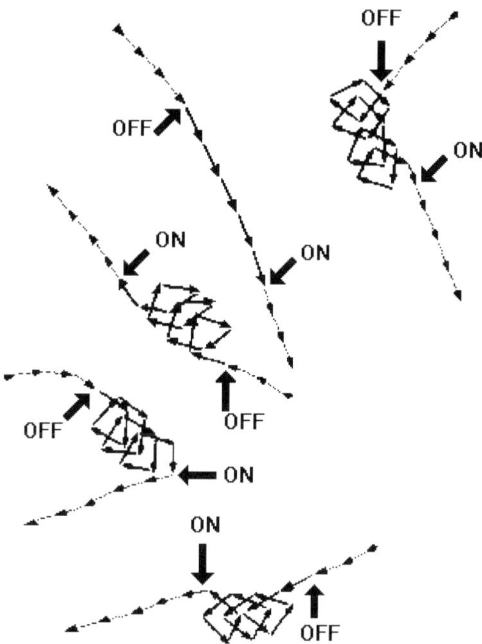

Figure 13.3 Step-down responses of *Euglena gracilis*, based on data from Barghigiani *et al.* (1979). These paths were patiently traced by hand on a transparent plastic sheet placed on the monitor screen by replaying the videotape frame by frame. As shown, a photophobic step-down reaction consists of a stop in forward movement upon switching off the light stimulus, followed by several rotations of 360° until normal swimming is restored.

performed within 1/60 s but thresholding and data transfer to the computer limited the rate to six pictures in 4.5 s, with a time span between each acquisition of 750 ms. The use of single-bit per pixel storage reduces the computer memory necessary to store the data, each byte corresponding to eight pixels. Each 'scene' contained six frames and this was enough to make quantitative measurements of phototactic responses of *H. salinarum*. At the end of the acquisition of a single scene, the computer stopped the videotape and started to track cells and count the number of reversals in the scene, sending a signal to the videotape to restart the playback at the end of the analysis. The computations to track cells were accomplished by bit operations for computing the correlation matrix between the pixels above threshold in the current frame and those of the cells identified in the previous one. As mentioned, this set-up was applied to the study of the photobehaviour of *H. salinarum*. Its first use was to investigate how the background light affected the responses to attractant stimuli, consisting of red–orange light. A full description of the method appeared in a later contribution (Takahashi, 1992). From a technical point of view, this work showed how it was possible to use a low-cost computer system equipped with an I/O device to produce an automatic off-line quantification of the behavioural responses in micro-organisms. A more recent version of this system is still in use to study the phototactic behaviour of both *H. salinarum* and unicellular algae (Takahashi *et al.*, 1991; Takahashi and Watanabe, 1993).

A system for automated off-line analysis of cell movement was also developed in Pisa in 1985 (Gualtieri *et al.*, 1985). This system used a minicomputer (PDP11) and the video signal was digitized by means of an image processor system. Its cost was therefore higher than that of the very compact system developed in 1982 by Takahashi. The same system was used until 1988 for studies on the photobehaviour of *Eu. gracilis* (Gualtieri *et al.*, 1988).

At the same time, a system for on-line image analysis was developed in Marburg (Häder and Lebert, 1985). This system was developed with the purpose of analysing the deviation of individuals in a population from a defined direction and was able to follow the movement of a single micro-organism on-line. The video signal was sent to a video digitizer working in real time (one frame every 40 ms), and the image stored in the RAM memory of the digitizer was analysed by the host computer (Z80, SD system). The goal of real-time analysis was achieved by programming in Z80 assembler language and by the strategy employed in the program. The program ran a loop in which the first step was the request for a new video frame sent to the digitizer. Starting from randomly chosen x and y coordinates, the program scanned the video image (horizontally or vertically) until a pixel above a predetermined threshold was found. If a pixel was found, the outline of the cell was traced using a standard algorithm for edge detection, and finally the centre of mass of the micro-organism was computed. Then a request for a new frame was sent to the digitizer (at almost equal time intervals) and the search for the *same cell* started from the position previously found. The reason why the program was capable of real-time analysis was because it was confined to finding a single micro-organism, thus avoiding the need to read most of the screen. This simple idea of obtaining information on a limited number of cells and following them for just the time required to measure their velocity has also been applied in more recent work of the Marburg group, as we report below. The program described above was used in the same paper to obtain polar diagrams of the distribution of cell velocity in studying the phototactic orientation of *Eu. gracilis*.

Here it is appropriate to recall that the distribution of cell velocities can be measured, with enviable statistics, by using a completely different method: heterodyne detection of the Doppler effect of light scattered by moving objects. Because of the Doppler effect, light scattered by a moving object has a frequency slightly different from that of the light falling on the sample, and this shift in frequency can be measured by a heterodyne technique. The spectral analysis of fluctuations in the photocurrent gives the distribution of a component of the cell velocity in the sample directly. The component measured is that along the bisector of the angle between the incident and the scattered beam. In measurements of phototaxis it is possible to align this bisector with the direction of the actinic light (Ascoli *et al.*, 1978). Moreover, an improved version of a heterodyne spectrometer allows measurement of the distribution of a component of single-cell velocity conserving the sign of the Doppler effect, which is usually lost in heterodyne detection (Cantatore *et al.*, 1989).

The heterodyne method is excellent for measuring phototaxis and in general for studying cell velocity and its dependence on physico-chemical factors. Its limitation is that it requires the constant presence of a trained and skilful experimenter, while cell tracking by image analysis is easier to master, once the program has been implemented and tested. Image analysis is now quite a widely available method, while heterodyne and other physical methods are less suitable for general use.

Coming back to the development of image analysis in connection with chemotaxis and phototaxis, it is remarkable that commercial general-purpose equipment for cell tracking (Expert Vision software and VP 110 digitizer, Motion Analysis Corporation (MAC), Santa Rosa, CA, USA) was already available in the same years in which the progress described above occurred (mid-1980s). At that time, its price was about $US60 000, too high for most scientific groups, particularly when compared to the cost of the home-made devices described above. The cost of the set-up described by Häder and Lebert (1985) can be evaluated, all in all, as about $US10 000.

A few groups used systems developed by MAC. In 1986 photoresponses of *H. salinarum* were monitored by making use of a MAC system (Sundberg *et al.*, 1986). The system was coupled to an electronic shutter controlling the delivery of the stimuli. The percentage of responding cells and the times of photoresponses were detected by the implemented program, running on a Sun2/120. All of the analysis was carried out using appropriate combinations of the operators defined in the MAC software package; video images were digitized frame by frame and transferred to the host computer, where a combination of searching operations was performed to identify the cells coordinates and to connect each single cell with its position in the successive frame. The time resolution was 67 ms (15 frames/s) and the program was set up to record 40 segments of 5 s data, merging the individual path data at the end, to form a single file containing information on hundreds of cells. The limit of 5 s long segments was not an absolute limitation and in subsequent work the system was used to track cells for a longer time, up to 90 s (McCain *et al.*, 1987).

Very important results on signal processing in *H. salinarum* were obtained through the use of this system over the years. As far as time resolution is concerned, the MAC system reaches the optimum within the limits of conventional video cameras. However, as reported by Takahashi (1992), the computation of centroids was a time-consuming process, requiring a waiting time of 20 min to analyse a sequence of video data digitized for 4 s at 5 frames/s. This long waiting time is actually a limit in a system for motion analysis, and below we describe home-made systems (with lower time resolution) in which the waiting time is considerably shorter. However the possibility of having high time resolution is relevant: for instance a resolution of 67 ms allowed measurement of the different minimum latencies in responses mediated by different pigments in *H. salinarum* (Sundberg *et al.*, 1986), an impossible task for a system with a lower time resolution. The technical work done in more recent years by several groups now makes it possible to reach the same temporal resolution using cheaper devices and PCs.

In the same years another paper appeared (Kuo and Koshland, 1989) that used a MAC system. This time, the bacterium under study was *E. coli* and the sample preparation was different. Cells were tethered to an *ad hoc* prepared glass slide by anti-flagellin antibodies and observed under darkfield illumination. Therefore, the cells were not swimming but rotating around the immobilized flagellar filament, and the problem was to detect the sense of rotation, its inversion and the time elapsed in changing from clockwise (CW) to counterclockwise (CCW). Previously this problem was solved either manually (Block *et al.*, 1983; Lapidus *et al.*, 1988) or optically (Kobayashi *et al.*, 1977) using linearly graded neutral density filters. The software package was intended for detection of the paths of freely swimming cells and was poorly suited for the analysis of rotating bacteria. The authors therefore developed their own analysis program in the C language. The analysis was performed off-line because the problem required a high temporal resolution (histograms of the distribution of time span between reversals with 17 ms per bin were obtained). This time resolution was attained by processing videotaped images at a 60 Hz video sampling rate off-line. The goal of this work was a detailed analysis of motor switching kinetics, a key problem in understanding the chemotactic behaviour of *E. coli*.

Later, the motion of cells was recorded on-line and frames were digitized at 10 or 15 frames/s for periods of 5–6 s by another group using a MAC system (Marwan and Oesterhelt, 1990; Marwan *et al.*, 1991). The program used was written in Pascal and allowed simultaneous display of each individual track and the parameters of motion. The motion analysis, made with good time resolution, allowed measurement of the distribution of pausing times which occur before the reversal takes place.

We conclude this historical survey of image analysis applications to cell motility and cell responsiveness by briefly describing two more recent studies carried out in Pisa and Marburg. The approaches used in these studies are different, as we will discuss in detail later, but the common feature is the attempt to produce real on-line analysis, with good statistics, using video digitizers with RAM memory to grab the image and PCs. The work in this direction started in Pisa in 1990 and was developed in two or three years (Coltelli and Gualtieri, 1990; Gualtieri and Coltelli, 1991, 1992). This approach is characterized by the use of the hardware possibilities of the frame grabber board. We discuss this approach further in Section 13.4.4. The device developed in Marburg (Häder, 1988, 1991; Häder and Vogel, 1991, 1992) performs two different tasks. In its original version the program was written to measure the distribution of the velocities in a sample, and this task was accomplished by splitting the 1024×1024 frame grabber memory into four quadrants and storing four successive (not necessarily consecutive) images in these four quadrants. The first image was thoroughly analysed in order to identify the cells, while the three successive ones were analysed only in the proximity of the initial position of the detected cells (this method greatly reduces the time needed to identify the cells). In this way, the velocities of all of the cells present in the first frame could be measured in a short time.

In order to track cells for a long time interval a different method was used: a small number of cells was identified in the first frame and then searched for in the following ones. This was done by following an approach similar to that developed by Häder and Lebert (1985). The program analyses a first frame, stopping when eight cells have been identified. Then it searches the following frame for the next position of the identified cells, reading and analysing pixels in the proximity of the previous position, again greatly reducing the time spent in cell identification. In fact, as we will discuss in more detail below, the time spent reading and transferring the pixel values from the frame grabber to the memory is the most important component of the time required for cell identification.

A great advantage of this program is its modular structure. We obtained the source code of this program from the authors and it was possible for us to change it and obtain long-term tracking of swimming *H. salinarum* cells (Lucia *et al.*, 1996, 1997). The time resolution was not very high, since we could analyse two or three frames per second, depending on the number of cells in the field, but this was enough to allow us to track many cells (in the order of 50 per field) for a long time. This was essential in the kind of experiments we performed, in which the effect of sequential stimuli (two or three sequential stimuli were used) was studied. An important point for obtaining long-term tracking by this alternative version of the Häder and Vogel program was the use of the program on cells that were represented in the grabbed image by a small number of pixels (usually the area of a cell was between 20 and 80 pixels).

In the next sections we discuss the hardware and software methods used in the past, or which could be used in future, in more technical detail, also taking into account the improved technical specifications of the most recent generation of frame grabbers. We also discuss recent work aimed at studying more rapid movements, such as the movements of eukaryotic flagella, by image analysis.

13.4 METHODS FOR STUDYING MICRO-ORGANISM MOTION BY IMAGE ANALYSIS

13.4.1 General tasks of a motility analysis system using image analysis

Here we give a schematic description of the tasks performed by a motility analysis system designed to study micro-organism motion, which is discussed in detail in the following sections. The first of these tasks, namely image acquisition, is performed by a video rate charge-coupled device (CCD) camera. Since data are to be manipulated by a computer, the information must be subsequently digitized by an analog/digital (A/D) converter (Chapter 2) and stored in the image analysis board plugged into the host PC. The image analysis board contains several components that can be controlled through the PC in order to obtain an image useful for the extraction of the relevant parameters. Once an image has been captured, the subsequent task is cell detection (or segmentation; see Chapter 6). Since the final aim is to observe cell motion, several successive frames must be acquired and the cell positions in different frames must be linked to form paths. The last step is to analyse the paths in order to detect singular events or to measure velocity distribution or any other relevant parameter in the cell population.

13.4.2 The basic set-up

In applications where image analysis is used to study the motility and motor responses of micro-organisms, the video camera is mounted on the microscope. The sample on the microscope stage is held in a chamber usually built from two glass slides and monitored by darkfield infrared illumination. The cells appear on the monitor as white particles on a rather dark background. The higher the contrast, the easier the image analysis. Therefore, out-of-focus cells must be reduced in number, since light scattered by them lowers the contrast. Therefore, thin samples should generally be used. Moreover, thin samples prevent the cells from going out of focus continuously. The probability of losing some cells because they fall below or rise above the plane of focus depends on the dimensions of the micro-organisms as well as on the way they move. If we want the apparent length of a cell on the video to be about 10 pixels (1/50 of the video width when a frame grabber of 512 × 512 is used), the absolute magnification will be very different, depending on whether we are dealing with bacteria (cell length in the 1 μm range) or with protozoa (cell length in the 100 μm range). This usually involves a greatly different depth of field in these two samples. Therefore, the probabilities of losing cells because they swim out of the focal plane are different. Thus it might seem that tracking protozoa should be simpler than tracking bacteria. However, bacteria are sometimes limited in their motion due to restricted space in the liquid medium between the two glass slides, and thus remain in focus most of the time. On the other hand, certain algae prefer not to swim in two dimensions for prolonged periods.

Cell motion can be observed on a monitor directly connected to the video camera; a second monitor is connected to an image analysis board plugged into the host computer, in recent years usually a PC (Figure 13.4). At a given magnification of the microscope, the resolution in the digitized image depends on the number of pixels. A grey level is assigned to each pixel by the A/D converter (Chapters 1 and 2). Most recent digitizers allow software control of the offset and gain of the digitizer. This makes it possible to optimize the digitization. As shown schematically in Figure 13.5, the best way to digitize a signal should take advantage of the whole range of possible values, putting the lowest level to zero and the highest, for instance, to 252 (this leaves the two least significant bits for special uses). In recent video digitizers,

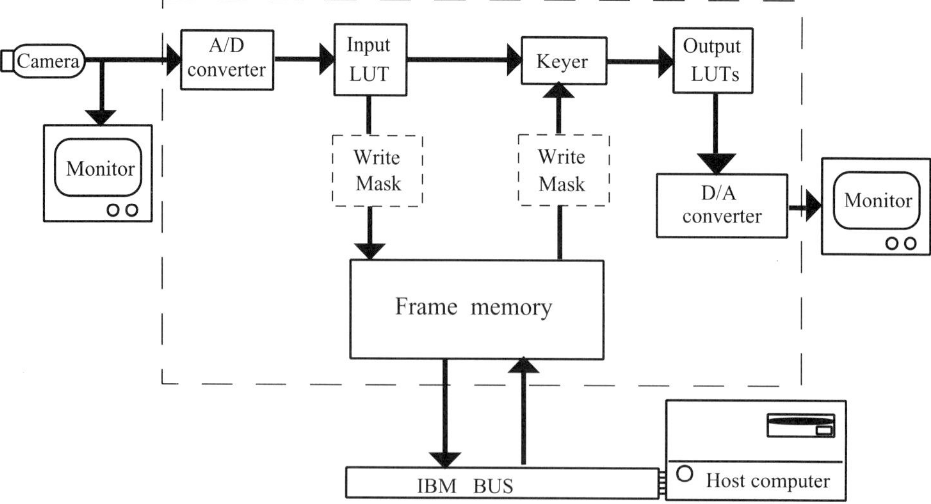

Figure 13.4 General scheme of a system for image analysis of motion of micro-organisms. All of the components in the dashed frame are within the image analysis board. Not all digitizers are equipped with write masks.

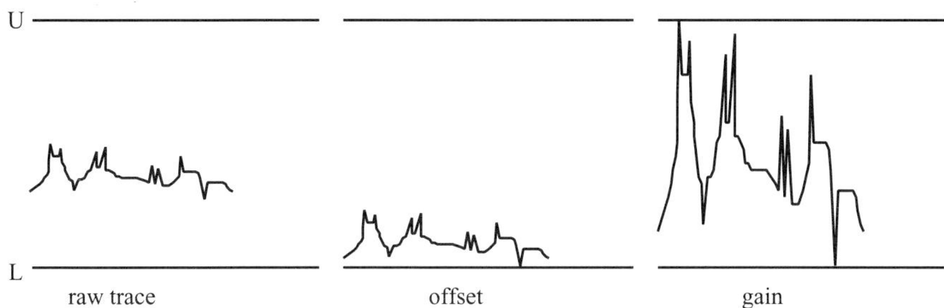

Figure 13.5 A schematic description of offset and gain adjustments. The available range goes from the lower (L) to the (U) upper limit.

the process of A/D conversion is performed quickly (digitizing a video image in 512 × 512 pixels in 40 ms requires at least a digitizer working at 6.6 MHz); the repetition rate is determined only by the video camera triggers (25 frames/s in the European standard, 30 frames/s in the US standard). It is important to realize that this is an intrinsic limit in the time resolution that can be achieved by using conventional video systems and frame grabbers. If the problem to be faced demands a resolution of better than a few tens of milliseconds, the use of a special-purpose video camera is required. Work is in progress to develop instrumentation which allows lowering of the limit of the time resolution of commercial video systems (Ishijima, 1995).

After digitization, the image is stored in numeric form in RAM on the frame grabber (Chapter 5). The problem of analysing such digitized images can be dealt with by adopting either of two distinct general strategies. In the first, the board is used to *digitize and memorize* many consecutive images (four in the Häder and Vogel program; up to 192 in that by Coltelli and Gualtieri (1990)), which will eventually be analysed after this global acquisition. In the second possible approach, the board is used *solely to digitize* the images and the analysis is carried out frame-by-frame by the computer. Using PCs this approach can be pursued only by limiting the time resolution of the analysis. These two approaches are very different, and show different limitations and advantages. It is immediately clear that in frame-by-frame analysis it is crucial to have as speedy as possible a routine for analysis, while in the hardware (or digitize and memorize) approach this is desirable but not crucial. In order to explain this point more clearly, let us recall the typical scheme of an experiment aimed at studying the motor responses of micro-organisms, which means that we are interested in getting information on the trend of a parameter (velocity, frequency of tumbles or of reversals etc.) as a function of the time elapsed since the stimulus was applied. Usually the results obtained in successive applications of the stimulus are collected. A certain time interval must elapse between one application and the next in order to let the sample come back to a stationary condition corresponding to the absence of stimulation (re-adaptation time). Thus the speed of the analysis is not critical, if it is carried out during such a re-adaptation time. If, on the other hand, frame-by-frame analysis is chosen, the speed of the routine performing the analysis becomes relevant. Similarly, even when the motility is studied *per se*, i.e. under stationary conditions, a waiting time of several seconds at the end of a set of acquisitions may be acceptable, whereas a worse time resolution between frames may not be tolerated in general.

Whichever strategy is adopted, the computer must access the image stored in the frame grabber and must identify the cells within it. Therefore, this part of the software is common to both methods. The next two sections are devoted to a discussion of the hardware and software methodologies that have been used in image analysis of the motility of micro-organisms.

13.4.3 Hardware potentialities

In order to understand what can be done by means of the hardware, i.e. by getting the most out of an image analysis board, we must discuss the kind of tasks that can currently be performed by a commercial board in more detail. There are many boards available commercially; for example, PIP 1024 (Matrox, Canada), VP 110 (Motion Analysis Corporation, USA) and FG100 AT (Image Technology, USA), just to mention those used in papers cited in this chapter. We shall illustrate their general functions with the help of the block diagram shown in Figure 13.4. The input signal from the video camera is digitized by the A/D converter, but before it is stored in the frame memory, it passes through a programmable input look-up table (LUT). This represents an important possibility for manipulating the grey values of the digitized pixels before they are stored in the memory in real time. The programmable input LUT allows replacement of each grey level by a preset value at the LUT position corresponding to the grey level. Thus, it is possible to invert an image by replacing the 0 value by 255, the maximum value 255 by 0 and the other values accordingly. It is also possible to expand part of the grey scale by assigning the 0 value, for instance, to all grey levels from 0 to 127, thus reserving the whole range to pixels with grey levels from 128 to 255.

The digitized image is usually displayed on a different monitor, in order to compare the grabbed image with that of the monitor directly connected with the video camera. As shown in Figure 13.4, the signal sent to the output monitor can be modified by output LUTs. Note that, while the output LUTs influence only the image as it appears on the monitor, the input LUT affects the signal to be stored (Chapter 5). A possibility offered by the scheme in Figure 13.4 is that represented by the 'keyer' block. If keying is enabled, it allows overlaying of the image contained in the frame grabber and that contained in the least significant bit(s) (LSB) on the video screen. This feature can be used, for example, to write messages. This is done by setting the LSB to 0 using the input LUT, and writing any kind of message in the LSB plane. The overlaying of the two images is made by the keyer, which sends the pixel value of the input image if the LSB is 0 and sends 1 otherwise.

The block labelled 'write mask' in Figure 13.4 is a most interesting function, which is present only in some boards. It allows 'freezing' of the value of some bits, so that it is possible to write in the grabber memory without changing the previous values of those bits. This reduces the dynamic range of the A/D converter (more precisely, it reduces the range available to the converted values) but in the meantime makes it possible to use part of the grabber memory to store information. This usage of the frame grabber memory allows storage of the entire course of an experiment, analysis of which is performed in a few seconds at the end of the acquisition.

This opportunity for tracking moving cells has been used by Coltelli and Gualtieri (1990; Gualtieri and Coltelli, 1991). We describe their method and then discuss the possibilities offered by more recent frame grabbers for the same purpose. The board used in their studies was equipped with a 12-bit converter and allowed: (i) freezing some bits to their preset value, and (ii) storage of the difference between two images. The six most significant bits of the memory were used to store the current image, whereas the six least significant bits were used to store the difference between the current image and the previously stored one. This was achieved by suitably programming the input LUT. Moreover, the use of zoom facilities allowed division of the entire frame grabber memory into many parts (up to 192), while maintaining a tolerable spatial resolution. The current frame was stored at the top-left corner of the frame grabber memory in the six least significant bit planes, while the difference images produced for the final analysis were stored at the proper level and position by shifting the x and y coordinates of their origin in real time by writing the appropriate commands in the corresponding address registers of the board. This procedure made it possible to store frames with a minimal delay of 40 ms, i.e. it was possible to store each and every frame, thus attaining a very good temporal resolution. When using the frame grabber memory to store many images by zoom operations using boards with only 1024×1024 bytes of memory, there is a price to be paid: spatial resolution. If the reduced images of the micro-organisms are to be large enough to be identified in the stored data, the initial magnification has to be higher (magnified by the zoom factor), and this reduces the number of trackable micro-organisms considerably. On the other hand, this limitation becomes less severe when working with more recent boards whose frame grabber is equipped with a greater memory capacity. In prospective terms, it will be more and more advantageous to use hardware memorization.

Some boards can be equipped with more memory space; e.g. boards designed for colour acquisition have three frame memories. Let us assume that each frame memory has 1024×1024 pixels, each with 8-bit planes, and that an input write mask can be used. Here, the write mask allows manipulation of the data bit by bit. For instance (see Figure 13.4), by applying a threshold to the signal and programming the input LUT, it is possible to reduce the image to a binary signal and, through the digitizer mask, the 24 1-bit planes can be used separately to store information. By combining this feature with the use of electronic zoom (for instance, by dividing the memory into only four parts, thus storing the video image in 512×512 pixels), it is possible to store 24×4 frames, i.e. the entire course of an experiment on sensory responses or adaptation. This discussion shows that the use of hardware capabilities can be very advantageous and will probably become more frequent in future.

Let us now briefly discuss the method, also used by Coltelli and Gualtieri (1990), of memorizing difference images. This is done with the aim of clearing the images of dust or debris of the same dimensions as the cells, and retaining only the moving objects, an operation that would otherwise be carried out by the software. At the end of the acquisition, it is possible either to analyse the subsequent images and identify the cells, or to skip this step and overlay the different images to look directly at the trajectories, whose shape appears very clearly in the final image. Although the use of difference images reduces the presence of debris etc., we cannot tell whether there is an advantage in doing this operation through the hardware at the time of acquisition, instead of doing it later by software analysis. When the acquisition ends, there is plenty of time for any kind of analysis (even for very busy people, waiting a few seconds is not a problem), and the difference method does not allow measurement of the percentage of non-motile cells, which can be an important parameter. For instance, people studying the effect of UV radiation on cell motility are certainly interested in it. As a conclusion, at least in some cases, it could be better simply to identify cells, and not to confine the information to moving objects.

13.4.4 Cell identification

We shall discuss the software methods by referring to the case in which each image is analysed between two subsequent acquisitions. The speed of the analysis routine is therefore of crucial importance. The first step is reading the data stored in the frame grabber. Just to give an idea, when using a Matrox PIP 1024 the time for sequential reading is 1.8 μs on average per pixel (pp. 5–16 of the PIP 1024 hardware manual), so that reading all of the pixels of the image is rather time-consuming (about one million bytes are stored in the frame grabber memory). This suggests that optimization of the software should reduce the useless information as soon as possible. Thus the first step in the program should be dedicated to the extraction of the important features of the image, in our case the identification of objects that are probably cells. As already mentioned, cells appear as bright objects against a dark background. Thus the first operation is to set a threshold value on the grey scale, in order to separate the bright pixels from the dark ones. Looking at the histogram of pixel values is of help in choosing the threshold value (Section 6.7.3). It is unfavourable to read all of the pixels stored in the frame grabber memory. For each image row, it can be sufficient to read one pixel out of n (and one row out of n) and analyse the part of the image around the pixels that happen to be above the threshold value in detail. The value of n depends on the area (in pixels) of a single cell (in a non-uniform sample, where cell growth is not synchronous, the reference area will be that of a small cell). When elongated cells are observed, the smallest diameter must be considered in choosing the value for n. This methodology to reduce the reading time of an image was adopted in the program written by Häder and Vogel.

One might consider first identifying the set of the pixels above the threshold value and subsequently finding which pixels belong to the same object (or connected component, Section 6.7.7). However, this is by no means optimal, as can readily be appreciated by considering the algorithms for cell identification (see next section). In the following we report a very schematic description of two types of algorithms implemented for this purpose. From the point of view given above, a cell is characterized by appearing as a connected set of pixels. In a similar way, we can say that a cell is a region with a closed contour. The two fundamental methods for identifying cells (generally referred to as 'region growing' and 'contour tracing' methods) are based on these two definitions of the objects that have to be found (see Section 6.9).

13.4.5 Description of algorithms

In this section we describe two algorithms used to identify cells (region growing and contour tracing) and also try to offer an evaluation of their relative speed. It is not easy to make a direct and absolute comparison between these two different algorithms. Coltelli and Gualtieri (1990) reported that, in order to analyse an image of 1024×1024 pixels (which in their work was a combination of various successive acquisitions), 10 s were needed, using a PC with a 386 processor and a program written in C that read each pixel. In the program written by Häder and Vogel (1991), which we have modified slightly in order to detect *H. salinarum* reversals (the basic routines are written in assembler language), the analysis of an image of 512×512 pixels takes about 0.5 s when around a hundred cells are tracked in the field (this time interval increases with the number of cells).

However, these two programs are written in different languages and have a different overall structure. If we try to compare the two algorithms on a common basis, an *a priori* comparison might be made by evaluating the number of operations required in either algorithm. However, such an evaluation is not easy since the time requirements of reading steps, numerical computations and conditional statements (decisions) weigh differently in determining the time intervals taken by each routine. Thus we tried a different approach by implementing two distinct test programs corresponding to each algorithm, and measuring their performances. The task of these test programs is to find the coordinates of the centre of mass of a connected set of pixels R starting from a pixel selected randomly from a subset A of R. A is a subset $n \times n$, n being the step in pixels that is supposed to be used in reading the pixels from the frame grabber, as in the Häder and Vogel program. In these test programs the pixel values are given *a priori* in a matrix in the computer memory, and consequently the time spent 'reading' them is practically zero. Thus, the time that would be spent in communications between the board and the computer is not taken into account. This has been done on purpose, in order to obtain a comparison independent of the technical characteristics of different boards for image analysis and also to simplify these unpretentious test programs.

Let us start by describing the contour tracing algorithm. In this kind of algorithm an identification strategy is adopted based on the detection of edges and closed contours. For a detailed description of the methods to be used in programming this algorithm in assembler language, reference can be made to the work of Häder and Vogel (1991, 1992). The basic scheme is shown in Figure 13.6. During the scanning of the image memory of the frame grabber, each pixel is compared with the threshold value C. When a pixel above the threshold is found the routine for edge detection starts. The first step is to move, for example, to the left until an edge pixel is found. Let us call the coordinates of the identified pixel on the outline of the object (x_0, y_0). The first question to be considered is whether (x_0, y_0) belongs to a previously identified object. It is simple to find out whether a pixel belongs to an already detected object: in this case, a pixel above threshold, connected with the pixel under analysis, will be found in the preceding row. Alternatively, and much more simply, the pixels of previously found objects are given a grey value not allowed by the input LUT (Häder and Vogel, 1991, 1992). In

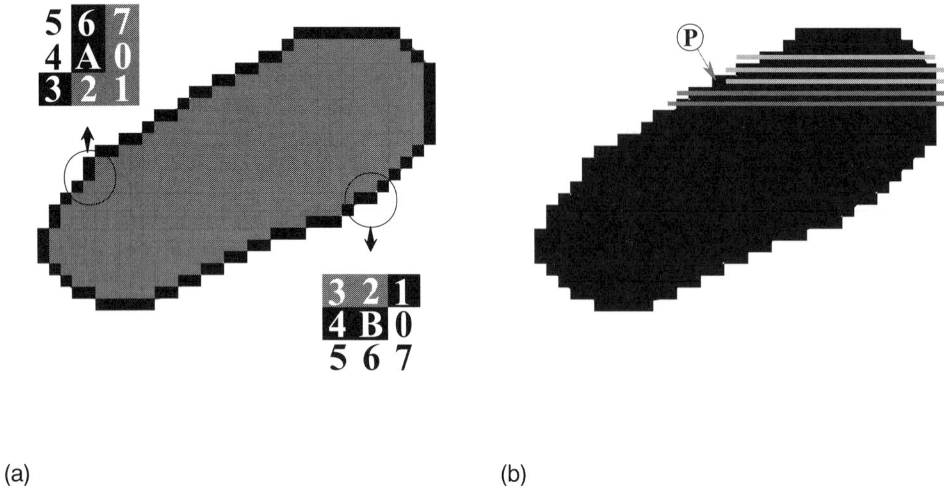

Figure 13.6 A scheme showing how contour tracing (CT) and region growing (RG) algorithms work. (a) CT (the relevant pixels are magnified in the circles): the test program moves clockwise if the 0 neighbour of the current pixel is above threshold and counterclockwise in the opposite case. Thus, starting from pixel A the next edge pixel is found at position 3, whereas from pixel B it is found at position 1. (b) RG: light grey rows are the first three swept in the upward direction, starting from pixel P, dark grey rows are the first two in the downward direction.

the case of a positive answer, the stepwise scanning of the frame buffer continues, otherwise the neighbours of pixel (x_0, y_0) are read. In our test program we chose to move clockwise around the current pixel (x_0, y_0) if its '0' neighbour is above threshold, counterclockwise if it is below threshold. Each pixel encountered in the clockwise (counterclockwise) path around (x_0, y_0) is compared to the threshold value until a pixel below (above) the threshold is found. Then the coordinates (x_1, y_1) of the last pixel above threshold (edge pixel) are stored in a small array containing the contour of the object. The process is repeated for all new contour pixels until an (x_i, y_i) is found which coincides with (x_0, y_0); the contour tracing routine stops, and area, perimeter and centre of mass coordinates can be computed. Finally, the coordinates of cells with appropriate outline and area are stored for further analysis.

For a description of the region growing algorithm, reference can be made to Sections 6.6.1 and 6.7.7. The task performed is again finding connected sets of foreground pixels, to evaluate their area, perimeter and centre of mass, and to store just this information, which represents the relevant data, permanently. The final purpose is in fact to store the positions of the objects that probably represent cells, at the end of scanning an image. Further identification controls can be done after this preliminary identification process, e.g. objects must have a suitable area and outline in order to be counted as cells, as it is obvious that the area and perimeter of a certain cell type must fall within a previously fixed range.

The use of a shape factor can also be of some help: the ratio of the square of the perimeter to the area is a parameter independent of scale factors and provides a measure of the indentation of the object outline (see Sections 7.5 and 9.2.2, equation (9.2)). This factor equals 4π for a circle, and $2\pi(a^2 + b^2)/ab > 4\pi$ for an elliptical shape with semi-axes a and b. This may help in evaluating the range of values that can be accepted as a shape factor in a given cell population. However, while the use of a shape factor is very useful in order to identify immobilized or non-motile cells (e.g. to differentiate between two different cell types), its usefulness for the analysis of moving cells is more limited. A cell with an ellipsoidal shape may appear circular while it swims, due to changes in its orientation, thus widening the range for the possible values of the shape factor. These control tests may be applied after a first screening for potential cells.

In Table 13.1 we report the time required to identify the centroids of some objects by both our test programs. It is obvious that the result of this comparison depends on the dimensions of the objects to be identified: increasing their dimensions should favour the algorithm based on the search for a closed contour, since the ratio of the area to the perimeter increases (note that, in computer programming, perimeter and area of an object are comparable quantities, both given in pixels). For the same reason, it is also clear that the results depend on the object shape: the circle has the smallest circumference compared to other shapes of the same area. For the sake of simplicity, we have used rectangles of various dimensions and proportions. The results of the comparison for squares of various dimensions are shown in Table 13.1. For each test program, the reported processing times were measured after iterating the procedure 100 000 times, and are normalized to match the computing times required to identify 100 cells. The data are given for both a Pentium processor (Intel, 100 MHz) and a 486 processor (66 MHz).

The data in Table 13.1 show that, for small objects, the two algorithms are equivalent (since both area and outline are roughly of the same size), but when dimensions increase, the contour tracing algorithm appears to be faster. Moreover, as expected, the contour tracing works faster for squares than for 2:1 rectangles of the same area (data not shown). However, the difference between the two routines never becomes very large. Moreover, their absolute computing times are always much lower than the required reading times. These in turn will be discussed in more detail in the following section.

Table 13.1 Processing times required for the identification of 100 cells, of the contour tracing and region growing algorithms for a 66 MHz 80486 and a 100 MHz Pentium.

	Contour tracing (ms)		Region growing (ms)	
Object dimensions (pixels)	80486	Pentium	80486	Pentium
6×6	16	6	16	6
9×9	25	10	24	9
17×17	49	19	51	20
34×34	103	40	154	60

13.4.6 The time required to identify cells

In order to assess the relevance of various factors determining the time required by the routine that identifies the cell centroids, we refer to a real instance, namely the program we used to study the photoresponses in *H. salinarum*. We obtained this program through the courtesy of Donat Häder and Kurt Vogel, and changed some features in the organization of the program, in order to conform it to our requirements, in particular, to track *H. salinarum* cells for a long time and to measure the percentage of reversals in response to sequential stimuli (Lucia *et al.*, 1996). Although the basic routines remained unchanged, we changed the main program, in order to perform a complete analysis of each frame and identify the coordinates of any potential cells. We used this routine iteratively, so the delay between successive acquisitions (the time spent in the analysis) was crucial in this application.

The program is written for the PIP 1024B board (Matrox, Canada) and the basic routines are written in assembler language. If we use the program to scan an empty field, reading every third pixel and every third row, the time spent by the routine is about 270 ms (about 9 μs per pixel). On the other hand, the program spends a total of 360 ms on average to identify the centres of mass of about 50 cells and produce output. The program uses the contour tracing algorithm, so that each cell entails the reading of $n \times 4$ pixels on average, n being the length in pixels of the average outline. The reason for the factor 4 becomes clear by reference to Figure 13.6: each boundary pixel has eight neighbours, which are tested sequentially; on average, the new next boundary point is found after four attempts. Let us consider a field containing about 50 cells, each with an outline 30 pixels long and an area of 45 pixels (which implies a ratio of approximately 2:1 between the two axes). This is rather similar to the real conditions of our experiments. For 50 cells with an average outline of 30 pixels the number of pixels to be read is about $50 \times 30 \times 4 = 6000$, which corresponds to a reading time of nearly 60 ms (at 9 μs per pixel).

The three components of the computing effort therefore contribute to the overall computing time as follows: 270 ms are spent reading the pixels during the scanning of the board, 60 ms reading the outline pixels of 50 cells, and about 10 ms on actual calculations (as in Table 13.1), which adds up to 340 ms, i.e. nearly the actual time required by the computer to display the result (360 ms). For the same 50 cells, the number of pixels used in region growing is $50 \times (45 + 30) = 3750$ (45 pixels in the whole cell area and 30 pixels in the outline, which are read twice), with a time for reading the cell pixels of 34 ms. From this point of view, as far as pixel reading is concerned, the better algorithm for small elongated cells is region growing, but yielding only a 26 ms, or 7%, improvement.

In summary, the clear (and predictable) outcome of these tests on algorithms is that the time required to analyse a frame is chiefly the time spent in reading the data. As a consequence, although the (small) amount of time required for actual calculations can be shortened noticeably by using a faster processor, this hardly affects the overall time required, since the (most important) reading time is a feature of the board-to-computer communication. Thus, when frame-by-frame analysis is adopted, the reading time of the board is critical in determining the overall duration of the analysis and therefore the time resolution of the system.

13.5 GENERAL STRUCTURE OF MOTION ANALYSIS PROGRAMS

We now consider the general structure of the image analysis programs for studying cell movements. Usually a program of this kind is organized by a menu, for ease of use during measurement. The values of several parameters are to be preset: it is necessary to adjust the offset and the gain, and assign a grey threshold for identification objects (potential cells), and this in turn requires the possibility of viewing the histogram of the pixel grey levels. In addition, it is helpful to have access to statistics of areas, perimeters and shape factors of the identified objects. Finally, the program must show the cells it has identified. There are many graphic operations that the program could do automatically or on request, and this obviously represents a burden for programs of motion analysis. However, the core of the program is dedicated to cell identification and to measuring the motion parameters of interest.

Motion analysis programs have been exploited for two different purposes: (i) measurements of the distribution of cell velocities, and (ii) detection of particular events, such as reversals or tumbles. The programs directed to detect such events can work by tracing the path of a cell and then detecting the occurrence of particular features in the trajectory of the cell. Once the coordinates of the objects identified in frame number i are stored in the computer memory, the problem to be faced is finding the corresponding positions in the next frame $(i + 1)$. A simple criterion to use is that based on distance. If the object a in frame i is near enough to object b in frame $i+1$, the program identifies both a and b as the same object (or cell). A 'bias' can also be included in this search, taking into account the direction of movement of each cell, evaluated from the previous positions in frame i and $i-1$. However, in our experience this is not strictly necessary (we typically work with 30–50 cells in a field), and may actually lead to underestimation of the number of reversal events.

After the reconstruction of the trajectories, the next problem is that of identifying reversals. Two different indicators are used by Sundberg *et al.* (1986) to determine whether and when a reversal occurs in a given path. The first indicator used is the rate-of-change-of-direction (RCDI), which has a maximum at the time of reversal. Occasionally other maxima are present, but the absolute values of the change in direction (given in degrees per second) are lower. Thus the criterion for identifying a reversal is the occurrence of a maximum in RCDI together with a threshold criterion. The second indicator is based on the measurement of the swimming velocity, which has a minimum when a reversal occurs. Again, a threshold criterion was used together with the search for a minimum. A reversal is identified when both the indicators coincide.

A different way to identify reversals was used by Marwan and Oesterhelt (1990), using the MAC system. Starting from frame n, a mean direction was identified in the previous six and the following six frames (corresponding to a time shift of ± 4 s); if the angle β between these two directions was higher than a predetermined value ($\cos \beta > 0.4$), the program assigned a reversal to the frame where $\cos \beta$ had a maximum. A similar criterion can also be used on image analysis programs working at a lower time resolution. Takahashi (1992) took the cell displacements, which were assigned eight possible values (e.g. W, NW, N, NE and so on), directly, and the program considered a cell as smooth swimming if the next displacement was in either of the two adjacent sectors; otherwise, it marked a reversal. Lucia *et al.* (1996) also used a criterion based on the angle between successive displacements: if the absolute value of the angle between two successive displacements (computed over five frames) was greater than 70°, a reversal was identified at the middle frame. Despite the fact that reversals are clear-cut events characterized by a sudden change in the direction, some errors, mainly false identifications, occur in each of the systems reported above, but this does not necessarily change the shape of light responses very much. However, there are instances where the occurrence of such errors can be severely misleading, e.g. when a mutant is tested for its total inability to reverse its motion, direct observation cannot be substituted by automated analysis (Rudolph and Oesterhelt, 1996d).

Different bacteria cannot be tracked and analysed with equal ease. For instance, the swimming velocity of *E. coli* during a run averages 20 μm/s, with peaks of 40 μm/s, while its cell dimensions are close to those of *H. salinarum*, which has a mean swimming velocity of 2 μm/s. It is therefore nearly impossible to track *E. coli* cells with the same program we used to track *H. salinarum*. For this task, a system with a good time resolution is required – either the MAC system or that developed by Coltelli and Gualtieri (1990). A similar difficulty is encountered in identifying the tumbling behaviour of other Bacteria, or the step-down reaction of protozoa and flagellated algae (see Figures 13.1 and 13.3). Several papers in the past three years have dealt with the identification of tumbles (see Lopez-de-Victoria *et al.*, 1995 and references therein). Together with the measurements of speed and of the RCDI, another parameter was considered, the net to gross displacement ratio (NGDR). This parameter reflects the degree of path linearity. Looped paths have origin and endpoint near each other, so the NGDR will be close to 0, whereas a straight path will have an NGDR value close to 1. Both RCDI and NGDR can be used to identify tumbles at a given frame in a path, but their average value can also be plotted against the concentration of chemical species to generate dose–response curves describing the sensitivity of bacterial populations to the chemical agents applied (Lopez-de-Victoria *et al.*, 1995).

As for the measurement of velocities of phototactic algae and protozoa, the use of motion analysis has been confined to the measurement of their distribution. The measurements of velocity distribution do not require tracking of cells for a long period of time. For instance, in the Häder and Vogel program, the measurement of cell velocity is accomplished by storing only four consecutive frames in the frame grabber memory: this makes it possible to measure the velocities and their distribution in the sample, but it does not allow tracking their time course after the onset of the orientating (attractant or repellent) light. The latter should be possible using the set-up of Coltelli and Gualtieri (1990), which (to date) has only been used to measure the distribution of the velocities in samples to which the orientating stimulus has already been applied. The measurement of time required by single micro-organisms to orientate themselves at the onset of phototaxis has not yet been considered in motion analysis studies. This would be a result of great interest, but this measurement requires tracking many cells for long periods of time. To achieve this, it is necessary to reconstruct the cell paths, which can probably be done for algae using the same approach, and encountering the same difficulties as in the case of bacteria. In fact, since flagellated algae travel at a speed of 100 μm/s, if we measure the velocity in cell-body lengths per second, the values for flagellated algae are not so different from those of *H. salinarum*. The apparent velocity in pixels per second is about the same if the magnification of the microscope is set to yield the same apparent area for the two types of cells. The choice of a suitable microscope magnification is certainly important, and widely different values have been used. Takahashi (1992) reports an area of a few pixels (>4) per cell, whereas from the work of Coltelli and Gualtieri (1990) and Häder and Vogel (1992) areas ranging from 500 to 2000 can be inferred. The larger the cells appear on the video, the shorter the time they remain in the field, and the longer the time needed for analysis. In our experience (Lucia *et al.*, 1996) a value of 30–50 pixels per cell is suitable for tracking of *H. salinarum* cells for long periods of time, i.e. up to 60 s. On the other hand, this also depends on the way the cells move; for example, it is difficult to track cells going up and down, i.e. frequently moving out of focus, for a long time, as discussed previously. This behaviour depends on the species of micro-organism.

13.6 HIGH-SPEED VIDEO ANALYSIS OF FLAGELLA AND CILIA MOVEMENTS

Improved video analysis techniques can in principle be applied to study the motion of cilia and flagella, whose beating frequencies reach 90 Hz. Information on flagellar beating can also be obtained using simpler methods. In flagellated algae, the analysis of the light scattered by swimming cells allows measurement of the beating frequencies of the flagella (Angelini *et al.*, 1986). Using this method, it has been shown that the flagellar beating frequency increases during a response to a light flash.

More direct measurements of flagellar beating in unicellular algae have been obtained by placing a light microspot on the region swept by one or both flagella of a cell sucked into a micropipette. The use of this method has given relevant insights into both phototactic behaviour and photophobic responses. In particular, it has been shown that, during a periodic stimulation of *Haematococcus pluvialis*, the frequency of the *cis* flagellum lowers in the dark, whereas that of the *trans* flagellum increases, opposite changes occurring in the light (Sineshchekov, 1991). Similar results have previously been obtained by Smyth and Berg (1982) on *Chlamydomonas* sp. using a different technique.

These examples serve to illustrate the point that there is no need for more complicated techniques when the sole aim of the study is the measurement of flagellar beating frequency or its rotation direction. However, when the interest is in specific features of the motion of these organelles, such as the curvature along the flagellum, the wave shape, the inclination of the flagellum axis with respect to the cell body etc., then the development of adequate image analysis methods is required because a precise definition of flagellar movements can be achieved only by looking at the shape of the flagellum and its variation during the beating cycle. In past years, attempts in this direction were made by high-speed microcinematography. Microcinematography implies the use of a very high light intensity, but its major drawback is in fact that it is not directly compatible with computers, and thus it is not easily amenable to quantitative measurements of the geometrical features of a beating flagellum. Brokaw (1990) managed to interface a microcomputer to a digitizing camera which analysed frames from a 35 mm film acquired at 120–150 Hz. Using this instrumentation he was able to analyse the shape of sperm flagella.

In the past two years several papers have appeared aiming to apply video recording and image analysis techniques to study the motion of cilia and flagella (Ishijima, 1995 and references cited therein). Approaches to the study of movements of flagella and cilia by video analysis first require the use of cameras able to detect quite rapid movements. The beating frequency of flagella is in the range from 20 to 90 Hz and several images should be obtained within one period to allow the features of the motion (not merely its frequency) to be detected (temporal aliasing: Section 7.2). Furthermore, it should be emphasized that the image has to be collected in a short enough time to get a sharp picture of the moving object.

For those readers who are not familiar with the working principles of CCD cameras, reference to Chapter 3 of this book may be helpful. Although the design and construction may differ in several details, every CCD camera has a sensor consisting of an array of pixels. There are many ways to transfer an image and many possible configurations of the pixel array. A relevant feature common to these kind of devices is that charge transfer from an element to another takes about 0.1 μs. Thus, the overall time required to sequentially transfer an image of 512 × 512 pixels is about 30 ms, a value of the order of the reciprocal of the standard frequency of video images. The way the pixels of a CCD array are configured and the way the charge is transferred towards the output (or outputs) determines the maximum read-out rate of a CCD camera. To increase the number of frames per second a camera can generate, the number of output channels can be increased. An alternative way around this problem is to reduce the number of pixels sent to the output; e.g. if only a window of the CCD array is sent out, a higher repetition rate can be achieved. Video cameras at 120, 240 and 500 frames/s are already commercially available. When the field rate is a multiple of the conventional value (for instance, 60 Hz; see Section 2.3.3, Table 2.1), it is possible to use a conventional video cassette recorder (VCR) and record every image on a tape, e.g. storing four consecutive images collected at 240 Hz field rate in a single frame. It is also possible to request such a modification from the manufacturer (Ishijima, 1995).

The criteria for a CCD camera to be suitable for the study of rapid movements are: (i) high repetition rate, (ii) synchronous charging periods for all of the pixels of the image, and (iii) possibility of determining the exposure time independently of the repetition rate. The problem of shortening exposure times can be solved by working with a camera equipped with a mechanical shutter, by working under stroboscopic illumination, or by using flashes synchronized by the video triggers. In the last case, the sensor is illuminated for a time interval as short as the flash duration (for instance, 1 ms), thus allowing study of movements such as cilia and flagella beating at frequencies of up to 90 Hz.

13.7 GENERAL COMMENTS AND PERSPECTIVES

Let us now summarize what can be learned from a technical point of view from the discussion of the different attempts to use image analysis in the study of swimming micro-organisms. Several conclusions can be drawn, which are useful for evaluation of video digitizers and give insights into the perspectives of this technique of analysis.

A point that is certainly relevant is the speed of the communication between board and host computer. As discussed above, the most important contribution to the time required to collect and store the cell coordinates in the computer is the time spent reading the pixels. Recent systems used in image analysis are able to perform many sophisticated tasks (for instance, contrast enhancement, Fourier transform and filtering, template matching, etc.). Nevertheless, if frame-by-frame analysis has been chosen, the fundamental technical feature of a board being applied to the study of the motion of unicellular organisms is certainly that of the speed of the communication with the host computer.

There are distinct timing parameters that can be gathered from the board manuals, or measured through simple test programs written for this specific purpose: (i) the time required for a sequential reading, (ii) the time for a DMA transfer of data from the board to the computer memory and (iii) the time required to access a particular pixel, i.e. the time for random access. The latter is obviously longer than the previous two, because the selection of a pixel implies the setting of the x and y addresses, which are updated automatically during sequential reading or DMA transfer. However, random access allows use of the trick of 'stepwise reading' introduced by Häder and Vogel (1991), which seems to be a useful method. As shown by the data reported in Section 13.4.6, this method was useful up to a few years ago, but since then a dramatic improvement in the hardware characteristics of both boards and PCs has occurred. The computing speed of recent PCs has been improved greatly and structural changes in the communication between CPU and peripherals has reduced the time for data transfer enormously. For instance, it can be determined from the user manual of a recent image analysis board (Matrox MAGIC) that the time for transferring a 512×512 byte frame by DMA is about 15 ms. As shown in Table 13.1, the time for the actual calculations is also 10–15 ms, which means that frame-by-frame analysis can be performed directly in the computer memory without losing the time resolution of the video camera (40 ms in Europe, 33 ms in the USA).

Let us now consider the alternative approach, i.e. using the board memory to store the entire course of the experiment, as discussed in Section 13.4.2. The amount of memory and the presence of write masks in the board block scheme (Figure 13.4) allow storage of many different images in the board, using distinct frame grabbers and bit-planes to store different images. Most recent boards (for use of colour images) already have three distinct memories of 1024×1024, and the memory of frame grabbers is expected to increase in the next few years, following the same trend as computer memory, disc size, etc. It is therefore a reliable prediction that the use of hardware memorization will become easier and easier. Most probably, hardware memorization will be very helpful in the study of particular problems, such as, for instance, the dynamics of flagellar beating. The reconstruction of the shape of a beating flagellum is indeed an expensive task in terms of computing time (Brokaw, 1990). The required time resolution can be maintained by exploiting the board memory and postponing the analysis until the end of data acquisition. Hardware memorization will also be very useful in those cases where the repetition rate of conventional video systems is limiting. For example, the reversal of ciliary movement in ciliated protozoa, e.g. *Stentor* or *Blepharisma*, appears to occur within 10 ms, a time interval shorter than the time resolution of a conventional motion analysis system. Recent developments of rapid motion analysis systems applied to the study of flagellar motility in spermatozoa and unicellular organisms are of high relevance in this context.

ACKNOWLEDGEMENT

We are deeply indebted to Dr Piero Alberto Benedetti for his very helpful discussion on the technical features of CCD cameras.

REFERENCES

Angelini, F., Ascoli, C., Frediani, C. and Petracchi, D. (1986) Transient photoresponses of a phototactic microorganism, *Haematococcus pluvialis*, revealed by light scattering. *Biophys. J.* **50**: 929–36.

Armitage, J.P. (1992) Behavioural responses in bacteria. *Ann. Rev. Physiol.* 54: 683–714.

Ascoli, C., Barbi, M., Frediani, C. and Murè, A. (1978) Measurements of *Euglena* motion parameters by laser light scattering techniques. *Biophys. J.* **24**: 585–99.

Barghigiani, C., Colombetti, C., Franchini, B. and Lenci, F. (1979) Photobehavior of *Euglena gracilis*: action spectrum for the step-down photophobic response of individual cells. *Photochem. Photobiol.* **29**: 1015–9.

Berg, H.C. (1971) How to track bacteria. *Rev. Sci. Instrum.* **42**: 868–71.

Berg, H.C. and Brown, D.A. (1972) Chemotaxis of *Escherichia coli* analysed by three-dimensional tracking. *Nature* **239**: 500–4.

Blair, D.F. (1995) How bacteria sense and swim. *Ann. Rev. Microbiol.* **49**: 489–522.

Block, S.M., Segall, J.E. and Berg, H.C. (1983) Adaptation kinetics in bacterial chemotaxis. *J. Bacteriol.* **154**: 312–23.

Brokaw, C.J. (1990) Computerized analysis of flagellar motility by digitization and fitting of film images with straight segments of equal length. *Cell Motil. Cytoskel.* **17**: 309–16.

Cantatore, G., Ascoli, C., Colombetti, G. and Frediani, C. (1989) Doppler velocity measurements of phototactic response in flagellated algae. *Biosci. Rep.* **9**: 475–80.

Colombetti, G. and Petracchi, D. (1989) Photoresponse mechanisms in flagellated algae. *CRC Crit. Rev. Plant Sci.* **8**: 309–55.

Coltelli, P. and Gualtieri, P. (1990) A real-time, automatic velocity determination of moving objects in microscope images. *J. Comput.-Assist. Microsc.* **2**: 47–52.

Ekelund, N.G.A. (1990) Effects of UV-B radiation on growth and motility of four phytoplankton species. *Physiol. Plant.* **78**: 590–4.

Gerber, S., Biggs, A. and Häder, D.P. (1996) A polychromatic action spectrum for the inhibition of motility in the flagellate *Euglena gracilis*. *Acta Protozool.* **35**: 161–5.

Ghetti, F. (1991) Photoreception and photomovements in *Blepharisma japonicum*. In *Biophysics of Photoreceptors and Photomovements in Microorganisms* (Lenci, F., Ghetti, F., Colombetti, G., Häder, D.P. and Song, P.S., eds), NATO ASI Series, Series A: Life Sciences, vol. 211, pp. 257–65. Plenum Press: New York.

Gualtieri, P. and Coltelli, P. (1991) A real-time automated system for the analysis of moving images. *J. Comput.-Assist. Microsc.* **3**: 15–21.

Gualtieri, P. and Coltelli, P. (1992) Image analysis of motile flagellates. In *Image Analysis in Biology* (Häder, D.P., ed.), pp. 343–51: CRC Press: Boca Raton, FL.

Gualtieri, P., Colombetti, G. and Lenci, F. (1985) Automatic analysis of the motion of microorganisms. *J. Microsc.* **139**: 57–62.

Gualtieri, P., Ghetti, F., Passarelli, V. and Barsanti, L. (1988) Microorganism track reconstruction: an image processing approach. *Comput. Biol. Med.* **18**: 57–63.

Häder, D.P. (1988) Computer-assisted image analysis in biological sciences. *Proc. Indian Acad. Sci. (Plant Sci.)* **98**: 227–49.

Häder, D.P. (1991) Phototaxis and gravitaxis in *Euglena gracilis*. In *Biophysics of Photoreceptors and Photomovements in Microorganisms* (Lenci, F., Ghetti, F., Colombetti, G., Häder, D.P. and Song, P.S., eds), NATO ASI Series, Series A: Life Sciences, vol. 211, pp. 203–21. Plenum Press: New York.

Häder, DP. (ed.) (1992) *Image Analysis in Biology*. CRC Press: Boca Raton, FL.

Häder, D.P. and Lebert, M. (1985) Real time computer-controlled tracking of motile microorganisms. *Photochem. Photobiol.* **42**: 509–14.

Häder, D.P. and Vogel, K. (1991) Simultaneous tracking of flagellates in real time by image analysis. *J. Math. Biol.* **30**: 63–72.

Häder, D.P. and Vogel, K. (1992) Real-time tracking of microorganisms. In *Image Analysis in Biology* (Häder, D.P., ed.), pp. 289–313. CRC Press: Boca Raton, FL.

Häder, D.P., Colombetti, G., Lenci, F. and Quaglia, M. (1981) Phototaxis in the flagellates, *Euglena gracilis* and *Ochromona danica*. *Arch. Microbiol.* **130**: 78–82.

Ishijima, S. (1995) High-speed video microscopy of flagella and cilia. In *Methods in Cell Biology* (Dentler, W. and Witman, G., eds), vol. 47, pp. 239–43. Academic Press: San Diego, CA.

Kobayashi, S., Maeda, K. and Imae, Y. (1977) Apparatus for detecting rate and direction of rotation of tethered bacterial cells. *Rev. Sci. Instrum.* **48**: 407–10.

Krohs, U. (1994) Sensitivity of *Halobacterium salinarium* to attractant light stimuli does not change periodically. *FEBS Lett.*, **351**: 133–6.

Krohs, U. (1995) Damped oscillations in photosensory transduction of *Halobacterium salinarium* induced by repellent light stimuli. *J. Bacteriol.* **177**: 3067–70.

Kuo, S.C. and Koshland, D.E. (1989) Multiple kinetic states for the flagellar motor switch. *J. Bacteriol.* **171**: 6279–87.

Lapidus, I.R., Welch, M. and Eisenbach, M. (1988) Pausing of flagellar rotation is a component of bacterial motility and chemotaxis. *J. Bacteriol.* **170**: 3627–31.

Lindes, D.A., Diehn, B. and Tollin, G. (1965) Phototaxigraph: recording instrument for determination of rate of response of phototactic microorganisms to light of controlled intensity and wavelength. *Rev. Sci. Instrum.* **36**: 1721–24.

Lucia, S., Ascoli, C. and Petracchi, D. (1992) Photobehavior of *Halobacterium halobium*: sinusoidal stimulation and a suppression effect of responses to flashes. *Biophys. J.* **61**: 1529–39.

Lucia, S., Ferraro, M., Cercignani, G. and Petracchi, D. (1996) Effects of sequential stimuli on *Halobacterium salinarium* photobehavior. *Biophys. J.* **71**: 1554–62.

Lucia, S., Cercignani, G. and Petracchi, D. (1997) Photosensory transduction in *Halobacterium salinarium*: evidence for a non linear network of cross-talking pathways. *Biochim. Biophys. Acta* **1334**: 5–8.

Lopez-de-Victoria, G., Zimmer-Faust, R.K., Lowell, C.R. (1995) Computer-assisted videomotion analysis: a powerful technique for investigating motility and chemotaxis. *J. Microbiol. Meth.* **23**: 329–41.

Marwan, W. and Oesterhelt, D. (1990) Quantitation of photochromism of sensory rhodopsin-I by computerized tracking of *Halobacterium halobium* cells. *J. Molec. Biol.* **215**: 277–85.

Marwan, W., Alam, M. and Oesterhelt, D. (1991) Rotation and switching of the flagellar motor assembly in *Halobacterium halobium*. *J. Bacteriol.* **173**: 1971–7.

McCain, A., Amici, L.A. and Spudich, J.L. (1987) Kinetically resolved states of the flagellar motor switch and modulation of the switch by sensory rhodopsin I. *J. Bacteriol.* **169**: 4750–8.

Noble, P.B. and Levine, M.D. (1986) *Computer-assisted Analyses of Cell Locomotion and Chemotaxis*. CRC Press: Boca Raton, FL.

Petracchi, D., Lucia, S. and Cercignani, G. (1994) Photobehaviour of *Halobacterium halobium*: proposed models for signal transduction and motor switching. *J. Photochem. Photobiol. B* **24**: 75–99.

Rudolph, J. and Oesterhelt, D. (1996) Deletion analysis of the Che operon in the archaeon *Halobacterium salinarium*. *J. Molec. Biol.* **258**: 548–54.

Schimz, A. and Hildebrand, E. (1987) Effects of cGMP, calcium and reversible methylation on sensory signal processing in halobacteria. *Biochim. Biophys. Acta* **923**: 222–32.

Sgarbossa, A., Lucia, S., Lenci, F., Gioffrè, D., Ghetti, F. and Checcucci, G. (1995) Effects of UV-B radiation on motility and photoresponsiveness of the coloured ciliate *Blepharisma japonicum*. *Photochem. Photobiol. B* **27**: 243–9.

Sineshchekov, O.A. (1991) Electrophysiology of photomovements in flagellated algae. In *Biophysics of Photoreceptors and Photomovements in Microorganisms* (Lenci, F., Ghetti, F., Colombetti, G., Häder, D.P. and Song, P.S., eds), NATO ASI Series, Series A: Life Sciences, vol. 211, pp. 191–202. Plenum Press: New York.

Smyth, R.D. and Berg, H.C. (1982) Change in flagellar beat frequency of *Chlamydomonas* in response to light. *Cell Motil. Suppl.* **1**: 211–15.

Sundberg, S.A., Alam, M. and Spudich, J.L. (1986) Excitation signal processing times in *Halobacterium halobium* phototaxis. *Biophys. J.* **50**: 895–900.

Takahashi, T. (1992) Automated measurement of movement responses in halobacteria. In *Image Analysis in Biology* (Häder DP, ed.), pp. 315–28. CRC Press: Boca Raton, FL.

Takahashi, T. and Kobatake, Y. (1982) Computer-linked method for measurement of the reversal frequency in phototaxis of *Halobacterium halobium*. *Cell Struct. Funct.* **7**: 183–92.

Takahashi, T. and Watanabe, M. (1993) Photosynthesis modulates the sign of phototaxis of wild-type *Chlamydomonas reinhardtii*. Effects of red background illumination and 3-(3′,4′-dichlorophenyl)-1,1-dimethylurea. *FEBS Lett.* **336**: 516–20.

Takahashi, T., Yoshihara, K., Watanabe, M., Kubota, M., Johnson, R., Derguini, F. and Nakanishi, K. (1991) Photoisomerization of retinal at 13-ene is important for phototaxis of *Chlamydomonas reinhardtii*: simultaneous measurements of phototactic and photophobic responses. *Biochem. Biophys. Res. Commun.* **178**: 1273–9.

Part III

Multicellular Organisms and Consortia

14

Quantitative Morphology of Filamentous Micro-organisms

Anders Spohr, Teit Agger, Morten Carlsen and Jens Nielsen
Technical University of Denmark, Lyngby, Denmark

14.1 INTRODUCTION

Filamentous fungi and Actinomycetes form two groups of industrially very important micro-organisms. Due to their ability to secrete large amounts of protein, filamentous fungi are used extensively for the production of industrial enzymes – a continuously increasing market. They are also very important in the production of various secondary metabolites including β-lactams, which constitute by far the largest group of antibiotics. Actinomycetes (mostly species of *Streptomyces*) are mainly of industrial interest for their production of secondary metabolites, which include both classic antibiotics in the group of β-lactams and new pharmaceuticals. Despite the large differences between these two groups of micro-organisms, they have many similarities with respect to their growth mechanisms. They both grow filamentously, i.e. as multicellular structures called hyphal elements, and in submerged cultures they may both form pellets, which are spherical agglomerates of one or more hyphal elements. Methods for quantification of the morphology of these organisms are therefore similar. Furthermore, the growth kinetics for different micro-organisms can often be expressed using the same mathematical equations. Finally, the effect of environmental conditions on the development in their morphology is in many cases similar. For example, the effect of shear forces on fragmentation of both hyphal elements and pellets is the same for filamentous fungi and Actinomycetes.

The morphology of filamentous micro-organisms has been classified into microscopic and macroscopic morphology (Nielsen, 1992), where the microscopic morphology characterises single hyphal elements or hyphal populations. The properties used to describe the microscopic morphology are typically the total hyphal length, the number of tips, the diameter of the hyphae and the diameter of spores. The macroscopic morphology describes the shape, structure and size of hyphal agglomerates or pellets. In this chapter we describe the morphology at both levels, after which we give an introduction to cellular differentiation, which is believed to have an influence on the product formation of filamentous

Digital Image Analysis of Microbes: Imaging, Morphometry, Fluorometry and Motility Techniques and Applications. Edited by M.H.F. Wilkinson and F. Schut.

micro-organisms. The focus in the description is on *quantitative morphology*, i.e. the combination of measurements of the morphology by image analysis and mathematical modelling of the mechanisms determining the developing morphology. For this reason, fractal analysis of hyphal growth is not included in the discussion, because of the difficulties in including it in data models. Readers interested in fractal analysis of microscopic images can refer to Donelly *et al.* (1995) and Jones *et al.* (1993, 1995).

14.2 MICROSCOPIC MORPHOLOGY

A hyphal element arises from the outgrowth of a single spore. Upon germination the growth of the spore rapidly becomes polarized, resulting in the formation of a germ tube which extends by growth at the tip. Eventually the germ tube develops into a hypha and when the hypha has exceeded a certain length, new tips (or branches) are formed, resulting in a hyphal element consisting of a branched network of hyphae (Figure 14.1). The process may continue until a densely branched hyphal element is formed. In a submerged culture several hyphal elements may agglomerate to form pellets, or a single hyphal element may develop into a pellet (Figure 14.1A).

14.2.1 Quantification of microscopic morphology

Quantification of microscopic morphology has previously been performed manually, e.g. from photographs (Trinci, 1970). These measurements were very labour intensive, and the numbers of measurements were relatively low, thus reducing the accuracy of the results. The development of computers with frame grabbers enabled the measurements to be performed semi-automatically, i.e. the operator had to delineate the structures of interest. Recently, the increase in computer performance has allowed automation of many of the tasks associated with

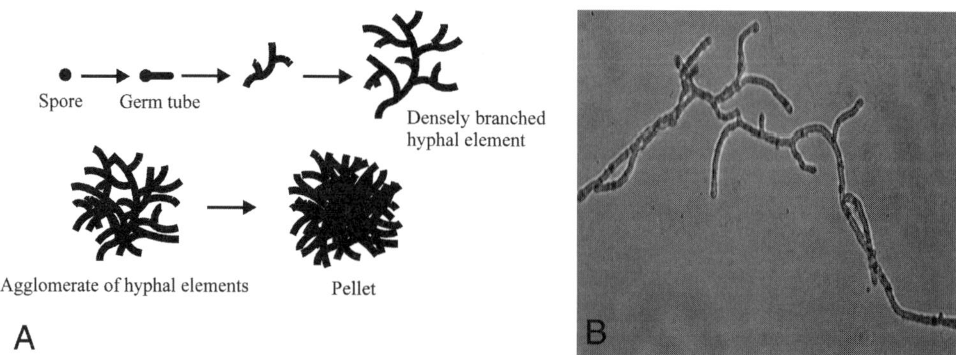

Figure 14.1 Morphology of filamentous fungi. (A) The different morphologies existing for a filamentous microorganism. (b) Structure of a hyphal element of *Aspergillus oryzae*, 12 hours old. Magnification 200×.

image analysis, and today many types of measurements can be performed completely automatically (Thomas and Paul, 1996).

Quantification of the morphology of individual hyphal elements and the population distribution in submerged cultures using image analysis has been a major research topic during the past five years, and several results have been published (Paul *et al.*, 1993; Durant *et al.*, 1994; Nielsen and Krabben, 1995; Vanhoutte *et al.*, 1995) and reviewed (Thomas, 1992; Nielsen, 1996). The measurements have been performed by taking samples from well-controlled bioreactors, which are then fixed using dyes such as lactophenol. For image analysis, a drop of a sample is transferred to a microscopic slide, overlaid by a cover slip and transferred to the microscope, where quantification takes place. For hyphal elements the total average length and the average number of tips are determined, and models describing the growth can be proposed based on these results.

14.2.2 Measurements of hyphal length and number of tips using image analysis

Before beginning processing and measurement of images of the micro-organisms in the microscopic field of view, various parameters must be specified. These are application-specific, e.g. to optimize the acquired image the brightness of the microscope lamp and the contrast and gain of the camera must be set. The image-processing phase of the program includes acquisition of the picture of interest. The picture is converted into a digital image, which is segmented by thresholding, i.e. all grey levels above a threshold value are treated as being of interest, and a masking binary image is defined, showing the selected objects. Since the images frequently contain objects other than micro-organisms, such as dust or media particles, further processing is necessary. A circularity test has been used by Packer and Thomas (1990) and Tucker *et al.* (1992) to eliminate these interfering objects since they are more spherical than typical filamentous micro-organisms. Objects with a circularity index smaller than a preset value are removed from the image. Because some entangled mycelia can also give roughly circular structures, it is also a criterion that a 'circular' object being considered for removal does not contain holes, i.e. regions of background within the object. Each object in the resulting image is then skeletonized, i.e. the pixels medial between the edges of the object are found (Section 1.6.1f). The skeletons are used to determine the length of the hyphal elements. Very small branches that appear due to debris touching the hyphal elements are removed by a pruning routine, i.e. removal of a predefined number of pixels from the tips of the skeleton. Accurate determination of the length of the hyphal structures requires more than just counting the number of pixels, since on a square pixel grid the distance between pixels that touch diagonally is greater by a factor of $\sqrt{2}$ than the distance between pixels that touch orthogonally. Counting the touching pairs as a function of direction and calculating the dimension accordingly produces a slight overestimate of the length. For lines that run in all directions across the square pixel grid Smeulders and Beckers (1989) found a correlation minimizing the mean error to 2.5% for the actual length. This is done by weighting the diagonal neighbours and the orthogonal neighbours differently. Depending on the image analyser used, these corrections must be implemented in order to obtain the true length of the structures of interest. It is possible to derive the number of tips from the skeleton, by counting the number of pixels in a skeleton with only one other neighbour, e.g. using mathematical morphology. Packer and Thomas (1990) and Tucker *et al.* (1992) estimated the number of tips by counting the number of branching points, i.e. the number of pixels in a skeleton with three or more neighbour pixels. The length and the number of tips of the hyphal elements are only two of several important variables in submerged cultures, since mycelium growing submerged also consists of clumps and in special cases pellets (for a description see Section 14.3). According to Packer and Thomas (1990) clumps can be identified because they contain holes within the binary image of the objects. The fractions of mycelium growing as freely dispersed hyphal elements and clumps can be quantified based on this identification.

In order to understand the basic physiology and hence the morphology of filamentous fungi, some kinetic expressions must be derived based on the image analysis data. Due to the growth mechanisms, i.e. growth like a binary tree, the most important variables describing the growth are the rates of tip extension and branching. Population average values of these two variables can only be obtained if every single element is measured. For *Penicillium chrysogenum* it has been shown that clumps appear late in a cultivation, and they consist mainly of large hyphal elements (Nielsen and Krabben, 1995). As a consequence of this it is possible to get a good estimate of average tip extension and branching rates early in a batch cultivation, since the distribution of the individual hyphal elements is shifted towards smaller values later in the cultivation.

Paul *et al.* (1993) used the circularity index for characterization of germination of fungal spores, since non-germinated spores are almost spherical. From a binary image containing germinated and non-germinated spores as well as debris the fraction of germinated spores has been extracted by a series of erosions and dilations. The first step is an erosion removing only two pixels in order to remove smaller debris. After the erosion the image is rebuilt to its original form, excluding those small extraneous objects. The large debris is removed by an erosion with a larger kernel size, where only the cores of the large objects of interest remain. After rebuilding the objects to their original size by a dilation routine, the resulting image is subtracted from the earlier image, leaving only the spores and germ tubes. The spores and germ tubes are then separated by a preset circularity parameter, using the spherical appearance of the non-germinated spores.

14.2.3 Models describing growth of populations of hyphal elements

14.2.3a Germination

Initiation of spore germination of filamentous fungi and Actinomycetes appears to be identical when observed under the microscope. Germination initiates when spores are placed in conditions favourable for growth, and the process is defined by a series of irreversible events, leading to the emergence of a germ tube (Griffin, 1994). A spore is considered to be fully germinated when the emerging germ tube has a length equal to or larger than one-half the largest dimension of the spore (Paul *et al.*, 1993). For filamentous fungi it has been shown that prior to germination the spore increases in size and there is a rapid increase in the content of the active cell material, e.g. the number of mitochondria and the size of the endoplasmic reticulum (Smith, 1975; Adunola *et al.*, 1976). During spore germination three basic structural changes can be recognized (Paul *et al.*, 1993): spore swelling, germ tube emergence and germ tube elongation. Initially, i.e. during spore swelling, the spore grows as a sphere, and new cell material is deposited uniformly over the entire inner surface of the spore.

Germination of spores in a submerged culture is non-synchronous, and the germination frequency is determined by many factors, such as the spore age and the medium composition. However, the overall process of spore germination may be quantified by five parameters (Nielsen and Krabben, 1995): (i) the fraction of viable spores (y_{viab}), (ii) the time t_{s} at which germination starts, (iii) the time t_{f} at which no more spores germinate, and (iv and v) the two parameters α and β of the B-distribution, which determine the steepness and skew, respectively, of the germination frequency. The parameter y_{viab} is determined as the fraction of germinated spores after no additional germination can be observed. With these parameters the germination frequency takes the form:

$$g(t) = e_{\mathrm{spore}} \cdot y_{\mathrm{viab}} \cdot \begin{cases} \dfrac{1}{B(\alpha, \beta)} \cdot \bar{t}^{\alpha - 1}(1 - \bar{t})^{\beta - 1} & \text{if } t_{\mathrm{s}} \leqslant t \leqslant t_{\mathrm{f}} \\ 0 & \text{otherwise} \end{cases} \tag{14.1}$$

with:

$$\bar{t} = \frac{t - t_{\mathrm{s}}}{t_{\mathrm{f}} - t_{\mathrm{s}}}$$

where e_{spore} is the concentration of spores after inoculation. The two form parameters α and β are estimated from a least squares fit, and are usually constant for individual species. For *P. chrysogenum* they were both found to be 3 (Figure 14.2 and Table 14.1) (Nielsen and Krabben, 1995).

The results in Table 14.1 clearly show that the germination frequency is determined by the spore quality and the environmental conditions.

14.2.3b Tip extension

A characteristic of filamentous micro-organisms is localized growth at the hyphal tips. This phenomenon was recognized as early as the 1890s. At that time it was correctly speculated that substrate is taken up throughout the hyphal element, processed to cell metabolites and sent toward the hyphal tip (Reinhardt, 1892). Despite the extensive research in filamentous growth since this early discovery, the mechanisms of tip extension have not yet been fully elucidated. It has, however, been shown for filamentous fungi that cells distal from the hyphal tip have the same rate of RNA and protein formation as new cells close to the hyphal tip (Zalokar, 1959). An important contribution to the elucidation of the actual mechanism of tip extension for filamentous fungi has been made by Bartnicki-Garcia (1973), who proposed that cell wall material, lytic enzymes and synthetic enzymes are transported to the tip section in vesicles. The mechanisms of vesicle supply to the

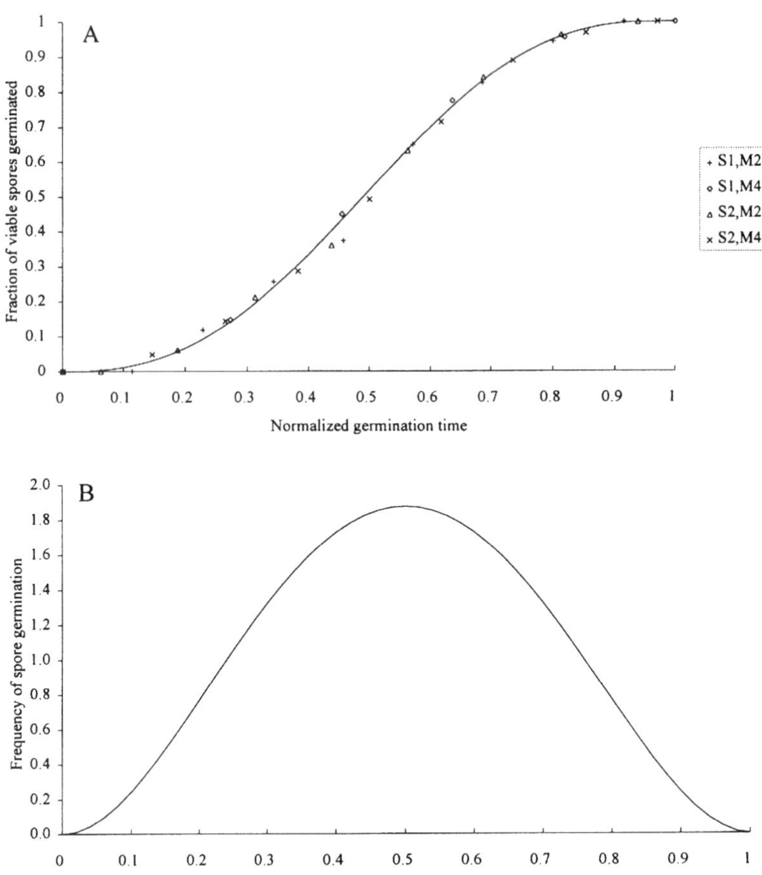

Figure 14.2 Germination of *Penicillium chrysogenum* spores in submerged batch cultivations. (A) The normalized fraction of germinated spores as a function of the normalized germination time $\bar{t}$ of equation (14.1). The parameters y_{viab}, t_s and t_f are listed in Table 14.1. The experimental data are taken from Paul (1992). S1, fresh spores; S2, old spores; M2, defined medium; M4, complex medium. The curve represents the accumulated frequency function $g(t)/(e_{\text{spore}} \cdot y_{\text{viab}})$ for spore germination given by equation (14.1) with the parameters $\alpha = \beta = 3$. (B) The frequency of spore germination $g(t)/(e_{\text{spore}} \cdot y_{\text{viab}})$.

Table 14.1 Parameters of the spore germination frequency function given in equation (14.1) for four different inoculum conditions reported by Paul (1992).

Experiment	y_{viab}	t_s (h)	t_f (h)	Comments
S1,M2	0.51	4	39	Fresh spores, defined medium
S1,M4	0.89	2	24	Fresh spores, complex medium
S2,M2	0.33	6	38	Old spores, defined medium
S2,M4	0.63	3	37	Old spores, complex medium

cell membrane at the hyphal tip are thought to be governed by a vesicle supply centre (VSC) from which vesicles are transported at an equal rate in all directions (Bartnicki-Garcia, 1990). The idea of a VSC is able to describe both spherical growth of spores and tip extension, if the VSC is displaced along the hypha with a fixed velocity. The final shape of the hyphal tip is determined by the velocity of the VSC and the number of vesicles released from the VSC per second. The VSC was shown to coincide with the *spitzenkörper* (Bartnicki-Garcia, 1990), which had long been believed to be a distribution centre for vesicles (Grove, 1978).

A similar hypothesis does not hold for Actinomycetes since they do not exhibit internal differentiation. In these organisms transport of building blocks needed for tip extension by molecular diffusion is, however, believed to be an important factor in the control of tip extension (Yang *et al.*, 1992a). In a submerged culture it is not possible to determine the kinetics of the single branches and an average population view is therefore required. By image analysis the average total hyphal length and the average number of tips can be quantified as a function of time. The results of such measurements for *Aspergillus oryzae* are shown in Figure 14.3A. Using these data the tip extension rate and the branching frequency can be estimated. From Figure 14.3A one can conclude that the total hyphal length increases exponentially, with a specific growth rate μ, i.e.

$$\frac{dl_{t,av}}{dt} = \mu \cdot l_{t,av} \tag{14.2}$$

Here $l_{t,av}$ is the average hyphal length for a population. The average tip extension rate q_{tip} can be described as the rate of increase in the average total hyphal length divided by the average number of tips. Combining this with equation (14.2) gives:

$$q_{tip}(l_{t,av}, \mathbf{z}) = \frac{dl_{t,av}}{dt} \cdot \frac{1}{n_{av}} = \mu \cdot \frac{l_{t,av}}{n_{av}} \tag{14.3}$$

where $\mathbf{z}$ a vector specifying the environmental conditions. The average tip extension rate can then be calculated from measurements of the average total hyphal length and number of tips using equation (14.3).

From analysis of the average tip extension kinetics of *P. chrysogenum*, *A. oryzae* and *Streptomyces anulatus* in submerged cultures, the following kinetic equation has been proposed for the dependence of the average tip extension rate on the average total hyphal length (see Figure 14.3B):

$$q_{tip}(l_{t,av}, \mathbf{z}) = k_{tip,av}(\mathbf{z}) \cdot \frac{l_{t,av}}{l_{t,av} + K_t} \tag{14.4}$$

Here $k_{tip,av}(\mathbf{z})$ is the average maximum tip extension for a population, and K_t is an average saturation constant for the population. The morphological parameters have been estimated using a Haynes plot and the results for the three strains are summarized in Table 14.2.

Equation (14.4) is derived empirically from measurements of average properties in submerged cultures, and it therefore does not necessarily give information about

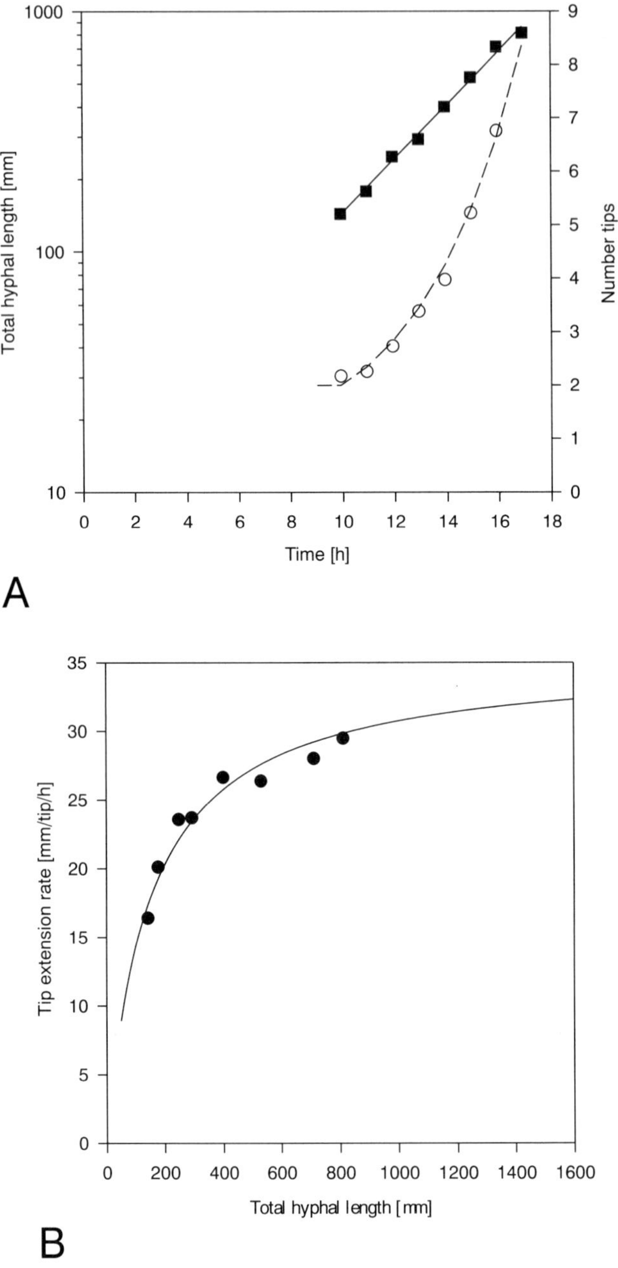

Figure 14.3 Morphological data for a wild-type strain of *Aspergillus oryzae*. (A) Profile of the average total hyphal length, and the number of tips as a function of time for *Aspergillus oryzae* during a batch cultivation on defined medium (Spohr *et al.*, 1997). The solid line represents exponential growth with a specific growth rate of $\mu = 0.26\ h^{-1}$. The broken line is model simulation using the kinetics given in equation (14.5) and $k_{bran} = 0.0023$ tip/μm/h. (B) The average tip extension rate as a function of the average total hyphal length. The line reresents the kinetics given by equation (14.6) with the parameters $k_{tip} = 35\ \mu$m/tip/h and $K_t = 148\ \mu$m.

Table 14.2 Morphological parameters for different filamentous microorganisms.

Strain	μ_{max} (h^{-1})	k_{tip} (μm/tip/h)	k_t (μm)	k_{bran} (tip/μm/h)	d_h (μm)	Reference
Streptomyces anulatus JH801	0.20	3.2	11	0.0164	0.6	Holmalahti *et al.* (1998)
Penicillium chrysogenum P8	0.16	10	60	0.0044	4	Nielsen and Krabben (1995)
Aspergillus oryzae A1560	0.27	35	148	0.0023	3.2	Spohr *et al.* (1997)

the underlying mechanisms. However, as mentioned above, it has been confirmed for several species of filamentous micro-organisms with different internal differentiation, and it may therefore be useful in average property models (Krabben and Nielsen, 1997), where the aim is to extract key kinetic parameters of growth from experimental data. Due to its empirical nature, both parameters are functions of the environmental conditions, e.g. both k_{tip} and K_t differ with growth on different media (Nielsen and Krabben, 1995).

14.2.3c *Branching*

Branching is the mechanism that allows exponential growth of filamentous micro-organisms, and together with tip extension it determines the overall specific growth rate and the morphology of freely dispersed hyphal elements. The actual mechanism of branching is poorly understood. A simple mechanistic explanation has been given by Trinci (1978), who suggested that branching occurs at positions where there is a high concentration of vesicles, e.g. at septa where reduced cytoplasmic streaming may result in an increase in the local concentration of vesicles. In the filamentous fungus *Geotrichum candidum*, for example, more than 70% of the branch points are in close proximity of septa. In *Aspergillus nidulans* branches are, however, distributed much more evenly, but this difference has been explained by a difference in the size of the septal pores in the two species (Trinci, 1984).

Since little is known about the branching mechanism, this process is often viewed as a more or less random process, for which the frequency is proportional to the total hyphal length (Yang *et al.*, 1992b; Lejeune *et al.*, 1995; Nielsen and Krabben, 1995), i.e.:

$$q_{bran}(l_t, \mathbf{z}) = \frac{dn_{av}}{dt} = \begin{cases} 0 & \text{if } l_t < l_{bran} \\ k_{bran}(\mathbf{z}) \cdot l_t & \text{if } l_t \geqslant l_{bran} \end{cases} \tag{14.5}$$

where q_{bran} is the average branching rate, k_{bran} is the branching constant [tips $(\mu m\ h)^{-1}$] and l_{bran} is a parameter (μm) which determines the lag phase for initiation of branching.

Since $l_{t,av}$ increases exponentially, integration of (14.5) from (l_{bran}, n_0) to $(l_{t,av}, n_{av})$ gives:

$$n_{av} = n_0 + \frac{k_{bran}}{\mu} \cdot (l_{t,av} - l_{bran}) \tag{14.6}$$

From equation (14.6), k_{bran} can be estimated from simple linear regression of n_{av} plotted as a function of $l_{t,av}$. It is important to notice that the parameter $k_{bran}(\mathbf{z})$ is a function of the environmental conditions, e.g. the medium composition. For *A. oryzae*, a computer simulation of equation (14.5) is illustrated in Figure 14.3A. The branching kinetics given in equation (14.5) has been used to describe branching for several micro-organisms, and the specific branching constants obtained are summarized in Table 14.2.

Comparison of the three strains with respect to morphology shows that *S. anulatus* is much smaller than the filamentous fungi. The diameter and the maximum tip extension rate are 5–10 times smaller, whereas the specific branching frequency is significantly higher. This indicates that *S. anulatus* is a more branched micro-organism than *P. chrysogenum* and *A. oryzae*. Despite the empirical nature of the simple expressions (14.4) and (14.5), they are valuable for extracting a few kinetic parameters that supply the investigator with valuable information about the growth kinetics.

Generally, the models described above give a good description of the average properties of the morphology, but the morphology of the average level may be quite different from that at the individual element level. It is only when the kinetics of individual elements are known that our understanding of the underlying mechanisms behind growth of filamentous micro-organisms will increase. Since it is not possible to control the culture conditions on solid media (the nutrient concentrations change during growth) surface cultures cannot provide the necessary information on growth kinetics. However, with novel techniques it is now possible to perform on-line image analysis of individual hyphal elements growing submerged (Reichl *et al.*, 1992a). In the following section we describe such a technique.

14.2.4 Experimental set-up using a flow-through cell

An experimental set-up for on-line image analysis of individual hyphal elements can be based on a small flow-through cell placed under a modified phase-contrast microscope with a scanning table (Reichert, Austria). A charge-coupled device (CCD) camera (Bischke CCD-5230P, Neumünster, Germany) was mounted on top of the microscope and connected to an image analyser (Quantimet 600, Leica Cambridge Ltd, Cambridge, UK) through a Panasonic NV-HS800EC video recorder. The set-up is illustrated in Figure 14.4. It was used to take series of images (every 10th min) of developing hyphal elements at magnifications of 40, 20, 10, 6 and 4, depending on the size of the elements. The pictures obtained were stored on an optical disc (One Technology, Karlsruhe, Germany) and the morphology was characterized by quantifying the images after the experiment.

Because the growth is followed using two-dimensional microscopy and image analysis, it is important to have growth in only two dimensions, i.e. the hyphal elements should not grow out of the plane of focus. Reichl *et al.* (1992a) showed that growth out of the focal plane constitutes a real problem in reliability using flow-through cells. It is therefore necessary to work with a chamber with a minimal height. The flow-through cell used in this study consists of a base unit of 102 mm × 49 mm made of stainless steel. Two steel tubes (1 mm in diameter) are built into the base unit to ensure flow of medium to the chamber (see Figure 14.4). A plastic cover slip (26 mm × 76 mm) with two transverse slits, used to ensure an evenly distributed flow of medium, is fixed in the bottom of the base unit. A second microscopic slide (26 mm × 76 mm) is used as a top window, and can be removed for cleaning of the inner chamber. Sealing is obtained by a thin Parafilm® membrane (American National Can, Neenah, WI). A stainless steel plate is fixed on the top of the base unit, resulting in compression of the chamber (see Figure 14.4). The height of the inner chamber becomes approximately 40 μm and its volume 79 μl. The construction of the chamber ensures an identical chamber size in each experiment.

The scanning table enables monitoring of 20–30 hyphal elements positioned at different places in the chamber during each experiment. The flow-through cell, microscope and substrate reservoir are placed in a thermostatic cupboard to ensure a constant temperature of 30 °C during the experiments. For sterilization, the flow-through cell is filled with ethanol (70%) and incubated at 30 °C for 2 h. The chamber is washed with sterile water followed by addition of 4 ml 0.1% poly-D-lysine (Sigma, MW > 300 000 kD) to facilitate fixation of spores. One millilitre of a spore suspension (10^3 spores/ml) is injected into the chamber. Finally, the flow-through cell is mounted on the microscope and connected to a reservoir via tubes. The medium in the reservoir is constantly aerated using a magnetic stirrer. A peristaltic pump is used to pump fresh medium through the cell at a constant flow rate of 0.025 ml/min.

14.2.5 Image analysis using a flow-through cell

Figure 14.5 shows a flow chart illustrating the procedure and the image transformation used. The routine starts by acquiring a picture. An 'enhance' function is then applied – this transform improves the definition of the edges by setting each pixel to the local minimum or maximum value of its neighbouring

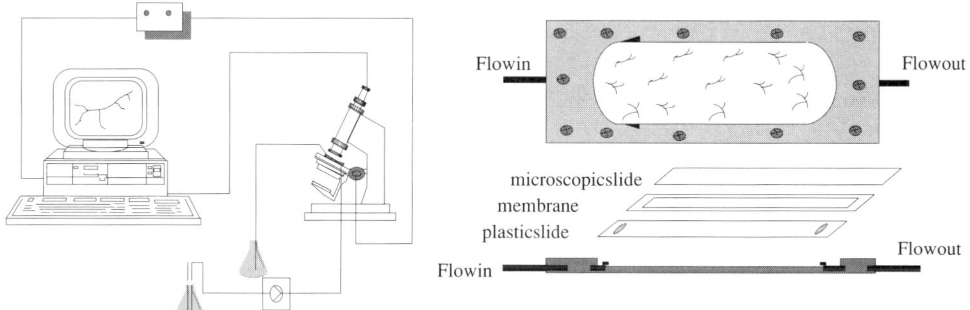

Figure 14.4 System for on-line image analysis. Left: the system consists of an image analyser (Quantimet Q600, Leica Ltd.), a Reichert microscope with an automated stage, and a flow-through chamber connected to a pump and a medium reservoir. The automated stage enables analysis of multiple hyphal elements during the same experiment. Right: construction of the flow-through chamber. The Figure illustrates the different parts of the flow-through cell and the chamber seen from the top.

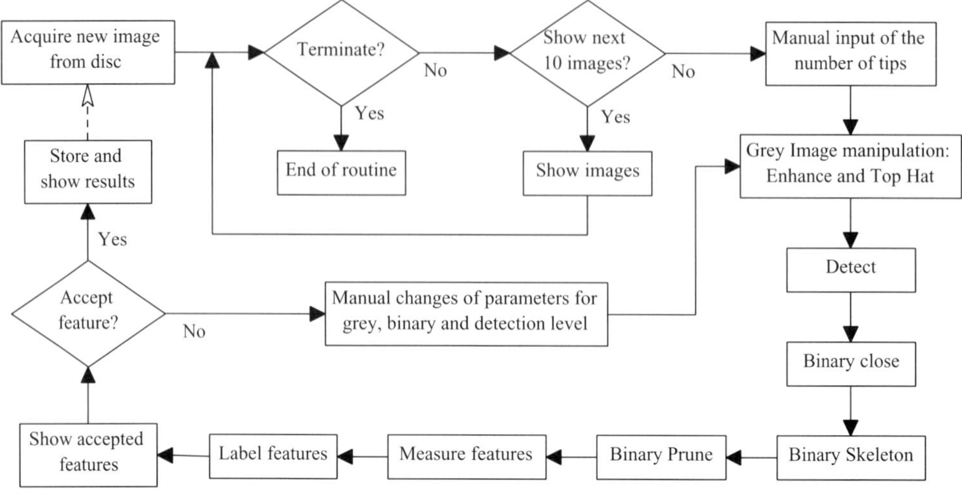

Figure 14.5 Flow chart of routine for measurements on hyphal elements. In the automatic measurements it is not necessary to determine the number of tips manually.

pixels (whichever is closer in value; see Section 6.7.2). This function is followed by a black top-hat transformation (Bright and Steel, 1987), where the background is closed and the original picture is subtracted (Sections 1.6.1c and 6.4.2). The black details in the original picture, i.e. the hyphal elements, appear as very clear white areas. The picture is segmented by thresholding (the 'detect' function in QUIPS), resulting in a binary image containing the hyphal elements and some debris. The threshold value used for segmentation can be adjusted during the measurements. The binary picture is transformed by a series of 'amend' functions in order to enhance the hyphal structure. For estimation of the hyphal length a routine similar to that of Packer and Thomas (1990) is used. Each object in the resulting image is skeletonized followed by pruning of three pixels in order to avoid small artificial branches, before the total hyphal length and the number of tips are measured. The measurements are performed by a feature measurement routine, which measures and labels each 'countable' object in the image. The picture contains objects other than micro-organisms, such as dust and media particles, and these objects will also be labelled. The label is allocated to the feature count point (X-FCP and Y-FCP), defined as the coordinate of the last detected pixel of the object (which is the pixel located in the lower right end of the object when using a two-dimensional system with an origin in the top left-hand corner of the image). The feature count point is used to identify the hyphal element of interest automatically, since the hyphal spore is placed at a fixed position inside the chamber, and because the scanning table returns to the same position each time, to acquire an image of the hyphal element developing in the proximity of this spore. Since the hyphal element grows, the feature count point might change due to a tip growing towards the lower right corner of the image. In order to account for this movement, the feature count point of interest must lie within a square of 40×40 pixels around the feature count point of the hyphal element in the previous picture, in order for the routine to proceed.

Since a typical branch does not grow more than approximately 10–20 pixels from one picture to the next, depending on the growth conditions, the hyphal element of interest can be detected automatically. Changing parameters, e.g. magnification, light intensity etc., cause the skeleton to break up into smaller parts, due to holes in the segmented image. This demands changes in the routine parameters, e.g. changing the threshold value for segmenting the image, introducing additional dilation steps etc. The automatic routine has been implemented with an 'accept' criterion which counters these problems. If the length of the hyphal element of interest is smaller than 80% of the same element in the previous image, no objects are accepted and the routine stops. After changing the image routine parameters, the measurements can be resumed by entering the label number of the hyphal element of interest.

During the growth experiment the structure of the hyphal elements becomes more complex, i.e. many branches are formed. The increased complexity leads to problems measuring the number of tips accurately because the tips might be hidden behind other parts of the hyphal structure. Sensitivity analysis of the model proposed by Nielsen and Krabben (1995) (equations (14.5–14.9)) shows that accurate tip estimations are necessary. To obtain the desired accuracy, a semi-automatic routine is used, where the number of tips must be entered manually. Figure 14.6 illustrates the same hyphal element after different image transformations used during the routine.

An automatic routine for measurements of individual hyphae has also been developed. The routine is identical to the one just described, except that an image with a white line positioned at the base of the new developing hypha is subtracted from the original image. This separates the hypha from the rest of the hyphal element, and thereby enables automatic measurements of single hypha.

14.2.6 Germination of single spores

The germination of *A. oryzae* has been studied using the growth chamber and the routine described above, and Figure 14.7 shows the circularity index and the estimated spore volume during swelling, together with the subsequent emergence and growth of the germ tube, all as functions of time. The circularity index C is defined as in Section 8.3.1:

$$C = \frac{P^2}{4 \cdot \pi \cdot A_i} \tag{14.7}$$

where P is the perimeter, and A_i is the measured projected area of the spores. The observed spore volume is determined from the measured projected area of the spore:

$$V_{\text{spore}} = \frac{1}{3 \cdot \sqrt{4 \cdot \pi}} \cdot A_i^{\frac{3}{2}} \tag{14.8}$$

Using the observed area, the swelling of spores has been correlated to the mechanisms of spore swelling proposed by Ekundayo and Carlile (1964) and by

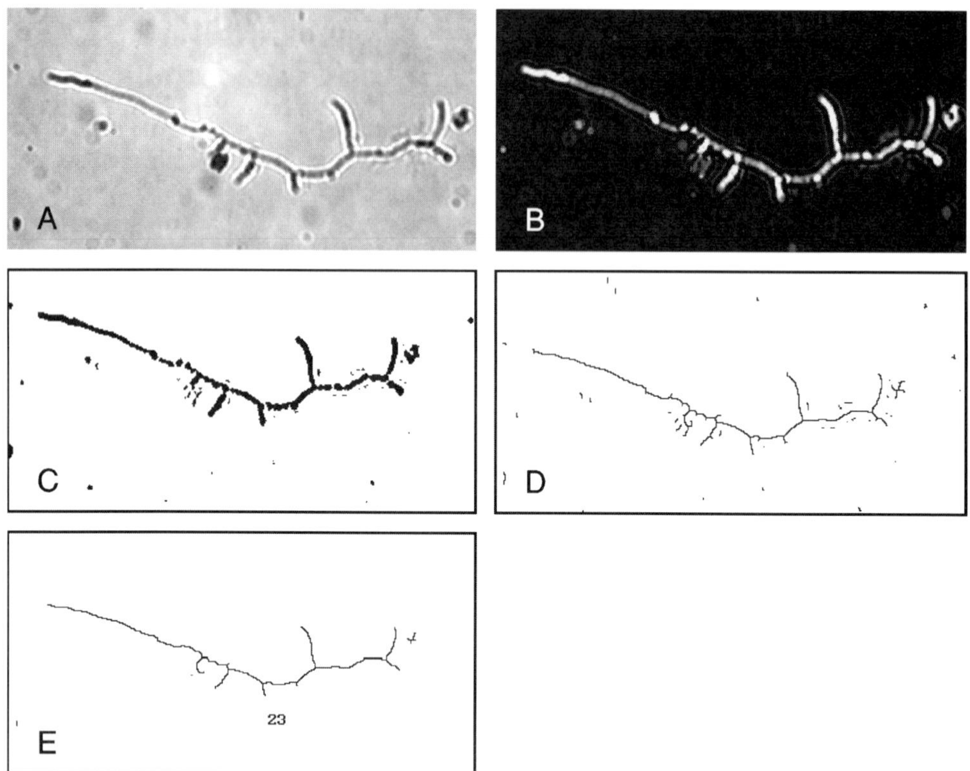

Figure 14.6 Different transformed images of the same hyphal element used in the routine for measuring the length of individual hyphal elements. (A) The grey image, (B) the grey image transformed with a black top-hat, (C) the detected binary image, (D) the skeleton and (E) the measured image after a pruning of image (D). Note the label used to identify the hyphal element.

Mandels and Darby (1953). The time course of the increase in spore diameter has been estimated by a linear (or log-linear) regression of the equivalent spore diameter, defined as (see Section 8.3.1, equation (8.2)):

$$d_{\mathrm{eq}} = \sqrt{\frac{4}{\pi} \cdot A_{\mathrm{i}}} \tag{14.9}$$

The results show that the spores' circularity is close to 1 during spore swelling, indicating that the spore is spherical, and it appears difficult to differentiate between exponential growth and a linear increase in the spore diameter (Figure 14.7A).

When the spore has attained a certain size, growth polarity is established and wall formation ceases for all but one or two areas, resulting in the outgrowth of a germ tube. Eventually, the germ tube develops into a hypha, as shown in Figure 14.7B. Just prior to the formation of the germ tube the metabolic activity is quite high inside the spore, which results in rapid growth (high specific growth rate) as soon as the germ tube is formed. As the germ tube increases in size, the activity

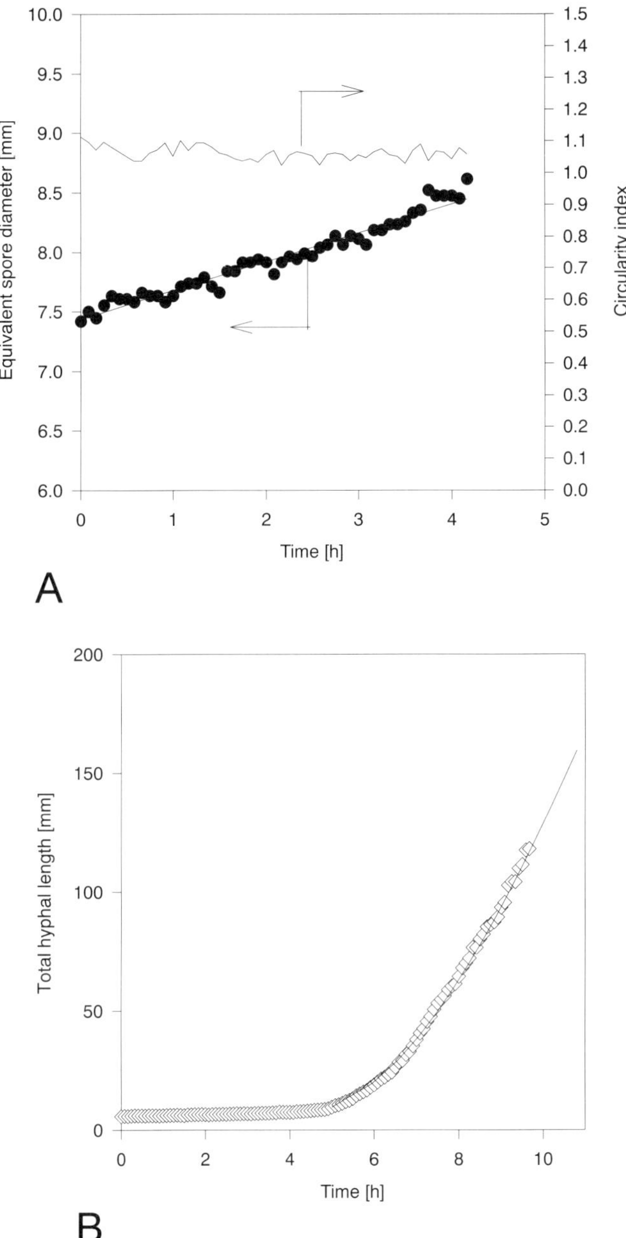

Figure 14.7 Spore germination. (A) (●) The equivalent spore diameter and the circularity index of an *Aspergillus oryzae* spore. (B) (◇) Increase in the total hyphal length of an *Aspergillus oryzae* germ tube. The solid line represents a simulation using equation (14.3) with $k_{germ} = 160\ \mu m/tip/h$ and $K_{germ} = 168.75\ \mu m$. Data from Spohr *et al.* (1998b).

inside the spore cannot keep up with the increasing demand for sustaining the same specific growth rate. Growth, i.e. incorporation of cell-wall components in the tip, must therefore be supported by the germ tube itself, and the rate of supply of wall material will thus become relatively smaller, which will result in a decrease in the specific growth rate (see also Figure 14.7B). Based on the postulated growth mechanism, it follows that the growth kinetics can be explained by simple saturation type kinetics:

$$\mu_h = \frac{q_{germ}}{l_{germ}} = k_{germ} \cdot \frac{l_{germ}}{l_{germ} + K_{germ}} \tag{14.10}$$

where q_{germ} is the tip extension rate of the germ tube, l_{germ} is the germ tube length, k_{germ} is the maximum tip extension rate of the germ tube, and K_{germ} is a saturation constant. Using equation (14.10), the germ tube growth was simulated and, as shown in Figure 14.7B, was found to describe the growth of a germ tube very well.

14.2.7 Growth of single hyphal elements

Using the flow-through cell, the length and number of tips of individual hyphal elements have been measured as a function of time, as shown in Figure 14.8; 26 branches formed in just under 14 hours after the start of the germination. Exponential growth can be observed after 8–10 hours, and the branching can be described by equation (14.5), indicating that the simple mechanism for branching postulated earlier does not only hold for population growth but also for individual hyphal elements. The average tip extension rate for the individual hyphal element was determined by dividing the growth rate (estimated using finite differentiation) of the hyphal element by the number of tips. As illustrated in Figure 14.8B, the observed tip extension rate for an individual hyphal element cannot be explained by simple saturation type kinetics with respect to total hyphal length, and it is therefore not possible to apply the empirical expression derived from measurements of average properties in submerged cultures in a search for a mechanistic basis for tip extension.

If the rate controlling process for tip extension is at the hyphal tip, e.g. vesicle fusion with the cell membrane or reorganization of vesicles at VSC, then the tip extension rate should be independent of the branch length. However, if biosynthesis of macromolecules is the limiting process for tip extension, then it depends on the length of the branch involved in the synthesis of macromolecules used for tip extension. Since there is a maximum tip extension rate, governed by mechanical constraints of the transport of vesicles, this dependence could be expressed by saturation kinetics with respect to the length of the individual branch, i.e.:

$$q_{tip}(l_{bran}, \mathbf{z}) = k_{tip}(\mathbf{z}) \cdot \frac{l_{bran}}{l_{bran} + K_{bran}} \tag{14.11}$$

where l_{bran} is the branch length, $k_{tip}(\mathbf{z})$ is the maximum tip extension rate and K_{bran} is a saturation constant. It is clear that the saturation constant K_{bran} is significantly smaller than K_{germ}, since the new tip in a hyphal element is formed in a

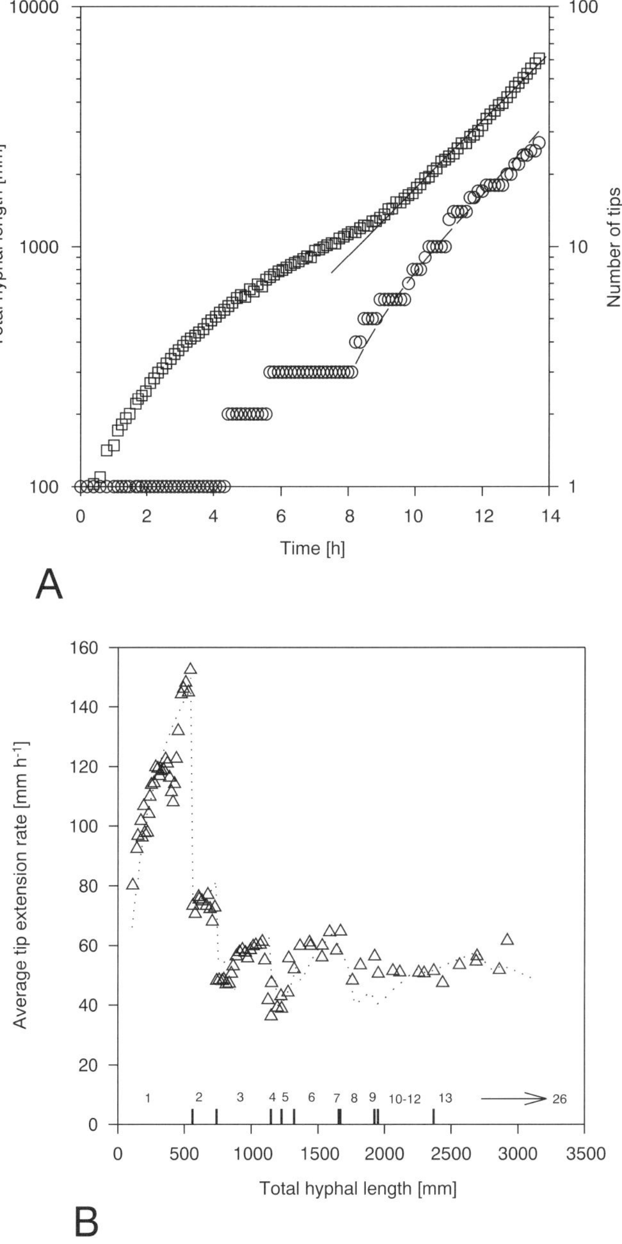

Figure 14.8 (A) The development of the total hyphal length (□) and the number of tips (○) of a single hyphal element growing in a defined medium with 250 mg/l glucose. The solid line represents exponential growth with a specific growth rate of $\mu_h = 0.33\ h^{-1}$. The broken line represents a simulation of equation (14.5) with $k_{bran} = 0.0018$ tips/μm/h. (B) The average tip extension rate as a function of the total hyphal length for the hypha of Figure A. (△) Experimental values. The broken line represents a simulation based on the summation of the tip extension rate for the individual branches using equation (14.11). Numbers above the x axis signify the number of tips for each interval in l_t.

compartment which is already active, and able to support tip growth, whereas in a developing germ tube the apparatus for cell wall production has to be synthesized first. Figure 14.9 illustrates that equation (14.11) describes the tip extension of an individual branch well, indicating that the biosynthesis of macromolecules is the rate controlling factor for tip extension.

The total rate of tip extension can be described as the sum of equation (14.11) for each branch formed, and subsequently the average tip extension rate is derived by dividing this total by the number of tips. Figure 14.8B illustrates the simulation of the average tip extension rate using the algorithm as described in Lejeune *et al.* (1995). Parameters $k_{\text{tip}}(\mathbf{z})$ and K_{bran} have been estimated for all individual branches using the automatic image analysis routine described earlier. The simulation fits the data well, indicating that quantification of individual branches is necessary in order to understand and model the mechanistic basis for tip extension.

14.2.8 The influence of substrate concentration on the morphology

In order to understand the underlying mechanisms for hyphal growth and hence the morphology, knowledge about the influence of the substrate is necessary. Here the flow-through cell is an excellent tool. Spohr *et al.* (1998b) studied the influence of the glucose concentration on the specific growth rate, the tip extension rate and the

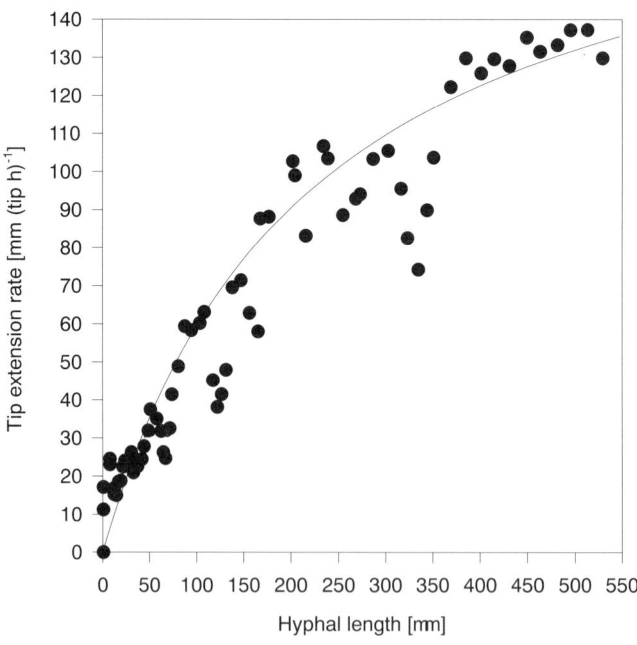

Figure 14.9 The tip extension rate (●) for the first branch formed. The solid line represents equation (14.11) with $k_{\text{tip}}(\mathbf{z}) = 185\ \mu\text{m/h}$ and $K_{\text{bran}} = 200\ \mu\text{m}$. The abrupt decrease in the tip extension rate at a length of 300 μm occurs when a new branch is formed on the branch studied. Data from Spohr *et al.* (1998b).

specific branching frequency for *A. oryzae*, and the results for tip extension are shown in Figure 14.10. At the same glucose concentration there is a significant variability in the developing morphology of the individual hyphal elements, which illustrates that individual hyphal elements grow differently under the same conditions, i.e. there is a natural variation in the growth of individual elements. From the data it was found that the specific growth rate, the maximum tip extension rate and the specific branching frequency all follow saturation type kinetics with respect to the glucose concentration, i.e.:

$$\mu_h = \mu_{max} \cdot \frac{s}{K_s + s} \tag{14.12}$$

$$k_{tip}(s) = k_{tip,max} \cdot \frac{s}{K_{s,tipext} + s} \tag{14.13}$$

$$k_{bran}(s) = k_{bran,max} \cdot \frac{s}{K_{s,bran} + s} \tag{14.14}$$

Here μ_{max}, $k_{tip,max}$ and $k_{bran,max}$ are the maximum specific growth rate, the maximum tip extension rate and the maximum specific branching frequency, respectively. $K_{s,i}$ are saturation constants with respect to glucose. The estimated

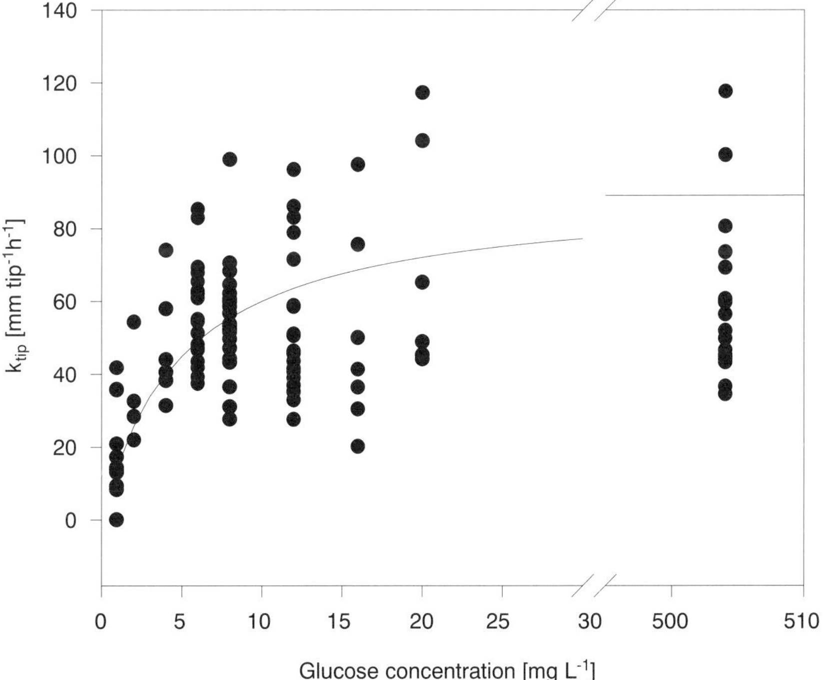

Figure 14.10 The maximum tip extension rate as a function of the glucose concentration. The line represents equation (14.9) with $k_{tip,max} = 90\ \mu m/tip/h$ and $K_s = 5$ mg/glucose.

parameters are summarized in Table 14.3. It is observed that glucose exerts the same control on both branching and tip extension, and the microscopic morphology will therefore not change with the glucose concentration.

14.2.9 Correlation between microscopic morphology and protein secretion

A possible correlation between the microscopic morphology and extracellular protein production in filamentous fungi has been shown by Wösten *et al.* (1991) and a qualitative model has been formulated by Peberdy (1994). Determination of the molecular threshold of a typical cell wall suggests that the size limit is around 20 kD, whereas many secreted proteins have molecular weights of up to 300 kD. Wösten *et al.* (1991) used immunogold labelling to localize the secretion of glucoamylase in *A. niger* to the tips of actively growing hyphae. The exact mechanisms behind transport of proteins through the cell wall have still to be uncovered. Chang and Trevithick (1974) (reviewed by Peberdy (1994)) suggest that the growing apical wall is more porous than the rigid, mature wall, and therefore permits a rapid diffusion of proteins.

Using the same growth system as Wösten *et al.* (1991), the protein secretion in *A. oryzae* has been studied. The growth system consisted of a minimal agar medium on which a sandwich of a polycarbonate membrane, a thin agarose gel (0.03 mm) and a second polycarbonate membrane were placed. Prior to the start of the experiment, spores of *A. oryzae* were placed in the agarose gel. Medium diffused through the carbonate membrane and into the agarose gel, resulting in germination of the spores. After 20 hours of cultivation the agarose gel was isolated, the α-amylase in the gel was fixed, and the gel was washed several times. After this the α-amylase levels could be detected in the agarose gel using rabbit antibodies raised against α-amylase, and secondary FITC-conjugated antibodies. The result is illustrated in Figure 14.11. The secretion of α-amylase is clearly located at the tips of growing hyphae.

These findings indicate that there is a correlation between protein secretion and the number of tips in the mycelium. Using different morphological mutants derived from the same strain, Spohr *et al.* (1998a) showed that a more branched mutant of a high yielding strain of *A. oryzae* has a slightly higher extracellular α-amylase production compared with the less branched mutant. These findings could indicate that the translocation through the tips could be a controlling factor in extracellular protein production.

Table 14.3 The morphological parameters determined as a function of the glucose concentration. Data from Spohr *et al.* (1998b).

	μ	k_{tip}	k_{bran}
Maximum value, $s > 500$ mg/l	$\mu_{max} = 0.38$ h^{-1}	$k_{tip,max} = 90$ μm/tip/h	$k_{bran,max} = 0.0023$ tip/μm/h
K_s	5 mg/l glucose	5 mg/l glucose	5 mg/l glucose

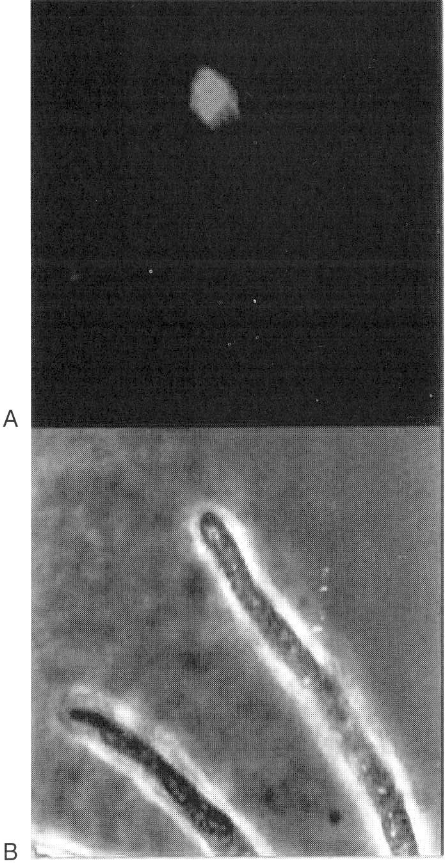

Figure 14.11 Localisation of α-amylase secretion in *Aspergillus oryzae* using antibodies directed against α-amylase and secondary FITC conjugated antibodies. (A) Two hyphae, using normal light microscopy. The one to the right is growing whereas the one to the left is not. (B) The same hyphae, using fluorescence microscopy. Here the FITC conjugated antibodies are located around the tip of the growing hypha.

14.2.10 Hyphal fragmentation

When growing in submerged cultures, filamentous micro-organisms fragment, i.e. the hyphae break up, resulting in the formation of two or more smaller hyphal elements from a large hyphal element. Fragmentation of the hyphal element occurs when the local shear rate becomes greater than the tensile strength of the hyphal wall, which is believed not to be the same throughout the hyphal element. The hyphal wall has been reported to be weaker at the septa (Savage and Van der Brook, 1946). Due to the above-mentioned mechanism for hyphal fragmentation, it is likely that there is an equilibrium effective hyphal length. Hyphae longer than this equilibrium length will fragment and those shorter than this equilibrium length will not be further reduced in size. Van Suijdam and Metz (1981a) derived a model describing hyphal

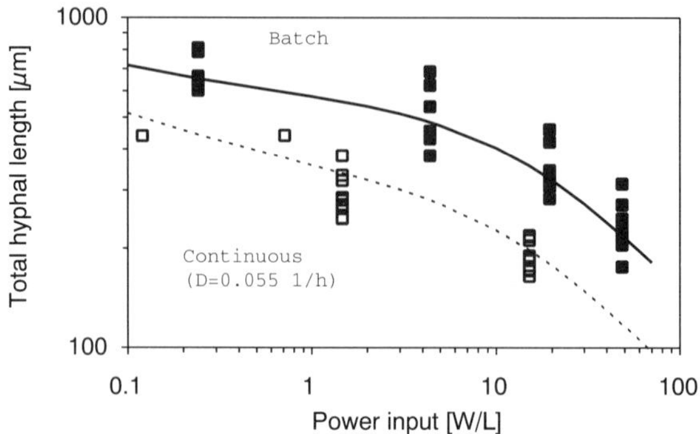

Figure 14.12 Effect of power input on the total hyphal length of *Penicillium chrysogenum*. The two sets of data are from a series of batch cultivations (batch) and a series of steady state continuous cultivations ($D = 0.055\ h^{-1}$). The lines are model simulations carried out by Nielsen and Krabben (1995). Data from Van Suijdam and Metz (1981a) and Metz (1976).

fragmentation based on this assumption. The rate of fragmentation is a function of the frequency with which the hyphal elements enter the zones of sufficiently high shear rates, and of the residence time distribution for the hyphal elements in these zones. Based on models of the liquid flow circulation in a stirred tank, and of the influence of the local energy input on the rate of fragmentation, Van Suijdam and Metz (1981a) derived a term describing the rate of fragmentation in a stirred tank as a function of the effective length and a specific fragmentation frequency, k_{frac}. The parameter k_{frac} is a function of the power input (correlated to the agitation rate), the environmental conditions and the geometry of the actual reactor vessel. Using data of Van Suijdam and Metz (1981a), it was later found that the fragmentation parameter k_{frac} is a linear function of the power input (Nielsen and Krabben, 1995), resulting in the influence of power input on the total average length as shown in Figure 14.12.

14.3 MACROSCOPIC MORPHOLOGY

Macroscopic morphology is more complex to describe than microscopic morphology, since there is no set of obvious morphological parameters that can be used for quantitative analysis. Nonetheless, image analysis is well suited to study the shapes and structures of pellets. Cox and Thomas (1992) used the solid core of mycelium within each pellet in their analysis whereas Reichl *et al.* (1992b) used an algorithm based on the notion that pellets are large, and their margins are less dense than their centres. The existence of a solid body can be ascertained by an ultimate skeletonization, i.e. the exhaustive removal of the outer pixel layers to attempt to identify a single central pixel. Pellets often appear as fluffy pellets, i.e. entangled lateral hyphae exists on the core. This form causes a failure of the solid core identification using the simple skeleton routine, since the hyphae appear as lines after skeletonization. Cox and Thomas (1992) avoided this problem by introducing an opening routine, i.e. an erosion followed by a dilation. The removal of pixel layers by erosion permanently removes edge features, which are not reconstructed by the subsequent dilation operation. If skeletonization after a preset number of opening cycles results in a single pixel the object is considered to be a pellet; if this is not the case, it is considered to be a clump. After identification of a pellet, the respective area, convex area, circularity and equivalent diameter can be determined. Additional variables can be determined. For example, Cox and Thomas (1992) quantified the optical density of the outer fluffy core in order to determine the compactness of the pellets. The results from measurements of the above variables are difficult (if not impossible) to correlate to other variables of the process. It is therefore not possible to set up a general model for the macroscopic morphology. However, the most important processes that influence the macroscopic morphology are pellet formation, pellet growth and pellet break-up and fragmentation (Figure 14.13).

14.3.1 Pellet formation

The mechanisms for pellet formation vary from strain to strain. However, pellet forming micro-organisms can be roughly divided into three groups:

(a) The spore coagulating type, where the spores coagulate, and upon germination of the agglomerated spores a pellet is formed. Many species of *Aspergillus* belong to this group (Galbraith and Smith, 1969; Nielsen and Carlsen, 1996).
(b) The non-coagulating type, where a single spore develops into a pellet. Some species of *Streptomyces* belong to this group (Yang *et al.*, 1992a).
(c) The hyphal element agglomerating type, where hyphal elements agglomerate and form a clump of hyphal elements that eventually develops into a pellet. *Penicillium chrysogenum* has been shown to belong to this group (Packer and Thomas, 1990; Tucker *et al.*, 1992; Nielsen *et al.*, 1995).

For each group the mechanisms behind pellet formation are complex and only in a few cases understood. Among the factors that influence pellet formation are the medium composition, spore concentration, pH and power input into the bioreactor.

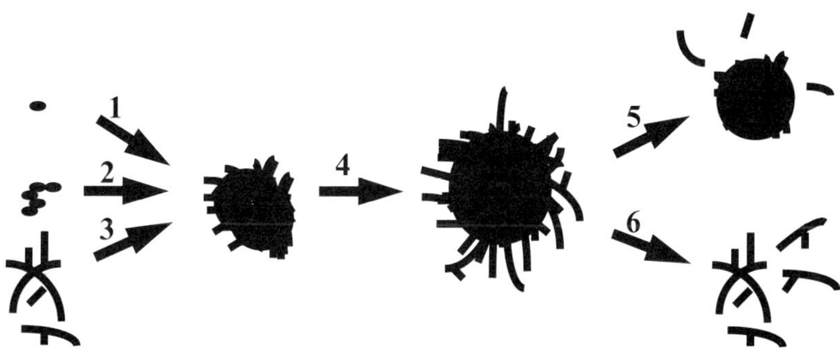

Figure 14.13 Processes affecting the macroscopic morphology. 1. Pellet formation from a single spore. 2. Pellet formation from coagulating spores. 3. Pellet formation due to agglomeration of hyphal elements. 4. Pellet growth. 5. Pellet fragmentation. 6. Pellet break-up. Adapted from Nielsen (1996).

For some systems qualitative models have been formulated that describe the influence of the various variables on pellet formation (Nielsen and Carlsen, 1996), but the complexity of these models clearly shows that much more information is required before reliable mathematical models can be formulated. However, one exception is the model of Yang *et al.* (1992b), which is able to describe the outgrowth of a single spore of *Streptomyces tendae* to a pellet.

14.3.2 Pellet growth

It was demonstrated early that the growth of fungal pellets is not exponential (Emerson, 1950; Marshall and Alexander, 1960; Trinci, 1970), but may be described by the so-called cube root law:

$$m^{1/3} = k_p t + m_0^{1/3} \tag{14.15}$$

where m is the pellet mass and m_0 is the pellet mass at time $t = 0$. The cube root law can easily be shown to be a result of mass transfer limitations, resulting in growth only in an outer shell of the pellet. In this case the kinetic parameter k_p can be specified as a function of the thickness of this shell, the density of the pellet and the specific growth rate of the cells within the pellet (Nielsen and Carlsen, 1996). However, the equation can also be derived by assuming a constant rate of increase in the pellet radius (Prosser and Tough, 1991; Nielsen and Carlsen, 1996). A linear increase in the pellet radius (see Figure 14.14), which is observed for many systems (Nielsen *et al.*, 1995; Carlsen *et al.*, 1996), does, however, not necessarily imply that

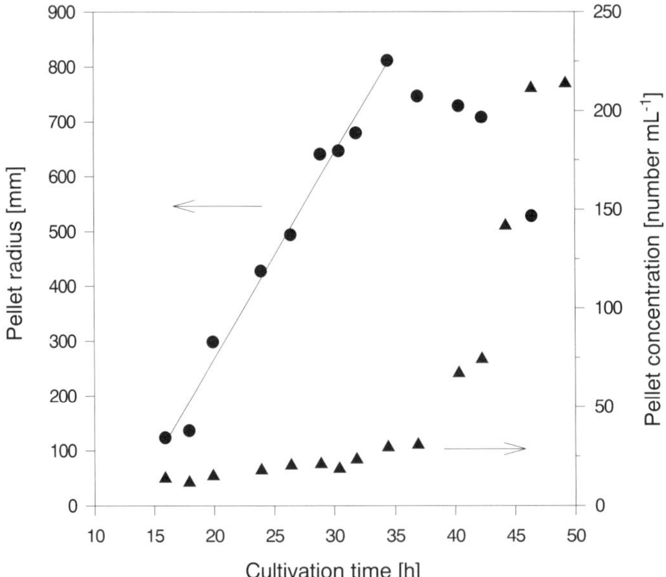

Figure 14.14 Pellet concentration and equivalent diameter during a batch culture of *Aspergillus oryzae*. Data from Carlsen *et al.* (1996).

the cube root law can be used to describe the growth of the pellets, since this equation only holds when there are mass transfer limitations. To evaluate whether there are mass transfer limitations, one can calculate the critical radius, which has been defined as the pellet radius where the substrate concentration is just zero at the centre of the pellet (Yano *et al.*, 1961). The critical radius is given as the ratio of the substrate diffusion and substrate consumption, i.e.:

$$R_{\text{crit}} = \sqrt{\frac{6D_{\text{eff}}c_{\text{s}}}{\rho_{\text{pel}}r(c_{\text{s}})}} \tag{14.16}$$

where D_{eff} is the effective diffusion coefficient, c_{s} is the substrate concentration outside the pellet, ρ_{pel} is the density of the pellet, and $r(c_{\text{s}})$ is the specific rate of substrate consumption. For a typical pellet the critical radius is found to be between 50 and 200 μm when oxygen is the limiting substrate (which is most often the case). For pellets with a radius larger than R_{crit} there will only be growth in an outer shell of the pellet, and consequently non-exponential growth. The thickness of this layer depends on the pellet radius. If the thickness of this layer is known, it is possible to calculate the effectiveness factor for the growth process, and hence the overall specific growth rate of the pellets can be evaluated. This approach is relatively straightforward, but a pitfall is that the calculations are quite sensitive to the pellet radius, and application of an average pellet radius in the calculations may therefore result in a wrong estimate of the overall specific growth rate of the biomass. To obtain a correct result one should take into account the pellet size distribution, for which data are only available for a few systems (Reichl *et al.*, 1992b; Nielsen *et al.*, 1995).

14.3.3 Pellet break-up and pellet fragmentation

Shear forces acting on pellets affect the pellet size (and its distribution) by fragmentation of hyphae at the pellet surface. This mechanism is referred to as pellet fragmentation. It has many similarities with fragmentation of hyphal elements, and has been modelled in a similar fashion (Van Suijdam and Metz, 1981b; Nielsen and Carlsen, 1996). This fragmentation not only influences pellet size, but also the pellet shape, which can be quantified by the core circularity (Cox and Thomas, 1992). In fed-batch cultures of *Penicillium chrysogenum* the core circularity of the pellet population approaches 1 as a result of fragmentation (Nielsen *et al.*, 1995). Another result of pellet fragmentation is the appearance of a skewed pellet size distribution, which can be described reasonably well with a log-normal distribution (Reichl *et al.*, 1992b).

Besides pellet fragmentation there may be complete pellet break-up. This process is most likely a result of mass transfer limitation, which leads to a non-growing zone at the centre of the pellet. Eventually there will be lysis of these non-growing cells, and this leads to structural destabilization of the pellet, which, when exposed to shear forces, may break up completely. With this mechanism, the break-up is dependent not only on the shear forces, but also on the history of the pellet, i.e. for how long there has been substrate depletion at the centre of the pellet, and how fast cell lysis occurs after substrate depletion. However, pellet break-up in a stirred bioreactor has been described as a first-order process in the pellet concentration, with the rate constant being a function of the pellet size and the agitation rate (Taguchi *et al.*, 1968; Taguchi, 1971).

14.4 QUANTIFICATION OF FUNGAL DIFFERENTIATION

The growth of filamentous micro-organisms results in a multicellular mycelium which is morphologically very heterogeneous. A time-dependent cellular differentiation of the hyphae occurs due to the flow of cytoplasm toward the tips, which leaves the older parts of the hyphae vacuolated and inactive (Zalokar, 1959; Smith and Berry, 1974).

One of the major problems in studying this cellular differentiation has been the lack of quantitative experimental techniques. However, the recent advancement of automated image analysis combined with application of various stains has provided a powerful technique for quantifying the differentiation of hyphae. This technique relies on special dyes, which are able to stain the regions of interest specifically. The staining produced by these dyes may be visible in a normal light microscope or may require the use of a fluorescence microscope. It is important to note that the result of measurements performed using automated image analysis often rely to a great extent on the dyes used for staining the hyphal elements, and these therefore have to be selected carefully to ensure that specific staining of the desired regions is obtained, and problems with, for example, fading of fluorescent dyes or unspecific background staining have to be dealt with. To exploit fully the capabilities of automated image analysis it is therefore important to choose dyes that are very specific, stable and easy to detect.

Different types of staining principles have been used for investigating the cellular differentiation in hyphae. Paul and Thomas (1996) used neutral red, a redox dye visible in a normal microscope, to stain the growing regions of *P. chrysogenum*. Vanhoutte *et al.* (1995) used a combination of methylene blue and Ziehl fuchsin, both visible in a normal microscope. Methylene blue is converted from a colourless, oxidized form when reduced by active mitochondria, hence the compound is an indicator of metabolic activity. Ziehl fuchsin, which dyes cytoplasmic components red–orange, was used to stain the parts of the hyphal elements containing cytoplasm, but not active mitochondria. Using colour image analysis, Vanhoutte *et al.* (1995) were able to distinguish between six different hyphal zones in *P. chrysogenum*, ranging from the active, apical regions to the most distal dead regions. Finally, Agger *et al.* (1998) used a combination of the fluorescent dyes 3,3′-dihexyloxocarbocyanin ($DiOC_6$) and Calcofluor White to study the cellular differentiation of *A. oryzae*. Here we describe the technique for staining hyphal elements with $DiOC_6$ and Calcofluor White, and thereafter discuss how measurements of cellular differentiation can be used to set up mathematical models.

14.4.1 $DiOC_6$ and Calcofluor White as fluorescent probes

The fluorochrome $DiOC_6$ is a positively charged lipophilic compound, consisting of a polycyclic fluorescent part with hydrocarbon groups attached (Figure 14.15). It has an optimal excitation wavelength of 478 nm and when fluorescing, the emitted light

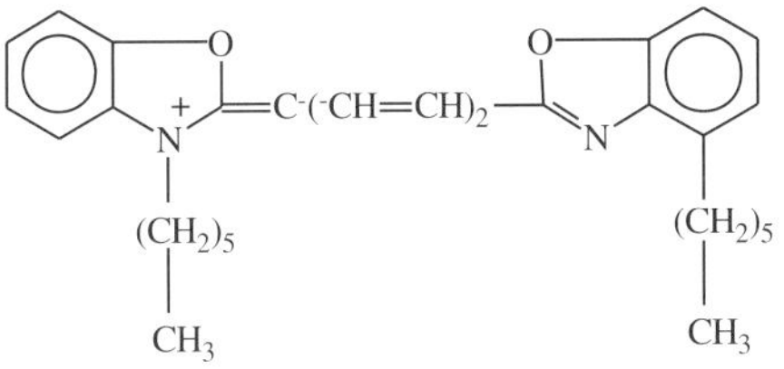

Figure 14.15 The structure of 3,3′-dihexyloxocarbocyanin ($DiOC_6$).

spectrum has a maximum at 496 nm (Kasten, 1993). The dye is visualized using a filter block with excitation passband and emission longpass cutoff of 450–490 nm and 520 nm, respectively. The dye is able to pass through the plasma membrane, following which it disperses into the lipid phase of intracellular membranes. According to Terasaki (1993), the dye does not seem to be chemically altered or metabolized by the cell in any way. When administered in low doses, the positive charge on the dye molecules will cause them to accumulate in the mitochondria due to the negative potential of the mitochondrial membrane, caused by the extrusion of protons from the mitochondrial matrix by the respiratory electron transfer system. At higher doses $DiOC_6$ accumulates in other membranous organelles as well, including the endoplasmic reticulum (Terasaki and Reese, 1992).

Figure 14.16 illustrates the use of $DiOC_6$ for staining organelles inside the hyphal elements of *A. oryzae*. The elongated appearance of what seem to be mitochondria can be seen. These are dynamic structures, and when observed they are frequently seen stretching, contracting and fusing with each other.

Calcofluor White is a stilbene compound which binds to chitin, β-glucans and cellulose fibres, thereby being able to stain the cell walls of plants and fungi (Kasten, 1993). The dye is visualized using a filter block with an excitation wavelength range of 330–380 nm and an emission longpass cutoff of 420 nm. Calcofluor White has an

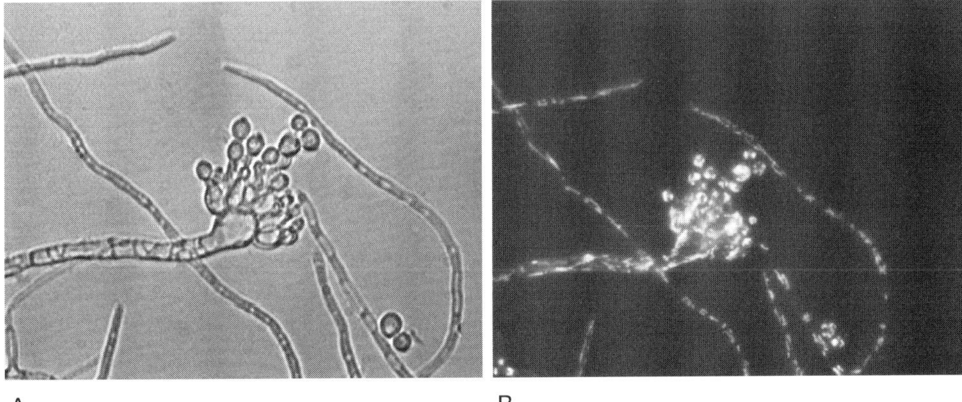

Figure 14.16 Images of hyphal elements as seen in (A) the normal light microscope and (B) the fluorescence microscope. Note the high degree of staining in the conidiophore in the centre of the images.

increased affinity for non-crystalline chitin, and has been used to stain growing tips of *Schizophyllum commune* selectively (Vermeulen and Wessels, 1986), among many other applications.

Agger *et al.* (1998) developed a method to incorporate the two fluorescent probes in one assay, thereby obtaining double-staining of the hyphae. Calcofluor White is used to optimize the measurement of the total hyphal area, a measurement otherwise performed by detection of the hyphae in images taken during normal light microscopy. The accuracy of the latter procedure is highly susceptible to vacuoles, as these structures normally appear bright in an otherwise dark hypha, as well as to the focus, illumination (contrast) and interfering dust particles. The effect of Calcofluor White on total area measurements is illustrated in Figure 14.17.

$DiOC_6$ has spectral characteristics different from those of Calcofluor White, thus making it possible to visualize specifically the parts stained by this probe in a sample containing the other probe as well, by using a filter block specially suited for $DiOC_6$.

14.4.1a Double-staining method

A sample, taken from the bioreactor, was diluted 13 times with a 0.1 M PBS buffer (pH 7.4) and stained with $DiOC_6$ (0.06–0.07 μg/ml) and Calcofluor White (0.023 μg/ml) for 10 minutes. The mixture was then transferred to a microscopic slide and placed under the fluorescence microscope. A number of hyphal elements were located with the fluorescence microscope, and images of both the Calcofluor White and the $DiOC_6$ staining were acquired by the image analyser. These images were processed by the computer as described in the next section, and the ratio of the areas stained by $DiOC_6$ and by Calcofluor White (the total hyphal area) was determined. The microscopic slide was then manually moved to a new position and another set of images was acquired, processed and measured. This procedure was repeated 60–100 times on each sample and the average value was then taken as a measure of the fraction of active cells. When performing measurements on chemostats in steady state, 10–20 samples were measured as described above and the average of these measurements was taken as the final result.

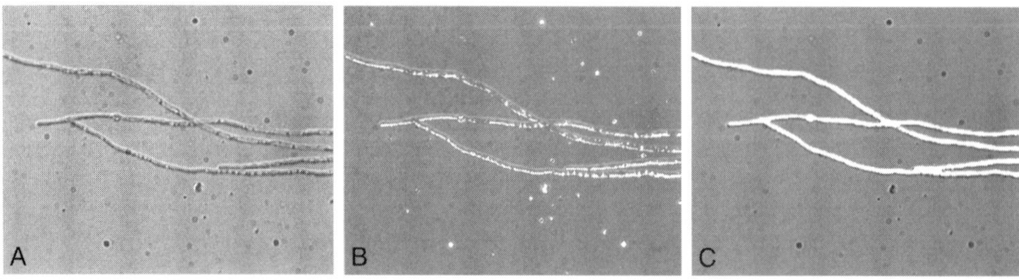

Figure 14.17 Illustration of the effect of Calcofluor White in total area measurements. A detection of the hyphae seen with normal light microscopy (A) has been performed on the normal light image, after increasing the contrast to a maximum (B) and on the Calcofluor White fluorescence image, the result of which is shown superimposed on the normal light image (C). The detected parts are white.

14.4.1b Detection by automated image analysis

The fluorescence microscope used (Nikon Optiphot 2) is connected via a monochrome video camera (Kappa CF 8/1 FMC) to a Leica Quantimet 600S Image Analyser, which is programmed with the Quips language for image processing. A Quips routine, illustrated in Figure 14.18, is used for the quantification of the images.

The routine is repeated a given number of times (usually 50–100), and during each cycle changing of filter blocks and adjustment of focus (if necessary) have to be performed manually. Between cycles the position of the microscope stage also has to be adjusted manually. The routine is started using the filter block for $DiOC_6$, and the program pauses while the desired image is found and adjusted on the microscope, after which the image is acquired by the computer as a grey-scale image. Following this acquisition, the filter blocks are switched and the Calcofluor White image is acquired. The excitation light is switched off to protect the sample from damage induced by the UV light during the subsequent image processing, which lasts a few seconds.

The processing of the $DiOC_6$ image starts with the application of the 'enhance' function (described in Section 14.2.5) to the grey-scale image, which enhances white detail in the image. The grey-scale image is then segmented into a binary image by the 'detect' function (global thresholding). When detecting bright structures during fluorescence microscopy, only pixels with a grey value above a given threshold value will be detected, as illustrated in Figure 14.19. Following the segmentation, the 'prune' function is used to erode a small number of pixels off the edges of the detected structures to compensate for the slight increase in area caused by the 'enhance' function. The reason for this increase is that the video camera is not able to perform an exact delimitation of the bright, fluorescent structures but creates a

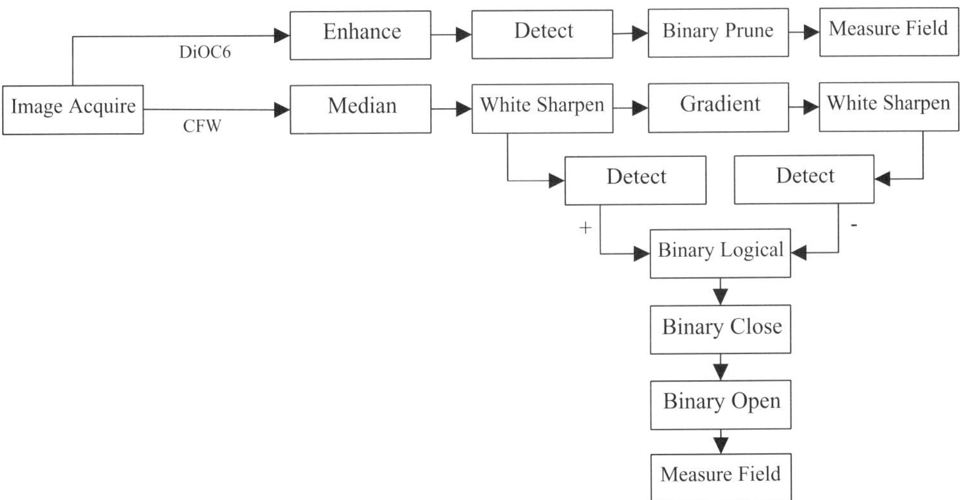

Figure 14.18 Diagram showing the main operations of the Quips routine used for the double-staining method.

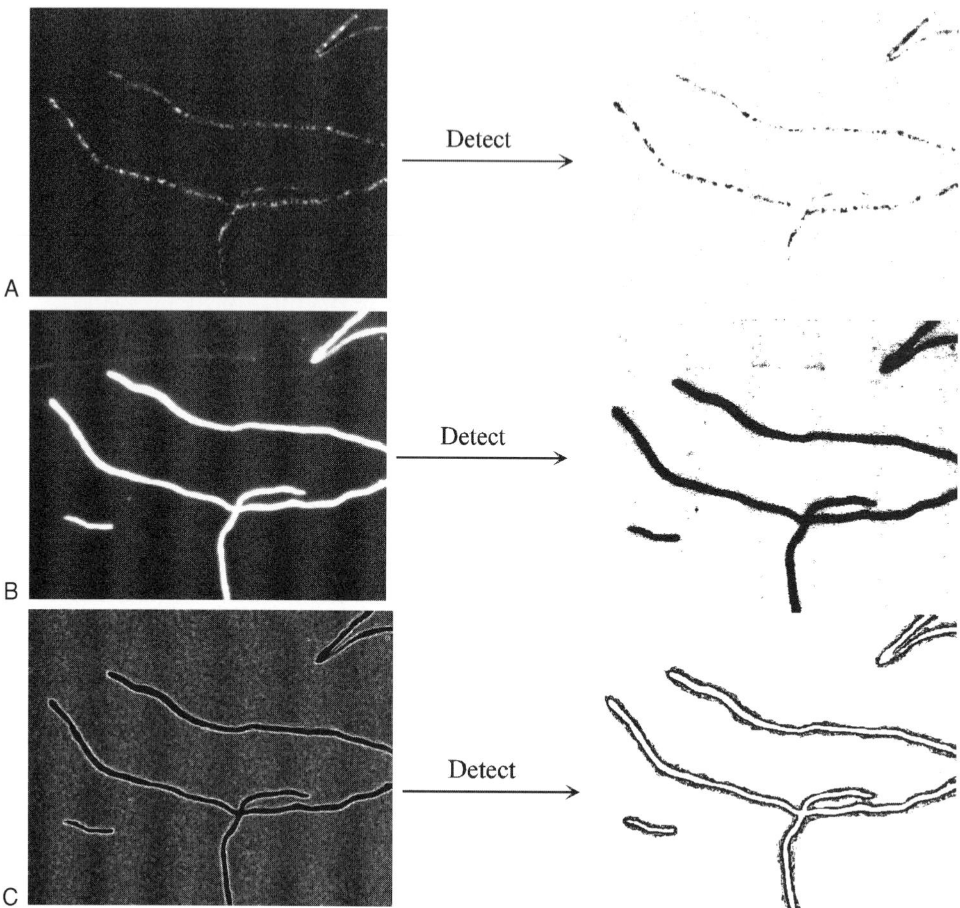

Figure 14.19 Effect of the 'detect' function on the enhanced $DiOC_6$ image (A), on the sharpened Calcofluor White image (B) and on the sharpened 'gradient' Calcofluor White image (C). The pixels detected are shown in black in the binary images (to the right).

gradient in intensity from these structures to the dark background. This results in the detection of some of the pixels in the gradient region following the 'enhance' function, and these are removed by the 'prune' function. Finally, the area of the pixels detected is measured.

The first function applied to the Calcofluor White image is a median filter. This function sets the grey value of each pixel equal to the median of the grey values of the neighbouring pixels, thereby being useful in reducing noise (see Section 1.4.2a). To enhance the contrast of the image and improve segmentation, a 'white sharpen' function is then used. This function emphasizes white detail by adding a white top-hat to the original image. The resulting image is then treated in two different ways. It is used to form a binary image directly, by using the 'detect' function as illustrated in Figure 14.19, but it is also submitted to a nonlinear gradient operator (Section 6.5.1). The latter function produces an image in which areas where gradients in grey level occur are enhanced, by setting the grey value of each pixel equal to the difference between the local maximum and minimum values. This is useful to remove the glare surrounding the hyphae in the Calcofluor White image which, as mentioned above, is something inherently created by the video camera when recording very bright objects on a completely dark background. The glare can be minimized by using a concentration of Calcofluor White as low as possible. The 'gradient' function will not affect images where no glow is present.

A 'white sharpen' function is used to enhance the contrast in the image produced by the 'gradient' function and the resulting image is thresholded to form a binary image, as shown in Figure 14.19. The binary image thus obtained is then subtracted from the binary image previously created from the Calcofluor White image by use of a binary logical operator, and the glare is removed from the image. The resulting binary image is then closed, the effect of which is to fill any small holes in the detected hyphae. To remove any small, unwanted structures in the surrounding image, an opening is applied next. Finally, the area of the detected structures in the binary image is measured, and the value thus obtained is used as the total hyphal area.

The routine calculates the ratio between the area stained by $DiOC_6$ and the total hyphal area. It also calculates the average of all the preceding measurements in order for the operator to know when a stable average value has been obtained. These data are stored on a hard disc drive, as well as the values of the total area and the $DiOC_6$ area. The routine is normally repeated 100 times or until the average of the preceding measurements reaches a stable value. The coefficient of variance of the double-staining method has been estimated at 12% by consecutive measurements on chemostats in the steady state.

14.4.2 Modelling cellular differentiation

In order to link cellular differentiation with the growth and production kinetics, mathematical models are very useful. Mathematical models describing cellular differentiation are referred to as morphologically structured models (Nielsen, 1993), and several of these models have been developed for a variety of fungal species, including *P. chrysogenum* (Nestaas and Wang, 1983; Paul and Thomas, 1996; Zangirolami *et al.*, 1997), *Aspergillus awamori* (Megee *et al.*, 1970), *A. oryzae* (Brown and Fitzpatrick, 1979; Agger *et al.*, submitted) and filamentous micro-organisms in general (Nielsen, 1993). However, only a few of these models have been validated through measurements of hyphal differentiation. The work of Megee *et al.* (1970) has formed the basis for many of the later models. In this model the hyphae were divided into five different compartments, each of which had different activities regarding substrate utilization and product formation. The actively growing regions of the hyphae were composed of an apical and a sub-apical compartment, whereas the regions responsible for conidia formation were divided into conidiophore developing hyphae, black spores and mature spores. Product formation was also included as a result of cellular differentiation, and the model was able to describe the complete life cycle of imperfect fungi. A drawback of the model was the large number of parameters that could not be validated experimentally.

The model of Nestaas and Wang (1983) is a much simpler model but with a similar structuring of the biomass. They divided the hyphae into three compartments: actively growing tips, non-growing penicillin producing regions and degenerated regions. However, the kinetics of the model included no influence of the substrate concentration, hence it was necessary to use one set of equations to describe the growth phase, and another set for the production phase, during fed-batch cultures with *P. chrysogenum*.

Nielsen (1993) constructed a general model for growth of filamentous organisms, based on the work of Megee *et al.* (1970). Like the model of Nestaas and Wang (1983) the biomass was divided into three compartments: apical, sub-apical and hyphal. The concept of an apical compartment was originally introduced by Fiddy and Trinci (1976), who defined it as consisting of the part of the hyphal elements from the hyphal apex to the first septum. The apical compartment is followed by the sub-apical compartment, and in the model of Nielsen (1993) these compartments are responsible for growth and uptake of substrate. The hyphal compartment was defined by Nielsen (1993) as consisting of the old parts of the hyphal elements, containing vacuolated and metabolically inactive cells. Cellular differentiation was described to occur during tip extension, when an apical cell is converted to a sub-apical cell, and during branching, when an apical cell is formed from sub-apical cell material. Also, when the sub-apical cells become sufficiently vacuolated, they turn into hyphal cells. The kinetics of all metamorphosis reactions, i.e. reactions where one morphological form is converted to another, was taken to be first order in the disappearing morphological form, and the formation of hyphal cells was assumed to be inhibited at high substrate concentrations. The growth of the apical and sub-apical compartments was described with simple Monod kinetics.

The model of Agger *et al.* (1998) is also based on a division of the hyphae into three compartments: an extension zone, an active compartment and a hyphal compartment. The extension zone consisted of the tip section, where cell wall synthesis takes place (i.e. the site of the vesicle supply centre, as described previously). The active compartment represents all metabolically active regions of the hyphae, whereas the hyphal compartment consists of the older, inactive and vacuolated regions. The model includes only two metamorphosis reactions: branching, in which new extension zones are formed from the active compartment, and formation of hyphal cells from active cells. The growth and cellular differentiation processes are described by the three rate equations (14.17–14.19), describing the rate of formation of new extension zones (by branching), hyphal cells and active cells (tip extension), respectively:

$$q_1 = \begin{cases} 0 & \text{if } \dfrac{x_a}{c_n} < \left(\dfrac{x_a}{c_n}\right)_0 \\[2ex] \dfrac{k_1 s}{a(s + K_{s1})} x_a & \text{if } \dfrac{x_a}{c_n} \geqslant \left(\dfrac{x_a}{c_n}\right)_0 \end{cases} \tag{14.17}$$

$$q_2 = k_2 x_a \tag{14.18}$$

$$q_3 = \frac{k_3 s}{s + K_{s3}} \frac{x_a / c_n}{x_a / c_n + K_3} a x_e \tag{14.19}$$

Similarly to equation (14.5), branching is described as initiating when the average active mass of a hyphal element exceeds a certain level, given as $(x_a/c_n)_0$, where c_n

is the concentration of hyphal elements in the culture. Once initiated, the rate of branching is assumed to be proportional to the concentration of active cells, but is also assumed to be inhibited at very low glucose concentrations. The parameter k_1 is the specific rate of branching (based on the mass of active compartment) and the parameter a represents the number of tips per unit mass of the extension zones. The rate of the differentiation process is assumed to be proportional to the concentration of active cells. The reasoning behind the expression describing the growth rate of the active cells (equation (14.19)) is as follows. As described previously, new cell wall material is synthesized in the active compartment and transported in vesicles toward the growing hyphal apex. Once a new hypha is formed, the ratio between the hyphal diameter and the hyphal length will decrease rapidly, thus increasing the rate of vesicle transport through a given cross-sectional area of the hypha. If it is assumed that this rate has an upper limit, governed by mechanical constraints, it is a reasonable assumption that the transport of vesicles toward the growing tip becomes the rate-controlling step in the elongation process. This assumption implies that the rate of tip extension for a single hypha increases in a saturation type fashion with respect to the length (or mass) of the active part of the hypha, as observed in Figure 14.8. It will eventually approach a maximal value, determined by the concentration of the limiting substrate in the surrounding medium. The growth rate of the total concentration of active cells is described as being equal to the growth rate of each tip multiplied by the concentration of tips, given as ax_e. The first part of the expression describes how the tip extension rate (based on the mass of active compartment) varies with the substrate concentration according to a Monod expression, as specified by equation (14.13).

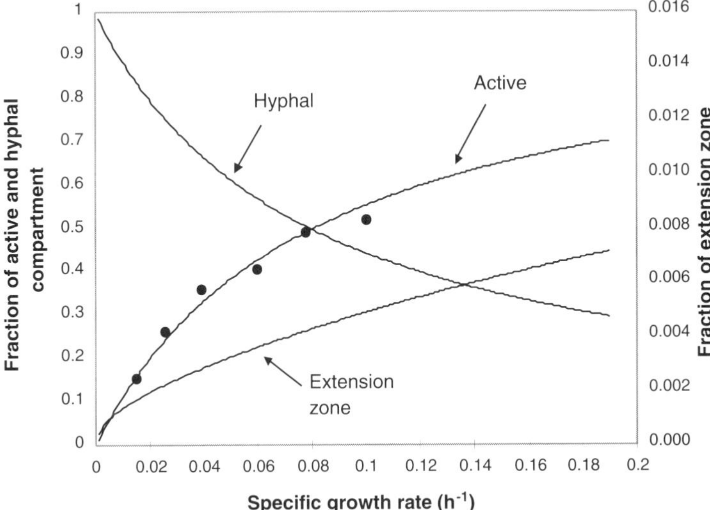

Figure 14.20 Experimentally determined values of the fraction of active compartment (●), determined using DiOC$_6$6, during steady state operation in continuous cultivations of *Aspergillus oryzae*. The lines represent model simulations. The data are taken from Agger *et al.* (1998).

Most of the model parameters were estimated from morphological data obtained from image analysis of hyphal elements growing in submerged cultures. For example, the maximum tip extension rate k_3, the specific branching frequency k_1, the saturation parameter for tip extension K_3 and the initiation parameter for branching $(x_a/c_n)_0$ were determined from the corresponding parameters found by using image analysis on submerged batch cultures of *A. oryzae*. This was done by converting the latter parameters from a length to a mass basis, using the density, water content and diameter of the hyphal elements. Furthermore, the model predictions of the fraction of active cells were confirmed by the measurements performed using fluorescent probes, as described above (see Figure 14.20).

The model also includes an expression describing the production of α-amylase, which is subject to glucose repression. The enzyme production was taken to be related to the active compartment, and with the model it was possible to describe α-amylase production both at transient operating conditions in continuous cultures and in fed-batch cultures (Agger *et al.*, submitted). Furthermore, the model can describe the developing morphology in a batch culture.

14.5 CONCLUDING REMARKS

As described above, mathematical models incorporating a certain structuring of the biomass can be quite successful in describing the growth and product formation of filamentous micro-organisms. For parameter estimation and to confirm the hypotheses on which the models are based, image analysis is an indispensable tool. The techniques developed using image analysis enable the investigator to obtain the large amount of data necessary in order to provide a reliable quantification of filamentous growth and cellular differentiation.

REFERENCES

Adunola, O.K., Anderson, J.G. and Smith, J.E. (1976) Control and autolysis of a spherical cell form of *Aspergillus niger*. *Trans. Br. Mycol. Soc.* **67**: 27–31.

Agger, T., Spohr, A.B., Carlsen, M. and Nielsen, J. (1998) Growth and product formation of *Aspergillus oryzae* during submerged cultivations: Verification of a morphologically structured model using fluorescent probes. *Biotechnol. Bioengng* **57**: 321–9.

Bartnicki-Garcia, S. (1973) Fundamental aspects of hyphal morphologenesis. *Symp. Soc. Gen. Microbiol.* **23**: 245–67.

Bartnicki-Garcia, S. (1990) Role of vesicles in apical growth and a new mathematical model of hyphal morphogenesis. In *Tip Growth* (Heath, I.B., ed.), pp. 211–32. Academic Press: San Diego, CA.

Brown, D.E. and Fitzpatrick, S.W. (1979) A structured model for the kinetics of fungal amylase production. *Biotechnol. Lett.* **1**: 3–8.

Carlsen, M., Spohr, A.B., Nielsen, J. and Villadsen, J. (1996) Morphology and physiology of an α-amylase producing strain of *Aspergillus oryzae* during batch cultivations. *Biotechnol. Bioengng* **49**: 266–76.

Chang, P.L.Y. and Trevithick, J.R. (1974) How important is secretion of exoenzymes through apical cell walls of fungi? *Arch. Microbiol.* **101**: 181–293.

Cox, P.W. and Thomas, C.R. (1992) Classification and measurement of fungal pellets by automated image analysis. *Biotechnol. Bioengng* **39**: 945–52.

Donelly, D.P., Wilkins, M.F. and Boddy, L. (1995) An integrated image analysis approach for determining biomass, radial extent and box-count fractal dimension of microscopic mycelial systems. *Binary Comput. Microbiol.* **7**: 19–28.

Durant, G., Cox, P.W., Formisyn, P. and Thomas, C.R. (1994) Improved image analysis algorithm for the characterisation of mycelial aggregates after staining. *Biotechnol. Tech.* **8**: 759–64.

Ekundayo, J.A. and Carlile, M.J. (1964) Germination of sporangiospores of *Rhizopus arrhizus*: spore swelling and germ-tube emergence. *J. Gen. Microbiol.* **35**: 261–9.

Emerson, S. (1950) The growth phase in *Neurospora* corresponding to the logarithmic phase in unicellular organisms. *J. Bacteriol.* **60**: 221–3.

Fiddy, C. and Trinci, A.P.J. (1976) Mitosis, septation, branching and the duplication cycle in *Aspergillus nidulans*. *J. Gen. Microbiol.* **97**: 169–84.

Galbraith, J.C. and Smith, J.E. (1969) Filamentous growth of *Aspergillus niger* in submerged shake cultures. *Trans. Br. Mycol. Soc.* **52**: 237–46.

Griffin, D.H. (1994) *Fungal Physiology*, 2nd edn. Wiley–Liss: New York.

Grove, S.N. (1978) The cytology of hyphal tip growth. In *The Filamentous Fungi* (Smith, J.E. and Berry, D.R., eds), pp. 28–50. Edward Arnold: London.

Holmalahti, J., Raatikainen, O., Von Wright, A., Laatsch, H., Spohr, A., Lyngberg, O. and Nielsen J. (1998) Production of dihydroabikoviromycin by *Streptomyces anulatus*. Production parameters and chemical characterization of geno-toxicity. *J. Appl. Bacteriol.* in press.

Jones, C.L., Lonergan, G.T. and Mainwaring, D.E. (1993) A rapid method for the fractal analysis of fungal growth using image processing. *Binary Comput. Microbiol.* **5**: 171–80.

Jones, C.L., Lonergan, G.T. and Mainwaring, D.E. (1995) Acid phosphatase positional correlations in solid surface fungal cultivations: a fractal interpretation of biochemical differentiation. *Biochem. Biophys. Res. Comm.* **208**: 1159–65.

Kasten, F.H. (1993) Introduction to fluorescent probes. In *Fluorescent and Luminescent Probes for Biological Activity* (Mason, W.T., ed.), pp. 12–31. Academic Press: London.

Krabben, P. and Nielsen, J. (1997) Modeling the mycelium morphology of *Penicillium* species in submerged cultures. *Adv. Biochem. Engng Biotechnol.* **60**: 125–52.

Lejeune, R., Nielsen, J. and Baron, G.V. (1995) Morphology of *Trichoderma reesei* QM 9414 in submerged cultures. *Biotechnol. Bioengng* **47**: 609–15.

Mandels, G.R. and Darby, R.T. (1953) A. rapid cell-volume assay for fungi toxicity using fungus spores. *J. Bacteriol.* **65**: 16–26.

Marshall, K.C. and Alexander, M. (1960) Growth characteristics of fungi and actinomycetes. *J. Bacteriol.* **80**: 412–6.

Megee, R.D., Kinoshita, S., Fredrickson, A.G. and Tsuchiya, H.M. (1970) Differentiation and product formation in molds. *Biotechnol. Bioengng* **12**: 771–801.

Metz, B. (1976) From pulp to pellet, PhD thesis, Delft University of Technology, Delft.

Nestaas, E. and Wang D.I.C. (1983) Computer control of the penicillin fermentation using the filtration probe in conjunction with a structured process model. *Biotechnol. Bioengng* **25**: 781–96.

Nielsen, J. (1992) Modelling the growth of filamentous fungi. *Adv. Biochem. Engng Biotechnol.* **46**: 187–223.

Nielsen, J. (1993) A. simple morphologically structured model describing the growth of filamentous microorganisms. *Biotechnol. Bioengng* **41**: 715–27.

Nielsen, J. (1996) Modelling the morphology of filamentous microorganisms. *TIBTECH* **14**: 409–46.

Nielsen, J. and Carlsen, M. (1996) Fungal pellets. In *Immobilised Living Cell Systems: Modelling and Experimental Methods* (Willaert, R.G., Baron, G.V. and De Backer, L., eds), pp. 273–93. Wiley: Chichester.

Nielsen, J. and Krabben, P. (1995) Hyphal growth and fragmentation of *Penicillium chrysogenum* in submerged cultures. *Biotechnol. Bioengng* **46**: 588–98.

Nielsen, J., Johansen, C.L., Jacobsen, M., Krabben, P. and Villadsen, J. (1995) Pellet formation and fragmentation in submerged cultures of *Penicillium chrysogenum* and its relation to penicillin production. *Biotechnol. Prog.* **11**: 93–8.

Packer, H.L. and Thomas, C.R. (1990) Morphological measurements on filamentous microorganisms by fully automatic image analysis. *Biotechnol. Bioengng* **35**: 870–81.

Paul, G.C. (1992) PhD thesis, University of Birmingham, Birmingham.

Paul, G.C. and Thomas, C.R. (1996) A. structured model for hyphal differentiation and penicillin production using *Penicillium chrysogenum*. *Biotechnol. Bioengng* **51**: 558–72.

Paul, G.C., Kent, C.A. and Thomas, C.R. (1993) Viability testing and characterization of germination of fungal spores by automatic image analysis. *Biotechnol. Bioengng* **42**: 11–23.

Peberdy, J.F. (1994) Protein secretion in filamentous fungi – trying to understand a highly productive black box. *Trends Biotechnol.* **12**: 50–7.

Prosser, J.I. and Tough, A.J. (1991) Growth mechanisms and growth kinetics of filamentous microorganisms. *Crit. Rev. Biotechnol.* **10**: 253–74.

Reichl, U., King, R. and Gilles, E.D. (1992a) Effect of temperature and medium composition on mycelial growth of *Streptomyces tendae*. *J. Basic Microbiol.* **32**: 193–200.

Reichl, U., King, R. and Gilles, E.D. (1992b) Characterization of pellet morphology during submerged growth of *Streptomyces tendae* by image analysis. *Biotechnol. Bioengng* **39**: 164–70.

Reinhardt, M.O. (1892). Das Wachstum der Pilzhyphen. Ein Beitrag zur Kenntniss des Flächenwachsthums vegetabilischer Zellmembranen. *Jahrb. Wiss. Botanik* **23**: 479–566.

Savage, G.M. and Van der Brook, M.J. (1946) The fragmentation of the mycelium of *Penicillium notatum* and *Penicillium chrysogenum* by a high-speed blender and the evaluation of blended speed. *J. Bacteriol.* **52**: 385–91.

Smeulders, A.W.M. and Beckers, A.D. (1989) Accurate image measurement methods (applied to 3D length and distance measurements). In Proceedings of the 1st International Conference on Confocal Microscopy. Academic Medisch Centrum: Amsterdam.

Smith, J.E. (1975) The structure and development of filamentous fungi. In *The Filamentous Fungi* (Smith, J.E. and Berry, D.R., eds), pp. 1–15. Edward Arnold: London.

Smith, J.E. and Berry, D.R. (1974) *Biochemistry of Fungal Development*. Academic Press: London.

Spohr, A., Carlsen, M., Nielsen, J. and Villadsen, J. (1997) Morphological characterization of recombinant strains of *Aspergillus oryzae* producing α-amylase during batch cultivations. *Biotechnol. Lett.* **19**: 257–61.

Spohr, A., Carlsen, M., Nielsen, J. and Villadsen, J. (1998a) α-Amylase production in recombinant *Aspergillus oryzae* during fed-batch and continuous cultivations. *J. Ferment. Bioengng* (in press).

Spohr, A., Dam-Mikkelsen, C., Carlsen, M., Nielsen, J. and Villadsen, J. (1998b) On-line study of fungal morphology during submerged growth using a small flow-through chamber. *Biotechnol. Bioengng.* (in press).

Taguchi, H. (1971) The nature of fermentation fluids. *Adv. Biochem. Engng* **1**: 1–30.

Taguchi, H., Yoshida, T., Tomita, Y. and Teramoto, S. (1968) The effects of agitation on disruption of the mycelial pellets in stirred fermentors. *J. Ferm. Technol.* **46**: 814–22.

Terasaki, M. (1993) Probes for the endoplasmic reticulum. In *Fluorescent and Luminescent Probes for Biological Activity* (Mason, W.T., ed.), pp. 120–3. Academic Press: London.

Terasaki, M. and Reese, T.S. (1992) Characterization of endoplasmic reticulum by co-localisation of BiP and dicarbocyanine dyes. *J. Cell Sci.* **101**: 315–22.

Thomas, C.R. (1992) Image analysis: putting filamentous microorganims in the picture. *TIBTECH* **10**: 343–8.

Thomas, C.R. and Paul, G.C. (1996) Applications of image analysis in cell technology. *Curr. Opin. Biotechnol.* **7**: 35–45.

Trinci, A.P.J. (1970) Kinetics of the growth of mycelial pellets of *Aspergillus nidulans*. *Arch. Microbiol.* **73**: 353–67.

Trinci, A.P.J. (1978) The duplication cycle and vegatative development in moulds. In *The Filamentous Fungi* (Smith, J.E. and Berry, D.R., eds), vol. III, pp. 134–63. Edward Arnold: London.

Trinci, A.P.J. (1984) Regulation of hyphal brancning and hyphal orientation. In *The Ecology and Physiology of the Fungal Mycelium* (Jennings, D.H. and Rayner A.D.M., eds), pp. 23–52. Cambridge University Press: Cambridge, UK.

Tucker, K.G., Kelly, T., Delgrazia, P. and Thomas, C.R. (1992) Fully-automatic measurement of mycelial morphology by image analysis. *Biotechnol. Prog.* **8**: 353–9.

Van Suijdam, J.C. and Metz, B. (1981a) Influence of engineering variables upon the morphology of filamentous molds. *Biotechnol. Bioengng* **23**: 111–48.

Van Suijdam, J.C. and Metz, B. (1981b) Fungal pellet breakup as a function of shear in a fermentor. *J. Ferm. Technol.* **59**: 329–33.

Vanhoutte, B., Pons, M.N., Thomas, C.R., Louvel, L. and Vivier, H. (1995) Characterization of *Penicillium chrysogenum* physiology in submerged cultures by color and monochrome image analysis. *Biotechnol. Bioengng* **48**: 1–11.

Vermeulen, C.A. and Wessels J.G.H. (1986) Chitin biosynthesis by a fungal membrane preparation. Evidence for a transient non-crystalline state of chitin. *Eur. J. Biochem.* **158**: 411–5.

Wösten, H.A.B., Moukha, S.M., Sietsma, J.H. and Wessels, J.G.H. (1991) Localization of growth and secretion of proteins in *Aspergillus niger*. *J. Gen. Microbiol.* **137**: 2017–23.

Yang, H., Reichl, U., King, R. and Gilles, E.D. (1992a) Measurement and simulation of the morphological development of filamentous microorganisms. *Biotechnol. Bioengng* **39**: 44–8.

Yang, H., King, R., Reichl, U. and Gilles, E.D. (1992b) Mathematical model for apical growth, septation, and branching of mycelial microorganisms. *Biotechnol. Bioengng* **39**: 49–58.

Yano, T., Kodama, T. and Yamada, K. (1961) Fundamental studies on the aerobic fermentation. Part VII. Oxygen transfer within mold pellet. *Agr. Biol. Chem.* **25**: 580–4.

Zalokar, M. (1959) Growth and differentiation of neurospora hyphae. *Am. J. Bot.* **46**: 602–10.

Zangirolami, T.C., Johansen, C.L., Nielsen, J. and Jørgensen, S.B. (1997) Simulation of penicillin production in fed-batch cultivations using a morphologically structured model. *Biotechnol. Bioengng* **56**: 593–604.

APPENDIX: NOMENCLATURE

Symbol	Unit	Definition
a	tips/g extension zone dry weight (DW)	Constant
A_I	μm^2	Measured projected area
$B(\alpha, \beta)$		B-distribution with parameters α, β
c_n	kg^{-1}	Concentration of hyphal elements
c_s	g/kg	Substrate concentration
d_{eq}	µm	Equivalent diameter
d_h	µm	Hyphal diameter
D_{eff}	m^2/h	Effective diffusion coefficient
e_{spore}	l^{-1}	Concentration of spores
$g(t)$	h^{-1}	Germination frequency
k_1	tips/g active DW/h	Specific branching frequency (active)
k_2	h^{-1}	Rate constant
k_3	g active DW/ tip/h	Maximal tip extension rate
$k_{bran}(\mathbf{z})$	tips/μm/h	Branching constant
$k_{bran,max}$	tips/μm/h	Maximal specific branching frequency
k_{germ}, $k_{tip}(\mathbf{z})$, $k_{tip.\,max}$	μm/tip/h	Maximal tip extension rates
k_p	kg/h	Kinetic parameter
K_{germ}, K_{bran}, K_t	μm	Saturation constants
K_{si}	g/l	Kinetic constants
l_{germ}	μm	Length of germ tube
l_{bran}	μm	Length of branch
l_t	μm	Total hyphal length
$l_{t,av}$	μm	Average length of hyphal elements
l_{bran}	μm	Branching initiation parameter
m	kg	Pellet mass
n_0		Number of tips at $l_{t,av} = l_{bran}$
n_{av}		Number of tips
q_i	g/kg/h	Reaction rates
q_{germ}, q_{tip}	μm/tip/h	Extension rates
$q_{bran}(l_t, \mathbf{z})$	tips/h	Average branching frequency
$r(c_s)$	g/g/h	Specific rate of substrate consumption
R_{crit}	μm	Critical pellet radius
s	g/l	Substrate concentration
t	h	Time
t_s	h	Time at which germination starts
t_f	h	Time at which germination ceases
$\bar{t}$		Normalized germination time
V_{spore}	μm^3	Estimated spore volume

x_{e}	g/kg	Concentration of extension zone cells
x_{a}	g/kg	Concentration of active cells
y_{viab}		Fraction of viable spores
$\mathbf{z}$		Vector of environmental conditions

Greek letters

α		Parameter in B-distribution
β		Parameter in B-distribution
μ	h^{-1}	Specific growth rate
ρ_{pel}	g/cm^3	Pellet density

15

The Study of Microbial Biofilms by Classical Fluorescence Microscopy

Ching-Tsan Huang, Philip S. Stewart and Gordon A. McFeters
Montana State University, Bozeman, Montana, USA

15.1 INTRODUCTION

In open environmental systems, the majority of microbial activity is associated with an interface within thin biological layers known as biofilms. Biofilms cause problems ranging from reducing transfer efficiency and deteriorating materials to increasing health risks. Conversely, biofilms are a set of immobilized cells that can facilitate cell-product separation, increase cell concentration and enhance overall productivity. Since these communities are highly heterogeneous in structure and in physiological activity, biofilm research relies on the understanding of the spatial information of composition and metabolic activity within biofilms. Traditional methods of studying these communities, which involve removing biofilm samples from the substratum followed by homogenization, and chemical and biological analyses, fail to provide spatial information about the structure and function of biofilms. To reveal spatial information within biofilms, fluorescent staining coupled with cryoembedding, cryosectioning and image analysis is proving useful in biofilm research. Using classical fluorescent microscopy, spatial patterns of biofilm thickness and species distribution, as well as physiological activity such as respiratory activity and growth rate can be visualized. These spatial patterns indicate that biofilms are highly heterogeneous in both structure and function.

15.1.1 Biofilms: general description

When solid surfaces are submerged in an aquatic environment, suspended microbial cells attach to the surface, the immobilized cells grow, replicate and secrete extracellular polymers that surround the cells within a gelatinous matrix. This constitutes the collective surface community referred to as a biofilm (Characklis and Marshall, 1990). Although the term is usually applied to bacterial cells and their extracellular polymers, any biologically active layer of cells (microbial, plant or mammalian) can be considered a biofilm (Bryers, 1987). There

Digital Image Analysis of Microbes: Imaging, Morphometry, Fluorometry and Motility Techniques and Applications. Edited by M.H.F. Wilkinson and F. Schut.

are several features unique to microbial biofilms (Hamilton, 1988):

(a) Biofilms are the principal sites of biological activity in many natural and man-made environments.
(b) Biofilms are heterogeneous, with discontinuities in both vertical and horizontal dimensions, involving organisms and biotic and abiotic components, creating physicochemical microenvironments.
(c) Biofilms are dynamic structures: their heterogeneities vary with time.

Biofilms consist of cells and their secreted insoluble extracellular polymers, which are primarily polysaccharides. Biofilm formation, illustrated in Figure 15.1, is the net result of suspended cell deposition, attached cell metabolism and biofilm removal processes (Bryers, 1987). Deposition of cells onto a substratum involves several individual processes: the macromolecular organic preconditioning of the target surface, cellular transport from the bulk phase to the solid substratum, and reversible and irreversible cell adhesion to the surface. Cellular growth, substrate conversion, endogenous decay and extracellular polymer production collectively constitute biofilm metabolic processes. Biofilms can be removed from the surface through chemical challenges, abrasion, shear-related detachment and sloughing. Deposition and biofilm detachment have been studied extensively and mathematical models have been established for each process during the past decade (Characklis, 1990).

15.1.1a *Importance of biofilms*

Biofilms have been implicated in microbiological problems associated with industrial and drinking water systems. Detrimental effects of biofilm formation

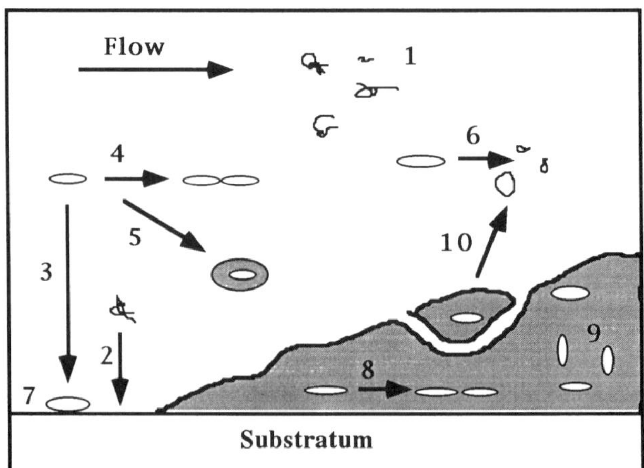

Figure 15.1 The mechanism of biofilm formation, in liquid phase: (1) dissolution of organic macromolecules, (2) pre-conditioning adhesion, (3) cell transport, (4) replication, (5) production of extracellular polymer, (6) death, (7) attachment to the substratum; within biofilm: (8) replication, (9) production of extracellular polymer and (10) biofilm detachment.

on system performance range from operational problems to economic losses. In industrial systems, biofilm formation and persistence create problems in engineered systems by increasing the resistance to mass, momentum and energy transfer, as well as by mediating chemical or biological reactions at the substratum. In drinking water systems, the results of a survey conducted in the US indicate that a relatively high proportion of potable water distribution systems occasionally experience excessive bacterial populations, in the absence of obvious defects within the distribution network (Smith *et al.*, 1990). Other studies have suggested that these unexplained occurrences of excessive heterotrophic and indicator bacteria result from microbial growth on pipe surfaces at the expense of nutrients in the bulk water (LeChevallier, 1990; Camper, 1994; LeChevallier *et al.*, 1996). Although limited direct experimental evidence is available that supports this hypothesis, results obtained from operating systems demonstrated increasing levels of coliforms with distance as water travelled through the distribution system, under conditions that would not allow proliferation of the bacteria within the bulk water (LeChevallier *et al.*, 1987). These reports are typical of a literature providing strong suggestive evidence associating the growth of coliforms and other heterotrophic bacteria within biofilms, in response to occasional physical and chemical conditions that are suitable for the multiplication of sessile bacteria on pipe surfaces. The subsequent release of these micro-organisms into the water column results in recalcitrant problems for the water utility in terms of meeting water quality standards and the erosion of consumer confidence.

Other microbiological problems have also been associated with biofilms in aquatic systems including drinking water networks. For example, biofilms containing sulphate-reducing bacteria (SRB) have been linked to biologically-enhanced corrosion of substrata containing iron (Little and Wagner, 1994). These biofilms require somewhat greater time to develop the necessary microzones and gradients that are found within thicker biofilms. Biocorrosion of other materials used in engineered aquatic systems is also promoted by attached bacteria. Nitrification is another biofilm-associated problem in drinking water systems (Wolfe *et al.*, 1990).

Apart from industrial and drinking water systems, biofilms also cause problems in dental plaque and device-related infections. Many microbiologists work in the area of oral health, and the clinical problems on which they focus increasingly involve microbial biofilms. As the problem of simple caries is partially solved by fluoride treatment and improved prophylaxis, microbiological attention is shifting to periodontal disease and to infections secondary to implantation and surgery. Periodontal disease is one of the most interesting and vexing biofilm diseases. It is clear that this disease is caused by a mixed species biofilm, some of whose microbial components are mobile cells that are only intermittently present in the biofilm itself (e.g. *Treponema*). The biofilm mode of growth in peridontitis promotes the survival of fastidious anaerobes, and the pathological cooperation of different microbial species, while modifying the interactions between the pathogens and the host tissues. Medical device-related bacterial infections have also been known to be caused by bacteria growing in biofilms (Khoury *et al.*, 1992). The surfaces of these devices are colonized by bacteria that are introduced at the time of surgery, or subsequently by haematogeneous spread, and these bacteria proliferate and produce exopolysaccharide slimes until mature biofilms are formed (Costerton and Anwar, 1994). These infections often develop slowly, as their causative biofilms withstand the sequential attacks of phagocytic cells (Jessen *et al.*, 1990) and antibodies (Costerton *et al.*, 1995), but most of them eventually serve as foci of acute disseminated infections that spread from the colonized surface and produce discernible symptoms.

15.1.1b Control of biofilms

The control of biofilms represents one of the most persistent challenges within engineered systems where these microbial communities are problematic. Mechanical cleaning and antimicrobial chemicals are the most commonly used methods of biofilm control. Mechanical cleaning is often impracticable and can be costly because it usually involves equipment downtime. Therefore, the main strategy of biofilm control relies on chemical biocides to kill the attached micro-organisms and/or remove them from the surface. However, biofilms are found to be hundreds of times more difficult to eradicate than their counterparts in suspended cultures. Bacteria within biofilms, such as those within drinking water distribution pipes, are notoriously difficult to control through the use of disinfectants. Concentrations of chlorine usually employed to disinfect potable water (i.e. 1–2 mg/l) are ineffective in controlling bacterial occurrences thought to be associated with biofilms (LeChevallier *et al.*, 1988). These and other studies have reported that surface-associated bacteria are as much as orders of magnitude less susceptible to the effects of antimicrobial agents (LeChevallier *et al.*, 1984; Wolfe *et al.*, 1990). Difficulties in formulating efficient control strategies are related to our incomplete understanding of biofilm processes and structure.

Mechanisms proposed to explain this enhanced resistance in bacteria within biofilms can be divided into two categories: transport limitation and physiological adaptation. Transport limitation is attributed to the neutralization of the antimicrobial agent in the biofilm faster than it can diffuse in to the community (Chen and Stewart, 1996; Stewart 1996; Xu *et al.*, 1996). Another explanation of biofilm resistance to chemical challenge is sought in physiological differences between biofilm and planktonic cells (Brown and Gilbert, 1993). It is quite reasonable to postulate that micro-organisms deep within a biofilm, where the chemical and physical microenvironment might be quite different from that in the bulk fluid, will be physiologically distinct and therefore less susceptible to disinfection. In particular, slow growing or starving micro-organisms in the interior of the biofilm are candidates for reduced susceptibility (Brown *et al.*, 1988; Tresse *et al.*, 1995).

Since biofilms are highly heterogeneous, traditional research methods that rely on the removal of biofilms from the substratum followed by disaggregation and enumeration provide little spatial information regarding the structure and physiological activity of individual bacteria within biofilms. To improve our understanding about the mechanisms of biofilm recalcitrance, and to develop more effective strategies of biofilm control, it is necessary to describe spatial patterns within biofilms. Using classical fluorescence microscopy in conjunction with cryoembedding, cryosectioning and image analysis, we are able to study the spatial patterns of biofilm structure (thickness variation and species distribution) and physiological activity (growth rate and respiratory activity).

15.2 APPROACHES TO REVEAL SPATIAL INFORMATION WITHIN BACTERIAL BIOFILMS

15.2.1 Introduction

Conventional microbiological approaches are inadequate for resolving spatial patterns of bacterial activity and species distribution as well as overall structure within biofilm communities. Traditional methods have relied on the quantitative removal of the biomass from the substratum using some type of shear force, such as scraping, followed by disaggregation to form a monodisperse bacterial suspension. This has usually been followed by identification and enumeration using colony formation or some other growth-dependent assay. This analytical approach is deficient for the analysis of biofilms because these communities are known to have established three-dimensional structural and chemical heterogeneities (Costerton *et al.*, 1995; Yang and Lewandowski, 1995) plus spatial and sometimes temporal gradients in physiological activities within individual cells (Huang *et al.*, 1995; McFeters *et al.*, 1995a; Wentland *et al.*, 1996). Hence, biofilm disaggregation obliterates the native community structure and any hope of gaining meaningful spatial information.

Cultural methods have also been criticized for their failure to detect bacteria under a number of circumstances. For example, natural bacterial communities are often composed of various autochthonous organisms, including some that have not been successfully cultivated (Ward *et al.*, 1992). Likewise, the detection and quantification of allochthonous bacteria in natural and some engineered systems can be problematic, because of reduced culturability under environmental conditions. Specifically, a variety of bacteria enter a viable but unculturable physiological state following aquatic exposure (Rozak and Colwell, 1987; see also Section 11.1.1). Some allochthonous bacteria can also become sublethally injured when exposed to sublethal conditions of antimicrobial treatment (McFeters, 1990; Singh and McFeters, 1990), and natural conditions can lead to bacteria that are both injured and viable but unculturable (Smith *et al.*, 1994) using established criteria. This is illustrated by results from controlled laboratory experiments using bacterial biofilm composed of both *Pseudomonas aeruginosa* and *Klebsiella pneumoniae* exposed to monochloramine for 2 hours (4 mg/l), where culturability decreased by orders of magnitude more than cellular respiration as well as net oxygen and glucose utilization (Stewart *et al.*, 1994). Under these circumstances, the bacteria become reversibly incapable of colony formation on certain media, while retaining metabolism and the ability to recover, grow and even initiate infections, in the case of pathogens, following a process of resuscitation (Singh and McFeters, 1990). This is significant since the detection and quantification of both autochthonous and allochthonous bacteria from environmental circumstances can be significantly underestimated when growth-dependent methodologies are used. These problems are further aggravated by the realization that the removal of biomass from the substratum as well as cellular disaggregation is seldom complete or quantitative.

15.2.2 Visualizing biofilm structure

A longstanding challenge in the study of bacterial biofilms has been the development of a method to visualize the structure of attached communities at the cellular level in their natural configuration without introduced artefacts. The commercial availability of the confocal laser scanning microscope (CLSM) provides this capability and allows the display of vertical sections through biofilms and the potential for two- and three-dimensional reconstruction of such communities (see Chapter 16; Lawrence *et al.*, 1991; Caldwell *et al.*, 1992a, b, 1993; Surman *et al.*, 1996). The cellular exclusion of low molecular weight fluorescent compounds has also been used to follow biofilm formation and structure, when coupled with CLSM and optical sectioning (Caldwell *et al.*, 1992b). However, problems have been encountered in the use of CSLM when studying thick or relatively opaque biofilms, and the instrumentation is still somewhat limited in terms of routine availability. A separate chapter in this volume (Chapter 16) is devoted to the use of CLSM to yield three-dimensional images of biofilms.

15.2.2a Cryosectioning and the study of bacterial biofilms

The examination of very thin biofilms can be accomplished by conventional microscopy coupled with conventional methodology to assess structure, viability and physiological activity (Yu *et al.*, 1993). Computer-enhanced darkfield microscopy has also been used in the quantitative analysis of bacterial growth and behaviour on surfaces (Lawrence *et al.*, 1989). However, biofilm communities are usually thicker than a monolayer and often exceed 100 μm in depth. Cryoembedding and cryosectioning coupled with fluorogenic stains and fluorescence microscopy are ideally suited to the examination of many biofilm types, irrespective of thickness, optical properties or substratum material. This approach has been extensively used in other biological as well as medical applications with great success, and was found practical for the detailed examination of biofilm structure, when used with fluorogenic dyes (Yu and McFeters, 1994a; Yu *et al.*, 1994). The method, illustrated in Figure 15.2, involves embedding the biofilm with a commercial medium (Tissue-Tech OCT, Miles Inc., Elkhart, IN, USA), and rapid freezing by placing the substratum on dry ice. The embedded biofilm is then easily removed from the substratum surface, turned over on the dry ice, and more OCT is added to surround the specimen fully. The resulting frozen block containing the embedded biofilm can then be sectioned with a cryostat to yield 5 μm slices of the community. This method requires minimal sample processing without prolonged fixation, and can be completed in only a few hours with minimal preparative artefacts. As an additional advantage, most of the equipment required for this procedure is readily available and the resulting images can be stored for later electronic processing.

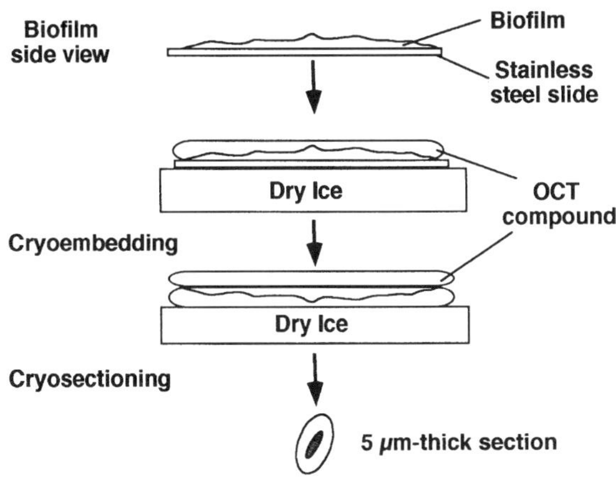

Figure 15.2 Cryoembedding and cryosectioning of biofilms.

Protocol 15.1 Cryoembedding and cryosectioning.

1 Gently dispense a layer of OCT to flood over biofilms accumulated on the substratum.
2 Place on a block of dry ice.
3 Wait until OCT turns white and is completely frozen.
4 Slightly bend the substratum and pop off the semi-embedded biofilms.
5 Turn the semi-embedded biofilms over and place on dry ice.
6 Dispense another layer of OCT.
7 After OCT is frozen, label the substratum side with marker.
8 Chop the embedded sample with Teflon-coated razor blade and mount on a cryostat stand.
9 Section the sample at thickness of 5 μm in −20 °C cryostat chamber.
10 Pick up the 5 μm thick section with a glass slide.

15.2.2b Physiological assessment of biofilm bacteria using specific fluorochromes

The challenge of accurately assessing the physiological status of individual bacteria is especially critical in the arena of biofilm research. Chemical and physical data have established the existence of structural complexity and chemical gradients within biofilms, suggesting the possibility of physiological heterogeneities within thicker biofilms (Costerton *et al.*, 1995; Stewart *et al.*, 1995). However, only limited experimental evidence has been available to test that hypothesis (Van Loosdrecht *et al.*, 1990; Marshall and Goodman, 1994). Such a task seems almost trivial from the perspective of pure culture studies of planktonic bacteria in the laboratory, but it becomes a significant methodological challenge when examining fixed three-dimensional communities like biofilms.

Fluorescent stains are used in a very wide range of biological applications where the response of individual cells can be observed microscopically (see Chapters 4 and 12; Mason, 1993; Haugland, 1996). Although fluorescent techniques are commonly used by microbiologists, including the acridine orange direct count (AODC) and fluorescent antibody detection techniques, there have been only a few microbiological applications of the numerous potentially useful physiological fluorochromes currently available (McFeters *et al.*, 1995a). However, recent studies have demonstrated the feasibility of describing various physiological parameters within biofilms, including growth rate, respiratory activity, membrane potential and adenylate energy charge with these stains. The combined application of fluorescent stains and cryosectioning is ideally suited for the task of examining spatial and temporal heterogeneities of these and other physiological activities in biofilms and other fixed bacterial communities.

15.2.3 Image analysis of fluorescent images

In order to distinguish specific components of biofilms (cells or extracellular polymers) or physiological activity, most fluorescent stains require counterstaining by a nucleic acid or protein dye. Specific fluorescent stains generally stain specific parts of biofilms, while counterstains provide overall staining of the sample. After counterstaining, positive cells alone and both positive and negative cells can be selected using different optical microscope filters. Once the images are grabbed, they can be digitized using a digital camera, saved in a grey scale TIFF file format, and converted into two-dimensional array numerical data for further analysis.

The image analysis system used at the Center for Biofilm Engineering at Montana State University includes an Olympus BH-2 Epi-Illumination ultraviolet (UV) upright microscope with 100 Watt mercury lamp and 50 Watt halogen lamp, Optronics OPDE-47OT cooled colour charge-coupled device (CCD) camera, with a $\frac{1}{2}$ in Sony ICX038AK chip with pixel resolution of 768(H) $\times$ 494(V), a Sony high resolution SVPVM1353 13 in RGB colour monitor, and a Sony UP-1200 thermal dye-sublimation video image printer. The image is processed by a Targa 64+ ADC (Truevision, Inc.) card, and handled by Image-Pro Plus 2.0 software (Media Cybernetics). The image is captured using different microscope filter cubic units and can be saved as either 8-bit grey scale or 24-bit colour image. Table 15.1 lists the specifications of microscope filters used at the Center for Biofilm Engineering. For quantitative analysis, the image is saved as a grey scale. Each pixel is assigned a number between 0 to 255 according to its intensity. An in-house software package, MARK, is used to analyse these digital image data. To avoid the error resulting from different image enhancement set-up, MARK is not used to enhance or sharpen images. The alignment of images is checked by placing a transparent sheet with grid on the monitor.

Table 15.1 Parameters of microscope filters used at the Center for Biofilm Engineering.

Filter type	Excitation filter	Dichroic mirror	Barrier filter
Olympus U	U (UG-1)	DM-400	L-420
Olympus B	B (BP-490)	DM-500	AFC + O-515
Olympus G	G (BP-545)	DM-570	O-590
Omega Optical 5716	485	505	535 ± 17.5
Omega Optical 5721	515	565	$\geqslant 590$

15.3 SPATIAL INFORMATION WITHIN BACTERIAL BIOFILMS

15.3.1 Thickness variability in biofilms

Microscopic examination of frozen biofilm sections affords high-resolution information about biofilm structure. Some of the physical properties that have or could be measured through image analysis of biofilm cross-sections are local biofilm thickness, nucleic acid density and extracellular polysaccharide (EPS) density (Kristensen and Christensen, 1982; Ganczarczyk and Zahid, 1994; Yu *et al.*, 1994; Zahid and Ganczarczyk, 1994; Zhang and Bishop, 1994; Stewart *et al.*, 1995).

Murga *et al.* (1995) used MARK image analysis software at the Center for Biofilm Engineering to measure biofilm thickness profiles along transects as long as 1 cm. This was done by catenating information from sequential, slightly overlapping, digitized images. Each image was a few hundred micrometres wide. Digitized images were imported into the custom software. This software allowed the user to draw a line at the location of the substratum and to trace, using the mouse, the outline of the biofilm surface. In their study, Murga *et al.* defined the biofilm surface as the outermost material stained by Gill II haematoxylin. The software then automatically calculated a thickness profile for the image by constructing perpendicular lines from substratum to the biofilm surface trace at defined lateral intervals of a few micrometres. An example of a thickness profile from a single image is reproduced in Figure 15.3. Thickness profiles from sequential images were subsequently catenated by eye in a spreadsheet.

The digitized images and thickness profiles obtained as described above reveal structural heterogeneity consistent with that observed by other techniques, such as CLSM and the Electroscan wet scanning electron microscopy (SEM). Features such

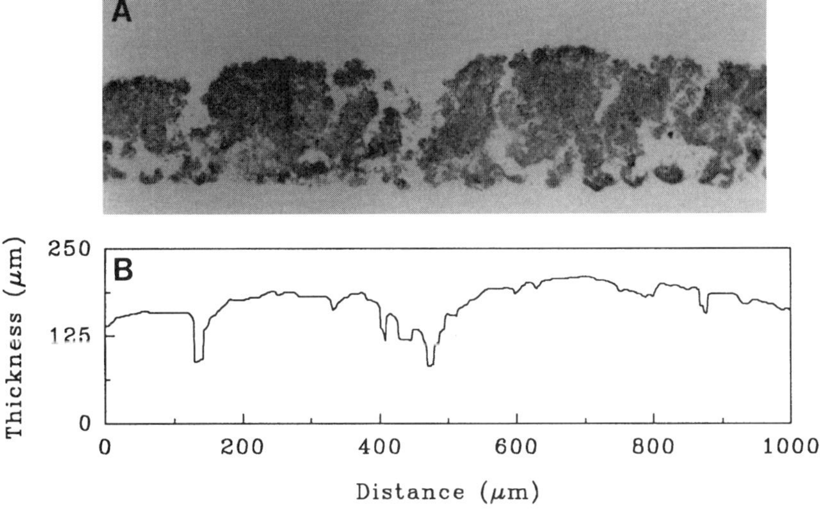

Figure 15.3 (A) Microscopic image of a frozen cross-section of a two-species (*K. pneumoniae*, *P. aeruginosa*) biofilm and (B) the thickness profile obtained from that image.

as thickness variation, void areas and putative water channels are visible. Although staining biofilm internal structure has not been quantified by image analysis, it should be straightforward to convert differences in relative staining intensity, for example with a DNA stain such as DAPI (4′,6-diamidino-2-phenylindole), into maps of relative cell density. An example of the potential for this type of analysis is presented in Plate 3, which compares the distribution of nucleic acids (as stained by ethidium bromide) with the distribution of EPS (as stained by Calcofluor White) in the same section. An overlay of the digitized patterns for the two stains indicates the presence of cell-free regions that nevertheless contain EPS.

15.3.2 Species distribution in biofilms

Spatial heterogeneity is a hallmark of biofilms. Spatial patterns of micro-organism distribution within biofilms have only recently begun to be investigated. To study species distribution within biofilms, a method is required to identify specific species while preserving biofilm spatial structure. The preservation of biofilm structure can be achieved by using cryoembedding techniques or CLSM. Recent advances in immunology and molecular biology have greatly expanded the array of tools that can be applied for speciation. Among these, immunofluorescent staining and oligonucleotide probing are the most frequently used methods in the investigation of species distribution within biofilms.

Various immunological techniques have been used successfully to study the distribution of methanogens in methanogenic granular sludge (Macario *et al.*, 1991; Visser *et al.*, 1991). Recently, Stewart *et al.* (1997) utilized cryoembedding and cryosectioning coupled with immunofluorescent staining to reveal the spatial distribution and coexistence of *K. pneumoniae* and *P. aeruginosa* in biofilms. They cultivated binary population biofilms using annular reactors by inoculating *K. pneumoniae* and *P. aeruginosa* at similar concentrations. Biofilm sections were stained with anti-*Klebsiella* fluorescein isothiocyanate (FITC) labelled monoclonal antibody followed by propidium iodide counterstaining. The spatial heterogeneity of *K. pneumoniae* and *P. aeruginosa* biofilm is shown in Plate 4A–C. They found that most of the biofilms fluoresce red only after staining (Plate 4A), indicating that *P. aeruginosa* dominated the biofilm. In some regions, both *K. pneumoniae and P. aeruginosa* were partially intermixed from the surface to the substratum (Plate 4B), while *K. pneumoniae* appeared as a distinct colony above a *P. aeruginosa* base film in other regions (Plate 4C). Although fluorescent antibodies have been used extensively in the study of microbial ecology, due to their high specificity, this approach has limitations. To prepare specific antibodies, it is necessary to isolate the target micro-organism and to cultivate it in pure culture prior to immunization of the animal. However, not all natural populations are amenable to pure-culture isolation.

In situ hybridization with fluorescently labelled oligonucleotide probes is another approach used in the detection and identification of micro-organisms in biofilms. Although DNA-based hybridization is also used (Holben *et al.*, 1988; Ogram and Saylor, 1988), rRNA-based oligonucleotide probes are most frequently used. Oligonucleotide probes targeting rRNA are attractive because rRNA is an abundant constituent of all living cells. The rRNA oligonucleotide probe has been successfully applied in the identification of ruminal fibrolytic bacteria (Odenyo *et al.*, 1994), sulphate-reducing bacteria (Amann *et al.*, 1992; Poulsen *et al.*, 1993; Ramsing *et al.*, 1993; Raskin *et al.*, 1996), methylotrophic bacteria (Tsien *et al.*, 1990), methanogenic bacteria (Raskin *et al.*, 1996) and *Pseudomonas fluorescens* (Boye *et al.*, 1995). There are some obstacles to general application of the rRNA approach. Difficulties in the application of the rRNA approach can occur in both sequence retrieval and probing of highly diverse samples (Amann *et al.*, 1995).

Although each method has its limitations, the combination of cryoembedding, cryosectioning, microscopy, immunofluorescent staining, molecular probing and image analysis should provide a better approach in the study of species distribution within biofilms.

Protocol 15.2 Immunostaining of *Klebsiella pneumoniae* and *Pseudomonas aeruginosa* biofilms.

1 Fix frozen sections with 1% formaldehyde at room temperature for 1 h.
2 Wash three times with TBS solution for 5 min. To prepare TBS solution, mix 4.5 g NaCl and 0.605 g Tris in 500 ml H_2O and adjust pH to 7.4.
3 Wash with blocking solution for 1 h at room temperature. To prepare blocking solution, add 0.3% Tween-80 and 2% skimmed milk in TBS solution.
4 Wash three times with TBST solution for 5 min. To prepare TBST solution, add 0.03% Tween-80 in TBS.
5 Add primary antibody solution and incubate at room temperature for 2 h. To prepare primary antibody solution, rat *K. pneumoniae* monoclonal antibody was diluted 500-fold by TBST.
6 Wash three times with TBST solution for 5 min.
7 Add secondary antibody solution and incubate at room temperature in darkness for 2 h. To prepare secondary antibody solution, goat anti-rat monoclonal antibody labelled with FITC was diluted 1000-fold by TBST.
8 Wash three times with TBST for 5 min in darkness.
9 Counterstain with 50 mg/ml propidium iodide for 2 in in darkness.
10 Wash once with TBS, shake in TBS and store in the freezer.

15.3.3 Physiological heterogeneities within biofilms

15.3.3a Growth rate

Only a few of the numerous fluorogenic stains with potential to indicate specific physiological and biochemical processes at the cellular level have been applied to microbiological studies (McFeters *et al.*, 1995a; see also Chapter 11). Among these, acridine orange (AO) has been widely used in the AODC technique for determining 'total' bacterial concentrations and has been interpreted with putative physiological implications. Some workers have used the staining reaction with AO to distinguish between active and inactive cells. While that practice has been openly questioned by some, it is based on the well-established differential spectrofluorescence of AO intercalated within single-stranded versus double-stranded nucleic acids and the relationship between the concentration of single-stranded (RNA) versus double-stranded (DNA) nucleic acids and growth rate in bacteria. Specifically, more rapidly growing enteric bacteria, which contain a higher proportion of single-stranded nucleic acids, emit red to orange fluorescence while inactive bacteria appear green due to the different modes of AO–nucleic acid intercalation. Despite these factors, and considerable disagreement, the physiological validity of the AO stain remained largely untested until a crucial study was done (McFeters *et al.*, 1991). This study examined the physiological interpretation of the AO stain in enteric bacteria, by using *Escherichia coli* grown under a range of controlled laboratory conditions. Both these bacteria and purified nucleic acids and a single-stranded bacteriophage were examined using a number of methodological variables of the AODC staining technique. The findings indicate that the fluorescent outcome of the AO stain is a potentially useful index of physiological activity under carefully controlled conditions with some bacteria. Hence, this concept has been employed successfully to examine growth rate heterogeneities within bacterial biofilms and colonies that were cryosectioned followed by microscopic examination and electronic quantitation by image analysis.

Wentland *et al.* (1996) stained colony biofilm frozen sections, pretreated with acidic fixative (5% acetic acid, 4% formaldehyde, 85% ethanol), using 4 mg/l acridine orange, and examined them using epifluorescence microscopy. Plate 4D shows the growth rate pattern of a real *K. pneumoniae* biofilm section stained with acridine orange. The biofilm was green in the interior with an orange band running along the biofilm–bulk fluid interface. In places where the biofilm was locally thin, the orange band extended to the substratum. On colony biofilms, they found that growth rates correlated with the orange:green intensity ratios. The orange and green images were captured using Omega Optical filter cubic units 5721 and 5716, respectively. The images were saved as grey-scale TIFF files and their intensities were calculated by MARK software. Figure 15.4 shows the comparison of average

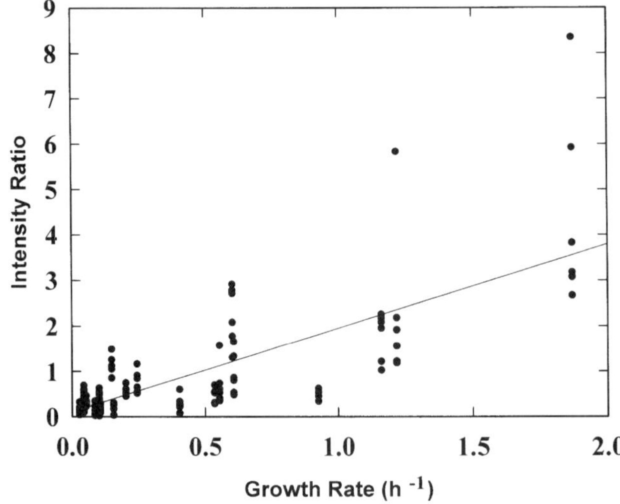

Figure 15.4 Comparison of average orange:green intensity ratio and colony average specific growth rate. The regressed line shown is given by $y = 1.84x + 0.107$ ($r^2 = 0.57$). Reproduced from Wentland *et al.* (1996), by permission of the American Institute of Chemical Engineers.

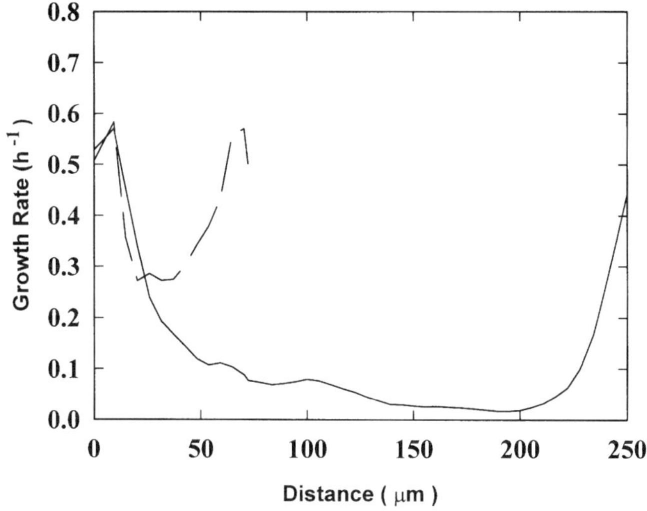

Figure 15.5 Estimated growth rate profiles in aerobically grown colonies. The colony ages and average specific growth rates were (solid line) 30.42 hours old, 0.086 h^{-1} and (broken line) 8.17 hours old, 0.56 h^{-1}. The membrane interface (agar side) was at approximately zero on the x axis. Reproduced from Wentland *et al.* (1996), by permission of the American Institute of Chemical Engineers.

orange:green fluorescent intensity ratios and average specific growth rates determined from cell count data. Using the linear regression of data in Figure 15.4 and profiles for the orange:green intensity ratio, spatial variations in the growth rate across a colony cross-section can be estimated (Figure 15.5). These profiles indicated that bacteria were growing rapidly near the air and agar interfaces, and more slowly in the centre of colonies. The correlation presented in Figure 15.5 is consistent with the differential staining of RNA and DNA by acridine orange and the known dependence of bacterial RNA:DNA ratio on specific growth rate (Moyer *et al.*, 1990; Berdalet and Dortch, 1991; Kerkhof and Ward, 1993).

These results indicate that orange colour corresponds to regions of rapid growth and green to regions of slow growth. The underlying mechanism for this approach for assessing actively growing bacteria has also been demonstrated through the use of fluorescent rRNA probes on planktonic cells (DeLong *et al.*, 1989) and bacteria within activated sludge flocs (Manz *et al.*, 1994).

Protocol 15.3 Acridine orange staining to distinguish growth rate.

1 Fix frozen sections mounted on glass slides in an acidic fixative at 4 °C for 10 min.
2 To prepare the acidic fixative, mix 5 ml 37% formaldehyde, 2.5 ml glacial acetic acid and 42.5 ml 95% ethanol.
3 Rinse the sections twice with 4 °C 85% ethanol.
4 Allow to air dry.
5 Add 5 μl of 4 μg/ml acridine orange on the section and leave in darkness for 1 min.
6 Blot excess acridine orange with tissue paper.

15.3.3b *Respiratory activity and membrane potential*

The microbiological adaptation of CTC (5-cyano-2,3-ditolyltetrazolium chloride) by Rodriguez *et al.* (1992) has enhanced the evaluation of respiratory activity of individual bacteria because of its fluorogenic properties. Although the mechanism of cellular CTC reduction is known (Smith and McFeters, 1997) and some limitations have been encountered in its use (Pyle *et al.*, 1995; Smith and McFeters, 1996), this compound remains useful in physiological studies of biofilms (McFeters *et al.*, 1995a; McFeters *et al.*, 1995b). When coupled with cryosectioning, this compound has allowed the visualization of spatial and temporal cellular respiratory activity in bacterial biofilms with disinfectant treatment (Huang *et al.*, 1995). CTC has also been used to determine that 5–35% of the bacteria within drinking water biofilms demonstrated active respiration and that starvation resulted in a rapid reduction in electron transport activity (Schaule *et al.*, 1993).

Fluorescent probes that reveal cellular bioenergetic properties have also been useful in the assessment of physiological activities within bacterial biofilms. Rhodamine 123 (Rh123), which signals membrane potential, is such a stain that has been applied to the determination of physiologically active bacterial populations in starved planktonic cultures of *Micrococcus luteus* (Kaprelyants *et al.*, 1996) and within biofilms (Yu and McFeters, 1994b). Enumeration results obtained after staining with Rh123 were comparable with data from CTC activity and an *in situ* direct viable count (DVC) method for enumeration of viable cells (Kogure *et al.*, 1979), but all three yielded significantly higher values than plate counts. Cellular energetic status was also evaluated within biofilm communities by determining spatial heterogeneities in adenylate energy charge, by cryosectioning followed by classical adenylate extraction and measurement (Kinniment and Wimpenny, 1992). Other membrane potential fluorogens will almost certainly be found to be useful in the description of cellular physiological heterogeneities within biofilms.

15.3.4 Physiological response of biofilms during disinfection

Accurate data describing the response of key cellular physiological processes within bacterial biofilms exposed to disinfectants is critical in the study of biofilm control. Such information is essential to (i) describe the mode of action of different disinfectants, (ii) help explain why bacteria within biofilms are less susceptible to antimicrobials and (iii) evaluate the efficacy of new and existing antimicrobial formulations. Therefore, the analytical approach utilizing fluorogenic stains for key physiological activities in combination with cryosectioning, microscopic examination and digital image analysis is ideal for gaining a greater understanding of biofilm control strategies.

Although more traditional methodologies such as DVC and fluorescence microscopy have been useful for describing the effects of biocides on very thin biofilms *in situ* (Yu *et al.*, 1993), alternative approaches are clearly needed when studying the disinfection of thicker biofilms, which are more typical of sessile communities in natural systems. Further, the method for assessing the cellular response to biocide exposure is optimal if it specifically addresses a key physiological process that is known to be compromised by the disinfectant under investigation. Chlorine-based biocides continue to be widely used as disinfectants in aquatic systems, and their coupling with CTC to assess the physiological response of the treated bacteria satisfies that experimental design criterion. Specifically, chlorine is known to damage bacterial respiratory activity, when used in a way that mimics the disinfection of drinking water (Camper and McFeters, 1979), and CTC provides a compatible microscopic method to evaluate the respiratory activity of the treated biofilm at the cellular level.

Studies were carried out to describe the spatial and temporal physiological response of bacterial biofilms to disinfection with chloramine (Huang *et al.*, 1995). Biofilms containing both *P. aeruginosa* and *K. pneumoniae* were grown on stainless steel coupons in a continuous-flow reactor, which were then treated with monochloramine (2 mg/l) for two hours. This resulted in very little removal of biofilm, but there was a linear one-log reduction of culturable bacteria belonging to both species. Coupons with control and treated biofilms were also examined at various times of disinfection exposure, to discriminate respiring and nonrespiring cells using CTC and DAPI, as a counterstain, followed by cryosectioning, microscopic examination and data acquisition by image analysis. Examination of these biofilms using epi-illuminated fluorescence microscopy clearly revealed uniform activity throughout the community in control samples that were not treated with biocide, but increasing gradients of respiratory activity, seen as striking images of active (orange) and inactive (green) cells, with time of exposure to disinfectant. The gradients in specific cellular respiratory activity were then quantified by calculating the ratio of reduced CTC to DAPI intensities as shown in Figure 15.6. The CTC and DAPI stained images were captured using Olympus G and U filters, respectively, and the intensities were measured by image analysis. These data, which resulted in a family of curves demonstrating respiratory activity

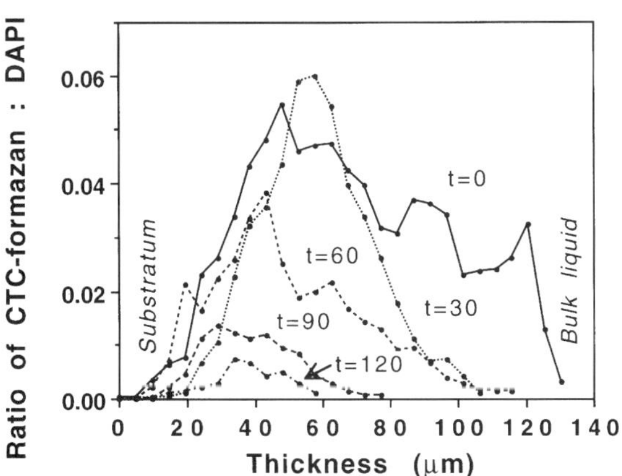

Figure 15.6 Ratio of intensities of CTC–formazan and DAPI obtained from 5 μm frozen sections following exposure to 2 mg monochloramine per litre. The biofilm thicknesses determined by the dimensions of the DAPI-stained region were 121 μm before treatment and 116, 106, 125 and 92 μm at 30, 60, 90 and 120 min, respectively. *t*, Time (min). Reproduced from Huang *et al.* (1995), by permission of the American Society for Microbiology.

versus position within the biofilm for each of the biocide exposure times, graphically revealed the spatial and temporal effects of monochloramine on bacterial cells within this biofilm community. Before application of chloramine, high levels of reduced CTC were seen throughout the entire 130 μm depth of the biofilm, indicating relatively uniform respiratory activity in the community. However, with disinfection, individual cells near the biofilm–bulk fluid interface lost respiratory activity first and after 120 minutes of exposure, only an approximately 60 μm thick region of the biofilm exhibited low levels of respiration. It is also of interest that the remaining active cells following disinfection were near the substratum and centrally located within structured cell clusters.

This study demonstrates the utility of using a specific fluorogenic stain to reveal the non-uniform physiological response of individual bacteria within a biofilm exposed to a disinfectant. The resulting data have provided a greater understanding of biofilm control, and allow the formulation of new hypotheses to stimulate further advances. This development in the area of biofilm research is mirrored by similar advances using flow cytometry technologies to evaluate the physiology of heterogeneous planktonic bacterial populations using some of the same fluorochromes (Davey and Kell, 1996).

ACKNOWLEDGEMENTS

This work was supported through cooperative agreement EEC-8907039 between the US National Science Foundation and Montana State University and by the industrial partners of the Center for Biofilm Engineering.

REFERENCES

Amann, R.I., Stromley, J., Deverux, R., Key, R. and Stahl, D.A. (1992) Molecular and microscopic identification of sulfate-reducing bacteria in multispecies biofilms. *Appl. Environ. Microbiol.* **58**: 614–23.

Amann, R.I., Ludwig, W. and Schleifer, K.-H. (1995) Phylogenetic identification and in situ detection of individual microbial cells without cultivation. *Microbiol. Rev.* **59**: 143–69.

Berdalet, E. and Dortch, Q. (1991) New double-staining technique for RNA and DNA measurement in marine phytoplankton. *Mar. Ecol. Prog. Ser.* **73**: 295–305.

Boye, M., Ahl, T. and Molin, S. (1995) Application of a strain-specific rRNA oligonucleotide probe targeting *Pseudomonas fluorescens* Ag1 in a mesocosm study of bacterial release into the environment. *Appl. Environ. Microbiol.* **61**: 1384–90.

Brown, M.R.W. and Gilbert, P. (1993) Sensitivity of biofilms to antimicrobial agents. *J. Appl. Bacteriol. Symp. Suppl.* **74**: 87S–97S.

Brown, M.R.W., Allison, D.G. and Gilbert, P. (1988) Resistance of bacterial biofilms to antibiotics: a growth-rate related effect? *J. Antimicrob. Chemother.* **22**: 777–83.

Bryers, J.D. (1987) Biologically active surfaces: processes governing the formation and persistence of biofilms. *Biotechnol. Prog.* **3**: 57–68.

Caldwell, D.E., Korber, D.R. and Lawrence, J.R. (1992a) Confocal laser microscopy and digital image analysis in microbial ecology. In *Advances in Microbial Ecology*, vol. 12, pp.1–67. Plenum Press: New York.

Caldwell, D.E., Korber, D.R. and Lawrence, J.R. (1992b) Imaging of bacteria cells by fluorescent exclusion using scanning confocal laser microscopy. *J. Microbiol. Meth.* **15**: 249–61.

Caldwell, D.E., Korber, D.R. and Lawrence, J.R. (1993) Analysis of biofilm formation using 2D and 3D digital imaging. *J. Appl. Bacteriol. Symp. Suppl.* **74**: 52S–66S.

Camper, A.K. (1994) Coliform regrowth and biofilm accumulation in drinking water systems: a review. In *Biofouling and Biocorrosion in Industrial Water Systems* (Geesey, G.G., Lewandowski, Z. and Flemming, H.-C., eds), pp. 91–105. CRC Press: Boca Raton, FL.

Camper, A.K. and McFeters, G.A. (1979) Chlorine injury and the enumeration of waterborne coliform bacteria. *Appl. Environ. Microbiol.* **37**: 633–41.

Chen, X. and Stewart, P.S. (1996) Chlorine penetration into artificial biofilm is limited by a reaction–diffusion interaction. *Environ. Sci. Technol.* **30**: 2078–83.

Characklis, W.G. (1990) Biofilm processes. In *Biofilms* (Characklis, W.G. and Marshall, K.C., eds), pp. 195–231. John Wiley & Sons: New York.

Characklis, W.G. and Marshall, K.C. (1990) Biofilms: a basis for an interdisciplinary approach. In *Biofilms* (Characklis, W.G. and Marshall, K.C., eds), pp. 3–15. John Wiley & Sons: New York.

Costerton, J.W. and Anwar, H. (1994) *Pseudomonas aeruginosa*: the microbe and the pathogen. In Pseudomonas aeruginosa *Infections and Treatments* (Baltch, A. and Smith, P., eds), pp. 1–18. Dekker: New York.

Costerton, J.W., Lewandowski, Z., Caldwell, D.E., Korber, D.R. and Lappin-Scott, H.M. (1995) Microbial biofilms. *Ann. Rev. Microbiol.* **49**: 711–45.

Davey, H.M. and Kell, D.B. (1996) Flow cytometry and cell sorting of heterogeneous microbial populations: the importance of single-cell analysis. *Microbiol. Rev.* **60**: 641–96.

DeLong, E.F., Wickman, G.S. and Pace, N.R. (1989) Phylogenetic stains: ribosomal RNA-based probes for the identification of single cells. *Science* **243**: 1360–3.

Ganczarczyk, J.J. and Zahid, W.M. (1994) Structure of RBC biofilms. *Wat. Environ. Res.* 66: 100–6.

Hamilton, A. (1988) Biofilms at the interface between microbiology and engineering. *Trends Biotechnol.* **7**: 19–20.

Haugland, R.P. (1996) *Handbook of Fluorescent Probes and Research Chemicals*. Molecular Probes: Eugene, OR.

Holben, W.E., Jansson, J.K., Chelm, B.K. and Tiedje, J.M. (1988) DNA probe method for the detection of specific micro-organisms in the soil bacterial community. *Appl. Environ. Microbiol.* **54**: 703–11.

Huang, C.-T., Yu, F.P., McFeters, G.A. and Stewart, P.S. (1995) Nonuniform spatial patterns of respiratory activity within biofilms during disinfection. *Appl. Environ. Microbiol.* **61**: 2252–6.

Jessen, E.T., Kharazmi, A. and Lam, K. (1990) Human polymorphonuclear leukocyte response to *Pseudomonas aeruginosa* biofilms. *Infect. Immun.* **58**: 2383–5.

Kaprelyants, A.S., Mukamolova, G.V., Davey, H.M. and Kell, D.B. (1996) Quantitative analysis of physiological heterogeneity within starved cultures of *M. luteus* by flow cytometry. *Appl. Environ. Microbiol.* **62**: 1311–6.

Kerkhof, L. and Ward, B.B. (1993) Comparison of nucleic acid hybridization and fluorometry for measurement of the relationship between RNA/DNA ratio and growth rate in a marine bacterium. *Appl. Environ. Microbiol.* **59**: 1303–9.

Khoury, A.E., Lam, K., Ellis, B. and Costerton, J.W. (1992) Prevention and control of bacterial infections associated with medical devices. *ASAIO J.* **38**: M174–8.

Kinniment, S.L. and Wimpenny, J.W.T. (1992) Measurement of the distribution of adenylate concentrations and adenylate energy charge across *P. aeruginosa* biofilms. *Appl. Environ. Microbiol.* **58**: 1629–35.

Kogure, K., Simidu, U. and Toga, N. (1979) A tentative direct microscopic method for counting living marine bacteria. *Can. J. Microbiol.* **25**: 415–20.

Kristensen, G.H. and Christensen, F.R. (1982) Application of cryo-cut method for measurements of biofilm thickness. *Wat. Res.* **16**: 1619–21.

Lawrence, J.R., Korber, D.R. and Caldwell, D.E. (1989) Computer enhanced darkfield microscopy for the quantitative analysis of bacterial growth and behavior on surfaces. *J. Microbiol. Meth.* **10**: 123–38.

Lawrence, J.R., Korber, D.R., Hoye, B.D., Costerton, J.W. and Caldwell, D.E. (1991) Optical sectioning of microbial biofilms. *J. Bacteriol.* **173**: 6558–67.

LeChevallier, M.W. (1990) Coliform regrowth in drinking water: a review. *J. Am. Water Works Assoc.* **82**(11): 74–86.

LeChevallier, M.W., Hassenauer, T.S., Camper, A.K. and McFeters, G.A. (1984) Disinfection of bacteria attached to granular activated carbon. *Appl. Environ. Microbiol.* **48**: 918–23.

LeChevallier, M.W., Babcock, T.M. and R.G. Lee (1987) Examination and characterization of distribution systems biofilms. *Appl. Environ. Microbiol.* **53**: 2714–24.

LeChevallier, M.W., Cawthon, C.D. and Lee, R.D. (1988) Inactivation of biofilm bacteria. *Appl. Environ. Microbiol.* **54**: 2492–9.

LeChevallier, M.L., Welch, N.J. and Smith, D.B. (1996) Full-scale studies of factors related to coliform regrowth in drinking water. *Appl. Environ. Microbiol.* **62**: 2201–11.

Little, B. and Wagner, P. (1994) Indicators for sulfate-reducing bacteria in microbiologically influenced corrosion. In *Biofouling and Biocorrosion in Industrial Water Systems* (Geesey, G.G., Lewandowski, Z. and Flemming, H.-C., eds), pp. 205–12. CRC Press: Boca Raton, FL.

Macario, A.J.L., Visser, F.A., van Lier, J.B. and Conway de Macario, E. (1991) Topography of methanogenic subpopulations in a microbial consortium adapting to thermophilic conditions. *J. Gen. Microbiol.* **137**: 2179–89.

Manz, W., Wagner, M., Amann, R. and Schliefer, K.-H. (1994) In situ characterization of the microbial consortia active in two wastewater treatment plants. *Water Res.* **28**: 1715–23.

Marshall, K.C. and Goodman, A.E. (1994) Effects of adhesion on microbial cell physiology. *Colloids Surf. B: Biointerfaces* **2**: 1–7.

Mason, W.T. (1993) *Fluorescent and Luminescent Probes for Biological Activity*. Academic Press: London.

McFeters, G.A. (1990) Enumeration, occurrence and significance of injured indicator bacteria in drinking water. In *Drinking Water Microbiology: Progress and Recent Developments* (McFeters, G.A., ed.), pp. 478–92. Springer-Verlag: New York.

McFeters, G.A., Singh, A., Byun, S., Callis, P.R. and Williams, S. (1991) Acridine orange staining reaction as an index of physiological activity in *Escherichia coli*. *J. Microbiol. Meth.* **13**: 87–97.

McFeters, G.A., Yu, F.P., Pyle, B.H. and Stewart, P.S. (1995a) Physiological assessment of bacteria using fluorochromes. *J. Microbiol. Meth.* **21**: 1–13.

McFeters, G.A., Yu, F.P., Pyle, B.H. and Stewart, P.S. (1995b) Physiological methods to study biofilm disinfection. *J. Ind. Microbiol.* **15**: 333–8.

Moyer, C.L., Mordy, C.W., Carlson, D.J. and Morita, R.Y. (1990) Ethidium homodimer used for the sensitive measurement of DNA and RNA of a psychrophilic marine bacterium grown at different growth rates during starvation-survival. *J. Microbiol. Meth.* **12**: 75–81.

Murga, R., Stewart, P.S. and Daly, D. (1995) Quantitative analysis of biofilm thickness variability. *Biotechnol. Bioengng* **45**: 503–10.

Odenyo, A.A., Mackie, R.I., Stahl, D.A. and White, B.A. (1994) The use of 16S rRNA-targeted oligonucleotide probes to study competition between ruminal fibrolytic bacteria: development of probes for Ruminococcus species and evidence for bacteriocin production. *Appl. Environ. Microbiol.* **60**: 3688–96.

Ogram, A.V. and Saylor, G.S. (1988) The use of gene probes in the rapid analysis of natural microbial communities. *J. Ind. Microbiol.* **3**: 281–92.

Poulsen, L.K., Ballard, G. and Stahl, D.A. (1993) Use of rRNA fluorescence in situ hybridization for measuring the activity of single cells in young and established biofilms. *Appl. Environ. Microbiol.* **59**: 1354–60.

Pyle, B.H., Broadaway, S.C. and McFeters, G.A. (1995) Factors affecting the determination of respiratory activity of the basis of CTC chloride reduction with membrane filtration. *Appl. Environ. Microbiol.* **61**: 4304–9.

Ramsing, N.B., Kühl, M. and Jørgensen, B.B. (1993) Distribution of sulfate-reducing bacteria, O_2 and H_2S in photosynthetic biofilms determined by oligonucleotide probes and microelectrodes. *Appl. Environ. Microbiol.* **59**: 3840–9.

Raskin, L., Rittmann, B.R. and Stahl, D.A. (1996) Competition and coexistence of sulfate-reducing and methanogenic populations in anaerobic biofilms. *Appl. Environ. Microbiol.* **62**: 3847–57.

Rodriguez, G.G., Phipps, D., Ishiguro, K. and Ridgway, H.F. (1992) Use of a fluorescent redox probe for direct visualization of actively respiring bacteria. *Appl. Environ. Microbiol.* **58**: 1801–8.

Rozak, D.B. and Colwell, R.R. (1987) Survival strategies of bacteria in the natural environment. *Microbiol. Rev.* **51**: 365–79.

Schaule, G., Flemming, H.-C. and Ridgway, H.F. (1993) Use of CTC for quantifying planktonic and sessile respiring bacteria in drinking water. *Appl. Environ. Microbiol.* **59**: 3950–7.

Singh, A. and McFeters, G.A. (1990) Injury of enteropathogenic bacteria in drinking water. In *Drinking Water Microbiology: Progress and Recent Developments* (McFeters, G.A., ed.), pp. 368–79. Springer-Verlag: New York.

Smith, D.B., Hess, A.F. and Hubbs, S.A. (1990) Survey of distribution system coliforms occurrences in the United States. In Proceedings of the Water Quality Technology Conference, American Water Works Association, San Diego, California, November, pp. 1103–16.

Smith, J.J. and McFeters, G.A. (1996) Effects of substrates and phosphate on INT and CTC reduction in *E. coli*. *J. Appl. Bacteriol.* **80**: 209–15.

Smith, J.J. and McFeters, G.A. (1997) Mechanisms of INT and CTC reduction in *E. coli* K-12. *J. Microbiol. Meth.* **29**: 161–75.

Smith, J.J., Howington, J.P. and McFeters, G.A. (1994) Survival, physiological response and recovery of enteric bacteria exposed to a polar marine environment. *Appl. Environ. Microbiol.* **60**: 2977–84.

Stewart, P.S. (1996) Theoretical aspects of antibiotic diffusion into microbial biofilms. *Antimicrob. Agents Chemother.* **40**: 2517–22.

Stewart, P.S., Griebe, T., Srinivasan, R., Chen, C.-I., Yu, F.P., de Beer, D. and McFeters, G.A. (1994) Comparison of respiratory activity and culturability during monochloramine disinfection of binary population biofilms. *Appl. Environ. Microbiol.* **60**: 1690–2.

Stewart, P.S., Murga, R., Srinivasan, R. and de Beer, D. (1995) Biofilm structural heterogeneity visualized by three microscopic methods. *Water Res.* **29**: 2006–9.

Stewart, P.S., Camper, A.K., Handran, S.D., Huang, C.-T. and Warnecke, M. (1997) Spatial distribution and coexistence of *Klebsiella pneumoniae* and *Pseudomonas aeruginosa* in biofilms. *Microb. Ecol.* **33**: 2–10.

Surman, S.B., Walker, J.T., Goddard, D.T., Morton, L.H.G., Keevil, C.W., Weaver, W., Skinner, A., Hanson, K., Caldwell, D. and Kurtz, J. (1996) Comparison of microscope techniques for the examination of biofilms. *J. Microbiol. Meth.* **25**: 57–70.

Tresse, O., Jouenne, T. and Junter, G.-A. (1995) The role of oxygen limitation in the resistance of agar-entrapped, sessile-like *Escherichia coli* to aminoglycoside and beta-lactam antibiotics. *J. Antimicrob. Chemother.* **36**: 521–6.

Tsien, H.C., Bratina, B.J., Tsuji, K. and Hanson, R.S. (1990) Use of oligodeoxynucleotide signature probes for identification of physiological groups of methylotrophic bacteria. *Appl. Environ. Microbiol.* **56**: 2858–65.

Van Loosdrecht, M.C.M., Lyklema, J., Norde, W. and Zhender, A.J.B. (1990) Influence of interfaces on microbial activity. *Microbiol. Rev.* **54**: 75–87.

Visser, F.A., van Lier, J.B., Macario, A.J.L. and Conway, de Macario, E. (1991) Diversity and population dynamics of methanogenic bacteria in a granular consortium. *Appl. Environ. Microbiol.* **57**: 1728–34.

Ward, D.M., Bateson, M.M., Weller, R. and Ruff-Roberts, A.L. (1992) Ribosomal analysis of microorganisms as they occur in nature. In *Advances in Microbial Ecology* (Marshall, K.C., ed.), vol. 12, pp. 219–86. Plenum Press: New York.

Wentland, E.J., Stewart, P.S., Huang, C.-T. and McFeters, G.A. (1996) Spatial variations in growth rate within *K. pneumoniae* colonies and biofilms. *Biotechnol. Prog.* **12**: 316–21.

Wolfe, R.L., Lieu, N.I., Izaguirre, G. and Means, E. (1990) Ammonia-oxidizing bacteria in a chloraminated distribution system: seasonal occurrence, distribution and disinfection resistance. *Appl. Environ. Microbiol.* **56**: 451–62.

Xu, S., Stewart, P.S. and Chen, X. (1996) Transport limitation of chlorine disinfection of *Pseudomonas aeruginosa* entrapped in alginate beads. *Biotechnol. Bioengng* **49**: 93–100.

Yang, S. and Lewandowski, Z. (1995) Measurement of local mass transfer coefficient in biofilms. *Biotechnol. Bioengng* **48**: 737–44.

Yu, F.P. and McFeters, G.A. (1994a) Rapid in situ assessment of physiological activities in bacterial biofilms using fluorescent probes. *J. Microbiol. Meth.* **20**: 1–10.

Yu, F.P. and McFeters, G.A. (1994b) Physiological responses of bacteria in biofilms to disinfection. *Appl. Environ. Microbiol.* **60**: 2462–6.

Yu, F.P., Pyle, B.H. and McFeters, G.A. (1993) A direct viable count method for the enumeration of attached bacteria and assessment of biofilm disinfection. *J. Microbiol. Meth.* **17**: 167–80.

Yu, F.P., Callis, G.M., Stewart, P.S., Grieb, T. and McFeters, G.A. (1994) Cryosectioning of biofilms for microscopic examination. *Biofouling* **8**: 85–91.

Zahid, W. and Ganczarczyk, J. (1994) A technique for a characterization of RBC biofilm surface. *Wat. Res.* **28**: 2229–31.

Zhang, T.C. and Bishop, P.L. (1994) Density, porosity, and pore structure of biofilms. *Wat. Res.* **28**: 2267–77.

16

The Study of Biofilms using Confocal Laser Scanning Microscopy

John R. Lawrence[1], Gideon M. Wolfaardt[2] and Thomas R. Neu[3]

[1]*National Hydrology Research Institute, Saskatoon, Saskatchewan, Canada*

[2]*University of Saskatchewan, Saskatoon, Saskatchewan, Canada*

[3]*UFZ, Magdeburg, Germany*

16.1 INTRODUCTION

The study of biofilms has been limited by the difficulty of relating structure and function in these highly organized microbial communities. Traditional approaches have often concentrated on disruption, fixation and other treatments to examine biofilms, thereby reducing or eliminating microbial relationships, complex structure and organization, which are the essential features of these systems. Various techniques have been applied to obtain non-destructive analysis of biofilms, including Fourier transformed infrared and nuclear magnetic resonance imaging (Geesey and White; 1990, Nivens *et al.*, 1995; Surman *et al.*, 1996). The examination of micro-organisms in their natural habitat may be achieved most effectively through the application of microscopic techniques (see Chapter 15). See, for example, the study of Surman *et al.* (1996) comparing a variety of microscopic techniques for biofilm studies. Among the most versatile and effective of these approaches is confocal laser scanning microscopy (CLSM).

Practical CLSM originated with the introduction of the first commercially available models in the early 1980s. Since that time CLSM technology has been applied in many fields, although medical applications have been the most extensive. CLSM allows detailed, non-destructive examination of thick, fully hydrated microbial biofilms. These capacities have been demonstrated for biofilms by Lawrence *et al.* (1991, 1994, 1996), Bott *et al.* (1997) and Neu and Lawrence (1997). The technique of optical sectioning removes out-of-focus information and leads to sharp, clear digital images with enhanced *xy* and *xz* resolution. As such, it is an ideal tool for studying the spatial distribution of cells in immobilized biofilm communities. The digital nature of the resulting images makes them amenable to image processing and analysis, allowing the user to obtain quantitative information, or to make three-dimensional reconstructions of the material under study (Wilson and Sheppard, 1984; Shotton and White, 1989; Caldwell *et al.*, 1992a). Concurrent

Digital Image Analysis of Microbes: Imaging, Morphometry, Fluorometry and Motility Techniques and Applications. Edited by M.H.F. Wilkinson and F. Schut.

with the development of CLSM technology has been the development of a wide range of fluorescent probes (Chapter 4) that have been very important for the application of CLSM in many fields, including microbial ecology and the study of biofilms. When CLSM is coupled with the large number of environmentally sensitive probes that are commercially available (Haugland, 1996), direct data may be acquired on diffusion (Lawrence *et al.*, 1994), redox, and pH and ion concentrations. Other probes reveal information on cell viability. CLSM used in conjunction with fluorescent antibodies or oligonucleotide probes provides vital information on the taxonomic affiliation (Assmus *et al.*, 1995) and growth rates (Poulsen *et al.*, 1993) of micro-organisms. Various studies have reshaped our concept of biofilms, emphasizing the presence of extensive structural complexity and heterogeneity within them (Lewandowski *et al.*, 1989, 1993; Lawrence *et al.*, 1991; Korber *et al.*, 1993; Stewart *et al.*, 1993; Wolfaardt *et al.*, 1994). Much of this change in our view of microbial biofilms arises from the development and application of CLSM and fluorescent molecular probes in biofilm studies.

The purpose of this chapter is to provide a basis for understanding CLSM and the approaches that may be used to apply high resolution digital microscopy to biofilm research. Additional reviews of digital imaging and CLSM for environmental microbiology applications are also available (Caldwell *et al.*, 1992a; Lawrence *et al.*, 1995, 1996).

16.2 CLSM DESIGN

16.2.1 The system

CLSM is a combination of traditional epifluorescence microscope hardware with a laser light source, specialized scanning equipment and computerized digital imaging. A CLSM system is equipped with a laser light source that generates a range of excitation wavelengths. A full range of commercially available lasers may be used in CLSM, including argon ion, helium/neon, argon/krypton, helium/cadmium and ultraviolet (UV) excimer lasers. Many commercial systems utilize the argon laser (25, 50 or 100 mW models with 5000–10 000 hour life spans) emitting two main lines at 488 nm (the blue line) and 514 nm (the green line) as well as minor lines ranging from 274 to 528 nm. The major disadvantages of the argon laser are the limited range of fluorochromes excited by the laser and the lack of separation of green and red fluorescence for dual labelling of samples. Increasingly, CLSM systems are equipped with additional helium/neon (543 or 633 nm) and mixed-gas krypton/argon lasers (488 nm blue, 568 nm yellow, 647 nm red lines). These lasers provide the advantage of a wider range of usable fluorochromes and simultaneous excitation of up to three fluorochromes with little spectral emission overlap. Helium/cadmium lasers are seldom used but can provide a strong 442 nm line. UV excimer lasers (157–351 nm) may now be obtained with commercially available CLSM systems. At present, UV systems exhibit problems with alignment, and are expensive to purchase and maintain. In most instances the lasers are now connected to a scanning head which is mounted on a standard epifluorescence microscope, by a fibre optic system. This greatly simplifies alignment and the use of multiple lasers. The scanning head consists of an integrated unit with galvanometric mirrors to scan the laser beam over the specimen, and filter sets and beam splitters to direct the selected wavelengths of light to specific photomultiplier tubes (PMTs). The range of PMTs offered in commercial CLSM systems is somewhat limited; however, potential exists for application of PMTs with various spectral ranges from 185 to 930 nm. The laser light source is used to illuminate the specimen in a point-by-point fashion through a conventional objective lens, thereby exciting the fluorescent probes that have been applied to the specimen. During this process the specimen is scanned over a defined area in a raster pattern, e.g. areas equal to 512 × 512, 512 × 768 or 1024 × 1024 pixels. The scan rate may be varied from 1/4 to 4 seconds. However, the more rapid the scan the lower the resolution of the image, whereas longer scan times may result in image degradation due to photobleaching. When a specimen is scanned, one or more confocal pinholes allow only those fluorescence signals that arise from a focused *xy* plane to be detected by the PMT (Figure 16.1).

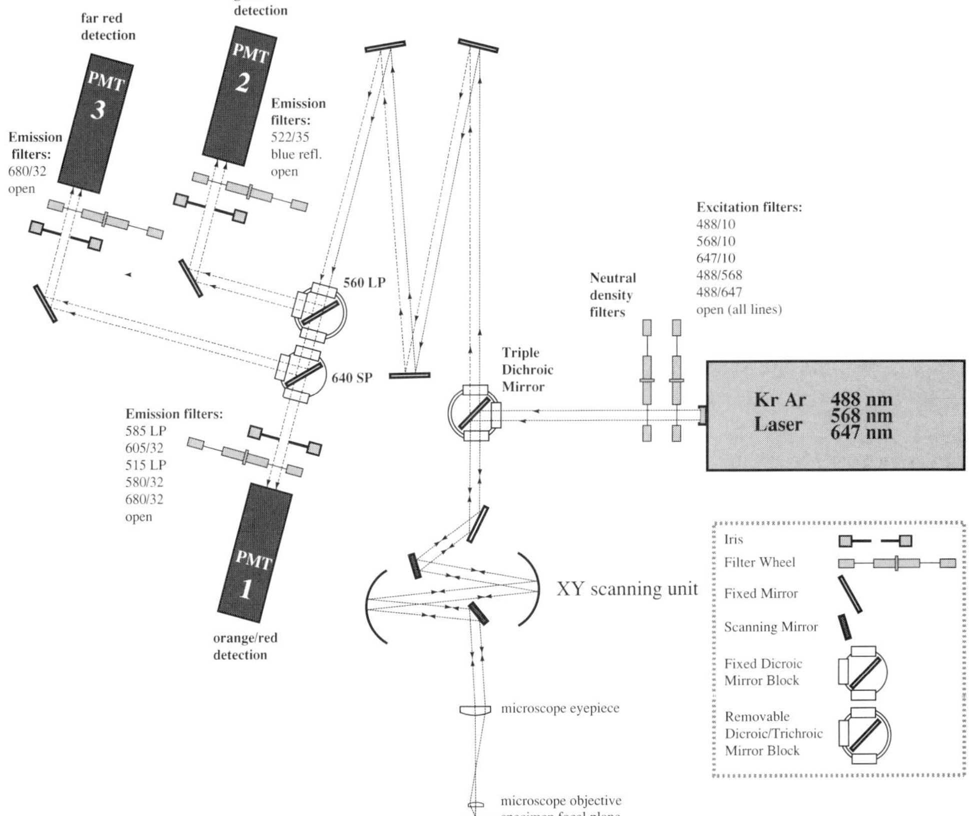

Figure 16.1 Schematic diagram of a confocal laser scanning microscope showing the positioning of mirrors, emission and excitation filters, photomultipliers, laser source and standard light microscope.

The pinholes are positioned confocally in front of the incident laser light and in front of the photomultiplier detector, preventing fluorescent signals originating from above, below or beside the point of focus from reaching the photodetector. In each case, the specific excitation and emission wavelengths are selected through insertion of filter sets and beam splitters in the light path. The PMT detects the emitted light at each point and converts this into a digital grey scale image. In addition, CLSM systems may incorporate either a fibre optic light guide below the condenser and connected to one of the PMTs, or a separate photodiode for imaging non-confocal transmitted laser images.

The major manufacturers provide a variety of configurations, e.g. standard or inverted microscope set-ups, multiple laser attachments, multi- or single-pinhole arrangements, and a selection of PMT sensitivities and wavelength selectivities. Figure 16.1 shows a representative configuration of a CLSM system with relative positions of the laser, the light path, scanning mirrors, pinholes (or irises), PMTs and conventional microscope indicated. Table 16.1 provides a quick guide to the lasers, fluorochromes and emission filters commonly used in CLSM.

Table 16.1 Laser types and filter sets for selected fluorescent dyes used in CLSM listed by increasing excitation/emission wavelength.

Laser (lines)	Dye	Excitation maximum (nm)	Emission maximum (nm)	Emission filter
Ar–Uv (351/363)	Hoechst 33258	346	460	455/30
Ar–Uv	Hoechst 33342	346	460	455/30
Ar–Uv	DAPI	359	461	455/30
Ar–Uv	Cascade blue	399	423	455/30
He–Cd (442)	Lucifer yellow	428	533	522/32
He–Cd	Fura red	436	640	640/40
He–Cd	BCECF (excitation ratio)	439	530	522/32
He–Cd	Chromomycin A3	458	590	605/32
Ar(488/514)	R-phycoerythrin	480	578	605/32
Kr–Ar(488/568/647)				
Ar/Kr–Ar	Acridine orange	487	520	522/32
Ar/Kr–Ar	SNARF	490	580	605/32
	(emission ratio)		630	640/40
Ar/Kr–Ar	BCECF (excitation ratio)	490	530	522/32
Ar/Kr–Ar	FITC	494	520	522/32
Ar/Kr–Ar	Biodipy FL	503	512	522/32
Ar/Kr–Ar	Rhodamine 123	505	533	522/32
Ar	Ethidium bromide	510	595	605/32
Ar	TOTO 1	514	533	522/32
Ar	Propidium iodide	536	617	605/32
GHeNe (543)/Kr–Ar	CY3	550	565	605/32
GHeNe/Kr–Ar	TRITC	554	576	605/32
Kr–Ar (488/568/647)	R-phycoerythrin	565	578	605/32
Kr–Ar/RHeNe (633)	Texas red	596	615	605/32
Kr–Ar/RHeNe	TOTO-3	642	660	680/32
Kr–Ar/RHeNe	CY5	650	670	680/32
Kr–Ar/RHeNe	Allophycocyanin	650	661	680/32

16.2.2 Objectives

Assuming that the system is properly set up and aligned, the next critical choice determining image quality is the nature of the objective lens used. In general, images cease to be confocal in nature when the numerical apperture (NA) of the lens is less than 0.5, with greater confocality obtained using high NA objective lenses (NA 1.4). The NA, in conjunction with the pinhole size, controls the thickness of the optical section. For example, when the NA is 0.2 the optical section is approximately 10 μm; for NA $\geqslant 0.6$, section thickness decreases to < 1 μm. The most frequently recommended objectives provide a large flat field and a high NA, such as 1.4 NA, 60× or 100× planapochromat lenses. In addition, chromatic correction between 350 nm and 650 nm is highly desirable. Water immersion lenses offered by manufacturers such as the Zeiss C-apochromat (60×, 1.2 NA, water immersion) with a relatively long working distance (220 μm) appear to be excellent lenses for CLSM applications, although they are very expensive. A 40×, 1.3 NA oil immersion lens will provide a high quality image at low cost and can benefit from the application of electronic zoom factors.

The nature of the recommended objectives is due to the predominant use of fixed, stained and mounted samples by many laser microscopists. However, these lens choices are often not appropriate for biofilm studies and can indeed reduce the advantages of examining fully hydrated biofilms and the dynamic processes within them. Most notably, high NA lenses are not suitable for collection of images in thick specimens. We have found that extra long working distance (ELWD) lenses and water immersible lenses such as the Zeiss 63×, 0.9 NA are extremely useful, eliminating the need for cover slips or fixation. In addition, these lenses provide high quality images of biofilm materials. Figure 16.2 shows a comparison of images of a biofilm obtained with a variety of objective lenses, including water immersible lenses. It should also be noted that images obtained with the lower magnification objectives can benefit from electronic zooming. For example, a 20×, 0.75 NA lens will increase in resolution up to approximately a 4× zoom. We have found that we can effectively section through 50–100 μm of biofilm materials using 60× 1.4 NA lenses, several hundred micrometres with 63× 0.9 NA water immersible lenses, and 1 mm and more when using 20× or 40× ELWD lenses or 40× water immersible lenses. The 40×, 1.0 NA oil immersion objectives have been reported to give working distances of 300–400 μm (Shotton, 1989). Lawrence *et al.* (1996) used water immersible lenses to examine biofilm development on sulphide minerals, while Bott *et al.* (1997) and Neu and Lawrence (1997) have applied them for studies of river biofilms. Shotton (1989), Pawley (1995) and Rost (1992a,b) provide additional information on the choice of objective lenses for fluorescence and CLSM applications.

16.2.3 Computers, monitors, printers, archiving and storage

CLSM systems are usually controlled by a PC, which may be DOS or Windows 95 based. Some manufacturers offer the option of a workstation, which can be beneficial. We have found it useful to have two computers linked in a network to allow one to be dedicated to image collection, while the other is dedicated to image

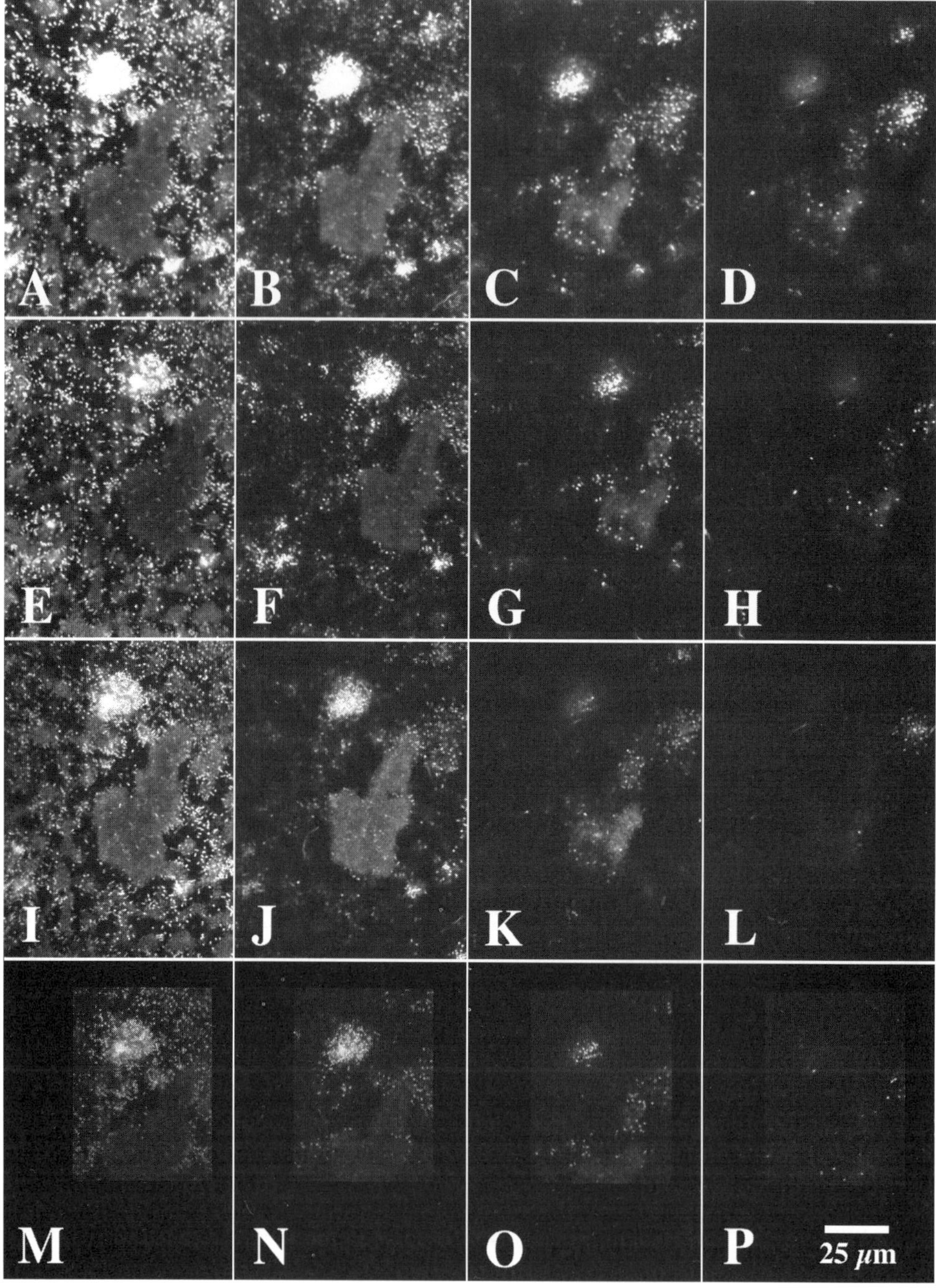

Figure 16.2 A series of images comparing the results obtained using: (A–D) 40×1.3 NA oil immersion lens, (E–H) 60×1.4 NA planapochromat oil immersion lens, (I–L) 63×0.9 NA water immersible lens and (M–P) 100× NA 1.3 oil immersion phase 4, Plan Neofluor lens. The biofilm community was stained using the nucleic acid stain SYTO 9.

processing and analyses. The system is usually operated on one monitor, with a second being used for image display. The monitors used for image display are RGB analog, high resolution. Typical imaging systems utilize a high-capacity hard drive (a minimum of 4 Gbytes) for rapid data collection, and are interfaced with optical devices for storage/archiving of selected information. Although a number of options exist for archiving images, including optical drives (write-once and rewritable formats), Bernoulli drives, Syquest and ZIP drives, recordable compact discs (CD-Rs) are the recommended medium. CD-Rs are a safe, inexpensive, versatile medium for storage, archiving and transfer of CLSM images. Images may be stored in a variety of formats, such as TIFF, GIF, RAW, PICT, EPS and BioRadTIFF (see Section 5.4). Each of these has advantages and disadvantages in terms of its use by specific manufacturers, its degree of image fidelity and its ability to compress images (i.e. JPEG). In general, tagged image file formats (TIFF) are the most universally used and will be opened by most software. After images have been analysed and processed they may be printed for publication using a variety of means, including video printers, dye sublimation printers and slide printers. Dedicated flat-screen, high-resolution monitors are also a useful addition for the production of slides and negatives. An often overlooked challenge in setting up a CLSM facility is the need for careful archiving of the accumulated images, and although some software packages are available, none truly meets the need (i.e. documenting and archiving more than 1 Gbyte of material per day).

16.2.4 Operating software/analytical packages

Operating software packages come with all commercially available CLSM instruments and should be assessed carefully as part of the overall instrument operation. The degree of user-friendliness is highly variable between manufacturers and while the overall image quality is similar, based on our personal experience with Bio-Rad (Hemel Hempstead, UK), Zeiss (Jena, Germany) and Leica (Heidelberg, Germany), the greatest variability is in terms of ease of use. Pawley (1995) provides a comprehensive overview of the available confocal microscopes for viewing biological specimens. All of the manufacturers offer supplementary analytical packages that work with their operating systems and image formats; none offers all required options. The Quantimet system offered by Leica has been used effectively to analyse CLSM images of bacteria (Bloem *et al.*, 1995). The Kontron (IBAS system, Kontron, Eching, Germany) is an effective analytical system. GTFS (Ultimage, GTFS Inc., Santa Rosa, CA, USA) offers a comprehensive package as well. PC-based three-dimensional imaging/image analysis systems include, for example, MicroVoxel (Indec Systems, Sunnyvale, CA), VoxelView (Vital Images, Fairfield, IA), or VoxBlast (VayTek, Inc., Fairfield, IA). Additional commercial software suppliers are listed in Pawley (1995). However, the most dynamic area in software development is in users' laboratories as evidenced by the range of software in the published literature (Morgan *et al.*, 1991; Dubuisson *et al.*, 1994; Møller *et al.*, 1995). Møller *et al.* (1995) used Cellstat, which is available for UNIX workstations (see http://www.lm.dtu.dk/cellstat/index.html). The program NIH-image for a Macintosh platform is available as freeware over the internet at http://rsb.info.nih.gov/nih-image/. This is a user-friendly, versatile analysis package, well suited to handling CLSM images. Some users will also be interested in the use of software to remove out-of-focus signals by mathematical deconvolution (Agard, 1984; Gorby, 1994). A UNIX-based package for this is available over the internet that may be applied to CLSM images (see http://ibc.wustl.edu/bcl/xcosm/xcosm.html).

16.3 CLSM ALIGNMENT AND OPERATION

The following protocol is general in nature but assumes that the user has a modern, commercially available CLSM such as the Bio-Rad, Zeiss or Leica system. The approach also concentrates on the use of fluorescence microscopy, although reflected light imaging and non-confocal transmitted light detection for imaging, phase contrast, differential interference contrast (DIC) and darkfield options are also discussed. These options allow comparative analysis of the same sample location using a variety of microscopic techniques.

Although the set-up and operating procedures of CLSM systems vary, there are a number of fundamental elements that should always be considered. It is good practice to check the alignment of the system periodically, during extended use, and always at the beginning of each session. The system should be checked when filter blocks are changed.

Protocol 16.1 CLSM operation.

1 Internal mirrors and optical elements of the unit should be aligned periodically by the manufacturer (annually).
2 Alignment must be confirmed prior to each usage. This may include centring the incident laser light with respect to the objective lens, and optimizing the emitted fluorescent/reflected light with respect to the photodetector. Alignment is particularly critical for work with bacteria, where a misalignment of 1 μm cannot be ignored.
3 Ensure that the correct filters are installed for the fluorochromes to be used. For example, the Bio-Rad MRC 600 with argon laser utilizes 488 nm excitation, 515 nm longpass emission filter for green excitation/emission, and 514 nm excitation, 550 nm longpass emission filter for red fluorescence excitation/emission.
4 Place a sample on the microscope and select a field for imaging using either standard epifluorescence or phase-contrast microscopy.
5 Switch to CLSM mode and scan the sample using a low zoom factor, ensuring that the PMT gain is set to manual and the pinhole aperture is about half open. Focus on a portion of the sample, which should provide an even signal response once the CLSM is aligned. Scanning in real-time at 1/4 frames/s, the gain of the PMT should be adjusted such that the image is about at half brightness (no part of the image should be saturated), but of reasonable quality. The mirrors should then be adjusted so that the image is evenly illuminated. Standard samples should be used for checking alignments; our sample of choice is Focal/Check fluorescent beads (Molecular Probes, Inc., Eugene, OR, USA). The brightness of

the image should be optimized by adjusting each mirror in the light path. If the system was misaligned, this operation will result in a significant increase in brightness and image clarity. The gain should be reduced to prevent saturation as image quality is enhanced. This process should be repeated with the PMT detector aperture (pinhole) reduced incrementally, until the detector aperture is at its smallest size. If dual-PMT detection is required, ensure that the correct filter blocks are in place and repeat the alignment procedures outlined above.

6 Ensure that the objective lens used for a particular analysis has sufficient magnification and resolving potential.
7 Filter sets and laser source must be matched to the excitation and emission spectra of the fluorochrome.
8 The intensity of the incident laser light should be just sufficient to provide a strong fluorescent signal while minimizing signal loss due to bleaching. Laser intensity is usually controlled using neutral density filters.
9 The size of the pinhole should be adjusted to achieve the smallest aperture possible. This may involve a sequence of adjustments with progressive reduction in aperture size (note that larger apertures produce thicker optical sections).
10 Adjustment of PMTs: the black level should be set at the point where the boundary between scans becomes invisible against the dark background. If the black level is set too high then detail may be lost in the image. Gain level: the optimum brightness in an image is the point at which only a few pixels of the image approach saturation (a grey level of 255 on a scale 0–255 where 0 = black and 255 = white). The gain and black level should be adjusted to utilize the entire grey-scale range. Most CLSM systems include a range indicator option, which displays over-saturated or under-saturated pixels in different colours, making these adjustments easier to do.
11 Adjustment of the computer monitor should be carried out periodically under working lighting. The monitor test image should be displayed and the brightness and contrast controls adjusted until all intensities are visible and well defined. It is also critical that the monitor image be centred and sized correctly.

The user will have to establish the optimum conditions for each specimen, stain etc. However, in general, the lowest intensity laser settings (usually 1–10% transmission) and the smallest detector pinhole aperture size should be used to minimize photobleaching or photodamage of the specimen while optimizing image quality.

16.4 CLSM IMAGING

After initial alignment the user has a number of imaging options, since it is possible to collect images in the *xy* or *xz* direction or a series of *xy* images creating an image stack for three-dimensional reconstruction. It is also possible to collect images using combinations of 1, 2 or 3 excitation and emission wavelengths. This may be done sequentially or simultaneously, depending on the CLSM system in use. Options also exist to collect in fluorescence, reflection and non-confocal transmission modes.

16.4.1 Fluorescence imaging

The most common mode of operation for CLSM is fluorescence imaging. This is due to the wide range, sensitivity and specificity of fluorescent probes available for macromolecules and micro-organisms. In general, once a sample is prepared and placed on the microscope stage, a suitable field is found using epifluorescence, and the operator switches to CLSM mode. The sample may also be explored using real-time scanning although the fluorescence will have to be resistant to bleaching in this case. Alignment may have to be checked again, and PMTs and pinhole adjusted as described in Protocol 16.1 to optimize the image and achieve a high quality *xy* image. The image may then be collected and stored. Image collection may occur as a direct input (single scan), or the signal-to-noise ratio (SNR) of the image may have to be enhanced through application of a running average or Kalman type mathematical filter. Kalman filtration averages sequentially collected images, thereby reducing the noise level of the image but maintaining edge features, and is extremely effective in reducing random noise and producing a clean image. The general equation is as follows:

$$P_j(x, y) = \frac{I(x, y)}{j} + P_{j-1}(x, y)\left\{1 - \frac{1}{j}\right\} \tag{16.1}$$

where $P_j(x, y)$ is the new pixel value, $I(x, y)$ the input value and $P_{j-1}(x, y)$ the previous pixel value.

Figure 16.3 shows the impact of Kalman filtration on the SNR of an image of a biofilm stained with a lectin (Sigma Chemical Co., St Louis, MO, USA). Note that considerable detail may be observed in the filtered image, such as the occurrence of capsules of extracellular polymeric substances (EPS) surrounding individual cells. The use of these filters may cause bleaching due to the high frequency of scanning and is not recommended during collection of images for applications such as ratiometric imaging. Other mathematical collection filters offered by the various manufacturers allow for image enhancement through summation of signals, collection to preset maximum signal, local contrast enhancement, median filters, linear smoothing filters and gradient filters (Pawley, 1995). Each of these filters attempts to reduce noise levels in images, smooth the data set and enhance the edges of objects within the image. The Bio-Rad systems also offer a low signal detection or photon counting mode that may be applied to weak signals.

A number of factors limit optical sectioning of biological materials. Most significant of these are the ability of the laser beam to penetrate the sample and the occurrence of quenching of fluorescence. In addition to the inherent quality of the objective lens, the selected staining method will also influence the maximum depth from which optical sections may be obtained (see Section 16.6.1).

The user may choose to collect a series of *xy* images, a *z* series or scan in the *z* dimension, creating an *xz* line scan or sagittal section of the biofilm material. The *xz*

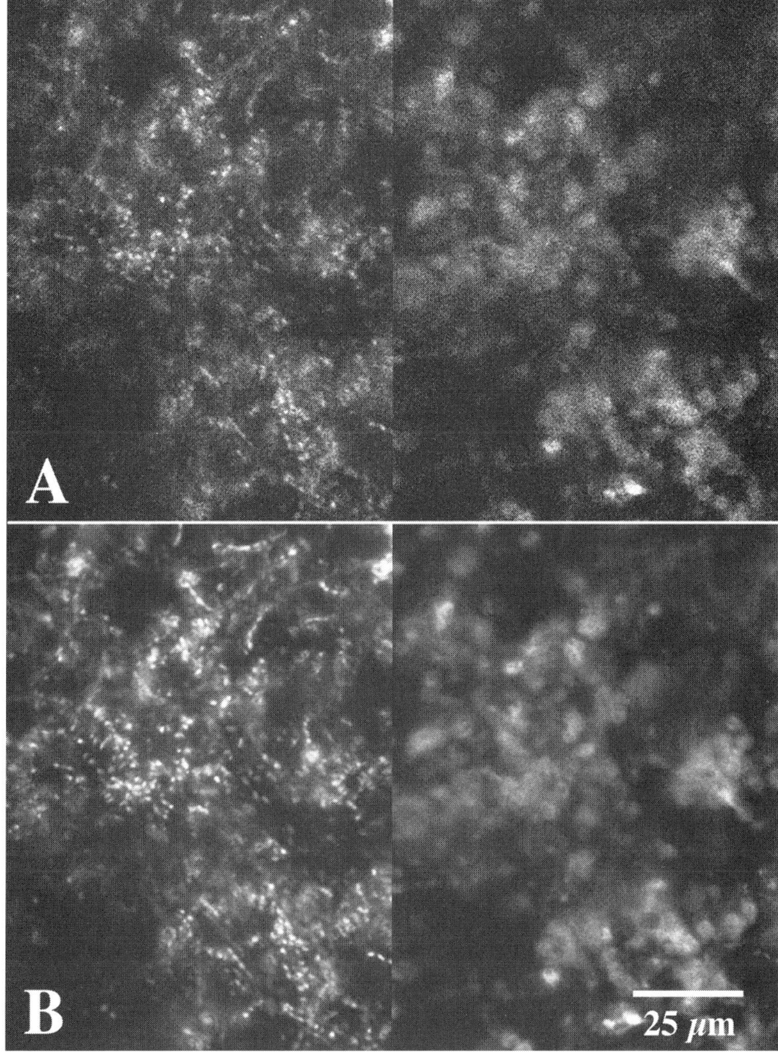

Figure 16.3 (A) Image collected without mathematical filtration, and (B) image of the same microscope field collected through a Kalman running average filter. Note the substantial increase in signal-to-noise ratio in the filtered image.

images lack the resolution of *xy* images due to the nature of the corrections in standard objectives and the point-spread function of the lens. Also, *xz* images may be calculated from a series of *xy* images. This can be advantageous since it reduces bleaching effects to the sample caused by multiple scanning. Figure 16.4 shows a series of *xy* images and both live-scanned *xz* and calculated *xz* images from the same location in a biofilm using fluorescein as a negative stain (see Section 16.6.1). Figure 16.5 illustrates the difference between positively and negatively stained images of the same optical section in a mixed species biofilm. Similar image series have been published in a number of publications, using a variety of positive and negative

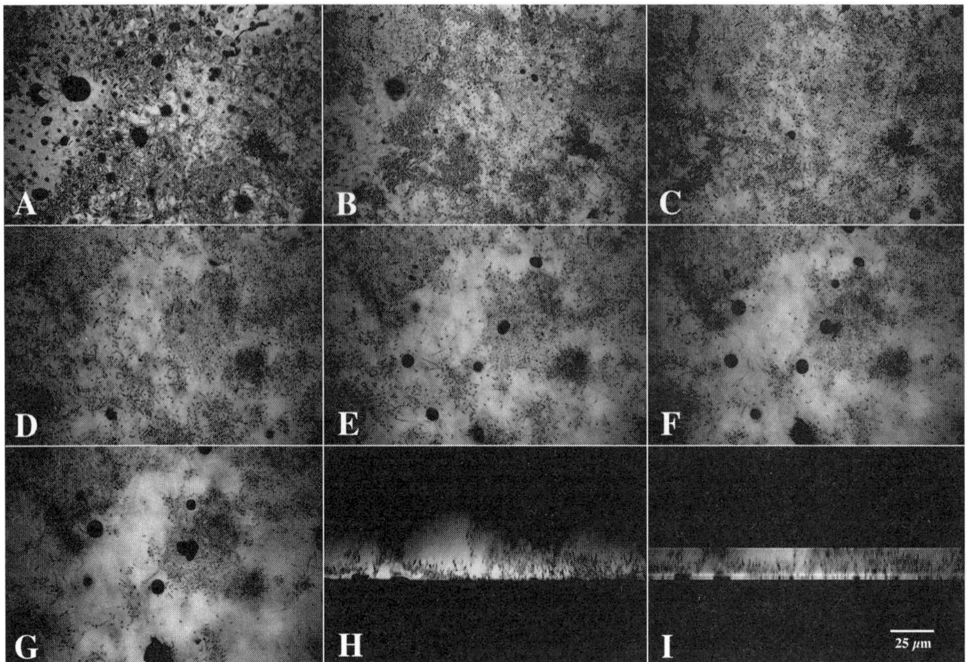

Figure 16.4 A series of CLSM images of a mixed species biofilm showing (A–G) the *xy* and (H) the *xz* scanning capacity of laser microscopes. (I) The production of an *xz* image through calculation from the *xy* series.

staining options (Lawrence *et al.*, 1991, 1994; Caldwell *et al.*, 1992a,b; Wolfaardt *et al.*, 1994; Amann *et al.*, 1995; Assmus *et al.*, 1995).

16.4.2 Three-dimensional imaging considerations

Besides improved resolution and contrast, another important property of confocal microscopy is its ability to form a three-dimensional image of an object with appreciable depth. This capability follows from the optical sectioning capacity and the creation of *z* series in perfect register.

When an ideal *z* series is collected for display or analytical purposes the sections are:

(a) collected non-destructively,
(b) adjacent without oversampling or undersampling error,
(c) homogeneous,
(d) aligned and in register,
(e) calibrated so that image intensity accurately represents a concentration or the presence of an object (deshading; see Section 2.4.7),
(f) isotropic such that the *xy* pixels correspond in size to the *z* dimension.

During collection of a *z* series or an *xz* image (Figure 16.5) it may be necessary to increase the gain to compensate for bleaching effects during scanning. Some

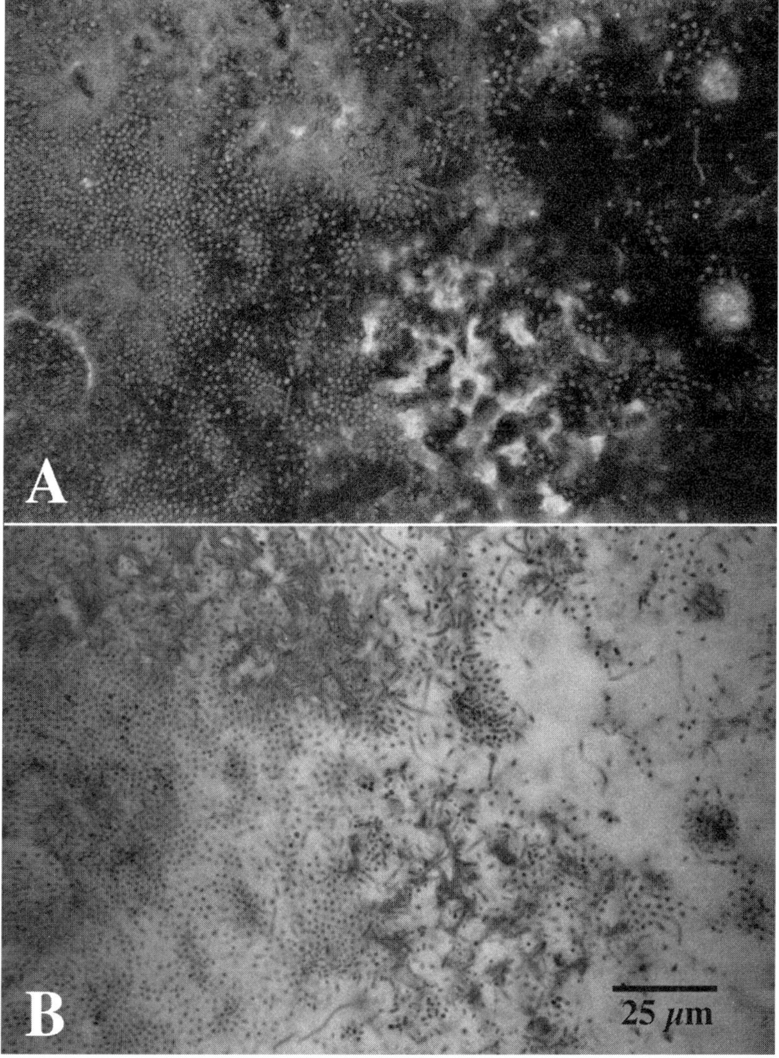

Figure 16.5 Single CLSM *xy* images showing (A) the positive (SYTO 9 stained) and (B) the negative fluorescein stained views of a degradative biofilm community.

instruments have features allowing this to be done automatically or manually and some software routines have been applied to correct for this type of signal loss. Another critical factor during collection of *xz* and *z* series is selection of the correct sectioning interval. Theoretically, with a high NA objective sections may be collected at intervals of 0.3 μm or less. In practice, optical sections are 0.5–1.5 μm in thickness and this must be reflected in the optical sectioning interval. If the section interval is incorrect, oversampling or undersampling errors may occur (Sections 2.2.1 and 7.2). These are most important when a series is analysed to extract depth information or during three-dimensional reconstruction. For example, Figure 16.6

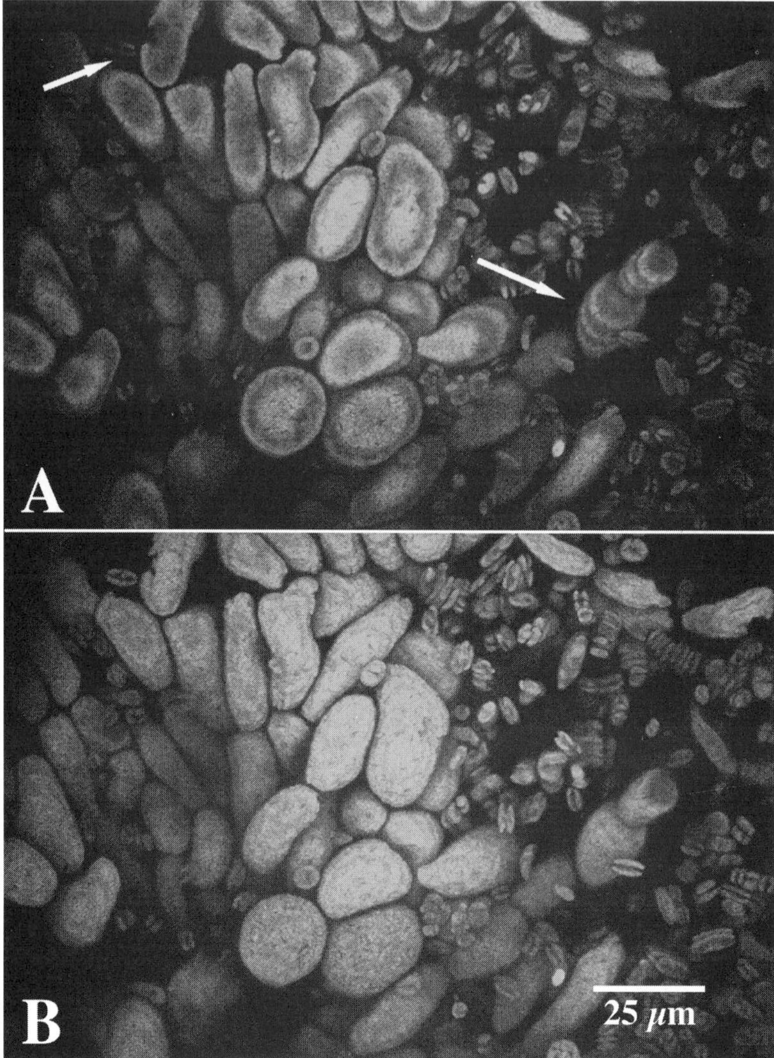

Figure 16.6 A series of xy images presented as a projection showing (A) undersampling error, i.e. evidence of individual sections and missing information (see arrows) and (B) where the sectioning interval was scaled to the thickness of the optical section.

shows the effect of undersampling error on the projection of a z series, i.e. the creation of a stepped image in which each section is apparent.

The user should also consider the concept of the total slice capacity (Stevens, 1994), i.e. a fixed number of sections that may be taken from a specimen before the effects of sampling (i.e. bleaching) create artefacts. Attenuation of the penetrating laser beam and the returning signal also creates sampling errors. In addition, the z series must be collected within the working distance of the objective. These sources of error will degrade both the appearance and the analytical usefulness of a z series.

However, mathematical solutions are available to some of these problems (Riguat and Vassy, 1991), although they may be somewhat complicated in nature. The user may assess the significance of the problems by varying laser beam intensity, z sampling frequency, number of scans and the angles of the scans through the biofilm. When comparing the sections and the resulting reconstructions of similar materials, those approaches that result in similar high quality views will be most effective in the system being studied.

During collection of any z series the user must always minimize damage to the specimen by the laser beam. The following suggestions are offered:

(a) minimize non-uniform bleaching by having the laser aligned (see Protocol 16.1),
(b) reduce bleaching through application of fade retardant solutions (see below),
(c) if necessary, apply image processing including, for example, normalization of the series between 0 and 255 grey values (contrast stretching; Section 1.4.1a), or data collection methods such as distributed bleaching (Stevens, 1994).
(d) change the fluorochrome in use or the staining approach (e.g. freshly prepared SYTO 9 (Molecular Probes, Inc.) or SYTO 17 are very stable fluorochromes, whereas rhodamine 123 is extremely unstable and bleaches readily) or switch from a positive to a negative stain (Caldwell *et al.*, 1992b).

16.4.3 Anti-fading reagents

When viewing mounted materials anti-fade solutions are essential. For many other applications where a fluorescent molecule is conjugated to a probe, the use of anti-fading reagents is highly recommended. When viewing living hydrated biofilms with a variety of stains and probes, we have found anti-fade reagents to be unnecessary. However, depending upon the stain, frequently only one z series can be taken from a specific site even with the use of anti-fading agents. The reader is referred to the following papers dealing with these reagents: Johnson and Nogueira Araujo (1981), Böck *et al.* (1985), Longin *et al.* (1993) and Florijn *et al.* (1995). It should also be noted that many of these mixtures are highly toxic and should be used with caution. We routinely purchase and use the anti-fade mixtures sold by Molecular Probes, Inc. and Citifluor (London, UK).

16.4.4 Multiple channel imaging

The application of more than one probe, each with different fluorescent reporter molecules, provides the option of either double-labelling or triple-labelling of samples, and in some instances four probes may be viewed. All of the commercial CLSM system manufacturers offer these options. Most commonly the excitation source is a krypton/argon laser offering three well-separated excitation lines (Table 16.1) although other combinations of lasers may be used to achieve similar results. Several factors become critical when collecting multi-channel images:

(a) the cutoffs for the emission and excitation filters,

(b) misalignments in the optical pathway, resulting in images from the different channels not being coincident, which cause offsets in the resulting dual- or triple-channel images,
(c) axial chromatic aberration in the objective lens, resulting in offset images,
(d) wavelength of light, resulting in sections collected in the far-red that are thicker than those collected in the red and green wavelengths due to the difference in chromatic correction of the objective at these wavelengths (see Cullander, 1994),
(e) field curvature and lateral chromatic aberration (Section 3.5), resulting in objects at the edges of images appearing in slightly different locations,
(f) loss of signal and resolution with increasing wavelength of emission, thus images originating in green and far red may be of considerably different quality.

The presence of these misalignments in the *xy* or *xz* may be checked through imaging Crimson Fluorophore (1 μm) or Focal/Check (15 μm) beads (Molecular Probes, Inc.) in dual- or triple-channel mode, and overlaying the resulting images. Any misalignment is evident in the overlay of the channels and may be corrected by following the alignment procedures outlined above. If the cause is in the objective lens realignment cannot correct the problem, although for most lenses this is only true at the edge of the image. Software solutions are also available that allow toggling of the digital images into alignment and cropping during the preparation of multi-channel overlays (Waggoner *et al.*, 1989). Independent software that allows this type of adjustment/alignment of images includes Adobe PhotoShop (Adobe Systems, Salinas, CA).

When three channels are not enough there are at least two approaches: (i) sequential addition of probes at a single location or (ii) the application of equimolar mixtures of probes labelled with different dyes. For example, combinations of red and green dyes result in yellow, green and blue, cyan and so on. Ried *et al.* (1992) displayed seven labels simultaneously in an RGB image after hybridization of DNA probes to isolated human chromosomes. Amann *et al.* (1996) used a similar approach to image bacteria in flocs, although this approach may be restricted to environments with very low background fluorescence. We have used multiple labelling with fluorochrome conjugated lectins and nucleic acid stains for multi-channel imaging. In addition, autofluorescence signals in the far red, red or green may provide useful information about a range of organisms. Plate 5A shows a three-channel image of a mixed species biofilm where algal cells were detected in the far red, exopolymer was stained using wheat germ agglutinin conjugated to tetramethyl rhodamine isothiocyanate and bacterial cells were stained with SYTO 9. Plate 5B shows the results of combined staining with SYTO 9 (green), *Ulex europeaus* lectin FITC (green), *Triticum vulgarus* lectin TRITC (red) and *Arachis hypogea* lectin CY5 (blue). A colour wheel is included to facilitate interpretation of the seven possible combinations of probes in the image.

16.4.5 Reflection imaging

Confocal reflection imaging may also be performed using CLSM systems. This has microbiological applications where differences in refractive index within the specimen result in the reflection of laser light, and has been used to visualize mineral and metal surfaces covered by microbial biofilms. Confocal reflection contrast images may be obtained using CLSM and can provide superior images to conventional interference microscopy. The combination of reflectance imaging with colloidal gold labelling has also been used in cell biology (Shotton, 1989) but not yet in the examination of biofilms.

16.4.6 Transmission imaging

It is also possible to equip the CLSM system with a non-confocal detector system for scanned transmitted light imaging. The detector may be either a photodiode or a

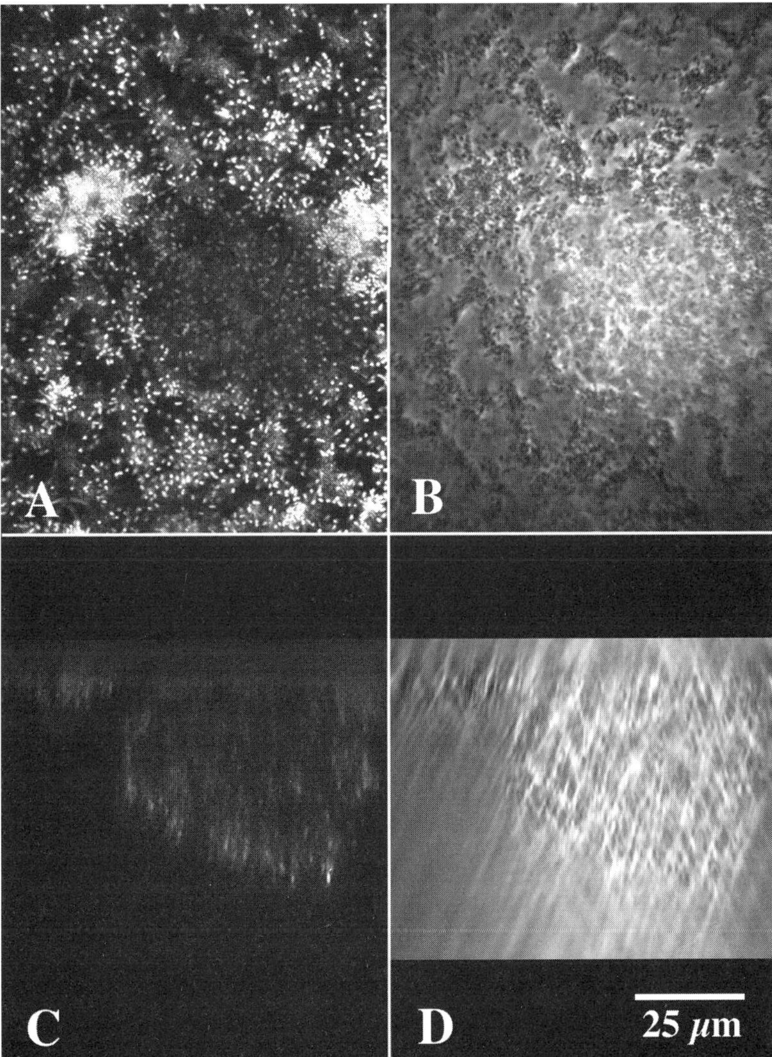

Figure 16.7 Photomicrographs illustrating the images obtained using: (A) fluorescent confocal microscopy, (B) transmission phase contrast and (C) the corresponding *xz* images by confocal fluorescence and (D) transmission of a herbicide degrading biofilm community.

fibre optic connection below the condenser. If the microscope is equipped for darkfield, phase-contrast and Nomarski DIC these images may be digitized images of the same microscope field. This provides the opportunity to combine confocal fluorescence (including multi-channel imaging), confocal reflectance, transmitted phase-contrast, darkfield, Nomarski and conventional epifluorescence imaging of the same sample materials. Figure 16.7 shows dual-channel images of the same location in a biofilm collected using phase-contrast and confocal fluorescence imaging.

16.5 THREE-DIMENSIONAL RECONSTRUCTION, PSEUDOCOLOUR AND DISPLAY

Collected *z* series may be used to create a variety of three-dimensional reconstructions. As few as 12 optical sections can be used to create a stereoscopic image, in which the individual planes are not evident, i.e. no undersampling is apparent (Figure 16.6). In contrast, the incorporation of too many sections will create a projection that is confusing for the human eye. Most commercial CLSM systems allow creation of extended focus or look-through images that present the summation of an entire series of *xy* images, thus presenting a single, in-focus image with increased depth of field. If one reconstructs two extended focus images through the same location in the specimen, and shifts one of the images sideways to create an oblique view, then these images may be aligned to create a stereo pair (see Figure 16.8). The lateral shift used during generation of stereo pairs is a critical factor: if the shift is too large the image will have poor contrast and a poor depth effect. A suggested total offset is in the range of 15 pixels for the entire image stack, or 5° offset. Stereo pairs have been presented to show development of *Vibrio parahaemolyticus* biofilms (Lawrence *et al.*, 1991), and colonization and weathering of sulphide minerals (Lawrence *et al.*, 1996). It is also possible to create a stereo effect through the addition of pseudocolour to the offset projections. If one is coloured red and the other green, and the resulting images are overlaid (a red–green anaglyph projection), then a three-dimensional effect is achieved when they are examined

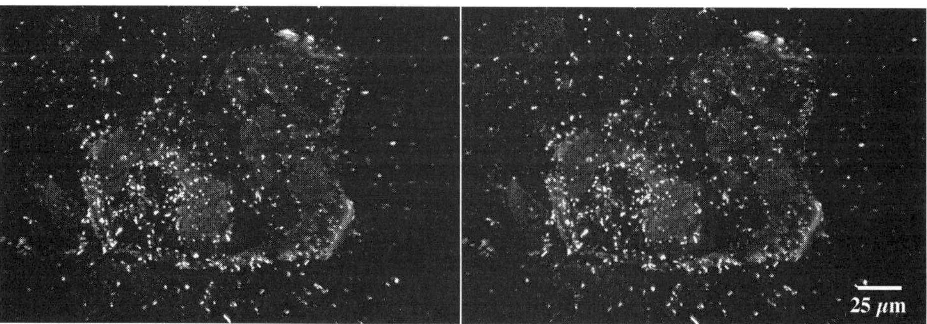

Figure 16.8 Stereo pair projection of a weathered pyrrhotite grain surrounded by pyrite and covered with *Thiobacillus ferrooxidans* and oxidation products. The equivalent red–green anaglyph projection is shown in Plate 6H. Reproduced from Lawrence *et al.* (1997), by permission of NRC Research Press.

using red and green glasses. Plate 6A and F–I are examples of red–green anaglyph projections of biofilm materials from a variety of habitats. Additional examples of this type of projection may be found in Amann *et al.* (1995), including a series of very good CLSM optical thin sections, and red–green anaglyph projections of rRNA probed materials. Similar effects may be achieved using polarized double projection, which also requires specialized glasses for observation. Another option for display of three-dimensional data sets is 'simulated fluorescence' whereby the material is viewed as though it were illuminated from an oblique angle and the surface layer were fluorescent. For some types of presentations the projection may be made as a solid body or surface projection, which is animated to show the entire data set.

Stereo projections may be colour-coded by depth, so that materials present at the same depth appear the same colour, an approach that can be very useful for interpretation of three-dimensional information. Colour coding may also be used to create three-colour stereo pairs, in which materials of the same type or binding the same fluorescent probe appear in the same colour. We have carried out this type of analysis, using a combination of nucleic acid stains, glycoconjugate staining (see below) and autofluorescence, to examine the structure of mixed-species biofilms in rivers (Lawrence and Neu, unpublished data). Plate 5A shows a typical result of this type of imaging approach.

Although confocal images have many advantages, they are not perfect, and the application of subsequent computational processing to remove residual blurring due to the limitations of all optical systems may be beneficial. Deconvolution sharpens an image through mathematical removal of out-of-focus information (see Section 16.2.4). The reader is referred to Agard (1984) and Gorby (1994) for examples of image enhancement through the application of deconvolution.

16.6 APPLICATIONS OF CLSM IN BIOFILM STUDIES

16.6.1 Imaging cells, cell types and biofilm architecture

One of the first applications of CLSM to biofilms was the study of the distribution of cells and biofilm architecture. General staining of biofilms falls into two categories: negative staining and positive staining. Negative staining has been used to yield high-resolution optical sections of biofilms (Caldwell *et al.*, 1992b) (Figures 16.4 and 16.5). Negative stains include fluorescein, resazurin and fluorochrome–dextran conjugates. In this procedure bacteria are imaged as dark objects against a bright background (Lawrence *et al.*, 1991; Korber *et al.*, 1993; Wolfaardt *et al.*, 1994). The major advantages of this approach are lack of fading, maintenance of contrast and so on. The drawback is significant quenching of excitation and emission signals by the biofilm. More commonly, bacterial cells in biofilms have been stained positively with acridine orange (AO), 4′,6-diamidino-2-phenylindole (DAPI), Nile red, fluorescein isothiocyanate (FITC), 5-(4,6-dichlorotriazin-2-yl)aminofluorescein (DTAF), tetramethyl rhodamine isothiocyanate (TRITC) other rhodamine stains and Texas Red. The SYTO™ series (Molecular Probes, Inc.) of nucleic acid stains, and particularly SYTO 9, SYTO 15 and SYTO 17, have worked well in our hands.

Both positive and negative staining have been used to study biofilm architecture (Lawrence *et al.*, 1991; Wolfaardt *et al.*, 1994; Massol-deya *et al.*, 1995). Figure 16.4 shows a series of *xy* images of a mixed species biofilm, using negative staining, in conjunction with *xz* images through the same biofilm region. Wells and Johnson (1994) also consider aspects of fluorochromes for CLSM applications.

Fluorochromes may also be coupled to molecules that recognize specific structures of the target micro-organisms. Lectins, antibodies and oligonucleotides may be labelled in this way. The principle conjugate fluorochromes are fluorescein (FLUOS, FITC), rhodamine derivatives (TRITC), the cyanins (CY3, CY5), aminomethylcoumarin (AMCA) and phycoerythrine (PE), which fluoresce in green and red wavelengths.

16.6.2 Viability

In addition to information on the presence, relative numbers and distribution of microbes in a biofilm, it is also important to know the metabolic condition of the cells, because the latter will ultimately determine the impact of the microbes on the environment. The development of fluorescent probe-based methods designed to demonstrate the presence of viable cells within heterogeneous communities has been discussed (Lloyd and Hayes 1995; McFeters *et al.*, 1995). These probes are based on aspects such as cytoplasmic redox potential, electron transport chain activity, enzymatic activity, cell membrane potential and membrane integrity (see also Chapters 4, 11 and 12). The major problems with these methods are: (i) the lack of a universal approach, (ii) limited application in biofilms to date, (iii) inability to correlate observations with cell viability (Black and Kroll, 1991), (iv) influence of environmental conditions, e.g. redox probes, and (v) the overall conditions such as pH, time, concentration and cause of death or injury to the cell. Thus, a 'ground truth' for the conditions studied must be established. However, various approaches have proven useful. Rodrigues *et al.* (1992) reported on a cyanoditolyltetrazolium chloride (CTC), Mason *et al.* (1995) indicated that bis-1,3-dibutylbarbituric acid and Calcofluor White were useful probes, ribosomal RNA probes can be used to assess the activity of individual cells (Ruimey *et al.*, 1994), combinations of rhodamine 123, propidium iodide and oxonol have been recommended (Lopez-Amoros *et al.*, 1995), as has the membrane potential sensitive dye 3,3′-dihexyloxacarbocyanine (Muller *et al.*, 1996). Korber *et al.* (1996) demonstrated that bacterial plasmolysis was a good indicator of viability. Additional staining methods with potential include the following: fluorogenic enzymatic substrates (fluorescein- and carboxy fluorescein-diacetate, calcein AM), probes sensitive to DNA damage (Hoechst 33342) and to membrane potential (rhodamine 123, RH-795 and carbocyanine derivatives) and commercial fluorescent viability staining kits (e.g. *Bac*lite Live/Dead stain, Molecular Probes, Inc.).

Plate 6B–D show the results obtained with the commercial *Bac*lite Live/Dead viability kit. In this experiment, the viability stain was used, as recommended by the manufacturer, to evaluate the effect of chlorine treatment on cell viability in a biofilm community cultivated in flow cells. Biofilm communities were cultivated in flow cells for 40 days on river water, before exposure to different concentrations of chlorine for 48 h. The kit uses a mixture of nucleic acid stains to distinguish between living cells with a functional plasma membrane, which fluoresce green, and dead cells with a damaged (compromised) membrane, which fluoresce red. Biofilms exposed to 4 mg/l chlorine contained exclusively red (dead) cells (Plate 6D), in contrast to biofilms that were exposed to a lower concentration of chlorine (0.5 mg/l), which contained a mixture of green and red cells (Plate 6C). Control biofilms contained the highest numbers of green (live) cells (Plate 6B). Conventional plate counts (Wolfaardt, Korber, Saftic and Caldwell, unpublished data) confirmed this pattern. In natural biofilms we have found that plant material, colloidal organics and algae autofluorescing in the red and far red make detection of red cells difficult.

16.6.3 Gram staining

When combined with CLSM, *in situ* Gram staining with fluorescent dyes allows visualization of Gram characteristics in three dimensions, thereby providing information about spatial organization in biofilms. Molecular Probes, Inc. offers a bacterial Gram stain kit (*Bac*light™) for use in living cells. This kit contains the green fluorescent SYTO 9 and red fluorescent hexidium iodide nucleic acid stains. Because of the different spectral characteristics of these stains, the kit can be used for dual-channel imaging with CLSM. SYTO 9 stains both Gram-positive and Gram-negative cells, while hexidium iodide preferentially stains Gram-positive cells. Because hexidium iodide displaces SYTO 9 in Gram-positive bacteria, they fluoresce red. Plate 6E shows a dual-channel image of the distribution of Gram-positive and Gram-negative cells in an optical thin section through a biofilm. The image is of a 2-week-old soil biofilm cultivated in continuous flow slide culture. Components A and B of the kit were combined at a ratio of 1:1 (v/v) and injected into the flow cell, followed by observations with CLSM after a 30-minute incubation at room temperature. The supplier recommends that different ratios of components A and B be evaluated for a particular application, to compensate for different rates at which the two stains fade due to photobleaching. In addition to photobleaching problems, we have observed variable staining, leakage of the green component out of cells, non-specific binding of the hexidium iodide, and interference from autofluorescent components of the biofilm. The use of lectins to differentiate Gram-negative and Gram-positive bacteria was explored by Sizemore *et al.* (1990).

16.6.4 Recognition of specific organisms

Two types of fluorescent probes have been demonstrated as effective tools to determine the distribution of species in complex biofilm communities. Fluorescent *in situ* hybridization (FISH) involves the hybridization of intact communities with fluorescently labelled rRNA probes to identify individual species. This technique is discussed in detail in Sections 10.5 and 17.2.4. The other type of probe to identify individual species in a community context is a fluorescently labelled antibody. Their use in the study of biofilms has been documented (Zambon *et al.*, 1984). Monoclonal as well as polyclonal antibodies are suitable for biofilm studies. To decide which type to use, aspects such as the desired specificity should be considered. Monoclonal antibodies are more specific, because a single determinant is recognized, while various determinants are recognized by a polyclonal antibody (Section 10.2.2). There is also a choice between fluorochrome-conjugated (direct immunofluorescence) and non-fluorescent antibodies (indirect immunofluorescence). When an antibody without a conjugated fluorochrome is used, an additional step involving staining with a fluorochrome-conjugated secondary antibody is required. This approach is particularly preferred when double or triple labelling is required since it leaves the researcher with greater flexibility in selecting excitation/emission spectra to detect antibody binding. A range of different fluorochrome-conjugated (e.g. FITC or TRITC-labelled anti-mouse, anti-rabbit etc.) secondary antibodies specific to the source of the primary antibody (e.g. mouse, rabbit, sheep) is available from a number of suppliers.

Protocol 16.2 Biofilm immunolabelling.

The staining protocol for using antibodies in flow cells is relatively simple.

1 It is not necessary to fix the biofilms prior to antibody staining. However, if desired, the biofilm can be fixed with 2% formaldehyde.
2 This is followed by blocking with 0.2% bovine serum albumin (BSA) for 30 minutes.
3 Next, a 1:100 dilution of the primary antibody in 0.2% BSA is used and incubated for 30 minutes.
4 This is followed by thorough washing, three times.
5 Finally, incubation with the secondary (fluorochrome-conjugated) antibody for 30 minutes is carried out. Excess secondary antibody should be washed out before CLSM visualization.

An excellent example of the immunolabelling approach to biofilm studies is provided by Rogers and Keevil (1992), who tracked the fate of *Legionella pneumophila* in aquatic biofilms. Schloter *et al.* (1993) used monoclonal antibodies and CLSM to localize *Azospirillum* sp. in the rhizosphere of wheat plants, while Schloter *et al.* (1995) and Dazzo and Wright (1996) provide more detailed evaluation of antibody applications in environmental microbiology.

16.7 EXTRACELLULAR POLYMERIC SUBSTANCES

So far, protocols for staining the microbial cells in biofilms have been described. Most of these fluorescent stains bind to nucleic acids, whereby the microbial cells become visible. However, microbial biofilms are defined as 'micro-organisms and their extracellular polymeric substances (EPS) associated with interfaces'. Thus EPS represent an essential part of microbial biofilms, as they attach the micro-organisms to the interface and to each other, thereby forming the biofilm structure. In addition, EPS represent a substantial amount of the biofilm weight. Biofilms are highly hydrated systems; roughly 95% of the biofilm may be water, the rest consists of up to 95% EPS, and only the small residual mass may represent cells and particles. The actual identity of EPS is defined in conflicting ways. Some authors define EPS as extracellular polysaccharides only. Other authors use a wider definition and discuss EPS as polysaccharides, proteins and nucleic acids. Nevertheless, a fluorescent probe that allows the visualization of the EPS matrix in biofilms would be of advantage. In the further discussion, we use the wider definition, and therefore have to distinguish probes specific for polysaccharides, proteins and nucleic acids. The free nucleic acids have already been discussed in a previous paragraph, together with the 'cell-specific' stains, which usually have nucleic acids as a target for staining.

There are different approaches to visualizing EPS compounds. Firstly, there are fluorescent stains that bind directly to the structure of interest, e.g. Hoechst 2495, which binds specifically to proteins, or Calcofluor White, which binds specifically to certain types of polysaccharides. Secondly, there is the option of producing antibodies against the EPS structure of interest. The antibodies can be labelled with a fluorescent marker, e.g. FITC, TRITC, BODIPY series or CY series, and then used as a highly specific probe. Antibodies against proteins may be produced easily, but polysaccharides may cause problems. Usually they have to be produced together with adjuvants, or chemically linked to an immuno-stimulans, to yield a substantial antibody titre (see immunology handbooks for details, e.g. Lam and Mutharia, 1994). Thirdly, for carbohydrate containing compounds, such as polysaccharides, there is a wide range of carbohydrate-specific lectins available. Lectins are proteins with a carbohydrate specificity (Liener *et al.*, 1986; Barondes, 1988; Sharon and Liss, 1989). They were initially employed for blood group typing and cell biology studies. Their specificity is usually not absolute for one carbohydrate unit, but covers several carbohydrates or di- and oligo-saccharides (Bog-Hansen, 1981; Reeke and Becker, 1988; Doyle and Slifkin, 1994). Despite this, lectins may be used for resolving the chemistry of the EPS matrix typically found in biofilms.

16.7.1 Probes for polysaccharides

For general polysaccharide staining two fluorescent dyes are typically in use. Calcofluor White M2R and Congo red are specific for $(1\rightarrow4)$- and $(1\rightarrow3)$-β-D-glucan polysaccharides (Wood, 1980). Both stains have been used in studies investigating growth and polysaccharide formation of bacteria and fungi. Calcofluor White M2R and Congo red inhibit bacterial cellulose formation and have an effect on fungal cell wall development (Colvin and Witter, 1983; Roncero and Duran, 1985). Calcofluor White M2R was also applied as a measure for exopolysaccharide production of *Rhizobium* and *Azospirillum* spp. (Doherty *et al.*, 1988; Del Gallo *et al.*, 1989). With respect to CLSM of microbial biofilms, one has to be aware that the application of Calcofluor White is dependent on the availability of a UV laser, as the excitation requires a laser line at 380 nm.

Congo red was employed for various purposes, such as general light microscopic staining for polysaccharides (Allison and Sutherland, 1984), indication of cellulose synthesis in bacteria (Zevenhuizen *et al.*, 1986), probing cell surface function (Arnold and Shimkets, 1988) and recording of hyphal growth (Matsuoka *et al.*, 1995). The Congo red emission (625 nm) may be recorded in the red channel of the CLSM which is normally used for Texas Red labelling.

The concentration of Calcofluor White and Congo red has to be tested and evaluated for each system. Calcofluor White may be applied at a final concentration of 25 μM. For Congo red 1 μM has been described as ideal for labelling.

16.7.2 Application of lectins

Lectins are commercially available (e.g. Sigma Chemical Co.) as purified proteins already labelled with different fluorescent and non-fluorescent markers. Both can be used for CLSM. The commercially available fluorescent markers include FITC and TRITC. In addition, the purified lectins may easily be labelled with other fluorescent markers, such as CY5. There are also commercial kits available for the labelling of a variety of proteins including lectins (Research Organics Inc., Cleveland, OH, USA). Furthermore, if the CLSM is operated in the reflection mode, lectins with non-fluorescent markers such as colloidal gold may be applied. However, the presence of organic and inorganic colloidal sized reflective materials in natural biofilms may make interpretation of the latter approach difficult.

Protocol 16.3 Lectin staining.

1 The sample of the substratum with the attached biofilm should be covered with original water. The biofilm sample may be kept in a moist chamber such as a Petri dish with a wet tissue. The staining does not require the fixation of the biofilm sample. The complete staining procedure of the biofilm is carried out in the liquid droplet covering the biofilm still attached to the substratum.
2 After removing the water from the biofilm carefully with a tissue, the staining solution with the lectin (1–5 mg lectin per ml) is applied with a microlitre pipette as a droplet to the biofilm. We have found that an incubation time of 10 minutes at room temperature works well.
3 After staining the sample is washed three times with filter-sterilized water. This is done by using a tissue to remove the staining solution, after which the biofilm is covered with water again. This sequence is repeated three times without any incubation.
4 Finally, the sample can be used for CLSM. To avoid any disturbance or destruction of the biofilm no cover slip is used. If a normal microscope is employed, a water immersible lens is ideal for direct observation in the water droplet covering the biofilm. The lens must be carefully immersed in the droplet to avoid shear forces. Then the lens can be lowered downwards, thereby slowly approaching the biofilm surface. If an inverted microscope is used, the substratum with the attached biofilm should be mounted upside down in a cover slip chamber using spacers. The same procedure as described above can be applied to this inverted set-up. Michael and Smith (1995) also outline procedures for lectin application and the use of positive and negative controls.

16.7.3 Multiple staining with lectins

With current CLSM systems, up to three different fluorescent labels can be recorded sequentially or even simultaneously. The labels used may have an excitation in the blue, yellow and red; they then may show an emission in the green, red and far red regions of the light spectrum. The three labels could be, for example, FITC, TRITC and CY5. Instead of these fluorescence markers, others with similar excitation/emission properties may be employed. With this approach several lectin binding sites can be stained, or a combination of cell labelling and EPS labelling can be applied to reveal the biofilm architecture. Plate 5B shows the results of combined staining with SYTO 9 (green), *Ulex europeaus* lectin FITC (green), *Triticum vulgarus* lectin TRITC (red) and *Arachis hypogea* lectin CY5 (blue). A colour wheel is provided to aid in the interpretation of regions where possible combinations of the probes occur.

With the more expensive CLSM instruments, having a UV laser available, the complete range of fluorescent probes and the complete spectrum of the visible light can be used. That means that four fluorescent probes and four channels can be employed to reveal the structural features of a microbial biofilm. UV may also be used for ratio imaging, or to release caged compounds (Haugland, 1996). However, at present UV laser systems have alignment problems, show chromatic errors, since the optics are still not fully corrected in this spectral region, and require new objectives calculated specially for CLSM and UV light (Bliton and Lechleiter, 1995).

16.7.4 Probes for proteins

As a general protein stain, Hoechst 2495 was applied to label 'footprints' of bacteria at interfaces (Paul and Jeffrey, 1985; Neu and Marshall, 1991). Similarly, this dye may be used to probe for proteins in biofilms. In addition, there is a whole range of reactive probes that can be linked to certain groups of amino acids, amines, proteins and thiols (e.g. Haugland, 1996). To our knowledge, these have not yet been applied directly to biofilms with subsequent examination by CLSM.

16.7.5 Probes for charge

It is useful to know more about the charge distribution in biofilms. Dextrans in combination with CLSM can be used for micro-scale determination of positively and negatively charged residues in biofilms. Because dextrans are relatively inert and have low toxicity, they are suitable for *in situ* determinations of charge and particle mobility in living biofilm systems. Furthermore, they are relatively resistant to microbial degradation, because of the presence of an α-1,6-polyglucose linkage. Fluorochrome-conjugated dextrans are commercially available from Molecular Probes, Inc. and Sigma Chemical Co. with molecular weights ranging between 3000 and 1.0×10^6 kD. Other charged probes that may be considered for biofilm studies include ficol and poly-L-lysine conjugates.

Application of the dextrans may be carried out as described above for lectins. Alternatively, if continuous-flow slide cultures are used, solutions may simply be pumped in and out of the system. We have observed extensive binding of cationic and polyanionic dextrans, with limited binding of dextrans with a net anionic charge.

16.7.6 Visualization of biofilm metabolism: gene expression

Fluorogenic substrates offer an alternative to conventional radioactive substrates to detect the expression of a variety of reporter genes, including β-galactosidase, β-glucuronidase, chloramphenicol acetyl transferase and luciferase. Davies *et al.* (1993) utilized a direct visualization approach to demonstrate that algC genes of *Pseudomonas aeruginosa* were upregulated by adhesion to the surface and growth in a biofilm, compared with planktonic cells in the liquid medium. In a subsequent

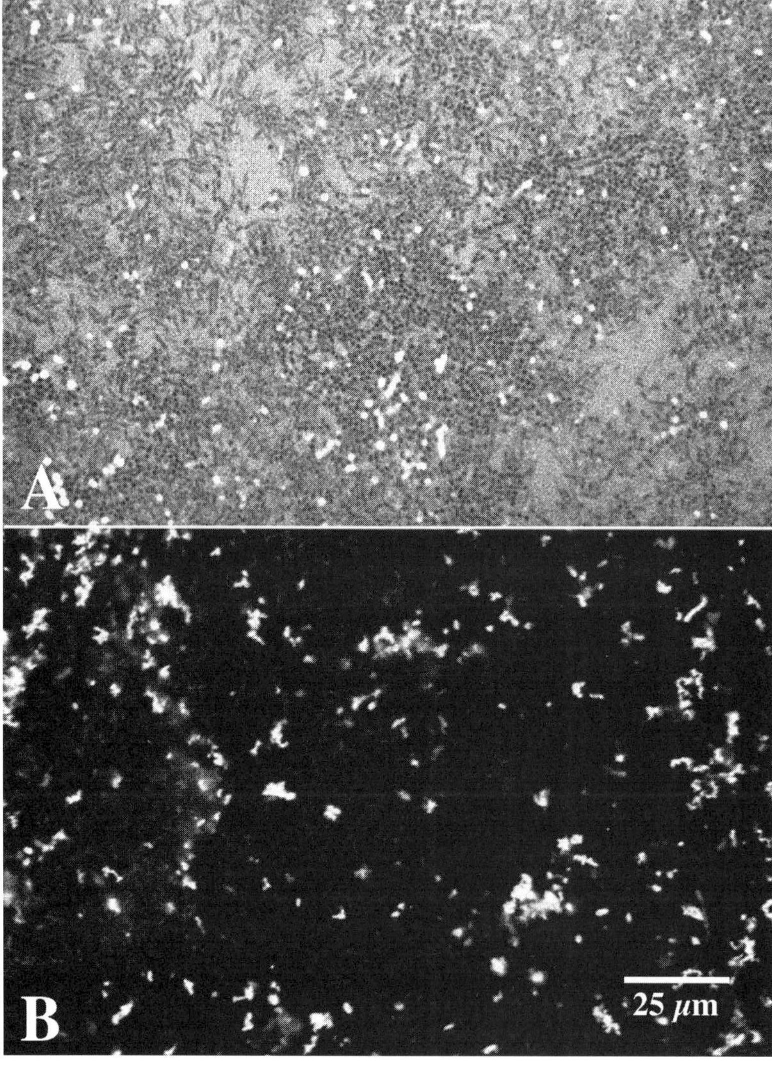

Figure 16.9 CLSM *xy* section showing the distribution of FDG labelled MK107 cells that have upregulated starvation genes; however, (A) the reporter molecule leaks out of the cells, resulting in negative staining. (B) A similar microscope field showing MK107 cells imaged using modified FDG, which does not leak out of the active cells.

study (Davies and Geesey, 1995), the authors demonstrated changes in expression of algC in *P. aeruginosa* following attachment to the substratum. The application of *in situ* polymerase chain reaction (PCR) (Hodson *et al.*, 1995) provides a powerful tool for visualization of genes and gene products. Fluorescein di-β-D-galactopyranoside (FDG) was the first fluorogenic substrate commercially available to detect expression of the *lacZ* gene in single living cells (Haugland, 1996). One problem is that the fluorescent enzymatic hydrolysis product (fluorescein) leaks out of the cells, with the result that the cells become less fluorescent and there is a simultaneous increase in background fluorescence (see Figure 16.9A). The FDG substrate was later modified by addition of fatty acyl chains, which yield fluorescent lipophilic products that are retained within the cell.

This approach has been used to monitor expression of starvation genes in *P. putida* strain MK107 (Kim *et al.*, 1995). Strain MK107 contains a mini-*Tn*5 *lacZ*1 transposon insert, which reports the expression of the starvation response genes. FDG and C_{12}-FDG, an analog of FDG containing a 12-carbon acyl chain, were used in combination with CLSM to study starvation gene expression by MK107 in biofilms.

Protocol 16.4 Monitoring gene expression.

1 MK107 was cultivated in flow cells for 48 hours in a minimal salts medium (Caldwell and Lawrence, 1986) plus 1 g/l glucose, before the irrigation solution was switched to different glucose concentrations varying between 1 and 0 g/l. Relatively small flow cells were used in these studies (dimensions of the flow cell channels were 30 mm $\times$ 3 mm $\times$ 0.2 mm) to minimize the amounts of FDG required.
2 Starvation gene expression was then determined after an additional 24 hours cultivation with the new glucose concentrations. We found that an incubation period of 45 minutes with FDG or C_{12}-FDG was sufficient.
3 A 1.9 mM solution was used for both FDG and C_{12}-FDG. FDG (5 mg) is first dissolved in a mixture of 0.4 ml EtOH and 0.4 ml DMSO before 3.2 ml ice-cold water is added to this solution. We used 0.2 ml of this working solution with 4 ml flow cell suspension, thus it was possible to do 20 analyses per vial.

Figure 16.9A and B show typical results when (C_{12}-)FDG is used to detect reporter gene expression. In Figure 16.9(A), much of the fluorescent hydrolysis product (fluorescein) leaked out of the cells when FDG was used as substrate, giving rise to a bright background fluorescence. In contrast, when biofilms cultivated under the same conditions and containing equivalent numbers of cells were incubated with C_{12}-FDG, there was no leakage of the fluorescent hydrolysis product from cells (Figure 16.9B).

Green fluorescent protein may also be used as a method to monitor gene expression in single cells by detection of fluorescence (Chalfie *et al.*, 1994; Webb *et al.*, 1995). This is an area of increasing potential, and a number of modified proteins with different fluorescence properties are under development.

Metabolism of biofilms may also be followed using two additional approaches that monitor the change in fluorescence within the biofilm. In one instance

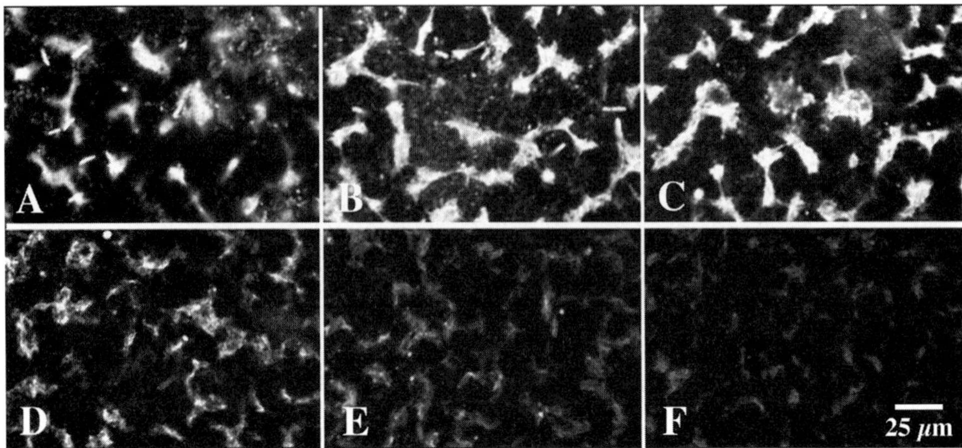

Figure 16.10 Application of CLSM fluorescence mode to monitor metabolization of the fluorescent herbicide diclofop methyl in a herbicide degrading biofilm community. Reproduced from Wolfaardt *et al.* (1995), by permission of NRC Research Press.

(Wolfaardt *et al.*, 1995), the metabolism of a fluorescent substrate, diclofop methyl, was followed by monitoring the decrease in fluorescence occurring as a result of ring cleavage of the molecule (Figure 16.10). Another alternative takes advantage of the effect of metabolic processes on the pH of the surrounding environment. Figure 16.11 shows a *Proteus mirabilis* biofilm where the pH has been shifted to pH 5 (time = 0). At this pH fluorescein is quenched, and thus cells appear as bright objects (internal pH 7) on a black background. Subsequently the activity of urease in the artificial urine medium results in the medium returning to pH > 8, and the background fluorescence undergoes a simultaneous increase in brightness. Images collected during these experiments may be subjected to image processing and analysis to obtain quantitative information (see Chapter 1; Lawrence *et al.*, 1996).

16.7.7 Monitoring the microenvironment in biofilms

Measurements using microelectrodes show significant gradients within microbial biofilms and bioaggregates (Lens *et al.*, 1993; Ramsing *et al.*, 1993; Costerton *et al.*, 1994; de Beer *et al.*, 1994). Probes such as fluorescein and 5,6-carboxyfluorescein have pH- and Eh-sensitive fluorescence, and thus potential for use in *in situ* measurements of microbially-associated chemical microenvironments. Some of these probes have been tried in biofilms, with the result that we have some intriguing images showing apparent pH gradients surrounding cells and cell clusters (Caldwell *et al.*, 1992a). The sales catalogue of Molecular Probes, Inc. (Haugland, 1996) provides a listing of a number of potential probes for biofilm studies (pp. 325–377). The cell biology literature contains examples of approaches that may find application in microbial systems (Tsien, 1989; Lattanzio, 1990; Hernandez-Cruz *et al.*, 1990; Tsien and Waggoner, 1995; Lattanzio and Bartschat, 1991). Other aspects of the microenvironment may be assessed using size-

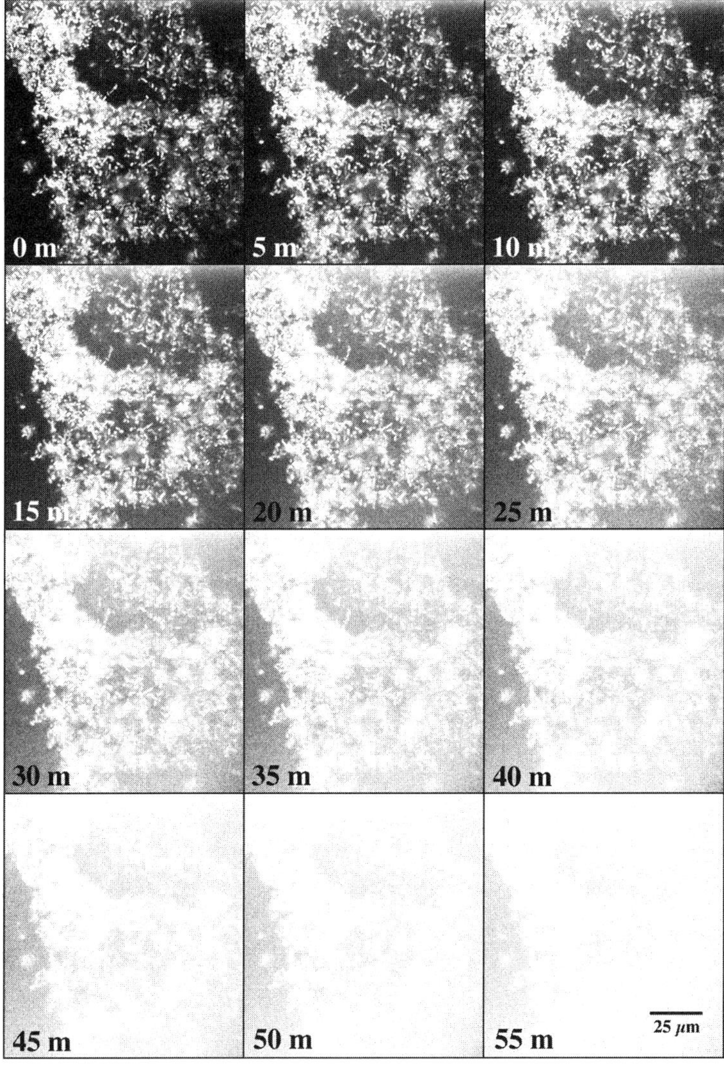

Figure 16.11 Monitoring the change in fluorescence of fluorescein as biofilm metabolism shifts the pH from pH 5 at 0 minutes where fluorescein is quenched except inside cells of *Proteus mirabilis* (pH approximately 7) to pH > 8 where the background pH matches or exceeds that of the cells, resulting in loss of contrast.

fractionated dextrans or ficols to examine diffusion within biofilms. An *in situ* monitoring approach was developed by Lawrence *et al.* (1994) to determine effective diffusion coefficients for biofilm systems. Figure 16.12 is a series of optical thin sections collected over time to monitor the diffusion of a 1.0×10^3 kD FITC conjugated dextran into a mixed-species biofilm. De Beer *et al.* (1997) have used microinjection and CSLM to determine diffusion coefficients in biofilm materials. The more standard FRAP (fluorescence recovery after photobleaching, Section 3.6.3)

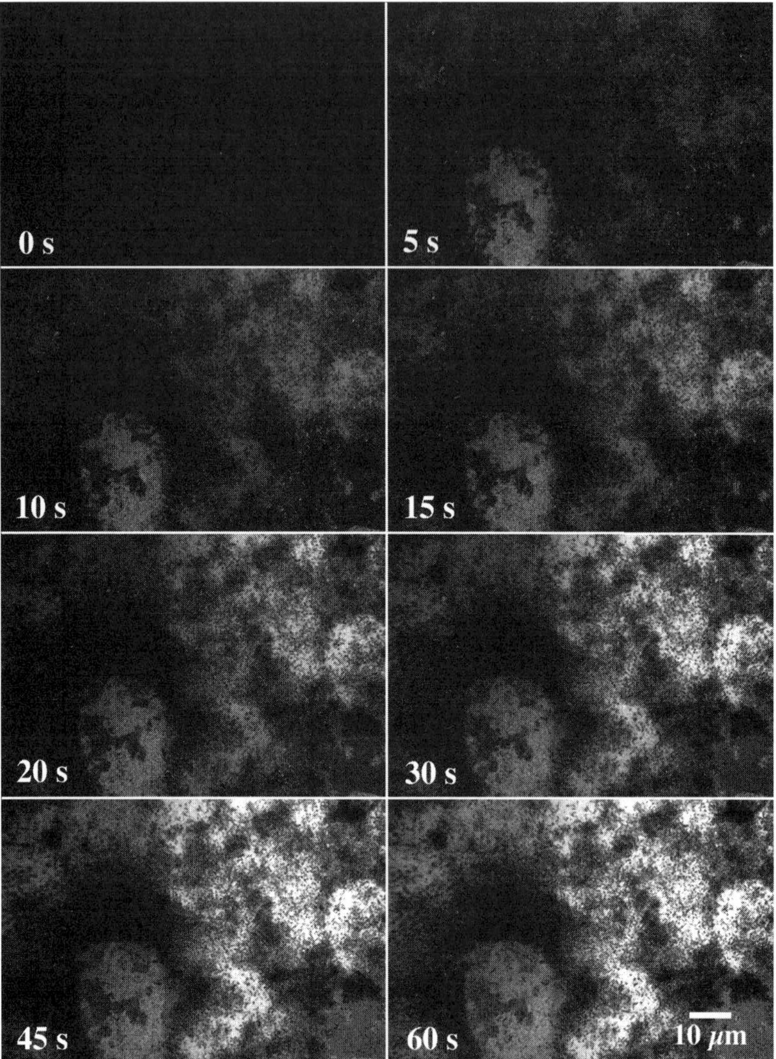

Figure 16.12 Monitoring the diffusion of a size fractionated fluorescein conjugated dextran (1×10^3 kD) at one depth or optical thin section in a herbicide degrading biofilm community. Data generated from image analysis can be used for the estimation of effective diffusion coefficients (see Lawrence *et al.* (1994) for additional details).

approach may also be applied to biofilms. Birmingham *et al.* (1995) used FRAP to monitor diffusion and binding of dextrans in oral biofilms. The details of the procedures are outlined in these and other papers (Axelrod *et al.*, 1976; Blonk *et al.*, 1993). Flow patterns in biofilms may be monitored easily using the approach described by Stoodley *et al.* (1994). This involves the introduction of fluorescent beads and collection of averaged or serial images in which the changing position of

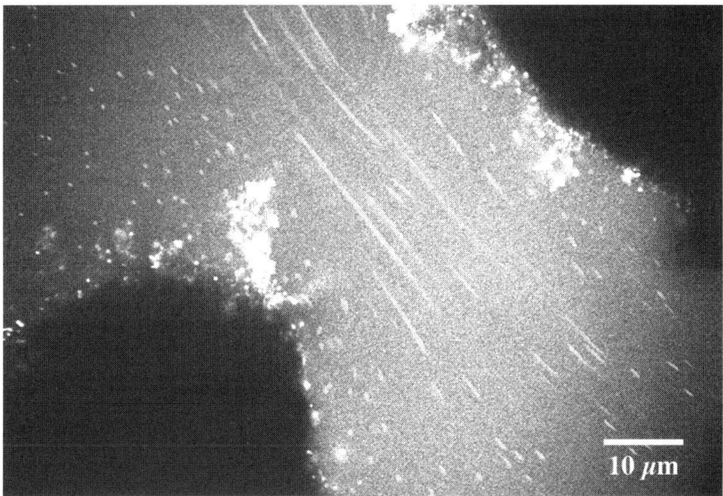

Figure 16.13 CLSM optical section of a *Klebsiella oxytoca* biofilm on sand grains and in pore throats. The image was collected as a summed image over time showing the changing position of 1 μm fluorescent beads. Images of this type may be analysed to determine the rates of water flow past or through biofilms.

the beads may be detected and measured. Figure 16.13 shows an example of this approach applied to a *Klebsiella oxytoca* biofilm developing on sand grains.

16.8 FUTURE REQUIREMENTS

Although CLSM allows the user to obtain extensive, detailed information on fully hydrated biological materials, there is a need for advances in a number of areas. Within the conventional design, major improvements are required in the areas of alignment, PMT sensitivity and response, scanning speed for image acquisition, and the range and nature of excitation/emission filters available. It is also possible to rethink completely the nature of confocal microscopy, developing new real-time approaches at lower cost (Juskaitis *et al.*, 1996). The application of real-time imaging is important for the study of microbial motion (Chapter 13). Although some advances have occurred in objective lens development, there is still a need for objectives with more favourable working distances, numerical aperture, and corrections for chromatic aberration encompassing 400–700 nm. The other major areas are the development of new fluorescent and reflective probes; specific areas include probes for specific structural components (EPS), and other environmentally sensitive probes for studies of the microenvironment. Software is also required to allow quantification of multi-channel (e.g. four) and multi-dimensional data sets, improved deconvolution software and in general more user-friendly approaches to both operating software and analytical packages. The past five years have seen a doubling of the number of manufacturers of CLSM systems in a variety of models and prices, and a vast increase in the range of software available. Thus, the trend is favourable for the increased availability and application of CLSM in microbial ecology.

REFERENCES

Agard, D.A. (1984) Optical sectioning microscopy: cellular architecture in three dimensions. *Ann. Rev. Biophys. Bioengng* **13**: 191–219.

Allison, D.G. and Sutherland, I.W. (1984) A staining technique for attached bacteria and its correlation to extracellular carbohydrate production. *J. Microbiol. Meth.* **2**: 93–9.

Amann, R., Ludwig, W. and Schleifer, K.H. (1995) Phylogenetic identification and in situ detection of individual microbial cells without cultivation. *Microbiol. Rev.* **59**: 143–69.

Amann, R., Snaid, R., Wagner, M., Ludwig, W. and Schleifer, K.H. (1996) In situ visualization of high genetic diversity in a natural microbial community. *J. Bacteriol.* **178**: 3496–3500.

Arnold, J.W. and Shimkets, L.J. (1988) Inhibition of cell–cell interactions in *Myxococcus xanthus* by Congo red. *J. Bacteriol.* **170**: 5765–70.

Assmus, B., Hutzler, P., Kirchhof, G., Amann, R., Lawrence, J.R. and Hartmann, A. (1995) In situ detection of *Azospirillum brasilense* in the rhizosphere of wheat using fluorescently labeled rRNA-targeted oligonucleotide probes and scanning confocal laser microscopy. *Appl. Environ. Microbiol.* **61**: 1013–19.

Axelrod, A., Koppel, D.E., Schlessinger, J., Elsen, E. and Webb, W.W. (1976) Mobility measurement by analysis of fluorescence photobleaching recovery kinetics. *Biophys. J.* **16**: 1055–69.

Barondes, S.H. (1988) Bifunctional properties of lectins: lectins redefined. *Trends Biochem. Sci.* **13**: 480–2.

Birmingham, J.J., Hughes, N.P. and Treloar, R. (1995) Diffusion and binding measurements within oral biofilms using fluorescence photobleaching recovery methods. *Phil. Trans. Roy. Soc. Lond. B, Biol. Sci.* **350**: 325–43.

Black, J.P. and Kroll, R.G. (1991) The differential fluorescence of bacteria stained with acridine orange and the effects of heat. *J. Appl. Bacteriol.* **71**: 51–8.

Bliton, A.C. and Lechleiter, J.D. (1995) Optical considerations at ultraviolet wavelengths in confocal microscopy. In *Handbook of Biological Confocal Microscopy* (Pawley, J.B., ed.), pp. 431–44. Plenum Press: New York.

Bloem, J., Veninga, M. and Sheperd, J. (1995) Fully automatic determination of soil bacterium numbers, cell volumes and frequencies of dividing cells by confocal laser scanning microscopy and image analysis. *Appl. Environ. Microbiol.* **61**: 926–36.

Blonk, J.C.G., Don, A., van Aalst, H. and Birmingham, J.J. (1993) Fluorescence photobleaching recovery in the confocal scanning laser microscope. *J. Microsc.* **169**: 363–74.

Böck, G., Hilchenbach, M., Schauenstein, K. and Wick, G. (1985) Photometric analysis of antifading reagents for immunofluorescence with laser and conventional illumination sources. *J. Histochem. Cytochem.* **33**: 699–705.

Bog-Hansen, T.C. (ed.) (1981) *Lectins – Biology, Biochemistry, Clinical Biochemistry*. Walter de Gruyter: Berlin.

Bott, T.L., Brock, J.T., Battrup, A., Chambers, P.A., Dodds, W.K., Himbeault, K., Lawrence, J.R., Planas, D., Snyder, E. and Wolfaardt, G.M. (1997) An evaluation of techniques for measuring periphyton metabolism in chambers. *Can. J. Fish. Aqu. Res.* **54**: 715–25.

Caldwell, D.E. and Lawrence, J.R. (1986) Bacterial growth kinetics in the hydrodynamic boundary layer of solid–liquid interfaces. *Microb. Ecol.* **12**: 299–312.

Caldwell, D.E., Korber, D.R. and Lawrence, J.R. (1992a) Confocal laser microscopy and digital image analysis in microbial ecology. *Adv. Microb. Ecol.* **12**: 1–67.

Caldwell, D.E., Korber, D.R. and Lawrence, J.R. (1992b) Imaging of bacterial cells by fluorescence exclusion using scanning confocal laser microscopy. *J. Microbiol. Meth.* **15**: 249–61.

Chalfie, M., Tu, Y., Euskirchen, G., Ward, W.W. and Prasher, D.C. (1994) Green fluorescent protein as a marker for gene expression. *Science* **263**: 802–5.

Colvin, J.R. and Witter, D.E. (1983) Congo red and Calcofluor White inhibition of *Acetobacter xylinum* cell growth and of bacterial cellulose microfibrill formation: isolation and properties of a transient, extracellular glucan related to cellulose. *Protoplasma* **116**: 34–40.

Costerton, J.W., Lewandowski, Z., de Beer, D., Caldwell, D., Korber, D. and James, G. (1994) Biofilms, the customized microniche. *J. Bacteriol.* **176**: 2137–42.

Cullander (1994) Imaging in the far-red with electronic light microscopy: requirements and limitations. *J. Microsc.* **176**: 281–6.

Davies, D.G. and Geesey, G.G. (1995) Regulation of the alginate biosynthesis gene algC in *Pseudomonas aeruginosa* during biofilm development in continuous culture. *Appl. Environ. Microbiol.* **61**: 860–7.

Davies, D.G., Chakrabarty, A. and Geesey, G.G. (1993) Exopolysaccharide production in biofilms: substratum activation of alginate gene expression by *Pseudomonas aeruginosa*. *Appl. Environ. Microbiol.* **59**: 1181–6.

Dazzo, F.B. and Wright, S.F. (1996) Production of anti-microbial antibodies and their use in immunofluorescence microscopy. *Molec. Microb. Ecol. Manual* **4.1.2**: 1–27.

de Beer, D., Stoodley, P., Roe, F. and Lewandowski, Z. (1994) Effects of biofilm structures on oxygen distribution and mass transfer. *Biotechnol. Bioengng* **43**: 1131–8.

de Beer, D., Stoodley, P. and Lewandowski, Z. (1997) Measurement of local diffusion coefficients in biofilms by microinjection and confocal microscopy. *Biotechnol. Bioengng* **53**: 151–8.

Del Gallo, M., Negi, M. and Neyra, C.A. (1989) Calcofluor- and lectin-binding exocellular polysaccharides of *Azospirillum brasilense* and *Azospirillum lipoferum*. *J. Bacteriol.* **171**: 3504–10.

Doherty, D., Leigh, J.A., Glazebrook, J. and Walker, G.C. (1988) *Rhizobium meliloti* mutants that overproduce the *R. meliloti* acidic Calcofluor-binding exopolysaccharide. *J. Bacteriol.* **170**: 4249–56.

Doyle, R.J. and Slifkin, M. (eds) (1994) *Lectin Microorganism Interaction*. Marcel Dekker: New York.

Dubuisson, M.P., Jain, A.K. and Jain, M.K. (1994) Segmentation and classification of bacterial culture images. *J. Microbiol. Meth.* **19**: 279–95.

Florijn, R.J., Slats, J., Tanke, H.J. and Raap, A.K. (1995) Analysis of antifading reagents for fluorescence microscopy. *Cytometry* **19**: 177–82.

Geesey, G.G. and White, D.C. (1990) Determination of bacterial growth and activity at solid–liquid interfaces. *Ann. Rev. Microbiol.* **44**: 579–602.

Gorby, G.L. (1994) Digital confocal microscopy allows measurements and three-dimensional multiple spectral reconstructions of *Neisseria gonorrhoe*/epithelial cell interactions in the human fallopian tube organ culture model. *J. Histochem. Cytochem.* **42**: 297–306.

Haugland, R.P. (1996) *Handbook of Fluorescent Probes and Research Chemicals*, 6th edn. Molecular Probes, Inc.: Eugene, OR.

Hernandez-Cruz, A., Sala, F. and Adams, P.R. (1990) Subcellular calcium transients visualized by confocal microscopy in a voltage-clamped vertebrate neuron. *Science* **247**: 858–62.

Hodson, R.E., Dustman, W.A., Garg, R.P. and Moran, M.A. (1995) In situ PCR for visualization of microscale distribution of specific genes and gene products in prokaryotic communities. *Appl. Environ. Microbiol.* **61**: 4074–82.

Johnson, G.D. and de C Nogueira Araujo, G.M. (1981) A simple method of reducing the fading of immunofluorescence during microscopy. *J. Immunol. Meth.* **43**: 349–50.

Juskaitis, R., Wilson, T., Neil, M.A.A. and Kozubek, M. (1996) Efficient real-time confocal microscopy with white light sources. *Nature* **383**: 804–6.

Kim, Y., Watrud, L.S. and Matin, A. (1995) A carbon starvation survival gene of *Pseudomonas putida* is regulated by σ^{54}. *J. Bacteriol.* **177**: 1850–9.

Korber, D.R., Lawrence, J.R., Hendry, M.J. and Caldwell, D.E. (1993) Analysis of spatial variability within mot^+ and mot^- *Pseudomonas fluorescens* biofilms using representative elements. *Biofouling* **7**: 339–58.

Korber, D.R., Choi, A., Wolfaardt, G.M. and Caldwell, D.E. (1996) Bacterial plasmolysis as a physical indicator of viability. *Appl. Environ. Microbiol.* **62**: 3939–47.

Lam, J.S. and Mutharia, L.M. (1994) Antigen–antibody reactions. In *Methods for General and Molecular Bacteriology* (Gerhardt P, ed.), pp. 104–32. American Society for Microbiology: Washington, DC.

Lattanzio, F.A. Jr (1990) The effects of pH and temperature on fluorescent calcium indicators as determined with Chelex-100 and EDTA buffer systems. *Biochem. Biophys. Res. Comm.* **171**: 102–8.

Lattanzio, F.A. Jr and Bartschat, D.K. (1991) The effect of pH on rate constants, ion selectivity and thermodynamic properties of fluorescent calcium and magnesium indicators. *Biochem. Biophys. Res. Comm.* **177**: 184–91.

Lawrence, J.R., Korber, D.R., Hoyle, B.D., Costerton, J.W. and Caldwell, D.E. (1991) Optical sectioning of microbial biofilms. *J. Bacteriol.* **173**: 6558–67.

Lawrence, J.R., Wolfaardt, G.M. and Korber, D.R. (1994) Monitoring diffusion in biofilm matrices using confocal laser microscopy. *Appl. Environ. Microbiol.* **60**: 1166–73.

Lawrence, J.R., Korber, D.R., Wolfaardt, G.M. and Caldwell, D.E. (1995) Bacterial behavioural strategies at interfaces. *Adv. Microbial. Ecol.* **14**: 1–75.

Lawrence, J.R., Korber, D.R., Wolfaardt, G.M. and Caldwell, D.E. (1996) Analytical imaging and microscopy techniques. In *Manual for Environmental Microbiology* (Hurst, C.J., Knudsen, G.R., McInerney, M.J., Stetzenbach, L.D. and Walker, M.V., eds), pp. 29–51. American Society for Microbiology Press, Washington, DC.

Lawrence, J.R., Kwong, Y.T.J. and Swerhone, G.D.W. (1997) Colonization and weathering of natural sulfide mineral assemblages by *Thiobacillus ferrooxidans*. *Can. J. Microbiol.* **43**: 178–88.

Lens, P.N.L., de Beer, D., Cronenberg, C.C.H., Houwen, F.P., Ottengraf, S.P.P. and Verstraete, W.H. (1993) Heterogeneous distribution of microbial activity in methanogenic aggregates: pH and glucose microprofiles. *Appl. Environ. Microbiol.* **59**: 3803–15.

Lewandowski, Z., Lee, W.C., Characklis, W.G. and Little, B. (1989) Dissolved oxygen and pH microelectrode measurements at water immersed metal surfaces. *Corrosion* **45**: 92–8.
Lewandowski, Z., Altobelli, S.A. and Fukushima, E. (1993) NMR and microelectrode studies of hydrodynamics and kinetics in biofilms. *Biotechnol. Prog.* **9**: 40–5.
Liener, I.E., Sharon, N. and Goldstein, I.J. (eds) (1986) *The Lectins*. Academic Press: New York.
Lloyd, D. and Hayes, A.J. (1995) Vigour, vitality and viability of microorganisms. *FEMS Microbiol. Lett.* **33**: 1–7.
Longin, A., Souchier, C., French, M. and Bryon, P.A. (1993) Comparison of anti-fading reagents used in fluorescence microscopy: image analysis and laser confocal microscopy study. *J. Histochem. Cytochem.* **41**: 1833–40.
Lopez-Amoros, R., Comas, J. and Vives-Rego, J. (1995) Flow cytometric assessment of *Escherichia coli* and *Salmonella typhimurium* starvation–survival in seawater using rhodamine 123, propidium iodide and oxonol. *Appl. Environ. Microbiol.* **61**: 2521–6.
Mason, D.J., Lopez-Amoros, R., Allman, R., Stark, J.M. and Lloyd, D. (1995) The ability of membrane potential dyes and Calcafluor White to distinguish between viable and non-viable bacteria. *J. Appl. Bacteriol.* **78**: 309–15.
Massol-deya, A.A., Whallon, J., Hickey, R.F. and Tiedje, J.M. (1995) Channel structures in aerobic biofilms of fixed-film reactors treating contaminated groundwater. *Appl. Environ. Microbiol.* **61**: 769–77.
Matsuoka, O., Takatori, K. and Kurata, H. (1995) Use of Congo red as a microscopic fluorescence indicator of hyphal growth. *Appl. Microbiol. Biotechnol.* **43**: 102–8.
McFeters, G.A., Yu, F.P., Pyle, B.H. and Stewart, P.S. (1995) Physiological assessment of bacteria using fluorochromes. *J. Microbiol. Meth.* **21**: 1–13.
Michael, T. and Smith, C.M. (1995) Lectins probe molecular films in biofouling: characterization of early films on non-living and living surfaces. *Mar. Ecol. Prog. Ser.* **119**: 229–36.
Møller, S., Kristensen, C.S., Poulsen, L.K., Carstensen, J.M. and Molin, S. (1995) Bacterial growth on surfaces: automated image analysis for quantification of growth rate-related parameters. *Appl. Environ. Microbiol.* **61**: 741–8.
Morgan, P., Cooper, C.J., Battersby, N.S., Lee, S.A., Lewis, S.T., Machin, T.M., Graham, S.C. and Watkinson, R.J. (1991) Automated image analysis method to determine fungal biomass in soils and on solid matrices. *Soil Biol. Biochem.* **23**: 609–16.
Muller, M., Loffhagen, N., Bley, T. and Babel, W. (1996) Membrane-potential-related fluorescence intensity indicates bacterial injury. *Microbiol. Res.* **151**: 127–31.
Neu, T.R. and Lawrence, J.R. (1997) Development and structure of microbial stream biofilms as studied by confocal laser scanning microscopy. *FEMS Microb. Ecol.* **24**: 11–25.
Neu, T.R. and Marshall, K.C. (1991) Microbial 'footprints' – a new approach to adhesive polymers. *Biofouling* **3**: 101–12.
Nivens, D.E., Palmer, Jr. R.J. and White, D.C. (1995) Continuous non-destructuve monitoring of microbial biofilms: a review of analytical techniques. *J. Ind. Microbiol.* **15**: 263–76.
Paul, J.H. and Jeffrey, W.H. (1985) Evidence for separate adhesion mechanisms for hydrophilic and hydrophobic surfaces in *Vibrio proteolytica*. *Appl. Environ. Microbiol.* **50**: 431–7.
Pawley, J.B. (ed.) (1995) *Handbook of Biological Confocal Microscopy*, 2nd edn. Plenum Press: New York.
Poulsen, L.K., Ballard, G. and Stahl, D.A. (1993) Use of rRNA fluorescence in situ hybridization for measuring the activity of single cells in young and established biofilms. *Appl. Environ. Microbiol.* **59**: 1354–60.
Ramsing, N.B., Kuhl, M. and Jorgensen, B.B. (1993) Distribution of sulfate reducing bacteria, O_2, and H_2S in photosynthetic biofilms determined with oligonucleotide probes and microelectrodes. *Appl. Environ. Microbiol.* **59**: 3840–9.
Reeke, G.N. and Becker, J.W. (1988) Carbohydrate-binding sites of plant lectins. *Cur. Topics Microbiol. Immunol.* **139**: 35–58.
Ried, T., Baldini, A., Rand, T.C. and Ward, D.C. (1992) Simultaneous visualization of seven different DNA probes by in situ hybridization using combinatorial fluorescence and digital imaging microscopy. *Proc. Natl. Acad. Sci. USA* **89**: 1388–92.
Riguat, J.P. and Vassy, J. (1991) High-resolution three dimensional images from confocal scanning laser microscopy. Quantitative study and mathematical correction of the effects from bleaching and fluorescence attenuation in depth. *Anal. Quant. Cytol. Histol.* **13**: 223–32.
Rodrigues, G.G., Phipps, D., Ishiguro, K. and Ridgway, H.F. (1992) Use of a fluorescent redox probe for direct visualization of actively respiring bacteria. *Appl. Environ. Microbiol.* **58**: 1801–8.
Rogers, J. and Keevil, C.W. (1992) Immunogold and fluorescein immunolabelling of *Legionella pneumophila* within an aquatic biofilm visualized by episcopic differential interference contrast microscopy. *Appl. Environ. Microbiol.* **58**: 2326–30.
Roncero, C. and Duran, A. (1985) Effect of Calcofluor White and Congo red on fungal cell wall morphogenesis: in vivo activation of chitin polymerization. *J. Bacteriol.* **163**: 1180–5.

Rost, F.W.D. (1992a) *Fluorescence Microscopy*, vol. I, pp. 253ff. Cambridge University Press: Cambridge.

Rost, F.W.D. (1992b) *Fluorescence Microscopy*, vol. II, pp. 457ff. Cambridge University Press: Cambridge.

Ruimy, R., Breittmayer, V., Boivin, V. and Christen, R. (1994) Assessment of the state of activity of individual bacterial cells by hybridization with a ribosomal RNA targeted fluorescently labelled oligonucleotide probe. *FEMS Microbiol. Ecol.* 15: 207–14.

Schloter, M., Borlinghaus, R., Bode, W. and Hartmann, A. (1993) Direct identification, and localization of *Azospirillum* in the rhizosphere of wheat using fluorescence-labelled monoclonal antibodies and confocal scanning laser microscopy. *J. Microsc.* **171**: 173–7.

Schloter, M., Assmus, B. and Hartmann, A. (1995) The use of immunological methods to detect and identify bacteria in the environment. *Biotechnol. Adv.* **13**: 75–90.

Sharon, N. and Lis, H. (1989) Lectins as cell recognition molecules. *Science* **246**: 227–34.

Shotton, D.M. (1989) Confocal scanning optical microscopy and its applications for biological specimens. *J. Cell Sci.* **94**: 175–206.

Shotton, D. and White, N. (1989) Confocal scanning microscopy: three-dimensional biological imaging. *Trends Biochem. Sci.* **14**: 435–9.

Sizemore, R.K., Caldwell, J.J. and Kendrick, A.S. (1990) Alternate Gram staining technique using a fluorescent lectin. *Appl. Environ. Microbiol.* **56**: 2245–7.

Stevens, J.K. (1994) Introduction to confocal three-dimensional volume investigation. In *Three-Dimensional Confocal Microscopy: Volume Investigation of Biological Systems* (Stevens, J.K., Mills, L.R. and Trogadis, J.E., eds), pp. 3–24. Academic Press: New York.

Stewart, P.S., Peyton, B.M., Dury, W.J. and Murga, R. (1993) Quantitative observations of heterogeneities in *Pseudomonas aeruginosa* biofilms. *Appl. Environ. Microbiol.* **59**: 327–9.

Stoodley, P., De Beer, D. and Lewandowski, Z. (1994) Liquid flow in biofilm systems. *Appl. Environ. Microbiol.* **60**: 2711–6.

Surman, S.B., Walker, J.T., Goddard, D.T., Morton, L.H.G., Keevil, C.W., Weaver, W., Skinner, A., Hanson, K., Caldwell, D. and Kurtz, J. (1996) Comparison of microscopic techniques for the examination of biofilms. *J. Microbiol. Meth.* **25**: 57–70.

Tsien, R.Y. (1989) Fluorescent indicators of ion concentrations. *Meth. Cell Biol.* **30**: 127–56.

Tsien, R.Y. and Waggoner, A. (1995) Fluorophores for confocal microscopy. In *Handbook of Confocal Microscopy* (Pawley, J.B., ed.), pp. 267–79. Plenum Press: New York.

Waggoner, A., DeBassio, R., Conrad, P., Bright, G.R., Ernst, L., Ryan, K., Nederlof, M. and Taylor, D. (1989) Multiple spectral parameter imaging. *Meth. Cell Biol.* **30**: 449–76.

Webb, C.D., Decatur, A., Teleman, A. and Losick, R. (1995) Use of green fluorescent protein for visualization of cell-specific gene expression and subcellular protein localization during sporulation in *Bacillus subtilis*. *J. Bacteriol.* **177**: 5906–11.

Wells, S. and Johnson, I. (1994) Fluorescent labels for confocal microscopy. In *Three-Dimensional Confocal Microscopy: Volume Investigation of Biological Specimens* (Stevens, J.K., Mills, L.R. and Trogadis, J.E., eds), pp. 101–29. Academic Press: London.

Wilson, T. and Sheppard, C. (1984) *Theory and Practice of Scanning Optical Microscopy*. Academic Press: London.

Wolfaardt, G.M., Lawrence, J.R., Robarts, R.D. and Caldwell, D.E. (1994) Multicellular organization in a degradative biofilm community. *Appl. Environ. Microbiol.* **60**: 434–46.

Wolfaardt, G.M., Lawrence, J.R., Robarts, R.D. and Caldwell, D.E. (1995) Bioaccumulation of the herbicide diclofop in extracellular polymers and its utilization by a biofilm community during starvation. *Appl. Environ. Microbiol.* **61**: 152–8.

Wood, P.J. (1980) Specificity in the interaction of direct dyes with polysaccharides. *Carbohydrate Res.* **85**: 271–87.

Zambon, J.J., Huber, P.S., Meyer, A.E., Slots, J., Fornalik, M.S. and Baier, R.E. (1984) In situ identification of bacterial species in marine microfouling films by using immunofluorescence technique. *Appl. Environ. Microbiol.* **4**: 1214–20.

Zevenhuizen, L.P.T.M., Bertochi, C. and van Neerven, A.R.W. (1986) Congo red adsorption and cellulose synthesis by Rhizobiaceae. *Antonie van Leeuwenhoek* **52**: 381–6.

17

Three-dimensional Analysis of Complex Microbial Communities by Combining Confocal Laser Scanning Microscopy and Fluorescence *In Situ* Hybridization

Michael Wagner[1], Peter Hutzler[2] and Rudolf Amann[3]

[1] *Technische Universität, München, Germany*

[2] *GSF-Forschungszentrum für Umwelt und Gesundheit, Oberschleissheim, Germany*

[3] *Max-Planck-Institut für marine Mikrobiologie, Bremen, Germany*

17.1 INTRODUCTION

In virtually all natural and engineered systems offering sufficient nutrient supply micro-organisms grow as spatially organized, matrix-enclosed multi-species communities in biofilms, aggregates or floccules rather than as single planktonic cells (Costerton *et al.*, 1995). In most ecosystems the number of bacterial cells present in biofilms exceeds the number of single microbial cells by one to several orders of magnitude. During the last decade it has become clear that biofilms have an inhomogeneous architecture. Numerous micro-environments exist which can be significantly different from each other and from the overall macro-environment (e.g. Caldwell *et al.*, 1992; Costerton *et al.*, 1994; Massol-Deyá *et al.*, 1995). Different bacterial species colonize the various micro-niches and most likely communicate with each other by various cell–cell interactions (Dworkin, 1991). Several studies suggest that distinct spatial arrangements of different microbial populations within biofilms exist, possibly reflecting syntrophic interactions (e.g. MacLeod *et al.*, 1990; Wolfaardt *et al.*, 1994; Mobarry *et al.*, 1996).

Traditional microbiological techniques are based on the isolation of bacterial pure cultures, mostly by the agar plate technique, followed by the identification and, if necessary, physiological characterization of the respective strains. To obtain clonal colonies originating from single cells, multicellular aggregates have to be effectively disrupted during the isolation procedure. At this step all information about the spatial organization of the bacteria relative to each other and within their ecological niche is already lost. Subculturing further selects for mutants of the isolated strain that have lost the ability to attach to surfaces. Recent studies have demonstrated that

Digital Image Analysis of Microbes: Imaging, Morphometry, Fluorometry and Motility Techniques and Applications. Edited by M.H.F. Wilkinson and F. Schut.

the gene expression pattern of a bacterial species is in part also dependent on the growth mode of the cells. Attached cells of *Pseudomonas aeruginosa* use a specific sigma factor to transcribe certain genes linked to attached growth (Deretic *et al.*, 1994). Consequently, physiological studies of bacterial isolates grown in suspension cultures provide us with valuable information about their capabilities in the planktonic growth mode, but do not necessarily fully inform us about their phenotypic properties during biofilm growth.

For the microbiologist interested in direct detection and identification of bacteria within their environment, two major techniques have been developed. Fluorescent antibodies (FA) and fluorescent *in situ* hybridization (FISH) with rRNA-targeted oligonucleotide probes can both be used for specific labelling of target cells present in complex environmental samples. In the following the advantages and limitations of both techniques are briefly summarized.

FA can be applied to stain both living or chemically fixed target bacteria. However, the production of FA requires the prior isolation of the target cells, which renders this technique unsuitable for those bacteria that are not yet culturable. The specificity of FA is in general restricted to the species or even subspecies level. Consequently, encompassing analysis of highly complex microbial communities typical of many natural samples requires the use of huge numbers of different FAs (e.g. Belser, 1979). Unspecific binding of FA to sheath material of some filamentous bacteria (M. Wagner, unpublished observation) and to fungal spores (Szwerinski *et al.*, 1985) results in false positive signals. An additional concern is that the expression of the antigen by the bacterial target cells can vary, depending on changes in environmental conditions or on the growth stage. In addition, the high molecular weight of FA causes penetration problems within thick biofilms, thereby hampering the applicability of this technique to many environmental samples.

Ribosomal RNA-targeted oligonucleotide probes can be designed and applied in a truly cultivation-independent way by performing a full-cycle rRNA analysis (Amann *et al.*, 1995). The specificity of these probes is almost freely adjustable to different phylogenetic levels, ranging from the domain to the species level (Amann *et al.*, 1995). For lower taxonomic levels, the specificity of FISH can be increased further by simultaneous application of up to three probes labelled with different fluorescent dye molecules. This approach also allows visualization of up to seven different bacterial populations simultaneously (Amann *et al.*, 1996). FISH requires prior permeabilization of microbial cell envelopes, which is achieved by chemical fixation of the sample. Since the first application of FISH in microbial ecology (DeLong *et al.*, 1989) different fixation protocols have been published. In general, for most, if not all, Gram-negative bacteria formaldehyde fixation is sufficient (Amann, 1995). Most Gram-positive bacteria are permeabilized by use of ethanol (Roller *et al.*, 1994). However, for certain Gram-positive species, especially for those with a hydrophobic cell surface, an additional enzymatic treatment is necessary (Beimfohr *et al.*, 1993; Schuppler *et al.*, 1998). The signal intensity of a single cell after FISH is directly linked to its ribosome content, which itself reflects the physiological potential of the cell. For a sulphate-reducing bacterium, quantification of the cellular signal intensity was successfully used to draw conclusions on its *in situ* growth rate (Poulsen *et al.*, 1993). However, it has been shown that the assumption of a strictly linear correlation between physiological activity and cellular ribosome content does not hold true for all bacteria. For example, some autotrophic ammonia-oxidizing bacteria keep a high cellular ribosome content for several days after complete inhibition of their physiological activity (Wagner *et al.*, 1995).

FISH with rRNA-targeted oligonucleotide probes offers two different routes for the analysis of the structure and dynamics of microbial consortia. The top-to-bottom approach offers the possibility of revealing the higher-level structure of microbial communities rapidly, by using a set of domain- and group-specific probes (Wagner *et al.*, 1993, 1994a; Wagner and Amann, 1997). Once numerically important bacterial groups have been identified, the analysis can be refined further by application of appropriate additional probes with more narrow specificities. The second, so-called bottom-up, approach is based on an initial direct 16S or 23S rDNA sequence retrieval from the sample. Comparative sequence analysis of the retrieved rDNA clones is used to identify their phylogenetic affiliation. In the next step oligonucleotide probes specific for selected clones (or group of clones) are designed and used for FISH. Such studies allow linking of the sequence (phylogenetic) information to a certain microbial morphotype in the original sample. In parallel, information about the abundance and spatial organization of the target organism can be obtained (Amann *et al.*, 1996; Schuppler *et al.*, 1998; Snaidr *et al.*, 1997). This approach is tedious but offers insights into the microbial community structure at unsurpassed resolution.

Direct comparison of microbial population structure analysis performed by classical cultivation-dependent techniques and FISH with rRNA-targeted oligonucleotide probes revealed three major advantages of FISH (Wagner *et al.*, 1993, 1994a; Manz *et al.*, 1994):

(a) Compared to cultivation-based techniques FISH allows the detection of one to three orders of magnitude more bacterial cells in environmental samples. For example, in activated sludge 60–90% of all cells present as detected with the DNA-intercalating dye 4′,6-diamidino-2-phenylindole (DAPI) can simultaneously be visualized with a bacterial probe. In contrast, even on optimized media only 1–19% of total bacterial cells form colonies. The EUB/DAPI ratio is lower for more oligotrophic environments such as lakes (29–64%; Glöckner *et al.*, 1996) or soil (30–50%; Zarda *et al.*, 1997). However, for these ecosystems this effect goes in parallel with a remarkable decrease in the cultivable fraction of bacterial cells to somewhere between 0.01 and 1%.

(b) While FISH allows study of the actual composition of microbial communities, cultivation leads to inevitable biases in community structure analysis. Bacterial genera or species well adapted to the applied cultivation conditions (nutrients, temperature, oxygen concentration etc.) are overestimated in number, while other bacterial populations are under-represented or even completely missed.

(c) FISH offers the possibility of studying the true three-dimensional arrangement of the bacteria (see below), while cultivation-dependent procedures wipe out all information about the spatial organization of the bacteria within their habitats.

In essence, keeping in mind the complexity of many environmental microbial consortia and the fact that according to recent estimates more than 90% of all micro-organisms have not been cultured yet, FISH is today's most powerful tool to study microbial community structure and dynamics in the environment. Detailed protocols for performing FISH using fluorescently labelled rRNA-targeted oligonucleotide probes have been published recently (Amann, 1995). The interested reader will find instructions for probe labelling, preparation of different fixatives and the hybridization procedure there.

17.2 THREE-DIMENSIONAL ANALYSIS OF MICROBIAL COMMUNITIES

The spatial organization of specifically labelled cells within bulky samples like biofilms can be analysed either by conventional epifluorescence microscopy in combination with cryosectioning or by confocal laser scanning microscopy (CLSM). Therefore the basic principles and practical tips for both techniques are given here.

Generally both techniques are realized at the same microscope stand and use the same lenses. Most technical requirements are similar. Both use top illumination through the imaging lens and a dichroic beamsplitter to separate the short wavelength excitation light from the Stokes-shifted longer wavelength fluorescence light.

17.2.1 Conventional epifluorescence microscopy

For fluorescence excitation an intense lamp with broad emission spectrum, such as a 50 or 100 Watt mercury short arc lamp, is used. It is worth looking at the illumination lens set-up. Whereas in transmission light microscopy a condenser with Köhler adjustment is standard, this is not the case at all in epifluorescence illumination. Only the latest models of expensive research microscopes offer a real field and aperture stops to optimize homogeneous illumination. This is most useful for quantitative work, e.g. when fluorescence intensities of cells at different locations within the image field are to be compared. Shading correction, available in image processing systems, is not an adequate substitute and should be used only for fine tuning.

The fluorescence filterset is of central importance in fluorescence microscopy. Besides the dichroic beamsplitter dividing the excitation and imaging paths, there is an excitation filter in the illumination path defining the spectral band of excitation, and an emission or stop filter in the imaging path selecting the spectral band contributing to the image. The choice of filterset depends on the application, whether a broadband visual inspection or a quantitative image data acquisition is wanted (see also Chapters 3 and 4).

For visual inspection two types of filtersets are in use. In cases where several fluorochromes are excitable by one excitation band, e.g. fluorescein-isothiocyanate (FITC) and propidium iodide (PI), a single-band excitation is combined with a longpass emission filter showing all fluorescence excited simultaneously. In other applications where different excitation bands are necessary, multiband filtersets are practicable (e.g. triple filter for DAPI, FITC and carboxytetramethyl-rhodamine-isothiocyanate). In these devices excitation and emission bands are cascaded in a highly sophisticated way such that cross-talk is minimized. For details see the product descriptions of suppliers like Omega Optical (Omega Optical, Brattleboro, VT, USA) or Chroma Technology (Chroma Technology, Brattleboro, VT, USA). Besides visual inspection these filtersets are also usable for taking colour images on photographic film or colour video.

For quantitative analysis of microbial populations, especially when using multi-fluorochrome labelling, the requirements are somewhat higher than for viewing only. In multi-fluorochrome labelling a fluorochrome stands for an event. An item is identified by the fluorescence intensity induced by a fluorochrome, or by the ratio of fluorescence intensities originating from two or more fluorochromes. To enable quantification, any cross-talk between fluorescence channels has to be avoided. For this purpose an excitation/emission band combination must be optimized for each fluorochrome, based on knowledge of the excitation and emission characteristics of the individual fluorochrome. General spectral data of fluorochromes are available from the suppliers, but should be checked after conjugation to the oligonucleotide in the relevant chemical environment. For simultaneous optimization of signal brightness and spectral separation filters from the high quality (HQ) class should be selected. These are multi-layer interference filters with very sharp edges of the transmitted spectral band.

The image data resulting from different fluorochromes are detected by sequentially applying the individual filtersets. Thus, a multi-parameter image data set is acquired. To avoid any image shift during sequential image acquisition, the following criteria have to be fulfilled:

(a) emission filters are free of optical wedge
(b) a rigid microscope stand
(c) no vibrations during change of filters
(d) a motorized and computer-controlled stand

Cooled, integrating charge-coupled device (CCD) cameras are best suited for detection of fluorescence images. Their image sensing chip should have a pixel size adapted to the optical resolution in the sensed image plane. A pixel size of 6–8 μm is a good approximation for most microscope lenses. A sensor frequently used is the Kodak KAF 1400 CCD chip with 1317 × 1035 square pixels, 6.8 μm each (see also Section 2.2.1).

17.2.2 Fluorochromes for quantitative fluorescence microscopy

During the past few years various fluorochromes have been used for labelling of oligonucleotide probes for FISH. Table 17.1 summarizes common fluorochromes for FISH in microbial ecology. In addition, several chemical compounds are available for DNA staining in whole microbial cells (e.g. acridine orange, Hoechst 33258, propidium iodide, DAPI). Of these compounds DAPI is particularly compatible for dual staining with FISH (e.g. Hicks *et al.*, 1992; Wagner *et al.*, 1994a).

Many of the commonly used fluorochromes show inherent problems, such as broad emission band (e.g. propidium iodide) and/or low photostability (e.g. FITC), and therefore are less suitable for quantitative multifluorescence microscopy. Recently, new cyanine-based fluorochromes have been developed

Table 17.1 Common fluorescent dyes for FISH in microbial ecology.

Fluorophore	Manufacturer (examples)	Colour of fluorescence	Absorption maximum (nm)	Fluorescence maximum (nm)	Extinction coefficient (M^{-1} cm^{-1})
FLUOS	Boehringer Mannheim	Green	494	518	75 000
Texas Red®	Molecular Probes	Red	589	615	85 000
Tetramethyl-rhodamine	Molecular Probes	Orange	555	580	80 000
Carboxytetra-methylrhodamine	Molecular Probes	Orange	550	576	93 000
Cy3	Nycomed Amersham	Orange	550	570	150 000
Cy5	Nycomed Amersham	Far-red	649	670	250 000

which overcome these problems. These fluorochromes show high photostability and narrow emission bands. Different types cover the emission range from blue to near infrared. Molecular Probes Inc. (Eugene, OR, USA) offers monomeric and dimeric cyanine nucleic acid stains (Hirons *et al.*, 1994), which are membrane-impermeant and therefore effectively label dead cells. Nycomed Amersham (Buckinghamshire, UK) provides monofunctional cyanine fluorochromes named CyDye, numbered from 2 (green emission) to 7 (near infrared emission), which can be successfully used for labelling of oligonucleotide probes for FISH (Poulsen *et al.*, 1994; Wagner *et al.*, 1995).

17.2.3 Sectioning and epifluorescence microscopy

Before the advent of CLSM the combination of sectioning and regular microscopy was the only possibility for analysing thicker samples in a satisfactory way. There are two much used ways for producing good sections, both originating in the field of clinical histochemistry. One relies on initial rapid freezing followed by the production of semi-thick cryosections (frequently 5–20 μm) in specially cooled microtomes. The other is based on initial embedding in paraffin, the sectioning of paraffin blocks to sections between 2 and 20 μm and the subsequent removal of the paraffin (deparaffinization) with xylene prior to hybridization. Both techniques have been successfully combined with FISH (cryosectioning plus FISH: Ramsing *et al.* 1993; Schramm *et al.* 1996; paraffin sectioning and FISH: Distel *et al.* 1991; Rothemund *et al.* 1996). It must, however, be pointed out that both procedures can be prone to artefacts. Paraffin sectioning requires complete dehydration, which is achieved more or less carefully by multiple changes of solvents. Rehydration is then the prerequisite for successful hybridization. Cryosectioning requires freezing, which itself could potentially alter the specimen under investigation. However, when applied carefully, both techniques are very helpful, and even today with CLSM technology in our hands, it is useful to have access to one or both techniques, since even the best CLSM is limited in depth penetration to some 100 μm. Anything that is thicker and needs to be analysed in the centre still requires sectioning prior to subsequent CLSM. Furthermore, it should be noted that the probes, whether antibody or oligonucleotide, still need to diffuse to the target molecules. If a matrix is dense and the length of diffusion wide, this can become impractical even for the smaller oligonucleotide probes. Sectioning to 10 or 20 μm prior to hybridization can circumvent this problem.

17.2.4 Confocal laser scanning microscopy

Conventional epifluorescence microscopy is adequate for thin samples (e.g. obtained by sectioning of bulky samples), where all information within the field of view is found in a single focus plane. In addition, epifluorescence microscopy can be combined with an integrating CCD camera, which is best suited for the detection of faint fluorescence signals. CLSM offers the third dimension by a technique known as optical sectioning of thick samples without the need for mechanical sample sectioning. Recently, CLSM has been introduced in the field of microbial ecology. Initial studies have combined CLSM with FISH for high resolution analysis of the spatial organization of probe-defined microbial populations in activated sludge (Wagner *et al.*, 1994b,c, 1995, in press; Amann *et al.*, 1996; Schuppler *et al.*, 1998), trickling filter biofilms (Schramm *et al.*, 1996), denitrifying sand filters (Neef *et al.*, 1996), epiphytic biofilms on plants (Wagner *et al.*, 1996), wetland particles (Ghiorse *et al.*, 1996), the rhizosphere of wheat (Assmus *et al.*, 1995), anaerobic granular sludge (Harmsen *et al.*, 1996) and a toluene-degrading multispecies biofilm (Møller *et al.*, 1996).

The basic principle of CLSM is illustrated in Figure 17.1. The specimen is illuminated through the imaging lens using the critical illumination scheme (Michel, 1962), i.e. the light source is imaged onto the sample. Using a laser light source, a diffraction-limited light spot is produced within the sample (Wilson and Sheppard, 1984). Re-emitted light, e.g. the Stokes-shifted fluorescence, is imaged by the microscope lens, deflected by a beamsplitter and detected by a photomultiplier. Thus, the image of one point in the sample is acquired. To generate a two-dimensional image, the focused spot is moved across the sample by an x–y light deflector. Thus, the image is detected sequentially, point by point and line by line. So far, the images from laser scan and conventional epifluorescence microscopy are quite similar and their difference is of more theoretical interest. The effect of optical sectioning is achieved by application of the confocal filter (Sheppard and Chaundhury, 1977; Sheppard and Wilson, 1978; Wilson, 1989). It is realized by a small pinhole in the image plane in front of the detector, adjusted in such a way that the light re-emitted from the illuminated spot is focused exactly onto the pinhole, and thus can pass with no reduction in intensity to the detector. However, light originating from layers above or below the focused plane is imaged to focus in front or behind the pinhole, and is thus weakened considerably. The effect of elimination of out-of-focus light depends on the numerical aperture of the microscope lens and the size of the pinhole. Under optimal conditions the axial (depth, or z direction) resolution is about three times weaker than the lateral resolution (d), which is itself dependent on the wavelength of the illuminating light (λ) and the numerical aperture (NA) of the lens ($d = \lambda/\text{NA}$) (Brakenhoff *et al.*, 1979; Wallén *et al.*, 1992).

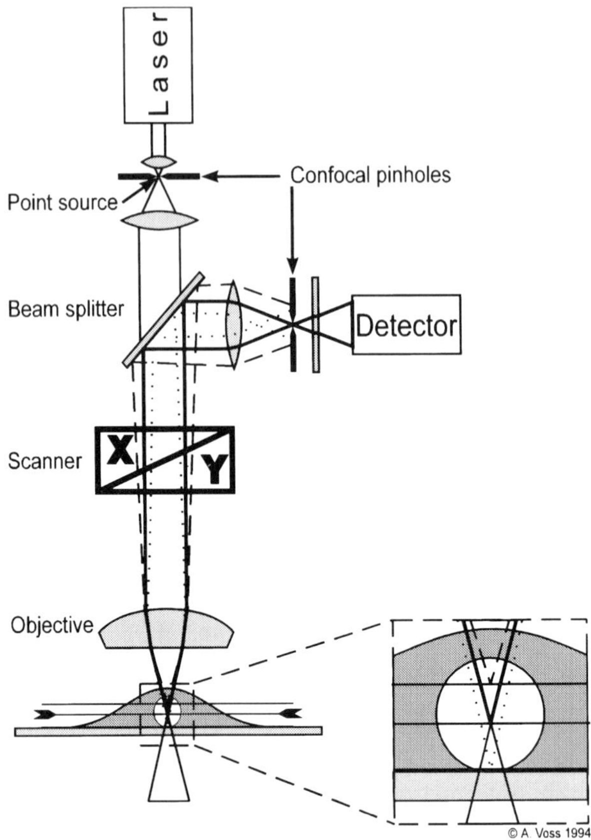

Figure 17.1 Flow diagram of a confocal laser scanning microscope.

17.2.4a Practical limits of axial resolution

Data in the literature on axial resolution of confocal imaging are based on the assumption of two-step perfect diffraction-limited imaging. First, an imaging lens of the infinity correction type should image a parallel laser light beam diffraction-limited onto the sample, utilizing the full nominal aperture of the lens (e.g. NA 1.2–1.4 for water or oil immersion lenses). Second, the light re-emitted from the sample should be imaged with the same quality to the image plane where the confocal pinhole is positioned. At the image plane an Airy pattern, the diffraction pattern of the limiting circular aperture, should be produced, and the pinhole selects the zero order of this pattern. These are a lot of assumptions, which in practice are difficult to meet at once. Even if the lens system were perfect, problems would arise from thick samples that have to be crossed twice by the light. Microscopic samples such as biological tissue or biofilm sections in general are not at all homogeneous media with respect to refractive index and absorption to allow perfect imaging through them.

A simple test of these general criteria for correct focusing might be done during pinhole alignment. A common way to do this is by moving the pinhole laterally and thus maximizing the intensity of the confocal image on the display while scanning continuously. During this alignment process the intensity should vary simultaneously all over the image. If there is a local shift of the image intensity or a movement of object structures within the image, then the pinhole is not positioned at the image plane, or the imaging quality is too poor to expect a reasonable axial resolution (see also Section 16.4.2).

For quantitative measurement of confocal resolution, slides containing fluorescent microspheres are best suited. These multifluorescently and homogeneously stained spheres have a diameter of about 5 μm and are excitable from deep blue to red. They are provided as standards, e.g. by Molecular Probes. A single isotropic *z* scan (meaning a depth scan along one line within the lateral *xy* plane, where the distance in depth between two adjacent lines is equal to the distance of two pixels within a line) through a sphere illustrates lateral versus axial resolution by comparing lateral versus axial signal half width. In multifluorescent imaging there is an additional requirement on the system, concerning colour correction, which means that light of different wavelengths irradiated from one volume element should be imaged onto a unique volume element in image space. A test is again performed easily by a *z* scan on multifluorescent microspheres. The *z* scans on a sphere are performed sequentially with different combinations of excitation laser line and emission filter under identical conditions, then an overlay of these scans on a colour display illustrates any shift by colour inhomogeneities. Microscope lenses with apochromatic correction mostly show a reasonable colour matching from blue to red excitation. However, extension of the excitation range to deep blue (e.g. by using a 364 nm ultraviolet (UV) laser) is sometimes problematic. Some lenses show a dramatic decrease of optical transmission at that range of wavelength; others, which have sufficient transmission, show considerable axial chromatic errors of up to several micrometres. Only the latest developments of microscope lenses, which are especially designed for confocal applications, meet the major requirements: intensity of light, image flatness and colour correction from deep blue to near infrared. However, these lenses are expensive (approximately $US6000 per lens) and their application requires some care. They are designed as water immersion lenses, and consequently require a medium with a refractive index of $n = 1.35$ on both sides of the cover slip. The thickness of the cover slip in use must be known, or measured exactly and adjusted on the correction wheel of the lens. When these conditions are met, high quality three-dimensional colour imaging over a depth range of 200 μm is possible, if the sample under examination is sufficiently translucent and homogeneous.

17.2.4b Image registration

Image registration covers the areas of (detector) system set-up, image data acquisition, data correction and data plus system parameter documentation. Prior to any image acquisition there should be an experimental concept. This includes the future destination of the data, which could be a coarse visual inspection, high quality images for presentation or quantitative analysis. The expense of work and the required storage space for the digital data strongly depend on these destinations.

Setting of detector parameters for CCD cameras as well as for laser scanning is done via a graphical user interface. Preferred settings might then be stored for reuse. In the case of the CCD camera there are two parameters of special interest: the binning factor and the dynamic data scaling.

The binning factor defines how many detector pixels in the x and y directions are summed up to one image pixel. Using a KAF 1400 sensor chip with a read out area of 1024×1024 pixels, a binning of 2, resulting in a 512×512 image, gives sufficient resolution in most applications, although there is some undersampling with regard to the theoretical optical resolution (Section 2.2.1). CCD cameras show a high dynamic range of 12 bits (4096 grey values) or even more. On the other hand, graphic displays and most of the popular file formats for raster images are limited to 8 bits (256 grey values). Therefore data must be scaled for display and storage. Most popular is the min/max scaling. The minimum and maximum signal values are detected, and the interval is imaged linearly onto the grey value range between 0 and 255. However, if artefacts, such as very bright spots, appear in the field of view, the grey value range for the signals of interest might be small and the signals will look dark. In such cases the disturbing pixels must be clipped, e.g. by a manual setting of the input range, prior to dynamic scaling. In addition, it is advisable to compare images obtained from positive and negative control experiments by using identical scaling intervals.

17.2.4c Setting of CLSM parameters

From the various parameters that can be modified in a confocal system, some of special interest will be discussed. Besides the numerical aperture of the lens, the size of the pinhole, the confocal filter at the image plane in front of the detector, is of major importance for the axial resolution, i.e. the thickness of an optical section (Wilson and Carlini, 1987; Wilson, 1989). For very large, 'infinite' circular diaphragms the images resulting from CLSM approach those from conventional fluorescence microscopy. Reducing the size of the pinhole increases depth resolution, but also reduces the detectable light intensity. The relationship between pinhole size and axial resolution is rather nonlinear, in that at small sizes a further decrease of pinhole size scarcely increases resolution. Keeping in mind that under ideal conditions the light distribution at the image plane should be like an Airy pattern (Section 2.2.1, Figure 2.1), it is a good compromise to let it pass to zero order, i.e. accept the centre peak, but not the first fringe. In practice, however, it is reasonable to determine the optimum pinhole size by making z scans on fluorescent microspheres as described above.

A frequent problem of confocal imaging, especially with respect to subsequent analysis, is given by its limited signal-to-noise ratio (SNR). Despite ongoing improvements, e.g. by applying cooled photomultipliers, the SNR quality of conventional epifluorescence microscopy based on cooled CCD cameras is still not reached by CLSM. To overcome this problem one option is to increase the pinhole size, accepting some loss in depth resolution. This is particularly suitable for analysis where only some representative optical sections are of interest. If real volumetric data are required, a significant improvement of the SNR can be obtained by averaging the signals from several scans (Kalman filtering,

Section 16.4.1). As the averaging procedure extends the excitation time of the fluorochrome per optical section, fluorochrome bleaching limits the frequency of averaging. Consequently, the use of high quality anti-fading reagents as mounting media (Florijn *et al.*, 1995) is critical for extended averaging. Theoretically, an identical increase of the SNR can be obtained by applying either a high number of averages at a high scan speed or by a lower number of averages using a slower scan speed. However, in general, we observed that averaging by the use of multiple high-speed scans is superior to the application of slow-speed scans with respect to bleaching of the fluorochrome.

Volumetric (voxel) image data are obtained by a sequence of optical sections. It is obvious that the axial distance of the individual sections is of importance for the quality of the volumetric analysis. From a theoretical point of view, an isotropic scan using equal lateral and axial distances of the voxels, where both distances have been adapted to the lateral optical resolution, would be optimal. However, such scans require extended excitation times, causing strong fluorochrome bleaching, and lead to the accumulation of huge amounts of digital data. In practice, these requirements might be considerably weakened, especially if the volumetric data are used exclusively for visualization. It is then sufficient when the smallest fluorescent structures of interest are not lost between two adjacent sections. They are certainly comprehended if they appear at two neighbouring sections.

17.2.4d Restoration of image volumes

Spatial resolution in microscopy is generally restricted by the limits given by wave optics, but also by imperfect corrections of lenses and many other technical constraints. According to Fourier optics (Castleman, 1979; Goodman, 1968) the image $\mu(\mathbf{r})$ of a self-luminous fluorescent specimen $\lambda(\mathbf{r})$ emitting incoherent light is expressed as:

$$\mu(\mathbf{r}) = \lambda(\mathbf{r})^{*}h(\mathbf{r}) \tag{17.1}$$

where $\mathbf{x} = (x, y, z)$ is the three-dimensional spatial variable where z stands for the axial variable, * denotes the three-dimensional convolution operation and $h(\mathbf{r})$ is the point-spread function (PSF) of the system. The PSF represents the three-dimensional image that the imaging system would produce from an ideal fluorescent point. From this point of view it should be possible to calculate the real three-dimensional fluorescence emission $\lambda(\mathbf{r})$ from the measured volume image $\mu(\mathbf{r})$ if the PSF $h(\mathbf{r})$ is known. The three-dimensional PSF can be measured by using small fluorescent microspheres (0.1–0.2 μm) following the instructions of Hiraoka *et al.* (1990).

A straightforward approach of image restoration, called inverse volume filtering or direct volume deconvolution, starts from equation (17.1). Both sides of (17.1) are three-dimensional Fourier transformed (FT), resulting in:

$$\mathrm{M}(\mathbf{f}) = \Lambda(\mathbf{f})\mathrm{H}(\mathbf{f}) \tag{17.2}$$

where $\mathbf{f} = (f_x, f_y, f_z)$ denotes the three-dimensional spatial frequency components and $\mathrm{H}(\mathbf{f})$, the FT of the PSF $h(\mathbf{r})$, is called the optical transfer function (OTF; see

Section 2.2.1). The convolution in space domain after FT results in a multiplication in the frequency domain. Starting from equation (17.2) the FT of the desired three-dimensional fluorescence data is obtained by:

$$\Lambda(\mathbf{f}) = \mathrm{M}(\mathbf{f})\mathrm{H_s}(\mathbf{f})/\mathrm{H}(\mathbf{f}) \tag{17.3}$$

where $\mathrm{H_s}(\mathbf{f})$ is a high frequency smoothing filter of Wiener type (Agard *et al.*, 1989; Section 1.5.3). The inverse filter reconstruction process subdivides into the following calculation steps:

(a) FT of image $\mu(\mathbf{r})$ and PSF $h(\mathbf{r})$
(b) preparation of reconstruction filter $\mathrm{H_r}(\mathbf{f}) = \mathrm{H_s}(\mathbf{f})/\mathrm{H}(\mathbf{f})$
(c) multiplication of $\mathrm{M}(\mathbf{f})$ with $\mathrm{H_r}(\mathbf{f})$
(d) inverse FT of $\Lambda(\mathbf{f})$ to obtain the reconstructed image $\lambda(\mathbf{r})$

A practical problem arises in a proper selection of the smoothing Wiener filter, which includes an empirical factor to avoid negative intensity values in reconstruction. More recent approaches for three-dimensional image reconstruction use more general mathematical techniques, and involve probability and statistics. They work iteratively. The application of nonlinear constraints is the main idea behind iterative constrained filters. In contrast to inverse filters, iterative filters do not resemble the mathematical equivalent of a convolution filter. The aim is to minimize the restoration error iteration by iteration. There is a variety of ways to achieve this, for example by:

(a) minimizing the squared restoration error (*least squares* restoration LSQ)
(b) maximizing the likelihood function (*maximum likelihood estimation,* MLE)
(c) maximizing the *entropy* function (ME)

All of these iterative techniques with nonlinear constraints use a PSF either measured or calculated from system parameters. An emerging variation of these iterative techniques is the blind deconvolution algorithm. With this method, the PSF of the system does not need to be known prior to the reconstruction.

A survey of all reconstruction techniques as well as detailed literature is given by Holmes and Liu (1992). A software package containing all of these three-dimensional reconstruction techniques is available from Carl Zeiss Vision (Eching, Germany) for MS-Windows 3.1 and NT.

All of these reconstruction techniques will correct errors induced by the imaging system more or less well. However, the PSF is not the only – and not the most important – limitation in confocal imaging. We have to keep in mind that light passes through the three-dimensional sample twice. As biological sample material is often optically inhomogeneous and more or less translucent, it refracts, scatters and absorbs light. Thus, object-dependent distortions are mainly responsible for loss of resolution with depth of scanning. Consequently, the reasonable depth of confocal scanning is limited mainly by the sample under study and less by the working distance of the imaging microscope lens.

17.2.4e Correction of geometric distortions

Related images taken through different fluorescence channels from the same microscopic field might suffer from various geometric distortions, such as translational and rotational image shifts and scaling differences. Most of these distortions are avoidable by proper optical alignment and by selection of high quality filtersets (to avoid pixel shift in conventional epifluorescence microscopy) and high quality lenses (to avoid scaling difference due to chromatic aberration). However, as already described above, one special distortion is common to most lenses used for confocal imaging. When changing from visible to UV excitation, an axial offset due to insufficient chromatic correction for short wavelengths cannot be avoided. Only some very recent lenses especially designed for confocal applications show a chromatic correction down to 360 nm. But this axial offset can also be measured and considered by a shift of the axial scanning positions. In the field of microbial ecology the problem of geometric distortions might become severe if we are interested in taking image data after different steps within the sequence of sample preparation; for example, if the morphology of a native biofilm is first measured in its aquatic environment and subsequently, after fixation and FISH, the specific fluorescence of the cells is detected. Under such circumstances we have to face possibly complex geometric changes of the biofilm caused by the fixation and hybridization procedure. As far as the volume deformations can be considered as being continuous, the well-known techniques of elastic modelling can be applied. Such reconstruction techniques are all based on a sufficient set of fiducial markers, i.e. fluorescent landmarks unique in both data sets to be matched.

17.2.4f Shading or flat field corrections

It is good practice in quantitative microscopy to check the homogeneity of illumination prior to image acquisition. This also holds true for conventional and confocal fluorescence microscopy, even though it is more troublesome. A prerequisite for an effective shading correction is a stable, non-bleaching fluorescence standard; the best suited is an inorganic material like uranyl glass. Such standards are available from the microscope manufacturers. For corrections it is essential to take the reference image (flat field) from the fluorescence standard under identical conditions as those images that need to be corrected. Examples for shading correction in bacterial fluorescence measurement are given by Wilkinson (1994).

Protocol 17.1 Acquiring CLSM images on FISH-stained environmental microbial consortia.

Oligonucleotide probe labelling with fluorescent dyes, as well as the fixation and hybridization procedure, has been described in detail by one of us elsewhere (Amann, 1995). Here we focus on practical tips for the application of CLSM for analysing FISH-stained samples. We tried to make the protocol as general as possible. However, as we use a Zeiss LSM 410 confocal laser scanning microscope in combination with the Zeiss LSM software (version 3.95), parts of the protocol are related to this particular equipment.

1 Fix the sample (formaldehyde, ethanol; see above) for FISH.
2 Fixed samples are then spotted on microscopic slides. We routinely use slides on which hydrophobic coating separates six glass surface windows (Paul Marienfeld KK, Bad Mergentheim, Germany). Due to the hydrophobic coating, mixing of probes applied to the different windows is prevented. Thus, up to six hybridization experiments can be performed on one slide.
3 Dehydration of the sample material by successive passages through 50, 80 and 98% ethanol washes (3 min each).
4 *in situ* hybridization of the sample. Oligonucleotide probes labelled with Cy3 and, to a lesser extent, Cy5 (Nycomed Amersham, Buckinghamshire, UK) are very photostable and therefore best suited for CLSM analysis. For simultaneous use of three differently labelled oligonucleotide probes, we also apply probes coupled to carboxyfluorescein-*N*-hydroxysuccinimidyl ester (FLUOS; Boehringer Mannheim, Mannheim, Germany). Note: FLUOS is sensitive to bleaching and its fluorescence intensity is pH dependent.
5 Mount slides in a glycerol/phosphate buffered saline (PBS) mountant with pH > 8.5 (e.g. Citifluor, Citifluor Ltd, Canterbury, UK).
6 Select a microscopic field by viewing the slide with the epifluorescence microscope equipped with suitable filtersets. We recommend the Zeiss filterset no. 09 for FLUOS and the Chroma HQ filtersets (Chroma Technology, Brattleboro, VT, USA) for Cy3 and Cy5. As the human eye is not able to detect the far-red fluorescence originating from Cy5 labelled probes, the use of an appropriate CCD camera (e.g. Photometrics CE250 camera equipped with a cooled KAF 1400 CCD; Photometrics, Tucson, AZ, USA) is recommended.
7 Record single optical sections by using a 633 nm HeNe laser for Cy5, a 543 nm HeNe laser for Cy3 and a 488 nm Ar ion laser for FLUOS to optimize the settings for pinhole size, contrast and brightness for each fluorescence channel independently. We do not recommend recording the different fluorescent labels simultaneously by simultaneous excitation with two or three lasers. This saves time but does not allow for adaptation of the various parameters to each fluorochrome (e.g. the selection of different numbers of passes for averaging for different fluorescent dyes). Pixel shift distortions between the different channels usually occur in the border regions of the image. Such distortions can be minimized by the selection of a high quality microscopic lens with low magnification (e.g. $40\times$ instead of $100\times$) and compensation of the low magnification with the electronic zoom function. In general, bright probe-derived signals will allow working with smaller pinhole sizes, thereby increasing the axial resolution. However, the minimum pinhole size is also dependent on the lens type (see above). A useful tool for defining the best settings is the so-called range indicator, which is implemented in the Zeiss LSM software. Once these adjustments have been made, the averaging function has to be activated to improve the SNR. In our experience 16 averages at a scan speed of 0.7 s/scan yield best results for the above-mentioned fluorochromes. For weak FLUOS signals it is reasonable to reduce the number of averages to

avoid bleaching. The best suited number of passes for averaging is also dependent on the sample thickness, which is linked to the number of required sections. Analysing thick samples with a high number of optical sections will require reducing the number of averages and/or the scanning time to protect the fluorescent dyes from bleaching.

8 Save configurations for each fluorescence channel.

9 Select a different microscopic field and start *z* sectioning by using the channel configurations determined before. First, the number of sections and the section thickness have to be defined. Both parameters are dependent on the thickness of the sample and the minimum size of the labelled target structures. Start recording with the longest wavelength (e.g. the 633 nm laser). Save the sequence of optical sections. Repeat recording and saving the sequence by sequentially using the shorter wavelength lasers (e.g. the 543 nm and 488 nm lasers).

10 Load the individual sequence to host memory. The series of optical sections can be visualized nicely as a gallery (Figure 17.2A). Calculate the two-dimensional (all-in-focus) projection (Figure 17.2B). Save the two-dimensional projection as a TIFF file. Repeat the procedure for the other two sequences. Load TIFF files in the red, green and blue channels of the RGB tool to obtain a superimposed, multicolour, all-in-focus image. Perfect alignment of the three channels is achieved by digital pixel shift compensation. Separate digital contrast enhancement for each channel might help to improve the image quality for visualization purposes but should not be used if quantification of signal intensities is of interest.

11 The spatial organization of the labelled bacteria can be viewed by calculating red–green stereo images, which should be viewed with red–green glasses, obtainable from Carl Zeiss, to recognize a black and white three-dimensional distribution (Plate 7A). Another option for three-dimensional reconstruction is to calculate coloured depth profiles (Plate 7B).

It is beyond the scope of this chapter to list all options that CLSM in combination with modern image analysis software packages offers (e.g. animation of optical section sequence data). For more detailed information the interested reader might refer to the respective software manuals.

17.2.4g Quantitative analysis

Quantitative analysis of specifically labelled microbial populations starts from properly registered two-dimensional and three-dimensional fluorescence image data sets. The fluorescence is induced by general cell stains like DNA intercalating fluorochromes and/or by more specific rRNA-targeted oligonucleotide probes using FISH. The registered signals are interpreted as monotonous mapping of the fluorochrome concentration within the sample. The effort necessary to analyse the data strongly depends on the signal quality, i.e. the amount of non-specific background fluorescence emitted from the sample.

Ramsing *et al.* (1996) describe a straightforward approach for quantitative analysis of bacteria present in aquatic ecosystems. They analysed the bacterial

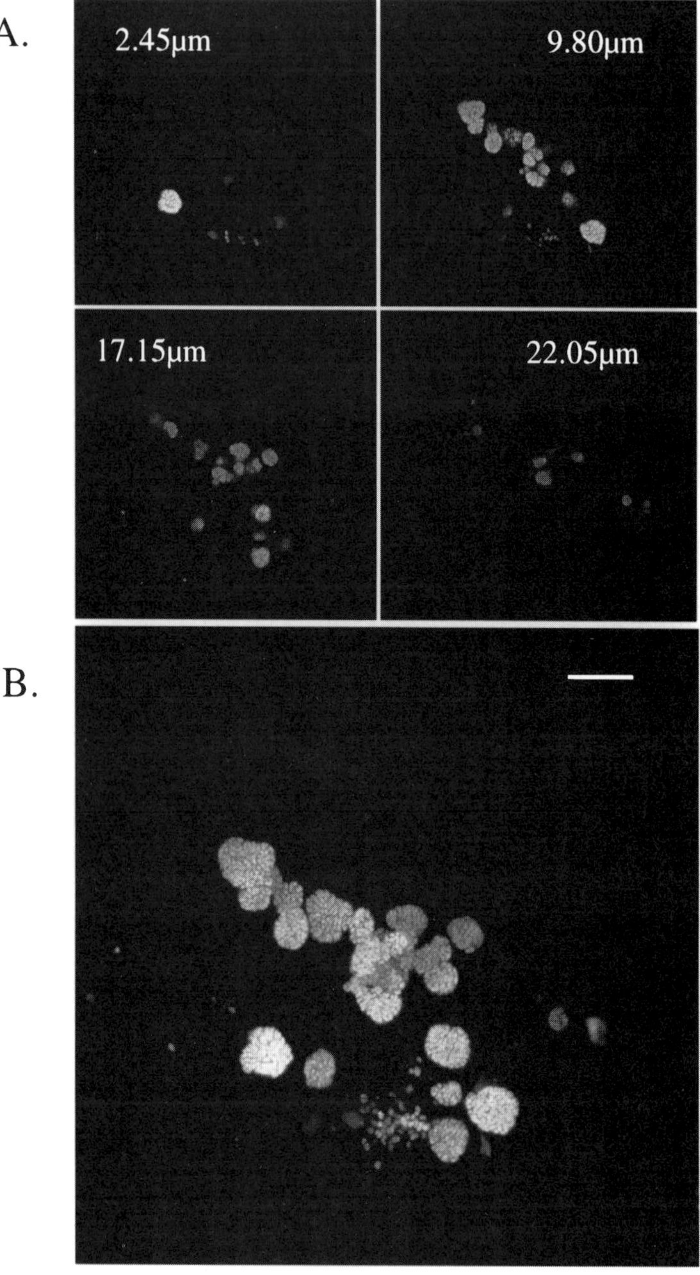

Figure 17.2 Analysing the spatial arrangement of specifically labelled bacteria in activated sludge. (A) Gallery of four representative optical sections selected from a series of 22 optical sections showing ammonia-oxidizing bacteria identified with Cy3-labelled probe NEU (Wagner *et al.*, 1995) in a 22 μm thick activated sludge floc. (B) Corresponding all-in-focus two-dimensional visualization. Scale bar = 10 μm and applies to both photomicrographs.

population occurring in the water column of a Danish fjord. Cells were prepared as flat samples by filtration and subsequently stained with ethidium bromide (EtBr) and fluorescently labelled specific oligonucleotide probes. Images were taken by conventional epifluorescence microscopy using a cooled CCD camera device. Cell demarcation was initiated by calculating a smoothed second-order derivative of the image applying a Marr–Hildreth operator (Marr and Hildreth, 1980). In the resulting data set zero crossings indicate locations with the steepest change of signal. The filigree pattern of zero crossings is then cleaned by binary opening (an erosion followed by a dilation). A further suppression of (noise induced) non-cell structures is obtained by thresholding the second derivative with values greater than zero (see Section 6.7.5). The authors calculated (estimated) cell volumes on the basis of the best filling ellipsoid with the assumption that the unknown vertical axis was equivalent to the measured minor axis. If the samples under study originate from ecosystems like activated sludge or soil, the analysis task becomes much more complicated due to the unavoidable unspecific background fluorescence.

Bloem *et al.* (1995) described an image analysis process working fully automatically on images from soil smears. After removal of detrital particles a diluted soil suspension was smeared on a microscope slide and stained with the protein stain DTAF. Fluorescence images were taken by a CLSM in extended focus mode (projection of a sequence of optical sections). For background removal the authors used a top-hat transform based on the method of Serra (1982). A top-hat transform is performed by first removing small peaks by opening. The remaining bigger peaks (background) are subtracted from the original image, and then only smaller peaks (white detail) remain. The size of the details retained depends on the size of the opening (see Section 6.7.2 for details).

To our knowledge all hitherto published papers presenting quantitative analysis of microbial populations (see above and, for example, Møller *et al.*, 1995) were restricted to two-dimensional image data. However, microbial cells present in microbial ecosystems like biofilms are distributed in three-dimensional space. If we want to understand the relative spatial arrangement of micro-organisms to each other and their distribution with respect to the architecture of the enclosing biofilm matrix, we have to tackle the three-dimensional problem. For this purpose, the elements of the image analysis toolbox must be expanded to three dimensions, where contours and areas become surfaces and volumes and most filtering operations are performed in frequency space (Section 6.10.2). A new problem arises from the need to visualize the intermediate and final results of the individual processing steps. Visualization is necessary at least in the design phase to judge the reliability of the operations. While in the two-dimensional case, relevant information might be displayed within a single frame, in the case of three dimensions, looking into the volume from various directions is desirable, and it would be convenient if the change of viewing direction could be managed interactively and if observation could be stereoscopic. Software that meets at least part of these requirements is on the way. For example Image Space (Molecular Dynamics), which runs on workstations from Silicon Graphics, contains some powerful analysis and visualization tools. The three-dimensional software package implemented in the KS 400 software distributed by Carl Zeiss Vision contains a sequence of performance-optimized three-dimensional analysis tools as well as a comprehensive collection of image reconstruction programs. KS 400 is currently available for Microsoft Windows 3.1(1) and Windows NT.

17.3 CONCLUSIONS AND FUTURE PERSPECTIVES

Three-dimensional analysis of complex microbial communities by combining CLSM and FISH is a key technique for a detailed understanding of microbial ecology. Coaggregation of different bacterial species, changes in the morphology and overall physiological activity of bacterial species in response to changes in their environment, and specific colonization of chemically defined micro-niches are just some topics that can now be studied *in situ* with the assistance of FISH and CLSM. Also, CLSM-based procedures will soon make it possible to quantify specifically labelled cells by semiautomatic measurements of their three-dimensional biovolume and might render tedious manual enumeration superfluous.

Future studies will focus on extending the *in situ* analysis on the structure, dynamics and spatial organization of environmental microbial populations to their function. Signal amplification techniques such as *in situ* polymerase chain reaction (PCR) or PRINS (e.g. Hodson *et al.*, 1995) might make it possible to improve the current sensitivity limits of the FISH technique for visualization of mRNA molecules in whole microbial cells. Combination of FISH with various fluorescent probes and/or microelectrodes will be used to characterize the chemical microenvironment of *in situ* identified bacteria. Such direct observation of bacteria will improve understanding of the complex microbiological processes in real ecosystems, and might contribute to the improvement of various biotechnological processes such as sewage treatment and bioremediation.

REFERENCES

Agard, D.A., Hiraoka, Y., Shaw, P. and Sedat, J.W. (1989) Fluorescence microscopy in three dimension. *Meth. Cell Biol.* **30**: 353–77.

Amann, R. (1995) *In situ* identification of micro-organisms by whole cell hybridization with rRNA-targeted nucleic acid probes. In *Molecular Microbial Ecology Manual* (Akkerman, A.D.L., van Elsas, J.D. and de Bruijn, F.J., eds), chapter 3.3.6, pp. 1–15. Kluwer: Dortrecht.

Amann, R.I., Ludwig, W. and Schleifer, K.-H. (1995) Phylogenetic identification and *in situ* detection of individual microbial cells without cultivation. *Microbiol. Rev.* **59**: 143–69.

Amann, R., Snaidr, J., Wagner, M., Ludwig, W. and Schleifer, K.-H. (1996) *In situ* visualization of high genetic diversity in a natural microbial community. *J. Bacteriol.* **178**: 3496–3500.

Assmus, B., Hutzler, P., Kirchhof, G., Amann, R., Lawrence, J.R. and Hartman, A. (1995) *In situ* localization of *Azospirillum brasiliense* in the rhizosphere of wheat with fluorescently labeled rDNA-targetted oligonucleotide probes and scanning confocal laser microscopy. *Appl. Environ. Microbiol.* **61**: 1013–9.

Belser, L.W. (1979) Population ecology of nitrifying bacteria. *Ann. Rev. Microbiol.* **33**: 309–33.

Beimfohr, C., Krause, A., Amann, R., Ludwig, W. and Schleifer, K.-H. (1993) *In situ* identification of Lactococci, Enterococci and Streptococci. *Syst. Appl. Microbiol.* **16**: 450–6.

Bloem, J., Veninga, M. and Shepherd, J. (1995) Fully automatic determination of soil bacterium numbers, cell volumes, and frequencies of dividing cells by confocal laser microscopy and image analysis, *Appl. Environ. Microbiol.* **61**: 926–36.

Brakenhoff, G.J., Blom, P. and Barends, P. (1979) Confocal scanning light microscopy with high aperture immersion lens, *J. Microsc.* **117**: 219–32.

Caldwell, D.E., Korber, D.R. and Lawrence, J.R. (1992) Confocal laser microscopy and digital image analysis in microbial ecology. In *Advances in Microbial Ecology* (Marshall, K.C. ed.), vol. 12, pp. 1–67. Plenum Press: New York.

Castleman, K.R. (1979) *Digital Image Processing*. Prentice Hall: Englewood Cliffs, NJ.
Costerton, J.W., Lewandowski, Z., de Beer, D., Caldwell, D.E., Korber, D.R. and James, G.A. (1994) Biofilms: the customized microniche. *J. Bacteriol.* **176**: 2137–42.
Costerton, J.W., Lewandowski, Z., Caldwell, D.E., Korber, D.R. and Lappin-Scott, H.M. (1995). Microbial biofilms. *Ann. Rev. Microbiol.* **49**: 711–45.
DeLong, E.F., Wickham, G.S. and Pace, N.R. (1989) Phylogenetic stains: ribosomal RNA-based probes for the identification of single microbial cells. *Science* **243**: 1360–3.
Deretic, V., Schurr, M.J., Boucher, J.C. and Martin, D.W. (1994) Conversion of *Pseudomonas aeruginosa* to mucoidy in cystic fibrosis: environmental stress and regulation of bacterial virulence by alternative sigma factors. *J. Bacteriol.* **176**: 2773–80.
Distel, D.L., DeLong, E.F. and Waterbury, J.B. (1991) Phylogenetic characterization and *in situ* localization of the bacterial symbiont of shipworms (Teredinidae: Bivalvia) by using 16S rRNA sequence analysis and oligodeoxynucleotide probe hybridization. *Appl. Environ. Microbiol.* **57**: 2376–82.
Dworkin, M. (1991) Introduction. In *Microbial Cell–Cell Interactions* (Dworkin M, ed.), pp. 1–6. American Society for Microbiology: Washington, DC.
Florijn, R.J., Slats, J., Tanke, H.J. and Raap, A.K. (1995) Analysis of antifading reagents for fluorescence microscopy. *Cytometry* **19**: 177–82.
Ghiorse, W.C., Miller, D.N., Sandoli, R.L. and Siering, P.L. (1996) Applications of laser scanning microscopy for analysis of aquatic microhabitats. *Microsc. Res. Technol.* **33**: 73–86.
Glöckner, F.O., Amann, R., Alfreider, A., Pernthaler, J., Psenner, R., Trebesius, K.H. and Schleifer, K.-H. (1996) An *in situ* hybridization protocol for the detection and identification of planctonic bacteria. *Syst. Appl. Microbiol.* **19**: 403–6.
Goodman, J.W. (1968) *Introduction to Fourier Optics*. McGraw-Hill: San Francisco, CA.
Harmsen, H.J.M., Akkermans, A.D.L., Stams, A.J.M. and De Vos, W.M. (1996) Population dynamics of propionate-oxidizing bacteria under methanogenic and sulfidogenic conditions in anaerobic granular sludge. *Appl. Environ. Microbiol.* **62**: 2163–8.
Hicks, R.E., Amann, R.I. and Stahl, D.A. (1992) Dual staining of natural bacterioplankton with 4′,6-diamidino-2-phenylindole and fluorescent oligonucleotide probes targeting kingdom-level 16S rRNA sequences. *Appl. Environ. Microbiol.* **58**: 2158–63.
Hiraoka, Y., Sedat, J.W. and Agard, D.A. (1990) Determination of three-dimensional imaging properties of a light microscope system. *Biophys. J.* **57**: 325–33.
Hirons, G.T., Fawcett, J.J. and Crissman, H.A. (1994) TOTO and YOYO: new very bright fluorochromes for DNA content analyses by flow cytometry. *Cytometry* **15**: 129–40.
Hodson, R.E., Dustman, W.A., Garg, R.P. and Moran, M.A. (1995) *in situ* PCR for visualization of microscale distribution of specific genes and gene products in prokaryotic communities. *Appl. Environ. Microbiol.* **61**: 4074–82.
Holmes, T.J. and Liu, Y.-H. (1992) Image restoration for 2-D and 3-D fluorescence microscopy. In *Visualization in Biomedical Microscopies* (Kriete, A., ed.), pp. 283–327. VCH Verlagsgesellschaft: Weinheim.
MacLeod, F.A., Guiot, S.R. and Costerton, J.W. (1990) Layered structure of bacterial aggregates produced in an upflow anaerobic sludge bed and filter reactor. *Appl. Environ. Microbiol.* **56**: 1598–607.
Manz, W., Wagner, M., Amann, R. and Schleifer, K.-H. (1994) *In situ* characterization of the microbial consortia active in two wastewater treatment plants. *Wat. Res.* **28**: 1715–23.
Marr, D. and Hildreth, E. (1980) Theory of edge detection. *Proc. R. Soc. Lond.* **207**: 187–217.
Massol-Deyá, A.A., Whallon, J., Hickey, R.F. and Tiedje, J.M. (1995) Channel structures in aerobic biofilms of fixed-film reactors treating contaminated groundwater. *Appl. Environ. Microbiol.* **61**: 769–77.
Michel, K. (1962) *Die Mikrophotographie*. Springer-Verlag: Vienna.
Mobarry, B.K., Wagner, M., Urbain, V., Rittmann, B.E. and Stahl, D.A. (1996) Phylogenetic probes for analysing abundance and spatial organization of nitrifying bacteria. *Appl. Environ. Microbiol.* **62**: 2156–62.
Möller, S., Kristensen, C.S., Poulsen, L.K., Carstensen, J.M. and Molin, S. (1995) Bacterial growth on surfaces: automated image analysis for quantification of growth related parameters. *Appl. Environ. Microbiol.* **61**: 741–8.
Møller, S., Pedersen, A.R., Poulsen, L.K., Arvin, E. and Molin, S. (1996) Activity and three-dimensional distribution of toluene-degrading *Pseudomonas putida* in a multispecies biofilm assessed by quantitative *in situ* hybridization and scanning confocal laser microscopy. *Appl. Environ. Microbiol.* **62**: 4632–40.
Neef, A., Zaglauer, A., Meier, H., Amann, R., Lemmer, H. and Schleifer, K.-H. (1996) Population analysis in a denitrifrying sand filter: conventional and *in situ* identification of *Paracoccus* spp. in methanol-fed biofilms. *Appl. Environ. Microbiol.* **62**: 4329–39.
Poulsen, L.K., Ballard, G. and Stahl, D.A. (1993) Use of rRNA fluorescence *in situ* hybridization for measuring the activity of single cells in young and established biofilms. *Appl. Environ. Microbiol.* **59**: 1354–60.
Poulsen, L.K., Lan, F., Kristensen, C.S., Hobolth, P., Molin, S. and Krogfelt, K.A. (1994) Spatial distribution of *Escherichia coli* in the mouse large intestine inferred from rRNA *in situ* hybridization. *Infect. Immunol.* **62**: 5191–4.

Ramsing, N.B., Kühl, M. and Jörgensen, B.B. (1993) Distribution of sulfate-reducing bacteria, O_2 and H_2S in photosynthetic biofilms determined by oligonucleotide probes and microelectrodes. *Appl. Environ. Microbiol.* **59**: 3820–49.
Ramsing, N.B., Fossing, H., Ferdelman, T.G., Andersen, F. and Thamdrup, B. (1996) Distribution of bacterial populations in a stratified fjord (Mariager Fjord, Denmark) quantified by *in situ* hybridization and related to chemical gradients in the water column, *Appl. Environ. Microbiol.* **62**: 1391–404.
Roller, C., Wagner, M., Amann, R., Ludwig, W. and Schleifer, K.-H. (1994) *In situ* probing of Gram-positive bacteria with high DNA G + C content using 23S rRNA-targeted oligonucleotides. *Microbiology* **140**: 2849–58.
Rothemund, C., Amann, R., Klugbauer, S., Manz, W., Bieber, C., Schleifer, K.-H. and Wilderer, P. (1996) Microflora of 2,4-dichlorophenoxyacetic acid degrading biofilms on gas permeable membranes. *Syst. Appl. Microbiol.* **19**: 608–15.
Schramm, A., Larsen, L.H., Revsbech, N.P., Ramsing, N.B., Amann, R. and Schleifer, K.-H. (1996) Structure and function of a nitrifying biofilm as determined by *in situ* hybridization and microelectrodes. *Appl. Environ. Microbiol.* **62**: 4641–7.
Schuppler, M., Wagner, M., Schön, G. and Göbel, U.B. (1998) *In situ* identification of nocardioform actinomycetes in activated sludge using fluorescent rRNA-targeted oligonucleotide probes and confocal laser scanning microscopy. *Microbiology* **144**: 249–59.
Serra, J. (1982) *Image Analysis and Mathematical Morphology*. Academic Press: London.
Sheppard, C.J.R. and Chaundhury, A. (1977) Image formation in the scanning microscope. *Opt. Acta* **24**: 1051–73.
Sheppard, C.J.R. and Wilson, T. (1978) Depth of field in the scanning microscope. *Opt. Lett.* **3**: 115–7.
Snaidr, J., Amann, R., Huber, I., Ludwig, W. and Schleifer, K.-H. (1997) Phylogenetic analysis and *in situ* identification of bacteria in activated sludge. *Appl. Environ. Microbiol.* **63**: 2884–96.
Szwerinski, H., Gaiser, S. and Bardtke, D. (1985) Immunofluorescence for the quantitative determination of nitrifying bacteria: interference of the test in biofilm reactors. *Appl. Microbiol. Biotechnol.* **21**: 125–8.
Wagner, M. and Amann, R. (1997) Molecular techniques for determining microbial community structures in activated sludge. In *Microbial Community Analysis: the Key to the Design of Biological Wastewater Treatment Systems. IAWQ Scientific Technical Report No. 5* (Cloete, T.E. and Muyima, N.Y.O., eds). Cambridge University Press: Cambridge.
Wagner, M., Amann, R., Lemmer, H. and Schleifer, K.-H. (1993) Probing activated sludge with oligonucleotides specific for proteobacteria: inadequacy of culture-dependent methods for describing microbial community structure. *Appl. Environ. Microbiol.* **59**: 1520–5.
Wagner, M., Erhart, R., Manz, W., Amann, R., Lemmer, H., Wedi, D. and Schleifer, K.-H. (1994a) Development of an rRNA-targeted oligonucleotide probe specific for the genus *Acinetobacter* and its application for *in situ* monitoring in activated sludge. *Appl. Environ. Microbiol.* **60**: 792–800.
Wagner, M., Amann, R., Kämpfer, P., Assmus, B., Hartmann, A., Hutzler, P., Springer, N. and Schleifer, K.-H. (1994b) Identification and *in situ* detection of Gram-negative filamentous bacteria in activated sludge. *Syst. Appl. Microbiol.* **17**: 405–17.
Wagner, M., Assmus, B., Hartmann, A., Hutzler, P. and Amann, R. (1994c) *In situ* analysis of microbial consortia in activated sludge using fluorescently labelled, rRNA-targeted oligonucleotide probes and confocal scanning laser microscopy. *J. Microsc.* **176**: 181–7.
Wagner, M., Rath, G., Amann, R., Koops, H.-P. and Schleifer, K.-H. (1995) *In situ* identification of ammonia-oxidizing bacteria. *Syst. Appl. Microbiol.* **17**: 251–64.
Wagner, M., Rath, G., Koops, H.-P., Flood, J. and Amann, R. (1996) *In situ* analysis of nitrifying bacteria in sewage treatment plants. *Wat. Sci. Technol.* **34**(1/2): 237–44.
Wagner, M., Noguera, D.R., Juretschko, S., Rath, G., Koops, H.-P. and Schleifer, K.-H. (in press) Combining fluorescent *in situ* hybridization (FISH) with cultivation and mathematical modeling to study population structure and function of ammonia-oxidizing bacteria in activated sludge. *Water Sci. Technol.*
Wallén, P., Carlsson, K. and Mosberg, K. (1992) Properties of confocal microscopy. In *Visualization in Biomedical Microscopies* (Kriete, A., ed.). VCH Verlagsgesellschaft: Weinheim.
Wilkinson, M.H.F. (1994) Shading correction and calibration in bacterial fluorescence measurement by image processing system. *Comput. Meth. Programs Biomed.* **44**: 61–7.
Wilson, T. (1989) Optical sectioning in confocal fluorescence microscopes. *J. Microsc.* **154**: 143–56.
Wilson, T. and Carlini, A.R. (1987) Size of the detector on confocal imaging systems. *Opt. Lett.* **12**: 227–9.
Wilson, T. and Sheppard, C.J.R. (1984) *Theory and Practice of Scanning Optical Microscopy*. Academic Press: London.
Wolfaardt, G.M., Lawrence, J.R., Robarts, R.D., Caldwell, S.J. and Caldwell, D.E. (1994) Multicellular organization in a degradative biofilm community. *Appl. Environ. Microbiol.* **60**: 434–46.
Zarda, B., Hahn, D., Chatzinotas, A., Schönhuber, W., Neef, A., Amann, R. and Zeyer, J. (1997) Analysis of bacterial community structure in bulk soil by *in situ* hybridization. *Arch. Microbiol.* **168**: 185–92.

18

Agar Plate Techniques

Adrian C. Peters[1], Linda V. Thomas[2] and Julian W.T. Wimpenny[2]
[1]*University of Wales Institute, Cardiff, UK*
[2]*Cardiff University, Cardiff, UK*

18.1 INTRODUCTION

Besides microscopes, the agar plate is the most familiar icon of microbiology. The use of agar as a gelling agent goes back to the time of Robert Koch, who originally used gelatine as a solidifying agent for his plates. This was of course unsatisfactory since many organisms produce proteases which liquefy it. It was Frau Fannie Hess, the wife of one of Koch's associates, who suggested the use of agar.

The most common use of agar is in Petri dishes, where individual organisms are isolated and recognized by the colonies that they form. As well as their routine use in the laboratory there has been considerable interest in the development of colonies for their own sake. Pirt (1967) examined the growth of *Escherichia coli* colonies, using a ruler to measure the diameter of projected images of colonies, and established that radial extension was linear in colonies visible to the naked eye (macro-colonies). This phenomenon has been reported for a range of bacteria by a number of other workers (Cooper *et al.* 1968; Palumbo *et al.*, 1971; Wimpenny, 1979). This linear increase is explained by exponential growth at the leading edge of the colony, and Pirt (1967) postulated that growth only occurred at the leading edge, the central zone remaining relatively flat with little or no growth occurring. It is now clear that colony height does increase during growth (Wimpenny, 1979; Lewis and Wimpenny, 1981) and a more complex conceptual model for colony growth has been suggested, where radial expansion is a consequence of leading edge growth at or near μ_{max} (maximum specific growth rate) and growth behind the leading edge occurs at lower rates, depending on the physiology of the cells and the local physico-chemical environment (Lewis and Wimpenny, 1981; Wimpenny, 1988).

Radial extension in very young colonies (micro-colonies) should be exponential (Pirt, 1967; Wimpenny, 1979) as growth is not limited by nutrient diffusion, and it is apparent that in older colonies there is a clear transition between linear radial extension and linear increase in area (Pirt, 1967).

Colonial growth of bacteria on food surfaces is of significant importance well before colonies become visible. The macro-colony work described above gives us insight into the behaviour of colonies, but more information is required concerning

Digital Image Analysis of Microbes: Imaging, Morphometry, Fluorometry and Motility Techniques and Applications. Edited by M.H.F. Wilkinson and F. Schut.

the initiation of growth and the early stages of micro-colony development. In Cardiff, image analysis has been used to study micro-colony development, with the aim of producing useful dynamic growth data. This work has yet to be published, and brief methods and results are presented. The method has been described previously with some images of micro-colonies (McKay and Peters, 1995).

Growth data of micro-colonies will be of great use in food microbiology, where much of our understanding of pathogen behaviour comes from liquid culture studies with large inocula and detection methods of limited sensitivity. The technique described should help with the construction and validation of predictive models, and yield data on microbial competition, where the distance between competing colonies can be measured, and, using specially constructed growth chambers, on the effect of modified atmospheres and water activity on pathogen growth.

As this chapter will show, image analysis has played an important part both in quantitating the *number* of colonies through dedicated colony counters or specific routines in image processing equipment, and in investigating colony form and growth dynamics in more detail.

A second major area of endeavour is the measurement of antimicrobial activity. Clearly, this has an essential role to play in the clinical world, where it is vital to determine the effectiveness of different antibiotics on clinical isolates from infected patients as soon as possible. The deployment of these agents as discs on the surface of agar plates leads to zones of inhibition around the disk. The diameter of these zones is a function of the sensitivity of the organisms. Further details on determining antibiotic susceptibility can be found in standard works (Baron and Finegold, 1994; Murray *et al.*, 1995). While zones are commonly measured manually with a ruler or calipers, there is obvious scope for automating the process using image analysis techniques.

There is an area of microbiology, a little more abstruse than the above, which investigates the effects of combinations of physico-chemical agents on the growth of and interactions between bacteria. The importance of understanding these relationships is most clearly felt in the food industry, where carefully selected combinations of agents, including pH, salt concentration and amount of preservative added, all serve to protect a product from food poisoning or food spoilage bacteria. While such experiments can be carried out using sets of tube cultures, a logistic problem soon arises if more than one or two combinations of protective agent are investigated. The development of the two-dimensional gradient plate has allowed such experiments to be set up easily. Recording growth at different positions on these agar plates has relied on the use of image processing equipment.

18.2 THE MICROBIAL COLONY

The collection of quantitative data from microbial populations is key to many microbiological investigations, and a large number of techniques have been developed to give some estimation of population size. The most enduring of these are methods that rely on the enumeration of microbial colonies on or within solid media. Such counts are used to estimate the level of contamination in a product at a given time and, by taking measurements over a period of time, to calculate dynamic growth data. Image analysis systems are now widely available to assist in data collection for a number of 'plate count' methods, and research techniques are growing in sophistication to allow the identification and selection of individual colonies with particular morphological attributes (Jones *et al.*, 1992). Computer based systems can also allow the development of 'paperless' operations where plate count data is automatically entered into a Laboratory Information Management System (LIMS) for data storage, handling and retrieval. This eases trend analysis of count data and can be of use in microbiological risk assessment.

It is also clear from the introduction to this chapter that the microbial colony, though ubiquitous in the laboratory environment, is poorly understood as a complex, heterogeneous structure dominated by physico-chemical gradients that determine an individual cell's physiology and growth or survival. That the colony is not just an artefact of the laboratory but a common growth form in many natural environments is now widely recognized (Wimpenny, 1992).

Given this evidence that the colony is a common growth form, it becomes clear that a greater understanding of colony development may give a valuable insight into the behaviour of microbes in the natural environment. A number of studies have examined aspects of colony development using a variety of techniques; image analysis would appear to be a potentially powerful tool for examining colony development, in terms of generating both precise dynamic growth and morphometric data.

18.3 COLONY COUNTING

The colony count, in all its guises, is the workhorse of analytical microbiology. In general, counts give an assessment of the number of cells in a sample that are capable of recovering and growing on the given medium. As it is impossible to determine whether visible colonies have developed from single cells, the term Colony Forming Unit (CFU) was coined to describe the undefined unit of cells from which a colony may have developed. All plate count methods suffer, to varying degrees, from inaccuracy and imprecision due to a number of factors (Table 18.1).

Although image analysis can do little to help the inherent inaccuracy of plate count methods, other than counting pinpoint colonies that may be overlooked during a manual count, it is used increasingly to (i) maximize precision by reducing operator variability and fatigue, (ii) increase objectivity by regulating the criteria governing whether a colony is included in a count, and (iii) reject or flag unreliable data automatically, for example plates with a low colony count.

18.3.1 Development of automatic counting protocols

Superficially the automation of colony counting is a simple matter. A digitized image of a plate is obtained using a lighting system and camera that gives maximum contrast between colony and medium. An algorithm can then be used to threshold the image to include all colonies above or below the threshold limit (see

Table 18.1 Factors affecting the accuracy and precision of plate count methods.

Factors affecting accuracy	Factors affecting precision
Variable recovery of cells	Operator variability
Choice of nutrient media	Operator fatigue
Cell dispersal	Subjective counting
Culture conditions	Inclusion of statistically unreliable data
Poor sensitivity	
Level of inoculum	

Section 6.7). The number of objects with a pixel value that lies within the threshold range are then counted. In reality there are a number of factors that need to be taken into account.

18.3.2 Light source

Maximizing contrast between colony and background depends on the nature of the nutrient medium used and on colony morphology.

18.3.2a Opaque media

Incident or surface illumination is necessary. This may be from above, but often illumination from the side will increase contrast. Surface illumination may affect the accuracy of any measurement of colony size, as colonies can cast shadows. Opaque media can only be used for surface count methods, including spread plates, spiral plates, drop plates and contact plates.

18.3.2b Transparent media

Highly transparent media such as Plate Count Agar, Nutrient Agar and Brain Heart Infusion Agar are best viewed by darkfield illumination, which highlights the colonies against a dark background. This can be used equally well with submerged colonies from pour plates as with surface counts.

18.3.2c Semi-transparent media

Certain media, including many selective agars such as violet red bile glucose agar, are insufficiently transparent for successful darkfield illumination, but are still often used as pour plates with an overlay of agar to reduce false positives. The presence of submerged colonies rules out the use of incident light, and instead transmitted light must be used. Transmitted light gives the poorest contrast between the colony and background, and it is important that the light source is as diffuse and even as possible.

18.3.3 Camera choice

Today's charge-coupled device (CCD) cameras have high resolution and good optical sensitivity, making them ideal for automated colony counting (see Chapter 2). With optimized lighting and optics it is possible to resolve and count colonies of around 150 μm diameter (Pover, 1996). The cameras are also compact, and images do not 'burn' into the light-sensitive chip – a significant problem with the earlier vidicon tube cameras. Monochrome imaging is the method of choice for most dedicated counting systems, because it maximizes processing speed and minimizes the cost of the camera, framestore and software. Aperture control, which may be manual, automatic or under software control, and high levels of illumination give the required depth of field to count all colonies developing in a pour plate, where colonies may be dispersed throughout a 5–6 mm depth of agar.

18.3.4 Image processing

The increase in computing power means that most PCs can now handle digitized video images. However, such systems are slow, and for most counting applications, where a high throughput of plates is necessary, a framestore or image board (see Section 5.2.4) with sufficient memory to allow on-board storage and processing of images will be essential.

A monochrome framestore will generally digitize and store an image as a 256 grey level (8 bit) image, with perhaps one or more 2-bit overlays to allow masking and annotation of images etc. The grey scale image may have a pseudocolour look-up table (LUT) applied to aid the visualization of small colonies during user thresholding of images. A 512 $\times$ 512 pixel, 8-bit image requires 256 kbytes of framestore memory, and a minimum of 1 Mbyte on-board memory is recommended.

The image of the plate will undergo a number of software manipulations to improve the reproducibility of the count. These processing steps can maximize contrast and reduce the effects of uneven illumination, medium variability, confluence of growth, irregular colonies and edge effects at the perimeter of the Petri dish.

18.3.5 Background subtraction

The subtraction of a background image will compensate for uneven illumination, which can cause problems during thresholding as background pixel values can vary across the plate. If it is desired to subtract the image of an uninoculated plate then care must be taken to ensure that plates are evenly poured with consistent depth and opacity or the subtraction process may introduce more problems. Subtracting a background is relatively simple, requiring the subtraction of the background pixel value from the image pixel value, but it is important that the system can cope with negative pixel values that may arise from such processing.

18.3.6 Ramping images

Unevenly poured plates can lead to noticeable variation in background pixel values across the image. This can cause the same thresholding problems seen with uneven illumination. By automatically measuring the changes in background pixel value a ramp can be applied which will smooth background variation.

18.3.7 Increasing image contrast by 'histogram stretching'

The contrast between microbial colonies and agar can be maximized by 'stretching' the image's pixel values across the full grey scale range, i.e. the darkest pixel in the image is set to a pixel value of zero and the brightest to a pixel value of 255; other pixel values are then interpolated (Section 1.4.1a).

18.3.8 Measurement frames

The stacking ring on Petri dishes and the agar meniscus at the edge of plates pose problems, and it is common to apply a standard frame (or region of interest (ROI); see Section 5.2.4f) just inside the Petri dish. Frames can be circular, square or rectangular. Counts are made within the frame and the count extrapolated to take account of the total plate area. It is also possible for the user to identify a particular count area; again, counts are extrapolated to account for the total area. Spiral count plates and surface drop plates also utilize different counting frames to allow sector counts for the former and individual counts for the latter. Annular counts for spiral count plates may also be possible.

18.3.9 Colony irregularities

Colonies rarely develop evenly distributed across a plate with no touching or overlapping. A count from an environmental sample will often include a variety of microbes with quite distinct colonial morphologies, from pinpoints to large effuse growths. Often, zones of confluent growth or a single spreading colony can present problems. A number of image processing approaches can be adopted to compensate for these problems.

Colonies that are touching can be separated either manually by using a cursor to mark the separation line or, in more sophisticated applications, by eroding (reverting to background level) the pixels around the edge of each colony (see Sections 6.8, 8.3.2c and 11.4.2f). This step can be reiterated until separation occurs. The opposite process, dilation, will add a layer of pixels to the perimeter of colonies. Overgrown areas of a plate can be excluded from a count on the basis of size or manually by marking the area with the cursor. Counts for the remaining plate area are then extrapolated to give a full count.

A particular problem occurs on overcrowded plates, where a large number of touching or pinpoint colonies makes automatic detection difficult, leading to underestimation. It is possible to compensate for this by applying a polynomial function to the counts derived from the relationship between manual and automated counts (Pover, 1990).

18.3.10 Thresholding

Thresholding allows the detection of all areas of an image with a pixel value above or below a particular value. A threshold pixel value is set and all pixels with a value below or above that value will be highlighted. In most systems a pseudocolour LUT will be applied so that the thresholded objects are highlighted in a vivid colour. This assists operators in determining whether all colonies have been successfully detected.

The threshold value can be user-defined interactively for each plate, or alternatively a series of threshold values can be recorded for different plating procedures. This latter approach minimizes user error and increases reproducibility, but not necessarily the accuracy of the count. Interactive thresholding enables counting of problematic plates.

18.3.11 Image analysis

A thresholded image allows rapid analysis of a number of parameters. Algorithms are applied that count for all thresholded objects (colonies): the size of each object in terms of pixel area, feret diameter or perimeter; the degree of circularity (Section 8.3.2a), which is a function of perimeter, area and diameter; and the presence of holes in the object, which may indicate non-colonial material. All dimensions are expressed in pixels by default, but it is possible to calibrate the system to report actual sizes.

It is possible to filter the count data by excluding objects that meet predetermined criteria, for example colonies below or above a particular size, or that do not conform to a certain degree of circularity. This can eliminate very small particles, which may be associated with the sample, other non-colonial material and large or irregular colonies.

18.3.12 Data handling

Raw count data will be converted to a count of CFU/unit sample (e.g. CFU/g). This requires information on the dilution factor used and the extrapolation of data from the counting frame to the full plate area assuming an initial even distribution of CFUs. Data will be transferable to other software packages such as spreadsheets, databases and fully integrated LIMS. Printed reports can then be produced automatically and complex trend analysis conducted.

18.4 IMAGE ANALYSIS IN COLONY GROWTH MEASUREMENT

The growth and morphology of developing colonies is not well understood, despite their ubiquity in the laboratory and detailed subjective descriptions used for diagnostic purposes. There is, however, increasing interest in the microbial colony as a dominant feature of many natural ecosystems (Wimpenny *et al.*, 1995a) and image analysis provides a potentially powerful tool for studying the nature of colony growth, and the morphology of both the colony and differentiation of cells within a colony. This section reviews some of this work and details the development of an image analysis system to measure growth rates of macro- and micro-colonies.

An image analysis application for characterizing bacterial colonies was described by Belyaev *et al.* (1992). Alkaliphilic strains of bacteria were grown on agar plates containing an indicator for cyclodextrin glucanotransferase (CGTase). Colonies producing this enzyme developed characteristic diffusion zones, which appeared as 'halos' around colonies. The image analysis system was used to obtain colour images of developing colonies at 4-hour intervals. By analysing the green component of the three-colour layer image an area measurement of the halo was obtained. The green component clearly indicated the edge of the halo but did not detect the colony. Analysing the blue component delineated the edge of the colony but not the halo and was used to measure colony area. Plots of colony area and halo area against time indicated increasing colony and halo sizes for all strains, but a plot of halo area/colony area clearly showed large dynamic differences between strains, dependent on their ability to produce CGTase. Data of this type clearly indicates the potential use of image analysis in objective screening and selection of bacterial colonies.

18.4.1 Determining colony growth kinetics using image analysis

Image analysis allows accurate, non-invasive measurement of colony size and thereby the determination of mean colony growth rates. The authors and co-workers have developed a system that generates growth data in three ways: (i) the increase in size of macro-colonies using transmitted light and changes in colony morphology with time; (ii) mapping of the optical density of a colony allowing measurement of changes in optical density over time; and (iii) the increase in size of micro-colonies viewed under phase-contrast microscopy. All three systems are reviewed here.

18.4.1a Image analysis system

A number of image analysis systems based around framestores and image processing software produced by Synoptics Ltd (Cambridge, UK) have been used by the authors to measure microbial growth on agar surfaces.

System 1 comprised a Synapse framestore (Synoptics) mounted in an IBM 80386 compatible PC with 80287 maths coprocessor running at 16 MHz (Dell Computers, Bracknell, UK). Images were displayed on a dedicated multisynch monitor (Taxan 770). System 2 utilized a Sprynt framestore (Synoptics) in an 80486 DX33 (Dell Computers) with images being displayed on a separate SVGA monitor (Sanyo). System 3 used the Synapse framestore mounted in an 80486 DX66 (Minstrel Computers Ltd). Systems 1 and 2 used a 1 in vidicon tube monochrome TV camera (WV1800, National Panasonic, Industrial Video Systems, Cwmbran, UK), while System 3 utilized a monochrome CCD camera (EOS Ltd, Barry, UK).

The Synapse framestore is used by the system for acquiring, processing and displaying images using IBM PC AT or XT microcomputers or compatibles fitted with at least 10 Mbytes hard disc storage and a maths coprocessor. Analog monochrome TV signals are digitized and stored in a 512 × 512 pixel framestore memory, which consists of 320 kbytes of video RAM split into an 8-bit image plane and two 2-bit overlay planes. The image plane stores a grey scale image with 256 possible grey levels while the overlay planes hold the cursor position, annotation, wire frame plots and contour maps. Image data pass through a look-up table allowing a grey scale or false colour to be displayed. The image is displayed as a 512 × 512 pixel array (maximum) using European standard 625 line 50 Hz interlaced composite video with a sampling frequency of 15 MHz, resulting in square pixels. The Sprynt framestore is more advanced, supporting full colour images and with a larger memory, allowing far more on-board processing and thus operating at greater speeds. For the applications described here, its increase in speed was the only significant difference. The system is shown in Figure 18.1.

18.4.1b Application software

Semper 6plus (Synoptics), a general-purpose high level image processing language, was used to manipulate the images. This software is a highly flexible language, which permits interactive use with immediate execution of commands. It provides a toolkit of image processing functions, which can be used interactively or, alternatively, sequences of commands can be combined to produce a complete

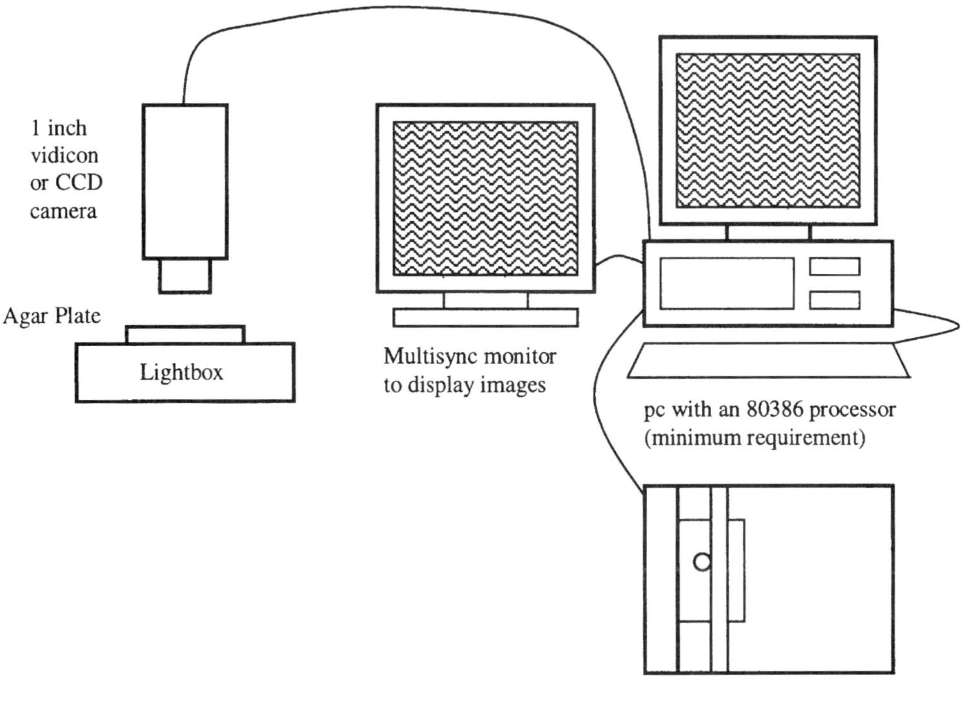

Figure 18.1 Image analysis system hardware. From Peters (1990), by permission of Academic Press Ltd.

applications package. Semper 6plus has a flexible picture filing system, which allows images to be stored in a variety of ways. The image held on the framestore is represented by a grey scale where pixel values of 0 and 255 correspond to black and white, respectively. This byte representation allows relatively compact data storage (256 kbytes disc space for a 512×512 pixel image). Semper 6plus can convert and store the data either as integer, floating point or complex point representation. In our image application programs, background subtraction sometimes generates negative pixel values, while the conversion of transmission data to optical density values requires a scale that approaches infinity. Neither of these data is supported by byte representation, so the image data is converted to floating point representation, which has operating limits of $\pm 1 \times 10^{-35}$ to $\pm 1 \times 10^{+35}$. Using Semper 6plus it is possible to access individual pixels, and arbitrary mathematical operations can be performed on either the whole image, part of an image or even a single pixel. The writing of menu-driven applications packages has allowed inexperienced operators to capture, process and store the images, and to present the data in a number of formats.

18.4.2 Macro-colony growth

Image analysis has been used in Cardiff to monitor the development of macro-colonies of *Bacillus cereus* (Wimpenny *et al.*, 1995b) and *Salmonella typhimurium* (McKay and Peters, 1995). Both studies used images of colonies on the surface of transparent agar using transmitted light.

18.4.2a Colony growth of Bacillus cereus

The size and a measure of the circularity of developing *B. cereus* colonies were followed over a 216-hour period. Nutrient Agar (NA, Oxoid) plates were inoculated with <10 cells to eliminate interference between colonies. Plates were incubated at 30 °C in a moist atmosphere to avoid drying out. Transmitted light images of five or six colonies on each plate were captured every 24 hours. A Semper 6plus program used particle analysis routines to determine feret diameters and coefficients of circularity. The latter (C) is defined as (see Section 8.3.2.1):

$$C = \frac{P}{\sqrt{4\pi A}}$$

where P is the total length of the perimeter and A is the area of the colony. C is 1.0 for a perfect circle.

The data show that colony diameter increases linearly over the period studied (Figure 18.2). The mean coefficient of circularity was determined from four colonies at each time interval. The coefficient of circularity decreased progressively with time as colonies became more irregular (Figure 18.3).

Changes in the coefficient of circularity, although of limited value in understanding the process of colony growth, confirm that older colonies become irregular due partly to a process known as 'roughening' and partly to nutrient depletion around the periphery of the colony (Wimpenny *et al.*, 1995b). More

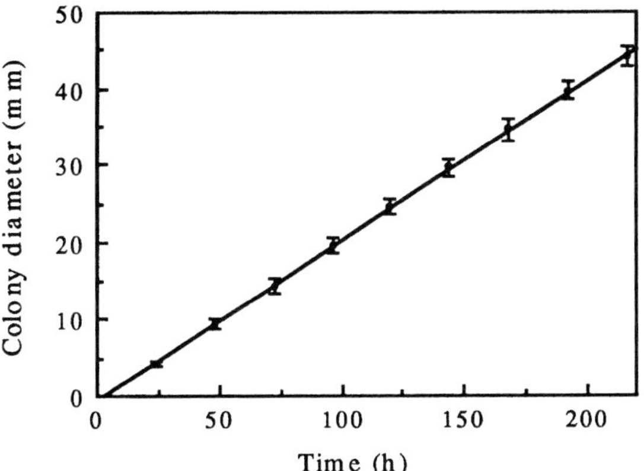

Figure 18.2 Increase in mean diameter of *B. cereus* colonies on nutrient agar ($n = 5$ or 6). Reproduced from Wimpenny *et al.* (1995b), by permission of Bioline.

detailed morphological analyses are possible using image analysis techniques, and these may open the way to rapid colony recognition and a greater understanding of the process of colony growth.

18.4.2b Growth of Salmonella typhimurium

A more detailed analysis of the growth of *S. typhimurium* colonies gave dynamic growth data of colonies in response to pH and sodium chloride (NaCl) concentration (McKay and Peters, 1995). Growth was followed from the time it was first discernible, when colonies had a radius of approximately 60 μm, for a total of up to 26 hours. By producing a calibration curve for viable cell number as a function of colony diameter it was possible to present data in terms of increase in biomass, and hence determine

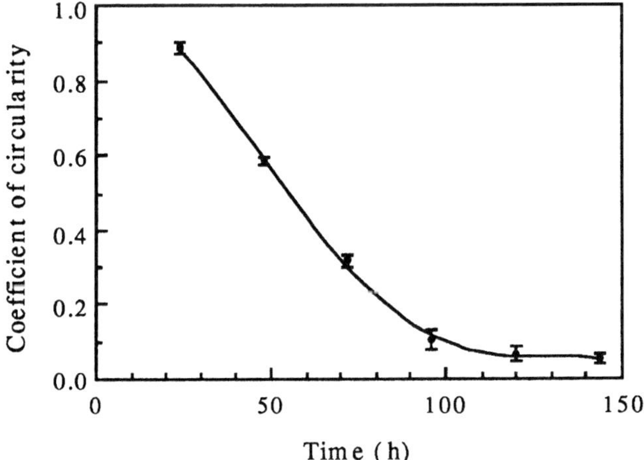

Figure 18.3 Change in mean coefficient of circularity over time for colonies of *B. cereus* on nutrient agar ($n = 4$). Reproduced from Wimpenny *et al.* (1995b), by permission of Bioline.

dynamic growth parameters. Colonies were grown on constant depth Brain Heart Infusion agar (BHIA, Oxoid, UK), pH adjusted to give a final pH of 5.0, 6.0 or 7.0 or supplemented with NaCl to give a total concentration of 0.5, 1.5, 2.5 or 3.5% w/v. Plates were inoculated with 20 μl of a 10^{-8} dilution of an overnight culture of the bacterium and incubated at 30 °C. This proved sufficient to ensure that typically a single colony would develop on each plate. Plates that showed no growth or supported more than one colony were discarded. This ensured that colony growth was not affected by the presence of neighbouring colonies.

Growth was monitored at 30-minute intervals after 13 hours incubation. Plates were removed from the incubator and positioned on a light box with opaque diffuser. Images were obtained using a CCD camera with macro zoom lens connected to image analysis system 3 (Section 18.4.1a). Plates were kept out of the incubator for a minimum of time during image acquisition (less than 1 minute).

Images were first calibrated and a background image subtracted to remove lighting and agar irregularities. An interactive threshold was applied to the image and particle analysis commands used to size each colony, recording a number of parameters in a log file.

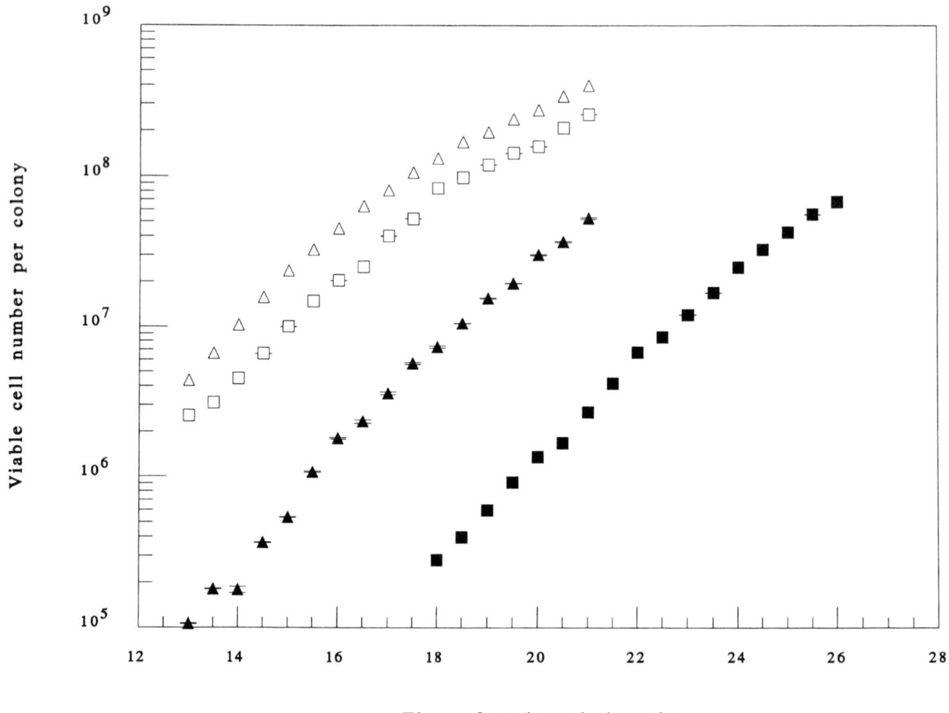

Figure 18.4 Growth of *S. typhimurium* colonies on NaCl supplemented BHIA at 30 °C. Error bars indicate ±3 standard errors of the mean (s.d./$\sqrt{n}$ where $n = 3$, 4 or 5). [NaCl] (w/v): open triangles, 0.5%; open squares, 1.5%; closed triangles, 2.5%; closed squares, 3.5%. From McKay and Peters (1995), by permission of Blackwell Science Ltd.

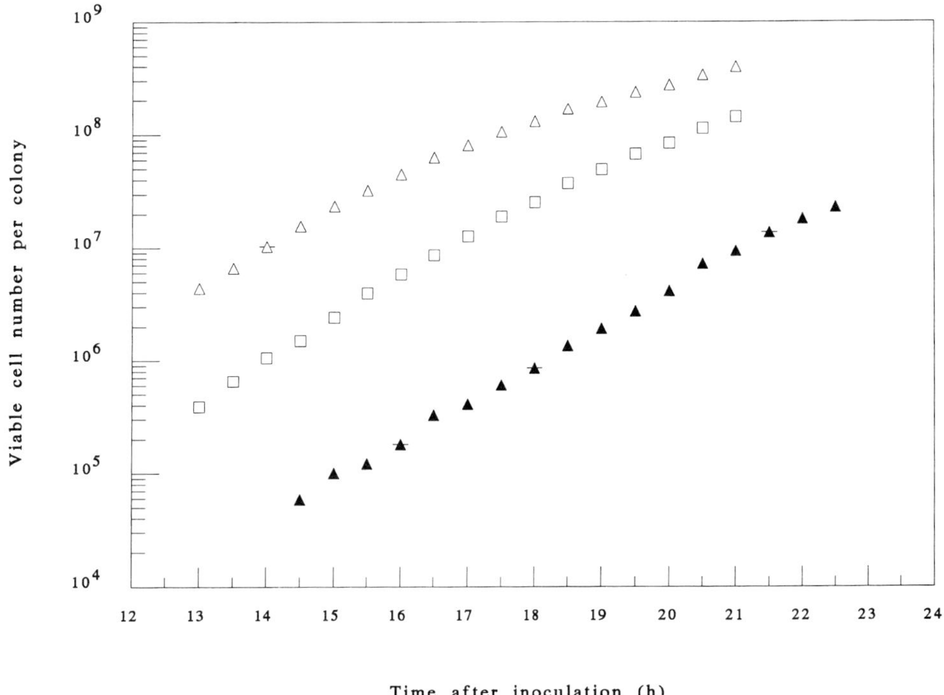

Figure 18.5 Growth of *S. typhimurium* colonies on pH adjusted BHIA at 30 °C. Error bars indicate ±3 standard errors of the mean (s.d./$\sqrt{n}$ where $n = 3$, 4 or 5). pH: open triangles, 7.0; open squares, 6.0; closed triangles, 5.0. From McKay and Peters (1995), by permission of Blackwell Science Ltd.

The average radius of each colony (mean of four feret diameters) was converted to viable cell number using a calibration curve relating colony size to the number of viable cells in a colony, and plotted to give the increase in viable cell number per colony over time. The calibration curve was prepared by growing a colony under the specified conditions, sizing it and then carefully removing the colony and a plug of agar beneath the colony using a sterilized cork borer, homogenizing this in maximum recovery diluent and performing a standard plate count. This was performed on a large number of colonies between 10 and 26 hours after inoculation. Figure 18.4 shows the effect of increasing NaCl on the growth of *S. typhimurium* colonies, while Figure 18.5 shows the effect of decreasing pH.

It is clear from the results that linear radial growth represented the exponential growth phase of the bacterium and that the smallest colonies visible using the system contained approximately 10^5 cells. Both increasing NaCl concentration and decreasing pH delayed the onset of exponential growth and the time taken for colonies to become visible, but not the specific growth rates. Exponential growth continued throughout the period studied, and it was possible to derive specific growth rate data from the curves.

This study showed for the first time that it was possible to obtain dynamic growth data in terms of increase in viable cell number using image analysis. The data was highly reproducible, and, as the technique is non-invasive, a large number of data points were collected for each curve. This is highly desirable in both the creation and validation of predictive models, which are of growing use in microbiological risk assessment of food safety control. Our understanding of the growth of foodborne pathogens is predominantly based on traditional liquid culture techniques, and a better understanding of the growth of colonies may lead to greater food safety.

18.4.2c Mapping optical density within colonies

Image analysis has been used by the authors to map the optical density of developing colonies. This has allowed the measurement of specific growth rates at different points within colonies, confirming conceptual models of colony development.

Using *B. cereus*, Wimpenny *et al.* (1995b) showed that most growth occurred at the leading edge of colonies during linear radial extension.

Nutrient agar plates inoculated at the centre were incubated at 30 °C, and a transmitted light image of colonial growth was recorded every day for 7 days. Image processing used a Semper 6plus application program, which required that the colony under investigation was centred in a square display window. An image was recorded using transmitted light, and data was then averaged down to a 30×30 pixel array. Data was then imported into Lotus 123 (Lotus Development UK Ltd., Windsor, UK).

The images were stored and processed in floating point format. Assuming that Beer's law (Section 3.3.1) is obeyed the pixel values recorded are approximately proportional to the optical density (OD) of the colony. The following expression was applied to each pixel:

$$\mathrm{OD} = 0.4343 \ln\left(\frac{255}{\text{pixel value}}\right)$$

Pixels with a value of zero would cause an error in this operation, so 1×10^{-5} was added to each pixel value to ensure that zero values were not encountered. This data was stored and transferred to Lotus 123, where three-dimensional wireframe representations of the optical density of colonies were produced using the add-on package 3D graphics (456 World Ltd, Colchester, UK). Further image processing allowed the subtraction of successive growth images. This process shows the amount of growth that has occurred over the time period in question. This data was also imported into Lotus 123 for plotting.

Optical density maps for a *B. cereus* colony growing over a 120-hour period are shown in Figure 18.6. It is clear that the younger colony (Figure 18.6a) has a clearly defined edge, while the older colony (Figure 18.6c) is less regular. This supports the coefficient of circularity data presented in Section 18.4.2a. The series of subtracted images presented in Figure 18.7 confirms that as a colony ages, growth is restricted to the periphery. The inner area of the colony has optical density (absorbance)

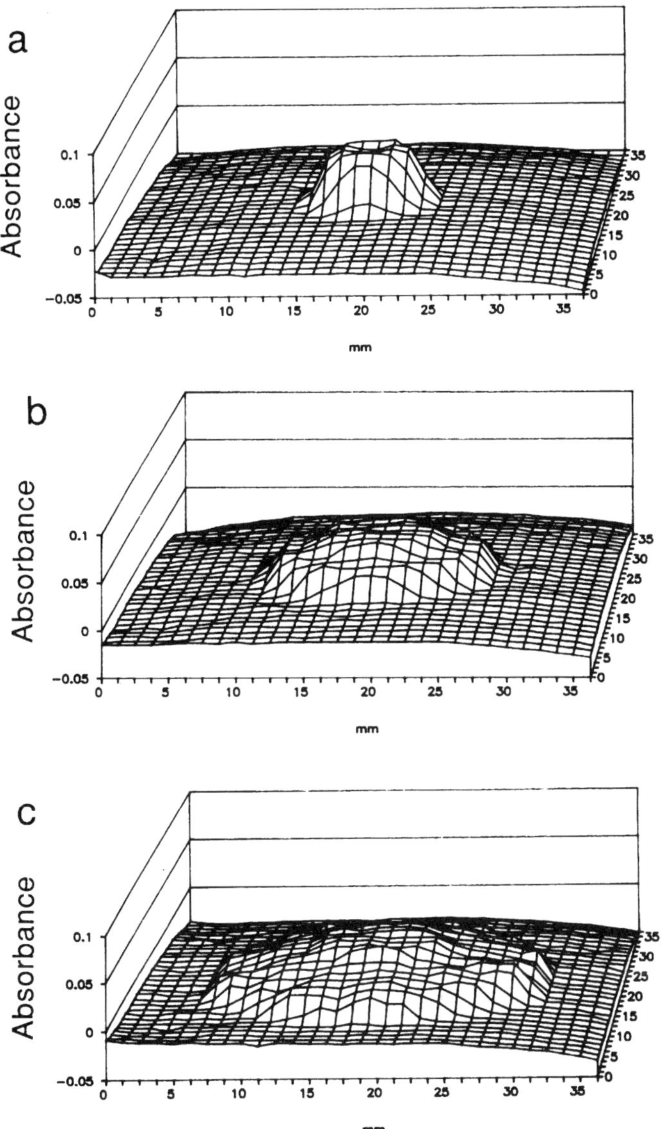

Figure 18.6 Wireframe diagrams showing increase in size of *B. cereus* colonies on nutrient agar. (a) 24 h, (b) 96 h and (c) 120 h. Reproduced from Wimpenny *et al.* (1995b), by permission of Bioline.

values similar to the background value, showing that little growth has occurred in the intervening period. This helps to confirm predictions first made by Pirt (1967), who identified the peripheral zones of colonies as the regions of most active growth. Wimpenny (1979), using profile measurements, agreed with this view from morphological considerations. The optical density data presented here confirms these views.

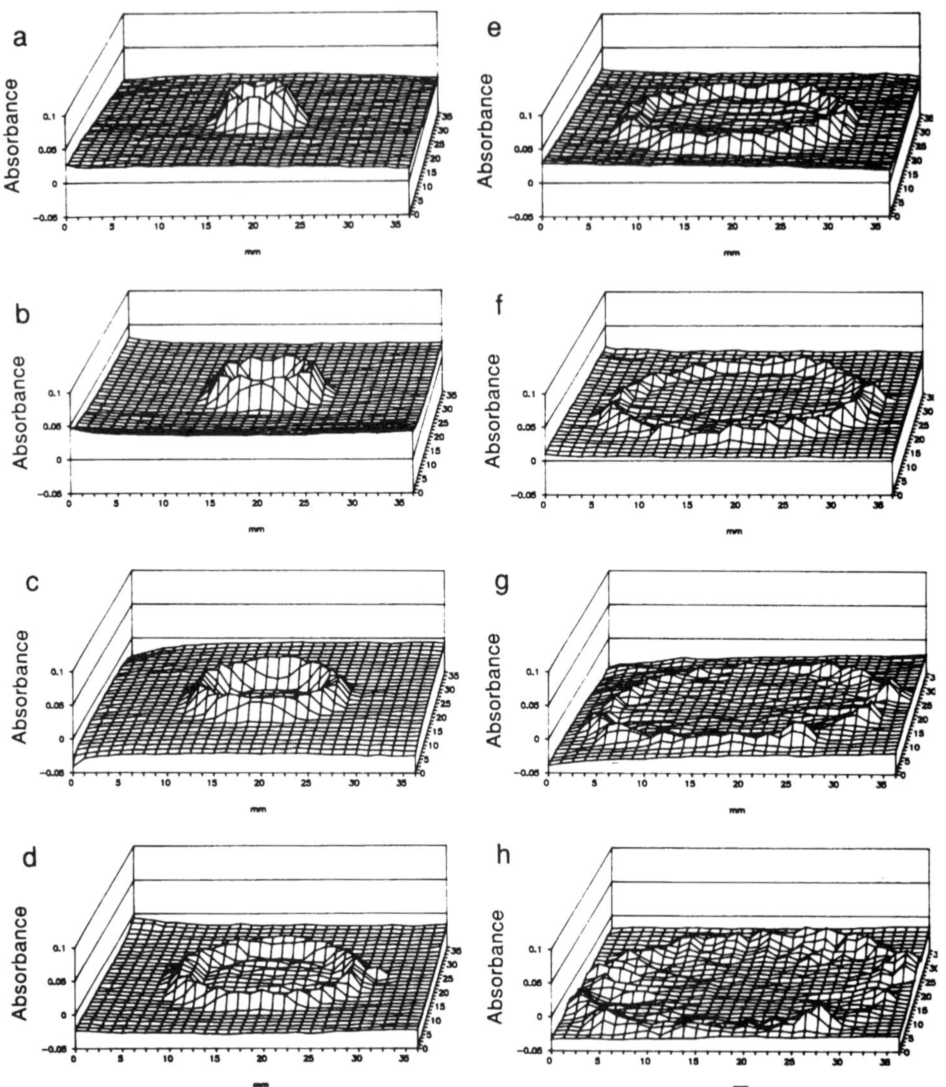

Figure 18.7 Series of subtracted images of *B. cereus* colonies on nutrient agar. (a) 48 h − 24 h, (b) 72 h − 48 h, (c) 96 h − 72 h, (d) 120 h − 96 h, (e) 144 h − 120 h, (f) 168 h − 144 h, (g) 192 h − 168 h and (h) 216 h − 192 h. Reproduced from Wimpenny *et al.* (1995b), by permission of Bioline.

In a further development of the method described above, McKay *et al.* (1997) presented specific growth rates measured at two points within a developing *S. typhimurium* colony. Optical density maps of *S. typhimurium* colonies were transferred to a Lotus 123 spreadsheet. The centre of each colony was determined and OD data in nine cells around this point were averaged. In addition OD data from a cluster of five cells at a point 200 μm from the centre point were averaged. Initially this point was at the very edge of the colony, but as the colony grew it became progressively more centralized. Average OD measurements were taken from the two points for successive images.

Assuming that the OD of the colonial material is proportional to the biomass in the colony, any increase in OD is the result of an increase in the population. It is possible to calculate 'apparent' specific growth rates (μ) by taking the natural log of the OD measurements, subtracting successive log values and dividing by the time interval between the readings. This only holds true if growth is in the exponential phase. Figure 18.8 shows the increase in OD at the centre and a point 200 μm from the centre of a *S. typhimurium* colony on BHIA at 30 °C. The data confirms that growth is exponential throughout the period studied, and that growth switches to a lower rate in the centre of the colony as it increases in size. The growth rates at each point converge as the points become progressively more centralized.

This study demonstrated the power of image analysis in deriving dynamic growth data from transmission images of colonies. The data support conceptual models of regional variation in specific growth rates within colonies, first proposed by Wimpenny (1988) following earlier studies on the form of colonial growth (Lewis and Wimpenny, 1981). These models described regional changes in cell physiology in growing bacterial colonies for strict aerobes, anaerobes and facultative anaerobes. In each case unrestricted growth could occur only at the leading edge of the colony. Growth behind the leading edge becomes oxygen- and/or nutrient-limited, while near the centre growth may cease and cell death and lysis occur. McKay *et al.* (1997) present a simplified conceptual model (Figure 18.9), which correlates with the growth data presented in Figure 18.8.

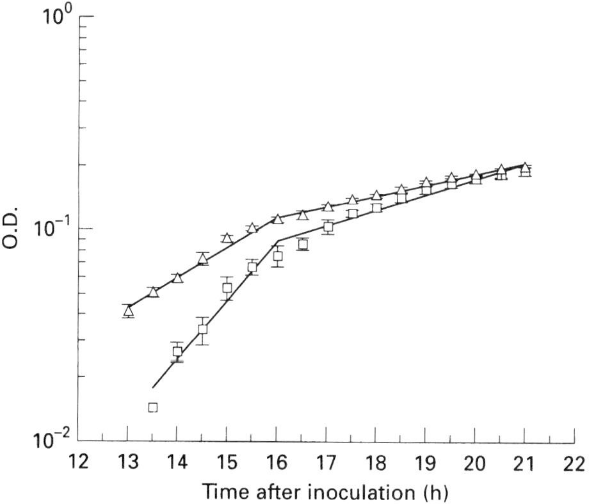

Figure 18.8 Increase in optical density (OD) at the centre (open triangle), and point 200 μm from the centre (open square) of an *S. typhimurium* colony grown on BHIA at 30 °C. From McKay *et al.* (1997), by permission of the Society for Applied Bacteriology.

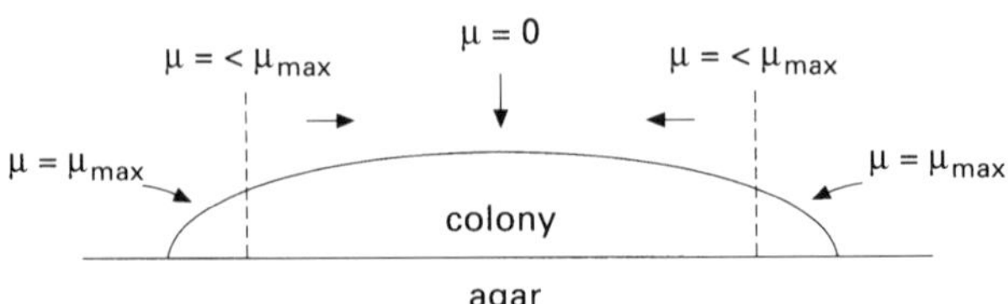

Figure 18.9 Simplified conceptual model illustrating regional variation in specific growth rate (μ) within a growing bacterial colony of a facultatively anaerobic organism. From McKay *et al.* (1997), by permission of the Society for Applied Bacteriology, developed from Wimpenny (1988), by permission of CRC Press.

18.4.3 Micro-colony growth

S. typhimurium was point-inoculated onto microscope slides coated with a consistent depth of BHIA. Slides were incubated under a water saturated atmosphere at 30 °C. After 4 hours incubation slides were examined microscopically (Olympus, Tokyo, Japan) by phase contrast.

The CCD camera used with image analysis system 3 was mounted via a C mount to a vertical eyepiece tube on the microscope using a 6.7× eyepiece and 40× objective. Images were displayed live at a resolution of approximately 0.1 μm per pixel. This allowed the slide to be scanned to identify developing micro-colonies. If a micro-colony could not be found within 15 minutes the slide was discarded as this was believed to have too large an impact on growth. With practice it was possible to identify micro-colonies within 1–2 minutes and, having noted their position on the slide using the vernier scales, return to them in a few seconds on subsequent imaging.

Image analysis was performed using a menu-driven, dedicated application program in Semper 6plus command language written by McKay. This package processed micro-colony images as follows:

(a) calibration of the system using stage graticule
(b) system kept in 'live' mode to enable the search for micro-colonies
(c) image of micro-colony grabbed and displayed
(d) operator outlines area of interest (AOI) using cursor
(e) AOI extracted
(f) AOI image saved to file
(g) operator sizes colonies and distance between colonies (if appropriate) using cursor. As many measurements as desired can be made to allow averaging
(h) size data converted to actual size (using calibration from step (a)) and output to a log file for further analysis.

The use of phase-contrast microscopy gave images that were poorly segmented by thresholding, as large areas of the agar had similar pixel values to the colonies. The edges of the colonies, and cells within the colonies could be defined clearly, and although edge detection algorithms were investigated it was determined that interactive operator measurement gave the most reproducible results.

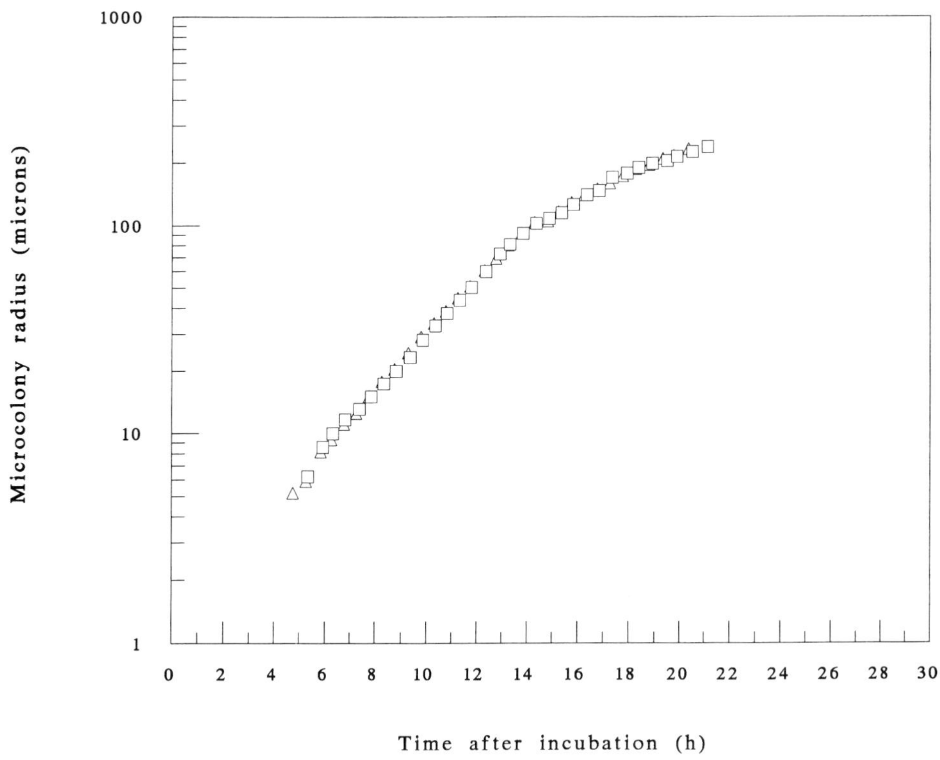

Figure 18.10 Radial growth of *S. typhimurium* micro-colonies on BHIA coated microscope slides at 30 °C ($n = 2$). McKay, personal communication, by permission.

Once micro-colonies reached about 60 μm in radius it was possible to image them using transmitted light and the macro zoom lens described previously. Measurements were still made using the application package described above. Results are presented as the log increase in colony radius with time (Figure 18.10). This shows that micro-colonies with a radius of 5 μm could be followed easily. These colonies contained a single layer of approximately 16 cells. At this point growth had clearly entered the exponential phase, and the data show an exponential increase in radius until about 14 hours after inoculation. At this point the radial extension of the colony slowed as growth became largely limited to the colony periphery, and as the colony reached 100 μm radius radial extension was linear.

Pirt (1967) predicted that young colonies would exhibit exponential radial extension, and this was supported by data from Hammonds and Galliford (1991). The results presented here correlate well with this prediction and show the transition from exponential to linear radial extension. Subsequent work allowed growth measurement from colonies containing just two cells, giving valuable insight into the lag period of cells subject to increased NaCl concentration (McKay, personal communication). One problem with the data is that it is not expressed in terms of biomass increase. However it is possible to count individual cells in small micro-colonies (<10 μm radius) and thus calculate actual specific growth rates. Image analysis routines are being written to allow automatic counting of cells. This will require erosion of images to separate closely associated cells.

18.5 THE DEVELOPMENT AND USES OF GRADIENT PLATES

Another way in which image analysis has been used in Cardiff is the recording of growth zones on square agar plates on which surface gradients of two factors are established. These are known as 'gradient plates' and have been used extensively to determine the response of microbes to a range of important growth determinants.

A wedge plate technique which created a concentration gradient in agar was first used in 1952 for the study of microbial resistance to antibiotics (Szybalski, 1952). Since then, the two most important areas of progress in gradient plate technology have been the development of two-dimensional plates, which has enabled simultaneous investigation of four environmental variables (Caldwell and Hirsch, 1973; Wimpenny and Waters, 1984, 1987), and the use of image analysis to record accurately the bacterial growth on the surface of the plates (Peters, 1990; Peters *et al.*, 1991b).

Early plates contained only a single gradient, e.g. antibiotic (Szybalski, 1952; Szybalski and Bryson, 1952; Kim and Anthony, 1983), pH (Sacks, 1956), or a 'double' gradient of antibiotic with metal ions (Weinberg, 1956). Two-dimensional plates have used the following combinations of gradients: pH/NaCl concentration, sodium nitrite/NaCl concentration; temperature/pH, nutrients, and light intensity/temperature (Bal'a and Marshall, 1996; Caldwell and Hirsch, 1973; Caldwell *et al.*, 1973; Halldal and French, 1958; McClure and Roberts, 1986; McClure *et al.*, 1989; Peters *et al.*, 1991b; Thomas *et al.*, 1991, 1992, 1993; Van Baalen and Edwards, 1973; Venables *et al.*, 1995; Waters and Lloyd, 1985; Wimpenny *et al.*, 1986; Wimpenny and Waters, 1984, 1987). Wimpenny and Waters (1987) introduced third and fourth dimensions by varying the incubation temperature for a series of plates and incorporating a third solute at different concentrations in each plate. These plates are useful for demonstrating possible synergy or antagonism between different chemicals acting on bacterial growth (particularly important in medical or food microbiology).

Gradient plates have been used for a diversity of purposes. Most gradient plate studies have been concerned with aspects of microbial ecology, food microbiology or medical microbiology (such as antibiotic resistance). In each case the technique was used to map the response of organisms to variations in environmental conditions. In most studies the bacterial response was measured by the area and/or density of visible growth on the surface of the plates. Information derived from one plate can be increased by using a mixture of strains for the inoculum. The boundaries of the growth area on such a plate are an indicator of the extremes of conditions limiting the growth of the tested strains (Thomas *et al.*, 1992). Bioluminescence, as well as growth area, was a bacterial response recorded by Waters and Lloyd (1985) when using gradient plates to investigate the effects of salt, pH and temperature on three species of luminescent bacteria (*Photobacterium leiognathi*, *P. phosphoreum* and *Vibrio fischeri*). The viability of organisms can be assessed by replica plating of gradient plates to non-gradient plates (Thomas and Wimpenny, 1996a). Additional areas on the gradient plate are thus revealed, where cells have remained viable but were unable to grow due to the combination of environmental conditions on that part of the plate. The replication method was also used to investigate competition between different species of bacteria (Thomas and Wimpenny, 1993, 1996b), and conjugal transfer of plasmids (Venables *et al.*, 1995).

18.5.1 Preparation of gradient plates

The procedure used in our laboratory for preparing square agar plates with gradients of pH and NaCl concentration is summarized in Figure 18.11 (Thomas *et al.*, 1994). A medium with a high buffering capacity should be used, e.g. a complex nutritious medium such as Brain Heart Infusion (Peters *et al.*, 1991a). Plates are poured as four separate layers in 10 cm square Petri dishes as follows:

(a) The Petri dish is placed onto a level surface and raised at one end by 3 mm. The first 15 ml layer of agar is poured, which contains an appropriate concentration of acid. This layer forms a wedge and is allowed to set.

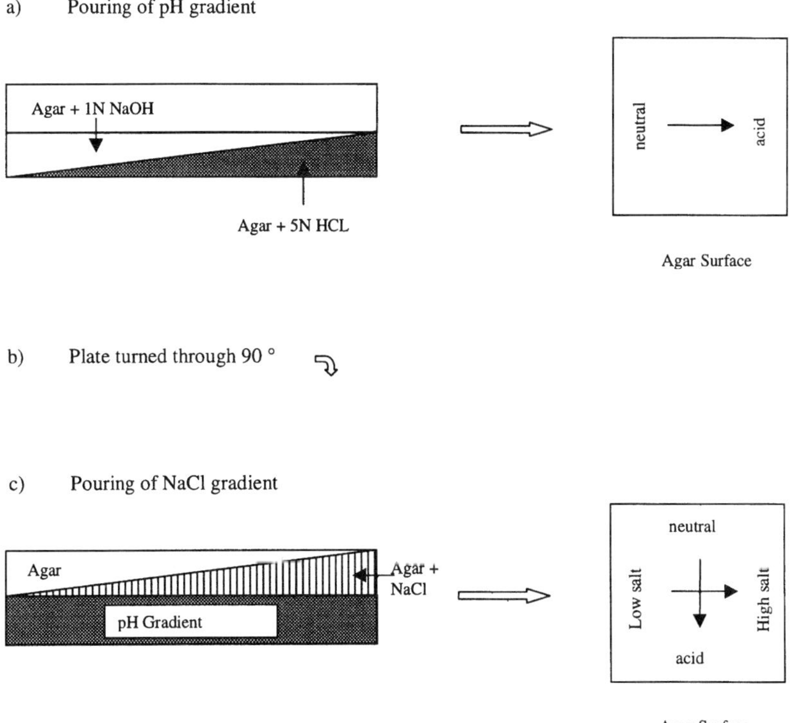

Figure 18.11 Sequence of operations for the preparation of pH/NaCl gradient plates. From Wimpenny *et al.* (1988), by permission of CRC Press.

(b) The Petri dish is placed on a level surface and the second 15 ml layer of agar, containing an appropriate concentration of alkali, is poured. This layer is allowed to set. These two layers comprise the pH gradient.
(c) The Petri dish is rotated through 90°, placed on a level surface, and one end raised by 3 mm. The third 15 ml layer of agar, containing 15–20% (w/v) NaCl, is poured. This is allowed to set.
(d) The Petri dish is placed on a level surface and the final 15 ml layer of agar, with no additions, is poured and allowed to set.
(e) The plates are kept overnight at ambient temperature to allow vertical diffusion of the NaCl and H^+ ions, which results in their dilution proportional to the ratio of the thickness of the agar layer and thus the formation of a uniform, horizontal concentration gradient.
(f) The gradients are measured the following day, when the plates are inoculated by flooding the surface with a broth culture.
(g) The growth on the plates is measured after 24 and 48 h. The incubation period of pH–NaCl gradient plates is limited to 48 h due to the decay of the gradients, in particular that of salt.

18.5.2 Temperature–pH gradient plates

The 'Gradiplate' (Biodata Oy, Helsinki, Finland) has been used to measure the temperature sensitivities of a range of bacteria (Junttila *et al.*, 1988; Niemela *et al.*, 1983; Peters *et al.*, 1991a; Thomas *et al.*, 1991, 1992). The system consists of a metal plate 10 cm wide by approximately 35 cm long, whose opposite edges are maintained at different temperatures by circulation of constant-temperature liquids from two thermostatically controlled water baths and pumps. A maximum of six agar or gradient plates poured in 92 mm square, flat glass dishes (Biodata), designed to make good thermal contact with the aluminium plate, can be incubated at one time. The use of such a steady-state temperature incubator in order to maintain a steady gradient for one of the two dimensions has the advantage of reducing the problem of gradient breakdown, and allows other variables, such as NaCl, antibiotics and other inhibitors, to be added to the plates at uniform and accurately measured concentrations. A continuous temperature gradient has a further advantage of measuring a greater range of temperature sensitivities rather than a few discrete incubation temperatures. Technical difficulties may be experienced when attempting to run the incubator over too great a temperature range. Condensation can build up and the agar has a tendency to crack at the hot end. This type of incubator may be more useful for investigations of growth at lower temperatures over longer periods, since the pH gradient remains stable for more than 48 h.

18.5.3 Recording and measurement of bacterial growth on gradient plates

Following incubation, the bacterial growth on gradient plates normally appears as a region of confluent growth whose area is bounded by the particular combinations of conditions produced by the gradients. In certain circumstances more than one area of confluent growth can appear and individual resistant colonies are often visible outside the region of confluent growth. Several methods have been used to record these growth areas. The area of visible growth can be measured by simply tracing its boundary, either directly or from a projected colour transparency. Tracings of this type can be digitized with a scanner, so that surface plots of the growth can be generated (Bal'a and Marshall, 1996). Quantitative data is harder to obtain. Wimpenny *et al.* (1988) cut grids into the plates, removed squares of agar, washed off the surface growth and measured the turbidity of the resulting suspension spectrophotometrically. McClure and Roberts (1986) scanned plates using a scanning laser densitometer. The optical density data were then transferred to a computer, compiled, interpolated and plotted. Both these methods are time consuming and both limit the amount of information potentially available on each plate. The authors have used image analysis to record and interpret gradient plate data using a dedicated, menu-driven, image processing application program written in Semper 6plus command language.

18.5.4 Data acquisition and processing

The sequence of operations for the purpose of obtaining growth data from a gradient plate is shown in Figure 18.12 (Peters *et al.*, 1991b). The plates are

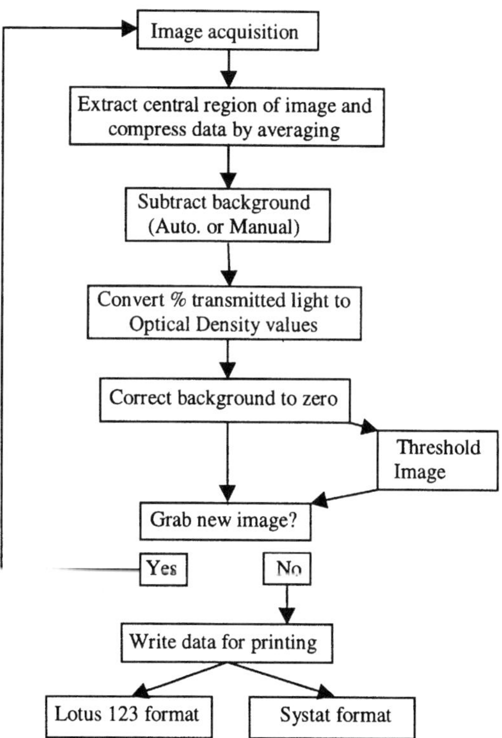

Figure 18.12 Sequence of events for the processing of gradient plate images. From Peters *et al.* (1991b), by permission of Academic Press Ltd.

photographed using transmitted light from a light box (Whatman, Maidstone, UK). The image from the camera is displayed on the computer screen to facilitate the alignment of the plate in the camera field of view; during positioning of the plate under the camera, guidelines generated in the overlay plane are superimposed on the image displayed on the screen. The grabbed image is stored in the frame store. This image is 512 × 512 pixels, with the 10 cm square plate occupying the central 411 × 411 pixels. The edge of the plate has an effect on the gradients, which in turn affects the growth zone, so only the central 9 cm square region is taken, which occupies the central 370 × 370 pixels. Averaging and extracting the data to give a final image of 37 × 37 pixels were necessary, not only to save storage space, but because this was the largest amount of data in one file which could be imported to Lotus 123 at the time the research was undertaken. Background subtraction, as described in Section 18.3.5, is necessary because the output from the light box may not be uniform.

After processing of the image, it is displayed on the screen (magnified with interpolation). Slight differences in the optical density of the plates and variation in camera output can result in areas of no growth having optical density values of 0 ± 0.01. This is corrected by outlining an area of the plate that has no growth, taking its average pixel value, setting this background value to zero and adjusting all other pixel values by the same factor. The image is then stored.

Figure 18.13 A PostScript image of the growth of *S. typhimurium* on a pH/NaCl gradient plate after 48 h incubation at 30 °C. The figure has been expanded with data interpolation from a 37 ×37 pixel image. The optical density is represented by a grey scale. The growth zone extends from the corner of the plate with low NaCl concentration and neutral pH, and is shown lighter (optical density >0) compared to the black (zero) background of the agar surface with no visible growth. From Peters *et al.* (1991b), by permission of Academic Press Ltd.

It is also possible manually to threshold an image, or set a fixed threshold. In this way the area of growth can be determined.

18.5.5 Image display

Using Semper 6plus, the processed image, with grey levels of 0 to 255, can be displayed on the monitor. This image, an example of which is shown in Figure 18.13, can be written to a PostScript file, enabling it to be printed on a laser printer which supports PostScript (Peters *et al.*, 1991b). Optical density is represented by a grey scale; the lighter the image, the higher the optical density. This type of output has limited use as it is not possible to label the axes (to show the gradients) or quantify the growth. Writing the data to file in ASCII format allows output in tabular form, as a wireframe three-dimensional plot or a contour map.

Figure 18.14 is a graphical representation of the data shown in Figure 18.13 (Peters *et al.*, 1991b). It was produced in the same way as Figure 18.6. Values for the pH and NaCl gradients and the optical density (OD) are now indicated. A contour map representation of the data in Figure 18.14, created using Systat, is shown in Figure 18.15, while Figure 18.16 shows the same data, now thresholded to give a binary image. The pH and NaCl gradients are indicated as before, while growth is shown as a single step. Figure 18.17 is an example of similar wire frame images obtained from temperature–pH gradient plates.

Once growth data has been stored in this way, further data analysis is possible. For instance, images can be subtracted from one another to reveal differences between growth zones. Figure 18.18a shows the growth zone of *S. typhimurium* under the same conditions as those described for Figure 18.14, but with the addition of an inhibitor, sodium nitrite. Subtracting the data shown in Figure 18.18a from the

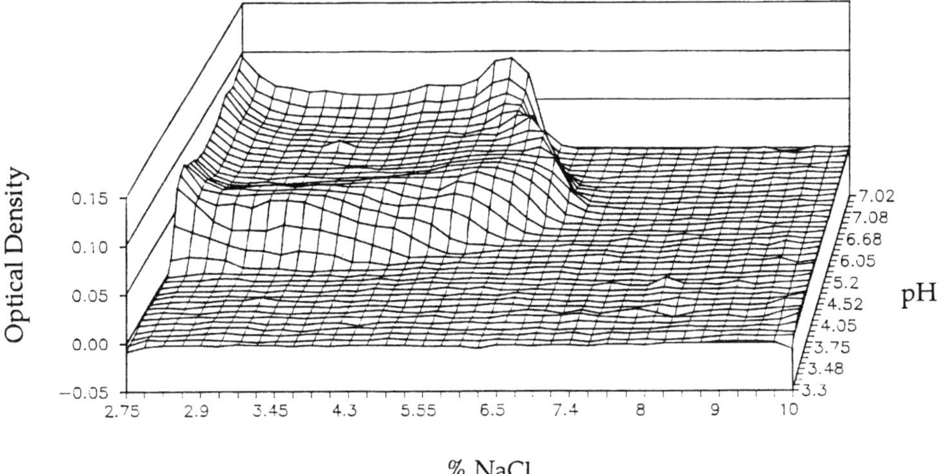

Figure 18.14 Wireframe representation of the growth of *S. typhimurium* as shown in Figure 18.13. The grey scale of Figure 18.13 has now been converted to optical density units, and the values for the gradients on the plate are shown as the *x* and *y* axes. The background has been forced to zero. From Peters *et al.* (1991b), by permission of Academic Press Ltd.

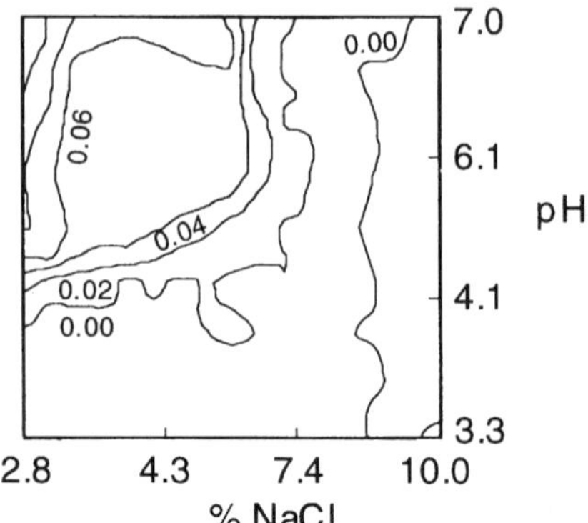

Figure 18.15 Contour map of the data from Figure 18.14. The contour lines represent lines of equal optical densities. From Peters *et al.* (1991b), by permission of Academic Press Ltd.

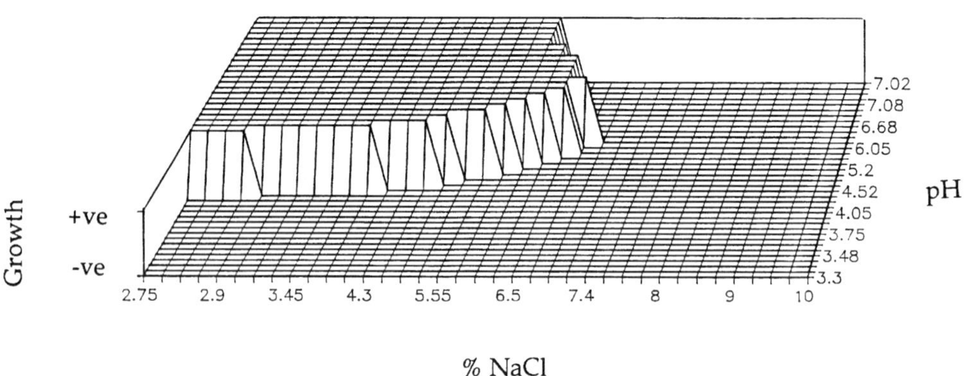

Figure 18.16 Data from Figure 18.14 after thresholding, producing a binary image. The background is zero, and all growth above the chosen threshold is recorded as 1. From Peters *et al.* (1991b), by permission of Academic Press Ltd.

larger growth area shown in Figure 18.14 gives Figure 18.18b, which reveals the zone of growth inhibited by the sodium nitrite. The subtraction of images was carried out in a Lotus 123 spreadsheet, but is also possible in Semper 6plus.

The growth on plates with different concentrations of an inhibitor and incubated at different temperatures can be shown as a 6 ×6 array of graphs which clearly illustrate the combined effects of several environmental variables (Thomas *et al.*, 1994). By combining data from several plates such as these, three-dimensional grids representing the growth of a particular organism over a range of values of pH, NaCl and a third variable such as temperature can be created (Thomas *et al.*, 1993). The effects of either different preservatives or different concentrations of the same preservative can be expressed in the same way. A program written by J.G. Davis (Thomas *et al.*, 1993) allows data files of this type to be created and displayed as sections using any two variables (e.g. pH and NaCl) as the x and y dimensions. Outlines defining the areas where growth exceeded a certain threshold value can

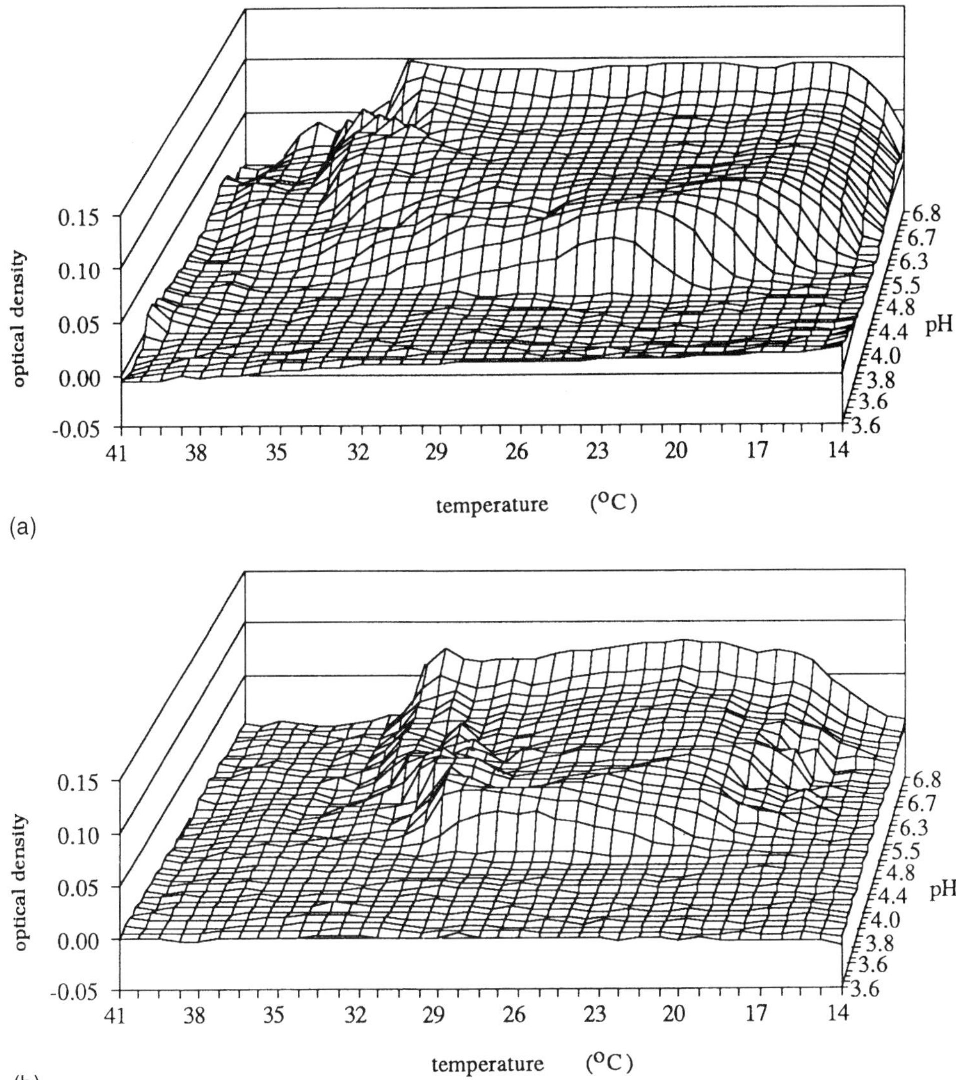

Figure 18.17 Wireframe representation of *S. typhimurium* grown on temperature/pH gradient plates: (a) with 0.5% (w/v) NaCl and (b) with 5% (w/v) NaCl. Reprinted from Peters A.C., Thomas L.V. and Wimpenny J.W.T., Effect of salt concentration on bacterial growth on plates with gradients of pH and temperature, *FEMS Microbiol. Lett.*, **77**, 309–14, (1991), with kind permission of Elsevier Science, Sara Burgerhartstraat 25, 1055 KV Amsterdam, The Netherlands.

then be added to each section and, by stacking sets of these outlines, the effects of three variables can be visualized on a single diagram. Three-dimensional graphics packages such as AutoCAD version 10 (AutoDesk Ltd., Cross Lanes, Guildford, Surrey, GU1 1UJ, UK), AcroSpin (Acrobits, PO Box 5563, Redwood City, California, USA) and PC3D (Jandel Scientific, 65 Koch Road, Corte Madera, California 94925,

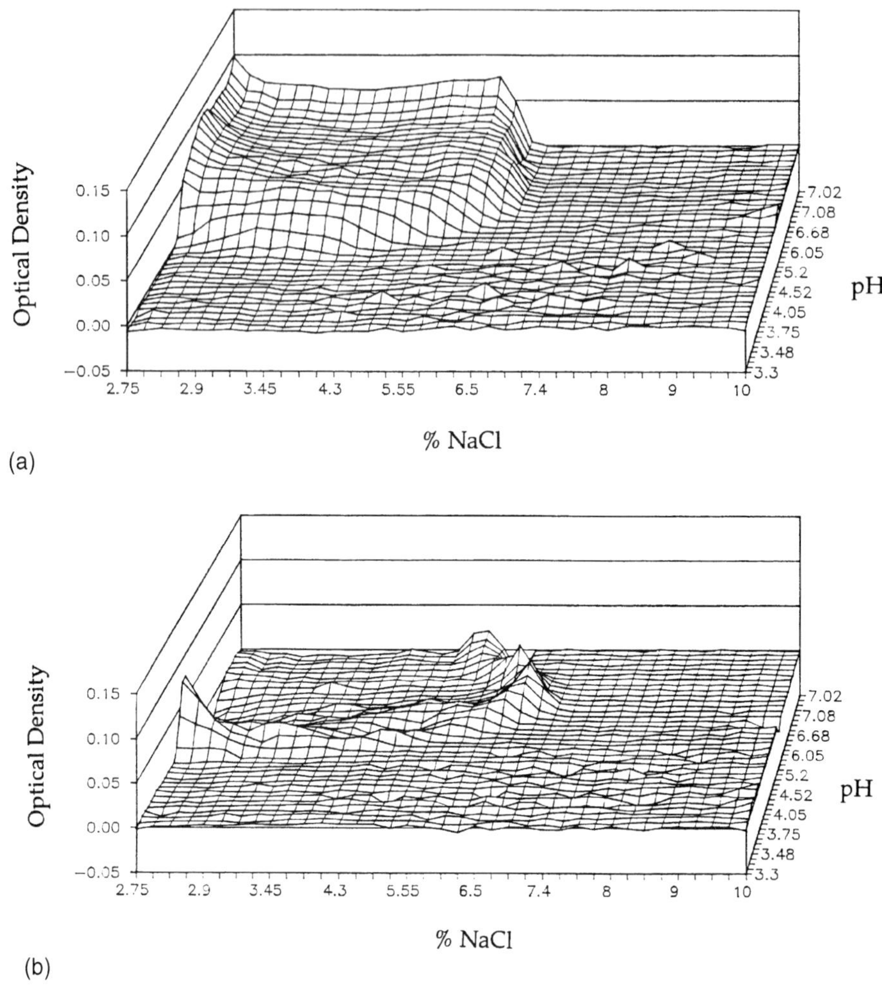

Figure 18.18 (a) Wireframe representation of the growth of *S. typhimurium* on a pH/NaCl gradient plate after 48 h incubation at 30 °C (as in Figure 18.14), but with the addition of an inhibitor, sodium nitrite, throughout the plate at a concentration of 0.2 g/l. (b) Wireframe representation of image data (a) subtracted from the image data of Figure 18.14. The band of growth represents the area of growth that is affected by the addition of 0.2 g/l sodium nitrite. From Peters *et al.* (1991b), by permission of Academic Press Ltd.

USA) can be used to display the diagrams from various viewpoints. For our particular purposes, a value of 0.01 optical density units was chosen as the most appropriate threshold value in the examples of stacked graphs shown in Figure 18.19 (showing the combined effects of temperature, pH and NaCl) and Figure 18.20 (showing the combined effects of pH, NaCl and different preservatives). For Figure 18.20 the gradients on the plates differed slightly due to the incorporation of the preservatives, and so the graphs were interpolated to give uniform *x* and *y* axes (pH and NaCl concentration), which allowed the data to be stacked. These graphs give a

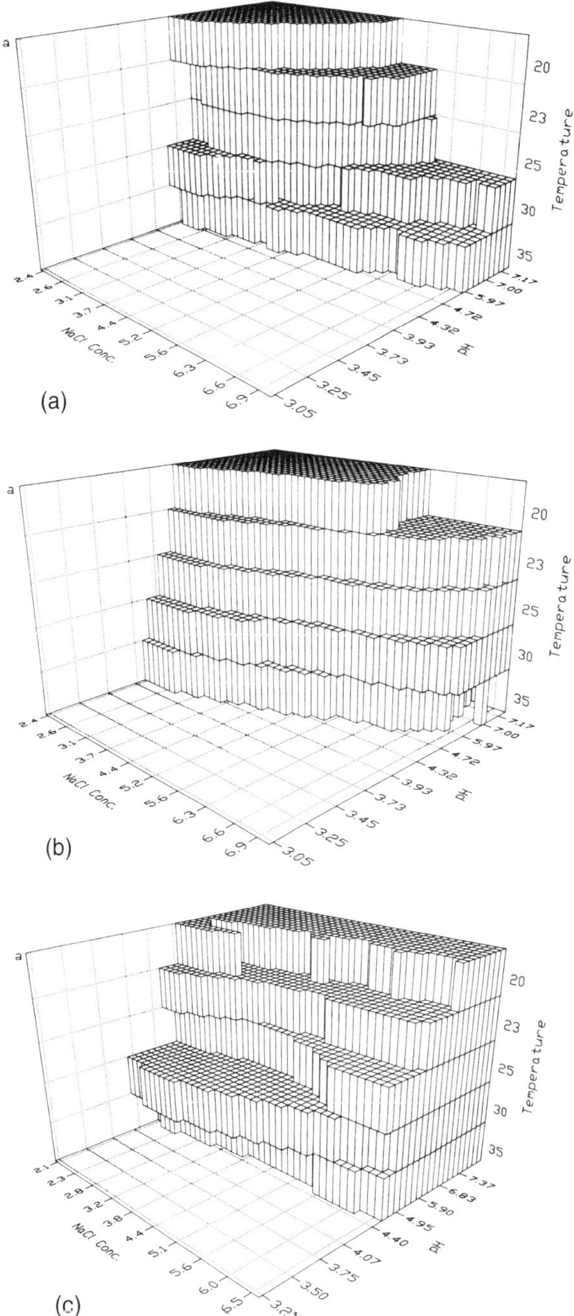

Figure 18.19 Stacking model of optical density data showing the effect of temperature, pH and NaCl concentration on foodborne pathogens grown for 48 h on pH/NaCl gradient plates: (a) *Bacillus cereus*, (b) Vero cytotoxigenic *Escherichia coli* O157 and (c) *Staphylococcus aureus*. Reprinted from Thomas, L.V., Wimpenny, J.W.T. and Davies, J.G., Effect of three preservatives on the growth of *Bacillus cereus*, Vero cytotoxigenic *Escherichia coli* and *Staphylococcus aureus* on plates with concentrations of pH and sodium chloride concentration, *Int. J. Food Microbiol.*, **17**, 309–14, (1993), with kind permission of Elsevier Science, Sara Burgerhartstraat 25, 1055 KV Amsterdam, The Netherlands.

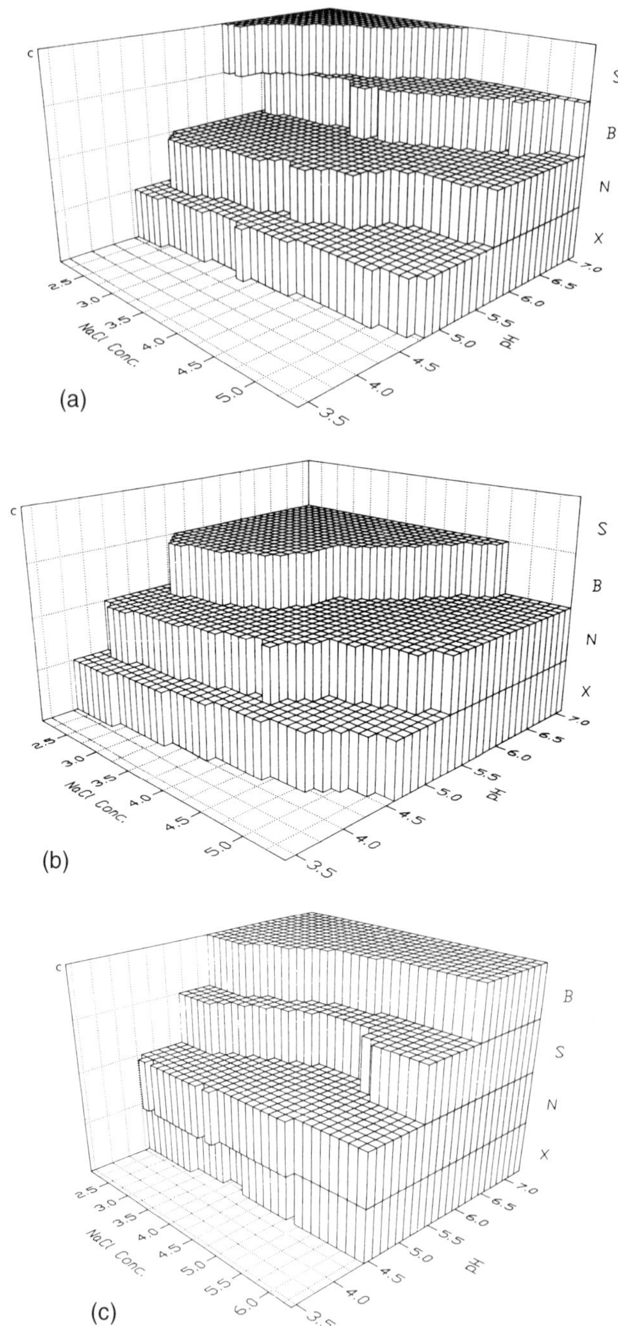

Figure 18.20 Stacking model of optical density data showing the effect of different preservatives, pH and NaCl concentration on foodborne pathogens grown for 48 h on pH/NaCl gradient plates at 35 °C. The preservatives were (X) control: no added preservative; (N) sodium nitrite (0.2 g/l); (B) sodium benzoate (1 g/l); (S) potassium sorbate (2 g/l). (a) *Bacillus cereus*, (b) Vero cytotoxigenic *Escherichia coli* O157 and (c) *Staphylococcus aureus*. Reprinted from Thomas, L.V., Wimpenny, J.W.T. and Davies, J.G., Effect of three preservatives on the growth of *Bacillus cereus*, Vero cytotoxigenic *Escherichia coli* and *Staphylococcus aureus* on plates with concentrations of pH and sodium chloride concentration, *Int. J. Food Microbiol.*, **17**, 309–14, (1993), with kind permission of Elsevier Science, Sara Burgerhartstraat 25, 1055 KV Amsterdam, The Netherlands.

clear representation of the effect of combined variables on the growth of each organism.

The stacks shown in this study are, of necessity, limited in that only three dimensions of the habitat domains of the organisms can be shown at one time. At any combination of the three coordinates one can establish whether there was growth or no growth of the test organism (as defined by the methodology of the experiment). Using another program, GRIDxD, written by J.G. Davis, the optical density data can be accessed at any combination of the four variables in order to determine growth/no growth under the terms of the experiment (i.e. ± optical density units of 0.01).

Testing four variables using two-dimensional gradient plates provides a large amount of data quickly and easily. One set of 36 plates (i.e. a 6 × 6 array) provides over 49 000 data points. To undertake such an experiment using individual test tubes or microtitre well cultures would be time consuming and difficult. However, presenting, understanding and analysing the quantity of data that can be obtained rapidly from gradient plates would have proved difficult, if not impossible, without the use of image analysis. This has enabled us to present or access our data in a variety of ways, facilitating the rapid comparison and prediction of the effect of combinations of up to four physico-chemical environmental variables on the growth of bacteria.

18.6 DETERMINATION OF MICROBIAL GROWTH RATES AS A FUNCTION OF ENVIRONMENTAL FACTORS, USING IMAGE ANALYSIS

The determination of microbial growth rates is a fundamental operation in quantitative microbiology, and numerous methods for doing this have been developed (Pirt, 1975). A two-dimensional gradient plate allows organisms to grow at all possible values of a subset of two factors varied simultaneously. This array of values can be determined as follows:

(a) Gradient plates were prepared as described in Section 18.5.1. The organism used in this study was *Staphylococcus aureus*. Plates were incubated at 30 °C and examined every hour from 5 to 24 hours after inoculation.
(b) Each plate used in these experiments, including an uninoculated blank, was examined by transmitted light using a 1 in videcon tube television camera. Frames were grabbed and processed as described earlier and the data reduced to a 37 × 37 array of values, which were incorporated into a spreadsheat. Two approaches were taken to determining growth rates. In the first, time-course data was taken from the plate to validate the second procedure.
(c) Absorbance data at each temperature were transferred to a spreadsheet on a Macintosh SE30 microcomputer. The data were then pasted into a graph plotting program (Cricket Graph) and plotted as conventional growth curves using a combination of Cricket Graph and MaCDraw II.

18.6.1 Automated determination of growth rates

Growth rate determination is as follows (Pirt, 1967). The standard growth equation:

$$\frac{dx}{dt} = \mu x$$

has the solution:

$$x = x_0 e^{\mu t}$$

The linearized form of this equation is as follows:

$$\ln x = \ln x_0 + \mu t$$

This is rearranged to solve for μ:

$$\mu = \frac{(\ln x - \ln x_0)}{t}$$

Natural logarithms of each absorbance data set were taken; the earlier data were then subtracted from the latter, and the result divided by the time interval to obtain a value for the growth rate. Despite the background correction procedures described above, there were problems still to be overcome. Some of the data were slightly negative. Zeros or negative values would prevent taking logs. Inspection suggested that the simplest procedure was to select the most negative value in the array, to make this positive and to add it plus a small additional amount to each of the image absorbance values. The lowest value was determined using:

= MIN(all cells in the array)

The natural log of the data was then taken after adding the correction factor:

= LN(V + (lowest value in array + 0.00011)

where V is the value in a given cell.

Each cell was then examined to see if the natural logarithm value was still very small, since such values would distort the axes of the wireframe diagrams. Values less than −5.0 were set to −5.0:

= IF(V < −5, −5, V)

The next step was to find the highest value for μ in each cell irrespective of plate age. To do this, the data set at each time interval was compared with its next older neighbour. An accumulator array was created the same size as the data sets. Values in corresponding cells were subtracted and divided by the time interval to give μ

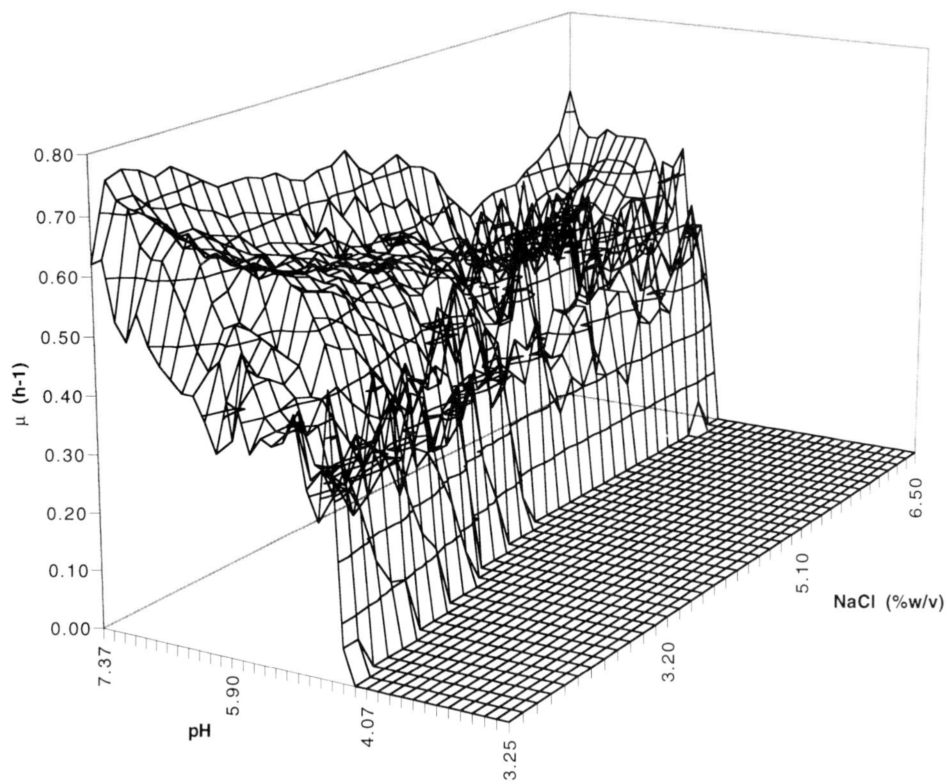

Figure 18.21 Specific growth rate for *Staphylococcus aureus* as a function of pH and NaCl concentration. For full details see text.

values. These were then compared to values in the accumulator. If the new value was larger than the value in the accumulator it replaced the latter; otherwise the accumulator remained the same:

$$= \text{IF}((V_{n+1} - V_n)/(T_{n+1} - T_n) > \text{ACC}, (V_{n+1} - V_n)/(T_{n+1} - T_n), \text{ACC}$$

T_{n+1} and T_n are the age of the (n + 1)th data set and the nth data set, respectively. Figure 18.21 shows specific growth rates for *S. aureus* growing on the surface of a pH–NaCl gradient plate.

18.6.2 Determination of lag periods using image processing

Data at each time interval were recorded as described above. The corrected image absorbance was used in the following way. Each data set was compared sequentially. An accumulator array the same size as the data arrays was created. Each cell in each data set is compared with its next older neighbour to see if any increase in absorbance above an arbitrary minimum value has taken place. If it has not then the age difference between the two is added to the value in the

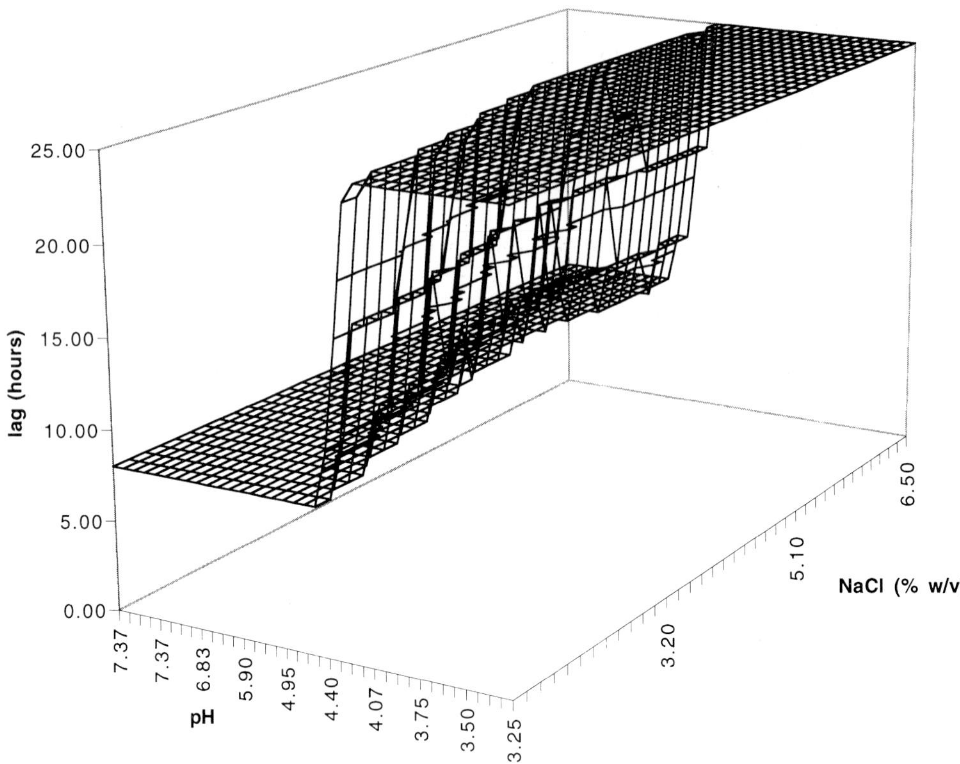

Figure 18.22 Lag period for *Staphylococcus aureus* as a function of pH and NaCl concentration. For full details see text.

accumulator. If the value exceeds the threshold a zero is added to the accumulator:

$$= \mathrm{IF}(\mathrm{V}_{n+1} - \mathrm{V}_n > 0.002, \mathrm{ACC} + 0, \mathrm{ACC} + ((\mathrm{T}_{n+1}) - \mathrm{T}_n)$$

Here V_{n+1} is the next older data set, V_n the earlier data set, ACC the corresponding cell in the lag accumulator array and T_{n+1} and T_n the corresponding times at which images were grabbed. This process is repeated for every time interval. Clearly, the zones where the lag is longest will have the larger numbers in them. Using this algorithm the longest possible lag is the total time of the experiment rather than infinity (Figure 18.22).

18.7 MEASUREMENTS OF THE ENZYMATIC ACTIVITY OF BACTERIA USING IMAGE ANALYSIS

An image analysis system was employed by Matchette *et al.* (1996) to determine β-galactosidase activity in cell suspensions of *Escherichia coli*. In these experiments the authors were testing the effect of UV laser light on the induction of lambda prophage containing the indicator gene, l-*lacZ*. The lysogen was incorporated in a thin agar layer on the plate. After irradiation an assay mixture containing 6-bromo-2-naphthol-β-D-galactoside plus fast blue RR salt in agar is poured over the previous layer. The 6-bromo-2-naphthol released is coupled with the fast blue diazonium salt to produce a blue colour. A CCD camera was used to capture images of the entire plate every 1–3 minutes and stored as 512×512 pixel images. Analysis was performed with IPLab Spectrum software (Signal Analytics Corp., Vienna, VA, USA).

Square regions were analysed around each spot of interest on the plate. Pixel values were corrected for background transmission and for residual values when the camera shutter was closed. Values were then converted to absorbance assuming that Lambert–Beer laws applied (see Sections 3.3.1 and 18.4.2c).

Two zones around each spot were measured, the second being roughly twice the area of the first. The sums of the OD values and the areas of each spot were recorded. The increased integrated OD of each was calculated as follows:

$$S_1 = A_1 B + D$$
$$S_2 = A_2 B + D$$

where S is the sum of the OD pixel values and A represents the area of the corresponding area of interest. B is the background pixel value while D is the integrated pixel value for the enzyme reaction. The solution for the integrated spot density is:

$$D = S_2 - \frac{S_2 - S_1}{A_2 - A_1}$$

Using this method, the dependence of reaction slopes versus enzyme concentration was nonlinear. A good linear fit was obtained if reaction rate was plotted against the logarithm of the enzyme concentration, and this can be used as a good calibration curve.

This rather careful work provides a useful general method for enzyme assays associated with biological samples as long as they can be determined on an agar surface.

18.8 THE DETERMINATION OF ANTIMICROBIAL ACTIVITY USING IMAGE ANALYSIS

The sensitivity of bacteria to antimicrobials is of paramount importance in medical microbiology, and many different techniques are available to measure this. Some rely on tube dilutions, others on the measurement of inhibition zones on agar plates. Image analysis has featured in automating these antibiotic test systems.

Hammonds and Adenwala (1990) and Hammonds and Galliford (1991) developed a method of measuring sensitivity by looking at the inhibition of micro-colony formation by the agents. Here 0.02 ml of a reasonably dense suspension of bacteria (approximately 10^5/ml) were spotted aseptically onto slides coated with Mueller–Hinton agar with or without added antibiotic. After incubation for 4 hours the plates were read using image analysis. Results were expressed as the total area of growth, expressed as pixels, in each spot. This procedure provides growth rate data; however, the authors appear not to have expressed their results in these terms. This system gave sensitivity results in 4 hours, which was significantly better than the time needed for disc diffusion tests to show. As they point out, it ought to be possible to reduce even this time, if a direct suspension from a growing colony were used as inoculum instead of an overnight stationary phase culture.

The measurement of antibiotic inhibition zones on agar plates has been a standard procedure for many years in microbiology and in clinical pathology departments. To make this method reasonably quantitative, the diameter of zones of inhibition is measured and related to minimum inhibitory concentration (MIC) values. Methods of automating these assays were obviously necessary, and a number of workers have turned to image analysis to do this.

Hejblom *et al.* (1993) used an Apple Macintosh Quadra 700 computer fitted with a Nubus image grabber board together with a CCD camera. Programs were written in Object-Oriented MPW Pascal. Organisms were grown on 165 mm square Petri dishes and 6 mm diameter antibiotic discs were deployed in a number of predetermined standard arrays, commonly 4×4. The authors used a radial profile analysis algorithm (RPAA) to determine inhibition zones. Firstly, to determine disc position, the data were converted to binary images and each identified area colour identified and filtered to ensure that they corresponded to actual discs in size and circularity. They were then matched with the predetermined disc pattern so that each area could be assigned to a given antibiotic disc. The algorithm could cope with missing discs or additional discs at positions outside the predetermined pattern. The second step built radial profiles around each disc. Thus a rectangular area was placed around each disc depending on the position of the disc. For central discs this was 35 mm square while side areas were half and corner zones one quarter of this so that edge effects could be ignored. Next, a radial profile of pixel values was constructed from the centre outwards. Three types of profile corresponded to information on the plate. Central pixels were bright and fell off quickly. These corresponded to the disc itself and were common to each profile. In type 1 profiles low pixel values followed, after which pixel values rose again. This corresponded to an inhibition zone that appeared dark, then brighter as zones of growth appeared. In type 2 profiles pixel value maintained a high plateau away from the edge of the disc. This represents bacterial growth, and hence no inhibition. Type 3 profiles were uniformly low, meaning that the inhibition zone was larger than the 35 mm^2 surrounding the disc. An algorithm was developed that distinguished between these patterns, and which drew a circular ring where the transition to higher pixel values was detected in type 1 profiles. The resulting estimate of the inhibition zone was automatically compared with conventional breakpoints for classifying the tested strain in one of the clinical categories of antibiotic susceptibility. When the system was tested against human readers correspondence of 95.5% resulted. Severe disagreements were seen in 5.6% of tests with staphylococci but only 0.3% of Gram-negative rods.

Schoevers *et al.* (1994) also used square Petri dishes with 4 × 4 arrays of discs. An IBAS (Zeiss/Kontron, Eching, Germany) image processing system was used in this study. Images of plates *or* of photographic negatives were captured for each plate, as 512 × 512 pixel images at 256 grey levels. To reduce noise, images were captured ten times and the data averaged. During processing all objects were replaced by interpolated grey scale data corresponding to the background. The whole background field was then subtracted from original data, and the results smoothed to generate a corrected image. Grey scales were enhanced using a DELIN filter (nonlinear edge enhancement, Section 6.7.2). Subsequently holes in the object were filled and the edges smoothed using a median filter. Inhibition zones were identified, and the area, diameter and circularity of each image were then recorded. Results from direct images, images from the photographic negative and images from human readers were well correlated statistically using plasma and urine as test systems for antibiotic levels.

Image analysis was also used by Gavoille *et al.* (1994) to measure inhibition zones. Once more, standard 4 × 4 arrays of discs were used on square Petri dishes. The image processing software was Magiscan 2 (Joyce-Loebl Ltd., UK). A black and white video camera (Bosch, Germany) was used to grab 512 × 512 images with 64 grey levels. A software library was used, allowing particular applications to be written easily in PASCAL. Petri dishes were put on a black background and illuminated with a reflected light source.

The algorithms applied included a step to find the actual position of each disc since this varied slightly. This algorithm was based on three hypotheses: (i) the size of the disc was constant and known; (ii) the real position was not very far from the predefined one and (iii) the disk was defined by high (white) pixel values. Taking these into account the program searched for the brightest circle around each predefined position. Once found, the centre of this circle corresponded to the centre of the disc. The algorithm needed for detecting the edge of the inhibition zone was more complex since it included the ability to detect zones that were incomplete or heterogeneous, or which overlapped with neighbouring zones. Two shape-defined circular or semicircular masks (depending on the position of the disc on the plate) were used. The masks were advanced outward from the centre and pixel grey levels and standard deviations computed for each mask at each position. Student's *t*-test was applied at each position. The point where this gave its highest value was regarded as the edge of the inhibition zone. This was marked on the image with a white circle. Kappa statistics were employed to compare results with those generated by a trained human and the data were shown to match extremely well.

An overlay inoculum susceptibility disc method was used by Chinwuba *et al.* (1991) to investigate synergy between antibiotic combinations. These workers used the commercially available Fisher and Lilly antibiotic zone reader. Finally, there are simpler computer-based techniques that involve human operators. Kath *et al.* (1993) developed a simple optical comparator, which measured the inhibition zone and stored the data in a computer. The image of the plate appears on a display and circles are superimposed manually over each zone. These are adjusted to fit the zone as closely as possible and the data are then recorded. The authors stress that this is quicker and more accurate than other manual methods and also cheaper than image analysis systems.

18.9 OTHER APPLICATIONS

Another interesting application was described by Watanabe *et al.* (1991). Using an image analysis system with an image intensifier, they followed synchronous photon emissions from *Photobacterium phosphoreum* as a colony developed from a single cell. Photon output was seen to decrease, undergo a prolonged lag period and then increase rapidly, as has been reported in liquid culture. They also measured periodicity in photon output during colony growth, which was attributed to cell cycle events.

Image analysis has also been used to identify recombinant/non-recombinant colonies of *E. coli* (Jones *et al.*, 1992). The image analysis system identified colonies in a three-stage process. Firstly, a morphological edge detection algorithm delineated objects that met the threshold criterion. Secondly, objects were screened against preset size and shape criteria. This excluded small dust particles, areas of agar imperfection and non-circular colonies (i.e. colonies that had merged or were touching). Finally, colonies were selected by grey level difference between the centre of the colony and the agar surrounding the colony. This discriminated between blue and white colonies for recombinant/non-recombinant selection in *E. coli*. Selected clones were picked and subcultured robotically.

18.10 CONCLUSIONS

Image analysis techniques examining surface growth of micro-organisms are now well established in both analytical microbiology and research. Routine plate counts and other analytical tests can be performed rapidly with enhanced precision and automatic data management. More specialized systems already allow automatic selection and robotic picking of colonies (e.g. Jones *et al.*, 1992) based on differences in colony morphology, and future developments for routine analytical tools should include automatic colony recognition for a range of common selective agars. This could be linked to robotic techniques for automatic picking and subculturing. Colony recognition routines could be based on expert systems, but it is the authors' opinion that the use of artificial neural networks in colony identification should be investigated.

The use of image analysis to monitor microbial growth on agar surfaces has proved a powerful research tool giving objective, detailed data on microbial behaviour. More specifically the research has provided:

(a) rapid screening of a microbe's response to multiple environmental factors and development of multidimensional habitat domain maps
(b) automatic determination of dynamic growth data including estimation of lag time and specific growth rates across the surface of gradient plates
(c) precise, non-invasive, measurement of growth rates of individual colonies with extremely high detection sensitivity
(d) determination of growth rates at different points within a colony

It is now acknowledged that microbial behaviour on gel surfaces and within gels may provide useful models of food environments, and much of the work reviewed here has concentrated on the growth of foodborne pathogens. There is also a growing recognition that safe food control depends on adequate microbiological risk assessment and more data concerning the growth of foodborne pathogens is needed to fulfil the food industry's needs, particularly in relation to the provision of predictive models (McClure *et al.*, 1994). Databases containing large, multidimensional habitat domain maps can give useful qualitative data concerning the ability of an organism to grow under a wide range of conditions, and may show important relationships between different effectors. The dynamic growth data generated by the systems described are suitable for use in both the development and validation of predictive models (Peters *et al.*, 1996), and the sensitivity of the technique can give an insight into the initiation of the growth of single cells and competition between individual cells in close proximity.

Work on the behaviour of foodborne pathogens is continuing, focusing on the effect of modified atmospheres and changes in water activity using a specially constructed growth chamber which, mounted on a microscope stage, allows imaging of single cells held under carefully controlled conditions. A preliminary study, using colour image analysis, has demonstrated that it is possible to map enzyme activity quantitatively within cryosectioned layers of a colony; building up a three-dimensional view of activity (Samantha Walker, personal communication).

This chapter demonstrates that image analysis has enormous potential for generating quantitative data on the development and activity of microbial growth on agar surfaces and we believe that this information will lead to a better understanding of the behaviour of microbes in the natural environment.

ACKNOWLEDGEMENTS

The authors acknowledge the support of MAFF, NERC and the University of Wales in undertaking the research presented, and the assistance of A.L. McKay, J. Griffiths and J.G. Davis. Work presented on the growth of *Salmonella typhimurium* colonies was undertaken by A.L. McKay during the tenure of her PhD Studentship.

REFERENCES

Bal'a, M.F.A. and Marshall, D.L. (1996) Use of double-gradient plates to study combined effects of salt, pH, monolaurin, and temperature on *Listeria monocytogenes*. *J. Food Protection* **59**: 601–7.

Baron, E.J. and Finegold, S.M. (1994) *Bailey and Scott's Diagnostic Microbiology*, 9th edn. C.V. Mosby: St Louis, MO.

Belyaev, N., Paavilainen, S. and Korpela, T. (1992) Characterization of bacterial growth on solid medium with image analysis. *J. Biochem. Biophys. Meth.* **25**: 125–32.

Caldwell, D.E. and Hirsch, P. (1973) Growth of micro-organisms in two-dimensional steady-state diffusion gradients. *Can. J. Microbiol.* **19**: 53–8.

Caldwell, D.E., Lai, S.H. and Tiedje, J.M. (1973) A two-dimensional steady-state diffusion gradient for ecological studies. *Bull. Ecol. Res. Communic. (Stockh)* **17**: 151–8.

Chinwuba, Z.G.N., Chiori, C.O., Ghobashy, A.A. and Okore, V.C. (1991) Determination of the synergy of antibiotic combinations by an overlay inoculum susceptibility disc method. *Arzneimittel-Forschung [Drug Research]* **41**: 148–50.

Cooper, A.L., Dean, A.C.R. and Hinschelwood, C. (1968) Factors affecting the growth of bacterial colonies on agar plates. *Proc. R. Soc. London* **B171**: 175–99.

Gavoille, A., Bardy, B. and Andremont, A. (1994) Measurement of inhibition zone diameter in disk susceptibility tests by computerized image analysis. *Comput. Biol. Med.* **24**: 179–88.

Halldal, P. and French, C.S. (1958) Algal growth in crossed gradients of light intensity and temperature. *Plant Physiol.* **33**: 249–52.

Hammonds, S.J. and Adenwala, F. (1990) Antibiotic sensitivity testing of bacteria by microcolony inhibition and image analysis. *Lett. Appl. Microbiol.* **10**: 27–9.

Hammonds, S.J. and Galliford, H.M. (1991) Assessment of the antimicrobial activity of antibiotics using microcolony areas measured by image analysis. *J. Microbiol. Meth.* **14**: 95–101.

Hejblom, G., Jarlier V., Grosset, J. and Aurengo, A. (1993) Automated interpretation of disk diffusion antibiotic susceptibility tests with the radial profile analysis algorithm. *J. Clin. Microbiol.* **31**: 2396–401.

Jones, P., Watson A., Davies, M. and Stubbings, S. (1992) Integration of image analysis and robotics into a fully automated colony picking and plate handling system. *Nucl. Acids Res.* **20**: 4599–606.

Junttila, J.R., Niemela, S.I. and Hirn, J. (1988) Minimum growth temperatures of *Listeria monocytogenes* and non haemolytic *Listeria. J. Appl. Bacteriol.* **65**: 321–7.

Kath G.S., McKeel, W.J. and Sundelof, J. (1993) Bacterial zone measurement apparatus. *Meas. Sci. Technol.* **4**: 231–3.

Kim, K.S. and Anthony, B.F. (1983) Use of penicillin-gradient and replicate plates for the demonstration of tolerance to penicillin in streptococci. *J. Infect. Dis.* **148**: 488–91.

Lewis, M.W.A. and Wimpenny, J.W.T. (1981) The influence of nutrition and temperature on the growth of colonies of *Escherichia coli* K12. *Can. J. Microbiol.* **27**: 679–84.

Matchette, L.S., Grossman, D.W., Hann, D.W. and Cooney, C. (1996) Induction of lambda prophage by 213 nm laser radiation: a quantitative comparison with 193 nm excimer radiation using image analysis. *Photochem. Photobiol.* **63**: 281–8.

McClure, P.J. and Roberts, T.A. (1986) Detection of growth of *Escherichia coli* on two-dimensional diffusion gradient plates. *Lett. Appl. Microbiol.* **3**: 53–6.

McClure, P.J., Roberts, T.A. and Oguru, O. (1989) Comparison of the effects of sodium chloride, pH and temperature on the growth of *Listeria monocytogenes* on gradient plates and in liquid medium. *Lett. Appl. Microbiol.* **9**: 95–9.

McClure, P.J., Blackburn, C.W., Cole, M.B., Curtis, P.S., Jones, J.E., Legan, J.D., Ogden, I.D., Peck M.W., Roberts, T.A., Sutherland, J.P. and Walker, S.J. (1994) Modelling the growth, survival and death of microorganisms in foods: the U.K. Food MicroModel approach. *Int. J. Food Microbiol.* **23**: 265–75.

McKay, A.L. and Peters, A.C. (1995) The effect of sodium chloride concentration and pH on the growth of *Salmonella typhimurium* colonies on solid medium. *J. Appl. Bacteriol.* **79**: 353–9.

McKay, A.L., Peters, A.C. and Wimpenny, J.W.T. (1997) Determining specific growth rates in different regions of *Salmonella typhimurium* colonies. *Lett. Appl. Microbiol.* **24**: 74–6.

Murray P.R.E., Baron, E.J., Pfaller, M.A., Tenover, F.C. and Yolken, R.H. (1995) *Manual of Clinical Microbiology*, 6th edn. American Society for Microbiology: Washington, DC.

Niemela, S.I., Mentu J., Vaatanen, P. and Lahti, K. (1983) Maximum growth temperature as a diagnostic character in Enterobacteriaceae. *INSERM* **114**: 619–27.

Palumbo, S.A., Johnson, M.G., Rieck, V.T. and Witter, L.D. (1971) Growth measurements on surface colonies of bacteria. *J. Gen. Microbiol.* **66**: 137–43.

Peters, A.C. (1990) Using image analysis to map bacterial growth on solid media. *Binary Comput. Microbiol.* **2**: 73–5.

Peters, A.C., Thomas, L.V. and Wimpenny, J.W.T. (1991a) Effect of salt concentration on bacterial growth on plates with gradients of pH and temperature. *FEMS Microbiol. Lett.* **77**: 309–14.

Peters, A.C., Thomas, L.V., Wimpenny, J.W.T. and Griffiths, J. (1991b) Mapping bacterial growth on gradient plates using image analysis. *Binary Comput. Microbiol.* **3**: 147–53.

Peters, A.C., Harrison W., McKay, A.L., Griffith, C. and Fielding, L. (1996) A comparison of experimentally generated microbial growth kinetics and rates predicted by the Food MicroModel and the Pathogen Modelling Program, Society of Applied Bacteriology Autumn Meeting, Colindale.

Pirt, S.J. (1967) A. kinetic study of the mode of growth of surface colonies of bacteria and fungi. *J. Gen. Microbiol.* **47**: 181–97.

Pirt, S.J. (1975) *Principles of Microbe and Cell Cultivation*. Blackwell Scientific: Oxford.

Pover, P.S. (1990) Colony counting and other Petri-dish applications of image analysis. *Binary Comput. Microbiol.* **2**: 77–9.

Pover, P.S. (1996) Automated colony counting in a contract testing laboratory. *Microbiol. Eur.* **4**: March/April, 2 pp.

Sacks, L.E. (1956) A pH gradient agar plate. *Nature* **178**: 269–70.

Schoevers, E.J., Terlou M., Pijpers, A. and Verheijden, J.H.M. (1994) An image analysis system: an objective and accurate alternative for reading the agar diffusion test. *J. Vet. Ther.* **17**: 38–42.

Szybalski, W. (1952) Gradient plates for the study of microbial resistance to antibiotics. *Bacteriol. Proc.* 38.

Szybalski, W. and Bryson, V. (1952) Genetics studies on microbial cross resistance to toxic agents. I. Cross resistance to *Escherichia coli* to fifteen antibiotics. *J. Bacteriol.* **64**: 489–99.

Thomas, L.V. and Wimpenny, J.W.T. (1993) Method for the investigation of competition between bacteria and a function of three environmental factors varied simultaneously. *Appl. Environ. Microbiol.* **59**: 1991–7.

Thomas, L.V. and Wimpenny, J.W.T. (1996a) Competition between *Salmonella* and *Pseudomonas* species growing in and on agar, as affected by pH, sodium chloride concentration and temperature. *Int. J. Food Microbiol.* **29**: 361–370.

Thomas, L.V. and Wimpenny, J.W.T. (1996b) Investigation of the effect of combined variations in temperature, pH and NaCl concentration on nisin inhibition of *Listeria monocytogenes* and *Staphylococcus aureus*. *Appl. Environ. Microbiol.* **62**: 2006–12.

Thomas, L.V., Wimpenny, J.W.T. and Peters, A.C. (1991) An investigation of the effects of four variables on the growth of *Salmonella typhimurium* using two types of gradient plates. *Int. J. Food Microbiol.* **14**: 261–75.

Thomas, L.V., Wimpenny, J.W.T. and Peters, A.C. (1992) Testing multiple variables on the growth of a mixed inoculum of *Salmonella* strains using gradient plates. *Int. J. Food Microbiol.* **15**: 165–75.

Thomas, L.V., Wimpenny, J.W.T. and Davies, J.G. (1993) Effect of three preservatives on the growth of *Bacillus cereus*, Vero cytotoxigenic *Escherichia coli* and *Staphylococcus aureus* on plates with concentrations of pH and sodium chloride concentration. *Int. J. Food Microbiol.* **17**: 289–301.

Thomas, L.V., Wimpenny, J.W.T. and Peters, A.C. (1994) Multifactorial control of the growth of food poisoning bacteria. A. report on contract N392 for MAFF.

Van Baalen, C. and Edwards, P. (1973) Light-temperature gradient plate. In *Handbook of Phycological Methods* (Stern, J.R., ed.), p. 267. Cambridge University Press: Cambridge.

Venables, W.A., Wimpenny J.W.T., Ayres A., Cook, S.M. and Thomas, L.V. (1995) The use of two-dimensional gradient plates to investigates the range of conditions under which conjugal plasmid transfer occur. *Microbiology* **141**: 2713–18.

Watanabe H., Suzuki S., Kobayashi M., Usa, M. and Inaba, H. (1991) Analysis of synchronous photon emissions from the bacterium *Photobacterium phosphoreum* during colony formation from a single cell. *J. Biolumin. Chemilumin.* **6**: 13–18.

Waters, P. and Lloyd, D. (1985) Salt, pH and temperature dependence of growth and bioluminescence of three species of luminous bacteria analysed on gradient plates. *J. Gen. Microbiol.* **131**: 2865–9.

Weinberg, E.D. (1956) Double agar plates. *Science* **125**: 196.

Wimpenny, J.W.T. (1979) The growth and form of bacterial colonies. *J. Gen. Microbiol.* **114**: 483–6.

Wimpenny, J.W.T. (1988) The bacterial colony. In *Handbook of Labarotory Model Systems for Microbial Ecosystems*, vol. 2 (Wimpenny J.W.T., ed.), pp. 101–139. CRC Press: Boca Raton, FL.

Wimpenny, J.W.T. (1992) Microbial systems: patterns in time and space. In *Advances in Microbial Ecology*, vol. 12 (Marshall, K.C., ed.), pp. 469–522. Plenum Press: New York.

Wimpenny, J.W.T. and Waters, P. (1984) Growth of microorganisms gel stabilized two-dimensional diffusion gradient systems. *J. Gen. Microbiol.* **130**: 2921–6.

Wimpenny, J.W.T. and Waters, P. (1987) The use of gel-stabilised gradient plates to map the responses of micro-organisms to three or four environmental factors varied simultaneously. *FEMS Microbiol. Lett.* **40**: 263–7.

Wimpenny J.W.T., Gest, H. and Favinger, J.L. (1986) The use of two dimensional gradient plates in determining the responses of non-sulphur purple bacteria to pH and NaCl concentration. *FEMS Microbiol. Lett.* **37**: 367–71.

Wimpenny J.W.T., Waters, P. and Peters, A.C. (1988) Gel plate methods in microbiology. In *Handbook of Laboratory Model Systems for Microbial Ecosystems*, vol. 1 (Wimpenny J.W.T., ed.), pp. 229–251. CRC Press: Boca Raton, FL.

Wimpenny J.W.T., Leistner L., Thomas, L.V., Mitchell A., Katsaras, K. and Peetz, P. (1995a) Submerged bacterial colonies within food and model systems: their growth, distribution and interactions. *Int. J. Food Microbiol.* **28**: 299–315.

Wimpenny J.W.T., Wilkinson, T. and Peters, A.C. (1995b) Monitoring microbial colony growth using image analysis techniques. *Binary Comput. Microbiol.* **7**: 14–18.

Appendix
List of Manufacturers

Here we list most of the manufacturers of equipment, software and chemicals that have been mentioned in this volume, or with which the authors or editors have some experience. The list is by no means a complete list of all manufacturers in this field. Others may be found using appropriate searches on the world-wide web (WWW). A good starting point is the list of manufacturers at:

http://www.kaker.com/mvd/list.html

Others starting points include:

http://corn.eng.buffalo.edu/www/CommercialInst/readme.html

and the Microscopy page:

http://www.ov.edu/research/electron/mirror

LIGHT MICROSCOPES (CONFOCAL AND CLASSICAL) AND ACCESSORIES

Bio-Rad Laboratories
1000 Alfred Nobel Drive,
Hercules, CA 94547, USA
http://www.biorad.com/

Carl Zeiss Jena GmbH
Division of Microscopy,
Tatzeadpromenade 1a, D-07745 Jena, Germany.
http://www.zeiss.de/

Chroma Technology Corporation
72 Cotton Mill Hill, A-9
Brattleboro, VT 05301, USA

Leica Mikroskopie und Systeme GmbH
PO Box 2040
D-35530 Wetzlar, Germany
http://www.leica.com/

Meridian Instruments, Inc.
2310 Science Parkway,
Okemos, MI 48864, USA
http://www.microscopy-online.com/Vendors/Meridian/

Molecular Dynamics
928 East Arques Avenue,
Sunnyvale, CA 94086-4520, USA
http://www.mdyn.com/

Nikon Corporation
Fuji Building, 2–3, 3-chome
Marunouchi, Chiyoda-ku, Tokyo 100, Japan.
http://www.klt.co.jp/Nikon/

NORAN Instruments Inc.
2551 West Beltline Highway,
Middleton, WI 53562-2697, USA
http://www.noran.com/

Olympus Optical Company, Ltd
43–2, Hatagaya 2-chome,
Shibuya-ku, Tokyo, Japan

Omega Optical, Inc.
PO Box 573, 3 Grove Street,
Brattleboro, VT 05302, USA
http://www.omegafilters.com/

Optiscan
27 Normanby Road,
Notting Hill, Victoria, Australia 3168

Technical Instruments Company
650 North Mary Avenue,
Sunnyvale, CA 94086, USA
http://www.technical.com/

SCANNING PROBE MICROSCOPES

Digital Instruments, Inc.
112 Robin Hill Road,
Santa Barabara, CA 93117, USA
http://www.di.com/

DME Danish Micro Engineering A/S
Transformervej 12,
DK-2730 Herlev, Denmark

TopoMetrix Corporation
5403 Betsy Ross Drive,
Santa Clara, CA 95054, USA
http://www.topometrix.com/

Park Scientific Instruments
1171 Borregas Avenue,
Sunnyvale, CA 94089, USA
http://www.park.com/

Surface Imaging Systems GmbH
Kaiserstrasse 100,
D-52134 Herzogenrath, Germany

IMAGE PROCESSING HARDWARE AND SOFTWARE

Carl Zeiss Vision GmbH
Oskar von Miller Str. 1,
D-85386 Eching, Germany
http://www.kontron.com/

Data Translation, Inc.
100 Locke Drive,
Marlboro, MA 01752-1192, USA
http://www.datx.com/

Foster Findlay Associates Ltd.
Newcastle Technopole, King's Manor,
Newcastle upon Tyne NE1 6PA, UK
http://www.demon.co.uk/ffaltd/

Improvision, Image Processing & Vision Co. Ltd
Barclays Venture Centre, University of Warwick Science Park,
Sir William Lyons Road, Coventry CV4 7EZ, UK
http://www.improvision.co.uk/

Leica Imaging Systems, Ltd
Clifton Road, Cambridge CB1 3QH, UK
http://www.leica.co.uk/ia/homepage.htm

Matrox Electronic Systems, Ltd
1055 St Regis Blvd,
Dorval, Quebec, H9P 2T4 Canada
http://www.matrox.com/

Media Cybernetics, Inc.
8484 Georgia Ave.,
Silver Spring, MD 20910, USA
http://www.mediacy.com/

Motion Analysis Corporation
3617 Westwind Blvd,
Santa Rosa, CA 95403, USA
http://www.motionanalysis.com/

Noesis Vision Inc.
Suite 200, 6800 Cote de Liesse,
St Laurent, H4T 2A7 Canada
http://www.noesisvision.com/

Optimas Corporation
19811 North Creek Parkway,
Bothwell, WA 98011, USA
http://www.optimas.com/

Synoptics Ltd
271 Cambridge Science Park,
Milton Road, Cambridge CB4 4WE, UK
http://www.synoptics.co.uk/

Truevision, Inc.
7340 Shadeland Station,
Indianapolis, IN 46256, USA
http://www.truevision.com/

IMAGE SENSORS

AstroCam Ltd
Innovation Centre, Cambridge Science Park,
Milton Road, Cambridge CB4 4GS, UK
http:/www.astrocam.co.uk/

Dage-MTI, Inc.
701 North Roeske Avenue,
Michigan City, IN 46360, USA
http://www.dagemti.com/

Eastman Kodak Company
343 State Street,
Rochester, NY 14650-1139, USA
http://www.kodak.com/

Hamamatsu Photonics K.K.
812 Joko-Cho,
Hamamatsu City, 431–32, Japan

Kappa Messtechnik GmbH
Kleinesfeld 6,
D-37130 Gleichen, Germany
http://www.kappa.de/

Optronics Engineering
175 Cremona Drive,
Goleta, CA 93117, USA
http://www.optronics.com/

Photometrics Ltd.
3440 East Britannia Drive,
Tucson, AZ 85706, USA
http://www.photomet.com/

Photonic Science Ltd
Milham, Mountfield,
Robertsbridge TN32 5LA, UK
http://www.photonic-science.ltd.uk/

CHEMICALS AND LABORATORY SUPPLIES

Biodata Oy
PO Box 75, SF-00371 Helsinki,
Finland

Boehringer Mannheim GmbH
Sandhofer Strasse 116,
D-68298 Mannheim, Germany
http://www.boehringer-mannheim.com/

Merck KGaA
Frankfurterstrasse 250
D-64293, Darmstadt, Germany
http://www.merck.com/

Microscreen BV
Zernikepark 8
9747 AM Groningen, The Netherlands
http://www.microscreen.com/

Miles, Inc., Diagnostics Division
PO Box 70,
Elkhart, IN 46515, USA

Molecular Probes Inc.
PO Box 22010
Eugene, OR 97402-0469, USA
http://www.probes.com/

Nycomed Amersham plc (formerly Amersham International plc)
Amersham Place, Little Chalfont,
Buckinghamshire HP7 9NA, UK
http://www.amersham.co.uk/

Pierce
PO Box 117
Rockford, IL 61105, USA
http://www.piercenet.com/

Promega Corporation
2800 Woods Hollow Road,
Madison, WI 53711, USA
htto://www.promega.com

Research Organics Inc.
4353 East 49th Street,
Cleveland, OH 44125, USA
http://www.resorg.com/

Sigma Chemical Co.
PO Box 14508,
St Louis, MO 63178-9916, USA
http://www.sigma.sial.com/sigma/

Vector Laboratories, Inc.
30 Ingold Road,
Burlingame, CA 94010, USA
http://www.vectorlabs.com/

Index

Note: Boldface page numbers refer to definitions of terms; italicized refer to illustrations.

CARDIFF UNIVERSITY
PRIFYSGOL CAERDYDD